APPLIED NUMERICAL ANALYSIS

SIXTH EDITION

Applied Numerical Analysis

SIXTH EDITION

Curtis F. Gerald
Patrick O. Wheatley

California Polytechnic State University
San Luis Obispo

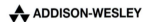 **ADDISON-WESLEY**

An imprint of Addison Wesley Longman, Inc.

Reading, Massachusetts • Menlo Park, California • New York • Harlow, England
Don Mills, Ontario • Sydney • Mexico City • Madrid • Amsterdam

Sponsoring Editors: Jennifer Albanese, Carolyn Lee-Davis
Project Manager: Rachel S. Reeve
Production Supervisor: Kim Ellwood
Production Services: Tobi Giannone, Michael Bass and Associates
Marketing Manager: Carter Fenton
Marketing Coordinator: Michael Boezi
Cover Designer: Barbara Atkinson
Manufacturing Manager: Ralph Mattivello
Manufacturing Buyer: Evelyn Beaton

Cover Photograph: Copyright © 1997 by PhotoDisc, Inc. The cover photograph is of the strings over the bridge of an electric guitar. The vibration of a taut string is an example of an hyperbolic partial-differential equation that can be solved through numerical methods.

DERIVE is a registered trademark of Soft Warehouse, Inc. *Mathematica* is a registered trademark of Wolfram Research, Inc. *Mathematica* is not associated with Mathematica, Inc., Mathematica Policy Research, Inc., or Math Tech, Inc. Maple is a registered trademark of Waterloo Maple Software. MATLAB is a registered trademark of The MathWorks, Inc.

Library of Congress Cataloging-in-Publication Data

Gerald, Curtis F.
    Applied numerical analysis / Curtis F. Gerald, Patrick O.
Wheatley. — 6th ed.
        p.   cm.
    Includes bibliographical references and index.
    ISBN 0-201-87072-X
    1. Numerical analysis.   I. Wheatley Patrick O.   II. Title.
QA297.G47   1999
519.4—dc21   98-3887
                                                        CIP

2 3 4 5 6 7 8 9 10 DOC 010099

Preface

The success of previous editions of *Applied Numerical Analysis* has encouraged us to retain the same features that have made the book popular: ease of reading so that the instructor does not have to "interpret the book" for the student, many illustrative examples that often solve the same problem with different procedures to clarify the comparison of methods, many exercises from which the instructor may choose appropriately for the class, more challenging problems and projects that show practical applications of the material, and the use of motivating examples to introduce most chapters. Many numerical methods are covered—we expect the instructor to encourage discussion of which ones would be preferred in different situations.

Although it may be controversial, we have put much of the theoretical discussions into a separate section of each chapter. We feel that this allows the student to concentrate on the various methods and to compare them before delving into the underlying theory. Then, when the theory is presented, there is an automatic review of the previous material.

We continue to include chapter summaries to remind the student to review that part of the chapter that is not fully understood. This, together with short descriptions of the contents at the beginning of each chapter, puts the content in clear perspective.

Our coverage of partial-differential equations is more complete than in most other texts on numerical procedures. We intend this to be an easily understood introduction to a most important application of the techniques of numerical analysis. We also lead the student carefully into the finite element method, but this is only sufficient to lay a foundation for further study.

Applied Numerical Analysis is written as a text for sophomores and juniors in engineering, science, mathematics, and computer science. It should be a valuable source book for practicing engineers. Because of its coverage of many numerical methods, the text can serve as a valuable reference.

Although we assume that the student has a good knowledge of calculus, appropriate topics are reviewed in the context of their use. An appendix gives a summary of the most important items that are needed to develop and analyze numerical procedures. We purposely keep the mathematical notation simple for clarity. Furthermore, the answers to exercises marked with a ▶ are found in the back of the text.

Changes in This Edition

There have been substantial changes from the fifth edition. We have followed the recommendations of several reviewers to move the last chapter on approximation of functions to follow Chapter 3 on interpolation and curve fitting; this is a natural sequence. The part of that chapter that described getting Fourier coefficients by numerical integration is now placed within the chapter on numerical integration.

The treatment of computer algebra systems now uses MATLAB as the primary example. This program is compared to Maple and *Mathematica* in several chapters. How programs can be written to extend the power of such software is illustrated. We also provide an introduction to the capabilities of two advanced models of calculators, the HP-48G and the TI-92, whose abilities to graph functions and to solve equations can assist the student.

The treatment of the finite element method that previously appeared in several chapters has been combined into a chapter of its own. Several reviewers have commented that this is a good introduction to a most important topic for applied mathematicians.

Instead of providing computer programs written in a variety of languages, we have used only the two most important computer languages for numerical analysts—Fortran 90 and C. The number of programs has been significantly reduced because not many of the users of the book will write their own programs. However, we continue to believe that students will benefit from some exposure to computer programs that are easy to read.

We have updated the appendices, modified some of the exercises, and added some projects. A few topics have been eliminated or minimized. The use of difference tables based on evenly spaced data is abbreviated. The multivalued method for solving ordinary differential equations has been omitted. We have streamlined the discussion of hyperbolic partial-differential equations. The method of characteristics has been dropped.

The changes made to this edition have produced a strong text which gives a broad coverage of the field of numerical analysis.

Supplements for the Instructor

- The *Instructor's Solutions Manual* gives the answers to nearly all the exercises and to some of the applied problems and projects. Hints for other problems or projects are included. In this *Solutions Manual*, we suggest how the instructor can select from the text when there is insufficient time to cover all of it. Such selection may be necessary because of the broad coverage that we give to the topics of numerical analysis.
- Instructors who adopt this book may obtain electronic copies of the programs in this book, in ASCII format, as well as programs that implement many of the algorithms. This disk is in IBM-compatible form on 3.5-inch diskettes.
- These same programs are also available via e-mail. Detailed instructions on accessing the programs via FTP follow.

Access to Programs in ASCII

The sample programs presented in this book are available on-line in ASCII form for your personal, noncommercial use. These files can be accessed using an Internet browser by visiting the Addison Wesley Longman Web site at http://hepg.awl.com and choosing the keyword GERALD. Alternatively, the program files are available on host pop.awl.com via anonymous FTP. To access the files from an Internet host, enter the following commands:

% ftp pop.awl.com

Name: anonymous

Password: ⟨your e-mail address⟩

cd aw.mathematics

cd gerald6

ls (to display files to select programs)

A 3.5-inch program disk is also available for professors who adopt this text. To order the disk, e-mail Addison Wesley Longman at math@awl.com.

Acknowledgments

We especially want to thank our many students whose feedback has helped us to improve over previous editions. Our wives have been supportive during this revision and have helped us with proofreading. Many instructors have given valuable suggestions and constructive criticism. In particular we thank our colleague Ed Garner at California Polytechnic State University.

In addition, we mention those whose thorough reviews have helped make this edition better:

Colin J. Aro, *University of California*

Neil Berger, *University of Illinois at Chicago*

Josef Brody, *Concordia University*

George Fadel, *Clemson University*

Mayer Humi, *Worcester Polytechnic Institute*

Seppo Korpela, *Ohio State University*

Zhi-Kui Ling, *Michigan Technological University*

Ramon E. Moore, *Ohio State University*

Daoqi Yang, *Wayne State University*

Kathie Yerion, *Gonzaga University*

We also want to express our thanks to those at Addison Wesley Longman who have worked extensively with us to ensure the publication of another quality edition: Greg Tobin, Jennifer Albanese, Carolyn Lee-Davis, Rachel S. Reeve, Kim Ellwood, Karen Guardino, Michael Boezi, Barbara Atkinson, and Joe Vetere.

Table of Contents

6   Numerical Solution of Ordinary Differential Equations   448

7   Boundary-Value Problems   525

8   Parabolic and Hyperbolic Partial-Differential Equations   600

9   The Finite Element Method   652

Appendixes

APPLIED NUMERICAL ANALYSIS

SIXTH EDITION

0

Numerical Computing and Computers

1

way that computers store numbers and do arithmetic, are discussed in some detail.

0.6 Theoretical Matters

Points out the theoretical aspects of the numerical methods of the chapter. The present chapter includes only one method, interval halving, but later chapters will cover several related methods.

0.7 Parallel and Distributed Computing

Discusses how doing numerical analysis can be speeded up by using a computer system that has many separate processing units operating in parallel on the same problem, and when one might need this potential increase in speed. Some of the special problems that are involved in parallel processing are mentioned.

Chapter Summary

Gives you a checklist against which you can measure your understanding of the topics of this chapter. You should then restudy those sections where your comprehension is not complete.

0.1 Introduction

Numerical analysis is the development and study of procedures for solving problems with a computer. The term "algorithm" is used for a systematic procedure that solves a problem. A numerical analyst often is interested in determining which of several algorithms that can solve the problem is, in some sense, the most efficient. Efficiency may be measured by the number of steps in the algorithm, the computer time and amount of memory that is required, or in other ways. A major advantage for numerical analysis is that a numerical answer can be obtained even when a problem has no "analytical" solution. For example, the following integral, which gives the length of one arch of the curve of $y = \sin(x)$, has no closed form solution:

$$\int_0^\pi \sqrt{1 + \cos^2(x)} \, dx. \tag{0.1}$$

Numerical analysis can compute the length of this curve by standard methods that apply to essentially any integrand; there is never a need to make special substitutions or to do integration by parts to get the result. Further, the only mathematical operations required are addition, subtraction, multiplication, and division plus the making of comparisons. Because these simple operations are exactly the functions that computers can perform, computers and numerical analysis make a perfect combination.

It is important to realize that a numerical analysis solution is always numerical. Analytical methods usually give a result in terms of mathematical functions that can then be evaluated for specific instances. There is thus an advantage to the analytical result, in that the behavior and properties of the function are often apparent; this is not the case for purely numerical results. However, numerical results can be plotted to show some of the behavior of the solution.

Another important distinction is that the result from numerical analysis is an approximation, but results can be made as accurate as desired. (There are limitations to the achievable level of accuracy, because of the way that computers do arithmetic; we will explain these limitations later.) To achieve high accuracy, very many separate operations must be carried out, but computers do them so rapidly without ever making mistakes that this is no significant problem. Actually, evaluating an analytical result to get the numerical answer for a specific application is subject to the same errors.

The analysis of computer errors and the other sources of error in numerical methods is a critically important part of the study of numerical analysis. This subject will occur often throughout this book.

Here are some of the operations that numerical analysis can do and that are described in this book:

Solve for the roots of a nonlinear equation.

Solve large systems of linear equations.

Get the solutions of a set of nonlinear equations.

Interpolate to find intermediate values within a table of data.

Find efficient and effective approximations of functions.

Approximate derivatives of any order for functions even when the function is known only as a table of values.

Integrate any function even when known only as a table of values. Multiple integrals can also be obtained.

Solve ordinary differential equations when given initial values for the variables. These can be of any order and complexity.

Solve boundary-value problems and determine eigenvalues and eigenvectors.

Obtain numerical solutions to all types of partial differential equations.

Fit curves to data by a variety of methods.

0.2   Using a Computer to Do Numerical Analysis

Numerical methods require such tedious and repetitive arithmetic operations that only when we have a computer to carry out these many separate operations is it practical to solve problems in this way. A human would make so many mistakes that there would be little confidence in the result. Besides, the manpower cost would be more than could normally be afforded. (Once upon a time military firing tables were computed by hand using desk calculators, but that was a special case of national emergency before computers were available.)

Of course, a computer is essentially dumb and must be given detailed and complete instructions for every single step it is to perform. In other words, a computer program must be written so the computer can do numerical analysis. As you study this book you will learn enough about the many numerical methods available that you will be able to write programs to implement them. The specific computer language used is not very important; programs can be written in BASIC (many dialects), FORTRAN, Pascal, C, C++, Java, and even assembly language. Presently C and FORTRAN seem to be the most widely used by numerical analysts. Java is becoming popular in some universities. In this book, most of the methods will be described fully through pseudocode in such a form that translating this code into a program is relatively straightforward. A number of example programs are given to illustrate the process. These programs have been designed to be easy to read rather than to be examples of professional caliber.

Section 0.4 shows a program that carries out a very simple numerical procedure, the bisection method, that finds the solution of the equation $f(x) = 0$. Such a solution is called a "zero of the function," also a "root of the equation."

Actually, writing programs is not always necessary. Numerical analysis is so important that extensive commercial software packages are available. The IMSL (International Mathematical and Statistical Library) MATH/LIBRARY has hundreds of routines, of efficient and of proven performance, written in FORTRAN and C that carry out the methods. Recently, LAPACK (Linear Algebra Package) has been made available at nominal cost. This package of FORTRAN programs incorporates the subroutines that were contained in the earlier packages of LINPACK and EISPACK. Appendix C of this book gives information on these and other programs. The bimonthly newsletter of the Society for Industrial and Applied Mathematics (*SIAM News*) contains discussions and advertisements on some of the latest packages. A set of books, *Numerical Recipes,* lists and discusses numerical analysis programs in a variety of languages: FORTRAN, Pascal, and C.

One important trend in computer operations is the use of several processors working in parallel to carry out procedures with greater speed than can be obtained with a single processor. Some numerical analysis procedures can be carried out this way. Special programming techniques are needed to utilize these fast computer systems.

An alternative to using a program written in one of the higher-level languages is to use a kind of software sometimes called a computer algebra system (CAS). (This name is not very standardized and not too descriptive.*) This kind of program mimics the way humans solve mathematical problems. Such a program is designed to recognize the type of function (polynomial, transcendental, etc.) presented and then to carry out requested mathematical operations on the function or expression. It does so by looking up in tables the new expressions that result from doing the operation or by using a set of built-in rules. For example, a program can use the ordinary rules for finding derivatives, employ tables of integrals to do integrations, and factor a polynomial or expand a set of factors. These are only a few of the capabilities. If an analytical answer cannot be given, most of these programs allow the user to get an answer by numerical methods.

In connection with numerical analysis, an important feature of many such programs is the ability to write utility files that are essentially macros: A sequence of the built-in operations is defined to perform a desired larger task or one not inherent in the program. A

---

*Such programs are also called symbolic algebra systems.

succession of operations, each of which uses the results of the previous one—a procedure called *iteration*—is also possible. Many numerical analysis procedures are iterative.

Many such computer algebra systems are available, including *Mathematica,* MATLAB, Maple, DERIVE, and MacSyma. In this chapter, we explain how one of these, the MATLAB program, can be very useful in doing numerical analysis as well as some of the more analytic steps that are preliminary to the numerical method. In later chapters we will explain the use of two other computer algebra systems.

One special feature of most of these programs is their ability to carry out many operations with exact arithmetic. An interesting example is to see π displayed to 100 decimal places. Ordinarily, we must be satisfied with a limited number of digits of precision when a normal computer program is employed.

Of particular importance in using such programs is that the plotting of functions, even functions of two independent variables (which require a three-dimensional plot), is built in. In *Mathematica* this graphical capability is especially well developed.

Computer algebra systems, with their ability to perform mathematics symbolically and to carry out numerical procedures with extreme precision, would seem to be almost a preferred tool for the numerical analyst. However, for the large "real-life" problems that a professional analyst often deals with, they do not have the necessary speed. They are good for "small problems" and are an excellent learning environment. However, in many "real world" situations, such as weather prediction or the computation of space vehicle trajectories, the scientist/engineer will employ programs written in FORTRAN or C. And he or she will almost always use the proven routines of IMSL or LAPACK.

Another alternative to writing a computer program to do numerical analysis is to employ a spreadsheet program.

Doing Numerical Procedures with a Calculator

There are other devices than computers that can carry out numerical procedures. Today, so-called "hand-held calculators" have been developed with remarkable power. Typical of these advanced calculators are the TI-92 from Texas Instruments and the HP-48G from Hewlett-Packard. These machines have much of the power of a personal computer to do mathematics. They have limited memory, but built into them are special facilities of interest to the numerical analyst. Programs that are coded in their Read Only Memory (ROM) can plot functions in 2- and 3-dimensions, solve for roots of a nonlinear equation, solve systems of linear equations, manipulate matrices, do interpolation, differentiate and integrate (both numerically and analytically), and solve ordinary differential equations as well as perform mathematical and statistical operations. Expressions can include terms like sine, cosine, and other mathematical functions. They not only handle numeric expressions; symbolic manipulations are also possible.

The number of operations that can be performed is so great that the buttons on their keyboards must do multiple things. To make this possible, extra shift keys are provided. For example, the HP-48G has three shift keys: a green left shift key that is marked with a curved left-pointing arrow, a purple right shift key with a curved right-pointing arrow, and a shift key marked with the Greek letter α (and there are two more shifted α-keys). These shift keys, when pressed before one of the regular keys, allow one to enter as many as four kinds of inputs, three shifted inputs besides the regular one. But even this is not enough! Many selections from the keyboard cause a menu of choices to be exhibited on the screen where entered values are normally displayed. In some cases additional "pages" of menu labels are displayed when the NXT key is pressed. Further choices are made from selection windows ("forms") when certain operations are requested. A set of four arrow keys lets the user move a cursor.

The TI-92 calculator makes similar multiple use of its keys. It differs from the HP-48G in having a regular QWERTY keyboard as well as special function keys. Eight keys, similar to the function keys on computer keyboards, cause menus to drop down; choices can be made from these menus to do various operations. A key marked "APPS" drops down another menu. Some of the menus have submenus as well. The TI-92 is unique in having a cursor pad that is something like a track ball. Using it allows one to move the cursor in four directions (and even in diagonal directions in the geometrical applications that are built into this calculator). To make it easy to use when held in the hand, it provides triplicate ENTER keys and duplicate shift keys.

Because such calculators are so easily carried and are so relatively inexpensive, it is important that the reader of this book be aware of them and know something about how they can be employed. It is also important to understand that they can be programmed to do calculations in special fields of applications such as business and finance, aviation, the sciences, electronic and electrical engineering, mechanical and civil engineering, and so forth. Special software packages are available so that one does not have to personally do the programming.

Later in this chapter we will explain how these modern computer-like calculators can be used to carry out some of the procedures and methods that are the content of applied numerical analysis. You will be introduced to plotting a function, finding the minimum point of a curve, getting a derivative, and solving a nonlinear equation.

A student of numerical analysis will find such calculators of great value when numerical procedures are examined and compared. Still, as is the case with program algebra systems, they are not sufficiently speedy or powerful to be of much use except as a learning device.

0.3 A Typical Example

We will introduce the subject of numerical analysis by showing a typical problem solved numerically. If you worked for a mining company, Example 0.1 might be a problem you would be asked to solve.

EXAMPLE 0.1   *The Ladder in the Mine.* Two intersecting mine shafts meet at an angle of 123°, as shown in Fig. 0.1(a). The straight shaft has a width of 7 ft, and the entrance shaft is 9 ft wide. What is the longest ladder that can negotiate the turn at the intersection of the two shafts? Neglect the thickness of the ladder members, and assume the ladder is not tipped as it is

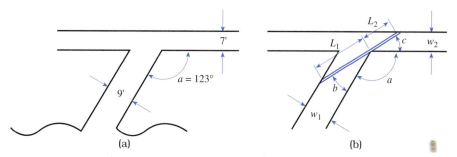

Figure 0.1

maneuvered around the corner. Provide for the general case in which the angle a is a variable as well as for the widths of the shafts.   ▲

Steps in Solving the Problem

Whenever a scientific or engineering problem is to be solved, there are four general steps to follow:

1. State the problem clearly, including any simplifying assumptions.
2. Develop a mathematical statement of the problem in a form that can be solved for a numerical answer. This process may involve, as in the present case, the use of calculus. In other situations, other mathematical procedures may be employed. When this statement is a differential equation, appropriate initial conditions and/or boundary conditions must be specified.
3. Solve the equation(s) that result from step 2. Sometimes the method will be algebraic, but frequently more advanced methods will be needed. This text may provide the method that is needed. The result of this step is a numerical answer or set of answers.
4. Interpret the numerical result to arrive at a decision. This will require experience and an understanding of the situation in which the problem is embedded. This interpretation is the hardest part of solving problems and must be learned on the job. This book will emphasize step 3 and will deal to some extent with steps 1 and 2, but step 4 cannot be meaningfully treated in the classroom.

The description of the problem has taken care of step 1. Now for step 2.

Here is one way to analyze our ladder problem. Visualize the ladder in successive locations as we carry it around the corner; there will be a critical position in which the two ends of the ladder touch the walls while a point along the ladder touches the corner where the two shafts intersect (see Fig. 0.1b). Let c be the angle between the ladder and the wall when in this critical position. It is usually preferable to solve problems in general terms, so we work with variables a, b, c, w_1, and w_2.

Consider a series of lines drawn in this critical position—their lengths vary with the angle c, and the following relations hold (angles are expressed in radian measure):

$$L_1 = \frac{w_1}{\sin(b)}, \qquad L_2 = \frac{w_2}{\sin(c)}$$

$$b = \pi - a - c, \tag{0.2}$$

$$L = L_1 + L_2 = \frac{w_1}{\sin(\pi - a - c)} + \frac{w_2}{\sin(c)}.$$

The maximum length of ladder that can negotiate the turn is the minimum of L as a function of the angle c. We hence set $dL/dc = 0$.

This is not a difficult derivative to do by hand, but MATLAB can do it for us. Let us see how this is done.

MATLAB is command-line driven, meaning that we type in commands to invoke operations. It is a large and powerful "computing environment," having nearly 500 separate commands or functions to manipulate objects, in addition to 43 mathematical operators and special characters. There are two "toolboxes" included in the Student Edition of MATLAB that manipulate expressions symbolically and for signal processing; these provide additional commands. Learning to use such a large system is not easy but is surely worth the effort. In this chapter we illustrate only a few features of MATLAB; later chapters will explore other aspects but we do not pretend to describe in this book all of its many commands that are useful for the student of numerical analysis.

It is easy to find the derivative of $L(c)$ in the ladder problem if we want to use the derivative to find the minimum through solving $dL/dc = 0$. We will get the derivative by first defining L as a function of c by entering, at the prompt,

```
L = 'w1/sin(pi − a − c) + w2/sin(c)'
```

MATLAB then echoes:

```
L =
w1/sin(pi − a − c) + w2/sin(c)
```

To get the derivative, we enter:

```
dL = diff(L,'c')
```

and see

```
dL =
−w1 / sin(a + c)^2 * cos(a + c) − w2 / sin(c)^2 * cos(c)
```

and now, entering "pretty" gives the answer as built-up fractions:

$$-\frac{w1\ \cos(a+c)}{\sin(a+c)^2} - \frac{w2\ \cos(c)}{\sin(c)^2}$$

It is important to realize that MATLAB is case-sensitive; L and l are different variables.

(We will not show MATLAB's prompt in our illustrations; with the educational version, it is EDU≫.)

With the derivative available, we could ask MATLAB to solve $dL/dc = 0$, but it is better to solve the problem through having it get the minimum of $L(c)$ directly. There is a built-in command to do this. First, however, we need to replace the constants in the expression with their numerical values. We do this with these three commands:

```
L = subs(L,9,'w1');
L = subs(L,7,'w2');
L = subs(L,2.147,'a');
```

where the trailing semicolons suppress printing the results. If we enter L to see what it now looks like, we see

```
L =
9 / sin (2147 / 1000 + c) + 7 / sin(c)
```

and we observe that MATLAB uses a ratio of integers and that it also does some simplification.

Now we can get the minimum point that we imagine must be about $c = 0.5$ (from looking at Fig. 0.1a) by using the MATLAB function fmin. However, we discover that the fmin function assumes that the independent variable is x, so we must make another substitution:

```
L = subs(L,'x','c');
```

Now we ask to search for the minimum point within the range $x = 0.3$ to $x = 0.6$:

```
Lmin = fmin (L,.3,.6)
```

and see

```
Lmin =
0.4676
```

and we know that the minimum point occurs at a value of $c = 0.4676$. We could substitute this into the equation for L to find the maximum length of the ladder, but we will instead use an alternative approach. We will ask MATLAB to draw the curve of L versus c and locate the minimum by visually inspecting that curve. So, we enter:

```
fplot (L,[.4 .5]); grid
```

and get Fig. 0.2. From this it is apparent that approximate values for L and c are 33.42 and 0.467, respectively. From the nature of the problem, we feel that this answer is of sufficient precision.

There are other approaches to solving the ladder problem that we do not consider here. The next chapter will cover the use of computer algebra systems to solve equations in more detail.

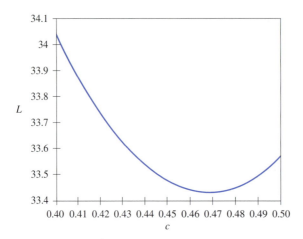

Figure 0.2

Getting the Solution with a Calculator

Both the TI-92 and the HP-48G can find the solution to the ladder problem. Our attack here in solving the problem will be similar to that with MATLAB. We will plot the equation for L as a function of c and find the coordinates of the minimum point. However, with these calculators, there is an operation to get this exactly, rather than getting an approximate answer by inspecting the graph.

The strategy is simple: (1) plot the curve for L versus c, and (2) find the values of c and L at the minimum point of this plot. (There are alternative ways to carry out the strategy.) First we use the TI-92. In the description, underlined commands are invoked through shifted keys.

1. Enter the equation for L and plot it. Here are the steps to do this:
    (a) Make sure that variable x is not defined with a value: x, ENTER
        If the display is just x, it is not defined with a value. If a number appears: VAR-LINK, x, F1, 1, ENTER, ESC and we are back at HOME with x deleted.
    (b) Clear the screen: F1, 8
    (c) Put into equation entry mode: Y = [we see y1(x) = on the entry line]
    (d) Enter the expression for L ($x = c$, $y = L$): 9 ÷ SIN(π − 2.147 − x) + 7 ÷ SIN(x), ENTER (be sure you are in radian mode)
    (e) Set window parameters: WINDOW, change values to xmin = 0, xmax = 1, xscl = 1, ymin = 0, ymax = 50, yscl = 1, xres = 2
    (f) Do the graph: GRAPH [we see a curve that is concave upward]
2. Find the minimum on the curve.
    (a) Turn on trace mode: F3
    (b) Prepare to get minimum: F5, 3
    (c) Set lower bound: Move cursor to left of low point of curve, press ENTER

  (d) Set upper bound: Move cursor to right of low point, press ENTER
  (e) Read the coordinates at the minimum: xc = .467607, yc = 33.4247

and we see that, at an angle of 0.467607 radians, the length of the ladder is 33.4247 feet. We note that such great precision is really not necessary in this example. (Who would care if the length of the ladder is known only as 33.5 feet with a possible error of an inch or two?) However, the more precise answer is found just as quickly; also, sometimes great accuracy is important. Actually, the TI-92 has found the answer to 12 digits of precision. The number of digits displayed is determined by the Display Digits setting in MODE.

  We now will repeat this same procedure, but instead will use the HP-48G calculator. As before, we plot the expression for L, the length of the ladder, against c, the angle that the ladder makes with the shaft. We then determine the values of L and c at the minimum of this curve through a built-in operation to find the "extremum" (here the minimum) of a plot.

  Here are the successive steps to solve the problem using the HP calculator:*

  1. Make sure that memory does not already hold a value for variable c. Press VAR and see if a variable named C appears in the menu. If it does, enter 'C', PURG and variable C is deleted.

  2. Open the plot application: PLOT. If Type is not Function, CHOOS, select Function, OK

  3. Move the cursor to the EQ: field and enter the expression that defines L: $9 \div \text{SIN}(\pi - 2.147 - c) + 7 \div \text{SIN}(c)$, then press ENTER (or OK) (be sure you are in radian mode)

  4. The INDEP: field will be highlighted; change the variable to c by entering c, press ENTER (or OK)

  5. Set H-VIEW to 0.3 0.6 by entering .3, ENTER, .6, ENTER

  6. Turn AUTOSCALE on by pressing CHK

  7. Erase any previous drawing with ERASE

  8. Draw the plot with DRAW

  9. The screen now shows a curve that is concave upward. Move the cursor (a small cross) with the arrow keys until it is near the minimum of the curve.

  10. Press FCN then press EXTR. The coordinates at the minimum point are displayed: (.4676, 33.4247)
    (The precision of the answer depends on the settings in MODES. This is with NUMBER FORMAT of FIX 4. The answer is actually obtained to 12 digits of precision, just as with the TI calculator.)

  11. Exit the application with CANCEL, NXT, CANCL to find the coordinates of the minimum point on the stack. If you want more precision, press VIEW and see all 12 digits!

---

*In the description of operations with the HP-48G, an underline indicates a command that is invoked with one of the shift keys; a double underline indicates a command from a menu on the screen or from a selection list on a form. Sometimes the NXT key is needed to see the desired menu command.

Getting the Answer from $dL/dc = 0$

Finding the minimum point from the plot of L versus c is probably the best way to solve the ladder problem, but both the TI-92 and the HP-48G can do it through solving $dL/dc = 0$. Here's how to do it this second way using the TI-92:

1. See if x is defined with a value by doing this: x, ENTER. If a number appears, x is so defined. If so, delete it by: VAR-LINK, x, F1, 1, ENTER, ESC
2. The equation for $L(c)$ should be in memory as y1(x). Ascertain that it is by: Y=. It should be there as y1 = ... If not, enter the equation again as y1(x) = 9 ÷ SIN(π − 2.147 − x) + 7 ÷ SIN(x), ENTER, HOME
3. Enter: d (y1(x),x), ENTER [See the derivative—and be sure to enter d from the keyboard, not typed in!]
4. Highlight the derivative expression, ENTER [Brings it to the command line]
5. Store this as d1 by entering: STO, d1, ENTER
6. Solve for dL/dc = 0: Enter nSolve(d1 = 0, x) | x > .3 and x < .6, ENTER [See answer: .4676068 . . .]
   (The last part of the command specifies the interval for x)
7. Store this value in x: Highlight the answer, ENTER, STO, x, ENTER
8. Find the value of L at the minimum point: Enter y1(x), ENTER [See 33.4247 . . .] and the problem is solved.

Here is how we do the same with the HP-48G:

1. Clear the memory of the value of c (gotten as the minimum point of the plot) by MEMORY, select 'c', PURG, OK (Alternatively, and quicker: c (from VAR menu), PURG)
2. Set up to get the derivative: SYMBOLIC, select Differentiate, OK
3. The equation for L as a function of c is already in memory from the plot operations. We can enter it into the EXPR field with CHOOS, select the equation, OK
4. Enter c as the differentiation variable in the VAR field.
5. Make sure that the RESULT field is SYMBOLIC. (If not, CHOOS, select SYMBOLIC, OK), press OK. We see the expression for dL/dc on the stack. (We can simplify this but that is not necessary.)
6. Put the expression for dL/dc into memory: Enter a name to the stack, say 'D', ENTER, STO
7. Set up to solve dL/dc = 0: SOLVE, select Solve equation, OK
8. Recall dL/dc from memory: CHOOS, select D:, OK
9. Enter an approximate value for c, say 0.5, move to the c: field, SOLVE. [See the solution on the screen, it is the same as before: .467606898925]
10. Put the answer on the stack with OK
11. Put the expression for L(c) on the stack: MEMORY, select expression for L(c), OK, RCL, OK (It is quicker to get it from the VAR menu.)
12. Evaluate with the value for c: →NUM [See the same value for L as before, 33.4247478486]

Conclusion: It is much easier to solve the ladder problem by finding the minimum point of the graph of $L(c)$.

0.4 Implementing Bisection

In Section 0.2 we mentioned the bisection method. In brief, this solves for a root of $f(x) = 0$ by starting with two values that enclose or bracket the root. (We know that a root is bracketed (enclosed) if the function changes sign at the endpoints.) We are sure that there is at least one root in the interval if $f(x)$ is continuous. We then halve the interval and test to see in which of these subintervals there is a sign change. The bisection method can be thought of as a kind of binary search. This is more efficient than a stepwise search, which can find where $f(x)$ changes signs by stepping through the enclosing interval by smaller and smaller steps.

We now demonstrate this technique in an example. Figure 0.3 shows graphically how successive values converge on a root of $f(x)$ when we begin with a pair of values that bracket the root. We see that x_3 is halfway between x_1 and x_2, and x_4 is halfway between x_2 and x_3. We always take the next x-value as the midpoint of the last pair that brackets the root: These values bracket the root when $f(x)$ changes sign at the two points. As the process is continued, it is clear that we will converge to the root.

We intend to write a computer program to carry out this algorithm. (*Algorithm* means a rule or prescription for carrying out a certain computation.) Before we do so, we will express the algorithm more specifically in *pseudocode,* a way to express the steps of a program in generic form; it describes the program logic in language-independent terms.

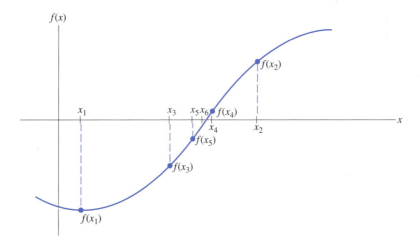

Figure 0.3

An Algorithm for Halving the Interval (Bisection)

To determine a root of $f(x) = 0$ that is accurate within a specified tolerance value, given values x_1 and x_2 such that $f(x_1) * f(x_2) < 0$,

REPEAT
    Set $x_3 = (x_1 + x_2)/2$.
    IF $f(x_3) * f(x_1) < 0$:
        SET $x_2 = x_3$.
    ELSE Set $x_1 = x_3$.
    ENDIF.
UNTIL $(|x_1 - x_2| < 2 *$ tolerance value) or $f(x_3) = 0$.

The final value of x_3 approximates the root; it is in error by not more than $(1/2)|x_1 - x_2|$.

Note. The method may give a false root if $f(x)$ is discontinuous on $[x_1, x_2]$.

It is not difficult to write a computer program when we have the algorithm expressed in pseudocode. Even though the BASIC programming language is not often used by numerical analysts, it is certainly easy to read. Figure 0.4 exhibits a BASIC program that does bisection for the ladder in the mine problem. (The version of BASIC is QuickBasic 4.5. This software is compatible with Microsoft's QBASIC.)* Figure 0.5 shows the output when the program is run. Observe that the root, 0.4677734, is near 0.4676, which we found before.

Knowing that the critical angle is 0.4678 radian (26.8°), we can compute the maximum length of ladder from Eq. 0.2. Using this relation, we find that $L = 33.42$ ft. It would appear that ladders up to this length will be adequate for use in the mine.

In the example program of Fig. 04, a function statement is used to define $f(x)$. It might be preferable to define $f(x)$ in a separate procedure and have the code for bisection call the procedure to compute values of $f(x)$. This step would make the code a general-purpose program; only the function defining procedure would have to be changed.

The program adds a test, although the pseudocode does not include it, to terminate the computations if more than a certain number of iterations have been done. It is always best to include this test to avoid an endless loop. (This program will be in an endless loop if the tolerance is set too small, an effect of precision in the computer, a topic that we will discuss later.)

We will show in later chapters how programs like this one for the bisection method can be written in computer algebra systems. One can thus supplement the routines for root finding that are built into them.

------

*See Exercise 40 in connection with this program.

```
`  USES BISECTION METHOD TO FIND A ROOT OF F(X) = 0
`  GETS X1, X2, TOLERANCE VALUE FROM USER
`  CHANGE DEF FNX() TO MATCH REQUIRED F(X)
DEF FNX (C) = 9 * COS(A + C) / SIN(A + C) ^ 2 + 7 * COS(C) / SIN(C) ^ 2
A = 123 * 3.14159 / 180
LIMIT = 50
INPUT "ENTER VALUES FOR X1, X2, TOLERANCE ", X1, X2, TOL
`  DO A HEADING
PRINT "ITER NO", "X1", "X2", "X3", "F(X3)": PRINT
ITER = 1
DO
   F1 = FNX(X1)
   F2 = FNX(X2)
   IF F1 * F2 > 0 THEN
      PRINT "VALUES DO NOT BRACKET A ROOT": EXIT DO
   END IF
   X3 = (X1 + X2) / 2
   F3 = FNX(X3)
   PRINT ITER, X1, X2, X3, F3
   IF F3 * F1 < 0 THEN X2 = X3 ELSE X1 = X3
   ITER = ITER + 1
LOOP UNTIL ABS(X1 - X2) / 2 < TOL OR F3 = 0 OR ITER > LIMIT
```

Figure 0.4 A program for bisection

ITER NO	X1	X2	X3	F(X3)
1	.4	.5	.45	4.661504
2	.45	.5	.475	−1.893359
3	.45	.475	.4625	1.364828
4	.4625	.475	.46875	−.2675005
5	.4625	.46875	.465625	.5476606
6	.465625	.46875	.4671875	.1398538
7	.4671875	.46875	.4679688	−6.388076E-02
8	.4671875	.4679688	.4675781	3.797662E-02
9	.4675781	.4679688	.4677734	−1.295546E-02

Figure 0.5 Program output for $x_1 = 0.4$, $x_2 = 0.5$, tolerance = 1E-4

0.5 Computer Arithmetic and Errors

We have mentioned the likelihood of errors due to the way that computers perform arithmetic computations. For example, the bisection program can go into an endless loop if the tolerance values are set very small. It is time that we discuss this most important subject of errors in numerical analysis. When we do numerical analysis there are several possible sources of error in addition to those due to the inexact arithmetic of the computer.

Truncation Error

The term *truncation error* refers to those errors caused by the method itself (the term originates from the fact that numerical methods can usually be compared to a truncated Taylor series). For instance, we may approximate e^x by the cubic

$$p_3(x) = 1 + \frac{x}{1!} + \frac{x^2}{2!} + \frac{x^3}{3!}.$$

However, we know that to compute e^x really requires an infinitely long series:

$$e^x = p_3(x) + \sum_{n=4}^{\infty} \frac{x^n}{n!}.$$

We see that approximating e^x with the cubic gives an inexact answer. The error is due to truncating the series and has nothing to do with the computer or calculator. For iterative methods, this error can usually be reduced by repeated iterations, but because life is finite and computer time is costly, we must be satisfied with an approximation to the exact analytical answer.

Round-Off Error

All computing devices represent numbers, except for integers, with some imprecision. (MATLAB and similar programs can work with integers and rational fractions to achieve results of higher precision.) Digital computers will nearly always use floating-point numbers of fixed word length; the true values are not expressed exactly by such representations. We call the error due to this computer imperfection the *round-off* error. If numbers are rounded when stored as floating-point numbers, the round-off error is less than if the trailing digits were simply chopped off. We discuss this in more detail later in the chapter.

Error in Original Data

Real-world problems, in which an existing or proposed physical situation is modeled by a mathematical equation, will nearly always have coefficients that are imperfectly known. The reason is that the problems often depend on measurements of doubtful accuracy. Further, the model itself may not reflect the behavior of the situation perfectly. We can do nothing to overcome such errors by any choice of method, but we need to be aware of such uncertainties; in particular, we may need to perform tests to see how sensitive the results are to changes in the input information. Because the reason for performing the computation is to reach some decision with validity in the real world, sensitivity analysis is of extreme importance. As Hamming says, "the purpose of computing is insight, not numbers."

Blunders

You will likely always use a computer or at least a programmable calculator in your professional use of numerical analysis. You will probably also use such computing tools extensively while learning the topics covered in this text. Such machines make mistakes very infrequently, but because humans are involved in programming, operation, input preparation, and output interpretation, blunders or gross errors do occur more frequently than we like to admit. The solution here is care, coupled with a careful examination of the results for reasonableness. Sometimes a test run with known results is worthwhile, but it is no guarantee of freedom from foolish error. When hand computation was more common, check sums were usually computed—they were designed to reveal the mistake and permit its correction.

Propagated Error

Propagated error is more subtle than the other errors. By *propagated error* we mean an error in the succeeding steps of a process due to an occurrence of an earlier error—such error is in addition to the local errors. It is somewhat analogous to errors in the initial conditions. Some root-finding methods find additional zeros by changing the function to remove the first root; this technique is called *reducing* or *deflating the equation.* Here the reduced equations reflect the errors in the previous stages. The solution, of course, is to confirm the later results with the original equation.

In examples of numerical methods treated in later chapters, propagated error is of critical importance. If errors are magnified continuously as the method continues, eventually they will overshadow the true value, destroying its validity; we call such a method *unstable.* For a *stable* method—the desirable kind—errors made at early points die out as the method continues. This issue will be covered more thoroughly in later chapters.

Each of these types of error, while interacting to a degree, may occur even in the absence of the other kinds. For example, round-off error can occur even if truncation error is absent, as in an analytical method. Likewise, truncation errors can cause inaccuracies even if we can attain perfect precision in the calculation. The usual error analysis of a numerical method treats the truncation error as though such perfect precision did exist.

Floating-Point Arithmetic

To examine round-off error in detail, we need to understand how numeric quantities are represented in computers. In nearly all cases, numbers are stored as floating-point quantities, which are very much like scientific notation.* For example, the fixed-point number

---

*Another name often used for floating-point numbers is *real numbers,* but here we reserve the term *real* for the continuous (and infinite) set of numbers on the "number line." When printed as a number with a decimal point, it is called *fixed point.* The essential concept is that these numbers are in contrast to integers.

13.524 is the same as the floating-point number .13524 * 10^2, which is often displayed as .13524E2. Another example: -0.0442 is the same as $-.442$E-1.

Different computers use slightly different techniques, but the general procedure is similar. In computers, floating-point numbers have three parts: the *sign* (which requires one bit); the *fraction part*—often called the *mantissa* but better characterized by the name *significand;* and the *exponent part*—often called the *characteristic*. The three parts of numbers have a fixed total length that is often 32 or 64 bits (sometimes even more). The fraction part uses most of these bits, perhaps 23 to as many as 52 bits, and that number determines the precision of the representation. The exponent part uses 7 to as many as 11 bits, and this number determines the range of the values.

Computers represent their floating-point numbers in the general form

$$\pm .d_1 d_2 d_3 \ldots d_p * B^e,$$

where the d_i's are digits or bits with values from zero to $B - 1$ and

B = the number base that is used, usually 2, 16, or 10.

p = the number of significand bits (digits), that is, the precision.

e = an integer exponent, ranging from E_{min} to E_{max}, with the values going from negative E_{min} to positive E_{max}.

The significand bits (digits) constitute the fractional part of the number. In almost all cases, numbers are normalized, meaning that the fraction digits are shifted and the exponent adjusted so that d_1 is nonzero. Zero is a special case; it usually has a fraction part with all zeros and a zero exponent. This kind of zero is not normalized and never can be.

In hand calculators, the base B is usually 10; in computers the base is often 2, but sometimes a base of 16 is used. Most computers permit two or even three types of numbers: *single precision,* which is equivalent to 6 to 7 significant decimal digits; *double precision,* equivalent to 13 to 14 significant decimal digits; and *extended precision,* which may be equivalent to 19 to 20 decimal digits.

The two advanced models of calculators that we discuss, the HP-48G and the TI-92, have precisions equivalent to double precision. The TI-92 stores numbers internally with 14 significant digits but the display is rounded to a maximum of 12. For the HP-48G, the precision is 12 significant digits. Both have an exponent range of ± 999.

Examples of Numbers as Represented in a Computer

Working with binary or hexadecimal digits is awkward, so we will begin with the more familiar base 10 in some examples. Suppose $B = 10$ and $p = 4$. Then these numbers would be represented as

$$27.39 \rightarrow +.2739 * 10^2;$$
$$-0.00124 \rightarrow -.1240 * 10^{-2};$$
$$37000 \rightarrow +.3700 * 10^5.$$

Observe that we have *normalized* the fractions—the first fraction digit is nonzero. If the base is 2, that means that the first fraction bit is always 1. Some systems take advantage of that fact and do not store the first bit, thus gaining one bit of precision. When this first bit is suppressed, it is referred to as a *hidden bit.*

Because the number of bits used for a floating-point number is fixed, there is a finite number of distinct values in the computer's number system—in great contrast to the real number system. Thus there are gaps within the computer's number system. To illustrate, let us use a greatly simplified case where $B = 2$, $p = 2$, and $-2 \le e \le 3$. For this system, all the normalized numbers would be of either of these forms:

$$\pm .10_2 * 2^e \quad \text{or} \quad \pm .11_2 * 2^e, \quad -2 \le e \le 3.$$

Because the binary fractions $.10_2 = \frac{1}{2}$ and $.11_2 = \frac{1}{2} + \frac{1}{4} = \frac{3}{4}$, these numbers range from $-.11_2 * 2^3 = -6$ to $+.11_2 * 2^3 = +6$. Here is a list of all the positive numbers:

$$.10_2 * 2^{-2} = \frac{1}{8}, \qquad\qquad .11_2 * 2^{-2} = \frac{3}{16},$$

$$.10_2 * 2^{-1} = \frac{1}{4}, \qquad\qquad .11_2 * 2^{-1} = \frac{3}{8},$$

$$.10_2 * 2^0 = \frac{1}{2}, \qquad\qquad .11_2 * 2^0 = \frac{3}{4},$$

$$.10_2 * 2^1 = 1, \qquad\qquad .11_2 * 2^1 = \frac{3}{2},$$

$$.10_2 * 2^2 = 2, \qquad\qquad .11_2 * 2^2 = 3,$$

$$.10_2 * 2^3 = 4, \qquad\qquad .11_2 * 2^3 = 6.$$

The following diagram shows the distribution of these positive values. The negative values will be similarly distributed. Gaps are present that are uneven in size.

These interior gaps are in addition to the underflow and overflow gaps that we will describe.

Observe that zero is not present. To represent zero, we define a special case: Zero is a value with all zeros in the fraction (it is not normalized) with a zero exponent. This corresponds to the standard in almost all computer systems.

Pay particular attention to the important consequence of these gaps in all computer number systems. In the tiny system illustrated in the preceding diagram, the value 2.3 will be stored as 2; to the computer, the two values 2.2 and 2.4 are precisely identical—this is true for all values greater than 2 and up to 3 if the rounding is done by chopping. This explains why the bisection program will loop indefinitely if the tolerance value is too small. (In one

Table 0.1   Comparative methods for storing floating-point numbers

Method	Total length	Bits in fraction	Bits in exponent	Bias value‡	Base	Max. exponent	Min. exponent	Largest number	Smallest number	Approx. precision, no. decimal digits
IEEE										
single	32	23*	8	127	2	127	−126	1.701E38	1.755E-38	7
double	64	52*	11	1023	2	1023	−1022	8.988E307	2.225E-308	16
extended	80	64	15	16383	2	16383	−16382	6E4931	3E-4931	19
VAX										
single	32	23	8	127	2	127	−127	1.701E38	5.877E-39	7
double-1	64	55	8	127	2	127	−127	1.701E38	5.877E-39	16
double-2	64	52	11	1023	2	1023	−1023	8.988E307	1.123E-308	15
extended	128	112	15	16383	2	16383	−16383	6E4931	1E-4931	33
IBM										
single	32	24	7	63	16	63	−64	7.237E75	8.636E-78	7
double	64	56	7	63	16	63	−64	7.237E75	8.636E-78	16
extended	128†	112	7	63	16	63	−64	7.237E75	8.636E-78	33

*Plus 1 "hidden bit."
†8 bits not used.
‡The bias value is added to the stored exponent so that negative exponents are stored as positive integers.

instance, after 21 iterations, $x_1 = 0.4677237$, $x_2 = 0.4677238$, and $x_3 = 0.4677238$; the computer could not distinguish between x_2 and x_3, so there can be no further approach to the true value of the root.)

Actual Computer Number Systems

In actual computer systems of floating-point numbers, the format is similar but not as easy to comprehend because the base is not 10. Table 0.1 compares the composition of three different systems. The fact that these systems are not identical means that different computers can give different results from the same set of computations, thus making programs not always readily transportable. Figure 0.6 illustrates the ranges of numbers representable in the IEEE standard, shown in Table 0.1.

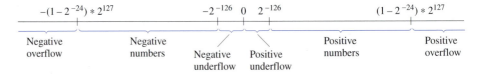

Figure 0.6 Numbers in the IEEE standard format

Cleve Moler, the chairman and co-founder of MathWorks Inc., the company that produces and distributes MATLAB, has a most informative article about floating-point numbers in the Fall 1996 issue of *MATLAB News and Notes.* The article not only tells about the IEEE standard for floating-point numbers and machine eps (which we define shortly) but gives some striking examples of how round-off can produce strange results for a problem.

Arithmetic Accuracy in Computers

We again avoid the confusion of thinking in number bases other than 10, by discussing accuracy of floating-point operations with examples using normalized base-10 numbers. The bases used in computers behave analogously. To keep it simple, we assume only three digits in the fraction part and one decimal digit for the exponent; $B = 10$, $p = 3$, $-9 \le e \le 9$. We compare rounding and chopping of the results.

When two floating-point numbers are added or subtracted, the digits in the number with the smaller exponent must be shifted to align the decimal points (normalization is forgotten in the adder). This shifting can lose some of the significant digits of one of the values. The result may also need to be shifted and the exponent adjusted to normalize it. Some computers automatically round the final answer, others just chop off extra digits beyond the precision of the system.

When multiplied (divided), the fractions are just multiplied (divided) and the exponents added (subtracted). The result is normalized.

Some examples follow.

EXAMPLE 0.2 Compute $1.37 + .0269 = .137 * 10^1 + .269 * 10^{-1}$.

$$
\left.\begin{array}{r}
.137 \quad * \ 10^1 \\
+ \ .00269 * \ 10^1
\end{array}\right\} \quad \text{Align decimal points.}
$$

$$.13969 * 10^1$$

$$
\begin{array}{ll}
\text{Chop} \rightarrow & .139 * 10^1. \\
\text{Round} \rightarrow & .140 * 10^1.
\end{array}
$$

▲

Rounding involves an extra operation, which may be done either in hardware or through a software routine.

EXAMPLE 0.3 Compute $4850 - 4820 = .485 * 10^4 - .482 * 10^4$.

$$
\begin{array}{r}
.485 * 10^4 \\
- \ .482 * 10^4 \\
\hline
.003 * 10^4 \\
.300 * 10^2 \quad \text{Normalize.}
\end{array}
$$

$$
\begin{array}{ll}
\text{Chop} \rightarrow & .300 * 10^2. \\
\text{Round} \rightarrow & .300 * 10^2.
\end{array}
$$

▲

In Example 0.3 observe that there is really only one digit of accuracy in the result even though the difference is represented as though the trailing zeros were significant. This loss of accuracy when two nearly equal numbers are subtracted is a major source of error in floating-point operations. In this case, rounding and chopping reach the same final answer.

EXAMPLE 0.4   Compute $3780 - .321 = .378 * 10^4 - .321 * 10^0$.

$$\left.\begin{array}{r} .378 * 10^4 \\ - .0000321 * 10^4 \end{array}\right\} \quad \text{Align decimal points.}$$

$$.3779679 * 10^4$$

$$\begin{array}{rl} \text{Chop}\rightarrow & .377 * 10^4. \\ \text{Round}\rightarrow & .378 * 10^4. \end{array}$$ ▲

In Example 0.4 shifting to align the decimal points has completely lost the significant digits of the subtrahend!

EXAMPLE 0.5   Compute $403000 * .0197 = .403 * 10^6 * .197 * 10^{-1}$.

$$\left.\begin{array}{r} .403 \\ * .197 \end{array}\right\} \quad \text{Multiply fractions.} \qquad \left.\begin{array}{r} 6 \\ -1 \end{array}\right\} \quad \text{Add exponents.}$$

$$.079391 \qquad\qquad\qquad 5$$

$$\begin{array}{ll} .079391 * 10^5 & \text{Combine fraction and exponent.} \\ .79391 * 10^4 & \text{Normalize.} \end{array}$$

$$\begin{array}{rl} \text{Chop}\rightarrow & .793 * 10^4. \\ \text{Round}\rightarrow & .794 * 10^4. \end{array}$$ ▲

In multiplying two n-digit numbers, a product with $2n$ digits results. Internally, double-length registers are used (but this is often accomplished by joining two single-length registers). The final result is truncated to the length of a single register.

EXAMPLE 0.6   Compute $.0356/1560 = .356 * 10^{-1}/.156 * 10^4$.

$$\left.\begin{array}{r} .356 \\ \div .156 \end{array}\right\} \quad \text{Divide fractions.} \qquad \left.\begin{array}{r} -1 \\ -4 \end{array}\right\} \quad \text{Subtract exponents.}$$

$$2.28205 \qquad\qquad\qquad -5$$

$$\begin{array}{ll} 2.28205 * 10^{-5} & \text{Combine fraction and exponent.} \\ .228205 * 10^{-4} & \text{Normalize.} \end{array}$$

$$\begin{array}{rl} \text{Chop}\rightarrow & .228 * 10^{-4} \\ \text{Round}\rightarrow & .228 * 10^{-4} \end{array}$$

Double-length registers are also involved in doing division. ▲

The time required to perform the different arithmetic operations will vary. Multiplication may be several times (2.5 to 10 times) slower than addition or subtraction. (Subtraction is equivalent to addition because it is done by adding two's-complements.) Floating-point division may be the slowest of all (4 to 25 times that for addition). These statements apply to computers of the 1980s. In some of these early computers, multiplication and division were performed through software routines—the timing differences were even greater in these machines.

In today's personal computers there is very little difference between the speeds for multiply/divide and add/subtract because of the very powerful math coprocessors that are built in. However, the historical difference in speeds that has been the case in the past has led numerical analysts to count only the number of multiply/divides in measuring the efficiency of an algorithm. It is more correct to count all arithmetic operations (including compares).

Errors in Converting Values

The numbers that are input to a computer are ordinarily base-10 values. Thus the input must be converted to the computer's internal number base, normally base 2. This conversion itself causes some errors. Terminating decimal fractions can be nonterminating in base 2: $(0.6)_{10} = (0.100110011001 \ldots)_2$. Because of the gaps in the computer's number system, we have to map the infinite set of real numbers into a finite set of computer numbers. For Examples 0.2 to 0.6, which have three decimal-digit fractions, there are only 900 different fractional values—all the mathematical numbers between 0.1 and 1.0 must be translated to one of these 900 values. In each decade, as represented by a constant value of the exponent, there are also only 900 values. The spacing between values in the different decades is therefore different, as we have said.

The representation of numbers smaller than $\frac{1}{8}$ in magnitude in the first base-2 example system is impossible. Such an input results in *exponent underflow.* This same phenomenon occurs with all computer number systems (see Fig. 0.6). The range of underflow values depends on the number of fraction bits and the number base. Some computers distinguish between *positive underflow* and *negative underflow.* In some systems, when this occurs, either from an attempt to input a value or from an intermediate computation, the value is replaced by zero (and, we hope, an error message is generated).

Similar things happen with number values larger than those that can be represented. Our simplified system has *exponent overflow* when values larger in magnitude than 6 are encountered. With actual computer number systems, much larger values can be accommodated, but there is always a limit. Some systems replace the number with the largest possible value when overflow occurs.

Machine eps

One important measure in computer arithmetic is how small a difference between two values the computer can recognize. This quantity is termed the *computer eps,* where "eps" is a shortening of the Greek letter epsilon. This measure of *machine accuracy* is standardized by finding the smallest floating-point number that, when added to floating-point 1.000, produces a result different from 1.000. In the exercises, you are asked to determine the size of the eps of machines available to you. This determination may even depend on the computer language in which a program is written, as some languages use only double precision for their floating-point numbers.

Peculiar things happen in floating-point arithmetic. For example, adding 0.001 one thousand times may not equal 1.0 exactly. In some instances, multiplying a number by unity does not reproduce the number. In many computations, changing the order of the calculations will produce different results.

Absolute Versus Relative Error, Significant Digits

The accuracy of any computation is always of great importance. There are two common ways to express the size of the error in a computed result: *absolute error* and *relative error*. The first is defined as

$$\text{absolute error} = |\text{true value} - \text{approximate value}|.$$

A given size of error is usually more serious when the magnitude of the true value is small. For example, 1036.52 ± 0.010 is accurate to five significant digits and may be of adequate precision, whereas 0.005 ± 0.010 is a clear disaster.

Using relative error is a way to compensate for this problem. Relative error is defined as

$$\text{relative error} = \frac{\text{absolute error}}{|\text{true value}|}$$

The relative error is more independent of the scale of the value, a desirable attribute. When the true value is zero, the relative error is undefined. It follows that the round-off error due to a finite length of the fractional part of floating-point numbers is more nearly constant when expressed as relative error than when expressed as absolute error. Most people define these errors in terms of magnitudes, in which case the error is always a positive quantity.

Another term that is commonly used to express accuracy is *significant digits,* that is, how many digits in the number have meaning. Extra digits that show up when numbers are shifted to normalize them are meaningless; this is a real problem when there are trailing zeros in a number. We may not know whether they are really zeros or just fillers.

A more formal definition of significant digits follows.

1. Let the true value have digits $d_1 d_2 \ldots d_n d_{n+1} \ldots d_p$.
2. Let the approximate value have $d_1 d_2 \ldots d_n e_{n+1} \ldots e_p$.

where $d_1 \neq 0$ and with the first difference in the digits occurring at the $(n + 1)$st digit. We then say that (1) and (2) agree to n significant digits if $|d_{n+1} - e_{n+1}| < 5$. Otherwise, we say they agree to $n - 1$ significant digits.

EXAMPLE 0.7   Let the true value = 10/3 and the approximate value = 3.333.

The absolute error is $0.000333 \ldots = 1/3000$.

The relative error is $(1/3000)/(10/3) = 1/10000$.

The number of significant digits is 4.   ▲

0.6 Theoretical Matters

Any user of a mathematical procedure should be concerned with its theoretical under-pinnings because they explain the limitations of the procedure and the conditions that must be true for the procedure to produce reliable results. Although this book mentions theory in the body of chapters, it will reserve a fuller discussion of theoretical matters for special sections like this one. In this chapter, only one method has been described—bisection (interval halving)—so this section is short. However, it introduces the kind of questions about theory that generally should be asked when a numerical method is employed. In addition, this section discusses some general points on how theory will be presented and some notations of importance.

When theory is to be discussed, two questions arise:

1. Where do we start—what background is assumed? Does every definition and postulate have to be stated before the pertinent theorems are developed? In this book we assume that the reader has a background in calculus and knows the definitions of that body of mathematics.
2. How are theorems presented? Is it better to use the rather condensed notations of mathematicians and their special symbols or to use language and style that is more accessible to the average person? We have opted for the latter, feeling that it appeals to the average student who is using this book as a text or reference.

Most of the methods of numerical analysis are iterative; approximate answers are obtained through a sequence of improved estimates. There are four questions we always ask about an iterative method:

1. Under what conditions does the method apply? For what kinds of functions does the method work, and how can we know that the conditions are satisfied?
2. Does the method converge? Do the successive approximations reach the true answer to a given accuracy?
3. What bounds can be placed on the error of each estimate? Can we know in advance the maximum size of the error after a certain number of iterations?
4. How rapidly do the errors of the successive estimates decrease? For example, do errors decrease proportionally to the number of iterations or is the accuracy improved more rapidly than linearly, a most desirable situation? How accurately do we know the error (or a bound to the error)?

When finding a root of $f(x)$ by interval halving, we require that $f(x)$ be continuous and of opposite sign at different values of x. This requirement guarantees that there is at least one root between the two x-values. The proof of this statement comes from visualizing the graph of $f(x)$—the function must cut the x-axis at some point(s) between the two values.

If we let x be the midpoint of the interval, $f(x)$ must either be zero at this new point (and the method is finished) or be nonzero, meaning that $f(x)$ at the new x-value is of sign opposite to one of the original points. From these three points, we then select the two points where $f(x)$ is of opposite sign as a new starting pair and subdivide again. Because the interval is cut in half at each step and a root lies in this last interval, the method obviously converges to the true value of a root. If we say that the midpoint of this last interval is our best estimate of the root, the error is certainly no greater than half the last interval. We see that the magnitude of this error estimate is precisely

$$\text{error} \leqq \frac{1}{2^n}|b - a|,$$

where n is the number of steps and a and b are the initial pair of x-values. Bisection or interval halving is one of the few methods of numerical analysis where we know in advance how many steps we must take to reduce the error to a prescribed size.

Be sure to recognize that such error analysis does not take machine errors into account. The theory of numerical analysis almost always assumes perfect computational precision. The effect of the machine is normally handled separately (and, more often than not, entirely neglected because so little can be said about it quantitatively).

As we consider the theory, it is instructive to ask the effect if the assumptions are violated. For example, why must $f(x)$ be continuous? Certainly, if there is jump discontinuity so that $f(x)$ changes sign without cutting the x-axis, there is no guarantee of a root in the starting interval. What if there is a single-jump discontinuity but the jump is not across the x-axis? The exercises will pose questions like this last one.

Measuring the Efficiency of a Procedure

Even though we have discussed just one method, later on we shall look at many algorithms that can accomplish the same thing. We need a measurement for saying which numerical technique is better in some sense. For this reason we introduce here a technique that is of great importance in the discussion of performance of algorithms—the *Big O* notation. Before giving a formal definition, let's look at an example or two. Suppose that an algorithm on n numbers takes

$$f(n) = 1 + 2 + 3 + \ldots + n \text{ additions.}$$

We would like to characterize the value of $f(n)$ more succinctly. It is easy to show that

$$f(n) = \frac{n(n + 1)}{2} = \frac{n^2}{2} + \frac{n}{2}.$$

In this case the more significant term in the expression is the first, and we say that $f(n) = O(n^2)$, because, for large n, doubling n increases $f(n)$ four times. (Try this for yourself, say with $n = 10,000$ and $n = 20,000$.)

Another example is one that will come up in Chapter 2. Suppose that an algorithm takes

$$g(n) = 1^2 + 2^2 + 3^2 + \ldots + n^2 \text{ multiplications.}$$

It can be shown that

$$g(n) = \frac{n(n + 1)(2n + 1)}{6} = \frac{2n^3}{6} + \text{(terms of lower degree).}$$

By the same reasoning as before for $f(n)$, we say that $g(n)$ is $O(n^3)$ or, less often, $O(n^3/3)$.

To define Big O more precisely, we make use of two additional symbols, $\Theta(\ldots)$ and $\Omega(\ldots)$. They are not used as frequently as the Big O notation, but for completeness we shall introduce them here. In this, we depart from our usual practice of avoiding a mathematician's style of definition.

Definition: A function, $f(n) = O(h(n))$, if and only if there exists an n_0 such that, for all $n > n_0$, we have $f(n) \leq K_1\, h(n)$ for some $K_1 > 0$. In the previous example, $f(n) = 1 + 2 + 3 + \ldots + n$ is $O(n^2)$ because $f(n) < n + n + n + \ldots + n = n^2$.

Similarly, we say that $f(n) = \Omega(h(n))$ if and only if for all $n >$ some n_0, we have $f(n) \geq K_2\, h(n)$ for some $K_2 > 0$. In our previous example of the function $f(n) = 1 + 2 + 3 + \ldots + n$, if we take the latter half of the terms in $f(n)$, we have

$$f(n) > n/2 + (n + 2)/2 + \ldots + (n + n)/2 > n/2 + n/2 + \ldots + n/2 > n^2/4$$

(for n even; otherwise, start with $n + 1$). From this, we conclude we have the inequality:

$$1/4\ n^2 \leq f(n) \leq n^2.$$

Definition: A function $f(n) = \Theta(h(n))$ if and only if $f(n) = \Omega(h(n)) = O(h(n))$; that is, there are positive numbers, K_1, K_2 so that we have

$$K_2\, h(n) \leq f(n) \leq K_1\, h(n) \qquad \text{for all } n \geq \text{ some given } n_0.$$

As a final example we shall now demonstrate that the summation:

$$g(n) = 1 + 2 + 2^2 + \ldots + 2^n$$

is $\Theta(2^n)$. Because $g(n) = 2^{n+1} - 1$, we have the following inequality:

$$2^n \leq g(n) = 2^{n+1} - 1 \leq 2 * 2^n$$

where, in this example, $K_2 = 1$ and $K_1 = 2$. This shows that $g(n)$ as defined is $\Theta(2^n)$. Using the limit definition from calculus, we can show that $f(n)$ is $O(h(n))$ if and only if

$$\lim_{n \to \infty} \frac{f(n)}{h(n)} = K > 0.$$

Be sure to note that throughout these discussions, we have assumed that the functions are all nonnegative because their application is in counting such things as the number of additions, function calls, and so on. The application of this definition will be found in Chapter 2 especially. It is also used in algorithms related to searching, sorting, and in the simplex method. Our use of it will be mainly in the discussion of the solution of a system of linear equations.

An Alternative Use of Big O

There is another different use of the Big O notation to indicate what happens when a variable gets very small. In this other usage, we say that some function, $f(x)$, is $\Theta(g(x))$ if and only if

$$\lim_{n \to \infty} \frac{|f(x)|}{|g(x)|} = K > 0.$$

There should be no confusion about this alternative use of the notation, because it will be very clear that, in the first usage, the variable, n, runs over the integers and in the second, x is a real number that is approaching 0.

For example, in the expansion of the exponential function, e^x, we have

$$e^x = 1 + \frac{x}{1!} + \frac{x^2}{2!} + O(x^3),$$

where $O(x^3) = Kx^3$. From the *extended mean value theorem,* we know that in this case,

$$O(x^3) = \frac{e^\xi x^3}{3!}, \qquad \text{where } 0 \le \xi \le x.$$

Except for Chapter 2, this latter definition of Big O is used throughout the book, especially when dealing with Taylor (Maclaurin) series expansions of functions.

0.7   Parallel and Distributed Computing

A computer normally runs its instructions sequentially—one after another—but another trend is emerging. Throughout the historical development of computers, faster and faster machines have been built, but today we have about reached the limits of speed improvements. The fastest machines now can operate with clock times of about 1 to 3 nanoseconds, giving on the order of 10^9 floating-point operations per second ("flops"). Machines with such high performance are today's very expensive supercomputers. However, for really large-scale problems, such as short-term weather forecasting, simulation to predict aerodynamics performance, image processing, and artificial intelligence, these speeds, although almost mind-boggling, are inadequate. Many of these applications involve the solution of very large sets of simultaneous equations by numerical methods.

One of the first techniques to increase the operating speed of a computer was "pipelining"—that is, performing a second instruction within the CPU before the previous instruction is completed. This technique takes advantage of the fact that doing a single "instruction" actually involves several micro-coded steps and that the initial micro-steps can be applied to an additional instruction even though the first sequence of micro-steps has not yet finished. Pipelining permits a speedup by a factor of two or more.

Another technique has been to build vector processing operations into the CPU. Because the individual steps required to solve sets of equations involve many multiplications of a vector by another vector, these machines offer significant speed improvements but only by a factor of 5 or 10, not by the factor of 10,000 that is really desired. Further, this feature increases the cost of mainframes considerably. The current trend is to use parallel processing, that is, to put several machines to work on a single problem, dividing the steps of the solution process into many steps that can be performed simultaneously. Not all problems permit such parallel operations, but many important problems of applied mathematics can be so structured. Obtaining many or even several supercomputers is outrageously costly, however. An alternative is to employ a massive number of low-cost microprocessors, of the order of a thousand (1024 is a practical number). Although the individual speed of a microprocessor is not equal to that of a supercomputer, the difference in speed is made up by the larger number of machines that are combined. Intel has been very active in the area of *massively parallel* computers. Their ASCI Red computer consists of over 9,000 Pentium Pros and can run at a peak speed of 1.3 teraflops. At the other extreme in this area is the *Beowulf-class* of supercomputers, which are PCs joined together to compete with the dedicated supercomputers. For a description and discussion of one of these, the Loki supercomputer, check the Web sites http://loki-www.lan1.gov/index.html and http://loki-www.lan1.gov.results/. The Loki computer consists of 16 Pentium Pro computers working together to create a modestly priced supercomputer. We can imagine a future of thousands of PCs working together and accessible through the Web.

Massively parallel computers are important in many applications. For example, as stated in the *Atlantic Monthly* for January 1998, "big parallel computers have proved useful for both global climate warming and detailed modeling of ocean circulation" to explain why Europe has winter temperatures about nine to eighteen degrees warmer than comparable latitudes elsewhere.

Classification of Computers

The most familiar grouping of computers has been into microcomputers, workstations, minicomputers, mainframes, and supercomputers. This breakdown ranks systems on the basis of cost, size, and speed; the peripherals connected to each group also vary similarly. The groupings are not precise. Microcomputers, better known as personal computers, have increased so much in power but with so little increase in price as to push minicomputers almost out of the market.

Mainframes and minis are nearly always used now as host machines for several to many terminals or as file servers. Often a number of interconnected personal computers use their local memories for much of their work, accessing files on the host for shared information or for copies of application programs. Supercomputers are needed when very large programs must be run or when rapid access to huge databases is required.

Another classification, a taxonomy suggested in 1966 for multiprocessing machines, subdivides computers on the basis of how instructions and data are processed. A sequential-instruction, single-data-stream machine (*SISD*) is the typical *von Neumann* computer that processes data through a single CPU, executing one instruction after another. Such a machine may involve pipelining.

A single-instruction, multiple-data-stream machine (*SIMD*) has several processing units, all supervised by a single control unit. Each processor receives the same instruction at any cycle but utilizes different data coming in separate data streams. A vector processor (also called an array processor) is of this type. A recent example of the application of the SIMD architecture has been to the *MMX* technology (multimedia extensions) for the latest Intel microprocessors at the beginning of 1997. Similar extensions have been made on the Sun SPARC and HP PA-RISC2.0 instruction set. The development of a multiple-instruction, single-data-stream machine (*MISD*), is possible but such a machine seems to be impractical, for no real computers of this type have ever been built.

Multiple-instruction, mutiple-data-stream machines (*MIMDs*) are the most promising of assemblies for parallel processing. MIMDs use many processing units, each executing instructions and utilizing data that are independent of the instructions and data used by the other CPUs.

Recently, much work and interest is in *distributed computing*. The basic idea here is to connect many different computers, which can work separately on their own tasks as well as in conjunction with each other. In the classification of parallel computers we implicitly assumed a single clock with all the parallel operations in step (synchronous), whereas with distributed computers each machine runs under its own clock; interrupts constantly occur throughout the system to coordinate the actions (asynchronous operations). Moreover, each machine has its separate memory, and the data can flow from one computer to another. Although this seems to complicate the whole business of parallel computing, there are good economic reasons for distributed computing. The hardware is not specialized. One can make use of what is already at hand. The major effort and expense is in software and in connecting the computers and this can be done in a variety of ways.

Two of the more recent and well-used techniques have been *PVM* (Parallel Virtual Machine) and *MPI* (Message Passing Interface). Walker and Dongarra* describe the latter model, which is to be a standard message-passing interface for concurrent computers with logically distributed memory, as opposed to parallelizing compilers. There are two very important aspects in parallel and distributed computing: namely, message-passing or moving of data/instructions and applying parallelism to the numerical algorithm. We shall just consider the latter in our discussions even though message-passing may be of equal or greater importance.

Special Problems in Parallel Computing

If parallel computing is to be used to solve a large problem rapidly, several new aspects come into play. Is the data stream provided from a single shared memory, or do the separate units have individual memories? If the memory is distributed, how is communication between the units accomplished? What type of bus provides the data channels to the separate units, and can separate units read and write data at the same time? What sort of intercommunication is there between the individual processors, and can they exchange data without going through memory?

Other questions remain. Do the units operate synchronously, with all controlled by a single clock, or do they run asynchronously? If operation is asynchronous, how does one unit know when to accept data from a prior operation of a different unit, or do all units operate "chaotically"? How can the loads for the separate processors be balanced—will some units sit idle while others are running at capacity? (It would be preferable for all units to run at full loading.) What about the programs for parallel processing? Does the programmer have to be concerned with synchronization and intercommunications? Is the code portable to other machines?

---

*Walker, David W., and Dongarra, Jack J. 1996. "MPI: A Standard Message-Passing Interface." *SIAM News* 29, no. 1 (January/February).

The questions about programming a parallel system are not yet settled. If it were possible to have the compiler recognize parallelism within a conventional program written for sequential operations and have it develop the changed code to be run on the parallel system, the task of programming would be much easier. On the other hand, writing code that specifically takes advantage of the parallel CPUs can be more efficient, but this task is tricky and complicated. It requires a skill that few programmers currently have. This mode would involve knowing exactly how the hardware is organized and what communications problems are involved. Further, it is likely that the best algorithm (solution procedure) for a parallel machine will not always be the optimum one for sequential processing.

Speedup and Efficiency

We do not intend to explain all these many aspects of parallel processing in this book. We must be content to show where parallelism exists for the various kinds of problems that we attack numerically. For example, here is a simple classical problem that exhibits the advantage of parallel processing. Suppose we are to add together n values. We can show the successive steps by a "directed acyclic graph" (dag), as shown in Fig. 0.7. Now imagine that we have many separate processors that can be applied to the job. Figure 0.8 shows that the number of time steps can be decreased from seven to three. In both Figs. 0.7 and 0.8, the "directed acyclic graphs" (dags) have steps that indicate the sequence of the operations in time. In both cases, step $i + 1$ cannot take place until step i is completed. The flow of operations is from the bottom to the top and one can characterize the dag as having a *height* of 7 in Fig. 0.7 and a height of 3 in Fig. 0.8. This is consistent with the definition of a "tree." At each level, indicated by a step n, we have the maximum number of processors used at the time. We see from Fig. 0.8 that we only need 4 and not 8 processors to speed up the addition of the eight numbers.

The term *speedup* is used to describe the increased performance of a parallel system compared to a single processor. It is the ratio of the execution time for the original sequential process, using a single processor, to the time for the same job using parallel processors. In the preceding simple example, the speedup is $7/3 = 2.333$. In computing the speedup

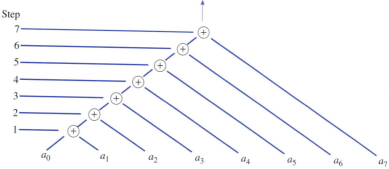

Figure 0.7 Adding eight numbers sequentially

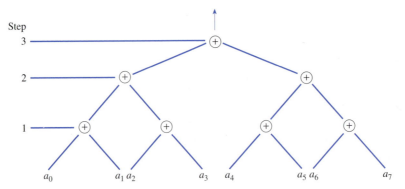

Figure 0.8   Adding eight numbers with parallel processors

for n data items, we use the time for the optimum sequential procedure (or for the best-known procedure if the actual optimum procedure is not known), $T_1(n)$, and for the best (known) parallel algorithm for p processors, $T_p(n)$. With these defined, we can now define speedup* as

$$S_p(n) = T_1(n)/T_p(n).$$

In our example we have $T_1(8) = 7$ and $T_4(8) = 3$ where 7 and 3 are the respective heights of the dags in Figs. 0.7 and 0.8. Another term, the *efficiency*, is based on how the speedup compares to the number of processors used, where

$$E_p(n) = T_1(n)/(pT_p(n)) = S_p(n)/p.$$

Theoretically, if we have p processors, we should be able to do the job p times as fast. In our example, however, $E_4(8) = 2.333/4 = 0.583$.

We have less than an efficiency of 1.00 because some of the processors are idle after the first step. Sometimes the speedup and efficiency are reduced because the size of the problem does not fit to the number of processors. For example, if we were to add only seven numbers in this example, we would still require four processors to get the sum in three steps, but now the speedup would be only $6/3 = 2.000$ and the efficiency would drop to 0.5. On the other hand, what if we were limited to four processors but had 15 numbers to add? We would subdivide the problem and, although trivial in this example, it would have a best solution that is not obvious.

You can see why a complete exposition of how parallel processing and numerical analysis interact will not be attempted in this book, whose major emphasis is on the numerical solution to applied problems.

---

*Bertsekas, Dimitri, and Tsitsiklis, John. 1989. *Parallel and Distributed Computation: Numerical Methods.* Englewood Cliffs, NJ: Prentice-Hall.

Chapter Summary

We will conclude each chapter with a summary of the important items of the chapter, as we do here.

After studying this chapter, you should be able to

1. Tell how a numerical solution differs from an analytical one, and know the advantages of each.

2. Describe the kinds of software that can be used to instruct a computer to carry out the procedures of numerical analysis.

3. Outline the four steps that should be followed in solving any scientific or engineering problem.

4. Show a fellow student the basic principles of bisection to find a root of a function, and explain how to get initial values that bracket the root.

5. Utilize MATLAB (or some other computer algebra system) to get the derivative of a function, to find a minimum, and to plot the function.

6. Plot a function and find a minimum point with a calculator such as the HP-48G or the TI-92.

7. Employ the program listed in the chapter to solve a nonlinear equation. Perhaps you can rewrite it in a different computer language and modify it so that the function evaluation is performed in a subprogram.

8. Distinguish among five types of errors and explain how each can be minimized.

9. Discuss the relative advantages of using relative error versus absolute error as a measure of accuracy.

10. Explain how floating-point numbers are stored in computers and tell what factors affect their accuracy and range. Be able to compute the eps value from the number of bits used to store the different parts of a floating-point number.

11. Tell why parallel processing is currently of great interest to numerical analysts, and describe some of the problems associated with this technique. Explain how a directed acyclic graph (dag) can show the difference in times required for sequential and parallel processing and how a dag can be used to compute the speedup and efficiency values.

Exercises*

Section 0.3

1. Use a computer algebra system to evaluate the following derivatives:

    a. $d/dx \, [x^2 \cos(2/x)]$
    b. $d^2/dy^2 \, [(x^2 + y) / (y^2 - x)]$
    c. $d/dx \, [e^{-x} \{\cos(bx^2)\}]$
    d. $d/dy \, [d/dx \, (x^2 - x/y)]$

---

*Answers are given at the end of the text for exercises marked by ▶.

2. Repeat Exercise 1 but use either the TI-92 or HP-48G calculator. Are the results the same as with a computer algebra system?

3. Use MATLAB to plot dL/dc in the "Ladder" problem of Section 0.3. Scale it so the portion between [.4, .5] occupies the entire screen. Can you find the value of the zero by visual inspection more accurately than getting the minimum from Fig. 0.2? Is using Fig. 0.2 preferred?

4. Repeat Exercise 3 but now use a graphing calculator.

5. Suppose the height of the inlet and straight shafts of Example 0.1 is 6 ft. If you are permitted to tip the ladder as the corner is negotiated, what maximum length of ladder can now be taken into the mine?

▶ 6. A circular well is 5.6 ft in diameter and is 14.3 ft deep, with a flat bottom. A ladder that is 17 in. wide (outside measurement) has side rails that measure 1 in. by 3 in. What is the longest length of ladder that can be placed in the well if its top is to be exactly even with the surface of the ground? Is a computer essential for solving this problem?

Section 0.4

7. Enter the program of Fig. 0.4 into your computer and run it. Do the results duplicate those of Fig. 0.5? If they do not, explain the difference. (You may have to modify the program if you use a different dialect of BASIC.)

8. Translate the program of Fig. 0.4 into some other computer language and run the program. If your results do not duplicate Fig. 0.5, explain the difference.

▶ 9. Modify the program of Fig. 0.4 to eliminate the test for maximum number of iterations. At what tolerance level does the program now enter an endless loop? How will you stop the program when this loop occurs?

10. Modify the program of Fig. 0.4 to print the result after the iterations terminate as:

        THE LENGTH OF THE LADDER IS
            XX.XXXX +/- .YYYY FT

   where .YYYY is the calculated error in the length.

11. Modify the program of Fig. 0.4 to obtain the roots of these equations.

   a. $\sin(x) - 2e^x = 0$, near $x = -3$, also near $x = -6$
   ▶ b. $x^3 + 2x - 1 = 0$ within [0, 1]
   c. $x \sin(2 + x^2/3) = 0$ within [1, 2], also within [3, 4]

12. Repeat Exercise 9 but now compare double-precision to single-precision arithmetic.

13. Solve the equations of Exercise 11 but use either the HP-48G or TI-92 calculator.

Section 0.5

14. Express the quantities below as floating-point numbers in the form

$$0.xxx \ldots xx \text{ E } yy$$

   where the first digit after the decimal point is nonzero and E stands for "times 10 to the power" and yy is an integer.

   a. 1234.5678   c. 1234567890
   b. -0.001020304 d. 0.000,000,001

15. Parallel the simplified examples of floating-point numbers in Section 0.5 with $B = 16$, $p = 4$, and $-4 \le e \le 5$.

   a. How many distinct numbers are there in this system?
   b. What is the largest positive number?
   c. What is the smallest (most negative) number?
   d. What is the positive number of least magnitude?
   e. What is the negative number of least magnitude?

▶ 16. Repeat Exercise 15 except for $B = 10$.

17. For some computer system available for your use, find the values of B, p, and E_{max} and E_{min}. Are these language-dependent (meaning does the answer vary with the computer language)? Answer the questions of Exercise 15 for this system. Your study should include the various precision levels that are provided (single, double, extended precision).

▶ 18. Using the simplified number system of Examples 0.2–0.6, determine the results with both rounding and chopping for each of the following.

   a. $12.3 + 0.0234$
   b. $-0.0321 + 0.000136$
   c. $12.3 - 0.0234$
   d. $-321 + 32.1$
   e. $132 * 0.987$
   f. $-2.14/0.000137$
   g. $(-.111 + .222) * .00111/999$ (in left to right order)

19. Write a computer program to determine experimentally the relative speeds of addition, subtraction, multiplication, and division on some computer system available to you. Be sure to do enough repetitions to overcome the effect of the clock ticks of your system and be sure to compensate for the loop overheads. Are these results language-dependent? Does the system have special hardware for doing arithmetic operations?

20. For some computer system available to you, find out what the machine does when there is exponent overflow and when there is underflow. Are the results language-dependent?

21. Write a computer program that determines the machine eps for some system available for your use. Is this experimentally determined value equal to what you would expect from the machine's number system?

22. Compute the absolute and relative errors of each result of Exercise 18.

23. Find the eps for the TI-92 and/or HP-48G calculators. Do these correspond to the stated precision of the machines?

▶ 24. Evaluate the following polynomial for $x = 1.07$, using both chopping and rounding to three digits, proceeding through the polynomial term by term from left to right. What are the absolute and relative errors of your results?

$$2.75x^3 - 2.95x^2 + 3.16x - 4.67$$

25. Repeat Exercise 24, except this time proceed from right to left, term by term.

26. Evaluating a polynomial in "nested form" is more efficient. The nested form of the polynomial in Exercise 24 is

$$((2.75x - 2.95)x + 3.16)x - 4.67$$

Repeat Exercise 24, but this time do it with the nested form.

27. Write a computer program that computes the sums that follow. Print out results that should equal 0.1, 0.2, 0.3, ... What are the absolute and relative errors of each final result?

a. 0.001 added 1000 times
b. 0.0001 added 10,000 times
c. 0.00001 added 100,000 times

▶ 28. The infinite series $1 + \frac{1}{2} + \frac{1}{3} + \frac{1}{4} + \ldots$ is divergent. Write a computer program to evaluate this sum. Is the computer series divergent? If not, why not?

29. Repeat Exercise 21 (that determines the machine eps for some system available to you), but do this with double and extended precision if your machine and language permit.

Section 0.6

30. Why is it necessary for $f(x)$ to be continuous on $[a, b]$ for bisection (interval halving) to be successful?

▶ 31. Suppose that $f(x)$ has five roots within $[a, b]$. Which of the roots will bisection find if it begins with $x = a$ and $x = b$?

32. In Section 0.5, the following expression is given for the error after n iterations, suggesting that the errors decrease steadily as n increases. Show that this is not always true.

$$\text{error} = (1/2^n)|b - a|$$

Section 0.7

33. Do any of the computer systems in your organization do pipelining? Do any of them have multiple CPUs?

34. Can bisection be speeded up by the use of parallel processing? If not, why not?

▶ 35. How can one recognize when parallel processing can speed up an iterative numerical procedure?

Applied Problems and Projects

36. Write a computer program (in C or FORTRAN) that finds a value of x that makes $f(x) = 0$ within the interval from $x = a$ to $x = b$ where $f(a)$ and $f(b)$ are of opposite signs. The program should step through the enclosing interval beginning at $x = a$ with steps equal to $(b - a)/10$ until a sign change is detected. Then it steps backward in steps one-tenth as large to more precisely isolate the root. The program continues the reversals of direction with smaller steps until an accuracy of $1 * 10^{-6}$ is attained.

37. Critique the procedure of Exercise 36. Consider these questions and others:
    a. What if $f(x)$ is discontinuous within $[a, b]$?
    b. What if there are multiple roots within the interval?
    c. How does the efficiency of this technique compare to the bisection method? How should "efficiency" be defined?

38. Can the bisection method be adapted to find the extrema of a function? How can you adapt the stepping technique (Exercise 36) to find extrema?

39. The IMSL library has a number of FORTRAN subroutines to find the zeros of a function. Do any of these employ bisection?

40. The BASIC program of Section 0.4 can be criticized on two points:

    (1) Computing the quantities f1 * f2 and f1 * f3 can cause overflow.

    (2) The computation x3 = (x1 + x2)/2 is more subject to round-off error than doing x3 = x1 + (x2 − x1)/2.

    a. How can (1) be overcome?

    b. Why is (2) correct?

41. The ABC Manufacturing Company currently ships a product in a cardboard box that measures $6 \times 7.5 \times 2.5$ in. The box is formed from a die-cut pattern using a piece of card stock that is 1/32-in. thick, 12.5-in. wide, and 17-in. high. After the card stock is cut, the unassembled box looks like the figure. Part T forms the top, part B is the bottom, and parts S make the sides. The

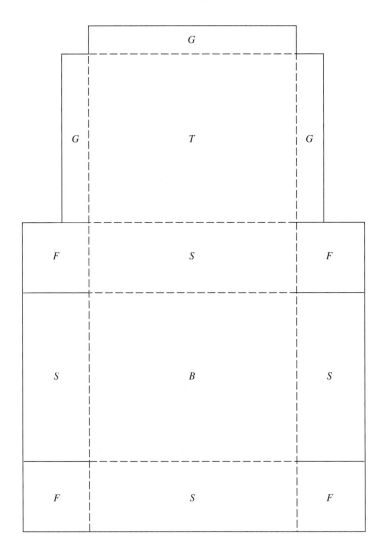

solid lines represent cuts and the dashed lines represent folds. Flaps *F* are folded inside the box and are glued to the side pieces. After the box has been filled, the top is folded over and flaps *G* (which are 1-in. wide) are folded and glued to the outside of the box to seal the box.

You have been asked to lay out the pattern for a new product. Folowing the same type of design as in the figure, draw the pattern for die-cutting the box material. The box is to have the largest volume that can be made from card stock that measures 15 × 20 in. Flaps *G* are still to be 1-in. wide.

Sketch the pattern for the new box. Show how you determined the dimensions to achieve the maximum volume for the box, proving that its volume is the maximum possible.

1

Solving Nonlinear Equations

Contents of This Chapter

Chapter 1 deals with a most important problem of applied mathematics: finding values of x to satisfy $f(x) = 0$. Such values are called the *roots* of the equation and also are known as the *zeros* of $f(x)$.

Some equations are easy to solve. If the relation is linear in x, such as $7x - 2 = 0$, we "solve" it by just rearranging to exhibit x by itself on the left side of the equation: $x = 2/7$. If the function is a second-degree polynomial (a *quadratic*), a formula that most students can recite in their sleep solves for x—the well-known quadratic formula. There is a formula to find the roots of a cubic, but it is fairly complicated and few of us know it. A fourth-degree polynomial can also be solved by a formula, but this process is even more complicated. It has been proven that no general formula exists for polynomials of degree greater than four (meaning that there is no way to exhibit the roots in terms of "ordinary" functions).

Many nonlinear relations involve sines, cosines, exponentials, and other "transcendental" functions. Usually such relations cannot be solved except by successive approximations; these techniques are precisely the subject of this chapter. One of the special advantages of numerical analysis is that the method used usually applies to any kind of relation. We describe the most important root-finding methods in this chapter as applied to functions of a single variable. Systems of nonlinear equations that involve two or more independent variables are discussed in the next chapter. These systems are harder to solve.

We show a total of ten root-finding procedures in this chapter: six that apply to any type of equation, plus four more that apply only to polynomials. Why so many? One reason is to show how the efficiency of a method that solves $f(x) = 0$ can be improved through various modifications. A second reason is to compare their rates of convergence. The central question in numerical analysis is: Which of the available procedures can achieve a desired level of accuracy most quickly (its rate of convergence), with greatest certainty (is it stable?), and with least trouble in starting it (when it is done by successive approximations, that is, iteratively). Even though this chapter gives a wide range of choices for selecting a root-finding procedure, we have little to say about which is best. That is because it depends on the specific problem. When discontinuities are absent, bisection is slow but certain and Newton's method is fast but may be unstable. Acton (1970) gives an exceptionally fine discussion.

Computer algebra systems are of special value for getting the zeros of a function. These have built-in functions that find the zeros. Even though, as we have said, they may be of limited use in large, real-world problems in general, they are still well adapted to the problems described in this chapter. Root finding can be thought of as a "small problem" when it is encountered in isolation. Actually, we really don't need a numerical procedure to solve $f(x) = 0$ approximately—we can merely draw the graph and see where it crosses the x-axis. And computer algebra systems do such graphing readily.

1.1 Pressure Drop for a Flowing Fluid

Describes how engineers compute the pressure required to force a fluid through a pipe. This may lead to solving a nonlinear equation.

1.2 Interval Halving (Bisection) Revisited

Reviews the method described in the previous chapter with greater emphasis on its error analysis. Its advantages and disadvantages are discussed, and times when it cannot be used to find a root are pointed out.

1.3 Linear Interpolation Methods

Describes two methods, the secant method and the method of false position, both of which are based on approximating $f(x)$ by a straight line in the vicinity of the root.

1.4 Newton's Method

Explains a method that is perhaps the most widely used technique for finding roots of functions. It is more rapidly convergent than almost all other methods, but there are pitfalls that you should know about. Complex-valued roots can be obtained by Newton's method if complex arithmetic is used in the computations.

1.5 Muller's Method

Shows that approximating $f(x)$ near a root with a quadratic polynomial (a parabola) significantly improves the rate of convergence over linear interpolation methods, but it still does not equal Newton's method in this regard.

1.6 Fixed-Point Iteration: $x = g(x)$ Method

Departs from the approach used in the previous methods by finding a re-arrangement of the form $x = g(x)$ that may converge to a root from a starting value x_0 if this is substituted into $g(x)$ and the resulting x is used as the argument of $g(x)$ and the process repeated. This method not only is feasible for root finding, but also it is used to establish important theoretical results.

1.7 Newton's Method for Polynomials

Tells about some special techniques that are useful when $f(x)$ is a polynomial. The section also illustrates how parallel processing can speed up certain operations.

1.8 Bairstow's Method for Quadratic Factors

Discusses a specialized method for finding complex-valued roots of polynomials without using complex arithmetic.

1.9 Other Methods for Polynomials

Describes four other methods that can be used to find the roots of polynomials. Three of these methods have the advantage of not needing starting values. The fourth is extremely rapid in convergence.

1.10 Multiple Roots

Shows some ways to cope with the problems that arise when $f(x)$ has multiple roots. This situation often gives the standard methods great difficulty.

1.11 Theoretical Matters

Examines the important subject of errors in the estimates of roots when the methods of this chapter are used. It develops the proofs for some statements about errors and convergence rates that are merely stated in earlier sections.

1.12 Using MATLAB

Gives information on this computer algebra system and ways that it can be used to solve a variety of equations of the form $f(x) = 0$.

Chapter Summary

Gives you a checklist against which you can measure your understanding of the topics of this chapter. You should then restudy those sections where your comprehension is not complete.

Computer Programs

Illustrates the process of writing programs to find roots.

1.1 Pressure Drop for a Flowing Fluid

An important problem for engineers is to determine the energy required to overcome the effects of friction when a fluid flows in a conduit. (There are other energy requirements but energy loss due to friction is one of the more difficult things to compute.) Usually a pump provides the energy required to cause the fluid to flow at the required rate, but if the source is elevated above the outlet, gravity may be sufficient to provide the energy needed.

Flow at low rates in a circular pipe without obstructions may be of a type called laminar; flow lines are straight and circular layers of fluid glide smoothly past adjacent layers. However, at the higher rates of flow that are more common in commercial applications, the flow is turbulent; eddy currents are spontaneously created within the fluid and more energy is required to produce the desired flow rate.

Osborne Reynolds, an Irish engineer, investigated this phenomenon in the 1880s and found that a dimensionless quantity that is termed the Reynolds number tells when the flow becomes turbulent. The Reynolds number is computed from

$$\text{RE} = (D * V)/v,$$

where D = the diameter of the pipe (ft), V = average velocity of flow (ft/s), and v = viscosity of the fluid (ft^2/s). (The values are often expressed in so-called SI units: meters, meters/s, and meters^2/s, but, because the quantity is dimensionless, any consistent set of units may be employed.)

If the Reynolds number is less than about 2100, the flow is laminar; if above about 3000, it is turbulent. For turbulent flow, the pressure difference between the inlet and outlet due solely to friction is given by

$$h_f = (f * L * V^2)/(2g * D)$$

where h_f is the "head loss," f is the so-called friction factor, L is the length of the pipe, V is the average velocity of the flowing fluid, g is the gravitational constant, and D is the diameter of the pipe. Consistent units must be used.

"Head loss" is a way of measuring pressure; it is the pressure at the bottom of a column of fluid h_f high. If h_f is multiplied by the density of the fluid, the result is pressure. (A column of fluid that is 10 feet high exerts a pressure of 500 lbs/ft^2 if the fluid has a density of 50 lb/ft^3, because 10 ft $*$ 50 lb/ft^3 = 500 lb/ft^2.)

For homogeneous fluids, f, the friction factor, is often read from a chart. If the fluid is not a homogeneous liquid but a suspension of particles (such as a slurry of fibers in paper production), an empirical equation for f, the friction factor, can be used:

$$1/\sqrt{f} = (1/k) \, [\ln(\text{RE} * \sqrt{f}) + 14 - 5.6/k],$$

in which k is a parameter known from previous experiments.

The methods of this chapter permit solving this equation for f, if we know the values for RE and k. (See Exercise 83.)

1.2 Interval Halving (Bisection) Revisited

This ancient but effective method for finding a zero of $f(x)$ was discussed in Chapter 0. Its strategy is to begin with two values of x—a and b—that bracket a root of $f(x) = 0$. It determines that the values $x = a$ and $x = b$ do bracket (enclose) a root by finding that $f(a) *$ $f(b) < 0$ (because they are of opposite signs). The method then successively divides the interval in half and replaces one endpoint with the midpoint so that again the root is bracketed. One knows in advance that the error in the estimate of the root must be less than $|(b - a) * (\frac{1}{2^n})|$, where n is the number of iterations performed. It is required that $f(x)$ be continuous in the interval.

This chapter describes several different methods for solving $f(x) = 0$. To compare them, we will solve the same equation by each method. This standard equation is

$$f(x) = x^3 + x^2 - 3x - 3 = 0.$$

One can almost see by inspection that a root is $\sqrt{3}$. We will then be able to see how quickly successive iterates converge on the value 1.732050808. Using the algorithm whose pseudocode was given in the previous chapter, the results of Table 1.1 are obtained.

The main advantage of interval halving is that it is guaranteed to work if $f(x)$ is continuous in $[a, b]$ and if the values $x = a$ and $x = b$ actually bracket a root. Another important advantage that few other root-finding methods share is that the number of interations to achieve a specified accuracy is known in advance. Because the interval $[a, b]$ is halved each time, the last value of x_3 differs from the true root by less than $\frac{1}{2}$ the last interval. So we can say with surety that

Table 1.1   Finding a root of $f(x) = x^3 + x^2 - 3x - 3$ starting with $a = 1$, $b = 2$, and tolerance of $1E - 4$ by interval halving

Iteration	x_1	x_2	x_3	$F(x_3)$	Maximum error	Actual error
1	1.000000	2.000000	1.500000	−1.875000	0.500000	−0.232051
2	1.500000	2.000000	1.750000	0.171875	0.250000	0.017949
3	1.500000	1.750000	1.625000	−0.943359	0.125000	−0.107051
4	1.625000	1.750000	1.687500	−0.409424	0.062500	−0.044551
5	1.687500	1.750000	1.718750	−0.124786	0.031250	−0.013301
6	1.718750	1.750000	1.734375	0.022030	0.015625	0.002324
7	1.718750	1.734375	1.726563	−0.051756	0.007813	−0.005488
8	1.726563	1.734375	1.730469	−0.014957	0.003906	−0.001582
9	1.730469	1.734375	1.732422	0.003512	0.001953	0.000371
10	1.730469	1.732422	1.731445	−0.005728	0.000977	−0.000605
11	1.731445	1.732422	1.731934	−0.001109	0.000488	−0.000117
12	1.731934	1.732422	1.732178	0.001202	0.000244	0.000127
13	1.731934	1.732178	1.732056	0.000046	0.000122	0.000005

Tolerance met

$$\text{error after } n \text{ iterations} < \left| \frac{(b-a)}{2^n} \right|.$$

The major objection of interval halving has been that it is slow to converge. Other methods require fewer iterations to achieve the same accuracy (but then we do not always know a bound on the accuracy).

Observe in Table 1.1 that the estimate of the root may be better at an earlier iteration than at later ones. (The second iterate is closer to the true root than are the next two; we are closer at iteration 6 than at iteration 7.) Of course, in this example we have the advantage of knowing the answer, which is never the case. However, the values of $f(x_3)$ themselves show that these better estimates are closer to the root. (This is not an absolute criterion—some functions may be nearly zero at points not so near the root, but, for smooth functions, a small value of the function is a good indicator that we are near to the root. This is especially true when we are quite close to the root.) The methods we consider in later sections use the values of $f(x)$ to find the root more rapidly.

With speedy computers so prevalent today the slowness of the bisection method is of less concern. When the values of Table 1.1 were computed from a program, the results were seen in less than a second.

When the roots of functions must be computed a great many times (this may be a requirement of some other program that does engineering analysis), the efficiency of interval halving may be inadequate. This will be particularly true if $f(x)$ is not given explicitly but, instead, is developed internally within the other program. In that case, finding values of x that bracket the root may also be a problem.

In spite of arguments that other methods find roots with fewer iterations, interval halving is an important tool in the applied mathematician's arsenal. Bisection is generally recommended for finding an approximate value for the root, and then this value is refined by more efficient methods. The reason is that most other root-finding methods require a starting value near to a root—lacking this, they may fail completely.

Do not overlook other techniques that may seem mundane for getting a first approximation to the root. Graphing the function is always helpful in showing where roots occur, and with programs like MATLAB (or a graphing calculator) that do plots so handily, getting the graph before beginning a root-finding routine is a good practice. Searching methods should also be considered as a preliminary step. Stepping through the interval $[-1, 1]$ and testing whether $f(x)$ changes sign will show whether there are roots in that interval. Roots of larger magnitude can be found by stepping through that same interval with x replaced by $1/y$, because the roots of this modified function are the reciprocals of the roots of the original function. Experience with the particular types of problems that are being solved may also suggest approximate values of roots. Even intuition can be a factor. Acton (1970) gives an especially interesting and illuminating discussion.

When there are multiple roots, interval halving may not be applicable, because the function may not change sign at points on either side of the roots. Here a graph will be most important to reveal the situation. In this case, we may be able to find the roots by working with $f'(x)$, which will be zero at a multiple root.

1.3 Linear Interpolation Methods

Although the bisection (interval halving) method is easy to compute and has simple error analysis, it is not very efficient. For most functions, we can improve the speed at which the root is approached through a different scheme. Almost every function can be approximated by a straight line over a small interval. We begin from a value—say, x_0—that is near to a root, r. (We would get x_0 from a graph or from a few applications of bisection.)

The Secant Method

Suppose we assume that $f(x)$ is linear in the vicinity of the root r. Now we choose another point, x_1, which is near x_0 and also near r (which we don't know yet), and we draw a straight line through the two points. Figure 1.1 illustrates the situation with the distances along the x-axis exaggerated. (Because we don't know the value of the root yet, the two points could be on opposite sides of the root or to the left of the root, rather than as shown.)

If $f(x)$ were truly linear, the straight line would intersect the x-axis at the root. But $f(x)$ will never be exactly linear because we would never use a root-finding method on a linear function! That means that the intersection of the line with the x-axis is not at $x = r$ but that it should be close to it. From the obvious similar triangles we can write

$$\frac{(x_1 - x_2)}{f(x_1)} = \frac{(x_0 - x_1)}{f(x_0) - f(x_1)}$$

and from this solve for x_2:

$$x_2 = x_1 - f(x_1)\frac{(x_0 - x_1)}{f(x_0) - f(x_1)}.$$

Because $f(x)$ is not exactly linear, x_2 is not equal to r, but it should be closer than either of the two points we began with. We can continue to get better estimates of the root if we do this repeatedly, always choosing the two x-values nearest to r for drawing the straight line.

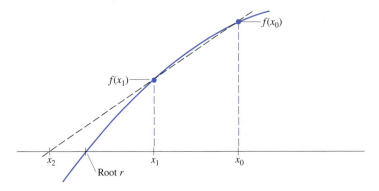

Figure 1.1

Because each newly computed value should be nearer to the root, we can do this easily after the second iterate has been computed, by always using the last two computed points. But after the *first* iteration there aren't "two last computed points." So we make sure to start with x_1 closer to the root than x_0 by testing $f(x_0)$ and $f(x_1)$ and swapping if the first function value is smaller.* The net effect of this rule is to set $x_0 = x_1$ and $x_1 = x_2$ after each iteration. The exceptions to this rule are pathological cases, which we consider next.

The technique we have described is known as the secant method because the line through two points on the curve is called the secant line. Here is pseudocode for the secant method algorithm:

An Algorithm for the Secant Method

To determine a root of $f(x) = 0$, given two values, x_0 and x_1, that are near the root,

IF $|f(x_0)| < |f(x_1)|$
    Swap x_0 with x_1.
REPEAT
    Set $x_2 = x_1 - f(x_1) * (x_0 - x_1)/[f(x_0) - f(x_1)]$.
    Set $x_0 = x_1$.
    Set $x_1 = x_2$.
UNTIL $|f(x_2)| <$ tolerance value.

Note: If $f(x)$ is not continuous, the method may fail.

An alternative stopping criterion for the secant method is when the pair of points being used are sufficiently close together.

An Example

Table 1.2 shows the results of the secant method on the function $f(x) = x^3 + x^2 - 3x - 3 = 0$ with starting values of $x = 1$ and $x = 2$. Values not very close to the root (which is at $x = 1.73205$) were deliberately chosen. Table 1.3 shows the results on a transcendental function $f(x) = 3x + \sin(x) - e^x$, starting from $x = 0$ and $x = 1$. This latter function has a zero at $x = 0.360421703$.

An objection is sometimes raised about the secant method. If the function is far from linear near the root, the successive iterates can fly off to points far from the root, as seen in Fig. 1.2.

If the method is being carried out by a program that displays the successive iterates, the user can interrupt the program should such improvident behavior be observed. Also, if the function was plotted before starting the method, it is unlikely that the problem will be

---

*$|f(x_0)| < |f(x_1)|$ does not always mean that x_0 is closer to the root, but that is often the case. When it is, the method is speeded up. In any case, the algorithm still converges to the root when $f(x)$ is continuous.

Table 1.2   Secant method on $f(x) = x^3 + x^2 - 3x - 3$

Iteration	x_0	x_1	x_2	$F(x_2)$
1	1	2	1.571429	−1.364432
2	2	1.571429	1.705411	−0.2477449
3	1.571429	1.705411	1.735136	2.925562E-02
4	1.705411	1.735136	1.731996	−5.147391E-04
5	1.735136	1.731996	1.732051	−1.422422E-06

At $x = 1.732051$, tolerance of .00001 met!

Table 1.3   Secant method on $f(x) = 3x + \sin(x) - e^x$

Iteration	x_0	x_1	x_2	$F(x_2)$
1	1	0	0.4709896	0.2651588
2	0	0.4709896	0.3722771	2.953367E-02
3	0.4709896	0.3722771	0.3599043	−1.294787E-03
4	0.3722771	0.3599043	0.3604239	5.552969E-06
5	0.3599043	0.3604239	0.3604217	3.554221E-08

At $x = .3604217$, tolerance of .0000001 met!

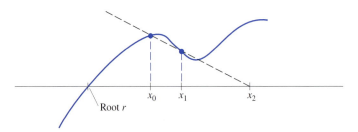

Figure 1.2   A pathological case for the secant method

encountered, because a better starting value would be used. There are times when this remedy is not possible: when the routine is being used within another program that needs to find a root before it can proceed.

Linear Interpolation (False Position)

A way to avoid such pathology is to ensure that the root is bracketed between the two starting values and remains between the successive pairs. When this is done, the method is known as *linear interpolation,* or, more often, as the *method of false position* (in Latin, *regula falsi*). This technique is similar to bisection except the next iterate is taken at the intersection of a line between the pair of x-values and the x-axis rather than at the midpoint. Doing so gives faster convergence than does bisection, but at the expense of a more complicated algorithm.

Here is the pseudocode for *regula falsi* (method of false position).

An Algorithm for the Method of False Position (*regula falsi*)

To determine a root of $f(x) = 0$, given values of x_0 and x_1 that bracket a root, that is, $f(x_0)$ and $f(x_1)$ are of opposite sign,

REPEAT
    Set $x_2 = x_1 - f(x_1) * (x_0 - x_1)/(f(x_0) - f(x_1))$.
    IF $f(x_2)$ of opposite sign to $f(x_0)$):
        Set $x_1 = x_2$.
    ELSE Set $x_0 = x_2$.
    ENDIF.
UNTIL $|f(x_2)| <$ tolerance value.

Note: The method may fail if $f(x)$ is discontinuous on the interval.

Table 1.4 compares the results of three methods—interval halving (bisection), linear interpolation, and the secant method—on $f(x) = 3x + \sin(x) - e^x = 0$. Observe that the speed of convergence is best for the secant method, poorest for interval halving, and intermediate for false position. Notice that false position converges to the root from only one side, slowing it down, especially if that end of the interval is farther from the root. There is a way to avoid this result, called modified linear interpolation. We omit the details of this method.

Table 1.4 Comparison of methods, $f(x) = 3x + \sin(x) - e^x = 0$, $x_0 = 0$, $x_1 = 1$

	Interval halving		False position		Secant method	
Iteration	x	$f(x)$	x	$f(x)$	x	$f(x)$
1	0.5	0.330704	0.470990	0.265160	0.470990	0.265160
2	0.25	-0.286621	0.372277	0.029533	0.372277	0.029533
3	0.375	0.036281	0.361598	$2.94 * 10^{-3}$	0.359904	$-1.29 * 10^{-3}$
4	0.3125	-0.121899	0.360538	$2.90 * 10^{-4}$	0.360424	$5.55 * 10^{-6}$
5	0.34375	-0.041956	0.360433	$2.93 * 10^{-5}$	0.360422	$3.55 * 10^{-7}$
Error after 5 iterations	0.01667		$-1.17 * 10^{-5}$		$< -1 * 10^{-7}$	

(Exact value of root is 0.360421703.)

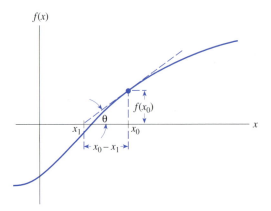

Figure 1.3

1.4 Newton's Method

One of the most widely used methods of solving equations is Newton's method.* Like the previous ones, this method is also based on a linear approximation of the function, but does so using a tangent to the curve. Figure 1.3 gives a graphical description. Starting from a single initial estimate, x_0, that is not too far from a root, we move along the tangent to its intersection with the x-axis, and take that as the next approximation. This is continued until either the successive x-values are sufficiently close or the value of the funciton is suffi- ciently near zero.†

The calculation scheme follows immediately from the right triangle shown in Fig. 1.3, which has the angle of inclination of the tangent line to the curve at $x = x_0$ as one of its acute angles:

$$\tan \theta = f'(x_0) = \frac{f(x_0)}{x_0 - x_1}, \qquad x_1 = x_0 - \frac{f(x_0)}{f'(x_0)}.$$

We continue the calculation scheme by computing

$$x_2 = x_1 - \frac{f(x_1)}{f'(x_1)},$$

or, in more general terms,

$$x_{n+1} = x_n - \frac{f(x_n)}{f'(x_n)}, \qquad n = 0, 1, 2, \dots.$$

---

*Newton did not publish an extensive discussion of this method, but he solved a cubic polynomial in *Principia* (1687). The version given here is considerably improved over his original example.
†Which criterion should be used often depends on the particular physical problem to which the equation applies. Customarily, agreement of successive x-values to a specified tolerance is required.

Newton's algorithm is widely used because, at least in the near neighborhood of a root, it is more rapidly convergent than any of the methods discussed so far. We show in a later section that the method is quadratically convergent, by which we mean that the error of each step approaches a constant K times the square of the error of the previous step. The net result of this is that the number of decimal places of accuracy nearly doubles at each iteration. However, offsetting this is the need for two function evaluations at each step, $f(x_n)$ and $f'(x_n)$.*

When Newton's method is applied to $f(x) = 3x + \sin x - e^x = 0$, we have the following calculations:

$$f(x) = 3x + \sin x - e^x,$$
$$f'(x) = 3 + \cos x - e^x$$

There is little need to use MATLAB to get this simple derivative, but, for practice, here is how to do it:

1. Define $f(x)$: fx = '3 * x+ sin(x) − exp(x)';
2. Get derivative: dfx = diff(fx)
   and see:

```
dfx =
3 + cos(x) - exp(x)
```

Observe that we use exp(x) to represent e^x and that the variable of differentiation is not required when it is the default value x.

If we begin with $x_0 = 0.0$, we have

$$x_1 = x_0 - \frac{f(x_0)}{f'(x_0)} = 0.0 - \frac{-1.0}{3.0} = 0.33333;$$

$$x_2 = x_1 - \frac{f(x_1)}{f'(x_1)} = 0.33333 - \frac{-0.068418}{2.54934} = 0.36017;$$

$$x_3 = x_2 - \frac{f(x_2)}{f'(x_2)} = 0.36017 - \frac{-6.279 \times 10^{-4}}{2.50226} = 0.3604217.$$

After three iterations, the root is correct to seven significant digits. Comparing this with the results in Table 1.4, we see that Newton's method converges considerably more rapidly than the previous methods. In comparing numerical methods, however, we usually count the number of times functions must be evaluated. Because Newton's method requires two function evaluations per step, the comparison is not as one-sided in favor of Newton's method as at first appears; the three iterations with Newton's method required six function evaluations. Five iterations with the previous methods also required six evaluations. If a difficult problem requires many iterations to converge, the number of function evaluations with Newton's method may be many more than with linear iteration methods because Newton always uses two per iteration whereas the others take only one (after the first step that takes two).

---

*Another problem with Newton's method is that finding $f'(x)$ may be difficult. Computer algebra systems can be a real help.

A more formal statement of the algorithm for Newton's method, suitable for implementation in a computer program, is shown here.

Newton's Method

To determine a root of $f(x) = 0$, given a value x_0 reasonably close to the root,

     Compute $f(x_0), f'(x_0)$.
   IF $(f(x_0) \neq 0)$ AND $(f'(x_0) \neq 0)$
     REPEAT
       Set $x_1 = x_0$.
       Set $x_0 = x_0 - f(x_0)/f'(x_0)$.
     UNTIL $(|x_0 - x_1| <$ tolerance value 1$)$ OR
          $(|f(x_0)| <$ tolerance value 2$)$.

Note: The method may converge to a root different from the expected one or diverge if the starting value is not close enough to the root.

When Newton's method is applied to polynomial functions, special techniques facilitate such application. We consider these in a later section of this chapter.

In some cases Newton's method will not converge. Figure 1.4 illustrates this situation. Starting with x_0, one never reaches the root r because $x_6 = x_1$ and we are in an endless loop. Observe also that if we should ever reach the minimum or maximum of the curve, we will fly off to infinity. We will develop the analytical condition for this in a later section and show that Newton's method is quadratically convergent in most cases.

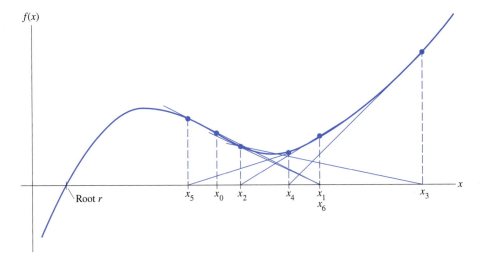

Figure 1.4

Relating Newton's Method to Other Methods

It is of interest to notice that the previous interpolation methods are closely related to Newton's method. For linear interpolation, whose algorithm we can write as

$$x_{n+1} = x_n - \frac{f(x_n)}{\dfrac{f(x_n) - f(x_{n-1})}{x_n - x_{n-1}}},$$

we see that the denominator of the fractional term is exactly the definition of the derivative except not taken to the limit as the two x-values approach each other. This *difference quotient* is an approximation to the derivative, as we will explain in detail in a later chapter. Because the denominator of the fractional term is an approximation to the derivative of f, we see the close resemblance to Newton's method.

The secant method has exactly this same resemblance to Newton's method because it is just linear interpolation without the requirement that the two x-values bracket the root. Because these two values usually are closer together than for linear interpolation, the approximation to the derivative is even better.

From this we see that there is an alternative way to get the derivative for Newton's method. If we compute $f(x)$ at two closely spaced values for x and divide the difference in the function values by the difference in x-values, we have the derivative (nearly) without having to differentiate. Although this sounds like spending an extra function evaluation, we avoid having to evaluate the derivative function and so it breaks even. (Convergence will not usually be as fast, however.)

Complex Roots

Newton's method works with complex roots if we give it a complex value for the starting value. Here is an example.

EXAMPLE 1.1 Use Newton's method on $f(x) = x^3 + 2x^2 - x + 5$.

Figure 1.5 shows the graph of $f(x)$. It has a real root at about $x = -3$, whereas the other two roots are complex because the x-axis is not crossed again.

If we begin Newton's method with $x_0 = 1 + i$ (we used this in the lack of knowledge about the complex root), we get these successive iterates:

1. $0.486238 + 1.04587i$
2. $0.448139 + 1.23665i$
3. $0.462720 + 1.22242i$
4. $0.462925 + 1.22253i$
5. $0.462925 + 1.22253i$

Because the fourth and fifth iterates agree to six significant figures, we are sure that we have an estimate good to at least that many figures. The second complex root is the conju-

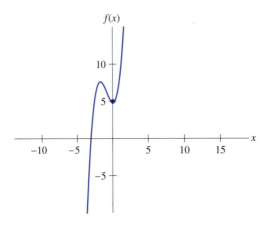

Figure 1.5 Plot of $f(x) = x^3 + 2x^2 - x + 5$

gate of this: $0.462925 - 1.22253i$. If we begin with $x_0 = 1 - i$, the method converges to the conjugate.

If we begin with a real starting value—say, $x_0 = -3$—we get convergence to the root at $x = -2.92585$. ▲

1.5 Muller's Method

Most of the root-finding methods that we have considered so far have approximated the function in the neighborhood of the root by a straight line. Obviously, this is never true; if the function were linear, finding the root would take practically no effort. Muller's method is based on approximating the function in the neighborhood of the root by a quadratic polynomial. This gives a much closer match to the actual curve.

A second-degree polynomial is made to fit three points near a root, $[x_0, f(x_0)]$, $[x_1, f(x_1)]$, $[x_2, f(x_2)]$. The proper zero of this quadratic, using the quadratic formula, is used as the improved estimate of the root. The process is then repeated using the set of three points nearest the root being evaluated.

The procedure for Muller's method is developed by writing a quadratic equation that fits through three points in the vicinity of a root, in the form $av^2 + bv + c$. (See Fig. 1.6.) The development is simplified if we transform axes to pass through the middle point, by letting $v = x - x_0$.

Let $h_1 = x_1 - x_0$ and $h_2 = x_0 - x_2$. We evaluate the coefficients by evaluating $p_2(v)$ at the three points:

$$v = 0: \quad a(0)^2 + b(0) + c = f_0;$$
$$v = h_1: \quad ah_1^2 + bh_1 + c = f_1;$$
$$v = -h_2: \quad ah_2^2 - bh_2 + c = f_2.$$

From the first equation, $c = f_0$. Letting $h_2/h_1 = \gamma$, we can solve the other two equations for a and b:

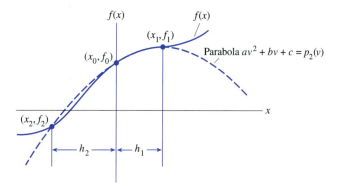

Figure 1.6

$$a = \frac{\gamma f_1 - f_0(1 + \gamma) + f_2}{\gamma h_1^2(1 + \gamma)}, \qquad b = \frac{f_1 - f_0 - ah_1^2}{h_1}.$$

After computing a, b, and c, we solve for the root of $av^2 + bv + c = 0$ by the quadratic formula, choosing the root nearest to the middle point x_0. This value is

$$\text{root} = x_0 - \frac{2c}{b \pm \sqrt{b^2 - 4ac}},$$

with the sign in the denominator taken to give the largest absolute value of the denominator (that is, if $b > 0$, choose plus; if $b < 0$, choose minus; if $b = 0$, choose either). The reason for using this somewhat unusual form of the quadratic formula is to make the next iterate closer to the root.

We take the root of the polynomial as one of a set of three points for the next approximation, taking the three points that are most closely spaced (that is, if the root is to the right of x_0, take x_0, x_1, and the root; if to the left, take x_0, x_2, and the root). We always reset the subscripts to make x_0 be the middle of the three values.

An algorithm for Muller's method is

Muller's Method

Given the points x_2, x_0, x_1 in increasing value,

1. Evaluate the corresponding function values f_2, f_0, f_1.
2. Find the coefficients of the parabola determined by the three points.
3. Compute the two roots of the parabolic equation.
4. Choose the root closest to x_0 and label it x_r.
5. IF $x_r > x_0$ THEN rearrange x_0, x_r, x_1 into x_2, x_0, x_1
   ELSE rearrange x_2, x_r, x_0 into x_2, x_0, x_1.
6. IF $|f(x_r)| <$ FTOL, THEN return (x_r)
   ELSE go to 1.

Muller's method, like Newton's, will find a complex root if given complex starting values. Of course, the computations must use complex arithmetic.

EXAMPLE 1.2   Find a root between 0 and 1 of the same transcendental function as before: $f(x) = 3x + \sin(x) - e^x$. Let

$$x_0 = 0.5, \quad f(x_0) = 0.330704 \qquad h_1 = 0.5,$$
$$x_1 = 1.0, \quad f(x_1) = 1.123189 \qquad h_2 = 0.5,$$
$$x_2 = 0.0, \quad f(x_2) = -1 \qquad\quad \gamma = 1.0.$$

Then

$$a = \frac{(1.0)(1.123189) - 0.330704(2.0) + (-1)}{1.0(0.5)^2(2.0)} = -1.07644,$$

$$b = \frac{1.123189 - 0.330704 - (-1.07644)(0.5)^2}{0.5} = 2.12319,$$

$$c = 0.330704,$$

and

$$\text{root} = 0.5 - \frac{2(0.330704)}{2.12319 + \sqrt{(2.12319)^2 - 4(-1.07644)(0.330704)}}$$

$$= 0.354914.$$

For the next iteration, we have

$$x_0 = 0.354914, \quad f(x_0) = -0.0138066 \qquad h_1 = 0.145086,$$
$$x_1 = 0.5, \qquad\quad f(x_1) = 0.330704 \qquad h_2 = 0.354914,$$
$$x_2 = 0, \qquad\qquad f(x_2) = -1 \qquad\qquad\quad \gamma = 2.44623.$$

Then

$$a = \frac{(2.44623)(0.330704) - (-0.0138066)(3.44623) + (-1)}{2.44623(0.145086)^2(3.44623)} = -0.808314,$$

$$b = \frac{0.330704 - (-0.0138066) - (-0.808314)(0.145086)^2}{0.145086} = 2.49180,$$

$$c = -0.0138066,$$

$$\text{root} = 0.354914 - \frac{2(-0.0138066)}{2.49180 - \sqrt{(2.49180)^2 - 4(-0.808314)(-0.0138066)}}$$

$$= 0.360465.$$

After a third iteration, we get 0.3604217 as the value for the root, which is identical to that from Newton's method after three iterations.   ▲

Experience shows that Muller's method converges at a rate that is similar to that for Newton's method.* It does not require the evaluation of derivatives, however, and (after we have obtained the starting values) needs only one function evaluation per iteration. There is an initial penalty in that one must evaluate the function three times, but this is frequently overcome by the time the required precision is attained.

1.6 Fixed-Point Iteration: $x = g(x)$ Method

The method known as *fixed-point iteration* (we also call it the $x = g(x)$ method) is a very useful way to get a root of $f(x) = 0$. This method is also the basis for some important theory. To use the method, we rearrange $f(x)$ into an equivalent form $x = g(x)$, which usually can be done in several ways. Observe that if $f(r) = 0$, where r is a root of $f(x)$, it follows that $r = g(r)$. Whenever we have $r = g(r)$, r is said to be a fixed point for the function g.

Under suitable conditions that we explain later, the iterative form

$$x_{n+1} = g(x_n) \qquad n = 0, 1, 2, 3, \dots ,$$

converges to the fixed point r, a root of $f(x)$.

Here is a simple example:

$$f(x) = x^2 - 2x - 3 = 0.$$

$f(x)$ is easy to factor to show roots at $x = -1$ and $x = 3$. (We pretend that we don't know this.)

Suppose we rearrange to give this equivalent form:

$$x = g_1(x) = \sqrt{2x + 3}.$$

If we start with $x = 4$ and iterate with the fixed-point algorithm, successive values of x are

$$x_0 = 4,$$
$$x_1 = \sqrt{11} = 3.31662,$$
$$x_2 = \sqrt{9.63325} = 3.10375,$$
$$x_3 = \sqrt{9.20750} = 3.03439,$$
$$x_4 = \sqrt{9.06877} = 3.01144,$$
$$x_5 = \sqrt{9.02288} = 3.00381,$$

and it appears that the values are converging on the root at $x = 3$.

---

* Atkinson (1978) shows that each error is about proportional to the previous error to the 1.85th power.

Other Rearrangements

Another rearrangement of $f(x)$ is

$$x = g_2(x) = \frac{3}{(x-2)}.$$

Let us start the iterations again with $x_0 = 4$. Successive values then are

$$x_0 = 4,$$
$$x_1 = 1.5,$$
$$x_2 = -6,$$
$$x_3 = -0.375,$$
$$x_4 = -1.263158,$$
$$x_5 = -0.919355,$$
$$x_6 = -1.02762,$$
$$x_7 = -0.990876,$$
$$x_8 = -1.00305,$$

and it seems that we now converge to the other root, at $x = -1$. We also see that the convergence is oscillatory rather than monotonic as we saw in the first case.

Consider a third rearrangement:

$$x = g_2(x) = \frac{(x^2 - 3)}{2}.$$

Starting again with $x_0 = 4$, we get

$$x_0 = 4,$$
$$x_1 = 6.5,$$
$$x_2 = 19.625,$$
$$x_3 = 191.070,$$

and the iterates are obviously diverging.

This difference in behavior of the three rearrangements is interesting and worth further study. First, though, let us look at the graphs of the three cases. The fixed point of $x = g(x)$ is the intersection of the line $y = x$ and the curve $y = g(x)$ plotted against x. Figure 1.7 shows the three cases.

Observe that we always get the successive iterates by this construction: Start on the x-axis at the initial x_0, go vertically to the curve, then horizontally to the line $y = x$, then vertically to the curve, and again horizontally to the line. Repeat this process until the points on the curve converge to a fixed point or else diverge. It appears that the different behaviors depend on whether the slope of the curve is greater, less, or of opposite sign to the slope of the line (which equals $+1$).

Here is pseudocode for the fixed-point ($x = g(x)$) method:

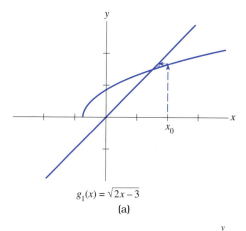

$g_1(x) = \sqrt{2x - 3}$

(a)

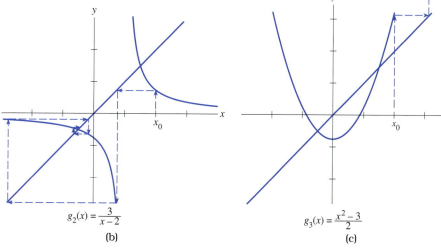

$g_2(x) = \dfrac{3}{x - 2}$

(b)

$g_3(x) = \dfrac{x^2 - 3}{2}$

(c)

Figure 1.7

Iteration with the Form $x = g(x)$

To determine a root of $f(x) = 0$, given a value x_1 reasonably close to the root,

Rearrange the equation to an equivalent form $x = g(x)$.

REPEAT
    Set $x_2 = x_1$.
    Set $x_1 = g(x_1)$.
UNTIL $|x_1 - x_2| <$ tolerance value.

Note: The method may converge to a root different from the expected one, or it may diverge. Different rearrangements will converge at different rates.

We will give the details of the proof of the following statement in a later section, but the statement is true.

> If $g(x)$ and $g'(x)$ are continuous on an interval about a root r of the equation $x = g(x)$, and if $|g'(x)| < 1$ for all x in the interval, then $x_{n+1} = g(x_n)$, $n = 0, 1, 2, \ldots$, will converge to the root $x = r$, provided that x_1 is chosen in the interval. Note that this is a sufficient condition only, because for some equations, convergence is secured even though not all the conditions hold.*

Examine again the three graphs of Fig. 1.7. Observe that in part (a) the slope of the curve $[g'(x)]$ is positive but less than 1. There will be monotonic convergence. In part (b), the slope of the curve is negative and less than 1 in value; there is oscillatory convergence. In part (c) the slope of the curve is greater than 1, so the iterates diverge. It is easy to see from similar graphs that convergence is faster if the slope of the curve is nearer to zero.

The preceding conditions for convergence are based on the fact that the error at each step of the iterations is

$$|e_{n+1}| = |g(\xi_n)| * |e_n|,$$

where ξ_n is a value between x_n and r, the fixed point, and e_i is the error of the ith iterate.

When we are near to r, the derivative of $g(x)$ is essentially constant. That is, each successive error is a fraction of the preceding one, or, in other words, the error at any step is approximately proportional to the previous one.

The fixed-point method is particularly easy to program. In later sections of this chapter we show how MATLAB can be used to do fixed-point iterations.

The major difficulty one finds with fixed-point iteration is determining a suitable $g(x)$ that converges to the desired root.

Accelerating Convergence

Even though the statement that each error is proportional to the previous one is only an approximation, assuming it to be true allows us to accelerate the convergence. This is known as *Aitken acceleration*.

To do such acceleration, we compute two values, D1 and D2, from the starting value and the first two iterates, x_0, x_1, and x_2. D1 $= x_0 - x_1$ and D2 is $x_2 - 2x_1 + x_0$. Then we compute

$$X = x_0 - (D1)^2/D2 = (x_0x_2 - x_1{}^2)/(x_2 - 2x_1 + x_0).$$

---

*The analytical test that $|g'(x)| < 1$ is often awkward to apply. A constructive test is merely to observe whether the successive x_i values converge. In a computer program it is worthwhile to determine whether $|x_3 - x_2| < |x_2 - x_1|$.

X is closer to the root than if we were to just apply fixed-point iteration a third time. After beginning again with X and iterating two more times, we can use these last three values to accelerate again.

We illustrate this with the values from the first example of the section, where

$$x_0 = 4, \; x_1 = 3.31662, \; x_2 = 3.10375$$

$$D1 = 4 - 3.31662 = 0.68338,$$

$$D2 = 3.10375 - 2(3.31662) + 4 = 0.47051,$$

$$X = 4 - \frac{(0.68338)^2}{0.47051} = 3.00744,$$

and we see that X is closer to the true root than x_4 of the example. We have jumped ahead by more than two iterations. Of course, $g(x)$ is so easy to evaluate in this example that the saving of effort is small. When $g(x)$ is costly to evaluate, Aitken acceleration is worthwhile.

You should realize that Aitken acceleration can be applied to any iterative process where the errors decrease proportionately, not only to fixed-point iteration.

See earlier editions of this book or Jones (1982) for a fuller explanation.

1.7 Newton's Method for Polynomials

Polynomial functions are of special importance. We will see throughout the remainder of this book that many valuable numerical procedures are based on polynomials. This important role of polynomial functions is due to their "nice" behavior: They are everywhere continuous, they are smooth, their derivatives are also continuous and smooth, and they are readily evaluated. Descartes' rule of signs (see Appendix A) lets us predict the number of positive roots. Polynomials are particularly well adapted to computers because the only mathematical operations they require for evaluation are addition, subtraction, and multiplication, all of which are speedy operations on computers.

Because of this special importance of polynomials, we now consider how our root-finding methods can be applied to them. For most of the methods previously discussed there is nothing new to say, but for Newton's method there are significant new ideas to consider. We begin on an historical note, with a procedure that saves time in hand computations. However, we will see that this same procedure is also the basis for computer calculations as well. We assume in this section that we are finding a simple root of the polynomial. Functions with multiple roots are discussed in another section.

In applying Newton's method to polynomials in a hand computation, it is most efficient to evaluate $f(x_n)$ and $f'(x_n)$ by use of synthetic division.* We illustrate this by the same

---

*The mechanics of synthetic division, whereby we divide a polynomial by the factor $x - x_i$, are explained in most algebra books. In the example, when $x^3 + x^2 - 3x - 3$ is divided by $(x - 2)$, the result is $x^2 + 3x + 3$, with a remainder of 3. In brief, as shown in the example, the coefficients of the polynomial are first written in a row. Suppose we are dividing by $(x - x_0)$. Below a line, the first coefficient is copied down. This is multiplied by x_0 and added to the second coefficient. The result is written below the line. This is then multiplied by x_0 and added to the next coefficient, and the process is repeated. The last result is equal to $f(x_0)$.

cubic polynomial that we used before, $x^3 + x^2 - 3x - 3 = 0$, which has a root at $x = \sqrt{3}$. We begin with the value $x = 2$. We utilize the remainder theorem* to evaluate $f(2)$, and evaluate $f'(2)$ as the remainder when the reduced polynomial (of degree 2 here) is divided by $(x - 2)$:

$$
\begin{array}{r|rrrr}
x_0 = 2 & 1 & 1 & -3 & -3 \\
 & & 2 & 6 & 6 \\
\hline
 & 1 & 3 & 3 & ③ \quad \longleftarrow \quad \text{Remainder} = f(2). \\
 & & 2 & 10 & \\
\hline
 & 1 & 5 & ⑬ & \quad \longleftarrow \quad \text{Second remainder} = f'(2).
\end{array}
$$

We use the values from the synthetic division—$f(2) = 3$ and $f'(2) = 13$—to get an improved estimate of the root by Newton's method:

$$x_1 = 2 - \frac{3}{13} = 1.76923. \ldots$$

Continuing,

$$
\begin{array}{r|rrrr}
x_1 = 1.76923 & 1 & 1 & -3 & -3 \\
 & & 1.76923 & 4.89940 & 3.36048 \\
\hline
 & 1 & 2.76923 & 1.89940 & 0.36048 \\
 & & 1.76923 & 8.02957 & \\
\hline
 & 1 & 4.53846 & 9.92897 &
\end{array}
$$

$$x_2 = 1.76923 - \frac{0.36048}{9.92897} = 1.73292.$$

Similarly,

$$x_3 = 1.73292 - \frac{0.00823}{9.47487} = 1.73205.$$

The value of x_3 is correct to five decimals. To observe the improvement in accuracy, consider the successive errors:

	Error	Number of correct figures
$x_0 = 2$	0.26895	1
$x_1 = 1.76923$	0.03718	2
$x_2 = 1.73292$	0.00087	4
$x_3 = 1.73205$	0.00000	6+

To compute with five decimal places, as in this example, we used a calculator. (If you have access to a calculator with storage for two or more values, you will find it especially well adapted to this method.)

---

*Why the first remainder equals $f(2)$ and the second remainder equals $f'(2)$ is developed next.

The initial value at which Newton's method is begun can make a considerable difference. For example, if this problem is started with $x = 1$, the following values result:

x	$f(x)$	$f'(x)$
$x_0 = 1$	-4	2
$x_1 = 3$	24	30
$x_2 = 2.2$	5.888	15.92
$x_3 = 1.83015$		

From here on, the convergence is rapid, for we are using iterates just about as near the root as in the previous example.

After a first root is found (as shown by a remainder that is very small), one normally proceeds to determine additional roots from the reduced polynomial (whose coefficients are in the third row of the synthetic-division tableau). This technique makes the computations somewhat shorter. In the example, the reduced equation is a quadratic, so the quadratic formula would be used, but if a higher-degree polynomial were being solved, Newton's method employing synthetic division would be employed to improve an initial estimate of a second root. The process is then repeated until the reduced equation is of second degree.

This technique of working with the reduced function can be used even if the function is not a polynomial. After a root r of $f(x) = 0$ has been found, the new function $F(x) = f(x)/(x - r)$ will have all the roots of $f(x)$ except the root r. This procedure is called *deflating the function*. One must remember that a discontinuity has been introduced at $x = r$, however. Figure 1.8 shows the example $P_3(x)$ and the deflated polynomial $P_2(x)$. We suggest that you explore how deflation works on nonpolynomial functions by graphing $f(x) = (x - 1)(e^x - \cos x)$, then $g(x) = f(x)/x$, $h(x) = f(x)/(x - 1)$, and comparing the graphs.

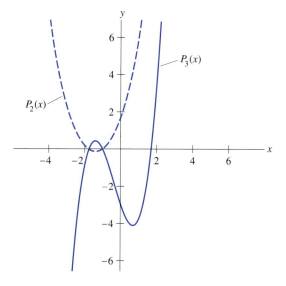

Figure 1.8 A polynomial and its deflated descendant

You may also wish to compare to $h(x)/x$ and to functions derived by deflating $f(x)$ with roots other than $x = 0$ and $x = 1$.

We used MATLAB to get the roots of $P_3(x)$ and the plot of Fig. 1.8 through the following:

1. Define the polynomial as an array of coefficients:

   ```
   P3 = [1 1 -3 -3];
   ```

2. Get the roots:

   ```
   roots (P3)
   ans =
   -1.7321
   1.7321
   -1.0000
   ```

3. Redefine P3 as an algebraic object:

   ```
   P3 = 'x^3 + x^2 - 3 * x - 3';
   ```

4. Define P2, also as an algebraic object:

   ```
   P2 = '(x^3 + x^2 - 3 * x - 3)/(x - 1.7321)';
   ```

5. Plot P3:

   ```
   fplot (P3, [-2.2 2.2])
   ```

   and see the graph of the cubic polynomial.

6. Plot P2, but don't erase the first plot:

   ```
   hold on; fplot (P2, [-2.2 2.2])
   ```

   and see Fig. 1.8.

From this sequence of commands we see that MATLAB uses the array definition of P3 to get the roots (all of them) but that in plotting with the fplot command, we must redefine it as an algebraic object. The 'hold on' command saves the previous plot so we can superimpose the next one. If we want to do a new plot on a blank screen, we can either say 'hold' (which toggles the hold operation) or 'hold off.'

It should be observed that using deflated functions can result in unexpected errors. If the first root is determined only approximately, the coefficients of the reduced equation are themselves not exact and the succeeding roots are subject not only to round-off errors and the errors that occur when iterations are terminated too soon, but also to inherited errors residing in the nonexact coefficients. This is an example of propagated error. Some functions are extremely sensitive, in that small changes in the value of the coefficients cause large differences in the roots.* Removing roots in order of increasing magnitude is said to minimize the difficulty, and the use of double-precision arithmetic will further help preserve accuracy.

---

*A classic example has been given by Wilkinson (1959). The 20th-degree polynomial with roots $-1, -2, \ldots, -20$ begins as $x^{20} + 210x^{19} + \cdots + 20! = 0$. If 2^{-23} is added to the coefficient of x^{19}, the roots change noticeably; five of them become complex with real parts that differ from the original roots by as much as 0.73.

The Synthetic Division Algorithm and the Remainder Theorem

We will now establish the remainder theorem and develop the synthetic division algorithm. The scheme is also the most efficient way to evaluate polynomials and their derivatives in a computer program.

Write the nth-degree polynomial as

$$P_n(x) = a_n x^n + a_{n-1} x^{n-1} + \cdots + a_1 x + a_0.$$

We wish to divide this by the factor $(x - x_1)$, giving a reduced polynomial $Q_{n-1}(x)$ of degree $n - 1$, and a remainder, R, which is a constant:

$$\frac{P_n(x)}{x - x_1} = Q_{n-1}(x) + \frac{R}{x - x_1}.$$

Rearranging yields

$$P_n(x) = (x - x_1)Q_{n-1}(x) + R.$$

Note that at $x = x_1$,

$$P_n(x_1) = (0)[Q_{n-1}(x_1)] + R = R,$$

which is the remainder theorem: The remainder on division by $(x - x_1)$ is the value of the polynomial at $x = x_1$, $P_n(x_1)$.

If we differentiate $P_n(x)$, we get

$$P'_n(x) = (x - x_1)Q'_{n-1}(x) + (1)Q_{n-1}(x) + 0.$$

Letting $x = x_1$, we have

$$P'_n(x_1) = Q_{n-1}(x_1).$$

We evaluate the Q-polynomial at x_1 by a second division whose remainder equals $Q_{n-1}(x_1)$. This verifies that the second remainder from synthetic division yields the value for the derivative of the polynomial.

We now develop the synthetic division algorithm, writing $Q_{n-1}(x)$ in form similar to $P_n(x)$:

$$\begin{aligned}
P_n(x) &= a_n x^n + a_{n-1}x^{n-1} + \cdots + a_1 x + a_0 \\
&= (x - x_1)Q_{n-1}(x) + R \\
&= (x - x_1)(b_{n-1}x^{n-1} + b_{n-2}x^{n-2} + \cdots + b_1 x + b_0) + R.
\end{aligned}$$

Multiplying out and equating coefficients of like terms in x, we get

$$
\left.
\begin{array}{ll}
\text{coef. of } x^n: & a_n = b_{n-1} \\
x^{n-1}: & a_{n-1} = b_{n-2} - x_1 b_{n-1} \\
x^{n-2}: & a_{n-2} = b_{n-3} - x_1 b_{n-2} \\
& \quad\vdots \\
x: & a_1 = b_0 - x_1 b_1 \\
\text{const: } & a_0 = R - x_1 b_0
\end{array}
\right\}
\quad \text{or} \quad
\left\{
\begin{array}{l}
b_{n-1} = a_n \\
b_{n-2} = a_{n-1} + x_1 b_{n-1} \\
b_{n-3} = a_{n-2} + x_1 b_{n-2} \\
\quad\vdots \\
b_0 = a_1 + x_1 b_1 \\
R = a_0 + x_1 b_0
\end{array}
\right.
$$

The general form is $b_i = a_{i+1} + x_1 b_{i+1}$, by which all the b's may be calculated, provided that we first set $b_n = 0$. If this is compared to the preceding synthetic divisions, it is seen to be identical, except that we now have a vertical array. The horizontal layout is easier for hand computation. For evaluation of the derivative, a set of c-values is computed from the b's in the same way in which the b's are computed from the a's.

Synthetic division is also known as the *nested multiplication method* of evaluating polynomials. Consider the fifth-degree polynomial, evaluated at $x = x_1$:

$$a_5 x_1^5 + a_4 x_1^4 + a_3 x_1^3 + a_2 x_1^2 + a_1 x_1 + a_0.$$

We can rewrite this as

$$((((a_5 x_1 + a_4) x_1 + a_3) x_1 + a_2) x_1 + a_1) x_1 + a_0.$$

In the original form, $5 + 4 + 3 + 2 + 1 = 15$ multiplications are required, plus five additions. In the nested form, only five multiplications are required, plus five additions; it is obviously the more efficient method.

Comparing this with the equations $b_{n-2} = a_{n-1} + x_1 b_{n-1}$ and $b_i = a_{i+1} + x_1 b_{i+1}$ for synthetic division, we see that the successive terms are formed in exactly the same way, so that synthetic division and nested multiplication are two names for the same thing.

Horner's Method

The historic name for synthetic division is Horner's method. It is not hard to show that for an nth-degree polynomial, the reduction in multiplies is from $n(n + 1)/2$ multiplies for the standard form to only n for Horner's method. There are n adds in either case.

We can write the algorithm for nested multiplication (Horner's method) in pseudocode as follows:

Algorithm for Horner's Method

To divide a given polynomial, $P_n(x)$, of degree n, by $(x - x_1)$, where

$$P_n(x) = a_n x^n + a_{n-1} x^{n-1} + \cdots + a_1 x + a_0,$$

    SET $b_n = 0$.
    SET $i = n - 1$.
    REPEAT
        SET $b_i = a_{i+1} + x_1 b_{i+1}$
        SET $i = i - 1$
    UNTIL $i < 0$.

The remainder, $R = a_0 + x_1 b_0$, is $P(x_1)$.

To divide two polynomials using MATLAB, we first define them as arrays of coefficients, then use the command 'deconv' (which really means to get the inverse of the convolution of two vectors, which is the equivalent to multiplying the polynomials). So, to divide $x^3 + x^2 - 3x - 3$ by $x - 2$, we do the following:

1. Define the two polynomials:

   ```
   N = [1 1 -3 -3]; D = [1 -2];
   ```

2. Divide them:

   ```
   [q, r] = deconv(N, D)
   q =
         1  3  3
   r =
         0  0  0  3
   ```

   which is MATLAB's way of telling us that $N/D = (x^2 + 3x + 3) + 3/(x - 2)$.

   There is no easy way to divide two polynomials if they are in algebraic form.

Parallel Processing

Horner's method for evaluating a polynomial is one of the classic examples where we can speed up a computation by using parallel processors. The directed acyclic graphs (dags) for the sequential and parallel algorithms are shown in Fig. 1.9. Although we have more operations (five multiplies and three adds) with the parallel scheme (compared to three multiplies and three adds), the time required to produce the result is reduced from six steps to four steps. The time savings comes from doing some operations in parallel rather than in succession, of course. Observe that the most efficient method for sequential processing (Horner's method) is not used in parallel processing.

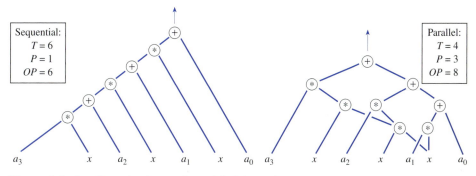

Figure 1.9 dags for evaluating a polynomial of degree 3

1.8 Bairstow's Method for Quadratic Factors

The methods considered so far are difficult to use to find a complex root of a polynomial. It is true that Newton's and Muller's methods work satisfactorily, provided that we begin with initial estimates that are complex-valued; however, in a hand computation, performing the multiplications and divisions of complex numbers is awkward. There is no problem in a computer program if complex arithmetic capabilities exist, but the execution is slower.

For polynomials, the complex roots occur in conjugate pairs if the coefficients are all real-valued. For this case, if we extract the quadratic factors that are the products of the pairs of complex roots, we can avoid complex arithmetic because such quadratic factors have real coefficients. We first develop the algorithm for synthetic division by a trial quadratic, $x^2 - rx - s$, which, hopefully, is near to the desired factor of the polynomial:

$$
\begin{aligned}
P_n(x) &= a_n x^n + a_{n-1} x^{n-1} + \cdots + a_0 \\
&= (x^2 - rx - s)Q_{n-2}(x) + \text{remainder} \\
&= (x^2 - rx - s)(b_n x^{n-2} + b_{n-1} x^{n-3} + \cdots + b_3 x + b_2) + \text{remainder}, \\
\text{remainder} &= b_1(x - r) + b_0.
\end{aligned}
$$

If $x^2 - rx - s$ is an exact factor of $P_n(x)$, then b_1 and b_0 will both be zero. The negative signs in the factor are conventional in developing the algorithm.

On multiplying out and equating coefficients of like powers of x, we get

$$
\left.
\begin{aligned}
a_n &= b_n \\
a_{n-1} &= b_{n-1} - rb_n \\
a_{n-2} &= b_{n-2} - rb_{n-1} - sb_n \\
a_{n-3} &= b_{n-3} - rb_{n-2} - sb_{n-1} \\
&\vdots \\
a_1 &= b_1 - rb_2 - sb_3 \\
a_0 &= b_0 - rb_1 - sb_2
\end{aligned}
\right\}
\quad \text{or} \quad
\left\{
\begin{aligned}
b_n &= a_n \\
b_{n-1} &= a_{n-1} + rb_n \\
b_{n-2} &= a_{n-2} + rb_{n-1} + sb_n \\
b_{n-3} &= a_{n-3} + rb_{n-2} + sb_{n-1} \\
&\vdots \\
b_1 &= a_1 + rb_2 + sb_3 \\
b_0 &= a_0 + rb_1 + sb_2
\end{aligned}
\right.
\quad \textbf{(1.1)}
$$

We would like both b_1 and b_0 to be zero, for that would show $x^2 - rx - s$ to be a quadratic factor of the polynomial (because the remainder is zero). This will normally not be so; if we properly change the values of r and s, we can make the remainder zero, or at least make its coefficients smaller. Obviously b_1 and b_0 are both functions of the two parameters r and s. Expanding these as a Taylor series for a function of two variables* in terms of $(r^* - r)$ and $(s^* - s)$, where $(r^* - r)$ and $(s^* - s)$ are presumed small so that

---

* Appendix A reviews this.

terms of higher order than the first are negligible, we obtain

$$b_1(r^*, s^*) = b_1(r, s) + \frac{\partial b_1}{\partial r} (r^* - r) + \frac{\partial b_1}{\partial s} (s^* - s) + \cdots,$$

$$b_0(r^*, s^*) = b_0(r, s) + \frac{\partial b_0}{\partial r} (r^* - r) + \frac{\partial b_0}{\partial s} (s^* - s) + \cdots.$$

Let us take (r^*, s^*) as the point at which the remainder is zero, and

$$r^* - r = \Delta r, \qquad s^* - s = \Delta s.$$

(Δr and Δs are increments to add to the original r and s to get the new values r^* and s^* for which the remainder is zero.) Then

$$b_1(r^*, s^*) = 0 \simeq b_1 + \frac{\partial b_1}{\partial r} \Delta r + \frac{\partial b_1}{\partial s} \Delta s,$$

$$b_0(r^*, s^*) = 0 \simeq b_0 + \frac{\partial b_0}{\partial r} \Delta r + \frac{\partial b_0}{\partial s} \Delta s.$$

All the terms on the right are to be evaluated at (r, s). We wish to solve these two equations simultaneously for the unknown Δr and Δs, so we need to evaluate the partial derivatives.

Bairstow showed that the required partial derivatives can be obtained from the b's by a second synthetic division by the factor $x^2 - rx - s$ in just the same way that the b's are obtained from the a's. (Observe that this is analogous to getting the derivative of a polynomial by dividing the reduced polynomial by $(x - x_1)$.)

Define a set of c's by the following relations shown at the left, and compare these to the partial derivatives in the right-hand columns:

$$c_n = b_n$$

$$\frac{\partial b_n}{\partial r} = \frac{\partial a_n}{\partial r} = 0 \qquad\qquad \frac{\partial b_n}{\partial s} = \frac{\partial a_n}{\partial s} = 0$$

$$c_{n-1} = b_{n-1} + rc_n$$

$$\frac{\partial b_{n-1}}{\partial r} = r\frac{\partial b_n}{\partial r} + b_n = b_n = c_n \qquad\qquad \frac{\partial b_{n-1}}{\partial s} = \frac{\partial a_{n-1}}{\partial s} + r\frac{\partial b_n}{\partial s} = 0$$

$$c_{n-2} = b_{n-2} + rc_{n-1} + sc_n$$

$$\frac{\partial b_{n-2}}{\partial r} = r\frac{\partial b_{n-1}}{\partial r} + b_{n-1} = c_{n-1} \qquad\qquad \frac{\partial b_{n-2}}{\partial s} = r\frac{\partial b_{n-1}}{\partial s} + s\frac{\partial b_n}{\partial s} = b_0$$

$$= b_n = c_n$$

$$c_{n-3} = b_{n-3} + rc_{n-2} + sc_{n-1}$$

$$\frac{\partial b_{n-3}}{\partial r} = r\frac{\partial b_{n-2}}{\partial r} + b_{n-2} + s\frac{\partial b_{n-1}}{\partial r} \qquad\qquad \frac{\partial b_{n-3}}{\partial s} = r\frac{\partial b_{n-2}}{\partial s} + s\frac{\partial b_{n-1}}{\partial s} + b_{n-1}$$

$$= b_{n-2} + rc_{n-1} + sc_n = c_{n-2} \qquad\qquad = b_{n-1} + rc_n = c_{n-1}$$

$$\vdots \qquad\qquad\qquad \vdots \qquad\qquad\qquad \vdots$$

$$c_1 = b_1 + rc_2 + sc_3$$

$$\frac{\partial b_1}{\partial r} = r\frac{\partial b_2}{\partial r} + b_2 + s\frac{\partial b_3}{\partial r} \qquad\qquad \frac{\partial b_1}{\partial s} = r\frac{\partial b_2}{\partial s} + s\frac{\partial b_3}{\partial s} + b_3$$

$$= b_2 + rc_3 + sc_4 \qquad\qquad = b_3 + rc_4 + sc_5$$

$$= c_2 \qquad\qquad\qquad = c_3$$

Hence the partial derivatives that we need are equal to the properly corresponding c's. Our simultaneous equations become, where Δr and Δs are unknowns to be solved for,

$$-b_1 = c_2\Delta r + c_3\Delta s,$$
$$-b_0 = c_1\Delta r + c_2\Delta s.$$

We express the solution as ratios of determinants, using Cramer's rule:

$$\Delta r = \frac{\begin{vmatrix} -b_1 & c_3 \\ -b_0 & c_2 \end{vmatrix}}{\begin{vmatrix} c_2 & c_3 \\ c_1 & c_2 \end{vmatrix}},$$

$$\Delta s = \frac{\begin{vmatrix} c_2 & -b_1 \\ c_1 & -b_0 \end{vmatrix}}{\begin{vmatrix} c_2 & c_3 \\ c_1 & c_2 \end{vmatrix}}.$$

As an exercise, the student should write the algorithm for this method.

EXAMPLE 1.3   Find the quadratic factors of

$$x^4 - 1.1x^3 + 2.3x^2 + 0.5x + 3.3 = 0.$$

Use $x^2 + x + 1$ as starting factor ($r = -1$, $s = -1$). (Frequently $r = s = 0$ are used as starting values if no information as to an approximate factor is known.) Equations (1.1) lead to a double synthetic-division scheme as follows:

	a_4	a_3	a_2	a_1	a_0
	1	-1.1	2.3	0.5	3.3
$r = -1$		-1.0	2.1	-3.4	0.8
$s = -1$		—	-1.0	2.1	-3.4
	1	-2.1	3.4	-0.8	0.7
		-1.0	3.1	-5.5	
		—	-1.0	3.1	
	1	-3.1	5.5	-3.2	b_1 b_0
		c_3	c_2	c_1	

Note that the equations for b_1 and c_1 have no term involving s. The dashes in the preceding tableau represent these missing factors. Then

$$\Delta r = \frac{\begin{vmatrix} 0.8 & -3.1 \\ -0.7 & 5.5 \end{vmatrix}}{\begin{vmatrix} 5.5 & -3.1 \\ -3.2 & 5.5 \end{vmatrix}} = \frac{2.23}{20.33} = 0.11, \qquad r^* = -1 + 0.11 = -0.89,$$

$$\Delta s = \frac{\begin{vmatrix} 5.5 & 0.8 \\ -3.2 & -0.7 \end{vmatrix}}{20.33} = \frac{-1.29}{20.33} = -0.06, \qquad s^* = -1 - 0.06 = -1.06.$$

The second trial yields

	1	−1.1	2.3	0.5	3.3	
−0.89		−0.89	1.77	−2.68	0.06	
−1.06		—	−1.06	2.11	−3.17	
	1	−1.99	3.01	−0.07	0.17	← b's
		−0.89	2.56	−4.01		
		—	−1.06	3.05		
		−2.88	4.51	−1.03		← c's

$$\Delta r = \frac{\begin{vmatrix} 0.07 & -2.88 \\ -0.17 & 4.51 \end{vmatrix}}{\begin{vmatrix} 4.51 & -2.88 \\ -1.03 & 4.51 \end{vmatrix}} = \frac{-0.175}{17.374} = -0.010, \qquad r^* = -0.89 - 0.010 = -0.900,$$

$$\Delta s = \frac{\begin{vmatrix} 4.51 & 0.07 \\ -1.03 & -0.17 \end{vmatrix}}{17.374} = \frac{-0.694}{17.374} = -0.040, \qquad s^* = -1.06 - 0.040 = -1.100.$$

The exact factors are $(x^2 + 0.9x + 1.1)(x^2 - 2x + 3)$.

Observe that the other factor has its coefficients in the row of b's. ▲

If we use MATLAB to divide a polynomial by a quadratic, as we do in Bairstow's method, we do not get the same remainder that Bairstow gives:

```
N = [1 -1.1 2.3 .5 3.3]; D = [1 1 1];
[q, r] = deconv(N, D)
```

results in

```
q =
      1.0000 -2.1000 3.4000
r =
      0   0   0   -.8000 -0.1000
```

We get a remainder that is $-0.8x - 0.1$, whereas Bairstow expresses this differently: $-0.8(x + 1) + 0.7$. These are really the same; if we expand the Bairstow expression we get $-0.8x - 0.1$.

Of course it is much better to ask MATLAB for the roots directly. Doing so gives all of them: $-0.45 \pm .9474i$ and $1 \pm 1.4124i$.

1.9 Other Methods for Polynomials*

Of the many other methods for finding the roots of polynomials, we discuss four in this section. Three methods—the QD algorithm, Graeffe's root-squaring method, and Lehmer's method—do not require that we start with a value near to a root. The fourth, Laguerre's method, does need a starting value, but it is remarkably efficient.

The QD Algorithm

The QD, or quotient-difference method, is quite efficient. We present it without elaboration.[†]

For the nth-degree polynomial

$$P_n(x) = a_n x^n + a_{n-1} x^{n-1} + \cdots + a_1 x + a_0,$$

we form an array of q and e terms, starting the tableau by calculating a first row of q's and a second row of e's:

$$q^{(0)} = -\frac{a_{n-1}}{a_n}. \quad \text{All other } q\text{'s are zero.}$$

$$e^{(i)} = \frac{a_{n-i-1}}{a_{n-i}}, \quad i = 1, 2, \ldots, n-1.$$

$$e^{(0)} = e^{(n)} = 0.$$

The start of the array is

$e^{(0)}$	$q^{(0)}$	$e^{(1)}$	$q^{(1)}$	$e^{(2)}$	$q^{(2)}$	$\cdots$	$e^{(n-1)}$	$q^{(n-1)}$	$e^{(n)}$
	$\dfrac{-a_{n-1}}{a_n}$		0		0	$\cdots$		0	
0		$\dfrac{a_{n-2}}{a_{n-1}}$		$\dfrac{a_{n-3}}{a_{n-2}}$		$\cdots$	$\dfrac{a_0}{a_1}$		0

A new row of q's is computed by the equation

$$\text{New } q^{(i)} = e^{(i+1)} - e^{(i)} + q^{(i)},$$

using terms from the e and q rows in the array. Note that this algorithm is "e to right minus e to left plus q above."

A new row of e's is now computed by the equation

$$\text{New } e^{(i)} = \left(\frac{q^{(i)}}{q^{(i-1)}} \right) e^{(i)};$$

"q to right over q to left times e above." The example in Table 1.5 isolates the roots of the quartic

$$P_4(x) = 128x^4 - 256x^3 + 160x^2 - 32x + 1$$

by continuing to compute rows of q's and then e's until all the e-values approach zero. When this occurs, the q-values assume the values of the roots. Because the method is slow to converge, it is generally used only to get approximate values, which are then improved by Newton's method.

Table 1.5   Example of QD method for $P(x) = 128x^4 - 256x^3 + 160x^2 - 32x + 1$

$e^{(0)}$	$q^{(0)}$	$e^{(1)}$	$q^{(1)}$	$e^{(2)}$	$q^{(2)}$	$e^{(3)}$	$q^{(3)}$	$e^{(4)}$
	2.000	0		0		0		
0		−0.625		−0.200		−0.031		0
	1.375		0.425		0.169		0.031	
0		−0.193		−0.079		−0.006		0
	1.182		0.539		0.242		0.037	
0		−0.088		−0.036		−0.001		0
	1.094		0.591		0.277		0.038	
0		−0.048		−0.017		−0.000		0
	1.046		0.622		0.294		0.038	
0		−0.028		−0.008		−0.000		0
	1.018		0.642		0.302		0.038	
0		−0.018		−0.004		−0.000		0
	1.000		0.656		0.304		0.038	
0		−0.012		−0.002		−0.000		0
	0.988		0.666		0.306		0.038	
0		−0.008		−0.001		−0.000		0
	0.980		0.673		0.307		0.038	
0		−0.005		−0.001		−0.000		0
	0.975		0.677		0.308		0.038*	

*The true values of the roots are 0.96194, 0.69134, 0.30866, and 0.03806.

What if there are complex roots? If the polynomial has a pair of conjugate complex roots, one of the e's will not approach zero but will fluctuate in value. The sum of the two q-values on either side of this e will approach r, and the product of the q above and to the left times the q below and to the right approaches $-s$ in the factor $x^2 - rx - s$. Two equal roots behave similarly.

Table 1.6 shows the result of the method for the polynomial

$$(x - 1)(x - 4)(x^2 - x + 3) = x^4 - 6x^3 + 12x^2 - 19x + 12.$$

For our example, the factors are $(x - 4)(x - 1)(x^2 - x + 3)$. We have:

$$q^{(0)} \text{ converging to } 4,$$

$$q^{(3)} \text{ converging to } 1.$$

Because $e^{(2)}$ does not approach zero, $q^{(1)}$ and $q^{(2)}$ represent a quadratic factor. We compute

$$r = q^{(1)} + q^{(2)} = 1.456 - 0.466 = 0.990;$$

$$s = -(-6.426)(-0.466) = -2.995.$$

This quadratic factor is $x^2 - rx - s = x^2 - 0.990x - (-2.995)$.

What if some a's are zero? Note that we cannot compute the first q and e rows if one of the coefficients in the polynomial is zero, for division by zero is undefined. In such a case, we change the variable to $y = x - 1$. (Subtracting 1 from the roots of the equation is an arbitrary choice, but this facilitates the reverse change of variable to get the roots of the original equation after the roots of the new equation in y have been found.)

Table 1.6   QD method with complex roots, for $P(x) = x^4 - 6x^3 + 12x^2 - 19x + 12$

$e^{(0)}$	$q^{(0)}$	$e^{(1)}$	$q^{(1)}$	$e^{(2)}$	$q^{(2)}$	$e^{(3)}$	$q^{(3)}$	$e^{(4)}$
	6.000	0		0		0		
0		-2.000		-1.583		-0.632		0
	4.000		0.417		0.951		0.632	
0		-0.208		-3.610		-0.420		0
	3.792		-2.985		4.141		1.052	
0		0.164		5.008		-0.107		0
	3.956		1.859		-0.974		1.159	
0		0.077		-2.624		0.127		0
	4.033		-0.842		1.777		1.032	
0		-0.016		5.538		0.074		0
	4.017		4.712		-3.687		0.958	
0		-0.019		-4.333		-0.019		0
	3.998		0.398		0.627		0.977	
0		-0.002		-6.826		-0.030		0
	4.000		-6.426		7.423		1.007	
0		0.003		7.885		-0.004		0
	4.003		1.456		-0.466		1.010	

For example, if $f(x) = x^4 - 2x^2 + x - 1 = 0$, we let $y = x - 1$ and use repeated synthetic division to determine the coefficients of $f(y) = 0$. The successive remainders on dividing by $x - 1$ are the coefficients of $f(y)$:

$$
\begin{array}{r|rrrrr}
\underline{1} & 1 & 0 & -2 & 1 & -1 \\
  &   & 1 & 1 & -1 & 0 \\
\hline
  & 1 & 1 & -1 & 0 & \boxed{-1} \\
  &   & 1 & 2 & 1 & \\
\hline
  & 1 & 2 & 1 & \boxed{1} & \\
  &   & 1 & 3 & & \\
\hline
  & 1 & 3 & \boxed{4} & & \\
  &   & 1 & & & \\
\hline
  & 1 & \boxed{4} & & & \\
\boxed{1} & & & & &
\end{array}
$$

Therefore,

$$f(y) = y^4 + 4y^3 + 4y^2 + y - 1.$$

We proceed to find the roots of $f(y) = 0$, and then get the roots of $f(x) = 0$ by adding 1.

Graeffe's Root-Squaring Method

Graeffe's method transforms $P_n(x)$ into another polynomial of the same degree but whose roots are the squares of the roots of the original polynomial. Thus if the roots of $P_n(x)$ are all real and distinct, the roots of the new polynomial are spread more widely apart than in the old polynomial. This is particularly so for roots greater than 1 in absolute value. Repeating this process until the roots are really far apart, we can compute the roots directly from the coefficients.

This simple example shows how to get the new polynomial when the initial polynomial is of the third degree. Let

$$P(x) = (x - 1)(x + 2)(x - 3).$$

Then

$$
\begin{aligned}
P(-x) &= (-x - 1)(-x + 2)(-x - 3) \\
&= (-1)^3(x + 1)(x - 2)(x + 3).
\end{aligned}
$$

Multiplying,

$$
\begin{aligned}
P(x) * P(-x) &= (-1)^3(x^2 - 1^2)(x^2 - 2^2)(x^2 - 3^2) \\
&= (-1)^3(x^2 - 1)(x^2 - 4)(x^2 - 9) \\
&= (-1)^3(z - 1)(z - 4)(z - 9),
\end{aligned}
$$

and we have a polynomial in $x^2 = z$ whose roots are indeed the squares of the roots of $P(x)$. [The power on the (-1) factor is always n, the degree of the polynomial.] Of course, we never have $P(x)$ in factored form, but the result is the same. Observe that the squaring process loses the signs of the roots so that "distinct" applies to the magnitudes of the roots.

Suppose we have squared the original k times. We then can estimate each of the roots with the 2^kth root of

$$\left| \frac{a_j}{a_{j+1}} \right| \qquad j = n - 1, n - 2, \ldots, 1, 0,$$

n being the degree of the polynomial.

EXAMPLE 1.4   Use root squaring on $x^3 - 3x^2 - 6x + 8$. The first three new polynomials (with x replacing x^2) are

$$k = 1: \quad x^3 - 21x^2 + 84x - 64,$$

$$k = 2: \quad x^3 - 273x^2 + 4368x - 4096,$$

$$k = 3: \quad x^3 - 65793x^2 + 16843008x - 16777216.$$

Estimates of the roots from the first new polynomial are

$$\sqrt{\frac{64}{84}} = 0.8729, \qquad \sqrt{\frac{84}{21}} = 2, \qquad \sqrt{\frac{21}{1}} = 4.5826.$$

From the second, we get

$$\sqrt[4]{\frac{4096}{4368}} = 0.9841, \qquad \sqrt[4]{\frac{4368}{273}} = 2, \qquad \sqrt[4]{\frac{273}{1}} = 4.0648.$$

From the third,

$$\sqrt[8]{\frac{16777216}{16843008}} = 0.9995, \qquad \sqrt[8]{\frac{16843008}{65793}} = 2, \qquad \sqrt[8]{\frac{65793}{1}} = 4.0020.$$

The exact values are 1, -2, 4. We must determine the signs of the roots from the estimates by substituting them into the original polynomial. If the sign of the root is positive, the result from the substitution is nearly zero; otherwise, the root is negative.   ▲

The method fails when there are roots of equal magnitude (and this will always be true for imaginary roots). Ralston (1965) shows how to overcome this difficulty.

Lehmer's Method

This method, also known as the Lehmer–Schur method, can be applied to find complex roots as well as real roots, and works with polynomials with complex coefficients. It does not require an initial approximation to a root. We will only outline the procedure. For details, Ralston (1965) is a good source.

We require that the polynomial not have a root at $x = 0$. (This will be obvious because there will then be no constant term in $P(x)$, so we reduce the polynomial until there is none.) We begin with a unit circle in the complex plane. Through a fairly complicated test procedure, we can determine whether the polynomial has any root(s) inside this circle. If it does, we test a circle half as large. If this circle also contains a root, we decrease the radius again and repeat until we find a circle that contains no roots. By this means we locate an annulus whose inner radius is one-half the outer radius that contains one or more roots. Now we draw a circle that covers at least $\frac{1}{6}$ of the annulus and test again for the presence of roots. We continue to test such smaller circles that move around the annulus by an amount equal to $\pi/3$ until, in at most six such trials, we locate a small circle that contains one or more roots. We apply the test on circles that shrink inside that circle to define another annulus, test still smaller circles that cover the annulus, and so on, until a root is located as precisely as desired. Of course, we continue to test within each annulus until it is entirely covered.

If we find no roots within the initial unit circle, instead of testing smaller circles, we enlarge the radius by a factor of two. If no roots are there, we enlarge again until we find the presence of a root. We now have a (larger) annulus that contains root(s), whereby we proceed analogously.

Although this method has wide applicability, it is less efficient than other root-finding procedures, even if the finding of initial values for, say, Newton's method, is by searching. After all, Newton's method works for complex roots if we are willing to do complex arithmetic (or we can use Bairstow if the polynomial has real coefficients).

Laguerre's Method

Suppose that $P_n(x) = (x - x_1)(x - x_2)(x - x_3) \cdots (x - x_n)$. By computing the derivative of $\ln|P_n|$, it is easy to show that

$$A = \frac{P_n'}{P_n} = \sum_{i=1}^{n} \frac{1}{(x - x_i)}.$$

By computing the second derivative of $\ln|P_n|$, we find that

$$B = \frac{-a^2(\ln|P_n|)}{ax^2} = \left(\frac{P_n'}{P_n}\right)^2 - \frac{P_n''}{P_n}$$

$$= \sum_{i=1}^{n} \frac{1}{(x - x_i)^2}.$$

We see that B is always positive.

Let x_1 be the root we want to determine. Assume that all other roots are distant from x_1 and are bunched closely together at some point $x = X$. (A pretty rash assumption, but it works!) Define $a = x - x_1$ and $b = x - X$. We can then rewrite the equations for A and B as

$$A = \frac{1}{a} + \frac{(n-1)}{b},$$

$$B = \frac{1}{a^2} + \frac{(n-1)}{b^2}.$$

From these two equations, eliminate b to get a:

$$a = \frac{n}{A \pm \sqrt{(n-1)(nB - A^2)}},$$

where we use plus if A is positive and minus if A is negative. We begin with a value for x_0 that is near to the desired root. Using this value, we compute A, B, and a. The next iterate is $x_1 = x_0 - a$. We repeat this until a is sufficiently small.

EXAMPLE 1.5

$$P(x) = x^3 - 8.6x^2 + 22.41x - 16.236$$

Take $x_0 = 1.0$: $P(1) = -1.426$, $P'(1) = 8.21$, $P''(1) = -11.2$. From these, $A = -5.7573$ and $B = 25.29306$. These give $a = -0.199973$, so $x_1 = -1.199973$. Continuing, we find the next $a = -2.800E-5$, so $x_2 = 1.200000$, which is exactly the smallest root of $P(x)$.

If we start with $x_0 = 5.0$, we get $a = 0.870960$; $x_1 = 4.12904$. Repeating, we get $a = 0.029025$ and $x_2 = 4.10000$, which is exactly the largest root. Because the sum of the roots is 8.6, the third root is at $x = 3.3$.   ▲

1.10 Multiple Roots

A function can have more than one root of the same value. When that is true, the graph will resemble Fig. 1.10. The curve on the left has a triple root at $x = -1$ [the function is $f(x) = (x + 1)^3$], whereas there is a double root at $x = 1$ for the curve on the right: $f(x) = (x - 2)^2$. If there were more than two or three roots, somewhat similar cuves would result, but they would be flatter near the x-axis and rise more steeply as the curve departs from the x-axis.

The methods we have studied do not work well with multiple roots. For example, Newton's method converges only linearly to a double root, as will be shown in the next section, whereas the method converges quadratically to a single (simple) root. (*Quadratic*

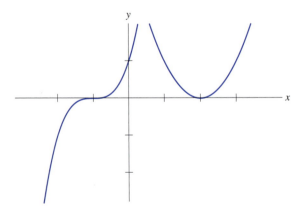

Figure 1.10 Functions with multiple roots

Table 1.7   Getting a double root, for $f(x) = (x - 1) * (e^{(x-1)} - 1)$

	Secant method	Newton's method	Muller's method
Estimate after 9 iterations	1.00331	1.00126	1.00058
Start value(s)	1.2, 1.5	2.0	0, 1.2, 1.5

convergence means that each error is proportional to the square of the preceding error. The net effect of quadratic convergence is that the number of significant figures in the estimates approximately doubles at each iteration.)

Still, our methods work to get a double root, except for bisection and linear iteration (because the function may not change sign at the root). Table 1.7 compares the convergence of Newton's method, the secant method, and Muller's method on $f(x) = (x - 1) *$ $(e^{(x-1)} - 1)$, which has a double root at $x = 1$.

From Table 1.7, we see that Newton's method is faster than the secant method, even though the latter starts with values nearer the root. Muller's method gets a more accurate estimate in nine iterations. (Comparing the closeness of starting values is not readily done.) Often Muller's method seems to converge more rapidly—but be careful! Some sets of starting values cause the method to fail, because the square root of a negative number is needed. For example, using starting values of 0, 0.5, and 1.5 with the function in Table 1.7 creates a parabola that does not cross the x-axis, so the parabola has complex roots.

Table 1.8 shows that convergence is linear with Newton's method, because each error is almost exactly half of the preceding error, especially as we get near the root. For a triple root, convergence will be even slower although all of our methods work, even those that require the function to change sign.

If we want MATLAB to plot $f(x) = (x - 1)(e^{(x-1)} - 1)$ over the interval $(-3, 4)$, we do the usual: first, define $f(x)$, then enter `fplot(f, [-3 4]` We see Fig. 1.11, which indicates that there is a double root at $x = 1$.

Table 1.8   Successive errors with Newton's method, for $f(x) = (x - 1) *$ $(e^{(x-1)} - 1)$

Iteration	Error	Iteration	Error
0	1.0	6	0.0199
1	0.5	7	0.0100
2	0.2798	8	0.0050
3	0.1494	9	0.0025
4	0.0775	10	0.00125
5	0.0395	11	0.000625

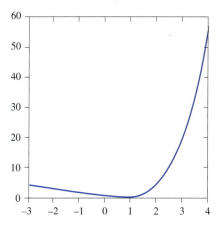

Figure 1.11   Plot of $(x - 1)(e^{(x-1)} - 1)$

Table 1.9   Getting a triple root, for $f(x) = (x + 1)^3 = 0$

	Bisection method	Newton's method	Muller's method
Estimate after 9 iterations	-0.99902	-1.01301	$-1.00000*$
Start value(s)	$-1.5, 0$	-1.5	$-1.5, -0.5, 0$

*Muller's method gave exact answer on first iteration!

Table 1.9 compares results from the bisection, Newton's, and Muller's methods on $f(x) = (x + 1)^3 = 0$, with a triple root at $x = -1$. Because this function is symmetrical about the root value, bisection will give immediate success if starting values of -2 and 0 are used. Remarkably, Muller's method converges immediately with many sets of starting values, but, again, a set such as $-1.5, -1.1, 0$ runs into a complex-valued estimate. Table 1.10 shows that Newton's method is again linearly convergent, because each error is $\frac{2}{3}$ times the previous one, and that convergence is slower than in the double-root example.

In addition to a slow convergence, there is another disadvantage to using these methods to find multiple roots: imprecision. Because the curve is "flat" in the neighborhood of the root—$f'(x)$ will always be zero at a multiple root, as is apparent from Fig. 1.10—there is a "neighborhood of uncertainty" around the root where values of $f(x)$ are very small. Thus the imprecise arithmetic of almost all computational devices will find $f(x)$ "equal" to zero throughout this neighborhood; that is, the program cannot distinguish which x-value is really the root. Using double precision will decrease the neighborhood of uncertainty. In fact, MATLAB's vpa command can give as much precision as desired, even to 100 significant figures, so this "neighborhood" can be very small.

It is interesting to see if our calculators give us an indication of multiple roots.

With the HP-48G, using <u>SOLVE</u> on $(x - 1)^2$ gives just one root: x = 1. (A plot shows a parabola tangent to the x-axis as expected.) However, if we use <u>Solve poly</u> on $x^2 - 2x + 1$, we get both: x = 1 and x = 1.

Table 1.10   Successive errors with Newton's method, for $f(x) = (x + 1)^3 = 0$

Iteration	Error	Iteration	Error
0	0.5	6	0.0439
1	0.3333	7	0.0293
2	0.2222	8	0.0195
3	0.1482	9	0.0130
4	0.0988	10	0.00867
5	0.0658		

With the TI-92, when we solve an equation such as $(x - 1)^2 = 0$ for x, we also find that only a single root, $x = 1$, is displayed. Even if we enter the quadratic in expanded form, $x^2 - 2x + 1 = 0$, we still get only the solution: $x = 1$. A simple cubic, $(x - 2)^3 = 0$, also gives only a single root value.

Remedies for Multiple Roots

Section 1.11 will examine the rate of convergence of Newton's method. We will see that, if there is a root of multiplicity k (there are k roots at $x = r$), we can restore quadratic convergence by modifying the algorithm to

$$x_{n+1} = x_n - k * \frac{f(x_n)}{f'(x_n)}.$$

Using this method to get the root of $f(x) = (x - 1) * (e^{(x-1)} - 1)$, we find that the third iterate is $x = 1.00088$ with $f(x) = 0.00000$. We also find that $e_{n+1} = 0.24 * e_n^2$, confirming quadratic convergence.

This algorithm would seem to solve the problem of multiple roots using Newton's method, but we don't know the multiplicity of the root in advance! (This objection is a little academic as the following argument shows.)

We might guess at the value for k and see whether we get quadratic convergence, or we could try several values and see what happens. Better yet, we could compare a graph of $f(x)$ with the plots of $(x - r)^k$, using an approximate value for r and various values for k. The "flatness" of the curves will be the same for $f(x)$ and the plot of equivalent multiplicity. We wonder, though, whether all such effort is justified—why not just live with the linear convergence? We will find the root with sufficient accuracy from that operation long before we complete the alternative explorations.

Another solution to multiple roots is tempting to consider. We can divide $f(x)$ by $(x - r)$ and deflate the function, reducing the multiplicity by one. The problem here is that we don't know r. However, dividing by $(x - s)$, where s is an approximation of r, does almost the same thing. We suggest that you might want to explore this idea. Be warned that the division creates an indeterminate form at $x = r$ and a strong discontinuity at $x = s$.

Using the Derivative Function

Because $f'(x) = 0$ at a multiple root, we can converge quadratically by applying Newton's method to $f'(x)$ to get a double root of $f(x)$. We need to apply this method to $f''(x)$ to get quadratic convergence to a triple root. The agony we sometimes experience in differentiating complicated functions is largely relieved if we let a computer algebra system do the work.

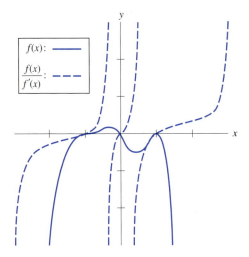

Figure 1.12 Plots of $f(x)/f'(x)$ when $f(x) = (x - 1)^3 \sin(x)$

A most tempting scheme is to compute the roots of $f(x)/f'(x)$. It is easy to show that, when $f(x)$ has a root of multiplicity n at $x = r$, $f'(x)$ has a root of multiplicity $n - 1$ at that point. Let

$$f(x) = (x - r)^n \, F(x), \qquad \text{and} \qquad F(r) \neq 0.$$

Then

$$
\begin{aligned}
f'(x) &= n(x - r)^{n-1}F(x) + (x - r)^n F'(x) \\
&= (x - r)^{n-1}[nF(x) + (x - r)F''(x)].
\end{aligned}
$$

Hence, if we divide $f(x)$ by $f'(x)$, we effectively deflate the function $(n - 1)$ times and we now work with a new function that has only a single root at $x = r$.

Some warnings are in order, however. This technique works fine with polynomials, but when $f(x)$ involves transcendentals, there may be difficulties. The deflated function will have infinite discontinuities at the maxima and minima of $f(x)$, which can distort the reduced function and make it difficult to converge to some single roots of $f(x)$. Figure 1.12 shows the plots of $f(x) = (x - 1)^3 \sin(x)$ and of $f(x)/f'(x)$. Observe the discontinuities at points where $f'(x) = 0$ and the distortions at the zeros of the sine function.

Nearly Multiple Roots

A problem related to multiple roots is a function that has two or more roots very close together. If these roots are all within the region of uncertainty (which is a function of the arithmetic precision we are using), they are effectively multiple roots because for all of them $f(x)$ is computationally equal to zero.

Newton's method is again essentially linearly convergent when we have nearly equal roots, provided that we start outside the interval that holds the roots. Unfortunately, modifying the method by considering them to be multiple roots doesn't work; often an infinite loop occurs. If we are so unlucky as to start between two almost equal roots, Newton's method can fly off to "outer space," as we previously observed.

Whenever we want to find roots that are near $f'(x) = 0$, we are in trouble. We strongly recommend that you graph the function, before jumping into a root-finding routine, to see in advance whether such problems will arise.

1.11 Theoretical Matters

We mentioned rates of convergence earlier in this chapter, so it is essential that we substantiate these statements. We also need to be more precise in how we express convergence rates.

We define the *order of convergence* for a method by the following:

If $|e_{n+1}|$ approaches $K * |e_n|^p$ as n becomes infinite,

we say that the method is of order p. (It is almost always true that the errors do not decrease this fast until we are near the root.) How do we know a method is of a certain order; why can we say, for example, "Newton's method is of order 2—its convergence is quadratic"?

One way to find the order of a method is to perform experiments with the method and see how the errors decrease. (We must be sure not to be too hasty in reaching the conclusion, because the order is attained only as a limit as the root is approached.) Figure 1.13 is

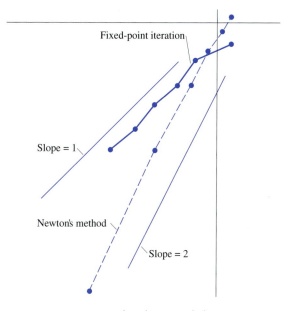

Fixed-point iteration

Slope = 1

Newton's method

Slope = 2

Figure 1.13 Plots of $\ln|e_{n+1}|$ versus $\ln|e_n|$ for Newton's method and for fixed-point iteration

a plot of $\ln|e_{n+1}|$ versus $\ln|e_n|$ for Newton's method on the cubic $x^3 - 2x - 3$ with a starting value of 5. A line of slope 2 is shown, and the points for Newton's method do seem to be about parallel. The figure also shows results when fixed-point iteration is used on the same function $[g(x) = \sqrt{2 + 3/x}]$. These points seem to parallel a line whose slope is 1.

Plotting the logarithms as we have done will give a straight line of slope p if the relation is true everywhere, not just in the limit. We see from Fig. 1.13 that although the points do not lie exactly on a line or even on a smooth curve, the slopes appear to match the orders that we claim for the two methods as the iterations proceed (points at the left end of the broken lines are from the later iterations).

The trouble with this experimental approach is that it is not general enough; we cannot conclude the order of a method from experiments with a few functions no matter how cleverly we chose them. That is why theoretical studies are so important; they give conclusions that are true for all functions. We must emphasize, though, that our analysis of errors in this theoretical discussion assumes perfect arithmetic. The errors are errors of the *method,* not errors resulting from imperfections of the computing device.

Fixed-Point Iteration

We begin with an analysis of fixed-point iteration (which we also call the $x = g(x)$ method). Recall that we iterate with

$$x_{n+1} = g(x_n).$$

If we let $x = r$ be a solution of $f(x) = 0$, then $f(r) = 0$ and $r = g(r)$. Subtracting, and then multiplying and dividing by $(x_n - r)$, we have

$$x_{n+1} - r = g(x_n) - g(r) = \frac{g(x_n) - g(r)}{(x_n - r)} (x_n - r).$$

If $g(x)$ and $g'(x)$ are continuous on the interval from r to x_n, the mean-value theorem* lets us write

$$x_{n+1} - r = g'(\xi_n) * (x_n - r),$$

where ξ_n lies between x_n and r. If we define the error of the ith iterate as $e_i = r - x_i$, we can write

$$e_{i+1} = g'(\xi_i) * e_i.$$

Taking absolute values (because the successive iterates may oscillate around the root), we get

$$|e_{i+1}| = |g'(\xi_i)| * |e_i|.$$

---

*Appendix A reviews certain principles of calculus, including this theorem.

Now suppose that $|g'(x)| < K < 1$ for all values of x in an interval of radius h about r. If x_0 is chosen in this interval, all succeeding iterates will be in this interval and the method will converge, because

$$|e_{n+1}| < K * |e_n| < K^2 * |e_{n-1}| < K^3 * |e_{n-2}| < \cdots < K^{(n+1)} * |e_0|.$$

We therefore conclude that fixed-point iteration is of order 1.

In summary, if $g(x)$ and $g'(x)$ are continuous on an interval about a root r of the equation $x = g(x)$, and if $|g(x)| < 1$ for all x in the interval, then $x_{n+1} = g(x_n)$, $n = 0, 1, 2, \ldots$, will converge to the root r, provided that x_0 is chosen in the interval. Note that this is a sufficient condition only, because for some equations, convergence is secured, even though not all the conditions hold.*

Ralston (1965) points out that measuring the worth of methods by their order is not enough. We should be concerned with efficiency by accounting for the "cost" of the iterations. We probably should determine the cost factor by totaling the number of operations required and the time for each; often, just counting the number of function evaluations is enough, because they occupy most of the computer's time.

Order of Convergence for Newton's Method

Newton's method uses the iterative scheme

$$x_{n+1} = x_n - \frac{f(x_n)}{f'(x_n)} = g(x_n),$$

from which we see that it is of the same form as fixed-point iteration. Hence, for this scheme we can use the preceding theory: Successive iterates converge if $|g'(x)| < 1$. We assume that the root at $x = r$ is simple. Differentiating $g(x)$ gives

$$g'(x) = 1 - \frac{f'(x) * f'(x) - f(x) * f''(x)}{[f'(x)]^2} = \frac{f(x) * f''(x)}{[f'(x)]^2}.$$

Hence, if

$$\frac{f(x) * f''(x)}{[f'(x)]^2} < 1$$

on an interval about the root r, the method will converge for any initial value x_0 in the interval. The condition is sufficient only; it requires the usual conditions of continuity and

---

*The analytical test that $|g'(x)| < 1$ is often awkward to apply. A constructive test is merely to observe whether the successive x_i values converge. In a computer program it is worthwhile to determine whether $|x_{n+1} - x_n| < |x_n - x_{n-1}|$ or, alternatively, whether the iterates "jump" far away.

existence for $f(x)$ and its derivatives. Note that $f'(x)$ must not be zero within the interval as it is for the pathological situation of Fig. 1.4. More specifically, $f'(r)$ must not be zero as it is for a multiple root at $x = r$. (See the next subsection).

Now we show that Newton's method is quadratically convergent. Because r is a root of $f(x) = 0$, $r = g(r)$. Because $x_{n+1} = g(x_n)$, we can write

$$x_{n+1} - r = g(x_n) - g(r).$$

Let us expand $g(x_n)$ as a Taylor series in terms of $(x_n - r)$, with the second derivative term as the remainder:

$$g(x_n) = g(r) + g'(r) * (x_n - r) + \frac{g''(\xi)}{2}(x_n - r)^2, \tag{1.2}$$

where ξ lies in the interval from x_n to r.

Because

$$g'(r) = \frac{f(r) * f''(r)}{[f'(r)]^2} = 0 \tag{1.3}$$

and because $f(r) = 0$ (r is a root), we have

$$g(x_n) = g(r) + \frac{g''(\xi)}{2}(x_n - r)^2.$$

Then, if we let $(x_n - r) = e_n$, we have

$$e_{n+1} = x_{n+1} - r = g(x_n) - g(r) = \frac{g''(\xi)}{2}e_n^2. \tag{1.4}$$

As the iterates approach r, so does ξ. Thus each error is (in the limit) proportional to the second power of the previous error, and Newton's method has convergence of order 2—it is quadratically convergent.

Newton's Method with Multiple Roots

If $f(x) = 0$ has k zeros at $x = r$, we say that r is a root of multiplicity k. Thus we can factor out $(x - r)$ k times to give

$$f(x) = (x - r)^k Q(x), \tag{1.5}$$

where $Q(x)$ has no root at $x = r$ [it is deflataed from $f(x)$]. That means that $Q'(r)$ is not zero. However, each of $f'(r), f''(r), \ldots, f^{(k-1)}(r)$ is zero, as is readily found by differentiating $f(x)$ repeatedly.

We see that Eq. (1.3) has a denominator equal to zero if the root at $x = r$ is multiple. Although the ratio is indeterminant, we cannot say that $g'(r) = 0$ and that Eq. (1.4) therefore does not apply. Because the $g'(r)$ term does not drop out of the Taylor series expansion, it must be true that Newton's method is linearly convergent if r is a multiple root.

We have stated that if the multiplicity k of the root is known, we can regain quadratic convergence by modifying Newton's method to read

$$x_{n+1} = x_n - k * \frac{f(x_n)}{f'(x_n)}. \tag{1.6}$$

We need to demonstrate that this modification does restore quadratic convergence.

As before, because $f(r) = 0$, $g(r) = r$. Here we have

$$g(x) = x - k * \frac{f(x)}{f'(x)}.$$

If we substitute $f(x) = (r - x)^k Q(x)$ and differentiate, we get

$$g'(x) = \frac{(r - x)\{k(r - x)QQ'' + Q'[2kQ - (k - 1)(r - x)Q']\}}{[(r - x)Q' + kQ]^2}$$

and we see that $g'(r) = 0$. From the preceding argument, then, the modified Newton's method now converges quadratically at a multiple root. (It also does so at a simple root with $k = 1$, of course.)

The only problem with this technique is that we may not know the proper value for k, although we gave some suggestions in Section 1.10.

Acton (1970) gives another technique by which we may obtain a multiple root with quadratic convergence. If $f(x)$ has a root of multiplicity k at $x = r$, we have $f(x) = (r - x)^k * Q(x)$. Let $S(x)$ be $f(x)/f'(x)$, so that

$$S(x) = \frac{(x - r)^k Q(x)}{k(x - r)^{k-1}Q(x) + (x - r)^k Q'(x)} = \frac{(x - r)Q(x)}{kQ(x) + (x - r)Q'(x)},$$

which has a simple root at $x = r$. When $S(x)$ is used in the Newton formula, we get

$$x_{n+1} = x_n - \frac{S(x_n)}{S'(x_n)} = x_n - \frac{f(x_n) * f'(x_n)}{[f'(x_n)]^2 - f(x_n) * f''(x_n)},$$

and we see that we need to evaluate three functions at each iteration: $f(x_n)$, $f'(x_n)$, and $f''(x_n)$. Acton also points out that there are nearly equal quantities being subtracted in the denominator, a source of arithmetic error in addition to the errors we are discussing here.

Convergence Order for Other Methods

Bisection (Interval Halving).   For this method, at each iteration we divide the original interval in half, confining the root in the new interval. If we take the midpoints of the successive intervals to be estimates of the root, obviously one-half the current interval is an upper bound to the error. Although we cannot say that $|e_{n+1}| = 0.5 * |e_n|$, if we are willing to call the upper bounds the estimates of the errors, we may say that bisection is linearly convergent. This does not mean, as we have seen, that each error is smaller than the previous one, even though the *error estimate* is smaller.

Linear Interpolation (False Position) and the Secant Method.   Both of these methods use iteration of the form

$$x_{n+1} = x_n - \frac{f(x_n)}{f(x_n) - f(x_{n-1})} (x_n - x_{n-1}) \tag{1.7}$$

to estimate r, a root of $f(x) = 0$. At each iteration, we replace one of the previous x-values with the new estimate. With false position, the replacement is such that the root is bracketed. It is usually found that one end of the interval does not change when this is done, so we can think of x_{n-1} as a constant in the formula. Thus Eq. (1.7) is of the form $x = g(x)$ and we can use the preceding development directly. We need to know whether $g'(r)$ is zero. We first write out $g(x)$:

$$g(x) = x - \frac{f(x)}{f(x) - F} (x - X),$$

where we use F and X to emphasize that they are fixed quantities. On differentiating we get

$$g'(x) = \frac{f'(x) * F * (x - X) - F * [f(x) - F]}{[f(x) - F]^2},$$

and we see that $g'(r)$ is not zero. Hence, because this is exactly the same as for fixed-point iteration, linear interpolation (false position) is linearly convergent.

The secant method differs in that neither of the two previous x-values is fixed. We must therefore consider the iterations to be

$$x_{n+1} = g(x_n, x_{n-1}),$$

so that g is a function of two independent variables. We can parallel the development for Newton's method, but now we must remember that

$$g'(x_n, x_{n-1}) = g_{x_n} + g_{x_{n-1}},$$

where the subscripts on g indicate partial derivatives.

The partial derivatives are

$$g_{x_n} = \frac{f(x_{n-1})[(x_n - x_{n-1})f'(x_n) - f(x_n) + f(x_{n-1})]}{[f(x_n) - f(x_{n-1})]^2},$$

$$g_{x_{n-1}} = \frac{f(x_n)[(x_n - x_{n-1})f'(x_{n-1}) - f(x_n) + f(x_{n-1})]}{[f(x_n) - f(x_{n-1})]^2},$$

and we observe that $g'(r, r)$ is zero because both partials are zero. This technique is similar to Newton's method, so we look at the next term in the Taylor series expansion of g. This process involves both variables and the second partial derivatives. We omit the details (the derivatives are very messy), but the partials are not zero at $x = r$. Although the secant method is similar to Newton's method, we cannot conclude that the secant method is quadratically convergent, because it turns out that

$$e_{n+1} = \frac{g''(\xi_1, \xi_2)}{2} (e_n)(e_{n-1}).$$

Because the error is proportional not to the square of a previous error but to the product of the two previous errors, we conclude that the secant method is better than the linear method but poorer than the quadratic method. Pizer (1975) shows that through solving an approximate difference equation, the order of the secant method is

$$\frac{1 + \sqrt{5}}{2} = 1.62.$$

Muller's Method.   Muller's method is more difficult to analyze for its rate of convergence. (The iterating function g is a function of three variables.) Ralston (1965) outlines a technique that can be used to show that Muller's method has a rate of convergence of order 1.84.

Aitken Acceleration.   If Aitken acceleration is applied after every second iteration of fixed-point iteration, and the improved estimate is used to begin each set of iterates, the accelerated iterates converge quadratically. This is known as Steffenson's method. Henrici (1964) should be consulted for the proof.

Special Methods for Polynomials

We omit the details, but it has been shown that Laguerre's method for polynomials is cubically convergent (Householder, 1970). The quotient-difference method, Lehmer's method, and Graeffe's method do not lend themselves to this kind of error analysis, as they are not iterative in the same sense.

1.12 Using MATLAB

MATLAB has facilities to solve equations of all types, but it is especially rich in operations on polynomials.

We have already seen that there are two ways to express a polynomial in MATLAB: as an array of coefficients or algebraically. We use the former technique to get the roots or to divide them. We also use the array definition to multiply two polynomials with 'conv,' the reverse operation to 'deconv.' We can also add or subtract two polynomials that are in array form but these operations require that the two arrays be of the same size; we can pad the lower degree polynomial with leading zeros to achieve this. We evaluate a polynomial that is expressed as an array with the command

```
v = polyval(p,x)
```

which stores the $p(x)$ in variable v. In a similar way, the derivative of $p(x)$ is stored in variable dp with

```
dp = polyder(p)
```

MATLAB even lets one get a polynomial with desired roots:

```
PR = poly([1.1 2.2 3.3])
```

gives

```
PR =
      1.0000 -6.6000 13.3100 -7.9860
```

In contrast to these operations on polynomials expressed as an array of coefficients, we use the algebraic form to plot a polynomial (or any function, for that matter). (The MATLAB manual calls the algebraic form a "symbolic string.") When the polynomial is in algebraic form, we can get the Horner representation of $P(x)$ with the command: horner(P). If the polynomial can be expressed as products of polynomial factors with integer coefficients, the command: factor(P) will find these. We can convert back to standard polynomial form with: expand( ). The command: pretty(P) displays the polynomial with the powers of x shown as exponents.

Solving Equations Other Than Polynomials

The 'solve' command will find the solution to almost any equation, including polynomials in algebraic form. If the parameter is an expression rather than an equation, the result is the solution when the expression is set equal to zero. Here is one example:

```
A = solve('x^2 - x + 3')


A =


[1/2+1/2*i*11^(1/2)]
[1/2-1/2*i*11^(1/2)]
```

The answer is not easy to comprehend because it is in symbolic form, but converting to numeric shows the two roots as we would ordinarily write them:

```
numeric(A)


ans =


   0.5000 + 1.6583i
   0.5000 - 1.6583i
```

Another example, with the format set to long, is:

```
fx = 'x^3 - exp(x) - 1';


solve (fx)


ans =


2.081116467461713
```

and we have found the root near $x = 2$. (A plot of fx shows clearly that there is a root there. In fact, if all we need is an approximate solution, doing the plot is the best way to find it; the graph also shows the behavior of the function within an interval.)

Another example:

```
solve('2 * x^2 = 3')

ans =

[ 1/2 * 6^(1/2)]
[-1/2 * 6^(1/2)]

numeric (ans)

ans =

    1.2247
   -1.2247
```

The solve command gets the solution with respect to any single variable in the expression or equation. For example,

```
fy = 'y^2 - 1/y = 3';

numeric(solve(fy))

ans =

  1.8794
 -1.5321
 -0.3473
```

If there is more than one variable, we can specify which one we want to solve for:

```
pretty(solve)('a * x^2 = 1/y','y'))
```

$$\frac{1}{ax^2}$$

A classical example is to get the quadratic formula:

```
solve('a * x^2 + b * x + c')

ans =

[1/2/a * (-b + (b^2 - 4 * a * c)^(1/2))]
[1/2/a * (-b - (b^2 - 4 * a * c)^(1/2))]
```

However,

```
solve ('a * x^2 + b * x + c','b')
```

gives

```
ans =
```

```
-(a * x^2 + c)/x
```

The MATLAB manual does not specify which algorithms are used by the solve function. If we want to employ one of the standard techniques for solving an equation, we ordinarily have to write a program. Programs must be stored in a file that MATLAB calls an "M-file" because they must be stored with an extension of m.

However, a fixed-point iteration is a simple enough procedure that we can do it directly. For example, to solve for the root near 2 of $f(x) = x^3 - e^x - 1$ through the rearrangement $x = (e^x + 1)/x^2$, we can do the following:

```
x = 2;
```

```
gx = '(exp(x) + 1)/x^2';
```

```
x = eval(gx)
```

```
x =
```

```
    2.0973
```

which is the first iterate. We get additional iterates by repeating:

```
x = eval(gx)
```

```
x =
```

```
    2.0789
```

```
x = eval(gx)
```

```
x =
```

```
    2.0815
```

and we are approaching the root that is near 2.08111. One nice thing about MATLAB is that previous commands can be recalled with the up-arrow key, so the commands that repeat the iteration do not have to be typed in.

Newton's method is actually a fixed-point iteration, so we can use it in a similar manner. We leave it to the reader to verify that repeating the command:

```
x = x - eval(fx)/eval(diff(fx))
```

will do Newton's method on fx starting with an initial value stored in variable x. You should find, after three iterations from $x = 2$, that the root is 2.081116.

With the 'solve' operation available and knowing how to use Newton's method to solve $f(x) = 0$, it is perhaps redundant to write a program to implement other methods. However, it is instructive to do this if only to illustrate programming in MATLAB. Here is a program to use the bisection method. We will approach this gradually, first doing bisection from the keyboard.

We begin by defining $f(x)$, setting starting x-values in xa and xb, and evaluating $f(xa)$ and $f(xb)$. Here is what MATLAB shows:

```
fx = 'x^3 + x^2 - 3 * x - 3'

fx =

x^3 + x^2 - 3 * x - 3

xa = 1, xb = 2

xa =

        1

xb =

        2
```

Now we set xc equal to the midpoint, evaluate $f(xc)$, see if there is a sign change in fx between xa and xc, setting xb equal to xc if there is or setting xa equal to xc if not. This we repeat as many times as we need to get a value of $f(xc)$ that is small, at which time xc approximates the root. In other words, we want to repeat these commands:

```
xc = (xa + xb)/2;
x = xc;fc = eval(fx);
if fc * fa < 0
   xb = xc;
else
   xa = xc;
end
xa,xb,xc,fa,fb,fc % This displays the results
```

and we expect MATLAB to display the successive commands. We could do this from the keyboarad (and the ability to recall commands helps) but it is better to put the repeated set of commands into an M-file and repeat them by entering the name of the M-file. Suppose we do this and name the file bis.m. Then when we enter "bis" successively, we see a set of results that match the entries in Table 1.1.

It would improve this if we could automatically repeat the steps n times, displaying the successive results all on one line. M-files in MATLAB can be a function as well as a simple succession of commands. Here is the M-file 'bisec.m.'

```
function rtn = bisec(fx,xa,xb,n)
% bisec does n bisections to approximate
% a root of fx
x = xa; fa = eval(fx);
x = xb; fb = eval(fx);
for i = 1:n
   xc = (xa + xb)/2; x = xc; fc = eval(fx);
   X = [i, xa, xb, xc, fc];
   disp(X)
   if fc * fa < 0
      xb = xc;
   else
      xa = xc;
   end  % of the if/else
end  % of the for loop
X = [n, xa, xb, xc, fx];
disp X  % repeat the last result
```

Then when we enter the command

```
bisec(fx,1,2,13)
```

we see Table 1.1.

Chapter Summary

If you have understood this chapter on solving nonlinear equations, you are now able to

1.   Recognize that a problem, even if it is disguised, may be one that does not require the powerful methods of this chapter.

2.   Use these methods to find solutions to $f(x) = 0$:

   Bisection (also called interval halving)

   Linear interpolation (also called false position)

   Secant method

   Newton's method

   Muller's method

   Fixed-point iteration [also called $x = g(x)$ method].

3.   Apply special methods to find roots of polynomials:

   Newton's method with synthetic division

   Bairstow's method

   QD algorithm

   Graeffe's root-squaring method

   Laguerre's method.

4. Explain how to use Aitken acceleration to speed the convergence of a method that is linearly convergent.

5. Show how Horner's method (synthetic division, nested multiplication) reduces the number of operations required to evaluate a polynomial.

6. Draw a directed acyclic graph (dag) to indicate the successive operations for a sequential procedure and a corresponding equivalent parallel processing procedure; show that the latter may run faster even though requiring more operations.

7. Realize that multiple roots cause problems in root finding, and know several techniques to overcome these difficulties.

8. Write the relation that indicates that the iterative method is of order p, and outline the proofs that substantiate these statements:

    Fixed-point iteration is linearly convergent.

    Newton's method converges with order 2 at a simple root.

    Newton's method is linearly convergent at a multiple root.

    The method of false position converges linearly.

9. For one or more of the methods of this chapter, make a plot that indicates the order of the method and explain why this is not the best way to determine the order.

10. Use MATLAB to evaluate $f(x)$ at $x = a$, to get the derivative of a function, to perform fixed-point iterations, and to carry out Newton's method. You can write a program in MATLAB.

Computer Programs

Beginning with this chapter, we conclude each chapter with example programs that implement algorithms of the chapter. We do not attempt to include code for every algorithm; this allows students to gain experience in writing their own programs. That skill is essential to the budding numerical analyst, even though prewritten programs will most often be used when doing numerical analysis professionally. Although these professional packages can supply the most useful algorithms, one cannot read these or decide which is best suited for the task at hand without some background in creating at least simple programs.

Even though a variety of computer languages can be used for programs that do numerical analysis, we will use only FORTRAN and C because these two languages are by far the more widely used. We do suggest in some of the exercises that some algorithms be implemented in another language just to emphasize the fact that the computer language is not of great importance. This can also help to hone the skill of the student in programming.

There are other ways to carry out numerical computations. Spreadsheets can be utilized to perform many procedures. (A good reference that shows a variety of such spreadsheet implementation is Orvis, 1987. See also Etter, 1993).

It is always good practice to incorporate informative comments within a computer program. We have commented liberally in our programs to make the logic easy to follow. When you write your programs, you should do the same.

Do not use our programs as models of good programming practice. Our programs are intended only to show clearly how a computer program can carry out the example algorithm. They may not always be written in a structured style, and we also do not incorporate the checks and error trapping that a good professional program should contain; we think this may make them easier to read.

There are just two programs in this chapter. The first is a FORTRAN program that uses Muller's method to get the root of an equation; the second is written in C and it carries out the Bairstow technique to find the quadratic factors of a polynomial.

The FORTRAN program solves Example 1.2, where a root between $x = 0$ and $x = 1$ is found. The definition of the equation and the required parameters are incorporated within the program.

The C program gets the coefficients of the polynomial and the necessary starting values from the user. The output represents the solution to Example 1.3.

Program 1.1 Muller's Method

Figure 1.14 is a listing of the FORTRAN program. It is written in FORTRAN 90 and should run correctly when compiled with any standard FORTRAN 90 compiler. It has been tested with elf90, F90, and xlf90 compilers. At the end of the listing is the output from an execution of the program, when Example 1.2 is solved.

```
Program PMuller
!
! !!!!!!!!!!!!!!!!!!!!!!!!!!!!!!!!!!!!!!!!!!!!!!!!!!!!!!!!!!!!!!!
! Chapter 1                 Muller's Method                    !
!                                                              !
! Gerald/Wheatley, APPLIED NUMERICAL ANALYSIS (Sixth Edition)  !
!                  Addison Wesley Longman, 1999                !
!                                                              !
! !!!!!!!!!!!!!!!!!!!!!!!!!!!!!!!!!!!!!!!!!!!!!!!!!!!!!!!!!!!!!!!
!
! This driver calls Subroutine Muller
!
! -----------------------------------------------------------
!
    Implicit None
    Real :: xr = 0.5
!
    Interface
       Function Fcn(x)
       Implicit None
       Real :: Fcn
       Real, Intent(In) :: x
       End Function Fcn
    End Interface
```

Figure 1.14

```
!
!
!
   Call Muller(Fcn,xr,0.2,0.0001,0.00001,50)
!
!
   Stop

 Contains
!
!----------------------------------------------------------
!
   Subroutine Muller(Fcn,xr,h,xtol,ftol,nlim)
!
!!!!!!!!!!!!!!!!!!!!!!!!!!!!!!!!!!!!!!!!!!!!!!!!!!!!!!!!!!!!!!
!                                                          !
!  Subroutine Muller:                                      !
!    This Subroutine finds the root of F(X) = 0 by         !
!    quadratic interpolation on three points - Muller's method. !
!                                                          !
!!!!!!!!!!!!!!!!!!!!!!!!!!!!!!!!!!!!!!!!!!!!!!!!!!!!!!!!!!!!!!
!                                                          !
!  PARAMETERS ARE :                                        !
!                                                          !
!  FCN      -function that computes values for f(x).       !
!  xr       -initial approximation to the root. It is used to !
!            begin iterations. Also returns the value of the root. !
!  h        -displacement from x used to begin calculations. The !
!            first quadratic is fitted at f(x), f(x+h), f(x-h). !
!  xtol,ftol -tolerance values for X, f(x) to terminate iterations. !
!                                                          !
!!!!!!!!!!!!!!!!!!!!!!!!!!!!!!!!!!!!!!!!!!!!!!!!!!!!!!!!!!!!!!
!
   Implicit None
   Interface
     Function Fcn(x)
     Implicit None
     Real :: Fcn
     Real, Intent(In)  :: x
     End Function Fcn
   End Interface
!
   Real, Intent(In) :: xtol, ftol, h
   Real, Intent(In Out) :: xr
   Real, Dimension(3) :: y
   Real  :: f1,f2,f3,h1,h2,g,a,b,c,disc,fr,delx
   Integer :: j = 0
   Integer, Intent(In) :: nlim
!
  Write(*,*) ' ****************************************** '
  Write(*,*) '              Output for pmullr.f90         '
  Write(*,'(/)')
!
```

Figure 1.14 *Continued*

```
! ---------------------------------------------------------
!
! Set initial values into Y array and evaluate F at these points
!
      y(1) = xr - h
      y(2) = xr
      y(3) = xr + h
      f1     = Fcn( y(1) )
      f2     = Fcn( y(2) )
      f3     = Fcn( y(3) )
      Write(*,199) j,xr,f2
!
! ---------------------------------------------------------
!
! Begin iterations
!
   Do  j = 1, nlim
      h1 = y(2) - y(1)
      h2 = y(3) - y(2)
      g = h1 / h2
      a = ( f3*g - f2*(1.0 + g) + f1 ) / ( h1*(h1 + h2))
      b = ( f3 - f2 - a*h2*h2 ) / h2
      c = f2
      Disc = Sqrt( b*b - 4.0*a*c )
      If ( b < 0.0 ) Disc = -Disc
!
! ---------------------------------------------------------
!
! Find root of quadratic : A * V**2 + B * V + C = 0
!
      Delx = -2.0 * c / ( b + Disc )
!
! Update xr
!
      xr = y(2) + delx
      fr = Fcn(xr)
      Write(*,199) j,xr,fr
!
! ---------------------------------------------------------
!
! Check stopping criteria
!
      If ( Abs(delx) <= xtol ) Then
        Write(*,202) j,xr,fr
        Return
      End If
!
      If (Abs(fr) <= ftol ) Then
        Write(*, 203) j,xr,fr
        Return
      End If
!
```

Figure 1.14 *Continued*

```
! -------------------------------------------------------
!
! Select the three points for the next iteration.
!       When delx > 0, choose Y(2), Y(3), & XR
!       When delx < 0, choose Y(1), Y(2), & XR.
!
! Enter the proper set into Y array so that they are in ascending
! order.
!
      If ( delx >= 0 ) Then
         y(1) = y(2)
         f1   = f2
         If ( delx > h2 ) Then
            y(2) = y(3)
            f2   = f3
            y(3) = xr
            f3   = fr
         Else
            y(2) = xr
            f2   = fr
         End If
      Else
         y(3) = y(2)
         f3   = f2
         If ( Abs(delx) > H1 ) Then
            y(2) = y(1)
            f2   = f1
            y(1) = xr
            f1   = fr
         Else
            y(2) = xr
            f2   = fr
         End If
!
      End If
!
   End Do
!
! -------------------------------------------------------
!
! When loop is normally terminated NLIM is exceeded.
!
         Write(*,200) nlim,xr,fr
!
! -------------------------------------------------------
!
 199  Format(' At iteration',I3,"   ",' x =',E12.5,"   ",'f(x) = ',
     &          E12.5)
 200  Format(/' Tolerance not met after ',I4,' iterations  x = ', &
     &          E12.5,' f(x) = ',E12.5)
 202  Format(/' tolerance met in ',I2,' iterations   x = ',E10.5, &
     &          '  f(x) = ',E10.5)
```

Figure 1.14 *Continued*

```
 203   Format(/' F tolerance met in ',I4,' iterations   x = ',E10.5, &
            '   f(x) = ',E10.5)
!
       Return
     End Subroutine Muller

   End Program PMuller
!
! ----------------------------------------------------------
!
     Function Fcn(x)
!
     Implicit None
     Real :: Fcn
     Real, Intent(In)  :: x
     Fcn = 3*x + Sin(x) - Exp(x)
     Return
     End Function Fcn

* * * * * * * * * * * * * * * Output for Pmullr Program  * * * * * * * * * * * * * * *

At iteration  0    x = 0.50000E+00    f(x) =  0.33070E+00
At iteration  1    x = 0.35995E+00    f(x) = -0.11837E-02
At iteration  2    x = 0.36042E+00    f(x) =  0.15497E-05

F tolerance met in   2 iterations  x = .36042E+00  f(x) = .15497E-05
```

Figure 1.14 *Continued*

Program 1.2 Bairstow's Method

Figure 1.15 lists the C program. The output, shown at the end of the program listing, is for the solution of Example 1.3.

```
/*
*  Chapter 1                    Bairstow's Method                  !
*
* Gerald/Wheatley, APPLIED NUMERICAL ANALYSIS (Sixth Edition)
*                  Addison Wesley Longman, 1999
*
* * * * * * * * * * * * * * * * * * * * * * * * * * * * * * * * * * * * * * * * * * * *
*                                                                    *
*    This program finds a quadratic factor of a polynomial          *
*    by Bairstow's method. The coefficients of the polynomial       *
*    are input by the user as well as the values for r and s in     *
*    the initial trial quadratic factor.                            *
*                                                                    *
```

Figure 1.15

```
*    After each iteration, the improved quadratic factor is        *
*    displayed. The user can request additional iterations,        *
*    terminating the program when the factor is obtained with      *
*    sufficient precision.                                         *
*                                                                  *
* * * * * * * * * * * * * * * * * * * * * * * * * * * * * * * * * * * * * * * * * * *

#include <stdio.h>
#include <math.h>

/*  Function to get determinant */
float det(a11, a12, a21, a22)
  float a11, a12, a21, a22;
{
  return (a11*a22 - a21*a12);
}                              /* end of function */

/* Here comes main */
void main ()
{
  float a[20], b[20], c[20], x, r, s, den;
  int nIt, nDg, i, quit;
  char inbuf[20];

/* Get degree and coefficients from user -- constant first */
  clrscr();
  printf ("What degree? \n");
  gets (inbuf);
  sscanf (inbuf, "%d" , &nDg);

  printf ("Enter coefficients, beginning with the constant term.\n");
  for (i= 0; i<=nDg; i++)
  {
    printf ("Coef of x[%2d]: ", i);
    gets (inbuf);
    sscanf (inbuf, "%f", &a[i]);
  }

/* Get r and s in quadratic factor: x^2 - rx - s from user */
  printf ("Enter r and s values: ");
  gets (inbuf);
  sscanf (inbuf, "%f %f", &r, &s);

/* Begin loop. To be stopped by user */

  nIt = 1;
  do                               /* do/while begins */
  {
 /* Do synthetic division */
    b[nDg+2] = b[nDg+1] = 0;
```

Figure 1.15 *Continued*

```
            for (i=nDg; i>=0; --i)
               b[i] = a[i] + r*b[i+1]+ s*b[i+2];
            c[nDg+2] = c[nDg+1] = 0;
            for (i=nDg; i>0; --i)
               c[i] = b[i] + r*c[i+1] + s*c[i+2];

          /* Get new r, s and print them */
               den = det(c[2], c[3], c[1], c[2]);
               r += det(-b[1], c[3], -b[0], c[2]) / den;
               s += det(c[2], -b[1], c[1], -b[0]) / den;

           printf ("At iteration %3d, r and s are : %7.4f, %7.4f\n",nIt, r, s);

        /* See if user wants another iteration */
            printf (" Enter 0 to quit, anything else to do another  ");
            gets (inbuf);
            sscanf (inbuf, "%d", &quit);
            nIt += 1;
          }
          while (quit != 0);      /* End of do/while loop */


          }                       /* End of main */


        * * * * * * * * * * * * * * *  Output for Bairstow Program  * * * * * * * * * * * * * * *

        What degree?
        4
        Enter coefficients, beginning with the constant term.
        Coef of x[ 0]: 3.3
        Coef of x[ 1]: 0.5
        Coef of x[ 2]: 2.3
        Coef of x[ 3]: -1.1
        Coef of x[ 4]: 1
        Enter r and s values: -1 -1
        At iteration   1, r and s are : -0.8903, -1.0635
         Enter 0 to quit, anything else to do another
        At iteration   2, r and s are : -0.9000, -1.1002
         Enter 0 to quit, anything else to do another
        At iteration   3, r and s are : -0.9000, -1.1000
         Enter 0 to quit, anything else to do another
```

Figure 1.15 *Continued*

Exercises

Section 1.2

1. $e^{-x} + 4x^3 - 5$ has a root at $x = 1.05151652$. Beginning with $[1, 2]$, use eight repetitions of bisection to approximate the root. Tabulate the error after each iteration and also tabulate the estimates of maximum error. Is the ac-

tual error always less than the maximum error estimate? Do the actual errors continuously decrease?

2. The quadratic $f(x) = (x - 0.4)(x - 0.6) = x^2 - x + 0.24$ has zeros at $x = 0.4$ and $x = 0.6$, of course. Observe that the endpoints of $[0, 1]$ are not satisfactory to begin bisection. Graph the function, and from this deduce the boundaries of intervals that will converge to each of the roots. If you start with $[0.5, 1]$, what is error bound after six iterations? What is the actual error after six applications of the method?

▶ **3.** Find where the graphs of $y = x^2 - 2$ and $y = e^x$ intersect (for $x < 0$), using interval halving. Get the intersection value correct to five decimal places. How many iterations are required?

4. Use bisection to find the smallest positive root of each of the following. Continue until the relative accuracy is 0.5%. Use graphs to find good starting values. (You probably will want to use either a graphing calculator or a computer algebra system to get the graphs.)

   a. $e^x = 2 - \sin(2x)$
   b. $x^4 - 2x - 1 = 0$
   c. $\cos(3x) + 1 = \exp(x^2)$
   d. $e^{x-1} = x^3 + 2$

▶ **5.** The function $f(x) = x \cos(x/(x - 2))$ has many zeros, especially near $x = 2$, where $f(x)$ is undefined. Graph the function. Determine the first four roots for $x > 0$ by bisection. Get each root correct to four significant digits.

Section 1.3

6. Repeat Exercise 4, this time using the secant method. Compare the number of iterations required with the number used in bisection.

7. Repeat Exercise 4, this time using the method of false position. Compare the number of iterations required with the number used in bisection and with the number used with the secant method (Exercise 6).

▶ **8.** Find where the cubic $y = x^3 - 2x^2 + x - 1$ intersects the parabola $y = 2x^2 + 3x + 1$. From a graph of the two equations, locate the intersection approximately, then use both *regula falsi* and the secant method to refine the value until you are sure that you have it correct to five significant digits. How many iterations are needed in each case?

9. Explain why the secant method usually converges to the root with a specified precision more rapidly than either bisection or the method of linear interpolation.

10. Implement the algorithms given in Section 1.3 with computer programs using any computer language that appeals to you.

11. Find the root near $x = 1$ of $f(x) = e^{x-1} - 5x^3$, beginning with $x = 1$. How accurate is the estimate after four iterations of Newton's method? How many iterations of

bisection does it take to achieve the same accuracy? Tabulate the number of correct digits at each iteration of Newton's method and see if these double each time.

12. Repeat Exercise 4, but this time use Newton's method. Compare the number of iterations required with bisection (Exercise 4), the secant method (Exercise 6), and *regula falsi* (Exercise 7).

▶ **13.** Use Newton's method on the equation $x^2 = N$ to derive the algorithm for obtaining the square root of N:

$$x_{i+1} = \frac{1}{2}\left(x_i + \frac{N}{x_i}\right).$$

14. Repeat Exercise 13, this time deriving algorithms for the third and fourth roots of N.

15. If the algorithm of Exercise 13 is applied twice, show that

$$N^{1/2} \simeq \frac{(A + B)}{4} + \frac{N}{(A + B)},$$

where $N = A * B$.

16. Show that the error of the algorithm in Exercise 15 is approximately

$$\frac{1}{8}\left(\frac{A - B}{A + B}\right)^4.$$

17. Expand $f(x)$ about the point $x = a$ in a Taylor series, and from this derive Newton's method. (See Appendix A if you have forgotten Taylor series.)

▶ **18.** $f(x) = (x + 1)^3 (x - 1)$ obviously has roots at $x = -1$ and $x = +1$. Using starting values that differ from the roots by 0.1, compare the number of iterations taken when Newton's method computes both of the roots until they are within 0.0001 of the correct values. (The problem that Newton's method has with multiple roots is discussed in Sections 1.10 and 1.11.)

19. Beginning with the interval $[0.9, 1.1]$, use the secant method on the root at $x = -1$ in Exercise 18. Can you explain why the secant method works better than Newton's method in this case?

▶ **20.** What starting values cause Newton's method to fail totally in Exercise 18?

21. Repeat Exercise 20 for each of the equations in Exercise 4 (if invalid starting values exist).

22. Each of the following has complex roots. Apply Newton's method to find them:

   a. $x^2 = -2$
   b. $x^3 - x^2 - 1 = 0$

c. $x^4 - 3x^2 + x + 1 = 0$

d. $x^2 = (e^{-2x} - 1)/x$

Section 1.5

23. Use Muller's method to solve the following equations. Apply five iterations and determine how the successive errors are related. Use starting values that differ by 0.2 from each other.

    a. $3x^3 + 2x^2 - x - 6 = 0$, root near 1.0

    b. $e^x - 2x^3 = 0$, root near 6. What are the other roots?

    c. $e^x = 3x^2$, root near 3.7

    d. $\tan(x) - x + 2 = 0$, root near 4.3

▶ 24. Muller's method is sometimes used in a "self-starting" form. Instead of specifying values near a root, the algorithm automatically begins with the values 0, 0.5, −0.5. Use these starting values on the equations of Exercise 4. We are supposed to get the root nearest the origin by this technique. Will this always be true?

25. An extension to the self-starting principle is the deflation of the function after finding a root. *Deflating* refers to the removal of a root by dividing by $(x - r)$, resulting in a new function that lacks the root $x = r$. (Of course, a discontinuity is introduced by the division, at $x = r$). Compare the graphs of

    $f_1(x) = x(x - 3)(x - 1)$, $f_2(x) = x(x - 3)$,

    $f_3(x) = x(x - 1)$,   and $f_4(x) = (x - 1)(x - 3)$

    to see that this is true. After beginning as in Exercise 24, continue with deflation to get all the roots of the following:

    a. $x^3 - 2x^2 - 1 = 0$

    b. $x^3 - 2x^2 - 1.185 = 0$

    c. $\exp(x^2 - 1) + 10 \sin(2x) - 5 = 0$

26. A problem with Muller's method is that the parabola that is created may not intersect the x-axis. When will you know that this occurs? What can you do to recover if this happens?

27. Because Muller's method subtracts function values that may be almost equal, there is a possibility of large relative errors due to the imprecise arithmetic of the computer. Investigate this problem by using starting values that are close together (but not near the root) on the equations of Exercise 23.

28. Muller's method finds complex roots, but we must use complex arithmetic. Find the complex roots of the equations of Exercise 22 by Muller's method.

Section 1.6

29. $f(x) = e^x - 3x^2 = 0$ has three real roots. An obvious rearrangement is

$$x = \pm \sqrt{\frac{e^x}{3}}.$$

Show that convergence is to the root near −0.5 if we begin with $x_0 = 0$ and use the negative value. Show also that convergence to a second root near 1.0 is obtained if $x_0 = 0$ and the positive value is used. Show, however, that this form does not converge to the third root near 4.0 even though a starting value very close to the root is used. Find a different rearrangement that will converge to the root near 4.0.

▶ 30. One root of the quadratic $x^2 + x - 1 = 0$ is at $x = 0.6180339$. The form $x = 1/(x + 1)$ converges to the root for $x_0 = 1$. How many iterations of the fixed-point method are required to get the root correct to five digits? If Aitken acceleration is employed whenever three values are available, how many iterations are needed?

31. Repeat Exercise 30, but now get the root accurate to seven digits.

32. For what range of starting values does the fixed-point operation of Exercise 30 converge to the root near 0.6? (Looking at a graph may help.) Do not stop with a division by zero, that is, for $x_0 = -1$, $x_1 = 1/0$. Use $x_3 = \lim_{x_2 \to \infty} (1/(x_2 + 1)) = 0$.

▶ 33. $2x^3 + 4x^2 - 4x - 6 = 0$ has a root near $x = 1.3$. Find at least three rearrangements that converge to this root.

34. The cubic of Exercise 33 has other roots near $x = -2.3$ and $x = -0.9$. Do the rearrangements you found in Exercise 33 converge to these roots? If not, find rearrangements that do.

Section 1.7

35. Do this exercise by hand computation: A fourth-degree polynomial is

$$P(x) = x^4 + 4.6x^3 + 6.6x^2 - 13x - 16.$$

Using synthetic division, find $P(-1)$. From the result, write $P(x)$ as the product of two polynomials, one of degree 3. Find the one positive root of the cubic polynomial by Newton's method, again using synthetic division. Finally, use the quadratic formula to get the last two roots.

▶ **36.** Do this exercise by hand computation: Using a calculator, find the two real negative roots of $P(x)$ correct to seven decimal places. Then determine the other roots from the deflated quadratic, again to seven places.

$$P(x) = 2x^4 + 7x^3 - 4x^2 + 29x + 14$$

37. Write a computer program in any language that implements the algorithm for Horner's method.

38. Working with deflated polynomials to get additional roots is a way to save computational effort, but any error in the first roots is propagated and causes the successive roots to be incorrect. Investigate this characteristic with the polynomial

$$P(x) = x^3 - 4x^2 + 3x + 1$$

by deliberately getting a first root imperfectly (say, correct to only 1% relative accuracy), and see how this affects the accuracy of the other roots.
    The roots of $P(x)$ are

$$x_1 = 1.44504, \qquad x_2 = 2.80193, \qquad x_3 = -0.246979.$$

Work this procedure three times, getting each individual root imperfectly, to see its effect on the others obtained from the quadratic formula. You may want to see how the amount of error in the first root affects the accuracy of the other roots.

▶ **39.** This exercise simulates the effect of inaccuracies such as those from imprecise experimental measurements. Suppose the coefficients of the polynomial in Exercise 36 are not perfectly known. Determine the influence of such imperfections by finding how much the roots change when any one of the five coefficients is varied by 1%, observing the effect when each coefficient is varied independently. Which coefficient causes the greatest change in the roots due to such variation?

40. Sometimes one or more of the coefficients of a polynomial are zero. How does this affect the speedup from parallel processing? In answering this, sketch the directed acyclic graphs for sequential and for parallel processing of a fourth-degree polynomial that has

a. One missing term,
b. Two missing terms.

Does it make a difference which term(s) are missing? Your graphs for parallel processing should maximize the speedup.

41. Synthetic division finds $P(a)$ by dividing $P(x)$ by $(x - a)$, then finds $P'(x)$ by a second synthetic division of the reduced polynomial. Does this mean that we can evaluate $P''(a)$, $P'''(a)$, ... by repeated divisions?

If not, what is the relation between the successive remainders and the successive derivative values?

Section 1.8

42. Using hand computations in Bairstow's method, beginning with the trial factor $x^2 + 2x - 4$, get the quadratic factors of

$$x^4 + 0.9x^3 - 3.36x^2 + 11.85x - 11.84.$$

From these, get the zeros of the polynomial.

▶ **43.** Use Bairstow's method to resolve into quadratic factors:

$$x^4 - 5.7x^3 + 26.7x^2 - 42.21x + 69.23.$$

You should find that all four zeros are complex.

44. Solve the equation of Exercise 35 using the Bairstow method.

▶ **45.** Use Bairstow's method to get the zeros of the polynomial in Exercise 36.

46. When the modulus of one pair of complex roots is the same as for another pair, Bairstow's method is slow to converge. Observe that this is true by using hand computation to get the quadratic factors of

$$x^4 - x^3 + x^2 - x + 1,$$

which factors into

$$(x^2 + 0.61803x + 1)(x^2 - 1.618034x + 1).$$

What is the modulus of the roots?

47. No formal algorithm is given in Section 1.8 for Bairstow's method, although the operations are shown clearly. Translate these operations into an algorithm of the nature used in previous sections of this chapter. Your algorithm should include a test to see whether additional repetitions of the procedure are required to get a quadratic factor with a specified degree of accuracy.

48. Write a computer program that implements your algorithm of Exercise 47. In the program, use $r = 0$, $s = 0$ as the starting values rather than have the user input them.

Section 1.9

49. Use the QD algorithm to get the roots of the following:

▶ a. $x^3 - x^2 - 2x + 1 = 0$
▶ b. $x^4 + 0.9x^3 - 3.6x^2 + 11.85x - 11.84$
  c. $x^4 + x^3 + 4.11x^2 + 5.45x + 18.86$

50. The QD algorithm also has difficulty with polynomials such as those in Exercise 46. Try your patience on that polynomial.

51. Write the algorithm for the QD method, and use this algorithm to write a computer program that carries it out.

52. Repeat Exercise 42, but his time use Graeffe's method. What happens when there are roots of equal magnitude?

▶ **53.** Repeat Exercise 42, this time using Laguerre's method. Is there a difficulty with roots of equal magnitude?

Section 1.10

54. This polynomial has a double root at $x = 1$ and a triple root at $x = 3$:

$$(x - 1)^2(x - 3)^3$$
$$= x^5 - 11x^4 + 46x^3 - 90x^2 + 81x - 27.$$

Can you get all the roots by bisection? Get the root at $x = 3$ starting with the interval $[2, 4]$.

55. Using the polynomial in Exercise 54, can you get all the roots by the secant method? Which root do you get starting with the interval $[0, 4]$? Try to predict this root in advance.

56. Using the polynomial in Exercise 54, start Newton's method at $x = 2$. Does it converge? If so, to which root? Can you predict this in advance? Tabulate the error of the successive iterates when you start with $x = 2.9$. Does the number of accurate digits about double each time?

▶ **57.** Using the polynomial in Exercise 54, use Bairstow's algorithm on the polynomial to get all the roots.

58. Using the polynomial in Exercise 54, try to determine the roots with Muller's method. Is it successful in finding both roots? If not, explain. Are there certain starting values that are unsuccessful?

59. Use the polynomial in Exercise 54 except now use the modified Newton's method:

$$x_{n+1} = x_n - k * \frac{f(x_n)}{f'(x_n)},$$

where k is the multiplicity of the root. Show from the successive iterates that convergence is now quadratic.

▶ **60.** Solve for the roots of the polynomial of Exercise 54 by applying Newton's method to $P(x)/P'(x)$. Show that quadratic convergence is secured for both roots.

61. This quadratic has nearly equal roots:

$$(x - 1.99)(x - 2.01) = x^2 - 4x + 3.9999.$$

a. Use Newton's method with a starting value of 2.1. Is there quadratic convergence?

b. Use Newton's method with a starting value of 1.9. Is there quadratic convergence?

c. Use Newton's method with a starting value of 2.0. What happens?

d. Repeat part (c), but use a quadratic that has roots at 2.01 and 2.03, starting with $x = 2.02$. If the results are different, explain.

Section 1.11

62. $f(x) = e^x \sin(x) - x^2$ has a zero near $x = 2.6$. By comparing the errors of successive iterates with Newton's method (beginning at $x = 3$), determine the order of convergence.

63. Repeat Exercise 62, but this time start at $x = 2.1$. Why is the first iterate so far from the zero of $f(x)$?

▶ **64.** Repeat Exercise 62, but this time use these starting values:

a. $x = 2.05$.

b. $x = 2.00$.

65. Repeat Exercise 62, this time using:

a. The *regula falsi* method (begin with $x = 2$, $x = 3$).

b. The secant method (begin with $x = 2$, $x = 3$).

c. Muller's method (begin with $x = 2, 2.5, 3$).

66. Make a sketch similar to Fig. 1.7 for a $g(x)$ where fixed-point iteration does converge even though not all the conditions given in Section 1.11 hold.

67. Show that Newton's method is not quadratically convergent by comparing errors for:

a. $e^{-x}(x^2 - 2x + 1) = 0$ (root at 1).

b. $1 - \cos(x) = 0$ (root at 2π).

▶ c. $(e^x - 1)(x)[\sin(x)]$ (root at zero).

68. Repeat Exercise 67, but this time use the modifications to Newton's method given in Section 1.10.

69. Do an experiment similar to that of Exercise 62, this time for Laguerre's method on some quadratic polynomial.

Section 1.12

The computer algebra system, DERIVE, mentioned in Exercises 71, 72, 75, and 76 is very easy to use because it is menu-driven.

70. Use MATLAB to get the smallest root of $\cos(x) = x \sin(x)$.

▶ **71.** Repeat Exercise 70 but now use DERIVE.

72. Use MATLAB and/or DERIVE to solve Exercise 30.

73. Modify the MATLAB program for bisection so that it uses the secant method.

74. Modify the MATLAB program for bisection so that it carries out the *regula falsi* method.

75. Use the ITERATES function of DERIVE to perform the secant method. Test it with several problems.

76. Use the ITERATE function of DERIVE to carry out *regula falsi*. How much change is required to have it use ITERATES instead?

77. Use the TI-92 or HP-48G calculator to solve Exercise 22.

▶ **78.** Use the HP-48G and/or the TI-92 calculator to solve Exercise 23 from the graph of the functions.

79. Use both the TI-92 and the HP-48G calculators to solve Exercise 4. Do this both from the graphs and directly from the equations.

Applied Problems and Projects

Beginning with this chapter and in each subsequent chapter, we present some problems that either are more involved or apply to the "real world."

80. Given are

$$x'' + x + 2y' + y = f(t), \qquad x'' - x + y = g(t), \qquad x(0) = x'(0) = y(0) = 0.$$

In solving this pair of simultaneous second-order differential equations by the Laplace transform method, it becomes necessary to factor the expression

$$(S^2 + 1)(S) - (2S + 1)(S^2 - 1) = -S^3 - S^2 + 3S + 1,$$

so that partial fractions can be used in getting the inverse transform. What are the factors?

81. DeSantis (1976) has derived a relationship for the compressibility factor of real gases of the form

$$z = \frac{1 + y + y^2 - y^3}{(1 - y)^3},$$

where $y = b/(4v)$, b being the van der Waals correction and v the molar volume. If $z = 0.892$, what is the value of y?

82. In studies of solar-energy collection by focusing a field of plane mirrors on a central collector, one researcher obtained this equation for the geometrical concentration factor C:

$$C = \frac{\pi(h/\cos A)^2 F}{0.5\pi D^2(1 + \sin A - 0.5 \cos A)},$$

where A is the rim angle of the field, F is the fractional coverage of the field with mirrors, D is the diameter of the collector, and h is the height of the collector. Find A if $h = 300$, $C = 1200$, $F = 0.8$, and $D = 14$.

83. Lee and Duffy (1976) relate the friction factor for flow of a suspension of fibrous particles to the Reynolds number by this empirical equation:

$$\frac{1}{\sqrt{f}} = \left(\frac{1}{k}\right) \ln(RE\sqrt{f}) + \left(14 - \frac{5.6}{k}\right).$$

In their relation, f is the friction factor, RE is the Reynolds number, and k is a constant determined by the concentration of the suspension. For a suspension with 0.08% concentration, $k = 0.28$. What is the value of f if RE $= 3750$?

84. Based on the work of Frank–Kamenetski in 1955, temperatures in the interior of a material with embedded heat sources can be determined if we solve this equation:

$$e^{-(1/2)t} \cosh^{-1}(e^{(1/2)t}) = \sqrt{\tfrac{1}{2}L_{cr}}.$$

Given that $L_{cr} = 0.088$, find t.

85. Suppose we have the 555 Timer Circuit

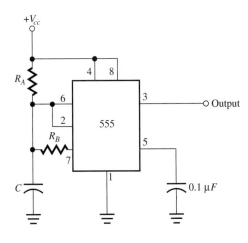

whose output waveform is

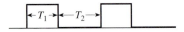

where

$$T_1 + T_2 = \frac{1}{f}$$

$$f = \text{frequency}$$

$$\text{Duty cycle} = \frac{T_1}{T_1 + T_2} \times 100\%.$$

It can be shown that

$$T_1 = R_A C \ln(2)$$

$$T_2 = \frac{R_A R_B C}{R_A + R_B} * \ln\left(\left|\frac{R_A - 2R_B}{2R_A - R_B}\right|\right).$$

Given that $R_A = 8670$, $C = 0.01 \times 10^{-6}$, $T_2 = 1.4 \times 10^{-4}$,

a. Find T_1, f, and the duty cycle.
b. Find R_B using any program you have written.
c. Select an f and duty cycle, then find T_1 and T_2.

86. The solution of boundary-value problems by an analytical (Fourier series) method often involves finding the roots of transcendental equations to evaluate the coefficients. For example,

$$y'' + \lambda y = 0, \qquad y(0) = 0, \qquad y(1) = y'(1),$$

involves solving $\tan z = z$. Find three values of z other than $z = 0$.

87. Find the max/min points of the function

$$f(x) = [\sin(x)]^6 * e^{20x} * \tan(1 - x)$$

on the interval $[0, 1]$. Compare your own root-finding program with the IMSL subroutine ZBRENT. [Note the disadvantage in trying to solve $f'(x) = 0$ using Newton's method.]

88. In Chapter 5, a particularly efficient method for numerical integration of a function, called *Gaussian quadrature,* is discussed. In the development of formulas for this method, it is necessary to evaluate the zeros of Legendre polynomials. Find the zeros of the Legendre polynomial of sixth order:

$$P_6(x) = \frac{1}{48}(693x^6 - 945x^4 + 315x^2 - 15).$$

(*Note:* All the zeros of the Legendre polynomials are less than one in magnitude and, for polynomials of even order, are symmetrical about the origin.)

89. The Legendre polynomials of Problem 88 are one set of a class of polynomials known as *orthogonal* polynomials. Another set are the *Laguerre* polynomials. Find the zeros of the following:

a. $L_3(x) = x^3 - 9x^2 + 18x - 6$
b. $L_4(x) = x^4 - 16x^3 + 72x^2 - 96x + 24$

90. Still another set of orthogonal polynomials are the *Chebyshev* polynomials. (We will use these in Chapter 4.) Find the roots of

$$T_6(x) = 32x^6 - 48x^4 + 18x^2 - 1 = 0.$$

(Note the symmetry of this function. All the roots of Chebyshev polynomials are also less than one in magnitude.)

91. A sphere of density d and radius r weighs $\frac{4}{3}\pi r^3 d$. The volume of a spherical segment is $\frac{1}{3}\pi(3rh^2 - h^3)$. Find the depth to which a sphere of density 0.6 sinks in water as a fraction of its radius. (See the accompanying figure.)

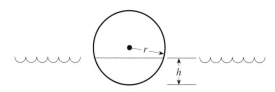

92. For several functions that have multiple roots, investigate whether Aitken acceleration improves the rate of convergence. Do this for several methods.

93. Make experimental comparisons of the rates of conversion for Newton's method, for Newton's method with the derivative estimated numerically, and for the secant method. Make a table that shows how the errors decrease for each method, then make a log plot of the errors.

94. Muller's method is said to converge with an order of convergence equal to 1.85. Verify this experimentally. Is it true if there is a multiple root?

95. Spreadsheet programs can perform iterated computations. Devise and test a spreadsheet program that implements fixed-point iterations.

96. Repeat Exercise 95 but for
(a) Bisection method,
(b) Newton's method.

97. We mentioned in Section 1.11 that Laguerre's method (Section 1.9) is cubically convergent. Show that this is true.

98. Fixed-point iterations sometimes converge (a) by "walking up a staircase (Fig. 1.7a) or (b) "spirally" (Fig. 1.7b), or they may diverge (Fig. 1.7c). The conditions for these cases are discussed in Sections 1.6 and 1.11, and we have shown that convergence is of order 1. This means that Aitken acceleration applies. Jones (1982) discusses other ways to accelerate the convergence; so does Acton (1970).

   Consider this problem: Where do the curves for $e^y + 1 = e^x$ and $x^2 + y^2 = 1$ cross? One intersection is near (0.9, 0.4). Find rearrangements of the form $x = g(x)$ that converge to this intersection and compare how fast they converge. Apply several acceleration techniques to this.

99. a. The rate of flow of water through a stream is often measured by installing a weir. This amounts to building a dam across the stream with a vee-shaped notch near the center (the point of the notch is down). If the upstream velocity is neglected, the flow Q (ft^3/sec) is related to distance h (ft) from the surface of the upstream water to the point of the vee and to the angle θ (degrees) between the sides of the notch by this formula:

$$Q = 0.59 * \left(\frac{8}{15}\right) * \tan\left(\frac{\theta}{2}\right) * \sqrt{(2g)} * h^{2.5},$$

   where g is the gravitational constant, 32.2 ft/sec^2. If $Q = 200$, make a table that shows how h is related to θ for values of θ between 20 and 130 degrees. Can this be done without using one of the methods of this chapter?

   b. More often, weirs have a rectangular notch. Look up formulas for this case, but now the formula should allow for the effects of the velocity of the incoming water, v. Make a table that shows how h varies with v for several values of the width of the notch. Can you do this without using a method of this chapter? (You might want to repeat this for other notch configurations.)

2

Solving Sets of Equations

Contents of This Chapter

This chapter covers the extremely important topic of how to solve a large system of equations. We emphasize sets of simultaneous linear equations, although we also treat less thoroughly the more difficult area of solving systems of nonlinear equations.

Linear systems are perhaps the most widely applied numerical procedure when real-world situations are to be simulated. Linear systems are used in statistical analysis in many applications. Methods for numerically solving ordinary and partial-differential equations depend on them.

2.1 Applications of Sets of Equations

Considers three very different engineering examples that are all solved the same way: by solving a system of linear equations.

2.2 Matrix Notation

Reviews the elementary concepts of matrices and vectors that are used extensively in this chapter.

2.3 The Elimination Method

Describes the classical elimination method that changes a system of linear equations so that back-substitution can find the solution.

2.4 The Gaussian Elimination and Gauss–Jordan Methods

Explains the *LU* method, which is a basis for solving systems of equations. The use of partial pivoting minimizes the effects of round-off error and may be required to avoid a division by zero. The "Big O" notation is used when a comparison of methods is made.

2.5 Other Direct Methods

Tells about additional *LU* decomposition methods that solve a system when applied to the coefficient matrix. The methods of this section and Section 2.4 are applied to solve a system with a series of right-hand sides. The section ends with an example of Gaussian elimination applied to a tridiagonal system.

2.6 Pathology in Linear Systems—Singular Matrices

Points out that there are times when an accurate solution to a system of equations is extremely difficult to obtain. Techniques are given to minimize the problem if a solution is possible.

2.7 Determinants and Matrix Inversion

Explains how the techniques of this chapter can get the determinant and the inverse of a matrix.

2.8 Norms

Extends the definition of "the magnitude of a number" to vectors and matrices. A norm is useful in analyzing the behavior of matrices when they are applied in solving problems.

2.9 Condition Numbers and Errors in Solutions

Shows that some systems have inherent problems associated with them. These are called *ill-conditioned* systems. The *condition number* is a measure of this ill-conditioning.

2.10 Iterative Methods

Explains these often preferred methods to solve the very large systems that occur frequently in modeling engineering and scientific problems. However, to use the iterative method, the coefficient matrix must meet certain conditions.

2.11 The Relaxation Method

Explains this technique for speeding the convergence of an iterative method. Although the original method is obsolete because it is not adaptable to computers, relaxation is a basis for an acceleration method that is widely used. The method has historical importance.

2.12 Systems of Nonlinear Equations

Shows that although these systems are very much harder to solve than linear systems, Newton's method from Chapter 1 combined with Gaussian elimination is one way to attack the problem.

2.13 Theoretical Matters

Presents some of the theory behind the elimination method. We see that Gaussian elimination can be expressed as the product of relatively simple operations. The importance of pivoting is shown.

2.14 Using Maple and MATLAB

Shows how systems of equations can be solved using both of these programs. In addition, either of these programs can do all the simple matrix operations, find determinants, norms, eigenvalues, eigenvectors, and solve systems of nonlinear equations.

2.15 Parallel Processing

Tells how parallel computing can be applied to the solution of linear systems. An algorithm is developed in several stages that allows a significant reduction in processing time.

Chapter Summary

Lists the topics you should understand after studying this chapter.

Computer Programs

Presents examples of computer programs that implement some of the algorithms of this chapter.

2.1 Applications of Sets of Equations

The methods of this chapter have such widespread application in engineering that we present in this section three very different engineering problems all of which are solved in the same way: by solving a system of linear equations.

(1) Finding the Steady-State Temperatures in a Plate

A flat thin plate is 2 ft by 2 ft by 1 in. The edges are kept at constant temperatures. We want to find the temperatures in the interior of this plate. We can estimate these temperatures if we make a grid of points with each point 0.5 ft apart, and let $u_1, u_2, \ldots, u_9$ be the temperatures at the grid points. The plate with these grid points is shown in Figure 2.1a. The temperature, $u(x, y)$, at each point satisfies a differential equation that can be simplified into a system of linear equations with constant coefficients. The following system of nine equations is the result.

$$
\begin{aligned}
-4u_1 + u_2 \qquad\quad + u_4 \qquad\qquad\qquad\qquad\qquad &= -50 \\
u_1 - 4u_2 + u_3 \qquad\qquad u_5 \qquad\qquad\qquad &= -50 \\
u_2 - 4u_3 \qquad\qquad + u_6 \qquad\qquad\qquad &= -150 \\
u_1 \qquad\qquad - 4u_4 + u_5 \qquad + u_7 \qquad\qquad &= 0 \\
u_2 \qquad + u_4 - 4u_5 + u_6 \qquad + u_8 \qquad &= 0 \\
u_3 \qquad + u_5 - 4u_6 \qquad\qquad + u_9 &= -100 \\
u_4 \qquad\qquad - 4u_7 + u_8 \qquad &= -50 \\
u_5 \qquad\qquad + u_7 - 4u_8 + u_9 &= -50 \\
u_6 \qquad\qquad + u_8 - 4u_9 &= -150
\end{aligned}
$$

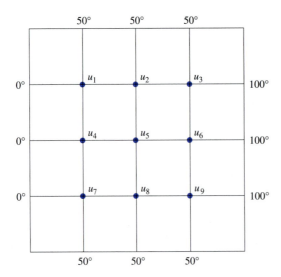

Figure 2.1a

Our problem is reduced to solving n linear equations where n depends on the size of the grid.

(2) Voltages and Currents in a Network

Figure 2.1b shows an electrical network with seven resistors. A voltage of 5.0 volts is applied to the network at nodes 1 and 6. We wish to know the current, i_{34}, that would flow between points 3 and 4, using the two well-known laws:

Kirchhoff's law: The sum of all the currents flowing into a node is zero.

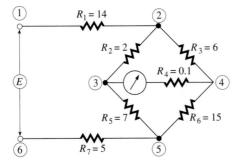

Figure 2.1b

Ohm's law: The current, i_{jk}, through a resistor equals the voltage across it divided by its resistance, $(V_j - V_k)/R_{jk}$.

We can set up eleven equations and from these equations solve for eleven unknown quantities (the four voltages and seven currents).

$$
\begin{aligned}
i_{12} - i_{23} - i_{24} &= 0 \\
i_{23} \quad\quad - i_{34} - i_{35} &= 0 \\
i_{24} + i_{34} \quad\quad - i_{45} &= 0 \\
i_{35} + i_{45} - i_{56} &= 0 \\
14i_{12} \quad\quad\quad\quad\quad + V_2 &= 5 \\
2i_{23} \quad\quad\quad\quad\quad - V_2 + V_3 &= 0 \\
6i_{24} \quad\quad\quad\quad\quad - V_2 \quad\quad + V_4 &= 0 \\
0.1i_{34} \quad\quad\quad\quad\quad - V_3 + V_4 &= 0 \\
7i_{35} \quad\quad\quad\quad - V_3 \quad\quad + V_5 &= 0 \\
15i_{45} \quad\quad\quad\quad - V_4 + V_5 &= 0 \\
5i_{56} \quad\quad\quad\quad - V_5 &= 0
\end{aligned}
$$

(3) Computing the Forces in a Planar Truss

A truss is a structure used for supporting loads such as in bridges or buildings. The truss is composed entirely of straight members arranged to form one or more triangles, giving the structure stability. Figure 2.1c shows a typical truss.

In this example, there are six pins (joints) and nine members. We resolve the force acting at each of the pins into its *x*- and *y*-components. The forces that act on the truss are pointed along each of the members. We consider the forces to act toward the center of the member, away from the joint. There are nine member-forces, F_i, $i = 1, \ldots, 9$. Setting the

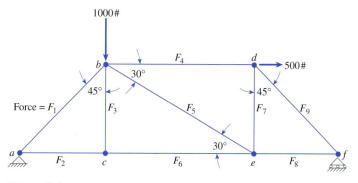

Figure 2.1c

sum of all the forces acting horizontally or vertically on each of the pins to zero, we can write the nine equations as follows:

$$
\begin{aligned}
\cos(45°) * F_1 - F_4 - \cos(30°) * F_5 &= 0 \\
\sin(45°) * F_1 + F_3 + \sin(30°) * F_5 &= -1000 \\
F_2 - F_6 &= 0 \\
- F_3 &= 0 \\
F_4 - \sin(45°) * F_9 &= 500 \\
F_7 + \cos(45°) * F_9 &= 0 \\
\cos(30°) * F_5 + F_6 - F_8 &= 0 \\
\sin(30°) * F_5 + F_7 &= 500 \\
F_8 + \cos(45°) * F_9 &= 0
\end{aligned}
\tag{2.1}
$$

Chapter 2 describes a variety of methods for solving sets of equations such as these. Our methods can be adapted to a computer. Because sometimes up to several hundred simultaneous linear equations must be solved, it is impractical to solve the system by hand.

2.2 Matrix Notation

Our discussion of methods to solve sets of linear equations will be facilitated by some of the concepts and notation of matrix algebra. Only the more elementary ideas will be needed.

A *matrix* is a rectangular array of numbers in which not only the value of the number is important but also its position in the array. The size of the matrix is described by the number of its rows and columns. A matrix of n rows and m columns is said to be $n \times m$. The elements of the matrix are generally enclosed in brackets, and double-subscripting is the common way of indexing the elements. The first subscript also denotes the row, and the second denotes the column in which the element occurs. Capital letters are used to refer to matrices. For example,

$$
A = \begin{bmatrix}
a_{11} & a_{12} & \cdots & a_{1m} \\
a_{21} & a_{22} & \cdots & a_{2m} \\
\vdots & & & \\
a_{n1} & a_{n2} & \cdots & a_{nm}
\end{bmatrix} = [a_{ij}], \quad i = 1, 2, \ldots, n, \quad j = 1, 2, \ldots, m.
$$

Enclosing the general element a_{ij} in brackets is another way of representing matrix A, as just shown. Sometimes we will enclose the name of the matrix in brackets, $[A]$, to emphasize that A is a matrix.

Two matrices of the same size may be added or subtracted. The sum of

$$A = [a_{ij}] \quad \text{and} \quad B = [b_{ij}]$$

is the matrix whose elements are the sum of the corresponding elements of A and B:

$$C = A + B = [a_{ij} + b_{ij}] = [c_{ij}].$$

Similarly, we get the *difference* of two equal-sized matrices by subtracting corresponding elements. If two matrices are not equal in size, they cannot be added or subtracted. Two matrices are equal if and only if each element of one is the same as the corresponding element of the other. Obviously, equal matrices must be of the same size. Some examples will help make this clear.

If

$$A = \begin{bmatrix} 4 & 7 & -5 \\ -4 & 2 & 12 \end{bmatrix} \quad \text{and} \quad B = \begin{bmatrix} 1 & 5 & 4 \\ 2 & -6 & 3 \end{bmatrix},$$

we say that A is 2×3 because it has two rows and three columns. B is also 2×3. Their sum C is also 2×3:

$$C = A + B = \begin{bmatrix} 5 & 12 & -1 \\ -2 & -4 & 15 \end{bmatrix}.$$

The difference D of A and B is

$$D = A - B = \begin{bmatrix} 3 & 2 & -9 \\ -6 & 8 & 9 \end{bmatrix}.$$

Multiplication of two matrices is defined as follows, when A is $n \times m$ and B is $m \times r$:

$$[a_{ij}][b_{ij}] = [c_{ij}]$$
$$= \begin{bmatrix} (a_{11}b_{11} + a_{12}b_{21} + \cdots + a_{1m}b_{m1}) \ldots (a_{11}b_{1r} + \cdots + a_{1m}b_{mr}) \\ (a_{21}b_{11} + a_{22}b_{21} + \cdots + a_{2m}b_{m1}) \ldots (a_{21}b_{1r} + \cdots + a_{2m}b_{mr}) \\ \vdots \\ (a_{n1}b_{11} + a_{n2}b_{21} + \cdots + a_{mn}b_{m1}) \ldots (a_{n1}b_{1r} + \cdots + a_{nm}b_{mr}) \end{bmatrix},$$

$$c_{ij} = \sum_{k=1}^{m} a_{ik}b_{kj}, \quad i = 1, 2, \ldots, n, \quad j = 1, 2, \ldots, r.$$

It is simplest to select the proper elements if we count across the rows of A with the left hand while counting down the columns of B with the right. Unless the number of columns of A equals the number of rows of B (so the counting comes out even), the matrices cannot be multiplied. Hence if A is $n \times m$, B must have m rows or else they are said to be "nonconformable for multiplication" and their product is undefined. In general $AB \neq BA$, so the order of factors must be preserved in matrix multiplication.

If a matrix is multiplied by a scalar (a pure number), the product is a matrix, each element of which is the scalar times the original element. We can write

$$\text{If } kA = C, \qquad c_{ij} = ka_{ij}.$$

A matrix with only one column, $n \times 1$ in size, is termed a *column vector,* and one of only one row, $1 \times m$ in size, is called a *row vector.* When the unqualified term *vector* is used, it nearly always means a *column* vector. Frequently the elements of vectors are only singly subscripted.

Some examples of matrix multiplication follow.

$$\text{Suppose } A = \begin{bmatrix} 3 & 7 & 1 \\ -2 & 1 & -3 \end{bmatrix}, \quad B = \begin{bmatrix} 5 & -2 \\ 0 & 3 \\ 1 & -1 \end{bmatrix}, \quad x = \begin{bmatrix} -3 \\ 1 \\ 4 \end{bmatrix}, \quad y = \begin{bmatrix} y_1 \\ y_2 \\ y_3 \end{bmatrix}.$$

$$A * B = \begin{bmatrix} 16 & 14 \\ -13 & 10 \end{bmatrix}; \quad B * A = \begin{bmatrix} 19 & 33 & 11 \\ -6 & 3 & -9 \\ 5 & 6 & 4 \end{bmatrix};$$

$$A * x = \begin{bmatrix} 2 \\ -5 \end{bmatrix}; \quad A * y = \begin{bmatrix} 3y_1 + 7y_2 + y_3 \\ -2y_1 + y_2 - 3y_3 \end{bmatrix}.$$

Because A is 2×3 and B is 3×2, they are conformable for multiplication and their product is 2×2. When we form the product of $B * A$, it is 3×3. Observe that not only is $AB \neq BA$; AB and BA are not even the same size. The product of A and the vector x (a 3×1 matrix) is another vector, one with two components. Similarly, Ay has two components. We cannot multiply B times x or B times y; they are nonconformable.

The product of the scalar number 2 and A is

$$2A = \begin{bmatrix} 6 & 14 & 2 \\ -4 & 2 & -6 \end{bmatrix}.$$

Because a vector is just a special case of a matrix, a column vector can be multiplied by a matrix, as long as they are conformable in that the number of columns of the matrix equals the number of elements (rows) in the vector. The product in this case will be another column vector. The size of a product of two matrices, the first $m \times n$ and the second $n \times r$, is $m \times r$. An $m \times n$ matrix times an $n \times 1$ vector gives an $m \times 1$ product.

The general relation for $Ax = b$ is

$$b_i = \sum_{k=1}^{\text{No. of cols.}} a_{ik}x_k, \qquad i = 1, 2, \ldots, \text{No. of rows.}$$

Two vectors, each with the same number of components, may be added or subtracted. Two vectors are equal if each component of one equals the corresponding component of the other.

This definition of matrix multiplication permits us to write the set of linear equations

$$a_{11}x_1 + a_{12}x_2 + \cdots + a_{1n}x_n = b_1,$$
$$a_{21}x_1 + a_{22}x_2 + \cdots + a_{2n}x_n = b_2,$$
$$\vdots$$
$$a_{n1}x_1 + a_{n2}x_2 + \cdots + a_{nn}x_n = b_n,$$

much more simply in matrix notation, as $\boxed{Ax = b,}$ where

$$A = \begin{bmatrix} a_{11} & a_{12} & \cdots & a_{1n} \\ a_{21} & a_{22} & \cdots & a_{2n} \\ \vdots & & & \\ a_{n1} & a_{n2} & \cdots & a_{nn} \end{bmatrix}, \qquad x = \begin{bmatrix} x_1 \\ x_2 \\ \vdots \\ x_n \end{bmatrix}, \qquad b = \begin{bmatrix} b_1 \\ b_2 \\ \vdots \\ b_n \end{bmatrix}.$$

For example,

$$\begin{bmatrix} 3 & 2 & 4 \\ 1 & -2 & 0 \\ -1 & 3 & 2 \end{bmatrix} * x = \begin{bmatrix} 14 \\ -7 \\ 2 \end{bmatrix}$$

is the same as the set of equations

$$3x_1 + 2x_2 + 4x_3 = 14,$$
$$x_1 - 2x_2 \qquad = -7,$$
$$-x_1 + 3x_2 + 2x_3 = 2.$$

A very important special case is the multiplication of two vectors. The first must be a row vector if the second is a column vector, and each must have the same number of components. For example,

$$[1 \quad 3 \quad -2] * \begin{bmatrix} 4 \\ -1 \\ 3 \end{bmatrix} = [-5]$$

gives a "matrix" of one row and one column. The result is a pure number, a scalar. This product is called the *scalar product* of the vectors, also called the *inner product*.

If we reverse the order of multiplication of these two vectors, we obtain

$$\begin{bmatrix} 4 \\ -1 \\ 3 \end{bmatrix} * \begin{bmatrix} 1 & 3 & -2 \end{bmatrix} = \begin{bmatrix} 4 & 12 & -8 \\ -1 & -3 & 2 \\ 3 & 9 & -6 \end{bmatrix}.$$

This product is called the *outer product.* Although not as well known as the inner product, the outer product is very important in nonlinear optimization problems.*

Special Properties of Matrices

Certain square matrices have special properties. The diagonal elements are the line of elements a_{ii} from upper left to lower right of the matrix. If only the diagonal terms are non-zero, the matrix is called a *diagonal matrix.* When the diagonal elements are each equal to unity while all off-diagonal elements are zero, the matrix is said to be the *identity matrix of order n.* The usual symbol for such a matrix is I_n, and it has properties similar to unity. For example, the order-4 identity matrix is

$$\begin{bmatrix} 1 & 0 & 0 & 0 \\ 0 & 1 & 0 & 0 \\ 0 & 0 & 1 & 0 \\ 0 & 0 & 0 & 1 \end{bmatrix} = I_4.$$

The subscript is omitted when the order is clear from the context.

An important property of an identity matrix, I, is that for any $n \times n$ matrix, A, it is always true that

$$I * A = A * I = A.$$

Another closely related matrix, P, is created by interchanging two rows (columns) of the identity matrix. P is called a *transposition matrix.* A matrix that is the product of several transposition matrices is called a *permutation matrix.* Example: Let

$$P_1 = \begin{bmatrix} 1 & 0 & 0 & 0 \\ 0 & 0 & 0 & 1 \\ 0 & 0 & 1 & 0 \\ 0 & 1 & 0 & 0 \end{bmatrix}, \quad A = \begin{bmatrix} 9 & 6 & 2 & 13 \\ 4 & 2 & 8 & 11 \\ 0 & 7 & 1 & 9 \\ 3 & 2 & 6 & 8 \end{bmatrix}.$$

Here, P_1 is derived from the identity matrix, I, by interchanging rows 2 and 4 of I. Consider the two products of P_1 and A: $P_1 * A$ and $A * P_1$.

The first of these, $P_1 * A$, interchanges rows 2 and 4 of A; the second, $A * P_1$, interchanges columns 2 and 4 of A. You should check this for yourself.

In the elementary case where P is a transposition matrix, we have $P * P = I$ because it is symmetric; in the more general case, it is an easy exercise to show that $P^T * P = I$ where

---

* Another important product of three-component vectors is the *vector product,* also known as the *cross product.*

P^T denotes the *transpose* of matrix P. More generally, the transpose of a matrix is the matrix that results when the rows are written as columns (or, alternatively, when the columns are written as rows). The symbol A^T is used for the transpose of A.

EXAMPLE 2.1

$$A = \begin{bmatrix} 3 & -1 & 4 \\ 0 & 2 & -3 \\ 1 & 1 & 2 \end{bmatrix}; \qquad A^T = \begin{bmatrix} 3 & 0 & 1 \\ -1 & 2 & 1 \\ 4 & -3 & 2 \end{bmatrix}.$$

▲

It should be obvious that the transpose of A^T, $(A^T)^T$, is just A itself. However, it is also true, but perhaps less obvious, that

$$(A * B)^T = B^T * A^T.$$

When a matrix is square, a quantity called its *trace* is defined. The trace of a square matrix is the sum of the elements on its main diagonal. For example, the traces of the previous matrices are

$$\text{tr}(A) = \text{tr}(A^T) = 3 + 2 + 2 = 7.$$

It should be obvious that the trace remains the same if a square matrix is transposed.

A vector whose length is one is called a *unit vector*. A vector that has all its elements equal to zero except one element, which has a value of unity, is called a *unit basis vector*. There are three distinct unit basis vectors for order-3 vectors:

$$\begin{bmatrix} 1 \\ 0 \\ 0 \end{bmatrix}, \qquad \begin{bmatrix} 0 \\ 1 \\ 0 \end{bmatrix}, \qquad \text{and} \qquad \begin{bmatrix} 0 \\ 0 \\ 1 \end{bmatrix}.$$

If all the elements above the diagonal are zero, a matrix is called *lower-triangular;* it is called *upper-triangular* when all the elements below the diagonal are zero. For example, these order-3 matrices are lower- and upper-triangular:

$$L = \begin{bmatrix} 1 & 0 & 0 \\ 4 & 6 & 0 \\ -2 & 1 & -4 \end{bmatrix}, \qquad U = \begin{bmatrix} 1 & -3 & 3 \\ 0 & -1 & 0 \\ 0 & 0 & 1 \end{bmatrix}.$$

Triangular matrices are of special importance, as will become apparent later in this chapter and in several other chapters.

Tridiagonal matrices are those that have nonzero elements only on the diagonal and in the positions adjacent to the diagonal; they will be of special importance in certain partial-differential equations. An example of a tridiagonal matrix is

$$\begin{bmatrix} -4 & 2 & 0 & 0 & 0 \\ 1 & -4 & 1 & 0 & 0 \\ 0 & 1 & -4 & 1 & 0 \\ 0 & 0 & 1 & -4 & 1 \\ 0 & 0 & 0 & 2 & -4 \end{bmatrix}.$$

For a tridiagonal matrix, only the nonzero values need to be recorded, and that means that the $n \times n$ matrix can be compressed into a matrix of 3 columns and n rows. For this example, we can write the matrix as

$$\begin{bmatrix} x & -4 & 2 \\ 1 & -4 & 1 \\ 1 & -4 & 1 \\ 1 & -4 & 1 \\ 2 & -4 & x \end{bmatrix}.$$

(The x entries are not normally used; they might be entered as zeros.)

Examples of Operations with Matrices

We present here some additional examples of arithmetic operations with matrices.

$$(3) * \begin{bmatrix} 1 & 2 \\ 3 & 4 \end{bmatrix} = \begin{bmatrix} 3 & 6 \\ 9 & 12 \end{bmatrix}.$$

$$\begin{bmatrix} 1 & 3 & 2 \\ -1 & 0 & 4 \end{bmatrix} + \begin{bmatrix} -1 & 0 & 2 \\ 4 & 1 & -3 \end{bmatrix} = \begin{bmatrix} 0 & 3 & 4 \\ 3 & 1 & 1 \end{bmatrix};$$

$$\begin{bmatrix} 2 & 1 \\ 0 & -4 \\ 7 & 2 \end{bmatrix} - \begin{bmatrix} 3 & -2 \\ 4 & 1 \\ 0 & -2 \end{bmatrix} = \begin{bmatrix} -1 & 3 \\ -4 & -5 \\ 7 & 4 \end{bmatrix}.$$

$$\begin{bmatrix} 2 & 0 & -1 \\ 3 & 2 & 6 \end{bmatrix} * \begin{bmatrix} -1 \\ 2 \\ 1 \end{bmatrix} = \begin{bmatrix} -3 \\ 7 \end{bmatrix} \quad \text{but} \quad \begin{bmatrix} 6 & 1 \\ 3 & -2 \end{bmatrix} * \begin{bmatrix} -1 \\ 2 \\ 1 \end{bmatrix} \text{ is not defined.}$$

$$\begin{bmatrix} 1 & 3 \\ 2 & -1 \end{bmatrix} * \begin{bmatrix} 0 & 3 \\ -1 & 1 \end{bmatrix} = \begin{bmatrix} -3 & 6 \\ 1 & 5 \end{bmatrix} \quad \text{but} \quad \begin{bmatrix} 0 & 3 \\ -1 & 1 \end{bmatrix} * \begin{bmatrix} 1 & 3 \\ 2 & -1 \end{bmatrix} = \begin{bmatrix} 6 & -3 \\ 1 & -4 \end{bmatrix}.$$

Division of a matrix by another matrix is not defined, but we will discuss the *inverse* of a matrix later in this chapter.

The *determinant* of a square matrix is a number. For a 2×2 matrix, the determinant is computed by subtracting the product of the elements on the minor diagonal (from upper right to lower left) from the product of terms on the major diagonal. For example,

$$A = \begin{bmatrix} -5 & 3 \\ 7 & 2 \end{bmatrix}, \quad \det(A) = (-5)(2) - (7)(3) = -31;$$

$\det(A)$ is the usual notation for the determinant of A. Sometimes the determinant is symbolized by writing the elements of the matrix between vertical lines (similar to representing the absolute value of a number).

For a 3×3 matrix, you may have learned a crisscross way of forming products of terms (we call it the "spaghetti rule") that probably should be forgotten, for it applies only to the special case of a 3×3 matrix; it won't work for larger systems. The general rule that applies in all cases is to expand in terms of the minors of some row or column. The *minor* of any term is the matrix of lower order formed by striking out the row and column in which the term is found. The determinant is found by adding the product of each term in any row or column by the determinant of its minor, with signs alternating $+$ and $-$. We expand each of the determinants of the minor until we reach 2×2 matrices. For example,

$$\text{Given } A = \begin{bmatrix} 3 & 0 & -1 & 2 \\ 4 & 1 & 3 & -2 \\ 0 & 2 & -1 & 3 \\ 1 & 0 & 1 & 4 \end{bmatrix},$$

$$\det(A) = 3 \begin{vmatrix} 1 & 3 & -2 \\ 2 & -1 & 3 \\ 0 & 1 & 4 \end{vmatrix} - 0 \begin{vmatrix} 4 & 3 & -2 \\ 0 & -1 & 3 \\ 1 & 1 & 4 \end{vmatrix}$$

$$+ (-1) \begin{vmatrix} 4 & 1 & -2 \\ 0 & 2 & 3 \\ 1 & 0 & 4 \end{vmatrix} - 2 \begin{vmatrix} 4 & 1 & 3 \\ 0 & 2 & -1 \\ 1 & 0 & 1 \end{vmatrix}$$

$$= 3 \left\{ (1) \begin{vmatrix} -1 & 3 \\ 1 & 4 \end{vmatrix} - (3) \begin{vmatrix} 2 & 3 \\ 0 & 4 \end{vmatrix} + (-2) \begin{vmatrix} 2 & -1 \\ 0 & 1 \end{vmatrix} \right\}$$

$$+ (-1) \left\{ (4) \begin{vmatrix} 2 & 3 \\ 0 & 4 \end{vmatrix} - (1) \begin{vmatrix} 0 & 3 \\ 1 & 4 \end{vmatrix} + (-2) \begin{vmatrix} 0 & 2 \\ 1 & 0 \end{vmatrix} \right\}$$

$$- 2 \left\{ (4) \begin{vmatrix} 2 & -1 \\ 0 & 1 \end{vmatrix} - (1) \begin{vmatrix} 0 & -1 \\ 1 & 1 \end{vmatrix} + (3) \begin{vmatrix} 0 & 2 \\ 1 & 0 \end{vmatrix} \right\}$$

$$= 3\{(1)(-7) - (3)(8) + (-2)(2)\} + (-1)\{(4)(8) - (1)(-3) + (-2)(-2)\}$$

$$- 2\{(4)(2) - (1)(1) + (3)(-2)\}$$

$$= 3(-7 - 24 - 4) + (-1)(32 + 3 + 4) - 2(8 - 1 - 6)$$

$$= 3(-35) + (-1)(39) - 2(1) = -146.$$

In computing the determinant, the expansion can be about the elements of any row or column. To get the signs, give the first term a plus sign if the sum of its column number and row number is even; give it a minus if the sum is odd, with alternating signs thereafter. (For example, in expanding about the elements of the third row we begin with a plus; the first element a_{31} has $3 + 1 = 4$, an even number.) Judicious selection of rows and columns with many zeros can hasten the process, but this method of calculating determinants is a lot of work if the matrix is of large size. Methods that triangularize a matrix, as described in Sections 2.4 and 2.5, are much better ways to get the determinant.

If a matrix, B, is triangular (either upper or lower), its determinant is just the product of the diagonal elements: $\det(B) = \Pi B_{ii}, i = 1, \ldots, n$. It is easy to show this if the determinant of the triangular matrix is expanded by minors. The following example illustrates this.

$$\det \begin{bmatrix} 4 & 0 & 0 \\ 6 & -2 & 0 \\ 1 & -3 & 5 \end{bmatrix} = 4 * \det \begin{bmatrix} -2 & 0 \\ -3 & 5 \end{bmatrix} + 0 + 0$$

$$= 4 * (-2) * |5| = 4 * (-2) * 5 = -40.$$

One special application of determinants is in the computation of the *characteristic poly-nomial* and the *eigenvalues* of a matrix. The eigenvalues are most important in applied mathematics. For a square matrix A, we define its characteristic polynomial as $p_A(\lambda) = |A - \lambda I| = \det(A - \lambda I)$.

For example, if

$$A = \begin{bmatrix} 1 & 3 \\ 4 & 5 \end{bmatrix},$$

then

$$p_A(\lambda) = |A - \lambda I| = \det \begin{bmatrix} (1 - \lambda) & 3 \\ 4 & (5 - \lambda) \end{bmatrix}$$

$$= (1 - \lambda)(5 - \lambda) - 12$$

$$= \lambda^2 - 6\lambda - 7.$$

(The characteristic polynomial is always of degree n if A is $n \times n$.) If we set the character-istic polynomial to zero and solve for the roots, we get the eigenvalues of A. For this ex-ample, these are $\lambda_1 = 7, \lambda_2 = -1$, or, in more symbolic mathematical notation,

$$\Lambda(A) = \{ 7, \quad -1 \}.$$

We also mention the notion of an *eigenvector* corresponding to an eigenvalue. The ei-genvector is a nonzero vector $\mathbf{w}$ such that

$$Aw = \lambda w, \qquad \text{that is,} \qquad (A - \lambda I)w = 0. \tag{2.2}$$

In the current example, the eigenvectors are

$$w_1 = \begin{bmatrix} 1 \\ 2 \end{bmatrix}, \qquad w_2 = \begin{bmatrix} -3 \\ 2 \end{bmatrix}.$$

We leave it as an exercise to show that these eigenvectors satisfy Eq. (2.2).

Observe that the trace of A is equal to the sum of the eigenvalues: $\text{tr}(A) = 1 + 5 = \lambda_1 + \lambda_2 = 7 + (-1) = 6$. This is true for any matrix: The sum of its eigenvalues equals its trace.

For now, we limit the finding of eigenvalues and eigenvectors to small matrices because getting these through the characteristic polynomial is difficult. In Chapter 7, we examine other, more efficient ways to get these important quantities.

Later in this chapter we examine two computer algebra systems. We show here that MATLAB comes up with the solutions.

In MATLAB, define matrix A by: A $=$ [1,3; 4,5]. The command

```
poly(A)
```

returns the coefficients of the characteristic polynomial as a vector, [1 $\quad$ −6 $\quad$ −7]. The commands det(A) and trace(A) return the values, −7 and 6, respectively. The eigen-values and eigenvectors of A come back a little differently. Here the command,

```
[W,L] = eig(A)
```

returns the eigenvectors/eigenvalues of A in matrices W and L respectively. For the matrix A of this example, the response from MATLAB is

```
W = -0.8321   -0.4472     L = -1   0
     0.5547   -0.8944          0   7
```

where the diagonals of L are the eigenvalues and the columns of W are the eigenvectors. These are "normalized" (the sum of the squares of the components equal one), which is different from our previous values. (However, they really are the same: If ω is an eigenvector corresponding to an eigenvalue, then so is $\alpha\omega$ for any nonzero scalar α.)

2.3 The Elimination Method

Suppose we have a system of equations that is of a special form, an *upper-triangular system,* such as

$$5x_1 + 3x_2 - 2x_3 = -3,$$
$$6x_2 + x_3 = -1,$$
$$2x_3 = 10.$$

Whenever a system has this special form, its solution is very easy to obtain. From the third equation we see that $x_3 = 5$. Substituting this value into the second equation quickly gives $x_2 = -1$. Then substituting both values into the first equation reveals that $x_1 = 2$; now we have the solution: $x_1 = 2, x_2 = -1, x_3 = 5$.

The first objective of the *elimination method* is to change the matrix of coefficients so that it is upper triangular. Consider this example of three equations:

$$4x_1 - 2x_2 + x_3 = 15,$$
$$-3x_1 - x_2 + 4x_3 = 8,$$
$$x_1 - x_2 + 3x_3 = 13.$$

Multiplying the first equation by 3 and the second by 4 and adding these will eliminate x_1 from the second equation. Similarly, multiplying the first by -1 and the third by 4 and adding eliminates x_1 from the third equation. (We prefer to multiply by the negatives and add, to avoid making mistakes when subtracting quantities of unlike sign.) The result is

$$4x_1 - 2x_2 + x_3 = 15,$$
$$-10x_2 + 19x_3 = 77,$$
$$-2x_2 + 11x_3 = 37.$$

We now eliminate x_2 from the third equation by multiplying the second by 2 and the third by -10 and adding to get

$$4x_1 - 2x_2 + x_3 = 15,$$
$$-10x_2 + 19x_3 = 77,$$
$$-72x_3 = -216.$$

Now we have a triangular system and the solution is readily obtained; obviously $x_3 = 3$ from the third equation, and *back-substitution* into the second equation gives $x_2 = -2$. We continue with back-substitution by substituting both x_2 and x_3 into the first equation to get $x_1 = 2$.

The essence of any elimination method is to reduce the coefficient matrix to a triangular matrix and then use back-substitution to get the solution.

We now present the same problem, solved in exactly the same way, in matrix notation:

$$\begin{bmatrix} 4 & -2 & 1 \\ -3 & -1 & 4 \\ 1 & -1 & 3 \end{bmatrix} \begin{bmatrix} x_1 \\ x_2 \\ x_3 \end{bmatrix} = \begin{bmatrix} 15 \\ 8 \\ 13 \end{bmatrix}.$$

The arithmetic operations that we have performed affect only the coefficients and the constant terms, so we work with the matrix of coefficients *augmented* with the right-hand side vector:

$$A|b = \begin{bmatrix} 4 & -2 & 1 & \vdots & 15 \\ -3 & -1 & 4 & \vdots & 8 \\ 1 & -1 & 3 & \vdots & 13 \end{bmatrix}$$

(The dashed line is usually omitted.)

We perform elementary row transformations* to convert A to *upper-triangular* form:

$$\begin{bmatrix} 4 & -2 & 1 & 15 \\ -3 & -1 & 4 & 8 \\ 1 & -1 & 3 & 13 \end{bmatrix}, \qquad \begin{matrix} 3R_1 + 4R_2 \to \\ (-1)R_1 + 4R_3 \to \end{matrix} \begin{bmatrix} 4 & -2 & 1 & 15 \\ 0 & -10 & 19 & 77 \\ 0 & -2 & 11 & 37 \end{bmatrix},$$

$$2R_2 - 10R_3 \to \begin{bmatrix} 4 & -2 & 1 & 15 \\ 0 & -10 & 19 & 77 \\ 0 & 0 & -72 & -216 \end{bmatrix}. \qquad \textbf{(2.3)}$$

The steps here are to add 3 times the first row to 4 times the second row and to add -1 times the first row to 4 times the second row. The next and final phase (in order to get a triangular system) is to add 2 times the second row to -10 times the third row.

The array in Eq. (2.3) represents the equations

$$\begin{aligned} 4x_1 - 2x_2 + x_3 &= 15, \\ -10x_2 + 19x_3 &= 77, \\ -72x_3 &= -216. \end{aligned} \qquad \textbf{(2.4)}$$

---

*Elementary row operations are arithmetic operations that obviously are valid rearrangements of a set of equations: (1) Any equation can be multiplied by a constant; (2) the order of the equations can be changed; (3) any equation can be replaced by its sum with another of the equations.

The back-substitution step can be performed quite mechanically by solving the equations of (2.4) in reverse order. That is,

$$x_3 = -216/-72 = 3,$$
$$x_2 = (77 - 3(19))/(-10) = -2,$$
$$x_1 = (15 - 1(3) - (-2)(-2))/4 = 2.$$

Thinking of the procedure in terms of matrix operations, we transform the augmented coefficient matrix by elementary row operations until a triangular matrix is created on the left. After back-substitution, the x-vector stands as the rightmost column.

These operations, which do not change the relationships represented by a set of equations, can be applied to an augmented matrix, because this is only a different notation for the equations. (We need to add one proviso: Because round-off error is related to the magnitude of the values when we express them in fixed-word-length computer representations, some of our previous operations may have an effect on the accuracy of the computed solution.)

One of the more essential features of the elimination method we have just presented is that we reduced the original problem to an upper-triangular system. In other words, we changed the original problem to this triangular system, shown next as both equations and in matrix form:

$$
\begin{aligned}
4x_1 - 2x_2 + x_3 &= 15 \\
-10x_2 + 19x_3 &= 77 \\
-72x_3 &= -216
\end{aligned}
\qquad \text{or} \qquad
\begin{bmatrix} 4 & -2 & 1 \\ 0 & -10 & 19 \\ 0 & 0 & -72 \end{bmatrix}
\begin{bmatrix} x_1 \\ x_2 \\ x_3 \end{bmatrix}
=
\begin{bmatrix} 15 \\ 77 \\ -216 \end{bmatrix}.
$$

Now we need only solve the variables in reverse order: x_3, then x_2, and finally x_1 for at most $1 + 2 + 3 = 6$ multiplications/divisions. For an upper-triangular system of n equations, we can expect at most $1 + 2 + \cdots + n = n(n + 1)/2$ multiplications/divisions to solve the system. For this reason, using the notation introduced earlier, we say that such a system requires at most $O(n^2)$ multiplications/divisions.

The same can be said for a lower-triangular system. For example, if we have this set of equations (or an augmented matrix):

$$
\begin{aligned}
6x_1 &= 18 \\
-2x_1 + 5x_2 &= 2 \\
3x_1 - 4x_2 + 3x_3 &= 11
\end{aligned}
\qquad \text{or} \qquad
\begin{bmatrix} 6 & 0 & 0 \\ -2 & 5 & 0 \\ 3 & -4 & 3 \end{bmatrix}
\begin{bmatrix} x_1 \\ x_2 \\ x_3 \end{bmatrix}
=
\begin{bmatrix} 18 \\ 2 \\ 11 \end{bmatrix}.
$$

Here we would solve for the variables in this order: x_1, then x_2, and finally x_3, with the same number of computations as in the case of a lower-triangular system. Both the lower- and upper-triangular systems play an important part in the development of algorithms in the following sections, because these systems require fewer multiplications/divisions than the general system. We shall also show that we can often write a general matrix A as the product LU, a lower-triangular matrix times an upper-triangular matrix.

Note that there exists the possibility that the set of equations has no solution, or that the prior procedure will fail to find it. During the triangularization step, if a zero is encountered on the diagonal, we cannot use that row to eliminate coefficients below that zero element. However, in that case, we can continue by interchanging rows and eventually achieve an upper-triangular matrix of coefficients. The real stumbling block is finding a zero on the diagonal after we have triangularized. If that occurs, the back-substitution fails, for we cannot divide by zero. It also means that the determinant is zero: There is no solution.

It is worthwhile to explain in more detail what we mean by the *elementary row operations* that we have used here, and to see why they can be used in solving a linear system. There are three of these operations:

1. We may multiply any row of the augmented coefficient matrix by a constant.
2. We can add a multiple of one row to a multiple of any other row.
3. We can interchange the order of any two rows (this was not used earlier).

The validity of these row operations is intuitively obvious if we think of them applied to a set of linear equations. Certainly, multiplying one equation through by a constant does not change the truth of the equality. Adding equal quantities to both sides of an equality results in an equality, and this is the equivalent of the second transformation. Obviously, the order of the set is arbitrary, so rule 3 is valid.

We conclude this section by returning to our original example to see how two well-known computer algebra systems can solve it. Maple and MATLAB are the two computer algebra systems that we shall discuss in detail in a later section. They solve the preceding example with the following simple commands:

Maple does it by

```
solve({ 4*x − 2*y + z = 15, −3*x − y + 4*z = 13, x − y + 3*z =
13}, { x,y,z});
```

which then returns the solution: $\{ x = 2, z = 3, y = -2 \}$. MATLAB does it somewhat differently:

```
[x,y,z] = solve('4*x − 2*y + z − 15', '−3*x − y + 4*z − 13',
'x − y + 3*z = 13')
```

and we then see the solution: $x = 2$, $y = -2$, $z = 3$. It may not be obvious, but a very important difference here is that in Maple we terminate the command line with a semicolon and in MATLAB we do not. Later, we shall see that the algorithms used to solve this example are the same as we have described.

2.4 The Gaussian Elimination and Gauss–Jordan Methods

Although the procedure of the previous section is satisfactory for hand calculations on small systems, there are several objections that we should eliminate before we write a computer program to perform Gaussian elimination. In a large set of equations, and that is the situation we must prepare for, the multiplications will give very large and unwieldy numbers that may overflow the computer's registers. We will therefore eliminate the first coefficient in the ith row by subtracting a_{i1}/a_{11} times the first equation from the ith equation.

(This is equivalent to making the leading coefficient 1 in the equation that retains that leading term.) We use similar ratios of coefficients in eliminating coefficients in the other columns.

We must also guard against dividing by zero. Observe that zeros may be created in the diagonal positions even if they are not present in the original matrix of coefficients. A useful strategy to avoid (if possible) such zero divisors is to rearrange the equations so as to put the coefficient of largest magnitude on the diagonal at each step. This is called *pivoting*. Complete pivoting may require both row and column interchanges. This is not frequently done. Partial pivoting, which places a coefficient of larger magnitude on the diagonal by row interchanges only, will guarantee a nonzero divisor if there is a solution to the set of equations, and will have the added advantage of giving improved arithmetic precision. The diagonal elements that result are called *pivot elements*. (When there are large differences in magnitude of coefficients in one equation compared to the other equations, we may need to *scale* the values; we consider this later.)

We repeat the example of the previous section, incorporating these ideas and carrying four significant digits in our work. We begin with the augmented matrix.

$$\left[\begin{array}{ccc|c} 4 & -2 & 1 & 15 \\ -3 & -1 & 4 & 8 \\ 1 & -1 & 3 & 13 \end{array}\right] \tag{2.5}$$

$$\begin{array}{c} (3/4)R_1 + R_2 \rightarrow \\ -(1/4)R_1 + R_3 \rightarrow \end{array} \left[\begin{array}{ccc|c} 4 & -2 & 1 & 15 \\ 0 & -2.5 & 4.75 & 19.25 \\ 0 & -0.5 & 2.75 & 9.25 \end{array}\right] \tag{2.6}$$

$$-(-0.5/-2.5)R_2 + R_3 \rightarrow \left[\begin{array}{ccc|c} 4 & -2 & 1 & 15 \\ 0 & -2.5 & 4.75 & 19.25 \\ 0 & 0.0 & 1.80 & 5.40 \end{array}\right]$$

The method we have just illustrated is called *Gaussian elimination*. (In this example, no pivoting was required to make the largest coefficients be on the diagonal.) Back-substitution, as presented with Eqs. (2.4), gives us, as before, $x_3 = 3, x_2 = -2, x_1 = 2$. We have come up with the exact answer to this problem. Often it will turn out that we shall obtain answers that are just close approximations to the exact answer because of round-off error. When there are many equations, the effects of round-off (the term is applied to the error due to chopping as well as when rounding is used) may cause large effects. In certain cases, the coefficients are such that the results are particularly sensitive to round-off; such systems are called *ill-conditioned*.

In the example just presented, the zeros below the main diagonal show that we have reduced the problem (Eq. (2.3)) to solving an upper-triangular system of equations as in Eqs. (2.4). However, at each stage, if we had stored the ratio of coefficients in place of zero (we show these in parentheses), our final form would have been

$$\left[\begin{array}{ccc|c} 4 & -2 & 1 & 15 \\ (-0.75) & -2.5 & 4.75 & 19.25 \\ (0.25) & (0.20) & 1.80 & 5.4 \end{array}\right].$$

Then, in addition to solving the problem as we have done, we find that the original matrix

$$A = \begin{bmatrix} 4 & -2 & 1 \\ -3 & -1 & 4 \\ 1 & -1 & 3 \end{bmatrix}$$

can be written as the product:

$$\underbrace{\begin{bmatrix} 1 & 0 & 0 \\ -0.75 & 1 & 0 \\ 0.25 & 0.20 & 1 \end{bmatrix}}_{L} * \underbrace{\begin{bmatrix} 4 & -2 & 1 \\ 0 & -2.5 & 4.75 \\ 0 & 0 & 1.80 \end{bmatrix}}_{U}. \tag{2.7}$$

This procedure is called a *LU decomposition* of A. In this case,

$$A = L * U,$$

where L is *lower-triangular* and U is *upper-triangular.* As we shall see in the next example, usually $L * U = A'$, where A' is just a permutation of the rows of A due to row interchange from pivoting.

Finally, because the determinant of two matrices, $B * C$, is the product of each of the determinants, for this example we have

$$\det(L * U) = \det(L) * \det(U) = \det(U),$$

because L is triangular and has only ones on its diagonal so that $\det(L) = 1$. Thus for the example given in Eqs. (2.7), we have

$$\det(A) = \det(U) = (4) * (-2.5) * (1.8) = -18,$$

because U is upper-triangular and its determinant is just the product of the diagonal elements.

From this example, we see that Gaussian elimination does the following:

1. It solves the system of equations.
2. It computes the determinant of a matrix very efficiently.
3. It can provide us with the *LU* decomposition of the matrix of coefficients, in the sense that the product of the two matrices, $L * U$, may give us a permutation of the rows of the original matrix.

With regard to item (2), when there are row interchanges,

$$\det(A) = (-1)^m * u_{11} * \cdots u_{nn}, \tag{2.8}$$

where the exponent m represents the number of row interchanges.

Let us summarize the operations of Gaussian elimination in a form that will facilitate the writing of a computer program. Note that in the actual implementation of the algorithm, the L and U matrices are actually stored in the space of the original matrix A.

Gaussian Elimination

To solve a system of n linear equations: $Ax = b$.

For $j = 1$ To $(n - 1)$
   pvt $= |a[j, j]|$
   pivot$[j] = j$
   ipvt_temp $= j$

   For $i = j + 1$ To n (Find pivot row)
      IF $|a[i, j]| >$ pvt   Then
         pvt $= |a[i, j]|$
         ipvt_temp $= i$
      End IF
   End For i

   (Switch rows if necessary)
   IF pivot$[j] <>$ ipvt_temp
     [switch_rows(rows j and ipvt_temp)]

   For $i = j + 1$ to n (Store multipliers)
     $a[i, j] = a[i, j]/a[j, j]$
   End For i

   (Create zeros below the main diagonal)
   For $i = j + 1$ To n
     For $k = j + 1$ To n
        $a[i, k] = a[i, k] - a[i, j] * a[j, k]$
     End For k
     $b[i] = b[i] - a[i, j] * b[j]$
   End For i

End For j;

(Back Substitution Part)
$x[n] = b[n]/a[n, n]$
For $j = n - 1$ Down To 1
   $x[j] = b[j]$
   For $k = n$ Down To $j + 1$
     $x[j] = x[j] - x[k] * a[j, k]$
   End For k
   $x[j] = x[j]/a[j, j]$

End For j.

Some computer programs do not actually move rows to interchange all the elements of the rows when pivoting. In these programs, one keeps track of the order in which the rows are to be used in a vector whose elements represent row order. (In the next example, we shall make use of such a vector.) When an interchange is indicated, only the elements of this ordering vector are changed. These numbers are then used to locate the positions of the elements in the matrix of coefficients that are to be operated on, during both the reduction step and the back-substitution. This method can reduce the computer time for large systems, but because it adds to the complexity of the program, we shall make the actual row interchanges.

The algorithm for Gaussian elimination will be clarified by an additional numerical example. Solve the following system of equations using Gaussian elimination. In addition, compute the determinant of the coefficient matrix and the *LU* decomposition of this matrix.

Given the system of equations, solve

$$
\begin{aligned}
2x_2 \qquad\qquad + \ x_4 &= \ \ \ 0, \\
2x_1 + 2x_2 + 3x_3 + 2x_4 &= -2, \\
4x_1 - 3x_2 \qquad\quad + \ x_4 &= -7, \\
6x_1 + \ x_2 - 6x_3 - 5x_4 &= \ \ \ 6.
\end{aligned}
\tag{2.9}
$$

The augmented coefficient matrix is

$$
\begin{bmatrix}
0 & 2 & 0 & 1 & 0 \\
2 & 2 & 3 & 2 & -2 \\
4 & -3 & 0 & 1 & -7 \\
6 & 1 & -6 & -5 & 6
\end{bmatrix}.
\tag{2.10}
$$

We cannot permit a zero in the a_{11} position because that element is the pivot in reducing the first column. We could interchange the first row with any of the other rows to avoid a zero divisor, but interchanging the first and fourth rows is our best choice. This gives

$$
\begin{bmatrix}
6 & 1 & -6 & -5 & 6 \\
2 & 2 & 3 & 2 & -2 \\
4 & -3 & 0 & 1 & -7 \\
0 & 2 & 0 & 1 & 0
\end{bmatrix}.
\tag{2.11}
$$

We make all the elements in the first column zero by subtracting the appropriate multiple of row one:

$$
\begin{bmatrix}
6 & 1 & -6 & -5 & 6 \\
0 & 1.6667 & 5 & 3.6667 & -4 \\
0 & -3.6667 & 4 & 4.3333 & -11 \\
0 & 2 & 0 & 1 & 0
\end{bmatrix}.
\tag{2.11a}
$$

We again interchange before reducing the second column, not because we have a zero divisor, but because we want to preserve accuracy.* Interchanging the second and third

---

*A numerical example that demonstrates the improved accuracy when partial pivoting is used will be found in Section 2.9.

rows puts the element of largest magnitude on the diagonal. (We could also interchange the fourth column with the second, giving an even larger diagonal element, but we do not do this.) After the interchange, we have

$$\begin{bmatrix} 6 & 1 & -6 & -5 & 6 \\ 0 & -3.6667 & 4 & 4.3333 & -11 \\ 0 & 1.6667 & 5 & 3.6667 & -4 \\ 0 & 2 & 0 & 1 & 0 \end{bmatrix}. \tag{2.12}$$

Now we reduce in the second column

$$\begin{bmatrix} 6 & 1 & -6 & -5 & 6 \\ 0 & -3.6667 & 4 & 4.3333 & -11 \\ 0 & 0 & 6.8182 & 5.6364 & -9.0001 \\ 0 & 0 & 2.1818 & 3.3636 & -5.9999 \end{bmatrix}. \tag{2.13}$$

No interchange is indicated in the third column. Reducing, we get

$$\begin{bmatrix} 6 & 1 & -6 & -5 & 6 \\ 0 & -3.6667 & 4 & 4.3333 & -11 \\ 0 & 0 & 6.8182 & 5.6364 & -9.0001 \\ 0 & 0 & 0 & 1.5600 & -3.1199 \end{bmatrix}. \tag{2.14}$$

Back-substitution gives

$$x_4 = \frac{-3.1199}{1.5600} = -1.9999,$$

$$x_3 = \frac{-9.0001 - 5.6364(-1.9999)}{6.8182} = 0.33325,$$

$$x_2 = \frac{-11 - 4.3333(-1.9999) - 4(0.33325)}{-3.6667} = 1.0000,$$

$$x_1 = \frac{6 - (-5)(-1.9999) - (-6)(0.33325) - (1)(1.0000)}{6} = -0.50000.$$

The correct answers are $-2, \frac{1}{3}, 1$, and $-\frac{1}{2}$ for x_4, x_3, x_2, and x_1. In this calculation we have carried five significant figures and rounded each calculation. Even so, we do not have five-digit accuracy in the answers. The discrepancy is due to round-off. The question of the accuracy of the computed solution to a set of equations is a most important one, and at several points in the following discussion we will discuss how to minimize the effects of round-off and avoid conditions that can cause round-off errors to be magnified.

In this example, if we had replaced the zeros below the main diagonal with the ratio of coefficients at each step, the resulting augmented matrix would be

$$\begin{bmatrix} 6 & 1 & -6 & -5 & 6 \\ (0.66667) & -3.6667 & 4 & 4.3333 & -11 \\ (0.33333) & (-0.45454) & 6.8182 & 5.6364 & -9.0001 \\ (0.0) & (-0.54545) & (0.32) & 1.5600 & -3.1199 \end{bmatrix}. \quad \textbf{(2.15)}$$

This gives a *LU* decomposition as

$$\begin{bmatrix} 1 & 0 & 0 & 0 \\ 0.66667 & 1 & 0 & 0 \\ 0.33333 & -0.45454 & 1 & 0 \\ 0.0 & -0.54545 & 0.32 & 1 \end{bmatrix} * \begin{bmatrix} 6 & 1 & -6 & -5 \\ 0 & -3.6667 & 4 & 4.3333 \\ 0 & 0 & 6.8182 & 5.6364 \\ 0 & 0 & 0 & 1.5600 \end{bmatrix}. \quad \textbf{(2.16)}$$

It should be noted that the product of the matrices in (2.16) produces a permutation of the original matrix, call it A', where

$$A' = \begin{bmatrix} 6 & 1 & -6 & -5 \\ 4 & -3 & 0 & 1 \\ 2 & 2 & 3 & 2 \\ 0 & 2 & 0 & 1 \end{bmatrix}, \quad \textbf{(2.16a)}$$

because rows 1 and 4 were interchanged in (2.11) and rows 2 and 3 in (2.12). The determinant of the original matrix of coefficients—the first four columns of (2.10)—can be easily computed from (2.14) or (2.15) according to the formula

$$\det(A) = (-1)^2 * (6) * (-3.6667) * (6.8182) * (1.5600) = -234.0028,$$

which is close to the exact solution: -234.* The exponent 2 is required, because there were two row interchanges in solving this system. To summarize, you should note that the Gaussian elimination method applied to (2.9) produces the following:

1. The solution to the four equations.
2. The determinant of the coefficient matrix

$$\begin{bmatrix} 0 & 2 & 0 & 1 \\ 2 & 2 & 3 & 2 \\ 4 & -3 & 0 & 1 \\ 6 & 1 & -6 & -5 \end{bmatrix}.$$

3. A *LU* decomposition of the matrix, A', which is just the original matrix, A, after we have interchanged its rows in the process.

---

*The difference is because the computed values have been rounded to four decimal places.

It is of interest to see whether MATLAB confirms the results of this example. We define matrix A by

```
A = [0, 2, 0, 1; 2, 2, 3, 2; 4, −3, 0, 1; 6, 1, −6, −5]
```

and call for the LU decomposition

```
[L, U, P] = lu(A).
```

MATLAB then returns the matrices L and U as seen in Eq. (2.16). In addition, MATLAB gives us the permutation matrix, P:

$$P = \begin{bmatrix} 0 & 0 & 0 & 1 \\ 0 & 0 & 1 & 0 \\ 0 & 1 & 0 & 0 \\ 1 & 0 & 0 & 0 \end{bmatrix},$$

which is the matrix that produces $PA = A'$ of Eq. (2.16a).

The efficiency of the Gaussian elimination algorithm is usually measured by counting the total number of multiplications and divisions involved. To get from a system like that of Eq. (2.10) to that of Eq. (2.15) takes at most $4^3/3 + 4^2/2 - 5 * 4/6$ multiplications and divisions. For a general system of n equations, the algorithm for both triangularization and back-substitution requires at most $n^3/3 + n^2 - n/3$ multiplications and divisions. In the terminology we described earlier in Chapter 0, this is $O(n^3)$ such operations.

The Gauss–Jordan Method

There are many variants to the Gaussian elimination scheme. The back-substitution step can be performed by eliminating the elements above the diagonal after the triangularization has been finished, using elementary row operations and proceeding upward from the last row. This technique is equivalent to the procedures described in the following example. The diagonal elements may all be made ones as a first step before creating zeros in their column; this performs the divisions of the back-substitution phase at an earlier time.

One variant that is sometimes used is the *Gauss–Jordan* scheme. In it, the elements above the diagonal are made zero at the *same time* that zeros are created below the diagonal. Usually the diagonal elements are made ones at the same time that the reduction is performed; this transforms the coefficient matrix into the identity matrix. When this has been accomplished, the column of right-hand sides has been transformed into the solution vector. Pivoting is normally employed to preserve arithmetic accuracy.

The previous example, solved by the Gauss–Jordan method, gives this succession of calculations. The original augmented matrix is

$$\begin{bmatrix} 0 & 2 & 0 & 1 & 0 \\ 2 & 2 & 3 & 2 & -2 \\ 4 & -3 & 0 & 1 & -7 \\ 6 & 1 & -6 & -5 & 6 \end{bmatrix}.$$

Interchanging rows one and four, dividing the first row by 6, and reducing the first column gives

$$
\begin{bmatrix}
1 & 0.16667 & -1 & -0.83335 & 1 \\
0 & 1.6667 & 5 & 3.6667 & -4 \\
0 & -3.6667 & 4 & 4.3334 & -11 \\
0 & 2 & 0 & 1 & 0
\end{bmatrix}.
$$

Interchanging rows two and three, dividing the second row by -3.6667, and reducing the second column (operating above the diagonal as well as below) gives

$$
\begin{bmatrix}
1 & 0 & -1.5000 & -1.2000 & 1.4000 \\
0 & 1 & 2.9999 & 2.2000 & -2.4000 \\
0 & 0 & 15.000 & 12.400 & -19.800 \\
0 & 0 & -5.9998 & -3.4000 & 4.8000
\end{bmatrix}.
$$

No interchanges are required for the next step. We divide the third row by 6.8182 and make the other elements in the third column into zeros:

$$
\begin{bmatrix}
1 & 0 & -0.8182 & -0.6364 & 0.5 \\
0 & 1 & -1.0909 & -1.1818 & 3 \\
0 & 0 & 6.8182 & 5.6364 & -9 \\
0 & 0 & 2.1818 & 3.3636 & -6
\end{bmatrix}.
$$

We now divide the fourth row by 1.5599 and create zeros above the diagonal in the fourth column:

$$
\begin{bmatrix}
1 & 0 & 0 & 0 & -0.49999 \\
0 & 1 & 0 & 0 & 1.0001 \\
0 & 0 & 1 & 0 & 0.33326 \\
0 & 0 & 0 & 1 & -1.9999
\end{bmatrix}.
$$

The solution is essentially the same as with the usual Gaussian method; round-off errors have created inaccuracies in a slightly different way than did the previous computation.

Although it might seem that the work to accomplish the Gauss–Jordan method is similar to that for Gaussian elimination, this assumption is not true. In fact, the Gauss–Jordan technique requires almost 50% more arithmetic operations (Exercise 19). It can be shown that the total number of multiplications/divisions to solve an $n \times n$ system through Gauss–Jordan elimination equals $n^3/2 + n^2 - 7n/2 + 2$, whereas as we have seen, Gaussian elimination takes $n^3/3 + n^2 - n/3$. (Although in both cases, these algorithms are of $O(n^3)$.)

The special merit of Gaussian elimination is that we get the LU decomposition of matrix A, which can be used in solving a system with multiple right-hand sides, and these right-hand sides can be input sequentially after getting the L and U matrices. This is because, if we have $LU = PA$, then $Ax = b$ can be written as $PAx = Pb = L(Ux)$, and solving $Pb = L(Ux)$ takes only $O(n^2)$ multiplications/divisions. In Section 2.5 we will consider an example where multiple right-hand sides are input in succession.

Of course, if the multiple right-hand sides are known at the start, we can augment matrix A (the coefficient matrix) with all of them and use either the Gaussian or Gauss–Jordan method to treat each column of the right-hand sides in the same way as with a single b-vector. Example 2.2 illustrates this situation.

EXAMPLE 2.2   Solve the system $Ax = b$, with multiple values of b, by Gaussian elimination:

$$A = \begin{bmatrix} 3 & 2 & -1 & 2 \\ 1 & 4 & 0 & 2 \\ 2 & 1 & 2 & -1 \\ 1 & 1 & -1 & 3 \end{bmatrix}, \quad b^{(1)} = \begin{bmatrix} 0 \\ 0 \\ 1 \\ 0 \end{bmatrix}, \quad b^{(2)} = \begin{bmatrix} -2 \\ 1 \\ 3 \\ 4 \end{bmatrix}, \quad b^{(3)} = \begin{bmatrix} 2 \\ 2 \\ 0 \\ 0 \end{bmatrix}.$$

We augment A with all of the b's, then triangularize:

$$\begin{bmatrix} 3 & 2 & -1 & 2 & 0 & -2 & 2 \\ 1 & 4 & 0 & 2 & 0 & 1 & 2 \\ 2 & 1 & 2 & -1 & 1 & 3 & 0 \\ 1 & 1 & -1 & 3 & 0 & 4 & 0 \end{bmatrix} \rightarrow$$

$$\begin{bmatrix} 3 & 2 & -1 & 2 & 0 & -2 & 2 \\ (0.333) & 3.333 & 0.333 & 1.333 & 0 & 1.667 & 1.333 \\ (0.667) & -0.333 & 2.667 & -2.333 & 1 & 4.333 & -1.333 \\ (0.333) & 0.333 & -0.667 & 2.333 & 0 & 4.667 & -0.667 \end{bmatrix} \rightarrow$$

$$\begin{bmatrix} 3 & 2 & -1 & 2 & 0 & -2 & 2 \\ (0.333) & 3.333 & 0.333 & 1.333 & 0 & 1.667 & 1.333 \\ (0.667) & (-0.100) & 2.700 & -2.200 & 1 & 4.500 & -1.200 \\ (0.333) & (0.100) & -0.700 & 2.200 & 0 & 4.500 & -0.800 \end{bmatrix} \rightarrow$$

$$\begin{bmatrix} 3 & 2 & -1 & 2 & 0 & -2 & 2 \\ (0.333) & 3.333 & 0.333 & 1.333 & 0 & 1.667 & 1.333 \\ (0.667) & (-0.100) & 2.700 & -2.200 & 1 & 4.500 & -1.200 \\ (0.333) & (0.100) & (0.259) & 1.630 & 0.259 & 5.667 & -1.111 \end{bmatrix}.$$

$$\underset{c^{(1)}}{\uparrow} \quad \underset{c^{(2)}}{\uparrow} \quad \underset{c^{(3)}}{\uparrow}$$

We obtain the three solution vectors by back-substitution, employing the proper b'-vector. (These are indicated above as $c^{(i)}$.)

$$x^{(1)} = \begin{bmatrix} 0.137 \\ -0.114 \\ 0.500 \\ 0.159 \end{bmatrix}, \quad x^{(2)} = \begin{bmatrix} -0.591 \\ -1.340 \\ 4.500 \\ 3.477 \end{bmatrix}, \quad x^{(3)} = \begin{bmatrix} 0.273 \\ 0.773 \\ -1.000 \\ -0.682 \end{bmatrix}. \quad \blacktriangle$$

Scaling

We have mentioned that the rows of the augmented coefficient matrix may need to be scaled before a proper choice of pivot element can be made. *Scaling* is the operation of adjusting the coefficients of a set of equations so that they are all of the same order of magnitude. In some instances, a set of equations may involve relationships between quantities measured in widely different units (microvolts versus kilovolts, for example, or nanoseconds versus years). This may result in some of the equations having very large numbers and others very small. If we select the pivot elements without scaling, pivoting may put numbers on the diagonal that are not large in comparison to others in their row; this can actually create the round-off errors that pivoting was supposed to avoid. An example will clarify the concept.

$$\text{Given} \begin{bmatrix} 3 & 2 & 100 \\ -1 & 3 & 100 \\ 1 & 2 & -1 \end{bmatrix} x = \begin{bmatrix} 105 \\ 102 \\ 2 \end{bmatrix}.$$

Carrying only three digits to emphasize round-off, and using partial pivoting, we find that the triangularized system is

$$\begin{bmatrix} 3 & 2 & 100 & 105 \\ 0 & 3.66 & 133 & 137 \\ 0 & 0 & -82.6 & -82.7 \end{bmatrix},$$

from which $x_3 = 1.00$, $x_2 = 1.09$, $x_1 = 0.94$; the exact solution vector should be [1.00, 1.00, 1.00].

If we scale the values before reduction by dividing each row by the magnitude of the largest coefficient, so that the system is

$$\begin{bmatrix} 0.03 & 0.02 & 1.00 \\ -0.01 & 0.03 & 1.00 \\ 0.50 & 1.00 & -0.50 \end{bmatrix} x = \begin{bmatrix} 1.05 \\ 1.02 \\ 1.00 \end{bmatrix},$$

we get, with the same arithmetic precision, the triangularized system

$$\begin{bmatrix} 0.50 & 1.00 & -0.50 & 1.00 \\ 0 & 0.05 & 0.99 & 1.04 \\ 0 & 0 & 1.82 & 1.82 \end{bmatrix},$$

from which $x_3 = 1.00$, $x_2 = 1.00$, $x_1 = 1.00$. The reason for the improvement is that rows are interchanged after scaling has been done. No interchanges are indicated in the unscaled equations.

Whenever the coefficients in one column are widely different from those in another column, scaling is beneficial. When all values are about the same order of magnitude, scaling should be avoided, for the additional round-off error incurred during the scaling operation itself may adversely affect the accuracy.* The usual way to scale is as we have done

---

*Actually, scaling can be done without incurring greater round-off error by doing *virtual scaling*. One scales to determine the correct pivot row but then reduces the matrix using the values as they are without dividing by the largest coefficient.

here, by dividing each row by the magnitude of the largest term. Some authorities recommend scaling so that the sum of the magnitudes of the coefficients in each row is the same. This is probably very slightly more economical to do in a computer program.

2.5 Other Direct Methods

The term *direct method* indicates a method that solves a set of equations by techniques such as the Gaussian elimination method of Section 2.4. These methods are in contrast to *iterative methods,* which we will describe in Section 2.10.

Although Gaussian elimination is the best known, there are other *LU* decomposition methods. In fact, a square matrix A can be resolved into the product of a lower- and an upper-triangular matrix in an infinite number of ways. Suppose that we have $A = L * U$. Take matrix D, any purely diagonal matrix with nonzero entries, and form $L' = L * D$ and $U' = D^{-1} * U$. It follows that $L' * U' = A$, and there are an infinite number of ways to choose matrix D.

We will examine one of the alternative *LU* methods, the *Crout reduction method.* It also transforms the coefficient matrix, A, into $L * U$, but in contrast to Gaussian elimination, matrix U has ones on its diagonal (in the Gauss method, L has ones on its diagonal). Another *LU* method, the *Cholesky method,* resembles Gaussian elimination in having ones on the diagonal of L.

We now develop the procedure for Crout reduction.

We get the rules for Crout reduction from the fact that $LU = A$. In the case of a 4×4 matrix:

$$
\begin{bmatrix}
\ell_{11} & 0 & 0 & 0 \\
\ell_{21} & \ell_{22} & 0 & 0 \\
\ell_{31} & \ell_{32} & \ell_{33} & 0 \\
\ell_{41} & \ell_{42} & \ell_{43} & \ell_{44}
\end{bmatrix}
\begin{bmatrix}
1 & u_{12} & u_{13} & u_{14} \\
0 & 1 & u_{23} & u_{24} \\
0 & 0 & 1 & u_{34} \\
0 & 0 & 0 & 1
\end{bmatrix}
=
\begin{bmatrix}
a_{11} & a_{12} & a_{13} & a_{14} \\
a_{21} & a_{22} & a_{23} & a_{24} \\
a_{31} & a_{32} & a_{33} & a_{34} \\
a_{41} & a_{42} & a_{43} & a_{44}
\end{bmatrix}.
$$

Multiplying the rows of L by the first column of U, we get $\ell_{11} = a_{11}, \ell_{21} = a_{21}, \ell_{31} = a_{31}, \ell_{41} = a_{41}$; the first column of L is the same as the first column of A.

We now multiply the first row of L by the columns of U:

$$\ell_{11}u_{12} = a_{12}, \qquad \ell_{11}u_{13} = a_{13}, \qquad \ell_{11}u_{14} = a_{14}, \tag{2.17}$$

from which

$$u_{12} = \frac{a_{12}}{\ell_{11}}, \qquad u_{13} = \frac{a_{13}}{\ell_{11}}, \qquad u_{14} = \frac{a_{14}}{\ell_{11}}. \tag{2.18}$$

Thus the first row of U is determined.

In this method we alternate between getting a column of L and a row of U, so we next get the equations for the second column of L by multiplying the rows of L by the second column of U:

$$\ell_{21}u_{12} + \ell_{22} = a_{22},$$

$$\ell_{31}u_{12} + \ell_{32} = a_{32},$$

$$\ell_{41}u_{12} + \ell_{42} = a_{42},$$

which gives

$$\ell_{22} = a_{22} - \ell_{21}u_{12},$$

$$\ell_{32} = a_{32} - \ell_{31}u_{12}, \qquad\qquad (2.19)$$

$$\ell_{42} = a_{42} - \ell_{41}u_{12}.$$

Proceeding in the same fashion, the equations we need are

$$u_{23} = \frac{a_{23} - \ell_{21}u_{13}}{\ell_{22}}, \qquad u_{24} = \frac{a_{24} - \ell_{21}u_{14}}{\ell_{22}},$$

$$\ell_{33} = a_{33} - \ell_{31}u_{13} - \ell_{32}u_{23}, \qquad \ell_{43} = a_{43} - \ell_{41}u_{13} - \ell_{42}u_{23},$$

$$u_{34} = \frac{a_{34} - \ell_{31}u_{14} - \ell_{32}u_{24}}{\ell_{33}},$$

$$\ell_{44} = a_{44} - \ell_{41}u_{14} - \ell_{42}u_{24} - \ell_{43}u_{34}.$$

The general formula for getting elements of L and U corresponding to the coefficient matrix for n simultaneous equations can be written

$$\ell_{ij} = a_{ij} - \sum_{k=1}^{j-1} \ell_{ik}u_{kj}, \qquad j \le i, \qquad i = 1, 2, \ldots, n, \qquad (2.20)$$

$$u_{ij} = \frac{a_{ij} - \sum_{k=1}^{i-1} \ell_{ik}u_{kj}}{\ell_{ii}}, \qquad i \le j, \qquad j = 2, 3, \ldots, n. \qquad (2.21)$$

(For $j = 1$, the rule for ℓ reduces to

$$\ell_{i1} = a_{i1}.$$

For $i = 1$, the rule for u reduces to

$$u_{1j} = \frac{a_{1j}}{\ell_{11}} = \frac{a_{1j}}{a_{11}}.)$$

The reason this method is popular in programs is that storage space may be economized. There is no need to store the zeros in either L or U, and the ones on the diagonal of U can also be omitted. (Because these values are always the same and are always known, it is redundant to record them.) One can then store the essential elements of U where the zeros appear in the L array. Examination of Eqs. (2.17) through (2.21) shows that, after any element of A, a_{ij}, is once used, it never again appears in the equations. Hence its place in the original $n \times n$ array A can be used to store an element of either L or U. In other words, the A array can be transformed by these equations and becomes

$$
\begin{bmatrix}
a_{11} & a_{12} & a_{13} & a_{14} \\
a_{21} & a_{22} & a_{23} & a_{24} \\
a_{31} & a_{32} & a_{33} & a_{34} \\
a_{41} & a_{42} & a_{43} & a_{44}
\end{bmatrix}
\rightarrow
\begin{bmatrix}
\ell_{11} & u_{12} & u_{13} & u_{14} \\
\ell_{21} & \ell_{22} & u_{23} & u_{24} \\
\ell_{31} & \ell_{32} & \ell_{33} & u_{34} \\
\ell_{41} & \ell_{42} & \ell_{43} & \ell_{44}
\end{bmatrix}.
$$

Because we can condense the L and U matrices into one array and store their elements in the space of A, this method is often called a *compact scheme*.

EXAMPLE 2.3 Consider the matrix A:

$$
A = \begin{bmatrix}
3 & -1 & 2 \\
1 & 2 & 3 \\
2 & -2 & -1
\end{bmatrix}.
$$

Applying the equations for the ℓ's and u's, we obtain

$$\ell_{11} = 3, \qquad \ell_{21} = 1, \qquad \ell_{31} = 2; \qquad u_{12} = -\tfrac{1}{3}, \qquad u_{13} = \tfrac{2}{3}.$$

$$\ell_{22} = 2 - (1)(-\tfrac{1}{3}) = \tfrac{7}{3}, \qquad \ell_{32} = -2 - (2)(-\tfrac{1}{3}) = -\tfrac{4}{3}.$$

$$u_{23} = \frac{3 - (1)(\tfrac{2}{3})}{\tfrac{7}{3}} = 1, \qquad \ell_{33} = -1 - (2)(\tfrac{2}{3}) - (-\tfrac{4}{3})(1) = -1.$$

$$
L = \begin{bmatrix}
3 & 0 & 0 \\
1 & \tfrac{7}{3} & 0 \\
2 & -\tfrac{4}{3} & -1
\end{bmatrix}, \qquad
U = \begin{bmatrix}
1 & -\tfrac{1}{3} & \tfrac{2}{3} \\
0 & 1 & 1 \\
0 & 0 & 1
\end{bmatrix}.
$$

If the quantities are written in the compact form as they are computed, we have

$$
LU = \begin{bmatrix}
3 & -\tfrac{1}{3} & \tfrac{2}{3} \\
1 & \tfrac{7}{3} & 1 \\
2 & -\tfrac{4}{3} & -1
\end{bmatrix}. \quad \begin{matrix} \leftarrow ② \\ \leftarrow ④ \\ \end{matrix}
$$
$$\uparrow \quad \uparrow \quad \uparrow$$
$$① \quad ③ \quad ⑤$$

The circled numbers show the order in which columns and rows of the new matrix are obtained. ▲

Here is an algorithm for LU decomposition. It does not compute the L and U matrices in place, but sets them up as separate matrices.

LU Decomposition—Crout Reduction

To transform an $n \times n$ matrix A into $L * U$:

For $i = 1$ To n
  $L[i, 1] = a[i, 1]$
End For i

For $j = 1$ to n
  $U[1, j] = a[1, j]/L[1, 1]$
End For j

For $j = 2$ to n

  For $i = j$ To n
    sum $= 0.0$
    For $k = 1$ To $j - 1$
      sum $=$ sum $+ L[i, k] * U[k, j]$
    End For k
    $L[i, j] = a[i, j] -$ sum
  End For i

  $U[j, j] = 1$

  For $i = j + 1$ To n
    sum $= 0.0$
    For $k = 1$ To $j - 1$
      sum $=$ sum $+ L[j, k] * U[k, i]$
    End For k
    $U[j, i] = (a[j, i] -$ sum$)/L[j, j]$
  End For i

End For j

The solution of the set of equations $Ax = b$ is readily obtained with the L and U matrices. Once the coefficient matrix has been converted to its LU equivalent, we are prepared to find the solution to the set of equations that corresponds to any given right-hand-side vector b. The L matrix is really a record of the operations required to make the coefficient matrix A into the upper-triangular matrix U. We apply these same transformations to the right-hand-side vector b, converting it to a new vector b'. If we augment b' to U and back-substitute, the solution appears.

The general equation for the reduction of b (it is exactly the same as the rule for forming the elements of U) is

$$b_1' = \frac{b_1}{\ell_{11}},$$

$$b_i' = \frac{b_i - \sum_{k=1}^{i-1} \ell_{ik} b_k'}{\ell_{ii}}, \qquad i = 2, 3, \ldots, n.$$

The equations for the back-substitution are

$$x_n = b_n',$$

$$x_j = b_j' - \sum_{k=j+1}^{n} u_{jk} x_k, \qquad j = n-1, n-2, \ldots, 1.$$

For example, if

$$A = \begin{bmatrix} 3 & -1 & 2 \\ 1 & 2 & 3 \\ 2 & -2 & -1 \end{bmatrix},$$

we get

$$L = \begin{bmatrix} 3 & 0 & 0 \\ 1 & \frac{7}{3} & 0 \\ 2 & -\frac{4}{3} & -1 \end{bmatrix} \quad \text{and} \quad U = \begin{bmatrix} 1 & -\frac{1}{3} & \frac{2}{3} \\ 0 & 1 & 1 \\ 0 & 0 & 1 \end{bmatrix}.$$

For $b = \begin{bmatrix} 12 \\ 11 \\ 2 \end{bmatrix}$,

$$b' = \begin{bmatrix} 4 \\ 3 \\ 2 \end{bmatrix},$$

because

$$b_1' = \tfrac{12}{3} = 4,$$

$$b_2' = \frac{11 - (1)(4)}{\frac{7}{3}} = 3,$$

$$b_3' = \frac{2 - (2)(4) - (-\frac{4}{3})(3)}{-1} = 2.$$

Augmenting b' to U and back-substituting

$$\begin{bmatrix} 1 & -\frac{1}{3} & \frac{2}{3} & \vdots & 4 \\ 0 & 1 & 1 & \vdots & 3 \\ 0 & 0 & 1 & \vdots & 2 \end{bmatrix},$$

we get

$$x_3 = 2,$$
$$x_2 = 3 - 1(2) = 1,$$
$$x_1 = 4 - \tfrac{2}{3}(2) - (-\tfrac{1}{3})(1) = 3.$$

Examination of these operations reveals that we can get b' by augmenting L with b and solving that triangular system (a kind of "forward" substitution):

$$L_{\mid}^{\mid}b = \begin{bmatrix} 3 & 0 & 0 & \vdots & 12 \\ 1 & \frac{7}{3} & 0 & \vdots & 11 \\ 2 & -\frac{4}{3} & -1 & \vdots & 2 \end{bmatrix}$$

and

$$b_1' = \tfrac{12}{3} = 4,$$

$$b_2' = \frac{11 - (1)(4)}{\frac{7}{3}} = 3,$$

$$b_3' = \frac{2 - (2)(4) - (-\tfrac{4}{3})(3)}{-1} = 2.$$

We do not write the algorithm for reducing the b-vector and back-substituting. This is left as an exercise for the student.

A special advantage of these LU methods is that we can accumulate the sums in double precision. This gives us greater accuracy by just using one or two double-precision variables. This is not easily done with the Gaussian elimination method of Section 2.4. Moreover, the LU method can be easily adapted to solve a system of new right-hand-side vectors with great economy of effort.* The number of arithmetic operations to get the solution corresponding to each b turns out to be exactly the same as to multiply an $n \times n$ matrix by an n-component vector.

Pivoting with the LU method is somewhat more complicated than with Gaussian elimination because we do not usually handle the right-hand-side vector simultaneously with our reduction of the A matrix. This means we must keep a record of any row interchanges made during the formation of L and U so that the elements of the right-hand-side vector

---

*A numerical example that illustrates using the LU method with multiple right-hand sides will be found in Section 2.7.

can be similarly interchanged. We do the interchanges immediately after computing each column of L, choosing the value to appear on the diagonal so as to have the one of largest magnitude. We illustrate with an example:

$$\text{Given } A = \begin{bmatrix} 0 & 2 & 1 \\ 1 & 0 & 0 \\ 3 & 0 & 1 \end{bmatrix}.$$

We will keep a record of row order in a vector: $O = [1, 2, 3]$, representing the original ordering.

The first column of L is

$$\begin{bmatrix} 0 \\ 1 \\ 3 \end{bmatrix};$$

we need to interchange rows 3 and 1. To keep track of this, we interchange the first and third elements of O, so O becomes $[3, 2, 1]$.

Interchange the rows of A and compute the first row of the U matrix. (We use the compact scheme.)

$$\begin{bmatrix} 3 & 0 & \frac{1}{3} \\ 1 & 0 & 0 \\ 0 & 2 & 1 \end{bmatrix}, \qquad \text{with } O = [3, 2, 1].$$

Now compute the second column of L; it is $\begin{bmatrix} 0 \\ 0 \\ 2 \end{bmatrix}$.

We must interchange again, the second row with the third, and O becomes $[3, 1, 2]$. Making the interchange of rows and computing the second row of U gives

$$\begin{bmatrix} 3 & 0 & \frac{1}{3} \\ 0 & 2 & \frac{1}{2} \\ 1 & 0 & 0 \end{bmatrix}, \qquad \text{with } O = [3, 1, 2].$$

Completing the reduction, we find $\ell_{33} = 0 - (1)(\frac{1}{3}) = -\frac{1}{3}$, giving

$$LU = \begin{bmatrix} 3 & 0 & \frac{1}{3} \\ 0 & 2 & \frac{1}{2} \\ 1 & 0 & -\frac{1}{3} \end{bmatrix}, \qquad \text{with } O = [3, 1, 2].$$

To solve the problem $Ax = b$, with $b^T = [5, -1, -2]$, we rearrange the elements of b in the order given by O and compute b':

$$L \vdots b = \begin{bmatrix} 3 & 0 & 0 & \vdots & -2 \\ 0 & 2 & 0 & \vdots & 5 \\ 1 & 0 & -\frac{1}{3} & \vdots & -1 \end{bmatrix}, \qquad b' = \begin{bmatrix} -\frac{2}{3} \\ \frac{5}{2} \\ 1 \end{bmatrix},$$

so

$$U \vert b' = \begin{bmatrix} 1 & 0 & \frac{1}{3} \\ 0 & 1 & \frac{1}{2} \\ 0 & 0 & 1 \end{bmatrix} \quad \begin{matrix} -\frac{2}{3} \\ \frac{5}{2} \\ 1 \end{matrix}, \qquad \text{giving } x = \begin{bmatrix} -1 \\ 2 \\ 1 \end{bmatrix}.$$

The Gaussian elimination method of Section 2.4 can be used in the same way to get the solution if we have multiple right-hand sides. Suppose we want to solve $Ax = b^{(1)}$, then $Ax = b^{(2)}$, where the matrices A are the same but the right-hand sides are different.

As an example, we will solve this system by Gaussian elimination:

$$\begin{aligned} 4x - 3y &= -7, \\ 2x + 2y + 3z &= -2, \\ 6x + y - 6z &= 6. \end{aligned} \qquad (2.22)$$

In augmented matrix form, we get

$$\begin{bmatrix} 4 & -3 & 0 & -7 \\ 2 & 2 & 3 & -2 \\ 6 & 1 & -6 & 6 \end{bmatrix}, \qquad \text{Row order vector } O = [1, 2, 3].$$

After using the Gaussian elimination method of the previous section, the system is

$$\begin{bmatrix} 6 & 1 & -6 & 6 \\ (0.6667) & -3.667 & 4 & -11 \\ (0.3333) & (-0.4545) & 6.8182 & -9 \end{bmatrix}, \qquad \text{Row order vector } O = [3, 1, 2].$$

After performing back-substitution, we find that the solution vector for this problem is

$$[x = -0.5800, \quad y = 1.5600, \quad z = -1.3200],$$

and the LU decomposition is

$$\begin{bmatrix} 1 & 0 & 0 \\ 0.6667 & 1 & 0 \\ 0.3333 & -0.4545 & 1 \end{bmatrix} * \begin{bmatrix} 6 & 1 & -6 \\ 0 & -3.6667 & 4 \\ 0 & 0 & 6.8182 \end{bmatrix},$$

which gives, for A',

$$\begin{bmatrix} 6 & 1 & -6 \\ 4 & -3 & 0 \\ 2 & 2 & 3 \end{bmatrix} = A',$$

agreeing with the order given in the vector: $O = [3, 1, 2]$.

In solving a system of n equations, the number of multiplications/divisions used is of the order $O(n^3/3)$. Suppose we wish to solve the same system (2.22) with a different right-hand side:

$$\begin{aligned} 4x - 3y &= 14, \\ 2x + 2y + 3z &= 9, \\ 6x + y - 6z &= -8. \end{aligned} \qquad (2.23)$$

We first rearrange this system of equations to fit the above form of $A'x = b'$, using the order vector. We now have the system

$$6x + y - 6z = -8,$$
$$4x - 3y = 14,$$
$$2x + 2y + 3z = 9,$$

which we can express as: $LUx = b'$ or $L(Ux) = b'$. Our solution method will be the following:

1. Solve $Ly = b'$, where the vector $y = Ux$ (which can be done with $O(n^2/2)$ multiplications/divisions).
2. Now solve $Ux = y$, where y is the solution from step 1 (this also takes $O(n^2/2)$ multiplications/divisions).

In our current example, we have the lower-triangular system

$$L * y = \begin{bmatrix} 1 & 0 & 0 \\ 0.6667 & 1 & 0 \\ 0.3333 & -0.4545 & 1 \end{bmatrix} * \begin{bmatrix} y_1 \\ y_2 \\ y_3 \end{bmatrix} = \begin{bmatrix} -8 \\ 14 \\ 9 \end{bmatrix},$$

which gives us the intermediate solution: $y_1 = -8$, $y_2 = 19.3333$, $y_3 = 20.4545$. We now solve the upper-triangular system:

$$U * x = \begin{bmatrix} 6 & 1 & -6 \\ 0 & -3.6667 & 4 \\ 0 & 0 & 6.9182 \end{bmatrix} * \begin{bmatrix} x_1 \\ x_2 \\ x_3 \end{bmatrix} = \begin{bmatrix} -8 \\ 19.3333 \\ 20.4545 \end{bmatrix}.$$

Solving this system gives us the solution to the original problem in (2.23): $x_1 = 2$, $x_2 = -2$, and $x_3 = 3$. Although these two intermediate steps seem rather awkward, the actual total number of multiplications/divisions for the LU system that we now have is $O(n^2)$, in comparison to the $O(n^3/3)$ operations for the original system when we did not know the LU decomposition of matrix A. This shows the power and usefulness of Gaussian elimination for solving a variety of linear systems, evaluating the determinant, and obtaining the LU equivalent of a square matrix. Finally, in Section 2.7, we shall find that we can use Gaussian elimination to evaluate the inverse of a square matrix.

The LU Decomposition of a Tridiagonal Matrix

We defined a tridiagonal matrix in Section 2.2 as one with nonzero entries only on its diagonal and the positions adjacent to the main diagonal. We repeat that example matrix with an added right-hand side to create a system of five equations, $Ax = b$:

$$-4x_1 + 2x_2 = 0,$$
$$x_1 - 4x_2 + x_3 = -4,$$
$$x_2 - 4x_3 + x_4 = -11,$$
$$x_3 - 4x_4 + x_5 = 5,$$
$$2x_4 - 4x_5 = 6.$$

We can store the augmented matrix of a tridiagonal system of n equations in an $n \times 4$ matrix. The fourth column holds the right-hand side. For our example, we have

$$\begin{bmatrix} 0 & -4 & 2 & 0 \\ 1 & -4 & 1 & -4 \\ 1 & -4 & 1 & -11 \\ 1 & -4 & 1 & 5 \\ 2 & -4 & 0 & 6 \end{bmatrix}.$$

Such tridiagonal systems occur very frequently in applied problems, as you will see in later chapters. This is a most fortunate situation because the equations can be economically stored and are also solved speedily. If Gaussian elimination is used, we get the LU decomposition of the coefficient matrix at a cost in multiplications/divisions that is only $O(n)$, two orders of magnitude less than for a general system of equations! Further, in the tridiagonal systems that are normally encountered, pivoting is unnecessary. They are also diagonally dominant.

Here is an algorithm for the Gaussian solution of a tridiagonal system:

Gaussian Elimination for a Tridiagonal System

Given the $n \times 4$ matrix that has the right-hand side as its fourth column,

(LU decomposition phase)

For $i = 2$ To n
    $A[i, 1] = A[i, 1]/A[i - 1, 2]$
    $A[i, 2] = A[i, 2] - A[i, 1] * A[i - 1, 3]$
    $A[i, 4] = A[i, 4] - A[i, 1] * A[i - 1, 4]$
End For i

(Back-substitution)

$A[n, 4] = A[n, 4]/A[n, 2]$
For $i = (n - 1)$ Down To 1
    $A[i, 4] = (A[i, 4] - A[i, 3] * A[i + 1, 4])/A[i, 2]$
End For i

After execution of the algorithm, our example matrix would look like this:

$$\begin{bmatrix} 0.0000 & -4.0000 & 1.0000 & 1.0000 \\ -0.2500 & -3.5000 & 1.0000 & 2.0000 \\ -0.2857 & -3.7143 & 1.0000 & 3.0000 \\ -0.2692 & -3.7308 & 1.0000 & -1.0000 \\ -0.5361 & -3.4639 & 0.0000 & -2.0000 \end{bmatrix}$$

The first three columns are really the *LU* equivalent of *A*, but compressed. If we write the matrices in the usual style for a 5 × 5 coefficient matrix, we have:

$$A = \begin{bmatrix} -4 & 2 & 0 & 0 & 0 \\ 1 & -4 & 1 & 0 & 0 \\ 0 & 1 & -4 & 1 & 0 \\ 0 & 0 & 1 & -4 & 1 \\ 0 & 0 & 0 & 2 & -4 \end{bmatrix}$$

$$L = \begin{bmatrix} 1 & 0 & 0 & 0 & 0 \\ -0.2500 & 1 & 0 & 0 & 0 \\ 0 & -0.2857 & 1 & 0 & 0 \\ 0 & 0 & -0.2692 & 1 & 0 \\ 0 & 0 & 0 & -0.5361 & 1 \end{bmatrix}$$

$$U = \begin{bmatrix} -4 & 2 & 0 & 0 & 0 \\ 0 & -3.5000 & 1 & 0 & 0 \\ 0 & 0 & -3.7143 & 1 & 0 \\ 0 & 0 & 0 & -3.7308 & 1 \\ 0 & 0 & 0 & 0 & -3.4639 \end{bmatrix}$$

2.6 Pathology in Linear Systems—Singular Matrices

When a real physical situation is modeled by a set of linear equations, we can anticipate that the set of equations will have a solution that matches the values of the quantities in the physical problem, at least as far as the equations truly do represent it.* Because of round-off errors, the solution vector that is calculated may imperfectly predict the physical quantity, but there is assurance that a solution exists, at least in principle. Consequently, it must always be theoretically possible to avoid divisions by zero when the set of equations has a solution.

An arbitrary set of equations may not have such a guaranteed solution, however. There are several such possible situations, which we term "pathological." In each case, there is *no unique solution* to the set of equations.

First, if the number of equations relating the variables is less than the number of *unknowns,* we certainly cannot solve for unique values of the unknown variables. It turns out, in this case, that there is an infinite set of solutions, for we may arrange the *n* equations with all but *n* of the variables on the right-hand sides, grouped with the constant terms. We may assign almost any desired values to these segregated variables (combining them with the constant terms) and then solve for the *n* remaining variables.[†] Assigning new values to

---

*There are certain problems for which values of interest are determined from a set of equations that do not have a unique solution; these are called *eigenvalue* problems and are discussed in Chapter 7.
[†] An important instance of this situation is the solution of a linear programming problem by the simplex method. In this method, the segregated variables are all assigned the value of zero.

the variables on the right-hand sides gives another set of values for the unknowns, and so on. For example,

$$\begin{cases} x_1 - 2x_2 + x_3 = 4, \\ x_1 - x_2 + x_3 = 5. \end{cases}$$

We rewrite this as

$$x_1 - 2x_2 = 4 - x_3,$$
$$x_1 - x_2 = 5 - x_3.$$

If $x_3 = 0$,

$$x_1 = 6, \qquad x_2 = 1.$$

If $x_3 = 1$,

$$x_1 = 5, \qquad x_2 = 1.$$

If $x_3 = -1$,

$$x_1 = 7, \qquad x_2 = 1, \qquad \text{and so on.}$$

A second situation where we might not expect a set of equations to have a solution is that in which the number of equations is greater than the number of unknowns. If there are n unknowns, we can normally find a subset of the equations that can be solved for the unknowns. There are two subcases to consider.

If the remaining equations are satisfied by the values of the unknowns we have just determined (we would say these equations are *consistent* with the others), there exists a unique solution to the set of equations. Really, of course, there are not truly more equations than unknowns in this case; the extra equations are *redundant*.

The other subcase is a pathological one. If the solution to the first n equations does not satisfy the remaining ones, the set is clearly *inconsistent*, and *no* solution exists that satisfies the system.

Realizing that an equation may be redundant in the prior situation makes us reexamine our more standard case of n equations in n unknowns. What if there is redundancy there? How can we recognize redundancy when it is present? In this example it is obvious:

$$x + y = 3, \qquad 2x + 2y = 6.$$

The second equation is clearly redundant and contains no information not already given by the first. This system will then have an *infinity* of values for x and y; it is an example of fewer equations than unknowns.

Inconsistency may also be present:

$$x + y = 3, \qquad 2x + 2y = 7.$$

In this case there is *no* solution.

If $n \times n$ systems do not have a unique solution, they have a (square) coefficient matrix that is called *singular*. If the coefficient matrix can be triangularized without having zeros on the diagonal (hence the set of equations has a solution), the matrix is said to be *nonsingular*.

Larger systems may have redundancy or inconsistency even though it is not obvious at a glance. Even in a 3×3 system, it is not easy to tell:

$$
\begin{aligned}
x_1 - 2x_2 + 3x_3 &= 5, \\
2x_1 + 4x_2 - x_3 &= 7, \\
-x_1 - 14x_2 + 11x_3 &= 2.
\end{aligned}
$$

Are these inconsistent or redundant? (In other words, is the coefficient matrix singular?) Or do they have a unique solution? (That is, is the matrix nonsingular?) Is there a rule that we can apply, especially one that works for large systems? The answer is yes, there is a rule, or rather, there are several tests we can apply. The standard response from mathematics is to determine the *rank* of the coefficient matrix. If this value is less than n, the number of equations, then no unique solution exists; the equations are either inconsistent or one or more are redundant, depending on the right-hand-side values.

But how does one determine the rank? One practical method is to triangularize by Gaussian elimination: If no zeros show up on the diagonal of the final triangularized coefficient matrix, the rank is equal to n (the matrix is said to be of *full rank*) and a unique solution exists. If, in spite of pivoting, one or more zeros occur on the final diagonal, there is no unique solution. The set will be consistent (and have redundancy) if back-substitution gives (0/0) indeterminate forms. When we would need to divide a nonzero term by zero in back-substituting, inconsistency occurs. Let us apply this to our earlier 3×3 example:

$$
\left[
\begin{array}{ccc|c}
1 & -2 & 3 & 5 \\
2 & 4 & -1 & 7 \\
-1 & -14 & 11 & 2
\end{array}
\right].
$$

On reduction we get

$$
\left[
\begin{array}{ccc|c}
1 & -2 & 3 & 5 \\
0 & 8 & -7 & -3 \\
0 & 0 & 0 & 1
\end{array}
\right].
$$

We see that there is no solution and that the equations are inconsistent. If the constant term in the third equation were 1 rather than 2, reduction would give

$$
\left[
\begin{array}{ccc|c}
1 & -2 & 3 & 5 \\
0 & 8 & -7 & -3 \\
0 & 0 & 0 & 0
\end{array}
\right].
$$

This system is redundant.

Another way to find whether a set of equations has a unique solution is to test the rows or columns of the coefficient matrix for *linear dependency*. Vectors are called linearly dependent if a linear combination of them can be found that equals the zero vector (one with all components equal to zero). (Of course the linear combination $a\bar{x} + b\bar{y} + c\bar{z}$ always equals zero if all the coefficients are zero; we rule out this possibility in our test for linear dependency. When $a = b = c = 0$, we say we have the *trivial case*.)

If the vectors are linearly independent, the only way a weighted sum of them can equal the zero vector is to weight each of them with a zero coefficient. Equivalently, we may say that when vectors are linearly independent none of them is a linear combination of the others.

Our singular 3×3 system has columns that form vectors that are linearly dependent:

$$(-10)\begin{bmatrix} 1 \\ 2 \\ -1 \end{bmatrix} + (7)\begin{bmatrix} -2 \\ 4 \\ -14 \end{bmatrix} + (8)\begin{bmatrix} 3 \\ -1 \\ 11 \end{bmatrix} = \begin{bmatrix} 0 \\ 0 \\ 0 \end{bmatrix}.$$

Similarly, the rows form linearly dependent vectors:

$$(-3)[1 \quad -2 \quad 3] + (2)[2 \quad 4 \quad -1] + (1)[-1 \quad -14 \quad 11] = [0 \quad 0 \quad 0].$$

In the general case, we say that vectors $\bar{x}_1, \bar{x}_2, \bar{x}_3, \ldots, \bar{x}_n$ are *linearly dependent* if we can find scalar coefficients, $a_1, a_2, \ldots, a_n$ (with not all the a_i simultaneously zero), for which

$$\sum_{i=1}^{n} a_i \bar{x}_i = \bar{0}. \tag{2.24}$$

If the only linear combination of the $\bar{x}_i$ that equals the zero vector requires that all the a_i be zero, the set of vectors is called linearly independent. It follows that, if a set of vectors is linearly dependent, at least one of the vectors can be written as a linear combination of the others. If the set is linearly independent, none of the vectors can be written as a linear combination of the others. As a practical matter, we do not usually test the columns (or rows) for linear dependency to determine whether a matrix is singular.

If we are interested in determining the coefficients $a_1, a_2, \ldots, a_n$ that appear in the linear combination of Eq. (2.24), it turns out that we have to solve a set of linear equations to obtain them.

It is worthwhile to summarize the concepts and terminology of this section. The following lists of terms are all equivalent expressions. If a square matrix can be shown to have one property, it has all the others.

Equivalent Properties of Singular or Nonsingular Matrices

The matrix is singular.	The matrix is nonsingular.
A set of equations with these coefficients has no unique solution.	A set of equations with these coefficients has a unique solution.
Gaussian elimination cannot avoid a zero on the diagonal.	Gaussian elimination proceeds without a zero on the diagonal.
The rank of the matrix is less than n.	The rank of the matrix equals n.
The rows form linearly dependent vectors.	The rows form linearly independent vectors.
The columns form linearly dependent vectors.	The columns form linearly independent vectors.

In the next section we will consider two other properties of the matrix: its determinant and its inverse. This adds two more attributes to our lists: A singular matrix has a zero determinant and a nonsingular matrix has a nonzero determinant. A singular matrix has no inverse and a nonsingular matrix does have an inverse.

2.7 Determinants and Matrix Inversion

You have perhaps wondered why there has been no reference so far in this chapter to the solution of linear equations by determinants (Cramer's rule). The reason is that, except for systems of only two or three equations, the determinant method is too inefficient. For example, for a set of 10 simultaneous equations, about 70,000,000 multiplications and subtractions are required if the usual method of expansion in terms of minors is used. A more efficient method of evaluating the determinants can reduce this to about 3000 multiplications, but even this is inefficient compared to Gaussian elimination, which would require about 380.

In fact, the evaluation of a determinant can perhaps best be done by adapting the Gaussian elimination procedure. Its utility derives from the fact that the determinant of a triangular matrix (either upper- or lower-triangular) is just the product of its diagonal elements. This is easily seen, in the case of an upper-triangular matrix, by expansion in terms of minors of the first column at each step. We repeat the argument of Section 2.2. For example,

$$
\begin{vmatrix}
a_{11} & a_{12} & a_{13} & a_{14} \\
0 & a_{22} & a_{23} & a_{24} \\
0 & 0 & a_{33} & a_{34} \\
0 & 0 & 0 & a_{44}
\end{vmatrix}
= a_{11}
\begin{vmatrix}
a_{22} & a_{23} & a_{24} \\
0 & a_{33} & a_{34} \\
0 & 0 & a_{44}
\end{vmatrix}
- 0 + 0 - 0
$$

$$
= a_{11}(a_{22}
\begin{vmatrix}
a_{33} & a_{34} \\
0 & a_{44}
\end{vmatrix}
- 0 + 0)
$$

$$
= a_{11}a_{22}(a_{33}a_{44} - 0) = a_{11}a_{22}a_{33}a_{44}.
$$

Adding a multiple of one row to another row of a matrix does not change the value of its determinant. The other row transformations change the value in predictable ways: Interchanging two rows changes its sign, and multiplying a row by a constant multiplies the value of the determinant by the same constant. If these changes are allowed for, using the procedure of Gaussian elimination to convert to upper-triangular is a simple way to evaluate the determinant.

EXAMPLE 2.4   Find the value of the determinant by using elementary row transformations to make it upper-triangular:

$$
\begin{vmatrix}
1 & 4 & -2 & 3 \\
2 & 2 & 0 & 4 \\
3 & 0 & -1 & 2 \\
1 & 2 & 2 & -3
\end{vmatrix}
=
\begin{vmatrix}
1 & 4 & -2 & 3 \\
0 & -6 & 4 & -2 \\
0 & -12 & 5 & -7 \\
0 & -2 & 4 & -6
\end{vmatrix}
=
\begin{vmatrix}
1 & 4 & -2 & 3 \\
0 & -6 & 4 & -2 \\
0 & 0 & -3 & -3 \\
0 & 0 & \frac{8}{3} & -\frac{16}{3}
\end{vmatrix}
$$

$$
=
\begin{vmatrix}
1 & 4 & -2 & 3 \\
0 & -6 & 4 & -2 \\
0 & 0 & -3 & -3 \\
0 & 0 & 0 & -8
\end{vmatrix}
= (1)(-6)(-3)(-8) = -144.
$$

For programming, an easy and very efficient method for computing the determinant is to use the algorithm in Section 2.4. Then the determinant of the matrix is just the product of the diagonal elements, with a reversed sign if there were an odd number of row interchanges: $\pm a_{11} * a_{22} * \cdots * a_{nn}$, where $+$ is used if there were 0 or an even number of row interchanges (otherwise we use $-$).

Applying this algorithm to the example, but using row interchanges, we see

$$\begin{bmatrix} 1 & 4 & -2 & 3 \\ 2 & 2 & 0 & 4 \\ 3 & 0 & -1 & 2 \\ 1 & 2 & 2 & -3 \end{bmatrix} \rightarrow \begin{bmatrix} 3 & 0 & -1 & 2 \\ (0.667) & 4 & -1.677 & 2.333 \\ (0.333) & (0.5) & 3.167 & -4.833 \\ (0.333) & (0.5) & (0.474) & 3.789 \end{bmatrix}.$$

Because $3 * 4 * 3.167 * 3.789 = 144$ and there were 3 row interchanges in the process, we have the determinant $= -144$. ▲

Although division of matrices is not defined, the matrix inverse gives the equivalent result. If the product of two square matrices is the identity matrix, the matrices are said to be inverses. If $AB = I$, we write $B = A^{-1}$; also $A = B^{-1}$. Inverses commute on multiplication, which is not true for matrices in general: here $AB = BA = I$. Not all square matrices have an inverse. Singular matrices do not have an inverse, and these are of extreme importance in connection with the coefficient matrix of a set of equations, as discussed earlier.

The inverse of a matrix can be defined in terms of the matrix of the minors of its determinant, but this is not a useful way to find an inverse. The Gauss–Jordan technique can be adapted to provide a practical way to invert a matrix. The procedure is to augment the given matrix with the identity matrix of the same order. One then reduces the original matrix to the identity matrix by elementary row transformations, performing the same operations on the augmentation columns. When the identity matrix stands as the left half of the augmented matrix, the inverse of the original stands as the right half. It should be apparent that this is equivalent to solving a set of equations with n different right-hand sides; each of the right-hand sides is a *unit basis vector,* in which the position of the element whose value is unity changes from row 1 to row 2 to row 3 . . . to row n.

EXAMPLE 2.5   Find the inverse of

$$A = \begin{bmatrix} 1 & -1 & 2 \\ 3 & 0 & 1 \\ 1 & 0 & 2 \end{bmatrix}.$$

Augment A with the identity matrix and then reduce:

$$\begin{bmatrix} 1 & -1 & 2 & 1 & 0 & 0 \\ 3 & 0 & 1 & 0 & 1 & 0 \\ 1 & 0 & 2 & 0 & 0 & 1 \end{bmatrix} \rightarrow \begin{bmatrix} 1 & -1 & 2 & 1 & 0 & 0 \\ 0 & 3 & -5 & -3 & 1 & 0 \\ 0 & 1 & 0 & -1 & 0 & 1 \end{bmatrix}$$

$$
\overset{(1)}{\rightarrow}
\begin{bmatrix}
1 & -1 & 2 & 1 & 0 & 0 \\
0 & 1 & 0 & -1 & 0 & 1 \\
0 & 0 & -5 & 0 & 1 & -3
\end{bmatrix}
\overset{(2)}{\rightarrow}
\begin{bmatrix}
1 & -1 & 0 & 1 & \frac{2}{5} & -\frac{6}{5} \\
0 & 1 & 0 & -1 & 0 & 1 \\
0 & 0 & 1 & 0 & -\frac{1}{5} & \frac{3}{5}
\end{bmatrix}
$$

$$
\rightarrow
\begin{bmatrix}
1 & 0 & 0 & 0 & \frac{2}{5} & -\frac{1}{5} \\
0 & 1 & 0 & -1 & 0 & 1 \\
0 & 0 & 1 & 0 & -\frac{1}{5} & \frac{3}{5}
\end{bmatrix}.
$$

[1] Interchange the third and second rows before eliminating from the third row.
[2] Divide the third row by -5 before eliminating from the first row.

We confirm the fact that we have found the inverse by multiplication:

$$
\begin{bmatrix}
1 & -1 & 2 \\
3 & 0 & 1 \\
1 & 0 & 2
\end{bmatrix}
\begin{bmatrix}
0 & \frac{2}{5} & -\frac{1}{5} \\
-1 & 0 & 1 \\
0 & -\frac{1}{5} & \frac{3}{5}
\end{bmatrix}
=
\begin{bmatrix}
1 & 0 & 0 \\
0 & 1 & 0 \\
0 & 0 & 1
\end{bmatrix}.
$$

▲

However, it is more efficient to use the Gaussian elimination algorithm of Section 2.4 by adding additional unit vectors to the augmented matrix.

Performing steps 2 through 4 gives us

$$
\begin{bmatrix}
1 & -1 & 2 & 1 & 0 & 0 \\
3 & 0 & 1 & 0 & 1 & 0 \\
1 & 0 & 2 & 0 & 0 & 1
\end{bmatrix}
\rightarrow
\begin{bmatrix}
3 & 0 & 1 & 0 & 1 & 0 \\
(0.333) & -1 & 1.667 & 1 & -0.333 & 0 \\
(0.333) & (0) & 1.667 & 0 & -0.333 & 1
\end{bmatrix}.
$$

Now applying back-substitution on the last three columns, we get

$$
\begin{bmatrix}
3 & 0 & 1 & 0 & 0.4 & -0.2 \\
(0.333) & -1 & 1.667 & -1 & 0 & 1 \\
(0.333) & (0) & 1.667 & 0 & -0.2 & 0.6
\end{bmatrix},
$$

where the last three columns store the inverse matrix. This method is actually more efficient than the Gauss–Jordan method, which takes about $3n^3/2$ versus $4n^3/3$ multiplications and/or divisions to compute the inverse of a nonsingular matrix. Moreover, in the Gaussian elimination method, we have also found the LU matrix, which we can use to solve the system for other right-hand sides.

The inverse of the coefficient matrix provides a way of solving the set of equations $Ax = b$ because, when we multiply both sides of the relation by A^{-1}, we get

$$
A^{-1}Ax = A^{-1}b,
$$
$$
x = A^{-1}b.
$$

The second equation follows because the product $A^{-1}A = I$, the identity matrix, and $Ix = x$. If we know the inverse of A, we can solve the system for any right-hand-side b simply by multiplying the b-vector by A^{-1}. This would seem like a good way to solve systems of equations, and one finds frequent references to it.

If we care about the efficiency of our method of solving the equations, however, this is not the preferred method, because solving the system with the LU decomposition of A, and doing the equivalent of two back-substitutions, requires exactly the same effort as multiplying b by the matrix. We compare the efficiency of the two schemes, then, by comparing the work needed to get the inverse and that to get the LU equivalent. Getting the inverse is more work, because it is the equivalent of solving the system with n right-hand sides, whereas getting the LU is the equivalent of doing only the reduction to triangular form.

Even though the inverse is not the most efficient way to solve a set of simultaneous equations, the inverse is very important for theoretical reasons and is essential to the understanding of many situations in applied mathematics. The use of the inverse concept and notation often simplifies the development of some fundamental relationships. We illustrate this by again considering the LU decomposition.

Find a pair of matrices such that $LU = A$.

Then $Ax = b$ can be written as

$$LUx = b.$$

Multiply both sides by L^{-1},

$$(L^{-1}L)Ux = L^{-1}b,$$

so $Ux = L^{-1}b$ because $L^{-1}L = I$.

We see that Ux (which is a vector because the product of a matrix times a vector always yields a vector) is equal to the vector formed by $L^{-1}b$. Call this vector b'. Then

$$Ux = b', \qquad b' = L^{-1}b, \qquad \text{or} \qquad Lb' = b.$$

We can get the vector b' by solving the system $Lb' = b$. This is particularly easy to do because L is triangular; all we need to do is the back-substitution phase (actually a forward-substitution because L is lower-triangular).

Once we have b', we can solve for x from the system $Ux = b'$. This is also easy because U is triangular.

Observe how using the concept and notation of inverses helps to clarify and prove the validity of the LU method.

2.8 Norms

When we discuss multicomponent entities like matrices and vectors, we frequently need a way to express their magnitude—some measure of "bigness" or "smallness." For ordinary numbers, the absolute value tells us how large the number is, but for a matrix there are many components, each of which may be large or small in magnitude. (We are not talking about the *size* of a matrix, meaning the number of elements it contains.)

Any good measure of the magnitude of a matrix (the technical term is *norm*) must have four properties that are intuitively essential:

1. The norm must always have a value greater than or equal to zero, and must be zero only when the matrix is the zero matrix (one with all elements equal to zero).
2. The norm must be multiplied by k if the matrix is multiplied by the scalar k.
3. The norm of the sum of two matrices must not exceed the sum of the norms.
4. The norm of the product of two matrices must not exceed the product of the norms.

More formally, we can state these conditions, using $\|A\|$ to represent the *norm of matrix A:*

1. $\|A\| \geq 0$ and $\|A\| = 0$ if and only if $A = 0$.
2. $\|kA\| = |k|\|A\|$.
3. $\|A + B\| \leq \|A\| + \|B\|$.   (2.25)
4. $\|AB\| \leq \|A\|\|B\|$.

The third relationship is called the *triangle inequality.* The fourth is important when we deal with the product of matrices.

For the special kind of matrices that we call vectors, our past experience can help us. For vectors in two- or three-space, the length satisfies all four requirements and is a good value to use for the norm of a vector. This norm is called the *Euclidean norm,* and is computed by $\sqrt{x_1^2 + x_2^2 + x_3^2}$.

We compute the Euclidean norm of vectors with more than three components by generalizing:

$$\|x\|_e = \sqrt{x_1^2 + x_2^2 + \cdots + x_n^2} = \left(\sum_{i=1}^{n} x_i^2\right)^{1/2}.$$

This is not the only way to compute a vector norm, however. The sum of the absolute values of the x_i can be used as a norm; the maximum value of the magnitudes of the x_i will also serve. These three norms can be interrelated by defining the *p*-norm as

$$\|x\|_p = \left(\sum_{i=1}^{n} |x_i|^p \right)^{1/p}.$$

From this it is readily seen that

$$\|x\|_1 = \sum_{i=1}^{n} |x_i| = \text{sum of magnitudes;}$$

$$\|x\|_2 = \left(\sum_{i=1}^{n} x_i^2 \right)^{1/2} = \text{Euclidean norm;}$$

$$\|x\|_\infty = \max_{1 \le i \le n} |x_i| = \text{maximum-magnitude norm.}$$

Which of these vector norms is best to use may depend on the problem. In most cases, satisfactory results are obtained with any of these measures of the "size" of a vector.

EXAMPLE 2.6   Compute the 1-, 2-, and ∞-norms of the vector x, if $x = (1.25, 0.02, -5.15, 0)$.

$$\|x\|_1 = |1.25| + |0.02| + |-5.15| + |0| = 6.42.$$
$$\|x\|_2 = [(1.25)^2 + (0.02)^2 + (-5.15)^2 + (0)^2]^{1/2} = 5.2996.$$
$$\|x\|_\infty = |-5.15| = 5.15.$$
▲

The norms of a matrix are developed by a correspondence to vector norms. Matrix norms that correspond to the above, for matrix A, can be shown to be

$$\|A\|_1 = \max_{1 \le j \le n} \sum_{i=1}^{n} |a_{ij}| = \text{maximum column sum;}$$

$$\|A\|_\infty = \max_{1 \le i \le n} \sum_{j=1}^{n} |a_{ij}| = \text{maximum row sum.}$$

The matrix norm $\|A\|_2$ that corresponds to the 2-norm of a vector is not readily computed. It is defined in terms of the eigenvalues of the matrix $A^T * A$. Suppose r is the largest eigenvalue of $A^T * A$. Then $\|A\|_2 = r^{1/2}$, the square root of r. This is called the *spectral norm* of A, and $\|A\|_2$ is always less than (or equal to) $\|A\|_1$ and $\|A\|_\infty$.

For an $m \times n$ matrix, the Frobenius norm is defined as

$$\|A\|_f = \left(\sum_{i=1}^{m} \sum_{j=1}^{n} a_{ij}^2 \right)^{1/2}.$$

EXAMPLE 2.7   Compute the Frobenius norms of A, B, and C, and the ∞-norms, given that

$$A = \begin{bmatrix} 5 & 9 \\ -2 & 1 \end{bmatrix}; \qquad B = \begin{bmatrix} 0.1 & 0 \\ 0.2 & -0.1 \end{bmatrix}; \qquad \text{and} \qquad C = \begin{bmatrix} 0.2 & 0.1 \\ 0.1 & 0 \end{bmatrix}.$$

$\|A\|_f = \sqrt{25 + 81 + 4 + 1} = \sqrt{111} = 10.54;$ $\|A\|_\infty = 14.0;$

$\|B\|_f = \sqrt{0.01 + 0 + 0.04 + 0.01} = \sqrt{0.06} = 0.2449;$ $\|B\|_\infty = 0.3;$

$\|C\|_f = \sqrt{0.04 + 0.01 + 0.01 + 0} = \sqrt{0.06} = 0.2449;$ $\|C\|_\infty = 0.3.$

The results of our examples look quite reasonable; certainly A is "larger" than B or C. Although $B \neq C$, both are equally "small." The Frobenius norm is a good measure of the magnitude of a matrix. ▲

    We see then that there are a number of ways that the norm of a matrix can be expressed. Which way is preferred? There are certainly differences in their cost; for example, some will require more extensive arithmetic than others. The spectral norm is usually the most "expensive." Which norm is best? The answer to this question depends in part on the use for the norm. In most instances, we want the norm that puts the smallest upper bound on the magnitude of the matrix. In this sense, the spectral norm is usually the "best." We observe, in the next example, that not all the norms give the same value for the magnitude of a matrix.

EXAMPLE 2.8

$$A = \begin{bmatrix} 5 & -5 & -7 \\ -4 & 2 & -4 \\ -7 & -4 & 5 \end{bmatrix}.$$

$\|A\|_f = \text{Frobenius norm} = 15;$

$\|A\|_\infty = 17;$

$\|A\|_1 = 16;$

$\|A\|_2 = \text{spectral norm} = 12.03$

If a matrix is a diagonal matrix, all p-norms have the same value, however. ▲

    These values for Example 2.8 are readily calculated in MATLAB. Using the basic command, norm(), we find that

    norm(A, 'fro') produces 15,

    norm(A, inf) produces 17,

    norm(A, 1) has value 16,

    and norm(A) or norm(A, 2) gives us 12.03.

    Why are norms important? For one thing, they let us express the accuracy of the solution to a set of equations in quantitative terms by stating the norm of the error vector (the true solution minus the approximate solution vector). Norms are also used to study quantitatively the convergence of iterative methods of solving linear systems (which we will cover in a later section).

2.9 Condition Numbers and Errors in Solutions

When we solve a set of linear equations, $Ax = b$, we hope that the calculated vector $\bar{x}$ is a close representation of the true solution vector x. Pivoting and iterative improvement (discussed next) will improve the numerical accuracy of the computed solution. Even so, some linear systems are extremely sensitive to round-off errors and the solution vector may be quite inaccurate. The system is *unstable;* it is said to be *ill-conditioned.* A measure of this instability is the *condition number.* We discuss ill-conditioned systems in this section.

Pivoting

Previous examples have shown how round-off can make the computed solution differ from the exact solution. Pivoting is a way to minimize this effect. If the computer had infinite precision, we would not need pivoting except to avoid a zero divisor in the reduction to triangular form. (An exercise asks you to see how much the accuracy can be improved by pivoting when using four-digit arithmetic.)

We can illustrate the power of pivoting even in a system of two equations that have this general form:

$$\epsilon x + By = C,$$
$$Dx + Ey = F,$$

where ϵ is a sufficiently small number.

Suppose we solve this system, reducing to triangular form by Gaussian elimination, without pivoting. The second equation becomes:

$$(\epsilon - (DB)/\epsilon)y = (F - (CD)/\epsilon).$$

Solving this for y gives

$$y = \frac{F - (CD)/\epsilon}{E - (DB)/\epsilon} \approx \frac{C}{B}, \qquad \text{(for } \epsilon \text{ very small).}$$

From this it follows that

$$x \approx \frac{C - B(C/B)}{\epsilon} = \frac{C - C}{\epsilon} = 0!$$

We find that $x = 0$ for any values of C and F without pivoting if ϵ is small enough (and correspondingly, $1/\epsilon$ is large enough).

We look now at a variation of the example. First, we do pivoting, putting the equation with the largest coefficient of x as the first equation:

$$Dx + Ey = F,$$

$$\epsilon x + By = C.$$

Now suppose that $F = D + E$ and $C = \epsilon + B$, giving

$$Dx + Ey = (D + E),$$

$$\epsilon x + By = (\epsilon + B).$$

The exact answer is clearly $x = 1$, $y = 1$. If we employ Gaussian elimination on these equations, we get the correct answer.

The solutions now are

$$y = \frac{C - (F\epsilon)/D}{B - (\epsilon E)/D} \approx \frac{C}{B} = 1 + \frac{\epsilon}{B} = 1,$$

and

$$x = \frac{F - E}{D} = \frac{D + E - E}{D} = 1!$$

This example shows that pivoting can have a very significant effect on accuracy.

Iterative Improvement

When the solution to the system $Ax = b$ has been computed, and, because of round-off error, we obtain the approximate solution vector $\bar{x}$, it is possible to apply iterative improvement to correct $\bar{x}$ so that it more closely agrees with x. Define $e = x - \bar{x}$. Define $r = b - A\bar{x}$.

$$Ae = r. \tag{2.26}$$

If we could solve this equation for e, we could apply this as a correction to $\bar{x}$. Furthermore, if $\|e\|/\|\bar{x}\|$ is small, it means that $\bar{x}$ should be close to x. In fact, if the value of $\|e\|/\|\bar{x}\|$ is 10^{-p}, we know that $\bar{x}$ is probably correct to p digits.

The process of iterative improvement is based on solving Eq. (2.26). Of course this is also subject to the same round-off error as the original solution of the system for $\bar{x}$, so we actually get $\bar{e}$, an approximation to the true error vector. Even so, unless the system is so ill-conditioned that $\bar{e}$ is not a reasonable approximation to e, we will get an improved estimate of x from $\bar{x} + \bar{e}$. One special caution is important to observe: The computation of the residual vector r must be as precise as possible. One always uses double-precision arithmetic; otherwise, iterative improvement will not be successful. An example will make this clear.

We are given

$$A = \begin{bmatrix} 4.23 & -1.06 & 2.11 \\ -2.53 & 6.77 & 0.98 \\ 1.85 & -2.11 & -2.32 \end{bmatrix}, \quad b = \begin{bmatrix} 5.28 \\ 5.22 \\ -2.58 \end{bmatrix},$$

whose true solution is

$$x = \begin{bmatrix} 1.000 \\ 1.000 \\ 1.000 \end{bmatrix}.$$

If three-digit chopped arithmetic is used, the approximate solution vector is $\bar{x} = (0.991,$ $0.997, 1.000)^T$. Using double precision, we compute $A\bar{x}$, storing this product in a register that holds six digits, then we get the residual.

$$A\bar{x} = \begin{bmatrix} 5.24511 \\ 5.22246 \\ -2.59032 \end{bmatrix}, \qquad r = \begin{bmatrix} 0.0349 \\ -0.00246 \\ 0.0103 \end{bmatrix}.$$

We now solve $A\bar{e} = r$, again using three-digit precision, and get

$$\bar{e} = \begin{bmatrix} 0.00822 \\ 0.00300 \\ -0.00000757 \end{bmatrix}.$$

Finally, correcting $\bar{x}$ with $\bar{x} + \bar{e}$ gives almost exactly the correct solution:

$$\bar{x} + \bar{e} = \begin{bmatrix} 0.999 \\ 1.000 \\ 1.000 \end{bmatrix}.$$

In the general case, the iterations are repeated until the corrections are negligible. Because we want to make the solution of Eq. (2.26) as economical as possible, we should use an LU method to solve the original system and apply the LU to Eq. (2.26).

Ill-Conditioned Systems

Even with the best available algorithm, the error due to round-off unfortunately is sometimes large, because the problem itself may be very sensitive to the effects of small errors in the matrix or in the right-hand sides. Such a system is said to be ill-conditioned. A property of ill-conditioned systems—that their solution is extremely sensitive to small changes in the coefficients—explains why they are also so sensitive to round-off errors. A similar problem exists in a certain class of ordinary differential equations and is referred to as instability.

We can get an appreciation of ill-conditioning by examining a simple system of just two equations in two unknowns. Consider the system $Ax = b$:

$$\begin{bmatrix} 1.01 & 0.99 \\ 0.99 & 1.01 \end{bmatrix} \begin{bmatrix} x \\ y \end{bmatrix} = \begin{bmatrix} 2.00 \\ 2.00 \end{bmatrix}.$$

It is obvious that the solution is $x = 1$, $y = 1$. Suppose we modify b, the right-hand side, just slightly to have

$$\begin{bmatrix} 1.01 & 0.99 \\ 0.99 & 1.01 \end{bmatrix} \begin{bmatrix} x \\ y \end{bmatrix} = \begin{bmatrix} 2.02 \\ 1.98 \end{bmatrix}.$$

Now the obvious solution is $x = 2$, $y = 0$. Finally, if there were another slightly different right-hand-side b-vector,

$$\begin{bmatrix} 1.01 & 0.99 \\ 0.99 & 1.01 \end{bmatrix} \begin{bmatrix} x \\ y \end{bmatrix} = \begin{bmatrix} 1.98 \\ 2.02 \end{bmatrix},$$

we would have $x = 0$, $y = 2$!

It will be helpful to think of the system, $Ax = b$, as a *linear system solver machine*. In this view, the preceding example can be considered as having *inputs* to the machine of the three right-hand sides, the b's. The *outputs* from the machine are the solutions, the x's, for a fixed set of coefficients, A.

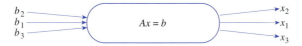

Even though the three inputs are "close together"—$b_1 = (2, 2)^T$, $b_2 = (2.02, 1.98)^T$, and $b_3 = (1.98, 2.02)^T$—we get very "distant" outputs—$x_1 = (1, 1)^T$, $x_2 = (2, 0)^T$, $x_3 = (0, 2)^T$. This modest example shows the basic idea of an ill-conditioned system: *For small changes in input, we get large changes in the output.*

In some situations, one can combat ill-conditioning by transforming the problem into an equivalent set of equations that are not ill-conditioned. The efficiency of this scheme is related to the relative amount of computation required for the transformation, compared to the cost of doing the calculations in higher precision.*

An interesting phenomenon of an ill-conditioned system is that we cannot test for the accuracy of the computed solution merely by substituting it into the equations to see whether the right-hand sides are reproduced. Consider again the ill-conditioned example we have previously examined:

$$A = \begin{bmatrix} 3.02 & -1.05 & 2.53 \\ 4.33 & 0.56 & -1.78 \\ -0.83 & -0.54 & 1.47 \end{bmatrix}, \qquad b = \begin{bmatrix} -1.61 \\ 7.23 \\ -3.38 \end{bmatrix}.$$

If we compute the vector Ax, using the exact solution $x = (1, 2, -1)^T$, we of course get

$$Ax = (-1.61, 7.23, -3.38)^T = b.$$

However, if we substitute a clearly erroneous vector

$$\overline{x} = (0.880, -2.34, -2.66)^T,$$

we get $A\overline{x} = (-1.6047, 7.2348, -3.3716)^T$, which is very close to b.

We define the residual of a solution vector as the difference between b and $A\overline{x}$, where $\overline{x}$ is the computed solution:

$$r = b - A\overline{x}.$$

---

*Double precision is not required throughout the computations. When the system is solved through *LU* decomposition, the accumulation of inner products in double precision is sufficient.

Our example shows that the norm of r is not a good measure of the norm of the error vector $(e = x - \bar{x})$ for an ill-conditioned system.

Condition Numbers

Because the degree of ill-condition of the coefficient matrix is so important in determining the magnitude of round-off effects, it is valuable to have a quantitative measure. The *condition number* is normally defined as the product of two matrix norms:

$$\text{Condition}(A) = \|A\|\|A^{-1}\|.$$

Unfortunately, this is not an inexpensive quantity to compute, for it requires us to invert A. Because inverting a matrix amounts to solving a linear system (solving it with n different right-hand sides, actually), and the computed solution for an ill-conditioned system may be inexact, we will not compute A^{-1} very accurately. This suggests that the condition number will not be computed very exactly either. Ordinarily this causes no great difficulty; if the condition number is large, we know we are in serious trouble. Observe that condition numbers will always be at least unity, which corresponds to the condition number of the identity matrix.

For an example, consider again the matrix

$$A = \begin{bmatrix} 3.02 & -1.05 & 2.53 \\ 4.33 & 0.56 & -1.78 \\ -0.83 & -0.54 & 1.47 \end{bmatrix}.$$

The inverse of A is found to be

$$A^{-1} = \begin{bmatrix} 5.661 & -7.273 & -18.55 \\ 200.5 & -268.3 & -669.9 \\ 76.85 & -102.6 & -255.9 \end{bmatrix}.$$

Using matrix ∞-norms, we find that the condition number is

$$\|A\|\|A^{-1}\| = (6.67)(1138.7) = 7595.$$

The elements of A^{-1} will be large relative to the elements of A when A is ill-conditioned. However, this can also be true when the elements of A are small, even in the absence of ill-conditioning. Multiplying the two norms has a normalizing effect, so the condition number is large only for an ill-conditioned system.

The condition number lets us relate the magnitude of the error in the computed solution to the magnitude of the residual. We use norms to express the magnitude of the vectors.

Let $e = x - \bar{x}$, where x is the exact solution to $Ax = b$ and $\bar{x}$ is an approximate solution. Let $r = b - A\bar{x}$, the residual. Because $Ax = b$, we have

$$r = b - A\bar{x} = Ax - A\bar{x} = A(x - \bar{x}) = Ae.$$

Hence,

$$e = A^{-1}r.$$

Taking norms and recalling Eq. (2.25), line 4, for a product, we write

$$\|e\| \le \|A^{-1}\|\|r\|. \tag{2.27}$$

From $r = Ae$, we also have $\|r\| \le \|A\|\|e\|$, which combines with Eq. (2.27) to give

$$\frac{\|r\|}{\|A\|} \le \|e\| \le \|A^{-1}\|\|r\|. \tag{2.28}$$

Applying the same reasoning to $Ax = b$ and $x = A^{-1}b$, we get

$$\frac{\|b\|}{\|A\|} \le \|x\| \le \|A^{-1}\|\|b\|. \tag{2.29}$$

Taking Eqs. (2.28) and (2.29) together, we reach a most important relationship:

$$\frac{1}{\|A\|\|A^{-1}\|}\frac{\|r\|}{\|b\|} \le \frac{\|e\|}{\|x\|} \le \|A\|\|A^{-1}\|\frac{\|r\|}{\|b\|},$$

or

$$\frac{1}{(\text{Condition no.})}\frac{\|r\|}{\|b\|} \le \frac{\|e\|}{\|x\|} \le (\text{Condition no.})\frac{\|r\|}{\|b\|}. \tag{2.30}$$

Equation (2.30) shows that the relative error in the computed solution vector $\bar{x}$ can be as great as the relative residual multiplied by the condition number. Of course it can also be as small as the relative residual divided by the condition number. Therefore, when the condition number is large, the residual gives little information about the accuracy of $\bar{x}$. Conversely, when the condition number is near unity, the relative residual is a good measure of the relative error of $\bar{x}$.

When we solve a linear system, we are normally doing so to determine values for a physical system for which the set of equations is a model. We use the measured values of the parameters of the physical system to evaluate the coefficients of the equations, so we expect these coefficients to be known only as precisely as the measurements. When these are in error, the solution of the equations will reflect these errors. We have already seen that an ill-conditioned system is extremely sensitive to small changes in the coefficients. The condition number lets us relate the change in the solution vector to such errors in the coefficients of the set of equations $Ax = b$.

Assume that the errors in measuring the parameters cause errors in the coefficients of A so that the actual set of equations being solved is $(A + E)\bar{x} = b$, where $\bar{x}$ represents the solution of the perturbed system and A represents the true (but unknown) coefficients. We let $\bar{A} = A + E$ represent the perturbed coefficient matrix. We desire to know how large $x - \bar{x}$ is.

Using $Ax = b$ and $\bar{A}\bar{x} = b$, we can write

$$x = A^{-1}b = A^{-1}(\bar{A}\bar{x}) = A^{-1}(A + \bar{A} - A)\bar{x}$$
$$= [I + A^{-1}(\bar{A} - A)]\bar{x}$$
$$= \bar{x} + A^{-1}(\bar{A} - A)\bar{x}.$$

Because $\bar{A} - A = E$, we have

$$x - \bar{x} = A^{-1}E\bar{x}.$$

Taking norms, we get

$$\|x - \bar{x}\| \le \|A^{-1}\|\|E\|\|\bar{x}\| = \|A^{-1}\|\|A\|\frac{\|E\|}{\|A\|}\|\bar{x}\|,$$

so that

$$\frac{\|x - \bar{x}\|}{\|\bar{x}\|} \le (\text{Condition no.})\frac{\|E\|}{\|A\|}.$$

This says that the error of the solution relative to the norm of the computed solution can be as large as the relative error in the coefficients of A multiplied by the condition number. The net effect is that, if the coefficients of A are known to only four-digit precision and the condition number is 1000, the computed vector x may have only one digit of accuracy.

2.10 Iterative Methods

As opposed to the direct method of solving a set of linear equations by elimination, we now discuss iterative methods. In certain cases, these methods are preferred over the direct methods—when the coefficient matrix is sparse (has many zeros) they may be more rapid. They may be more economical in memory requirements of a computer. For hand computation they have the distinct advantage that they are self-correcting if an error is made; they may sometimes be used to reduce round-off error in the solutions computed by direct methods, as discussed earlier. They can also be applied to sets of nonlinear equations.

The two methods for solving $Ax = b$ that we shall discuss in this section are the *Jacobi method* and the *Gauss–Seidel method*. These methods not only can solve a system of equations but they are also the basis for other accelerated methods that we shall introduce in later chapters of this book. When the system of equations can be ordered so that each diagonal entry of the coefficient matrix is larger in magnitude than the sum of the magnitudes of the other coefficients in that row—such a system is called diagonally dominant—

the iteration will converge for any starting values. Formally, we say that an $n \times n$ matrix A is diagonally dominant if and only if for each $i = 1, 2, \ldots, n$,

$$|a_{i,j}| > \sum_{\substack{j=1 \\ j \neq i}}^{n} |a_{i,j}|, \qquad i = 1, 2, \ldots, n.$$

Although this may seem like a very restrictive condition, it turns out that there are very many applied problems that have this property (steady-state and transient heat transfer are two). Our approach is illustrated with the following simple example of a system.

$$6x_1 - 2x_2 + x_3 = 11,$$
$$x_1 + 2x_2 - 5x_3 = -1,$$
$$-2x_1 + 7x_2 + 2x_3 = 5.$$

The solution is $x_1 = 2, x_2 = 1, x_3 = 1$. However, before we begin our iterative scheme we must first reorder the equations so that the coefficient matrix is diagonally dominant.

$$6x_1 - 2x_2 + x_3 = 11,$$
$$-2x_1 + 7x_2 + 2x_3 = 5, \qquad \text{(2.31)}$$
$$x_1 + 2x_2 - 5x_3 = -1.$$

The iterative methods depend on the rearrangement of the equations in this manner:

$$x_i = \frac{b_i}{a_{i,i}} - \sum_{\substack{j=1 \\ j \neq i}}^{n} \frac{a_{ij}}{a_{ii}} x_j, \qquad i = 1, 2, \ldots, n.$$

Each equation is now solved for the variables in succession:

$$x_1 = 1.8333 \qquad\quad + 0.3333x_2 - 0.1667x_3,$$
$$x_2 = 0.7143 + 0.2857x_1 \qquad\quad - 0.2857x_3,$$
$$x_3 = 0.2000 + 0.2000x_1 + 0.4000x_2.$$

We begin with some initial approximation to the value of the variables. (Each component might be taken equal to zero if no better initial estimates are at hand.) Substituting these approximations into the right-hand sides of the set of equations generates new approximations that, we hope, are closer to the true value. The new values are substituted in the right-hand sides to generate a second approximation, and the process is repeated until successive values of each of the variables are sufficiently alike. We indicate the iterative process on Eqs. (2.31), as follows, by putting superscripts on variables to indicate successive iterates. Thus our set of equations becomes

$$x_1^{(n+1)} = 1.8333 + 0.3333x_2^{(n)} - 0.1667x_3^{(n)},$$
$$x_2^{(n+1)} = 0.7143 + 0.2857x_1^{(n)} - 0.2857x_3^{(n)}, \qquad \text{(2.32)}$$
$$x_3^{(n+1)} = 0.2000 + 0.2000x_1^{(n)} + 0.4000x_2^{(n)}.$$

Starting with an initial vector of: $x^{(0)} = (0, 0, 0)$, we get:

Successive estimates of solution (Jacobi method)

	First	Second	Third	Fourth	Fifth	Sixth	...	Ninth
x_1	0	1.833	2.038	2.085	2.004	1.994	...	2.000
x_2	0	0.714	1.181	1.053	1.001	0.990	...	1.000
x_3	0	0.200	0.852	1.080	1.038	1.001	...	1.000

Note that this method is exactly the same as the method of fixed-point iteration for a single equation that was discussed in Chapter 1, but it is now applied to a set of equations; we see this if we write Eq. (2.32) in the form of

$$x^{(n+1)} = G(x^{(n)}) = b' - Bx^{(n)},$$

which is identical to $x_{n+1} = g(x_n)$ as used in Chapter 1.

In the present context, of course, $x^{(n)}$ and $x^{(n+1)}$ refer to the nth and $(n + 1)$st iterates of a vector rather than a simple variable, and g is a linear transformation rather than a non-linear function. For the preceding example, we restate Eq. (2.31) in matrix form:

$$Ax = b, \qquad \begin{bmatrix} 6 & -2 & 1 \\ -2 & 7 & 2 \\ 1 & 2 & -5 \end{bmatrix} \begin{bmatrix} x_1 \\ x_2 \\ x_3 \end{bmatrix} = \begin{bmatrix} 11 \\ 5 \\ -1 \end{bmatrix}. \qquad (2.33)$$

Now, let $A = L + D + U$, where

$$L = \begin{bmatrix} 0 & 0 & 0 \\ -2 & 0 & 0 \\ 1 & 2 & 0 \end{bmatrix}, \qquad D = \begin{bmatrix} 6 & 0 & 0 \\ 0 & 7 & 0 \\ 0 & 0 & -5 \end{bmatrix}, \qquad U = \begin{bmatrix} 0 & -2 & 1 \\ 0 & 0 & 2 \\ 0 & 0 & 0 \end{bmatrix}.$$

Then Eq. (2.33) can be rewritten as

$$Ax = (L + D + U)x = b, \qquad \text{or}$$

$$Dx = -(L + U)x + b, \qquad \text{which gives}$$

$$x = -D^{-1}(L + U)x + D^{-1}b.$$

From this we have, identifying x on the left as the new iterate,

$$x^{(n+1)} = -D^{-1}(L + U)x^{(n)} + D^{-1}b. \qquad (2.34)$$

In Eqs. (2.32) we see that

$$b' = D^{-1}b = \begin{bmatrix} 1.8333 \\ 0.7143 \\ 0.2000 \end{bmatrix},$$

$$B = D^{-1}(L + U) = \begin{bmatrix} 0 & -0.3333 & 0.1667 \\ -0.2857 & 0 & 0.2857 \\ -0.2000 & -0.4000 & 0 \end{bmatrix}.$$

The procedure we have just described is known as the *Jacobi method,* also called "the method of simultaneous displacements" because each of the equations is simultaneously changed by using the most recent set of *x*-values.

We can write the algorithm for the Jacobi iterative method as follows:

Algorithm for Jacobi Iteration

We assume that the system $Ax = b$ has been rearranged so that the matrix A is diagonally dominant. That is, for each row of A:

$$|a_{i,i}| > \sum_{\substack{j=1 \\ j \neq i}}^{n} |a_{i,j}|, \qquad i = 1, 2, \ldots, n.$$

This is a sufficient condition for convergence both for this method and for the one that we discuss next. We begin with an initial approximation to the solution vector, which we store in the vector: *old_x.*

> For *i* = 1 To *n*
>   *b*[*i*] = *b*[*i*]/*a*[*i*, *i*]
>   *new_x*[*i*] = *old_x*[*i*]
>   *a*[*i*, *j*] = *a*[*i*, *j*]/*a*[*i*, *i*]; *j* = 1 . . . *n* and *i* <> *j*
> End For *i*
>
> REPEAT
>
>   For *i* = 1 To *n*
>   *old_x*[*i*] = *new_x*[*i*]
>   *new_x*[*i*] = *b*[*i*]
>   End For *i*
>
>   For *i* = 1 To *n*
>   For *j* = 1 To *n*
>   If (*j* <> *i*) Then
>   *new_x*[*i*] = *new_x*[*i*] − *a*[*i*, *j*] * *old_x*[*j*]
>   End For *j*
>   End For *i*
>
> UNTIL *new_x* and *old_x* converge to each other.

Actually, the *x*-values of the next trial (*new_x*) are not used, even in part, until we have first found all its components. For instance, in the Jacobi method, even though we have computed the *new_x*[1], we still do not use this value in computing *new_x*[2], even though in nearly all cases the new values are better than the old, and should be used in preference to the poorer values. When this is done, the procedure is known as the *Gauss−Seidel method.* In this method our first step is to rearrange the set of equations by solving each equation for one of the variables in terms of the others, exactly as we have done in the Jacobi method. We then proceed to improve each *x*-value in turn, always using the most recent approximations to the values of the other variables. The rate of convergence is more rapid, as shown by reworking our earlier example (Eq. 2.31).

Successive estimates of solution (Gauss–Seidel method)

	First	Second	Third	Fourth	Fifth	Sixth
x_1	0	1.833	2.069	1.998	1.999	2.000
x_2	0	1.238	1.002	0.995	1.000	1.000
x_3	0	1.062	1.015	0.998	1.000	1.000

These values were computed by using this iterative scheme:

$$x_1^{(n+1)} = 1.8333 + 0.3333x_2^{(n)} - 0.1667x_3^{(n)},$$
$$x_2^{(n+1)} = 0.7143 + 0.2857x_1^{(n+1)} - 0.2857x_3^{(n)},$$
$$x_3^{(n+1)} = 0.2000 + 0.2000x_1^{(n+1)} - 0.4000x_2^{(n+1)},$$

beginning with $x^{(1)} = (0, 0, 0)^T$.

The algorithm for the Gauss–Seidel iteration is as follows:

Algorithm for Gauss–Seidel Iteration

We assume as we did in the previous algorithm that the system $Ax = b$ has been rearranged so that the coefficient matrix, *A,* is diagonally dominant. As before, we begin with an initial approximation to the solution vector, which we store in the vector: *x.*

```
For i = 1 To n
    b[i] = b[i]/a[i, i]
    a[i, j] = a[i, j]/a[i, i]; j = 1 . . . n and i <> j
End For i;

While Not (yet convergent) Do
    For i = 1 To n
        x[i] = b[i];
        For j = 1 To n
            If (j <> i) Then
                x[i] = x[i] − a[i, j] * x[j]
        End For j
    End For i
End While
```

The matrix formulation for the Gauss–Seidel method is almost the same as the one given in Eq. (2.34). For Gauss–Seidel, $Ax = b$ can be rewritten as

$$(L + D)x = -Ux + b, \tag{2.35}$$

and from this we get

$$x^{(n+1)} = -(L + D)^{-1}Ux^{(n)} + (L + D)^{-1}b. \tag{2.36}$$

The usefulness of this matrix notation will become apparent in Chapter 7, where the eigenvalues of matrices $D^{-1}(L + U)$ of Eq. (2.34) and $(L + D)^{-1}U$ of Eq. (2.36) will be studied. The eigenvalues of the two matrices indicate how fast the iterations will converge. We emphasize, however, that without diagonal dominance, neither Jacobi nor Gauss–Seidel is sure to converge. (Some authors use the term *row diagonal dominance* for our term *diagonal dominance*. Their term is perhaps more accurate.)

There are some instances of the system $Ax = b$ where the coefficient matrix does not have (row) diagonal dominance but still both Jacobi and Gauss–Seidel methods do converge. It can be shown that, if the coefficient matrix, A, is symmetric and positive definite,* the Gauss–Seidel method will converge from any starting vector. In another class of problems, where matrix A has diagonal elements that are all positive and off-diagonal elements that are all negative, both Jacobi and Gauss–Seidel methods will either converge or diverge. When both methods converge, the Gauss–Seidel method converges faster. Datta (1995) discusses this and gives examples.

For a general coefficient matrix, there is little that can be said. In fact, there are examples where Jacobi converges and Gauss–Seidel diverges from the same starting vector! Still, returning to the focus of this section, we can say that, given row diagonal dominance in the coefficient matrix, the Gauss–Seidel method is often the better choice. Having said that, we may still prefer the Jacobi method if we are running the program on parallel processors because all n equations can be solved simultaneously at each iteration.

2.11   The Relaxation Method

There is an iteration method that is more rapidly convergent than Gauss–Seidel and that can be used to advantage for hand calculations. Unfortunately, it is not well adapted to computer application. The method is due to a British engineer, Richard Southwell, and has been applied to a wide variety of problems. (Allen, 1954, is an excellent reference.) We discuss the method because of its historical importance and because it leads to an important acceleration technique called *overrelaxation.*

If we consider the Gauss–Seidel scheme, we realize that the order in which the equations are used is important. We should improve the x that is most in error, because, in the rearranged form, that variable does not appear on the right, and hence its own error will not affect the next iterate. By using that equation, then, we introduce lesser errors into the

---

*Matrix A is symmetric if $A = A^T$. It is positive definite if $x^TAx > 0$ for all nonzero vectors x.

computation of the next iterate. The *method of relaxation* is a scheme that permits one to select the best equation to be used for maximum rate of convergence.

We illustrate the method by the following example. The original equations are

$$8x_1 + x_2 - x_3 = 8,$$
$$2x_1 + x_2 + 9x_3 = 12,$$
$$x_1 - 7x_2 + 2x_3 = -4.$$

We again begin by a rearrangement of the equations, but different from that for the Gauss–Seidel or Jacobi methods. We transpose all the terms to one side, and then divide each equation by the negative of its largest coefficient:

$$-x_1 - 0.125x_2 + 0.125x_3 + 1 \quad\ = 0,$$
$$-0.222x_1 - 0.111x_2 - \quad\ x_3 + 1.333 = 0, \qquad (2.37)$$
$$0.143x_1 - \quad\ x_2 + 0.286x_3 + 0.571 = 0.$$

If we begin with some initial set of values and substitute in Eqs. (2.37), the equations will not be satisfied (unless, by chance, we have stumbled onto the solution); the left sides will not be zero, but some other value that we call the *residual* and denote by R_i. It is also convenient to reorder the equation so the -1 coefficients are on the diagonal. Equations (2.37) become, with these rearrangements,

$$-x_1 - 0.125x_2 + 0.125x_3 + 1 \quad\ = R_1,$$
$$0.143x_1 - \quad\ x_2 + 0.286x_3 + 0.571 = R_2,$$
$$-0.222x_1 - 0.111x_2 - \quad\ x_3 + 1.333 = R_3.$$

For example, with $x_1 = 0$, $x_2 = 0$, $x_3 = 0$, we have

$$R_1 = 1, \qquad R_2 = 0.571, \qquad R_3 = 1.333.$$

The largest residual in magnitude, R_3, tells us that the third equation is most in error and should be improved first. The method gets its name "relaxation" from the fact that we make a change in x_3 to relax R_3 (the greatest residual) so as to make it zero. Observing the coefficients of the various equations, we see that increasing the value of x_3 by one, say, will decrease R_3 by one, will increase R_1 by 0.125, and increase R_2 by 0.286. To change R_3 from its initial value of 1.333 to zero, we should increase x_3 by that same amount.

We then select the new residual of greatest magnitude, and relax it to zero. We continue until all residuals are zero, and when this is true, the values of the x's will be at the exact solution. In implementing this method, there are some modifications that make the work easier. We illustrate in Fig. 2.2.

We make three double columns, one for each variable and for the residual of the equation in which that variable appears with -1 coefficient. The initial x values and the initial residuals are entered as the first row of the table. It is convenient to work entirely with integers by multiplying the initial x-values and residuals by 1000, and then to scale down the solution by dividing by 1000 at the end of the computations. We avoid fractions; if a fractional change in a variable is needed to relax to zero, we only relax to near zero.

In Fig. 2.2, we set down the increments to the x's but record the cumulative effect on the residuals. (The old values of the residuals are crossed out when replaced by a new value.)

Eq. No.	x_1	x_2	x_3
1	-1	-0.125	0.125
2	0.143	-1	0.286
3	-0.222	-0.111	-1

x_1	R_1	x_2	R_2	x_3	R_3
0	~~1000~~	0	~~571~~	0	~~1333~~
	+167		+381		
	~~1167~~		~~952~~	+1333	~~0~~
			+167		−259
+1167	~~0~~		~~1119~~		~~−259~~
	−140				−124
	~~−140~~	+1119	~~0~~		~~−383~~
	−48		−109		
	~~−189~~		~~−109~~	−383	~~0~~
			−27		+42
−189	~~0~~		~~−136~~		~~42~~
	+17				+15
	~~17~~	−136	~~0~~		~~57~~
	+7		+16		
	~~24~~		~~16~~	+57	~~0~~
			+3		−5
+24	~~0~~		~~19~~		~~−5~~
	−2				−2
	~~−2~~	+19	~~0~~		~~−7~~
	−1		−2		
	~~−3~~		~~−2~~	−7	~~0~~
			0		+1
−3	~~0~~		~~−2~~		~~1~~
	0				0
	~~0~~	−2	~~0~~		~~1~~
	0		0	+1	0
999		1000		1000	
Check residuals:	1		−1		0

Figure 2.2 Solving a set of linear equations by relaxation

When the residuals are zero, we add the various increments to the initial value to get the final value. In this example, round-off errors cause an error of one in the third decimal.

It is important to make a final check by recomputing residuals at the end of the calculation to check for mistakes in arithmetic. The method is not usually programmed because searching on the computer for the largest residual is slow, adding enough execution time that the acceleration gives no net benefit. The search can be done rapidly by scanning the residuals in a hand calculation, however.

Southwell and his co-workers observed, for many situations, that relaxing the residuals to zero was less efficient than relaxing beyond zero (*overrelaxing*) or relaxing short of zero (*underrelaxing*). The reason this strategy is an improved one is that a zero residual doesn't

Table 2.1 Accelerated solution of linear equations by relaxation

Eq. No.	x_1	x_2	x_3
1	-1	-0.125	0.125
2	0.143	-1	0.286
3	-0.222	-0.111	-1

x_1	R_1	x_2	R_2	x_3	R_3
0	~~1000~~	0	~~571~~	0	~~1333~~
	+125		+286		
	~~1125~~		~~857~~	+1000	~~333~~
			+144		-225
+1013	~~112~~		~~1001~~		~~108~~
	-125				-111
	~~13~~	+1001	~~0~~		~~3~~
			-2		+3
-12	~~1~~		~~2~~		~~0~~
	+0				+0
	~~1~~	-2	~~0~~		~~0~~
			+0		+0
-1	0		0		0
1000		999		1000	

stay zero; relaxing the residual of another equation affects the first residual, so it is appropriate to anticipate and allow for this by an appropriate under- or overrelaxation.

Table 2.1 shows that a significant improvement in the speed of convergence is obtained if R_1 is underrelaxed by 10% and R_3 is underrelaxed by 25%. Unfortunately, the optimum degree of under- or overrelaxation is not easily determined. In many problems, acceleration is obtained by overrelaxing rather than underrelaxing.

Even though Southwell's relaxation method is not often used today, there is one aspect of it that has influence on the iterative solution of linear equations by computer. In using the Gauss–Seidel method, we can speed up the convergence by "overrelaxation," that is, by making the residuals go to the other side of zero instead of just relaxing to zero as in the first example. We can apply this technique to Gauss–Seidel iteration by modifying the algorithm.

The standard relationship for Gauss–Seidel iteration for the set of equations $Ax = b$, for variable x_i, can be written

$$ x_i^{(k+1)} = \frac{1}{a_{ii}} \left(b_i - \sum_{j=1}^{i-1} a_{ij} x_j^{(k+1)} - \sum_{j=i+1}^{n} a_{ij} x_j^{(k)} \right), \qquad (2.38) $$

where the superscript $(k+1)$ indicates that this is the $(k+1)$st iterate. On the right side we use the most recent estimates of the x_j, which will be either $x_j^{(k)}$ or $x_j^{(k+1)}$.

An algebraically equivalent form for Eq. (2.38) is

$$x_j^{(k+1)} = x_i^{(k)} + \frac{1}{a_{ii}}\left(b_i - \sum_{j=1}^{i-1} a_{ij}x_j^{(k+1)} - \sum_{j=i}^{n} a_{ij}x_j^{(k)}\right),$$

because $x_i^{(k)}$ is both added to and subtracted from the right side. In this form, we see that Gauss–Seidel and Southwell's relaxation can have identical arithmetic: The term we add to $x_i^{(k)}$ to get $x_i^{(k+1)}$ is exactly the increment that relaxes the residual to zero. (Of course, we apply the relaxation to the x_i's in a different sequence in the two methods.) Overrelaxation can be applied to Gauss–Seidel if we will add to $x_i^{(k)}$ some multiple of the second term. It can be shown that this multiple should never be more than 2 in magnitude (to avoid divergence), and the optimum overrelaxation factor lies between 1.0 and 2.0. Our iteration equations take this form, where w is the *overrelaxation factor:*

$$x_i^{(k+1)} = x_i^{(k)} + \frac{w}{a_{ii}}\left(b_i - \sum_{j=1}^{i-1} a_{ij}x_j^{(k+1)} - \sum_{j=i}^{n} a_{ij}x_j^{(k)}\right).$$

Table 2.2 shows how the convergence rate is influenced by the value of w for the system

$$\begin{bmatrix} -4 & 1 & 1 & 1 \\ 1 & -4 & 1 & 1 \\ 1 & 1 & -4 & 1 \\ 1 & 1 & 1 & -4 \end{bmatrix} x = \begin{bmatrix} 1 \\ 1 \\ 1 \\ 1 \end{bmatrix},$$

starting with an initial estimate of $x = 0$. The exact solution is

$$x_1 = -1, \qquad x_2 = -1, \qquad x_3 = -1, \qquad x_4 = -1.$$

Table 2.2   Acceleration of convergence of Gauss–Seidel iteration

w, the overrelaxation factor	Number of iterations to reach error $<1 \times 10^{-5}$
1.0	24
1.1	18
1.2	13
1.3	11←Minimum
1.4	14 of iterations
1.5	18
1.6	24
1.7	35
1.8	55
1.9	100+

We see that the optimum value for the overrelaxation factor is about $w = 1.3$ for this example. The optimum value will vary between 1.0 and 2.0, depending on the size of the coefficient matrix and the values of the coefficients. Overrelaxation is considered further in Chapter 7 in connection with methods to solve partial-differential equations. We also discuss there the question of finding w_{opt}.

2.12 Systems of Nonlinear Equations

As mentioned previously, the problem of finding the solution of a set of nonlinear equations is much more difficult than for linear equations. (In fact, some sets have no real solutions.) Consider the example of a pair of nonlinear equations:

$$x^2 + y^2 = 4,$$

$$e^x + y = 1. \tag{2.39}$$

Graphically, the solution to this system is represented by the intersections of the circle $x^2 + y^2 = 4$ with the curve $y = 1 - e^x$. Figure 2.3 shows that these are near $(-1.8, 0.8)$ and $(1, -1.7)$.

Newton's method can be applied to systems as well as to a single nonlinear equation. We begin with the forms

$$f(x, y) = 0,$$

$$g(x, y) = 0.$$

Let $x = r$, $y = s$ be a root, and expand both functions as a Taylor series about the point (x_i, y_i) in terms of $(r - x_i)$, $(s - y_i)$, where (x_i, y_i) is a point near the root:

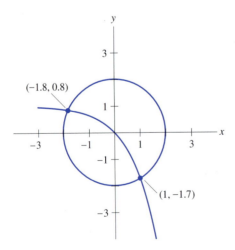

Figure 2.3

$$f(r,\ s) = 0 = f(x_i,\ y_i) + f_x(x_i,\ y_i)(r - x_i)$$
$$+ f_y(x_i,\ y_i)(s - y_i) + \cdots ,$$
$$g(r,\ s) = 0 = g(x_i,\ y_i) + g_x(x_i,\ y_i)(r - x_i)$$
$$+ g_y(x_i,\ y_i)(s - y_i) + \cdots .$$

Truncating the series gives

$$\begin{bmatrix} 0 \\ 0 \end{bmatrix} = \begin{bmatrix} f(x_i,\ y_i) \\ g(x_i,\ y_i) \end{bmatrix} + \begin{bmatrix} f_x(x_i,\ y_i) & f_y(x_i,\ y_i) \\ g_x(x_i,\ y_i) & g_y(x_i,\ y_i) \end{bmatrix} \begin{bmatrix} r - x_i \\ s - y_i \end{bmatrix}.$$

We rewrite this to solve as the system of equations

$$\begin{bmatrix} f_x(x_i,\ y_i) & f_y(x_i,\ y_i) \\ g_x(x_i,\ y_i) & g_y(x_i,\ y_i) \end{bmatrix} \begin{bmatrix} \Delta x_i \\ \Delta y_i \end{bmatrix} = - \begin{bmatrix} f(x_i,\ y_i) \\ g(x_i,\ y_i) \end{bmatrix}, \qquad \textbf{(2.40)}$$

where $\Delta x_i = r - x_i$, and $\Delta y_i = s - y_i$.

We solve (2.40) by Gaussian elimination and then, if we set

$$\begin{bmatrix} x_{i+1} \\ y_{i+1} \end{bmatrix} = \begin{bmatrix} x_i \\ y_i \end{bmatrix} + \begin{bmatrix} \Delta x_i \\ \Delta y_i \end{bmatrix},$$

we get an improved estimate of the root, $(r,\ s)$. We repeat this process with i replaced by $i + 1$ until f and g are close to 0. The extension to more than two simultaneous equations is straightforward.

We illustrate by repeating the previous example:

$$f(x,\ y) = 4 - x^2 - y^2 = 0,$$
$$g(x,\ y) = 1 - e^x - y = 0.$$

The partials are

$$f_x = -2x, \qquad f_y = -2y,$$
$$g_x = -e^x, \qquad g_y = -1.$$

Beginning at $x_0 = 1$, $y_0 = -1.7$, we solve the equations

$$\begin{bmatrix} -2 & 3.4 \\ -2.7183 & -1.0 \end{bmatrix} \begin{bmatrix} \Delta x_0 \\ \Delta y_0 \end{bmatrix} = - \begin{bmatrix} 0.1100 \\ -0.0183 \end{bmatrix}, \qquad \textbf{(2.41)}$$

to get $\Delta x_0 = 0.0043$, $\Delta y_0 = -0.0298$. This then gives us $x_1 = 1.0043$, $y_1 = -1.7298$. The results already agree with the true value of the root within two in the fourth decimal place.

Repeating the process once more produces $x_2 = 1.004169$, $y_2 = -1.729637$. The function values at (x_2, y_2) are approximately -0.0000001 and -0.00000001, respectively.

We can write Newton's method for a system of n equations by expanding Eq. (2.40). Thus we have

$$
\begin{bmatrix}
f_{1x} & f_{1y} & f_{1z} & \cdots \\
f_{2x} & f_{2y} & f_{2z} & \cdots \\
f_{3x} & f_{3y} & f_{3z} & \cdots \\
\vdots & \vdots & \vdots & \\
f_{nx} & f_{ny} & f_{nz} & \cdots
\end{bmatrix}
\begin{bmatrix}
\Delta x_i \\
\Delta y_i \\
\Delta z_i \\
\vdots \\
\;
\end{bmatrix}
= -
\begin{bmatrix}
f_1 \\
f_2 \\
f_3 \\
\vdots \\
f_n
\end{bmatrix}
\tag{2.42}
$$

evaluated at $(x_i, y_i, z_i, \dots)$. Solving this, we compute

$$
x_{i+1} = x_i + \Delta x_i, \qquad y_{i+1} = y_i + \Delta y_i, \qquad z_{i+1} = z_i + \Delta z_i, \dots\,.
$$

In a computer program, it is awkward to introduce each of the partial-derivative functions (which often must be developed by hand, unless one has access to a computer algebra system or an advanced calculator) to use in Eq. (2.40). An alternative technique is to approximate these partials by recalculating the function with a small perturbation to each of the variables in turn:

$$
(f_1)_x \doteq \frac{f_1(x + \delta, y, z, \dots) - f_1(x, y, z, \dots)}{\delta},
$$

$$
(f_1)_y \doteq \frac{f_1(x, y + \delta, z, \dots) - f_1(x, y, z, \dots)}{\delta},
$$

$$
\vdots
$$

$$
(f_i)_{x_j} \doteq \frac{f_i(x, y, z, \dots, x_j + \delta, \dots) - f_i(x, y, z, \dots, x_j, \dots)}{\delta}.
$$

Similar relations are used for each variable in each function.* A computer program at the end of this chapter exploits this idea.

It is interesting to observe that Newton's method, as applied to a set of nonlinear equations, reduces the problem to solving a set of *linear equations* in order to determine the values that improve the accuracy of the estimates. This points out quite dramatically how linear and nonlinear problems vary in difficulty.

Newton's method has the advantage of converging quadratically, at least when we are near a root, but it is expensive in terms of function evaluations. For the preceding 2×2 system, there are six function evaluations at each step, whereas for a 3×3 system, there

---

*Approximation of derivatives by such difference quotients is discussed in Chapter 5. If the limiting value (as $\delta \to 0$) of this ratio were used, we would have exactly the definition of a derivative. Because limits are not used, we have approximate values, as indicated by the dots over the equal signs.

are twelve. For n simultaneous equations, the number of function evaluations is $n^2 + n$. We can see why this is rarely applied to large systems. It is good strategy, in all cases of simultaneous nonlinear equations, to reduce the number of equations as much as possible by solving for one variable in terms of the others and eliminating that one by substituting for it in the other equations. For example, we could attack the previous example as follows:

$$\begin{cases} 4 - x^2 - y^2 = 0 \\ 1 - e^x - y = 0 \end{cases} \quad \text{Solve for } y: \ y = 1 - e^x.$$

Substituting in the first equation, we get

$$4 - x^2 - (1 - e^x)^2 = 0,$$
$$3 - x^2 + 2e^x - e^{2x} = 0.$$

We then use the methods of Chapter 1.

When we must solve a larger system of nonlinear equations, a modification of Newton's method is often used. It converges less than quadratically but usually faster than linearly. Unfortunately, it may *diverge* unless we start fairly near to a root. In this method we do not recompute the matrix of partials at each step. Rather, we use the same matrix for several steps before recomputing it again. For a system of n equations we would recompute the matrix after every n steps. In this way we need only n function evaluations at each step, except when we occasionally have to update the matrix of partials, which then adds an additional n^2 function evaluations.

We first illustrate the method on our earlier 2×2 system. Reworking Eq. (2.41) produces

$$\begin{bmatrix} \Delta x_0 \\ \Delta y_0 \end{bmatrix} = - \begin{bmatrix} -2 & 3.4 \\ -2.7183 & -1.0 \end{bmatrix}^{-1} \begin{bmatrix} 0.110 \\ -0.0183 \end{bmatrix} = \begin{bmatrix} 0.0043 \\ -0.0298 \end{bmatrix};$$

$$\begin{bmatrix} x_1 \\ y_1 \end{bmatrix} = \begin{bmatrix} 1.0 \\ -1.7 \end{bmatrix} + \begin{bmatrix} 0.0043 \\ -0.0298 \end{bmatrix} = \begin{bmatrix} 1.0043 \\ -1.7298 \end{bmatrix};$$

$$\begin{bmatrix} \Delta x_1 \\ \Delta y_1 \end{bmatrix} = - \begin{bmatrix} -2 & 3.4 \\ -2.7183 & -1.0 \end{bmatrix}^{-1} \begin{bmatrix} 0.000827 \\ 0.000196 \end{bmatrix} = \begin{bmatrix} -0.000133 \\ 0.000165 \end{bmatrix};$$

$$x_2 = 1.004167, \quad y_2 = -1.729635;$$

$$f(x_2, y_2) = -0.000011, \quad g(x_2, y_2) = -0.000002.$$

In addition, we consider a different example. Here we have

$$e^x - y = 0,$$
$$xy - e^x = 0.$$

Equation (2.40) for this system becomes

$$\begin{bmatrix} e^{x_i} & -1 \\ y_i - e^{x_i} & -x_i \end{bmatrix} \begin{bmatrix} \Delta x_i \\ \Delta y_i \end{bmatrix} = - \begin{bmatrix} f(x_i, y_i) \\ g(x_i, y_i) \end{bmatrix}.$$

Let $x_0 = 0.95$, $y_0 = 2.7$. Then solving the preceding with the inverse matrix

$$\begin{bmatrix} \Delta x_0 \\ \Delta y_0 \end{bmatrix} = - \begin{bmatrix} 2.5857 & -1 \\ 0.1143 & 0.95 \end{bmatrix}^{-1} \begin{bmatrix} -0.1143 \\ -0.0207 \end{bmatrix},$$

we get

$$x_1 = 1.00029, \qquad y_1 = 2.71575;$$

$$\begin{bmatrix} \Delta x_1 \\ \Delta y_1 \end{bmatrix} = - \begin{bmatrix} 2.5857 & -1 \\ 0.1143 & 0.95 \end{bmatrix}^{-1} \begin{bmatrix} 0.003325 \\ -0.002533 \end{bmatrix},$$

we get

$$x_2 = 1.000048, \qquad y_2 = 2.718445.$$

Because $n = 2$, we update our matrix of partials so that

$$\begin{bmatrix} \Delta x_2 \\ \Delta y_2 \end{bmatrix} = - \begin{bmatrix} 2.7184 & -1 \\ 0.00003 & 1.0000 \end{bmatrix}^{-1} \begin{bmatrix} -0.000033 \\ -0.000163 \end{bmatrix},$$

we get

$$x_3 = 1.000000, \qquad y_3 = 2.718282.$$

This agrees with the exact solution: $(1, e)$.

2.13 Theoretical Matters

In this section, we shall look a little deeper at the algorithm for Gaussian elimination of Section 2.4 for solving a system of n equations in n unknowns. We shall show that the algorithm is just a series of simple matrix multiplications that are important on their own. We will also show that there is a very big difference between a nearly singular matrix and an ill-conditioned matrix.

Interchanging Rows of a Matrix

In Section 2.2, we defined a *transposition matrix* and a *permutation matrix*. A transposition matrix, P, can be created by interchanging exactly two rows (or columns) of the identity matrix; and a matrix that is the product of several transposition matrices is called a permutation matrix. The idea becomes clear if we examine this definition for a 3×3 matrix. Define

$$P = \begin{bmatrix} 1 & 0 & 0 \\ 0 & 0 & 1 \\ 0 & 1 & 0 \end{bmatrix}, \qquad B = \begin{bmatrix} a & b & c \\ d & e & f \\ g & h & i \end{bmatrix}.$$

Observe that multiplying P and B interchanges rows 2 and 3 of B:

$$P * B = \begin{bmatrix} a & b & c \\ g & h & i \\ d & e & f \end{bmatrix}.$$

Conversely, if we change the order of multiplication, we get

$$B * P = \begin{bmatrix} a & c & b \\ d & f & e \\ g & i & h \end{bmatrix},$$

where now column 2 and column 3 of B have been interchanged. The identity matrix can be considered a permutation matrix that leaves all the rows and columns alone. Moreover, for any permutation matrix, P, we have

$$P^{-1} = P^T$$

and in the special case of a transposition matrix, P,

$$P^T = P.$$

Zeroing Elements Below the Diagonal

In Gaussian elimination, when we zeroed out the elements below the pivot element, we were actually doing the following:

$$H_1 B = \begin{bmatrix} 1 & 0 & 0 \\ -\frac{d}{a} & 1 & 0 \\ -\frac{g}{a} & 0 & 1 \end{bmatrix} * B = \begin{bmatrix} a & b & c \\ 0 & u & v \\ 0 & x & y \end{bmatrix}.$$

Similarly, if we set

$$H_2 = \begin{bmatrix} 1 & 0 & 0 \\ 0 & 1 & 0 \\ 0 & -\frac{x}{u} & 1 \end{bmatrix},$$

then it turns out that

$$H_2(H_1 B) = \begin{bmatrix} a & b & c \\ 0 & u & v \\ 0 & 0 & w \end{bmatrix}.$$

We use the symbol H_i because this is really a simple *Householder transformation* on the ith column of B. The inverses to these transformations are also very simple. For instance, in the example above,

$$H_1^{-1} = \begin{bmatrix} 1 & 0 & 0 \\ \frac{d}{a} & 1 & 0 \\ \frac{g}{a} & 0 & 1 \end{bmatrix}.$$

We shall now apply our discussion to the system of equations in Section 2.4, $Ax = b$ (Eq. 2.9), where

$$A = \begin{bmatrix} 0 & 2 & 0 & 1 \\ 2 & 2 & 3 & 2 \\ 4 & -3 & 0 & 1 \\ 6 & 1 & -6 & -5 \end{bmatrix}, \quad b = \begin{bmatrix} 0 \\ -2 \\ -7 \\ 6 \end{bmatrix}.$$

Letting P_1 be the permutation that interchanges rows 1 and 4 and letting

$$H_1 = \begin{bmatrix} 1 & 0 & 0 & 0 \\ -0.3333 & 1 & 0 & 0 \\ -0.6667 & 0 & 1 & 0 \\ 0 & 0 & 0 & 1 \end{bmatrix},$$

we get

$$(H_1 P_1)A = \begin{bmatrix} 6 & 1 & -6 & -5 \\ 0 & 1.6667 & 5 & 3.6667 \\ 0 & -3.6667 & 4 & 4.3333 \\ 0 & 2 & 0 & 1 \end{bmatrix}, \tag{2.43}$$

which is found in Eq. (2.11a).

Now let P_2 be the permutation that interchanges rows 2 and 3 of Eq. (2.43), and

$$H_2 = \begin{bmatrix} 1 & 0 & 0 & 0 \\ 0 & 1 & 0 & 0 \\ 0 & 0.4545 & 0 & 0 \\ 0 & 0.5455 & 0 & 0 \end{bmatrix},$$

then it is easy to see that

$$H_2(P_2 H_1 P_2^T)(P_2 P_1 A) = \begin{bmatrix} 6 & 1 & -6 & -5 \\ 0 & -3.6667 & 4 & 4.3333 \\ 0 & 0 & 6.8182 & 5.6364 \\ 0 & 0 & 2.1818 & 3.3636 \end{bmatrix}.$$

Finally, if we let

$$H_3 = \begin{bmatrix} 1 & 0 & 0 & 0 \\ 0 & 1 & 0 & 0 \\ 0 & 0 & 1 & 0 \\ 0 & 0 & -0.3200 & 0 \end{bmatrix}$$

and P_3 be the identity matrix, because there are no row interchanges, we finally get

$$H_3(P_3 H_2 P_3^T)(P_3 P_2 H_1 P_2^T P_3^T)(P_3 P_2 P_1 A) = \begin{bmatrix} 6 & 1 & -6 & -5 \\ 0 & -3.6667 & 4 & 4.3333 \\ 0 & 0 & 6.8182 & 5.6364 \\ 0 & 0 & 0 & 1.5600 \end{bmatrix},$$

which is the coefficient matrix part of Eq. (2.14).

To summarize: We wish to solve the system $Ax = b$. However, in order to help in understanding the derivation we need to point out that our computations are based on the two following facts:

Fact 1: $P_i^T P_i = I$, the identity matrix, for any permutation matrix. In what follows, $P_i^T = P_i$, because we are just considering elementary transpositions of two rows (or columns).

Fact 2: For $i > k$, we have $P_i^T H_k P_i = H_k'$; that is, we still have a *Householder transformation* because only elements below the main diagonal in column k of H_k are exchanged.

Using a sequence of elementary matrices, we have

$$H_3 H_2' H_1' (P_3 P_2 P_1) Ax = H_3 H_2' H_1' (P_3 P_2 P_1) b.$$

It should be noted that the right-hand side of this equation corresponds to the last column in the augmented matrix of Eq. (2.14). It can be shown for this example, but it is true even in the general case, that we can rearrange the order of the simple transformations to get

$$H_3 (P_3 H_2 P_3)(P_3 P_2 H_1 P_2 P_3)(P_3 P_2 P_1 A) = H_3 H_2 H_1 (P_3 P_2 P_1) A = U, \qquad \text{(2.44)}$$

where U is the upper-triangular matrix of Eq. (2.43). Then by multiplying both sides of the equation by the appropriate inverses we get

$$A' = P_3 P_2 P_1 A = (H_3' H_2' H_1')^{-1} U = LU, \qquad \text{(2.45)}$$

which for this 4×4 example is

$$\begin{bmatrix} 1 & 0 & 0 & 0 \\ 0.6667 & 1 & 0 & 0 \\ 0.3333 & -0.4545 & 1 & 0 \\ 0 & -0.5454 & 0.32 & 1 \end{bmatrix} * \begin{bmatrix} 6 & 1 & -6 & -5 \\ 0 & -3.6667 & 4 & 4.3333 \\ 0 & 0 & 6.8182 & 5.6364 \\ 0 & 0 & 0 & 1.5600 \end{bmatrix}.$$

In conclusion, the statements in Eqs. (2.44) and (2.45) are true for a general system of equations on which we use the Gaussian elimination of Section 2.4.

It also follows that

$$\det(A) = \det(P_n * \cdots * P_1) \det(L) \det(U)$$

$$= (-1)^k U(1, 1) * \cdots * U(n, n),$$

where k = the number of row interchanges, the $U(i, i)$'s are the diagonal elements of U, and $\det(L) = 1$, because its diagonal elements are all 1.

Matrix Norms

In Section 2.8, we introduced the notion of vector and matrix norms. Although the vector norms are more familiar to most students, the definition of $\|x\|_\infty$ is just derived from our general definition of $\|x\|_p$ because we can define it as

$$\|x\|_\infty = \max\{|x_i|\} = \lim \sqrt[p]{\sum_i |x_i|^p} \qquad \text{as } p \to \infty.$$

If we assume $\max_i\{|x_i|\} = |x_1|$, say, then we see first that

$$\sqrt[p]{\sum_i |x_i|^p} = |x_1| \sqrt[p]{1 + \sum_{i>1} |x_i|^p/|x_1|^p} \to |x_1| \qquad \text{as } p \to \infty,$$

Because we need only to observe that

$$\lim \sqrt[p]{n} = 1 \qquad \text{as } p \to \infty.$$

From our definition of norms on vectors, we proceed to define the norms of matrices as an extension of the vector norms. There are several equivalent ways to do this, but for our present purposes, we define

$$\|A\|_p = \max\|Ax\|_p \quad \text{for all } x \text{ where } \|x\|_p = 1.$$

Now let us consider our previous definition of $\|A\|_\infty$. We have

$$\|Ax\|_\infty = \max_i \left\{ \left| \sum_j A_{i,j} x_j \right| \right\} \le \max_i \left\{ \sum_j |A_{i,j}||x_j| \right\} \le \max_i \left\{ \sum_j |A_{i,j}| \right\} = \|A\|_\infty,$$

because the $|x_j| \le 1$ for $j = 1, \ldots, n$. Now suppose $\|A\|_\infty$ is achieved on the first row of A. Then all one has to do is to define a vector so that each component, y_j, is either a plus 1 or a minus 1. We choose values so that $A_{1,j} y_j = |A_{1,j}|$. Then $\|y\|_\infty = 1$ and $\|Ay\|_\infty = \|A\|_\infty$.

For the case of $\|A\|_2$, the definition is

$$\|A\|_2 = \max\|Ax\|_2 = \sqrt{x^T A^T A x}.$$

$A^T A$ is a *positive definite* matrix, so all its eigenvalues are nonnegative, say

$$\lambda_1 \ge \lambda_2 \ge \cdots \ge \lambda_n \ge 0$$

and have corresponding eigenvectors w_i that are *orthonormal* to one another and that form a *basis* for R^n. This means that $\|w_i\|_2 = 1$, and $w_i^T w_j = 0$ for $i \ne j$.

In addition, because every vector x can be written as

$$x = \alpha_1 w_1 + \alpha_2 w_2 + \cdots + \alpha_n w_n,$$

we have

$$\|Ax\|_2^2 = \sum_i \alpha_i^2 \lambda_i \le \lambda_1 \quad \text{for all } x, \|x\|_2 = 1.$$

Let $x = w_1$, the eigenvector for λ_1. Then it follows that

$$\|A\|_2 = \sqrt{\lambda_1}.$$

However, our chief purpose here is to emphasize that the definition of a matrix norm is just an extension of the one for a vector norm.

Ill-Conditioned Systems

In Section 2.9, we encountered the notion of an ill-conditioned system, $Ax = b$. This notion is quite different from that of a singular or near singular system. For instance, if we had 10 equations in 10 unknowns in the form

$$0.1x_i = 1, \quad \text{for } i = 1, \ldots, 10,$$

then the coefficient matrix, A, would just be a 10×10 matrix whose diagonal elements are all 0.1 for which $\det(A) = 10^{-10}$! This would indicate a near singular matrix, but actually its condition number is 1 and the solution is obvious.

As indicated in Section 2.9, the condition number of a system is a magnification factor for errors in the input matrix A or in the right-hand side b. We saw how much of the answers changed for just very small changes in the right-hand side b. In that example we were given

$$\begin{bmatrix} 1.01 & 0.99 \\ 0.99 & 1.01 \end{bmatrix} \begin{bmatrix} x \\ y \end{bmatrix} = \begin{bmatrix} 2.00 \\ 2.00 \end{bmatrix} \tag{2.46}$$

We thought of the system, $Ax = b$, as a linear system solver machine. In the example our inputs were just the right-hand sides, b. The outputs from the machine were the solutions to the system for the fixed coefficient matrix, A, and we drew a diagram:

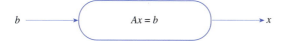

For three very "close" inputs—$b^{(1)} = (2, 2)^T$, $b^{(2)} = (2.02, 1.98)^T$, $b^{(3)} = (1.98, 2.02)^T$—we have in this very small example very "distant" outputs—namely, $x^{(1)} = (1, 1)^T$, $x^{(2)} = (2, 0)^T$, $x^{(3)} = (0, 2)^T$. We considered all this in Section 2.9. However, we did not complete our discussion of the condition number as a magnification factor. In this example we see that

$$\|b^{(1)} - b^{(2)}\| = 0.02,$$

and $\quad \text{cond}(A) = \|A\|_\infty * \|A^{-1}\|_\infty = 100.$

Observe that the norm of $(x^{(1)} - x^{(2)}) = \|x^{(1)} - x^{(2)}\|_\infty = 1$ is of the order of the product of the magnification factor (which is $\text{cond}(A) = 100$), and the norm of $(b_1^{(1)} - b_2^{(2)}) = 0.02$. A similar statement can be made for $b^{(1)}, b^{(3)}, x^{(1)}, x^{(3)}$.

Pivoting

Finally, a brief comment on pivoting. Gaussian elimination involves partial pivoting, *partial* meaning that at any time we are looking for the maximum magnitude in a particular column only. In contrast, full pivoting seeks the largest magnitude in the whole array at each stage and is more easily implemented in the Gauss–Jordan method. In any case, it should be pointed out that we would not need to use any pivoting if the computer could do arithmetic in infinite precision.

2.14 Using Maple and MATLAB

We described the MATLAB program in the previous chapter. Another important computer algebra system is Maple. Both are very powerful mathematical packages that allow the user to solve mathematical problems both numerically and through symbolic methods; they provide powerful graphics in two and three dimensions. Maple and MATLAB are very easy to use and each has its own script files for executing a sequence of several commands. There is a close relationship between MATLAB and Maple in that when MATLAB's symbolic operations are called for, they are actually done through calls to Maple routines.

The *Maple Technical Newsletter* published by Springer-Verlag and MATLAB *News & Notes* from The Math Works Inc. are newsletters that are devoted to applications of Maple and MATLAB, respectively. Each package offers a student edition.

Defining Vectors and Matrices

Before applying Maple to linear algebra problems, we must first call the package: `with(linalg)`. Vectors can be defined in a number of ways. A vector with four components is declared as follows:

```
v2 := array([1, 3, -2, 5]);
```

Once the vectors have been defined, they may be added, subtracted, or multiplied. The dot product can be done by the function, `dotprod(v1,v2)`. For Maple commands, note that the semicolon is required at the end of the expression.

Matrices can be defined in a number of ways. Entering either

```
A := array([[1, 2, 3], [3, 2, 1], [1, 0, 3]]);
```

or

```
A := array(3, 3, [1, 2, 3, 3, 2, 1, 1, 0, 3]);
```

will result in the following display:

```
1   2   3
3   2   1
1   0   3
```

We have already introduced MATLAB for defining vectors and matrices. Here, the vector `v2` would be defined as:

```
v2 = [1, 3, -2, 5]
```

Because the "single quote" is used to denote the transpose, all we would need for the dot product is `v1*v2'`. If we define `v3 = [1; 3; -2; 5]`, we get v3 as a column vector. Here of course, `v3 = v2'`.

As mentioned earlier, matrices are easily defined; for example,

```
A = [1, 2, 3; 3, 2, 1; 1, 0, 3]
```

Remember that leaving out the trailing semicolon in MATLAB gives the display of matrix *A*, and including it suppresses the display.

Determinants and Matrix Operations

Once a square matrix has been defined, Maple and MATLAB compute the determinant of the matrix *M* through the det(*M*) function. However, they use different commands for the simple matrix operations. Two matrices can be added or subtracted and, if conformable, can be multiplied. We present a partial list of a few matrix operations, on two conformable matrices, *A, B,* and with vector *v* to illustrate how the programs differ:

Math Operation	Maple	MATLAB
$\det(A)$	`det(A)`	`det(A)`
$A + B$	`add(A,B), evalm(A+B)`	`A + B`
$A * B$	`multiply(A,B)`	`A*B`
$A * v$	`multiply(A,v)`	`A*v (A*v')`
A^{-1}	`inverse(A), evalm(1/A)`	`inv(A), A^-1`
A^n	`evalm(A^n)`	`A^n`

It is remarkable how fast the 50th power of a matrix can be computed considering that it is done in exact arithmetic!

Solving Linear Systems

The basic commands for solving a system of linear equations were given at the end of Section 2.3. Both MATLAB and Maple will find the solution of a system of linear equations. One first enters the elements of the augmented matrix (multiple right-hand sides are permitted). We then get the solution by using the reduced-row echelon function `rref(A)` where A is the augmented matrix.

Here is an example. Assume we have defined an augmented matrix,

$$A = \begin{bmatrix} 1 & 2 & 3 & -2 & 4 \\ 3 & 2 & 1 & 3 & -1 \\ 2 & 0 & 2 & 1 & 2 \end{bmatrix}.$$

Then in both systems

```
rref(A);
```

produces this result

```
1  0  0    3/2   −3/4
0  1  0   −1/4   −1/4
0  0  1    −1     7/4
```

Observe that the `rref` function uses the Gauss–Jordan technique to reduce the coefficient matrix to the identity matrix, and at the same time, reduces the right-hand side column(s) so that they become the solution vector(s). Maple does have a `linsolve(A,b)` procedure. In Section 2.4 we already pointed out that MATLAB can give the LU decomposition of a matrix, A, as well as the permutation matrix, P, so that $LU = PA$.

Nonlinear Systems

It is easy to solve a system of equations in Maple. For instance,

```
fsolve({x^2 − y = 1, x + y^2 = 2},{x,y});
```

results in the following display of one of the answers as

```
{x = −1.701644095, y = 1.926303220}
```

One could also predefine the previous functions in `fsolve`. We first write

```
f := x^2 − y − 1 = 0;
g := x + y^2 − 2 = 0;
fsolve({f,g},{x,y});
```

All this generates the same result. Now if we rewrite the second equation,

```
h := y − sqrt(2 − x) = 0;
```

and then follow with

```
fsolve({f,h},{x,y},{x = 0..2, y = 0..2});
```

we get the other solution to the problem:

```
{x = 1.345089393, y = .809265474}
```

With MATLAB one can make use of the same function that was used in solving linear equations. In MATLAB, define

```
[x,y] = solve('x^2 − y = 1', 'x + y^2 = 2')
```

The solutions are very complicated, but they may be simplified by giving the command:

```
double(x), double(y)
```

The answer is converted to a recognizable form, and we get:

```
x: 0.1828 + 0.6334i, 0.1828 − 0.6334i, 1.3451, −1.7106
y: −1.3678 + 0.2315i, −1.3678 − 0.2315i, 0.8093, 1.9263
```

The plot procedures allow one to see the graphs of the different functions for quick starting values. In the example just given, we need only do the following to get the graph of the functions we want to solve:

```
f := x^2 - 1;
h := sqrt(2 - x);
with(plots);
plot({f,h}, x = -2..2);
```

Characteristic Polynomial

Both Maple and MATLAB have built-in functions to evaluate the characteristic polynomial of a matrix. Assume we have input the following matrix:

$$A = \begin{bmatrix} 1 & 3 & 4 & 2 \\ 3 & 2 & 4 & 1 \\ 4 & 2 & 1 & 3 \\ 2 & 1 & 4 & 3 \end{bmatrix}.$$

With Maple, the command `charpoly(A,x)` produces

```
x⁴ - 7x³ - 33x² + 21x + 90
```

whereas, with MATLAB

```
poly(A)
```

produces the coefficient vector:

```
1  -7  -33  21  90
```

Similarly,

```
eigenvals(A)
```

has the output $10, -3, 3^{1/2}, -3^{1/2}$ in Maple, while

```
eig(A)
```

in MATLAB produces the results in decimal form: $10, -3, -1.7321, 1.7321$. Of course there are variations of these commands that can give us much more information on the eigenvectors as well. Also, we have already indicated that in MATLAB, the command `norm(A)` will give us the (largest eigenvalue of $AA^T)^{1/2} = |A|_2$. Both Maple and MATLAB offer far more power in solving problems, especially with the programming capabilities offered in the packages.

2.15 Parallel Processing

We have mentioned that the operation of many numerical methods can be speeded up by the proper use of parallel processing or distributed systems. In this section we describe how vector/matrix operations, Gaussian elimination, and Jacobi iteration can be efficiently performed in a parallel or distributed processing environment. We shall show how much the performance can be improved depending on the topology of the network in each case and pay special attention to the implementation of Gaussian elimination.

Vector/Matrix Operations

For inner products, a very elementary case, we assume we have two vectors, v, u, of length n, and an equal number of parallel processors, proc(i), $i = 1 \ldots n$, where each proc(i) contains the components, v_i, u_i. Then the multiplication of all the $v_i * u_i$ can be done in parallel in one time unit. In Section 0.7 we found that if the processors are connected suitably we can actually do the addition part in $\log(n)$ time units. Thus we can estimate the time for an inner product as $1 + \log(n) = O(\log(n))$.

This assumes a high degree of connectivity between the processors. There has been much study of such connectivity. These different designs are referred to as the topologies of the systems. In our present example, we assume the topology of a hypercube. However, before we describe that design, we shall introduce the simpler topology of the linear array. Suppose our processors were only connected as a linear array in which the communication send/receive is just between two adjacent processors:

$$P_1 \leftrightarrow P_2 \leftrightarrow \cdots \leftrightarrow P_{n-1} \leftrightarrow P_n.$$

Then our addition of the n elements would be $n/2$ time steps, because we could do an addition at each end in parallel and proceed to the middle.

The n-dimensional hypercube is a graph with 2^n vertices in which each vertex has n edges (is connected to n other vertices). This graph can be easily defined recursively because there is an easy algorithm to determine the order in which the vertices are connected.

0-dimensional hypercube:

$\bullet\ P$

1-dimensional hypercube:

2-dimensional hypercube:

3-dimensional hypercube:

Two vertices are *adjacent* if and only if they differ in exactly one bit. We can get to the $(n + 1)$-dimensional hypercube by making two copies of the n-dimensional hypercube and then adding a zero to the leftmost bit of the first n-dimensional cube and then doing the same with a 1 to the second cube. It was this kind of connectivity that allowed us to make the addition of n numbers in $\log(n)$ time-steps in Section 0.7. There are many other designs for connecting the processors. Such designs include names like star, ring, torus, mesh, and others. However, for the rest of this section, we shall assume that we have the processors optimally connected.

For the *matrix/vector product, Ax,* we assume that processor proc(i) contains the ith row of A as well as the vector, x. Because one processor is performing this dot product of row i of A and of vector x, this could be done in $2n$ units of time. However, because the other processors are proceeding in parallel, the whole operation will only take $O(n)$ units of time. We would have had a less efficient algorithm had we made use of the inner product algorithm above on the individual rows of A and the vector x.

For a linear array of processors, Bertsekas and Tsitsiklis give the following time for our more simplified case. The time is

$$\alpha n + (n - 1)(\beta + \gamma),$$

where α is the time for an addition or multiplication, β is the time for a transmission of the product along the link, and γ is a positive constant.

For the *matrix/matrix product, AB,* for two $n \times n$ matrices, we suppose that we have n^2 processors. Before we start our computations, each processor, $P_{i,j}$, will have received the values for row i of A and column j of B. Based on our previous discussion, we can expect the time to be $O(n)$. For n^3 processors, this can be reduced to just $O(\log(n))$ time. Here each processor, proc(i,k,j), would store the elements, $A_{i,k}$, $B_{k,j}$. Then all the multiplications can be done in parallel, and the additional $(\log(n))$ time units are for the additions.

Gaussian Elimination

Recall how we achieve a solution to a system of linear equations through Gaussian elimination. We first perform a sequence of row reductions on the augmented matrix $(A : b)$ until the coefficient matrix A is in upper-triangular form. Then we employ back-substitution to find the solution.

To see how this can be done in a parallel processing environment, we must examine the row-reduction phase and the back-substitution phase in some detail.

We begin with the row-reduction phase: Consider the following example of the first stage of row reduction of a 4×4 system with one right-hand side:

$$\begin{bmatrix} 1 & 2 & 1 & 3 & 4 \\ 2 & 5 & 4 & 3 & 4 \\ 1 & 4 & 2 & 3 & 3 \\ 3 & 2 & 4 & 1 & 8 \end{bmatrix} \begin{array}{l} \\ R_2 - (2/1) * R_1 \\ R_3 - (1/1) * R_1 \\ R_4 - (3/1) * R_1 \end{array} \rightarrow \begin{bmatrix} 1 & 2 & 1 & 3 & 4 \\ 0 & 1 & 2 & -3 & -4 \\ 0 & 2 & 1 & 0 & -1 \\ 0 & -4 & 1 & -8 & -4 \end{bmatrix}.$$

Though each of these row reductions depends on the elements of row 1, they are completely independent of one another. For example, the elements of rows 2 and 4 play no part in the row operations performed on row 3. Thus the row reductions on rows 2, 3, and 4 can be computed simultaneously.

If we are computing in a parallel processing environment, we can take advantage of this independence by assigning each row-reduction task to a different processor:

$$
\begin{bmatrix}
1 & 2 & 1 & 3 & 4 \\
2 & 5 & 4 & 3 & 4 \\
1 & 4 & 2 & 3 & 3 \\
3 & 2 & 4 & 1 & 8
\end{bmatrix}
\begin{array}{l}
\\
\rightarrow \text{Processor 1: } R_2 - (2/1) * R_1 \\
\rightarrow \text{Processor 2: } R_3 - (1/1) * R_1 \rightarrow \dots \\
\rightarrow \text{Processor 3: } R_4 - (3/1) * R_1
\end{array}
$$

Suppose each row assignment statement requires 4 time units, one for each element in a row. Then the sequential algorithm performs this stage of the row reduction in 12 time units, whereas we need only 4 time units for the parallel algorithm. This example of parallel processing on the first stage of row reduction of a 4×4 system matrix generalizes to any row reduction in stage j of an $n \times n$ system matrix.

Recall that there are $n - 1$ row-reduction stages in Gaussian elimination, one for each of the n columns of the coefficient matrix except for the last column. This suggests that we need $n - 1$ processors to do the reduction in parallel.* Also recall that each row-reduction stage j creates zeros in every cell below the diagonal in the jth column. The following two pseudocodes compare the use of a single processor with the use of n processors to perform the entire row-reduction phase of Gaussian elimination.

Algorithms for Row Reduction in Gaussian Elimination

Sequential Processing (without pivoting)

```
For j = 1 To (n − 1)
    For i = (j + 1) To n
        For k = j To (n + 1)
            a[i, k] = a[i, k] − a[i, j]/a[j, j] * a[j, k]
        End For k
    End For i
End For j
```

Parallel Processing

```
For i = 1 To (n − 1)        (Counts stages = columns)
    For k = i To (n + 1)        (On Processor j = (i + 1) To n)
        a[i, k] = a[i, k] − a[i, j]/a[j, j] * a[j, k]
    End For k
End For i
```

If we total the arithmetic operations to carry out the reduction of an $n \times n$ coefficient matrix to upper-triangular form, we find that the sequential algorithm requires $O(n^3)$ successive operations and that the parallel algorithm with n processors accomplishes the same task in $O(n^2)$ successive operations.

---

*Even so, we will need n processors in the final algorithm, as will be seen.

What happens if we have more than n processors? As indicated in our earlier discussion, we can speed up the reduction process even more. Suppose we increase the number of processors from n to, say, $n^2 + n$. We can effectively use this extra power for Gaussian elimination just as we did for the matrix/vector operations earlier. The complete algorithm for the row-reduction phase of Gaussian elimination on these processors runs only $O(n)$ successive steps, and each step requires just three time steps for one subtraction, one division, and one multiplication. If, as before, we label each processor as proc(i,j), $i = 1, \ldots, n$, $j = 1, \ldots, n + 1$, proc(i,j) is responsible for each element a_{ij} of the matrix $[A : b]$. We can now rewrite the algorithm for parallel processing to reflect this improvement:

For $i = 2$ To n
  {On Processor (j,k)}
    $a[j, k] = a[j, k] - a[j, i]/a[i, i] * a[j, k]$
End For i

The row-reduction phase of Gaussian elimination leaves us with an upper-triangular coefficient matrix and an appropriately adjusted right-hand side. In the sequential algorithm we now find a solution using back-substitution. Before we consider parallelization of back-substitution, let us examine the activity of the processors during row reduction in greater detail.

As we have observed, the processors responsible for computations on the elements of row 1 sit idle during row reduction, because those elements of the matrix never change. In addition, after the first row-reduction stage in which zeros are placed in the first column, the processors for row 2 also sit idle. In fact, each stage of row reduction frees $n + 1$ processors.

It is natural to wonder if these idle processors could be employed in our algorithm. Indeed, they can. We use them to perform row reductions above the diagonal at the same time that corresponding row reductions occur below the diagonal. Thus at each stage j of the reduction, zeros appear in all but the diagonal element of the jth column.

This diagram illustrates our improved procedure, continuing the simple 4×4 example examined before and doing stages 3 and 4:

$$\rightarrow \begin{bmatrix} 1 & 0 & -3 & 9 & 12 \\ 0 & 1 & 2 & -3 & -4 \\ 0 & 0 & -3 & 6 & 7 \\ 0 & 0 & 9 & -20 & -20 \end{bmatrix} \begin{matrix} R_1 - (3/3) * R_3 \\ R_2 - (-2/3) * R_3 \rightarrow \\ \\ R_4 - (9/-3) * R_3 \end{matrix} \begin{bmatrix} 1 & 0 & 0 & 3 & 5 \\ 0 & 1 & 0 & 1 & \frac{2}{3} \\ 0 & 0 & -3 & 6 & 7 \\ 0 & 0 & 0 & -2 & 1 \end{bmatrix}$$

$$\begin{matrix} R_1 - (3/-2) * R_4 \\ R_2 - (1/-2) * R_4 \\ R_3 - (6/-2) * R_4 \end{matrix} \begin{bmatrix} 1 & 0 & 0 & 0 & \frac{13}{2} \\ 0 & 1 & 0 & 0 & \frac{7}{6} \\ 0 & 0 & -3 & 0 & 10 \\ 0 & 0 & 0 & -2 & 1 \end{bmatrix} \rightarrow x = \begin{bmatrix} \frac{13}{2} \\ \frac{7}{6} \\ -\frac{10}{3} \\ -\frac{1}{2} \end{bmatrix}.$$

The result of n such reductions—one for each column of the coefficient matrix A—is $[D : b']$, where D is a diagonal matrix and b' is an appropriately adjusted right-hand-side vector.

Specifically, the solution x for $Dx = b'$ also satisfies $Ax = b$. This solution is the vector x whose elements are $x_i = b'_i/d_{ii}$ for $i = 1, \ldots, n$. We can use n processors to perform these n divisions simultaneously. Notice that the back-substitution phase of Gaussian elimination is no longer necessary! We find that the Gauss–Jordan procedure is preferred when doing parallel processing!

The parallel algorithm for n^2 processors required n time units for row reduction, and one additional time unit for division. Recall that the sequential algorithm required $O(n^3)$ time units. To understand the magnitude of the improvement in running time, consider that a solution achieved in 10 seconds via the parallel algorithm would require around 15 minutes via the sequential algorithm.*

Our final parallel algorithm for solving a system of linear equations more closely resembles the Gauss–Jacobi solution technique than it does Gaussian elimination. This is not surprising. It is not uncommon for good parallel algorithms to differ dramatically from their speediest sequential counterparts.

Problems in Using Parallel Processors

It is essential to mention some important concerns that have been neglected in the preceding discussion. When we actually implement this parallel algorithm, we must worry about four issues.

1. The algorithm described here does not pivot. Thus our solution may not be as numerically stable as one obtained via a sequential algorithm with partial pivoting. In fact, if a zero appears on the diagonal at any stage of the reduction, we are in big trouble. Bertsekas and Tsitsiklis observe that Gaussian elimination with pivoting can have an upper bound of $O(n \log(n))$ time when $n^2 + n$ processors are used, and still $O(n^2)$ time in the case of n processors.
2. The coefficient matrix A is assumed to be nonsingular. It is easy to check for singularity at each stage of the row reduction, but such error-handling will more than double the running time of the algorithm.
3. We have ignored the communication and overhead time costs that are involved in parallelization. Because of these costs, it is probably more efficient to solve small systems of equations using a sequential algorithm.
4. Other, perhaps faster, parallel algorithms exist for solving systems of linear equations. One technique, which is easily derived from ours, involves computing A^{-1} via row operations and simply multiplying the right-hand side to get the solution $x = A^{-1}b$. Another technique requires computing the coefficients of the characteristic polynomial and then applying these coefficients in building A^{-1} from powers of A. This method finds a solution in only $[2 \log_2 n + O(\log n)]$ time units, but it requires $n^4/2$ processors to do so. In addition, it often leads to numeric instability.[†]

---

*This neglects the overhead of interprocessor communications.
[†] JaJa (1992) describes these alternative algorithms in some detail.

Despite these concerns, our algorithm is an effective approach to solving systems of linear equations in a parallel environment.

Iterative Solutions—The Jacobi Method

The method of simultaneous displacements (the Jacobi method) that was discussed in Section 2.10 is adapted very simply to a parallel environment. Recall that at each iteration of the algorithm a new solution vector $x^{(n+1)}$ is computed using only the elements of the solution vector from the previous iteration, $x^{(n)}$. In fact, the elements of the vector $x^{(n)}$ can be considered fixed with respect to the $(n + 1)$st iteration. Thus, though each element $x_i^{(n+1)}$ in the vector $x^{(n+1)}$ depends on the elements in $x^{(n)}$, these $x_i^{(n+1)}$ are independent of one another and can be computed simultaneously.

Suppose the solution vector x has m elements. Then each iteration of the Jacobi algorithm in a sequential environment requires m assignment statements. If we have m processors in parallel, these m assignment statements can be performed simultaneously, thereby reducing the running time of the algorithm by a factor of m.

Notice also that each assignment statement is a summation over approximately m terms. As demonstrated in Section 0.7, this summation can be performed in $\log_2 m$ time units with m parallel processors, compared to m time units for sequential addition. If m^2 processors are available, we can employ both of these parallelizations and reduce the time for each iteration of the Jacobi algorithm to $\log_2 m$ time units. This is a significant speedup over the sequential algorithm, which requires m^2 time units per iteration.

As seen in Section 2.10, the actual running time of the algorithm (the number of iterations) depends on the degree of diagonal dominance of the coefficient matrix. Parallelization decreases only the time required for each iteration.

Because Gauss–Seidel iteration requires that the new iterates for each variable be used after they have been obtained, this method cannot be speeded up by parallel processing. Again, the preferred algorithm for sequential processing is not the best for parallel processing.

Chapter Summary

After working through this chapter, you should be able to

1. Handle the basic operations of matrices and vectors. You know the definitions for triangular matrix, tridiagonal matrix, transpose, determinant, characteristic polynomial, and matrix inverse.

2. Use Gaussian elimination with partial pivoting and back-substitution to solve a system of equations, compute the determinant of the coefficient matrix, and find the LU decomposition. You are able to compare this method with the Gauss–Jordan method.

3. Compute the number of additions, multiplications, and divisions required in the implementation of several of the algorithms in this chapter.

4. Be acquainted with another *LU* method for solving a system of equations. You should know how to use the *LU* decomposition to solve a system efficiently when there are multiple right-hand sides.

5. Understand the several ways for computing the determinant of a square matrix.

6. Know more than one way to find the inverse of a matrix and why using Gaussian elimination can be more efficient than the standard Gauss–Jordan technique.

7. Understand the concept of an ill-conditioned system and compare this to singular and near-singular systems.

8. Compute the various norms for vectors and matrices and evaluate the condition number of a matrix.

9. Solve certain specialized systems by iterative methods and explain why the use of parallel processors may make some methods preferred that would not be in sequential processing. You should know what is meant by *diagonal dominance*.

10. Use Newton's method to solve a system of nonlinear equations and tell how the number of function evaluations may be reduced.

11. Explain what is meant by a *parallel algorithm.*

12. Express the Gaussian elimination method as a product of simple matrices.

13. Use an example to show the importance of pivoting in getting a more accurate solution.

14. Utilize MATLAB and/or Maple to solve systems of equations, compute norms of vectors and matrices, perform matrix operations, and find eigenvalues and eigenvectors.

Computer Programs

The two programs of this chapter are (1) a C program that does Gaussian elimination to solve a system of linear equations, and (2) a Fortran 90 program that solves a nonlinear system.

Program 2.1 Muller's Method

Figure 2.4a is the listing of a "driver" program; the output follows this. Then, in Figure 2.4b, are the listings of the two procedures that are called to obtain the final result. The example that is solved is the same as that in Section 2.4.

Program 2.2 Solving a Nonlinear System

The listing of this program, written in Fortran 90, is given in Figure 2.5. The output from the program follows the listing. This execution of the program solves a pair of equations; these are given in subroutine Fcn, which is at the end of the listing.

```
/* * * * * * * * * * * * * * * * * * * * * * * * * * * * * * * * * * * * * * * * * * * * * *
 *                                                                                        *
 *       Chapter 2             Gaussian Elimination                                       *
 *                                                                                        *
 *       Gerald/Wheatley, APPLIED NUMERICAL ANALYSIS (sixth edition)                      *
 *                       Addison Wesley Longman, 1999                                     *
 *                                                                                        *
 * * * * * * * * * * * * * * * * * * * * * * * * * * * * * * * * * * * * * * * * * * * * * *
 *                                                                                        *
 *     This program solves a system of linear equations by Gaussian                      *
 *     elimination with partial pivoting and back substitution. The                      *
 *     algorithm of Section 2.4 is used.                                                  *
 *                                                                                        *
 *     The program is implemented through two procedures:                                 *
 *                 ELIM and SOLVE.                                                        *
 *     The first of these returns the LU decomposition of the                            *
 *     coefficient matrix; the second solves the system using both                       *
 *     forward and back substitution.                                                     *
 *                                                                                        *
 *     The particular example that is solved is taken from                               *
 *     Section 2.4. The matrix of coefficients as well as the                            *
 *     vector of right-hand sides are defined within the driver                          *
 *     program.                                                                           *
 *                                                                                        *
 * * * * * * * * * * * * * * * * * * * * * * * * * * * * * * * * * * * * * * * * * * * * * */

#include <stdio.h>
#include <math.h>
#include "b:\2_1b.c"

float a[11][11],        /* the matrix of coefficients */
      b[11];            /* the right hand side of the system */
int   ipvt[11];         /* vector to keep track of row changes */
int   n, i, j;

/*
   This procedure displays the coefficient matrix A and the
   right-hand side vector, b.
*/
printing(a, b)
float a[][11], b[];

{
int i, j;

printf("\n");
for ( i = 1; i <= n; i++ )
  {
  for ( j = 1; j <= n; j++ )
    printf("%7.4f ", a[i][j]);
```

Figure 2.4a

```
      printf("          %7.4f   ", b[i]);
      printf("\n");
      }                     /* end of for i */
   }                      /* end of printing procedure */

void main()          /* main program begins */
{

/*
     Fill matrix a and b with zeroes.
*/
   for (i = 1; i <= 10; i++)
   {
     b[i] = 0.0;
     for ( j = 1; j <= 10; j++ )
       a[i][j] = 0.0;
   }                        /* end of for i */

/*
     Define the number of equations, matrix a, and vector b.
     The values are those of the example of Section 2.4.
*/
   n = 4;
   a[1][1] = 6; a[1][2] =  1; a[1][3] = -6; a[1][4] = -5;
   a[2][1] = 2; a[2][2] =  2; a[2][3] =  3; a[2][4] =  2;
   a[3][1] = 4; a[3][2] = -3;              a[3][4] =  1;
                a[4][2] =  2;              a[4][4] =  1;
   b[1] = 6; b[2] = -2; b[3] = -7;

/*
     Write out matrix a and vector b.
*/
   clrscr();
   printf("\n\n\n");
   printf("        THE MATRIX A              RIGHT HAND SIDE\n");
   printing(a,b);
   printf("\n\n");

*/
     Find the LU decomposition of matrix a, using procedure elim.
*/
   elim(a, n, ipvt);

*/
     Procedure solve now gets the solution.
*/
   solve(a, n, ipvt, b);

/*
     Display the LU decomposition matrix.
*/
   printf("\n   THE LU DECOMPOSITION OF A       SOLUTION VECTOR\n");
```

Figure 2.4a *Continued*

```
   printing(a, b);
   printf("\n\n");
   getch();

}            /* end of main */


* * * * * * * * * * * * * * * * * * * * * * * * * * * * * * * * * * * * * * * * * * * *

               OUTPUT FOR PDEC.C

         THE MATRIX A                    RIGHT HAND SIDE

    6.0000    1.0000   -6.0000   -5.0000      6.0000
    2.0000    2.0000    3.0000    2.0000     -2.0000
    4.0000   -3.0000    0.0000    1.0000     -7.0000
    0.0000    2.0000    0.0000    1.0000      0.0000



          THE LU DECOMPOSITION OF A         SOLUTION VECTOR

    6.0000    1.0000   -6.0000   -5.0000     -0.5000
    0.6667   -3.6667    4.0000    4.3333      1.0000
    0.3333   -0.4545    6.8182    5.6364      0.3333
    0.0000   -0.5455    0.3200    1.5600     -2.0000
```

Figure 2.4a *Continued*

```
/* * * * * * * * * * * * * * * * * * * * * * * * * * * * * * * * * * * * * * * * * * *
 *                                                                          *
 *     ELIM: this procedure solves a set of linear equations and           *
 *     gives an LU decomposition of the coefficient matrix. The            *
 *     Gaussian elimination method is used, with partial pivoting          *
 *     and back substitution.                                              *
 *                                                                          *
 *     INPUT:   a  - the coefficient matrix                                *
 *              n  - the number of equations                               *
 *                                                                          *
 *     OUTPUT:  a  - the LU decomposition of the matrix a                  *
 *              ipvt - a vector containing the order of the rows           *
 *                     of the rearranged matrix due to pivoting.           *
 *                                                                          *
 * * * * * * * * * * * * * * * * * * * * * * * * * * * * * * * * * * * * * * * * * */
void elim(a, n, ipvt)                      /* Procedure elim begins */
  float a[][11];
  int ipvt[], n;

{
```

Figure 2.4b

```
   float save, ratio, value, det;
   int i, ipvtemp, nMinus1, iPlus1, j, l, kcol, jcol, jrow, tempipvt;


/*
   Begin the LU decomposition
*/
  det = 1.0;
  nMinus1 = n-1;
  for (i = 1; i <= n; i++)
    ipvt[i] = i;
  for (i = 1; i <= nMinus1; i++)
    {
      iPlus1 = i+1; ipvtemp = i;
      for (j = iPlus1; j <= n; j++)                    /* find the */
        if (fabs(a[ipvtemp][i]) < fabs(a[j][i]))        /*   pivot   */
          ipvtemp = j;                                   /*   row     */

      if (ipvtemp != i)
        {
          tempipvt = ipvt[i];                       /* interchange */
          ipvt[i] = ipvt[ipvtemp];                  /* rows if a[i,i] */
          for (jcol = 1; jcol <= n; jcol++)  /* is not the max */
            {                                       /* in column.    */
              save = a[i][jcol];
              a[i][jcol] = a[ipvtemp][jcol];
              a[ipvtemp][jcol] = save;
            }              /* end of for jcol loop */
          ipvt[ipvtemp] = tempipvt;
          det = -det;
        }              /* end of if statement */

/*
   Now reduce all elements below the i'th row
*/
  for ( jrow = iPlus1; jrow <= n; jrow++ )
    if (a[jrow][i] != 0.0)
      {
        a[jrow][i] = a[jrow][i]/a[i][i];
        for ( kcol = iPlus1; kcol <= n; kcol++ )
          a[jrow][kcol] = a[jrow][kcol] - a[jrow][i]*a[i][kcol];
      }                          /* end of if statement */
  }                              /* end of for i loop */
}                              /* end of elim procedure */

/* * * * * * * * * * * * * * * * * * * * * * * * * * * * * * * * * * * * * * * * * * *
 *                                                                               *
 *    This is procedure SOLVE.                                                   *
 *    Making use of the LU decomposition of the matrix A, this                   *
 *    procedure solves the system by FORWARD and BACK substitution.              *
 *                                                                               *
 *    INPUT  a - LU matrix from elim                                             *
 *           n - number of equations                                            *
```

Figure 2.4b *Continued*

```
*            ipvt - a record of the rearrangement of the rows of      *
*                    a from procedure ELIM                            *
*            b - right hand side of the system of equations           *
*                                                                     *
*     OUTPUT b - the solution vector                                  *
*                                                                     *
* * * * * * * * * * * * * * * * * * * * * * * * * * * * * * * * * * * * * * * * * * * * * * * * * */

solve(a, n, ipvt, b)                    /* procedure solve begins */
  float a[][11], b[];
  int    n, ipvt[];

{
  int irow, jcol, i;
  float sum;
  float x[11];

/*
    Reduce the elements of the b vector, storing in the x vector
*/
  for (i = 1; i <= n; i++)
     x[i] = b[ipvt[i]];

/*
    Solve using forward substitution: Ly = b
*/
    for (irow = 2; irow <= n; irow++)
      {
        sum = x[irow];
        for (jcol = 1; jcol <= (irow-1); jcol++)
          sum = sum - a[irow][jcol]*x[jcol];
        x[irow] = sum;
      }         /* end of for irow */

/*
    Solve by back substitution: Ux = y
*/
    b[n] == x[n]/a[n][n];
    for ( irow = (n-1); irow >= 1; irow-- )
      {
        sum = x[irow];
        for ( jcol = (irow+1); jcol <= n; jcol++ )
          sum = sum - a[irow][jcol]*b[jcol];
        b[irow] = sum/a[irow][irow];
      }                       /* end of for irow */
}                             /* end of solve procedure */
```

Figure 2.4b *Continued*

```
Program Nlsys
!
!!!!!!!!!!!!!!!!!!!!!!!!!!!!!!!!!!!!!!!!!!!!!!!!!!!!!!!!!!!!
!                                                          !
!                      Chapter 2                           !
!                                                          !
!  Gerald/Wheatley, APPLIED NUMERICAL ANALYSIS(Sixth Edition)  !
!                 Addison Wesley Longman, 1999             !
!                                                          !
!!!!!!!!!!!!!!!!!!!!!!!!!!!!!!!!!!!!!!!!!!!!!!!!!!!!!!!!!!!!
!                                                          !
!    This Program Solves the following pair of equations   !
!                                                          !
!        f(x,y) = x*x + y*y - 5 = 0                        !
!        g(x,y) = y - exp(x) -1 = 0                        !
!                                                          !
!    using either numerical partials for the Jacobian      !
!    matrix (Method = 0) or analytic partials              !
!    (Method = 1). In the latter case,                     !
!    Subroutine Analytic_partials is modified.             !
!                                                          !
!!!!!!!!!!!!!!!!!!!!!!!!!!!!!!!!!!!!!!!!!!!!!!!!!!!!!!!!!!!!
!
      Implicit None
      Real(Kind = 8), Dimension(10) :: X,F
      Integer :: N = 2,i
      X = (/3,2,0,0,0,0,0,0,0,0/)
!
      Write(*,*) '  ************************************************  '
      Write(*,*) '              Output for nlsys.f90                '
      Write(*,'(/)')
!
      Call NL_System_Solver(n,50,X,F,0.125E-4,0.1E-6,0.1E-6,0)
      Write(*,100)
100   Format(///'   The X-values are:'/)
      Write(*,200) (x(i), i = 1,n)
200   Format(T6,F10.6)
      Stop
!
      Contains
!
!
      Subroutine
NL_System_Solver(N,Max_iterations,X,F,Delta,Xtol,Ftol,Method)
!   ---------------------------------------------------------
!
!    Subroutine NL_System_Solver :
!                      This subroutine solves a system of N non-
!    linear equations by Newton's Method. the partial derivatives of
!    the functions are estimated by difference quotients when a
!    variable is perturbed by an amount equal to Delta ( Delta is
```

Figure 2.5

```
!     added ). This is done for each variable in each function.
!     increments to improve the estimates for the X-values are
!     computed from a system of equations using Subroutines
!     LU_Decomp and Solve_LU_Eq_B
!
!     -------------------------------------------------------
!
!     parameters are :
!
!     Fcn     - Subroutine that computes values of the functions.
!
!     N       - the number of equations
!     Max_iterations - limit to the number of iterations
!                       that will be used
!     X       - array to hold the X-values. initially This array holds
!               the initial guesses. It returns the final values.
!     F       - an array that holds values of the functions
!     Delta   - a small value used to perturb the X-values so partial
!               derivatives can be computed by difference quotient.
!     Xtol    - tolerance value for change in X-values to stop itera-
!               tions. When the largest change in any X meets Xtol,
!               the Subroutine terminates.
!     Ftol    - tolerance value on F to terminate. when the largest F-
!               value is less Than Ftol, Subroutine terminates.
!     Method - allows one to compute the Jacobian matrix.
!
!       0       implies numerical partials are to be computed in
!               Subroutine Numeric_Partials.
!       1       implies analytic partials. This requires the user to
!               input these partial derivatives in Subroutine
!               Analytic_Partials
!
!     -------------------------------------------------------
!
      Implicit None
      Real(Kind = 0), Intent(In Out) :: X(:),F(:)
      Real(Kind = 8), Dimension(10,10) ::A
      Real(Kind = 8), Dimension(10) :: Xsave,B
      Real(Kind = 8) :: det
      Real, Intent(In) :: Delta,Xtol,Ftol
      Integer, Intent(In) :: Method, n, Max_iterations
      Integer, Dimension(10) :: Ipvt
      Integer :: iteration,i,itest,Irow
!
!     -------------------------------------------------------
!
!     Check that value of N is between 2 and 10.
!
        If ( N < 2 .OR. N > 10 ) Then
          Write(*,*) 1004, N
          Return
        End IF
```

Figure 2.5 *Continued*

```
!
!    ------------------------------------------------------
!
!    Begin iterations - save X-values, Then get F-values
!
        Do iteration = 1, Max_iterations
          Call Fcn(X,F)
!
!    ------------------------------------------------------
!
!    Test F values and Save TheM
!
          itest = 0
          Do i = 1,n
              If ( Abs(F(i)) > Ftol ) itest = itest + 1
              End Do
          Write(*,1000) iteration, ( x(i), i = 1,n)
          Write(*,1001) (F(i), i = 1,n)
!
!    ------------------------------------------------------
!
!    Check whether Ftol is met. If not, continue.
!
          If   ( itest == 0 ) Return
!
!    ------------------------------------------------------
!
!    ------------------------------------------------------
!
          If (Method == 0) Then
              Call Numeric_partials(A,n,Xsave,B,X,F, Delta)
            Else
              Call Analytic_partials(A,n,Xsave, B,X,F)
            End If
!
!
          Call LU_Decomp(A,n,Ipvt,det)
          If (Abs(det) < 1.0E-6) Then
            Write(*,*) 'matrix is Singular'
            Stop
            End If
          Call Solve_LU_Eq_B(A,n,Ipvt,B)
!
!
!    Check that the coefficient matrix is not ill-conditioned
!
            Do Irow = 1,n
              If ( Abs(A(Irow,Irow)) <= 1.0E-6 ) Then
                Write(*,1003)
                Return
                End If
              End Do
```

Figure 2.5 *Continued*

```
!
!     -----------------------------------------------------
!
!    Update the X-values, and check whether Xtol is met.
!
          itest = 0
          Do i = 1,n
            X(i) = Xsave(i) + B(i)
            If ( Abs(B(i)) > Xtol ) itest = itest + 1
            End Do
!
!     -----------------------------------------------------
!
!    If Xtol is met, print latest values and return
!    Otherwise, do another iteration.
!
          If ( itest == 0 ) Then
            Write(*,1002) iteration, (X(i), I = 1,n)
            Return
            End IF
  End Do
!
!     -----------------------------------------------------
!
!    When we have done the maximum number of iterations, Return
!
          Write(*,1005)
          Return
!
1000 Format(/' After iteration number',I3,' X and F values are' &
            //10F13.5)
1001 Format(10F13.5/)
1002 Format(/' After iteration number',I3,' X values (meeting', &
            ' Xtol) are '//10F13.5)
1003 Format(/' Cannot Solve_LU_Eq_B system. matrix nearly singular.')
1005 Format(/' Maximum number of iterations met without convergence')
!
      End Subroutine NL_system_Solver
!
!            Subroutine LU_Decomp
!
      Subroutine LU_Decomp(A,n,Ipvt_vector,det)
!
!     -----------------------------------------------------
!
!      Subroutine LU_Decomp: This Subroutine returns the LU
!    decomposition of the coefficient matrix. The method
!    is based on the algorithm presented in Section 2.4
!
!    INPUT:  a  -  the coefficient matrix
!            N  -  the number of equations
!
```

Figure 2.5   *Continued*

```
!    OUTPUT: a  -  the LU decomposition of the matrix A
!                    the original matrix a is lost
!            Ipvt_vector  -  a vector containing the order of the rows
!                             of the rearranged matrix due to pivoting
!            det  -  the determinant of the matrix. It is set to
!                     0 If any pivot element is less than 0.00001.
!
!    ------------------------------------------------------------
!
     Implicit None
     Real(Kind = 8), Dimension(:,:), Intent(In Out) :: A
     Real (Kind = 8), Dimension(10) :: Save
     Real(Kind = 8), Intent(Out) :: det
     Integer, Dimension(:), Intent(Out) :: Ipvt_vector
     Integer, Intent(In) :: n
     Integer :: i, nless1, iplus1, j,temp_pivot
     Integer :: jrow, tmpvt   ! kcol
!
     det = 1.0
     Nless1 = n—1
         Do   i = 1,n
              Ipvt_vector(i) = i
              End Do
!
         Do i = 1,nless1
             iPLUS1 = i+1
             temp_pivot = i
!
!        Find pivot row
!
             Do j = iPLUS1,n
                If (Abs(A(temp_pivot,I)) < Abs(A(J,I))) temp_pivot = j
                End Do
!
!    ------------------------------------------------------------
!
!    Check for small pivot element
!
!    ------------------------------------------------------------
!
         If (Abs(A(temp_pivot,I)) < 1.0E-05) Then
             det = 0.0
             Write(*,*) '(//)'
             Write(*,*) '    matrix is singular or near-singular '
             Write(*,*) '(//)'
             Return
             End IF
!
!        Interchange rows if necessary
!
         If (temp_pivot /= i) Then
             Save(1:n) = A(i, 1:n)
             A(i,1:n) = A(temp_pivot, 1:n)
```

Figure 2.5 *Continued*

```
                         A(temp_pivot,1:n) = Save(1:n)
                         tmpvt = Ipvt_vector(i)
                         Ipvt_vector(i) = temp_pivot
                         Ipvt_vector(temp_pivot) = tmpvt
                         det = -det
                    End IF
!
!          Reduce all elements below the i'th row
!
                    Do   jrow = iPLUS1,n
                         If (A(jrow,I) /= 0.0) Then
                             A(jrow,I) = A(jrow,I)/A(I,I)
                    A(jrow,iPLUS1:n) = A(jrow,iPLUS1:n) - &
                    A (jrow,i)*A(i,iPLUS1:n)
                             End If
                         End Do
!
           End Do
!
           If (Abs(A(n,n)) < 1.0E-5 ) Then
                Write(*,*) '(//)'
                Write(*,*) ' matrix is singular or near-singular '
                Write(*,*) '(//)'
                det = 0.0
                Return
           End IF
!
!     -----------------------------------------------------------
!
!    Compute the determinant of the matrix
!
!     -----------------------------------------------------------
!
     Do   I = 1,n
           det = det * A(I,I)
           End Do
!
     Return
     End Subroutine LU_Decomp
!
!     -----------------------------------------------------------
!
!               Subroutine Solve_LU_Eq_B
!
!     -----------------------------------------------------------
!
     Subroutine Solve_LU_Eq_B(A,n,Ipvt_vector,B)
!
!     -----------------------------------------------------------
!
! Making use of the LU decomposition of the matrix A, this sub-
! routine Solve_LU_Eq_BS the system by forward- and back-substitution.
!
```

Figure 2.5 *Continued*

```
!    INPUT:   a - LU matrix from Subroutine LU_Decomp
!             N - number of equations
!             Ipvt_vector - a record of the rearrangement of the
!                           rows from Subroutine LU_Decomp
!             B - right-hand side of the system of equations
!
!    OUTPUT:  B - the solution vector
!
!    ------------------------------------------------------------
!
     Implicit None
     Real(Kind = 8), Dimension(:,:), Intent(In) :: A
     Real(Kind = 8), Dimension(:), Intent(In Out) :: B
     Real(Kind = 8), Dimension(10) :: X
     Real(Kind = 8) :: sum
     Integer, Dimension(:),Intent(In) :: Ipvt_vector
     Integer, Intent(IN) :: n
     Integer :: Irow,jcol,i
!
!    ------------------------------------------------------------
!
!    Rearrange the elements of the B vector. Store them in the
!    X vector.
!
     Do I = 1,n
         x(i) = B(Ipvt_vector(I))
         End Do
!
!    ------------------------------------------------------------
!
!    Solve_LU_Eq_B using forward-substitution--LY = B
!
!    ------------------------------------------------------------
!
     Do Irow = 2,N
     sum = X(Irow)
     Do   jcol = 1,(Irow-1)
         sum = sum—A(Irow,jcol)*X(jcol)
         End Do
     X(Irow) = sum
     End Do
!
!    ------------------------------------------------------------
!
!    Solve_LU_Eq_B by back-substitution—LY = B, UX = Y
!
!    ------------------------------------------------------------
!
     B(n) = X(n)/A(n,n)
     Do   Irow = (N-1),1,-1
         sum = X(Irow)
```

Figure 2.5 *Continued*

```
                Do    jcol = (Irow+1),N
                    sum = sum - A(Irow,jcol)*B(jcol)
                    End Do
                B(Irow) = sum/A(Irow,Irow)
                End Do

        Return
        End Subroutine Solve_LU_Eq_B
!
!   -------------------------------------------------------
!
!   Subroutine Numeric_partials:
!       Here the matrix of partials are computed:
!       method:   0 - implies numerical partials
!                 1 - implies analytic partials
!                      which must be SUPPLIED.
!
!   -------------------------------------------------------
!
    Subroutine Numeric_partials(A,n,Xsave,B,X,F, Delta)
    Real(Kind = 8), Dimension(:), Intent(In Out) :: X, F
    Real, Intent(In) :: Delta
    Real(Kind = 8), Dimension(:,:), Intent(Out) :: A
    Real(Kind = 8), Dimension(:), Intent(Out) :: Xsave, B
    Real(Kind = 8), Dimension(10) :: Fsave
    Integer, Intent(In) :: N
    Integer :: Irow, jcol,i !,np
!
!     NP = NP + 1
!
    Do i = 1,n
      Fsave(i) = F(i)
      Xsave(i) = X(i)
    End Do
!
!
!                  Compute numerical partials
!
!   This double loop computes the partial derivatives of each function
!   for each variable and stores them in a coefficient array.
!
        Do jcol = 1,n
            X(jcol) = Xsave(jcol) + Delta
            Call Fcn(X,F)
            Do Irow = 1,n
                A(Irow,jcol) = (F(Irow)—Fsave(Irow)) / Delta
                End Do
!
!   Reset X-values for the next column of partials
!
            X(jcol) = Xsave(jcol)
            End Do
!
```

Figure 2.5 *Continued*

```
!
!    ------------------------------------------------------
!
!    Now we put negative of F-values as right-hand sides
!
            B(1:n) = -Fsave(1:n)
      Return
            End Subroutine Numeric_partials

        Subroutine analytic_partials(A,n,Xsave,B,X,F)
        Real(Kind = 8), Dimension(:), Intent(In Out) :: X, F
        Real(Kind = 8), Dimension(:,:), Intent(Out) :: A
        Real(Kind = 8), Dimension(:), Intent(Out) :: Xsave, B
        Integer, Intent(In) :: n
        Xsave(1:n) = X(1:n)
!
!                       compute analytic partials
!
        A(1,1) = 2.0*X(1)
        A(2,1) = -EXP(X(1))
        A(1,2) = 2.0*X(2)
        A(2,2) = 1.0
!
!
!
!    ------------------------------------------------------
!
!    Now we put negative of F-values as right-hand sides
!
            Do Irow = 1 , N
                B(Irow) = -Fsave(Irow)
                End Do
            B(1:n) = -F(1:n)
      Return
      End Subroutine analytic_partials
!
!    ------------------------------------------------------
!
!    Subroutine Fcn:
!       The two nonlinear functions are defined
!       The program can handle up to 10 functions.
!
!    ------------------------------------------------------
!
      Subroutine Fcn(X,F)
      Implicit None
      Real(Kind = 8), Intent(In) :: X(:)
      Real(Kind = 8), Intent(Out) :: F(:)

      F(1) = X(1)*X(1) + X(2)*X(2)-5.0
      F(2) = X(2)-Exp(X(1))-1.0
      Return
      End Subroutine Fcn
```

Figure 2.5 *Continued*

```
   End Program Nlsys

********************Output for Nlsys Program********************

After iteration number   1 X- and F-values are

     3.00000      2.00000
     8.00000    -19.08554

After iteration number   2 X- and F-values are

     2.02317      1.46525
     1.24017     -7.09701

After iteration number   3 X- and F-values are

     1.18227      2.20314
     1.25159     -2.05864

After iteration number   4 X- and F-values are

     0.56552      2.25006
     0.38259     -0.51030

After iteration number   5 X- and F-values are

     0.26959      2.23942
     0.08769     -0.07001

After iteration number   6 X- and F-values are

     0.20694      2.22739
     0.00407     -0.00252

After iteration number   7 X- and F-values are

     0.20434      2.22671
     0.00001      0.00000

After iteration number   8 X- and F-values are

     0.20434      2.22671
     0.00000      0.00000



  The X-values are:

     0.204337
     2.226712
```

Figure 2.5 *Continued*

Exercises

1. Given the matrices A, B, and the vectors x, y,

$$A = \begin{bmatrix} 3 & 0 & 2 & -4 \\ 4 & -3 & 1 & 2 \\ 5 & 1 & -1 & 2 \end{bmatrix}, \quad x = \begin{bmatrix} 2 \\ -3 \\ 0 \\ 1 \end{bmatrix},$$

$$B = \begin{bmatrix} 6 & 9 & 2 & -1 \\ 0 & 2 & 1 & 3 \\ 2 & 1 & -2 & 6 \end{bmatrix}, \quad y = \begin{bmatrix} 0 \\ 4 \\ 2 \\ 6 \end{bmatrix},$$

a. Find $3A$, $2A + 4B$, $2x - 3y$.
▶ b. Find $A - B$, Ax, By.
▶ c. Find x^Ty, xy^T.
d. Find B^T.

2. Given the matrices

$$A = \begin{bmatrix} -3 & 1 & -2 \\ 2 & 3 & 0 \\ -1 & 2 & 3 \end{bmatrix}, \quad B = \begin{bmatrix} 3 & -2 & 5 \\ 2 & -4 & 1 \\ -4 & 1 & 6 \end{bmatrix},$$

▶ a. Find BA, B^3, AA^T.
b. Find $\det(A)$, $\det(B)$.
c. A square matrix can always be expressed as a sum of an upper-triangular matrix U and a lower-triangular matrix L. Find two different combinations of L's and U's such that $A = L + U$.

3. Given the matrices

$$A = \begin{bmatrix} 1 & -2 & 2 \\ 3 & 1 & 1 \\ 2 & 0 & 1 \end{bmatrix}, \quad B = \begin{bmatrix} -1 & -2 & 4 \\ 1 & 3 & -5 \\ 2 & 4 & -7 \end{bmatrix},$$

$$C = \begin{bmatrix} 1 & 0 & 2 \\ 4 & 2 & -1 \\ 2 & 3 & 1 \end{bmatrix}.$$

a. Show that $AB = BA = I$, where I is the 3×3 identity matrix. We shall later identify B as the inverse of A.
b. Show that $AI = IA = A$.
c. Show that $AC \neq CA$ and also that $BC \neq CB$. In general, matrices do not commute under multiplication.
d. A square matrix can be expressed also as the sum of an upper-triangular matrix, a diagonal matrix, and a lower-triangular matrix. Express A as $L + D + U$.

4. Let

$$A = \begin{bmatrix} 9 & 2 \\ -10 & -3 \end{bmatrix} \quad B = \begin{bmatrix} 6 & 1 & -1 \\ 1 & 4 & 3 \\ -1 & 3 & -2 \end{bmatrix}.$$

a. Find the characteristic polynomials of both A and B.
▶ b. Find the eigenvalues of both A and B.

5. Write as a set of equations:

$$\begin{bmatrix} 2 & 4 & -1 & -2 \\ 4 & 0 & 2 & 1 \\ 1 & 3 & -2 & 0 \\ 3 & 2 & 0 & 5 \end{bmatrix} \begin{bmatrix} x_1 \\ x_2 \\ x_3 \\ x_4 \end{bmatrix} = \begin{bmatrix} 10 \\ 7 \\ 3 \\ 2 \end{bmatrix}.$$

▶ **6.** Write in matrix form:

$$\begin{aligned}
2x - 6y + z &= 11, \\
-5x + y - 2z &= -12, \\
x + 2y + 7z &= 20.
\end{aligned}$$

7. a. Solve by back-substitution:

$$\begin{aligned}
2x_1 - 3x_2 + x_3 &= -11, \\
4x_2 - 3x_3 &= -10, \\
2x_3 &= 4.
\end{aligned}$$

▶ b. Solve by forward-substitution:

$$\begin{aligned}
5x_3 &= 10, \\
3x_2 - 3x_3 &= 3, \\
2x_1 - x_2 + 2x_3 &= 7.
\end{aligned}$$

8. Solve the set of equations in Exercise 5.

9. Solve the set of equations in Exercise 6.

10. Solve the following (given as the augmented matrix):

$$\begin{bmatrix} 1 & 1 & -2 & \vdots & 3 \\ 4 & -2 & 1 & \vdots & 5 \\ 3 & -1 & 3 & \vdots & 8 \end{bmatrix}.$$

▶ **11.** Show that the following does not have a solution:

$$\begin{aligned}
3x_1 + 2x_2 - x_3 - 4x_4 &= 10, \\
x_1 - x_2 + 3x_3 - x_4 &= -4, \\
2x_1 + x_2 - 3x_3 &= 16, \\
-x_2 + 8x_3 - 5x_4 &= 3.
\end{aligned}$$

12. If the right-hand side of Exercise 11 is $(2, 3, 1, 3)^T$, show that there are an infinite number of solutions.

▶ **13.** Show that the set of left-hand sides (the rows of the coefficient matrix) in Exercise 11 are not independent vectors.

Section 2.4

14. Using Gaussian elimination with partial pivoting and back-substitution,

   a. Solve the equations of Exercise 5.

   b. Using part (a), find the determinant of the coefficient matrix.

   c. What is the LU decomposition of the coefficient matrix? (Rows may have been interchanged.)

▶ **15.** Using Gaussian elimination with partial pivoting and back-substitution,

   a. Solve the equations of Exercise 10.

   b. Using part (a), find the determinant of the coefficient matrix.

   c. What is the LU decomposition of the coefficient matrix? (Rows may have been interchanged.)

▶ **16.** a. Solve the system

$$2.51x_1 + 1.48x_2 + 4.53x_3 = 0.05,$$
$$1.48x_1 + 0.93x_2 - 1.30x_3 = 1.03,$$
$$2.68x_1 + 3.04x_2 - 1.48x_3 = -0.53,$$

   by Gaussian elimination, carrying just three significant digits and chopping. Do not interchange rows. Observe that there is a small divisor in reducing the third equation.

   b. Repeat part (a), but now use partial pivoting. Observe that there are no small divisors.

   c. Substitute each set of answers into the original equations and observe that the left- and right-hand sides match much better with the answers to part (b). The solution, correct to six digits, is

$$x_1 = 1.45310, \quad x_2 = -1.58919, \quad x_3 = -0.27489.$$

17. Solve the systems of Exercises 5, 6, and 10 by the Gauss–Jordan method.

18. Augment the coefficient matrix with all three of the right-hand sides and get all three solutions simultaneously, given

$$A = \begin{bmatrix} 4 & 2 & 1 & -3 \\ 1 & 2 & -1 & 0 \\ 3 & -1 & 2 & 4 \\ 0 & 2 & 4 & 3 \end{bmatrix}, \quad b_1 = \begin{bmatrix} 4 \\ 2 \\ 8 \\ 9 \end{bmatrix},$$

$$b_2 = \begin{bmatrix} 9 \\ 1 \\ 8 \\ 4 \end{bmatrix}, \quad b_3 = \begin{bmatrix} 4 \\ 2 \\ -7 \\ 5 \end{bmatrix}.$$

19. a. For a general $n \times n$ matrix, show that steps 1–4 of Gaussian elimination take at most $n(n-1)(2n-1)/6 + n(n-1)$ multiplications/divisions. You will need to know that

$$1 + 2 + 3 + \cdots + n = \frac{n(n+1)}{2},$$

$$1^2 + 2^2 + 3^2 + \cdots + n^2 = \frac{n(2n+1)(n+1)}{6}.$$

   b. Show also that the back-substitution part of Gaussian elimination takes $n(n+1)/2$ multiplications/divisions.

▶ c. Verify that Gauss–Jordan takes about 50% more operations than Gaussian elimination for the case of three equations. In this, add the number of adds, subtracts, multiplies, and divides.

▶ **20.** Suppose we want to solve the system $Az = b$, where the a_{ij}, z_i, and b_i are complex numbers.

   a. Show that this can be done using only real arithmetic. (*Hint: A* can be written as $B + Ci$.)

   b. If one solves the system in a computer using a language that permits complex numbers, compare the amount of storage space needed compared to the amount if done as in part (a).

21. a. Show that the system

$$\begin{bmatrix} 3 + i & 1 + 2i \\ -3i & 2 + i \end{bmatrix} \begin{bmatrix} z_1 \\ z_2 \end{bmatrix} = \begin{bmatrix} 6 + 2i \\ 1 - i \end{bmatrix}$$

   can be written as

$$\begin{bmatrix} 3 & 1 & -1 & -2 \\ 0 & 2 & 3 & -1 \\ 1 & 2 & 3 & 1 \\ -3 & 1 & 0 & 2 \end{bmatrix} \begin{bmatrix} x_1 \\ x_2 \\ y_1 \\ y_2 \end{bmatrix} = \begin{bmatrix} 6 \\ 1 \\ 2 \\ -1 \end{bmatrix}.$$

▶ b. Solve the system of part (a), then find z_1 and z_2.

Section 2.5

▶ **22.** Use Crout reduction to solve Exercise 10.

23. Use Crout reduction to solve Exercise 5.

24. Show where we can use double precision economically in Crout reduction when sums are being accumulated.

25. Suppose that we do not know all of the three right-hand sides of Exercise 18 in advance.

   a. Solve $Ax = b_1$ by Gaussian elimination, getting the LU decomposition. Then use the LU to solve with the other two right-hand sides.

   b. Repeat part (a), this time using Crout reduction.

26. a. Solve this tridiagonal system:

$$\begin{bmatrix} 4 & -1 & 0 & 0 & 0 \\ -1 & 4 & -1 & 0 & 0 \\ 0 & -1 & 4 & -1 & 0 \\ 0 & 0 & -1 & 4 & -1 \\ 0 & 0 & 0 & -1 & 4 \end{bmatrix} x = \begin{bmatrix} 100 \\ 200 \\ 200 \\ 200 \\ 100 \end{bmatrix}$$

by the special algorithm that stores the augmented matrix in a 5×4 array.

b. Compute the maximum number of multiplications/divisions and additions/subtractions needed in solving a general tridiagonal system in which pivoting is not done.

Section 2.6

27. Which of these matrices are singular?

a.
$$A = \begin{bmatrix} -2 & 3 & -2 \\ 1 & -4 & 7 \\ -1 & -6 & 17 \end{bmatrix}.$$

b.
$$B = \begin{bmatrix} 2 & 1 & 3 & 1 \\ 3 & 1 & 4 & 1 \\ 0 & -7 & -4 & -2 \\ -1 & 0 & 2 & -2 \end{bmatrix}.$$

c.
$$C = \begin{bmatrix} 2 & 1 & 3 & 1 \\ 1 & 3 & 4 & 1 \\ 0 & -7 & -4 & -2 \\ -1 & 0 & 2 & -1 \end{bmatrix}.$$

28. a. Find values of x and y that make A singular:
$$A = \begin{bmatrix} 5 & 3 & 1 \\ 3 & -1 & 2 \\ x & y & -1 \end{bmatrix}.$$

b. Find values for x and y that make A nonsingular.

29. a. Matrix A in Exercise 27 is singular. Do its rows form linearly independent vectors? Find the values for the a_i in Eq. (2.24).

▶ b. Repeat part (a) for the elements of A considered as column vectors.

30. Do these sets of equations have a solution? Find a solution if it exists.

a.
$$\begin{cases} 3x - 2y + z = 2, \\ x - 3y + z = 5, \\ x + y - z = -5, \\ 3x + z = 0. \end{cases}$$

▶ b.
$$\begin{bmatrix} 1 & 1 & 0 \\ 0 & 1 & 1 \\ 1 & 0 & 1 \\ 1 & 1 & 1 \end{bmatrix} x = \begin{bmatrix} 1 \\ -2 \\ 0 \\ 4 \end{bmatrix}.$$

c.
$$\begin{bmatrix} 2 & 1 & 3 \\ -1 & 0 & 2 \\ 6 & 2 & 2 \end{bmatrix} x = \begin{bmatrix} 1 \\ 0 \\ 0 \end{bmatrix}.$$

d.
$$\begin{bmatrix} 2 & 1 & 3 \\ -1 & 0 & 2 \\ 6 & 2 & 2 \end{bmatrix} x = \begin{bmatrix} 1 \\ 0 \\ 2 \end{bmatrix}.$$

▶ 31. The Hilbert matrix is a classic case of the pathological situation called "ill-conditioning." The 4×4 Hilbert matrix is

$$H_4 = \begin{bmatrix} 1 & \frac{1}{2} & \frac{1}{3} & \frac{1}{4} \\ \frac{1}{2} & \frac{1}{3} & \frac{1}{4} & \frac{1}{5} \\ \frac{1}{3} & \frac{1}{4} & \frac{1}{5} & \frac{1}{6} \\ \frac{1}{4} & \frac{1}{5} & \frac{1}{6} & \frac{1}{7} \end{bmatrix}.$$

For the system $Hx = b$, with $b^T = [25/12, 77/60, 57/60, 319/420]$, the exact solution is $x^T = [1, 1, 1, 1]$.

a. Show that the matrix is ill-conditioned by showing that it is nearly singular.

b. Using only three significant digits (chopped) in your arithmetic, find the solution to $Hx = b$. Explain why the answers are so poor.

c. Using only three significant digits, but rounding, again find the solution and compare it to that obtained in part (b).

Section 2.7

32. Find the determinant of the matrix
$$\begin{bmatrix} 2 & 5 & -1 \\ 1 & 6 & 4 \\ 7 & -4 & 2 \end{bmatrix}$$

by row operations to make it

a. upper-triangular.

b. lower-triangular.

33. Find the determinant of the matrix
$$\begin{bmatrix} 3 & -2 & 8 & 2 \\ -1 & 3 & 2 & -6 \\ -5 & -1 & 3 & -9 \\ 2 & 3 & -8 & 1 \end{bmatrix}.$$

34. Invert the coefficient matrix in Exercise 5, and then use the inverse to generate the solution.

35. If the constant vector in Exercise 5 is changed to one with components $[-6, -5, 1, 11]$, what is now the solution? Observe that the inverse obtained in Exercise 34 gives the answer readily.

36. Attempt to find the inverse of the coefficient matrix in Exercise 11. Note that a singular matrix has no inverse.

37. a. Find the determinant of the Hilbert matrix in Exercise 31. A small value of the determinant (when

the matrix has elements of the order of unity) indicates ill-conditioning.

▶ b. Find the inverse of the Hilbert matrix in Exercise 31. The inverse of an ill-conditioned matrix has some very large elements in comparison to the elements of the original matrix.

▶ **38.** Both Gaussian elimination and the Gauss–Jordan method can be adapted to invert a matrix. In Section 2.7, we say "it is more efficient to use the Gaussian elimination algorithm." Verify this for the specific case of a 3×3 matrix by counting arithmetic operations for each method.

Section 2.8

39. For the vectors in (a) and (b), evaluate the norms $\| * \|_p$, $p = 1, 2, \infty$. For the matrices in (c) and (d), evaluate the norms for $p = 1, f$, and ∞.

a. $x = [2.15, -3.1, 10.0, 2.2]$.
b. $y = [-4, -5, 0, 3, -7]$.
c.
$$A = \begin{bmatrix} -9 & 5 & -9 \\ -2 & 7 & 5 \\ 5 & 1 & 8 \end{bmatrix}.$$
d.
$$B = \begin{bmatrix} 8 & -2 & 1 \\ -2 & 2 & -1 \\ -2 & 4 & -3 \end{bmatrix}.$$
e. Find the norms of $B^2, A + B, AB$.
f. Does the triangle inequality of Eq. (2.25) hold for $A + B$, for $x + y$?

▶ **40.** Find the ∞-norm of the Hilbert matrix of Exercise 31.

41. Find the ∞-norm of the inverse of the Hilbert matrix of Exercise 31.

Section 2.9 *

42. Consider the system $Ax = b$, where
$$A = \begin{bmatrix} 3.01 & 6.03 & 1.99 \\ 1.27 & 4.16 & -1.23 \\ 0.987 & -4.81 & 9.34 \end{bmatrix}, \quad b = \begin{bmatrix} 1 \\ 1 \\ 1 \end{bmatrix}.$$

▶ a. Using double precision (or a calculator with 10 or more digits of accuracy), solve for x.
b. Solve the system using three-digit (chopped) arithmetic for each arithmetic operation; call this solution $\bar{x}$.

c. Compare x and $\bar{x}$, and compute $e = x - \bar{x}$. What is $\|e\|_2$?
d. Is the system ill-conditioned? What evidence is there to support your conclusion?

43. Repeat Exercise 42, but change the element a_{33} to -9.34.

▶ **44.** Suppose, in Exercise 42, that uncertainties of measurement give slight changes in some of the elements of A. Specifically, suppose a_{11} is 3.00 instead of 3.01 and a_{31} is 0.99 instead of 0.987. What change does this cause in the solution vector (using precise arithmetic)?

45. Compute the residuals for the imperfect solutions in 42(b) and 43(b). Use double precision in this computation.

46. What are the condition numbers of the coefficient matrices in Exercises 42 and 43? Use the 1-norms.

47. Verify Eq. (2.28), using the results in Exercises 42 and 43.

▶ **48.** Verify Eq. (2.30), using the results in Exercises 43, 45, and 46.

49. Verify Eq. (2.28) for the results of Exercise 44.

50. Apply iterative improvement to the imperfect solution of Exercise 42.

▶ **51.** Apply iterative improvement to the imperfect solution of Exercise 43.

Section 2.10

52. Solve Exercise 6 by the Jacobi method, beginning with the initial vector $(0, 0, 0)$. Compare the rate of convergence when Gauss–Seidel is used with the same starting vector.

53. Solve Exercise 26(a) by Gauss–Seidel iteration, beginning with approximate solution $[50, 100, 100, 100, 50]$.

▶ **54.** The pair of equations
$$x_1 - 2x_2 = -1,$$
$$3x_1 + x_2 = 4,$$
can be rearranged to give $x_1 = -1 + 2x_2, x_2 = 4 - 3x_1$. Apply the Jacobi method to this rearrangement, beginning with a vector very close to the solution $x^{(1)} =$

---

*In certain exercises (42, 43, 44, 50, 51), imperfect solutions will result because low-precision arithmetic is used when the condition number is large. This exaggerates the condition number problem.

$(1.01, 1.01)^T$, and observe divergence. Now apply Gauss–Seidel. Which method diverges more rapidly?

▶ **55.** Solve the system

$$9x + 4y + \ z = -17,$$
$$x - 2y - 6z = \ \ 14,$$
$$x + 6y \qquad = \ \ \ 4,$$

a. Using the Jacobi method.
b. Using the Gauss–Seidel method. How much faster is the convergence than in part (a)?

Section 2.11

▶ **56.** Beginning with $(0, 0, 0)$, use relaxation to solve the system

$$6x_1 - 3x_2 + \ x_3 = \ \ 11,$$
$$2x_1 + \ x_2 - 8x_3 = -15,$$
$$x_1 - 7x_2 + \ x_3 = \ \ 10.$$

57. Solve the system in Exercise 55 by relaxation.

58. Relaxation is expecially well suited to solve problems like Exercise 26(a). Solve it by relaxation, starting with the vector [50, 100, 100, 100, 50], which can be obtained by inspection.

Section 2.12

▶ **59.** Find the two intersections nearest the origin of the two curves $x^2 + x - y^2 = 1$ and $y - \sin x^2 = 0$.

60. Solve the system

$$x^2 + y^2 + z^2 = 9,$$
$$xyz = 1,$$
$$x + y - z^2 = 0,$$

by Newton's method to obtain the solution near (2.5, 0.2, 1.6).

▶ **61.** Solve by using Newton's method:

$$x^3 + 3y^2 = 21,$$
$$x^2 + 2y + 2 = \ \ 0.$$

Make sketches of the graphs to locate approximate values of the intersections.

62. Apply Eq. (2.42) to compute partials and solve this system by Newton's method:

$$xyz - x^2 + y^2 = 1.34,$$
$$xy - z^2 = 0.09,$$
$$e^x - e^y + z = 0.41.$$

There should be a solution near (1, 1, 1).

63. At the end of Section 2.12, it is suggested that it would be more efficient to avoid recomputing the partials at each step of Newton's method for a nonlinear system, doing it only after each nth step when there are n equations. Redo Exercises 59 and 62 using this modification. Compare the rate of convergence with that when the partials are recomputed at each step.

Section 2.13

64. Given matrix A, write the permutation matrix that does the following interchanges.

$$A = \begin{bmatrix} 4 & 7 & 3 & -2 \\ 0 & 0 & 2 & 1 \\ -6 & 1 & 1 & 0 \\ 1 & 0 & 1 & 8 \end{bmatrix}.$$

a. Row 3 with row 1
▶ b. Row 1 with row 4
▶ c. Column 2 with column 1
▶ d. Row 2 with row 4 and column 4 with column 2 simultaneously

65. Confirm that $P^{-1} = P$ for each transposition matrix of Exercise 64.

66. Confirm Eq. (2.43) by first computing $H = H_3 P_3 H_2 P_2 H_1 P_1$ and then multiplying this and A.

67. Repeat Exercise 66, this time using

$$H = H_3 H_2 H_1 P_3 P_2 P_1.$$

Section 2.14

(Note that for Exercises 68–76, the matrix A can indicate either the coefficient matrix or the augmented matrix.)

68. Use MATLAB or Maple to solve Exercise 10.

69. Use Maple or MATLAB to solve

a. Exercise 1.
b. Exercises 2(a) and 2(b).

70. Solve Exercises 3(a), (b), and (c) with MATLAB and/or Maple.

71. Use Maple to solve the system of Exercise 10,

a. through reduced row echelon, `rref(A)`.
b. using the inverse of A.
c. with `linsolve(A,b)`.

72. Use MATLAB to solve Exercise 4.

73. Use MATLAB to
   a. Solve the 4×4 Hilbert matrix of Exercise 31.
   b. Compute the condition number of that matrix.
   c. Compute the condition number of the 10×10 Hilbert matrix.

74. Solve Exercise 37 with Maple.

75. Plot the curves of the functions of Exercise 59 with Maple, then use `fsolve` to get the solution. Then use similar routines in MATLAB to do the same.

76. Solve Exercise 61 with Maple.

Section 2.15

77. Show that solving an $n \times n$ system by Gaussian elimination requires these numbers of steps:
Making upper-triangular:

   $(2n^3 + 3n^2 - 5n)/6$ multiplications/divisions

   $(n^3 - n)/3$ additions/subtractions

Back-substitution:

   $(n^2 + n)/2$ multiplications/divisions

   $(n^2 - n)/2$ additions/subtractions

For a total of

   $(n^3 + 3n^2 - n)/3$ multiplications/divisions

   $(2n^3 + 3n^2 - 5n)/6$ additions/subtractions

78. Find the equivalent number of operations (as in Exercise 77) for the Gauss–Jordan method.

79. Develop an algorithm for inverting an $n \times n$ nonsingular matrix by parallel processing using approximately n^2 processors.

80. The final algorithm developed in Section 2.15 used $n^2 + n$ processors. Show how this can further be improved so that only $(n + 1)(n - 1) = n^2 - 1$ processors are needed.

81. Develop an algorithm for solving a system of n linear equations by Jacobi iteration using n^2 processors.

Applied Problems and Projects

82. In considering the movement of space vehicles, it is frequently necessary to transform coordinate systems. The standard inertial coordinate system has the N-axis pointed north, the E-axis pointed east, and the D-axis pointed toward the center of the earth. A second system is the vehicle's local coordinate system (with the i-axis straight ahead of the vehicle, the j-axis to the right, and the k-axis downward). We can transform the vector whose local coordinates are (i, j, k) to the inertial system by multiplying by transformation matrices:

$$\begin{bmatrix} n \\ e \\ d \end{bmatrix} = \begin{bmatrix} \cos a & -\sin a & 0 \\ \sin a & \cos a & 0 \\ 0 & 0 & 1 \end{bmatrix} \begin{bmatrix} \cos b & 0 & \sin b \\ 0 & 1 & 0 \\ -\sin b & 0 & \cos b \end{bmatrix} \begin{bmatrix} 1 & 0 & 0 \\ 0 & \cos c & -\sin c \\ 0 & \sin c & \cos c \end{bmatrix} \begin{bmatrix} i \\ j \\ k \end{bmatrix}.$$

Transform the vector $[2.06, -2.44, -0.47]^T$ to the inertial system if $a = 27°, b = 5°, c = 72°$.

83. Exercise 31 shows the pattern for a 4×4 Hilbert matrix. The $n \times n$ Hilbert matrix can be defined more formally as:

$$H_n = (1/(i + j + 1)), \quad i, j = 0, 1, \ldots, n - 1.$$

   a. Show that the Hilbert matrices are derived from solving the following problem:
   Let $f(x)$ be a continuous function on $[0, 1]$. Find the polynomial $p_{n-1} = \Sigma a_i x_i, i = 0, 1, \ldots, n - 1$ of degree $n - 1$, so that the following is minimized:

   $$I(a_0, a_1, \ldots, a_{n-1}) = \int_0^1 (f(x) - a_0 - a_1 x - \ldots - a_{n-1} x^{n-1})^2 \, dx.$$

   b. Write out the 9×9 Hilbert matrix, H_9. What is its condition number?
   c. Solve $H_9 x = [1, 1, 1, 1, 1, 1, 1, 1, 1]^T$. Then increase the first component of the right-hand side by 1%. Which component of the solution vector is most changed?

84. Electrical engineers often must find the currents flowing and voltages existing in a complex resistor network. Here is a typical problem.

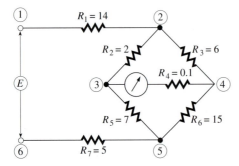

Figure 2.6

Seven resistors are connected as shown, and voltage is applied to the circuit at points 1 and 6 (see Fig. 2.6). You may recognize the network as a variation on a Wheatstone bridge.

Although we are especially interested in finding the current that flows through the ammeter, the computational method can give the voltages at each numbered point (these are called *nodes*) and the current through each of the branches of the circuit. Two laws are involved:

Kirchhoff's law: The sum of all currents flowing into a node is zero.

Ohm's law: The current through a resistor equals the voltage across it divided by its resistance.

We can set up eleven equations using these laws and from these solve for eleven unknown quantities (the four voltages and seven currents). If $V_1 = 5$ volts and $V_6 = 0$ volts, set up the eleven equations and solve to find the voltage at each other node and the currents flowing in each branch of the circuit.

85. Mass spectrometry analysis gives a series of peak height readings for various ion masses. For each peak, the height h_j is contributed to by the various constituents. These make different contributions c_{ij} per unit concentration p_i so that the relation

$$h_j = \sum_{i=1}^{n} c_{ij} p_i$$

holds, with n being the number of components present. Carnahan (1964) gives the values shown in Table 2.3 for c_{ij}.

Table 2.3

Peak number	Component				
	CH_4	C_2H_4	C_2H_6	C_3H_6	C_3H_8
1	0.165	0.202	0.317	0.234	0.182
2	27.7	0.862	0.062	0.073	0.131
3		22.35	13.05	4.420	6.001
4			11.28	0	1.110
5				9.850	1.684
6					15.94

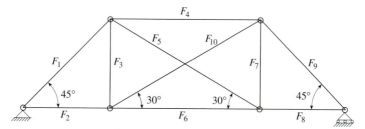

Figure 2.7

If a sample had measured peak heights of $h_1 = 5.20, h_2 = 61.7, h_3 = 149.2, h_4 = 79.4, h_5 = 89.3$, and $h_6 = 69.3$, calculate the values of p_i for each component. The total of all the p_i values was 21.53.

86. The truss in Section 2.1 is called *statically determinate* because nine linearly independent equations can be established to relate the nine unknown values of the tensions in the members. If an additional cross brace is added, as sketched in Fig. 2.7, we have ten unknowns but still only nine equations can be written; we now have a statically *indeterminate* system. Consideration of the stretching or compression of the members permits a solution, however. We need to solve a set of equations that gives the displacements x of each pin, which is of the form $ASA^Tx = P$. We then get the tensions f by matrix multiplication: $SA^Tx = f$. The necessary matrices and vectors are

$$A = \begin{bmatrix} 0.7071 & 0 & 0 & -1 & -0.8660 & 0 & 0 & 0 & 0 & 0 \\ 0.7071 & 0 & 1 & 0 & 0.5 & 0 & 0 & 0 & 0 & 0 \\ 0 & 1 & 0 & 0 & 0 & -1 & 0 & 0 & 0 & -0.8660 \\ 0 & 0 & -1 & 0 & 0 & 0 & 0 & 0 & 0 & -0.5 \\ 0 & 0 & 0 & 0 & 0 & 0 & 1 & 0 & 0.7071 & 0.5 \\ 0 & 0 & 0 & 1 & 0 & 0 & 0 & 0 & -0.7071 & 0.8660 \\ 0 & 0 & 0 & 0 & 0.8660 & 1 & 0 & -1 & 0 & 0 \\ 0 & 0 & 0 & 0 & -0.5 & 0 & -1 & 0 & 0 & 0 \\ 0 & 0 & 0 & 0 & 0 & 0 & 0 & 1 & 0.7071 & 0 \end{bmatrix}.$$

S is a diagonal matrix with values (from upper left to lower right) of

$$4255, \quad 6000, \quad 6000, \quad 3670, \quad 3000,$$
$$3670, \quad 6000, \quad 6000, \quad 4255, \quad 3000.$$

(These quantities are the values of aE/L, where a is the cross-sectional area of a member, E is the Young's modulus for the material, and L is the length.)

Solve the system of equations to determine the values of f for each of three loading vectors:

$$P_1 = [0, -1000, 0, 0, 500, 0, 0, -500, 0]^T,$$
$$P_2 = [1000, 0, 0, -500, 0, 1000, 0, -500, 0]^T,$$
$$P_3 = [0, 0, 0, -500, 0, 0, 0, -500, 0]^T.$$

87. For turbulent flow of fluids in an interconnected network (see Fig. 2.8), the flow rate V from one node to another is about proportional to the square root of the difference in pressures at the nodes. (Thus fluid flow differs from flow of electrical current in a network in that nonlinear equations result.) For the conduits in Fig. 2.8, find the pressure at each node. The values of b represent conductance factors in the relation $v_{ij} = b_{ij}(p_i - p_j)^{1/2}$.

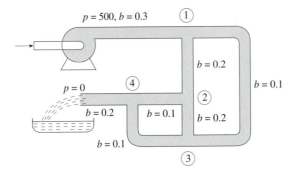

$p = 500, b = 0.3$ (1)

$b = 0.2$

$p = 0$ (4) $b = 0.1$

(2)

$b = 0.2$ $b = 0.1$ $b = 0.2$

$b = 0.1$

(3)

Figure 2.8

These equations can be set up for the pressures at each node:

$$\text{At node 1: } 0.3\sqrt{500 - p_1} = 0.2\sqrt{p_1 - p_2} + 0.1\sqrt{p_1 - p_3};$$
$$\text{node 2: } 0.2\sqrt{p_1 - p_2} = 0.1\sqrt{p_2 - p_4} + 0.2\sqrt{p_2 - p_3};$$
$$\text{node 3: } 0.1\sqrt{p_1 - p_3} + 0.2\sqrt{p_2 - p_3} = 0.1\sqrt{p_3 - p_4};$$
$$\text{node 4: } 0.1\sqrt{p_2 - p_4} + 0.1\sqrt{p_3 - p_4} = 0.2\sqrt{p_4 - 0}.$$

88. Translate each of the programs of this chapter to another computer language.

89. a. The elements in the matrix equation $Ax = B$ may be complex-valued. Write a program to do Gaussian elimination in some computer language that permits complex values and solve a few examples. How will you determine the proper pivot rows in this program? Does the coefficient matrix have an inverse? If so, multiply this inverse by the original coefficient matrix but, before doing this, try to predict the result.

b. It is not necessary to use complex arithmetic to solve a system that has complex-valued elements. How can this be done? Solve the examples that you used in part (a) in this way. You should get the same solutions; do you?

90. Rewrite the C program that solves a system by Gaussian elimination so that it includes scaling.

91. Suppose that matrix M is composed of four submatrices, A, B, C, and D, where A and D are square. These are arranged with A and B above C and D. Find the inverse of M in terms of A, B, C, and D. You will find that the inverse of A is involved, so A cannot be singular. What other conditions must hold for M to have an inverse?

92. a. The Hilbert matrix (see Exercise 31) is well known to be badly ill-conditioned. If the Hilbert matrix of order n has the solution $[1, 1, \ldots, 1]$, what must the right-hand-side vector be?

b. Use a program that solves linear systems using single-precision arithmetic and solve the $n \times n$ system of part (a) for increasing values of n. Record the magnitude of the largest error as a function of n. Is the element of the solution vector that is most in error always the same one?

93. Electrical circuits always have some capacitance and inductance in addition to resistance. Suppose that a 500 μF capacitor is added to the network of Exercise 84 between nodes 1 and 2 and a 4 mH inductance is added between nodes 5 and 6. Of course, if the voltage source E is a direct current source, no current will flow after the capacitance becomes saturated, but if E is an alternating voltage source, there will be continuous (though fluctuating) current in the network. Set up the equations that can be solved for the voltages at the nodes and the currents in each branch of the network. You may need to consult a reference to handle this mixture of resistors, capacitors, and inductance.

94. We have shown how a tridiagonal system is especially advantageous in that it can be solved with fewer arithmetic operations than a full $n \times n$ system. A banded matrix is similarly advantageous, and this is particularly true if the coefficient matrix is symmetric. What are the number of multiplies and divides for a symmetrical system of n equations that has m elements to the right and to the left of the diagonal? Your answer should be expressed in terms of n and m.

3

Interpolation and Curve Fitting

Contents of This Chapter

Interpolation, the computing of values for a tabulated function at points not in the table, is historically a most important task. Many famous mathematicians have their names associated with procedures for interpolation: Gauss, Newton, Bessel, Stirling. The need to interpolate began with the early studies of astronomy when the motion of heavenly bodies was to be determined from periodic observations.

Today, students rarely have to interpolate for values of sines, logarithms, and other such nonalgebraic functions from tables. Their calculators and computers compute the values using techniques that are described in the next chapter. Why then do we devote a lengthy chapter to a subject that seems almost obsolete? There are four reasons. First, interpolation methods are the basis for many other procedures that we will study: numerical differentiation and integration and solution methods for ordinary and partial-differential equations. Second, these methods demonstrate some important theory about polynomials and the accuracy of numerical methods. Third, interpolating with polynomials serves as an excellent introduction to some techniques for drawing smooth curves. And, finally, history itself has a special fascination. This chapter compares several ways of doing interpolation and contrasts these procedures with several ways for fitting imprecise data and for drawing smooth curves.

3.1 An Interpolation Problem

Describes a typical situation when a professional numerical analyst would need to do interpolation.

3.2 Lagrangian Polynomials

Presents the Lagrangian polynomial—a straightforward, but computationally awkward, way to construct an interpolating polynomial. The cost of determining the accuracy of the interpolated value is reduced by a variant, Neville's method.

3.3 Divided Differences

Describes these more efficient methods for constructing the interpolating polynomials and getting interpolated values. They readily allow changing the degree of the interpolating polynomial.

3.4 Interpolating with a Cubic Spline

Introduces spline curves, a newer and most important way to do interpolation. At the expense of added computations to find the polynomials, splines overcome some important problems that we find with ordinary interpolating polynomials.

3.5 Bezier Curves and B-Spline Curves

Describes modern techniques for constructing smooth curves that are widely used in computer graphics. They are not interpolating polynomials but are closely related to them.

3.6 Polynomial Approximation of Surfaces

Applies the techniques of the chapter to the more difficult problem of more than one independent variable. The section does not cover this topic in great detail.

3.7 Least-Squares Approximations

Considers the fitting of polynomials and other functions to data that are inexact. This classic problem must be faced when experimental data are interpreted.

3.8 Theoretical Matters

Presents the supporting theory behind many of the methods of the chapter.

3.9 Using MATLAB and *Mathematica*

Shows how these computer algebra systems can get interpolating polynomials and fit data by least squares. Both of these programs can fit a cubic spline to a set of data pairs, and *Mathematica* can compute the points on a Bezier curve.

Chapter Summary

Contains the usual review to check your understanding.

Computer Programs

Gives examples of programs that implement some of the methods of the chapter.

3.1 An Interpolation Problem

Rita Laski and Ed Baker were at lunch in the cafeteria of Ruscon Engineering. Both had just started summer jobs and were excited at the prospect of finding out how their college training really applied in industry. Rita was explaining her first assignment.

"They have huge amounts of data on the performance of that new rocket. Telemetry signals are received every 10 sec, giving the position of the rocket as well as other information. My boss has asked me to look into how we can determine the position at intermediate times. In essence, it's a kind of interpolation problem."

"I see," said Ed, "something like what we did when we studied log tables in algebra. There we calculated the logs for intermediate values by assuming that a straight line went between the two values from the table."

"Exactly," Rita replied, "except that it isn't appropriate to assume that the positions are linear with time. Typically, the points look something like this when they are plotted." She drew on a paper napkin to illustrate her point. "As you can see, sometimes a signal is missed, like on the third point."

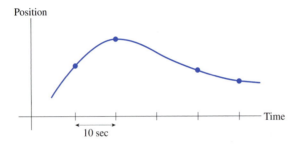

"You can see that a straight line is just impossible if we want to fit the data. And we can't just draw curves like I've done here because we want to get the intermediate values in the computer, and a computer can't look at a graph to find the values. What really bothers me is that Mr. Johnson, my boss, told me that we also must be able to get the intermediate values even when there is a maximum or minimum to the curve. Besides all this, we need a method of good efficiency because there are so many cases to handle."

In this chapter we will explore efficient techniques to interpolate, particularly for those situations where the data are far from linear. The principle that will be used is to fit a polynomial curve to the points. The problem is one of interest to many applications. Much of the development comes from work done by Newton and Kepler as they analyzed data on the positions of stars and planets.

In Chapters 1 and 2, we examined the question: "Given an explicit function of the independent variable x, what is the value of x corresponding to a certain value of the function?" In Chapter 1, x was a simple variable. In Chapter 2, x was a vector. We now want to consider a question somewhat the reverse of this: "Given values of an unknown function corresponding to certain values of x, what is the behavior of the function?" We would like to answer the question "What is the function?" but this is always impossible to determine with a limited amount of data.

Our purpose in determining the behavior of a function, as evidenced by the sample of data pairs $[x, f(x)]$, is severalfold. We will wish to approximate other values of the function at values of x not tabulated (interpolation or extrapolation) and to estimate the integral of $f(x)$ and its derivative. The latter objectives will lead us into ways of solving ordinary- and partial-differential equations.

The strategy we will use in approximating unknown values of the function is straight-forward. We will find a polynomial that fits a selected set of points $(x_i, f(x_i))$ and assume that the polynomial and the function behave nearly the same over the interval in question. Values of the polynomial then should be reasonable estimates of the values of the unknown function. When the polynomial is of the first degree, this leads to the familiar linear inter-polation. We will be interested in polynomials of degree higher than the first, so we can approximate functions that are far from linear, or so we can get good values from a table with wider spacing.

However, there are problems with interpolating polynomials when the data are not "smooth," meaning that there are local irregularities. In such cases, a polynomial of high degree would be required to follow the irregularities, but we will find that such polyno-mials, while fitting to the irregularity, deviate widely at other regions where the function is smooth. One solution is to fit subregions of the data with different polynomials, but this method too is problematic in that the joins of the different polynomials are not continuous in their slope. To remedy this problem, special types of polynomials, called splines, are useful.

The study of splines leads to some other special forms of polynomials (Bezier curves and B-spline curves) that do not interpolate (they do not pass exactly through all of the points) but are very useful for sketching smooth curves.

We do not always want to find a polynomial that fits exactly to the data. Often the values we wish to fit are not exact, or they may come from a set of experimental measurements that are subject to error. Fitting the approximating polynomial in this instance fits to the errors in the data and we do not want to do that.

A technique called *least squares* is normally used in such cases. Based on statistical theory, this method finds a polynomial (or some other kind of approximating function) that is more likely to approximate the true values. In this chapter we will cover this method as well as methods for fitting exactly to the points.

3.2 Lagrangian Polynomials

In this section and the next two we assume that the given data are exact and represent values of some unknown function. If we desire to find a polynomial that passes through the same points as our unknown function, we could set up a system of equations involving the co-efficients of the polynomial. For example, suppose we want to fit a cubic to these data:

x	$f(x)$
3.2	22.0
2.7	17.8
1.0	14.2
4.8	38.3
5.6	51.7

First, we need to select four points to determine our polynomial. (The maximum degree of the polynomial is always one less than the number of points.) Suppose we choose the first four points. If the cubic is $ax^3 + bx^2 + cx + d$, we can write four equations involving the unknown coefficients a, b, c, and d:

$$\text{when } x = 3.2: a(3.2)^3 + b(3.2)^2 + c(3.2) + d = 22.0,$$
$$\text{if } x = 2.7: a(2.7)^3 + b(2.7)^2 + c(2.7) + d = 17.8,$$
$$\text{if } x = 1.0: a(1.0)^3 + b(1.0)^2 + c(1.0) + d = 14.2,$$
$$\text{if } x = 4.8: a(4.8)^3 + b(4.8)^2 + c(4.8) + d = 38.3.$$

Solving these equations by the methods of the previous chapter gives us the polynomial. We can then estimate the values of the function at some value of x—say, $x = 3.0$—by substituting 3.0 for x in the polynomial.

For this example, the set of equations gives

$$a = -0.5275,$$
$$b = 6.4952,$$
$$c = -16.1177,$$
$$d = 24.3499,$$

and our polynomial is

$$-0.5275x^3 + 6.4952x^2 - 16.1177x + 24.3499.$$

At $x = 3.0$, the estimated value is 20.21.

We seek a better and simpler way of finding such interpolating polynomials. This procedure is awkward, especially if we want a new polynomial that is also made to fit at the point (5.6, 51.7), or if we want to see what difference it would make to use a quadratic instead of a cubic. Furthermore, this technique leads to an ill-conditioned system of equations.*

We will first look at one very straightforward approach—the Lagrangian polynomial. The Lagrangian polynomial is perhaps the simplest way to exhibit the existence of a polynomial for interpolation with unevenly spaced data. Data where the x-values are not equi-spaced often occur as the result of experimental observations or when historical data are examined.

Suppose we have a table of data with four pairs of x- and $f(x)$-values, with x_i indexed by variable i:

i	x	$f(x)$
0	x_0	f_0
1	x_1	f_1
2	x_2	f_2
3	x_3	f_3

*For this example, the system of four equations has a condition number of approximately 2700. Adding the last point to solve for the coefficients of a quartic polynomial causes the condition number to jump to near 100,000!

Here we do not assume uniform spacing between the x-values, nor do we need the x-values arranged in a particular order. The x-values must all be distinct, however. Through these four data pairs we can pass a cubic. The Lagrangian form for this is

$$P_3(x) = \frac{(x - x_1)(x - x_2)(x - x_3)}{(x_0 - x_1)(x_0 - x_2)(x_0 - x_3)} f_0 + \frac{(x - x_0)(x - x_2)(x - x_3)}{(x_1 - x_0)(x_1 - x_2)(x_1 - x_3)} f_1$$
$$+ \frac{(x - x_0)(x - x_1)(x - x_3)}{(x_2 - x_0)(x_2 - x_1)(x_2 - x_3)} f_2 + \frac{(x - x_0)(x - x_1)(x - x_2)}{(x_3 - x_0)(x_3 - x_1)(x_3 - x_2)} f_3$$

(3.1)

Note that this equation is made up of four terms, each of which is a cubic in x; hence the sum is a cubic. The pattern of each term is to form the numerator as a product of linear factors of the form $(x - x_i)$, omitting one x_i in each term, the omitted value being used to form the denominator by replacing x in each of the numerator factors. In each term, we multiply by the f_i corresponding to the x_i omitted in the numerator factors. The Lagrangian polynomial for other degrees of interpolating polynomials employs this same pattern of forming a sum of polynomials all of the desired degree; it will have $n + 1$ terms when the degree is n.

It is easy to see that the Lagrangian polynomial does in fact pass through each of the points used in its construction. For example, in the preceding equation for $P_3(x)$, let $x = x_2$. All terms but the third vanish because of a zero numerator, while the third term becomes just $(1) * f_2$. Hence $P_3(x_2) = f_2$. Similarly, $P_3(x_i) = f_i$ for $i = 0, 1, 3$.

EXAMPLE 3.1   Fit a cubic through the first four points of the preceding table and use it to find the interpolated value for $x = 3.0$.

$$P_3(3.0) = \frac{(3.0 - 2.7)(3.0 - 1.0)(3.0 - 4.8)}{(3.2 - 2.7)(3.2 - 1.0)(3.2 - 4.8)}(22.0)$$
$$+ \frac{(3.0 - 3.2)(3.0 - 1.0)(3.0 - 4.8)}{(2.7 - 3.2)(2.7 - 1.0)(2.7 - 4.8)}(17.8)$$
$$+ \frac{(3.0 - 3.2)(3.0 - 2.7)(3.0 - 4.8)}{(1.0 - 3.2)(1.0 - 2.7)(1.0 - 4.8)}(14.2)$$
$$+ \frac{(3.0 - 3.2)(3.0 - 2.7)(3.0 - 1.0)}{(4.8 - 3.2)(4.8 - 2.7)(4.8 - 1.0)}(38.3).$$

Carrying out the arithmetic, $P_3(3.0) = 20.21$.   ▲

Observe that we get the same result as before. The arithmetic in this method is tedious, although hand calculators are convenient for this type of computation. Writing a computer program that implements the method is not hard to do. Both MATLAB and *Mathematica* can get interpolating polynomials of any degree (but high degrees are usually undesirable).

An interpolating polynomial, although passing through the points used in its construction, does not, in general, give exactly correct values when used for interpolation. The reason is that the underlying relationship is often not a polynomial of the same degree. We are therefore interested in the error of interpolation.

As shown in a later section of this chapter, the error term is given by the expression

$$E(x) = (x - x_0)(x - x_1) \cdots (x - x_n)\frac{f^{(n+1)}(\xi)}{(n+1)!} \qquad \textbf{(3.2)}$$

with ξ in the smallest interval that contains $\{x, x_0, x_1, \ldots, x_n\}$.

The expression for error given in Eq. (3.2) is interesting but is not always extremely useful. This is because the actual function that generates the x_i, f_i values is often unknown; we obviously then do not know its $(n+1)$st derivative. We can conclude, however, that if the function is "smooth," a low-degree polynomial should work satisfactorily. (The smaller the higher derivatives of a function, the smoother it is. For example, for a straight line, all derivatives above the first are zero.) On the other hand, a "rough" function can be expected to have larger errors when interpolated. We can also conclude that extrapolation (applying the interpolating polynomial outside the range of x-values employed to construct it) will have larger errors than for interpolation. It also follows that the error is smaller if x is centered within the x_i, because this makes the product of the $(x - x_i)$ terms smaller.

Here is an algorithm for interpolation with a Lagrangian polynomial of degree N.

To interpolate for $f(x)$, given x and a set of $N + 1$ data pairs, (x_i, f_i), $i = 0, \ldots, N$:

Set SUM = 0.
DO FOR $I = 0$ to N
    SET $P = 1$.
    DO FOR $J = 0$ to N
        IF $J \neq I$
            SET $P = P * (x - x(J))/(x(I) - x(J))$.
    ENDDO (J).
    SET SUM = SUM + $P * f_i$
ENDDO (I).
SUM is the interpolated value.

Neville's Method

The trouble with the standard Lagrangian polynomial technique is that we do not know which degree of polynomial to use. If the degree is too low, the interpolating polynomial does not give good estimates of $f(x)$. If the degree is too high, undesirable oscillations in polynomial values can occur. (More on this later, in the section on spline curves.)

Neville's method can overcome this difficulty. It essentially computes the interpolated value with polynomials of successively higher degree, stopping when the successive values are close together.* The successive approximations are actually computed by linear interpolation from the intermediate values. The Lagrange formula for linear interpolation to get $f(x)$ from two data pairs, (x_1, f_1) and (x_2, f_2), is

$$f(x) = \frac{(x - x_2)}{(x_1 - x_2)} f_1 + \frac{(x - x_1)}{(x_2 - x_1)} f_2,$$

which can be written more compactly as

$$f(x) = \frac{(x - x_2) * f_1 + (x_1 - x) * f_2}{x_1 - x_2}. \tag{3.3}$$

We will use Eq. (3.3) in Neville's method.

If we examine Eq. (3.2) for the error term of Lagrangian interpolation, we see that the smallest error results when we use data pairs where the x_i's are closest to the x-value we are interpolating. Neville's method begins by arranging the given data pairs, (x_i, f_i), so the successive values are in order of the closeness of the x_i to x.

EXAMPLE 3.2 Suppose we are given these data:

x	$f(x)$
10.1	0.17537
22.2	0.37784
32.0	0.52992
41.6	0.66393
50.5	0.63608

and we want to interpolate for $x = 27.5$. We first rearrange the data pairs in order of closeness to $x = 27.5$:

i	$\lvert x - x_i \rvert$	x_i	$f_i = P_{i0}$
0	4.5	32.0	0.52992
1	5.3	22.2	0.37784
2	14.1	41.6	0.66393
3	17.4	10.1	0.17537
4	23.0	50.5	0.63608

Neville's method begins by renaming the f_i as P_{i0}. We build a table by first interpolating linearly between pairs of values for $i = 0, 1, i = 1, 2, i = 2, 3$, and so on. These values are written in a column to the right of the first P of each pair. The next column of the table is

---

*Neville's method is not the most efficient method to compute an interpolated value. It is better to obtain the interpolating polynomial by the procedures of the next section and then evaluate it for the desired x-value.

created by linearly interpolating from the previous column for $i = 0, 2, i = 1, 3, i = 2, 4$, and so on. The next column after this uses values for $i = 0, 3, i = 0, 4, \ldots$, and continues until we run out of data pairs.

Here is the Neville table for the preceding data:

i	x_i	P_{i0}	P_{i1}	P_{i2}	P_{i3}	P_{i4}
0	32.0	0.52992	0.46009	0.46200	0.46174	0.45754
1	22.2	0.37784	0.45600	0.46071	0.47901	
2	41.6	0.66393	0.44524	0.55843		
3	10.1	0.17537	0.37379			
4	50.5	0.63608				

The general formula for computing entries into the table is

$$P_{i,j} = \frac{(x - x_i) * P_{i+1,j-1} + (x_{i+j} - x) * P_{i,j-1}}{x_{i+j} - x_i}. \qquad (3.4)$$

Thus the values of P_{01} and P_{11} are computed by

$$P_{01} = \frac{(27.5 - 32.0) * 0.37784 + (22.2 - 27.5) * 0.52992}{22.2 - 32.0} = 0.46009,$$

$$P_{11} = \frac{(27.5 - 22.2) * 0.66393 + (41.6 - 27.5) * 0.37784}{41.6 - 22.2} = 0.45600.$$

Once we have the column of P_{i1}'s, we compute the next column. For example,

$$P_{22} = \frac{(27.5 - 41.6) * 0.37379 + (50.5 - 27.5) * 0.44524}{50.5 - 41.6} = 0.55843.$$

The remaining columns are computed similarly by using Eq. (3.4).

The top line of the table represents Lagrangian interpolates at $x = 27.5$ using polynomials of degree equal to the second subscript of the P's. Each of these polynomials uses the required number of data pairs, taking them as a set starting from the top of the table. (An exercise asks you to prove that the top line does represent Lagrangian interpolates with polynomials of increasing degree.)

The preceding data are for sines of angles in degrees and the correct value for $x = 27.5$ is 0.46175. Observe that the top line values get better and better until the last, when it diverges. This divergence becomes apparent when we notice that the successive values get closer to a constant value until the last one. (If the table is not arranged to center the x-value within the x_i, the convergence is not as quick.) ▲

If we instead do this computation by hand, we can save computing time by computing, not the entire table, but only as much as required to get convergence to the desired number of decimal places. We therefore do only the computations needed to compute those top row values that are required. In a computer program it is hardly worth the added programming complications because the entire table is computed so quickly.

Parallel Processing

The several terms of a Lagrange polynomial, as shown in Eq. (3.1), can all be computed simultaneously with parallel processing. Each entry in the successive columns of the table for Neville's method can be computed simultaneously. An exercise asks you to determine the number of time steps that are saved.

3.3 Divided Differences

There are two disadvantages to using the Lagrangian polynomial or Neville's method for interpolation. First, it involves more arithmetic operations than does the divided-difference method we now discuss. Second, and more importantly, if we desire to add or subtract a point from the set used to construct the polynomial, we essentially have to start over in the computations. Both the Lagrangian polynomials and Neville's method also must repeat all of the arithmetic if we must interpolate at a new x-value. The divided-difference method avoids all of this computation.

Actually, we will not get a polynomial different from that obtained by Lagrange's technique. As we will show in a later section, every nth-degree polynomial that passes through the same $n + 1$ points is identical. Only the way that the polynomial is expressed is different.

Our treatment of divided-difference tables assumes that function, $f(x)$, is known at several values for x:

$$x_0 \quad f_0$$
$$x_1 \quad f_1$$
$$x_2 \quad f_2$$
$$x_3 \quad f_3$$

We do not assume that the x's are evenly spaced or even that the values are arranged in any particular order (but some ordering may be advantageous).

Consider the nth-degree polynomial written in a special way:

$$P_n(x) = a_0 + (x - x_0)a_1 + (x - x_0)(x - x_1)a_2 + \cdots$$
$$+ (x - x_0)(x - x_1) \cdots (x - x_{n-1})a_n. \tag{3.5}$$

If we chose the a_i so that $P_n(x) = f(x)$ at the $n + 1$ known points, (x_i, f_i), $i = 0, \ldots, n$, then $P_n(x)$ is an interpolating polynomial. We will show that the a_i are readily determined by using what are called the *divided differences of the tabulated values*.

A special standard notation for divided differences is

$$f[x_0, x_1] = \frac{f_1 - f_0}{x_1 - x_0}, \qquad \text{with } f[x_0] = f_0 = f(x_0),$$

called the *first divided difference between x_0 and x_1*. The function

$$f[x_1, x_2] = \frac{f_2 - f_1}{x_2 - x_1}$$

is the first divided difference between x_1 and x_2.

In general,

$$f[x_s, x_t] = \frac{f_t - f_s}{x_t - x_s}$$

is the first divided difference between x_s and x_t. (Observe that the order of the points is immaterial:

$$f[x_s, x_t] = \frac{f_t - f_s}{x_t - x_s} = \frac{f_s - f_t}{x_s - x_t} = f[x_t, x_s].)$$

Second- and higher-order differences are defined in terms of lower-order differences. For example,

$$f[x_0, x_1, x_2] = \frac{f[x_1, x_2] - f[x_0, x_1]}{x_2 - x_0};$$

$$f[x_0, x_1, \ldots, x_n] = \frac{f[x_1, x_2, \ldots, x_n] - f[x_0, x_1, \ldots, x_{n-1}]}{x_n - x_0}$$

The concept is even extended to a zero-order difference:

$$f[x_s] = f_s.$$

Using the standard notation, a divided-difference table is shown in symbolic form in Table 3.1. Table 3.2 shows specific numerical values. (These data are the same as in the table of Section 3.2.)

Table 3.1

x_i	f_i	$f[x_i, x_{i+1}]$	$f[x_i, x_{i+1}, x_{i+2}]$	$f[x_i, x_{i+1}, x_{i+2}, x_{i+3}]$
x_0	f_0	$f[x_0, x_1]$	$f[x_0, x_1, x_2]$	$f[x_0, x_1, x_2, x_3]$
x_1	f_1	$f[x_1, x_2]$	$f[x_1, x_2, x_3]$	$f[x_1, x_2, x_3, x_4]$
x_2	f_2	$f[x_2, x_3]$	$f[x_2, x_3, x_4]$	
x_3	f_3	$f[x_3, x_4]$		
x_4	f_4			

Table 3.2

x_i	f_i	$f[x_i, x_{i+1}]$	$f[x_i, \ldots, x_{i+2}]$	$f[x_i, \ldots, x_{i+3}]$	$f[x_i, \ldots, x_{i+4}]$
3.2	22.0	8.400	2.856	−0.528	0.256
2.7	17.8	2.118	2.012	0.0865	
1.0	14.2	6.342	2.263		
4.8	38.3	16.750			
5.6	51.7				

We are now ready to establish that the a_i of Eq. (3.5) are given by these divided differences. We write Eq. (3.5) with x set equal to $x_0, x_1, \ldots, x_n$ in succession, giving

$$x = x_0: \qquad P_n(x_0) = a_0,$$

$$x = x_1: \qquad P_n(x_1) = a_0 + (x_1 - x_0)a_1,$$

$$x = x_2: \qquad P_n(x_2) = a_0 + (x_2 - x_0)a_1 + (x_2 - x_0)(x_2 - x_1)a_2,$$

$$\vdots$$

$$x = x_n: \qquad P_n(x_n) = a_0 + (x_n - x_0)a_1 + (x_n - x_0)(x_n - x_1)a_2 + \cdots$$
$$+ (x_n - x_0) \ldots (x_n - x_{n-1})a_n.$$

If $P_n(x)$ is to be an interpolating polynomial, it must match the table for all $n + 1$ entries:

$$P_n(x_i) = f_i \qquad \text{for } i = 0, 1, 2, \ldots, n.$$

If the $P_n(x_i)$ in each equation is replaced by f_i, we get a triangular system, and each a_i can be computed in turn.

From the first equation,

$$a_0 = f_0 = f[x_0] \qquad \text{makes} \qquad P_n(x_0) = f_0.$$

If $a_1 = f[x_0, x_1]$, then

$$P_n(x_1) = f_0 + (x_1 - x_0)\frac{f_1 - f_0}{x_1 - x_0} = f_1.$$

If $a_2 = f[x_0, x_1, x_2]$, then

$$P_n(x_2) = f_0 + (x_2 - x_0)\frac{f_1 - f_0}{x_1 - x_0}$$

$$+ (x_2 - x_0)(x_2 - x_1)\frac{(f_2 - f_1)/(x_2 - x_1) - (f_1 - f_0)/(x_1 - x_0)}{x_2 - x_0}$$

$$= f_2.$$

One can show in similar fashion that each $P_n(x_i)$ will equal f_i if $a_i = f[x_0, x_1, \ldots, x_i]$.

The notation for divided difference, particularly for those of higher order, is pretty complicated. In many instances from now on we will abbreviate with a nonstandard notation:

Standard Notation	Our Abbreviation
$f[x_0]$	$f_0^{[0]}$
$f[x_0, x_1]$	$f_0^{[1]}$
$f[x_1, x_2]$	$f_1^{[1]}$
$f[x_0, x_1, x_2]$	$f_0^{[2]}$
$f[x_0, x_1, \ldots, x_n]$	$f_0^{[n]}$

Observe that the subscript in our nonstandard notation indicates the first x-value that is used and the superscript indicates the order. Using this abbreviated notation, the interpolating polynomial that fits a divided difference table at $x = x_0, x_1, x_2, \ldots, x_n$ is

$$P_n(x) = f_0^{[0]} + (x - x_0)f_0^{[1]} + (x - x_0)(x - x_1)f_0^{[2]}$$
$$+ (x - x_0)(x - x_1)(x - x_3)f_0^{[3]} + \cdots \qquad (3.6)$$
$$+ (x - x_0)(x - x_1) \cdots (x - x_{n-1})f_0^{[n]}.$$

EXAMPLE 3.3   Write the interpolating polynomial of degree 3 that fits the data of Table 3.1 at all points from $x_0 = 3.2$ to $x_3 = 4.8$.

$$P_3(x) = 22.0 + 8.400(x - 3.2) + 2.856(x - 3.2)(x - 2.7)$$
$$- 0.528(x - 3.2)(x - 2.7)(x - 1.0).$$

What is the fourth-degree polynomial that fits at all five points? We only have to add one more term to $P_3(x)$:

$$P_4(x) = P_3(x) + 0.256(x - 3.2)(x - 2.7)(x - 1.0)(x - 4.8).$$

When this method is used for interpolation, we observe that nested multiplication can be used to cut down on the number of arithmetic operations, for example, for $x = 3$:

$$P_3(3) = \{[-0.528(3 - 1.0) + 2.586](3 - 2.7) + 8.400\}(3 - 3.2) + 22.0. \quad \blacktriangle$$

If we compute the interpolated value at $x = 3.0$ for each of the third-degree polynomials in Sections 3.2 and 3.3, we get the same result: $P_3(3.0) = 20.21$. This is not surprising, because all third-degree polynomials that pass through the same four points are identical. They may look different but they can all be reduced to the same form. The section on theory proves this assertion.

An algorithm for constructing a divided difference table is

Given a set of $N + 1$ data pairs, (x_i, f_i), $i = 0, \ldots, n$,
  DO FOR $i = 1$ to n
    DO FOR $j = 0$ to $n - i$
      Compute $f_j^{[i]} = (f_{j+1}^{[i-1]} - f_j^{[i-1]})/(x_{j+i} - x_j)$
      and enter into column i of the table.
    ENDDO (j).
  ENDDO (i).

Observe that parallel processing can compute all entries in the successive columns simultaneously. If there are $N + 1$ data pairs and a full table is constructed, the number of time steps equals the number of new columns, N. Sequential processing would require $N(N + 1)/2$ steps.

Divided Difference for $f(x)$ a Polynomial

It is of interest to look at the divided differences for $f(x) = P_n(x)$. Suppose that $f(x)$ is the cubic

$$f(x) = 2x^3 - x^2 + x - 1.$$

Here is its divided-difference table:

x_i	$f(x_i)$	$f_i^{[1]}$	$f_i^{[2]}$	$f_i^{[3]}$	$f_i^{[4]}$	$f_i^{[5]}$
0.30	-0.7360	2.4800	3.0000	2.0000	0.0000	0.0000
1.00	1.0000	3.6800	3.6000	2.0000	0.0000	
0.70	-0.1040	2.2400	5.4000	2.0000		
0.60	-0.3280	8.7200	8.2000			
1.90	11.0080	21.0200				
2.10	15.2120					

Observe that the third divided differences are all the same. (It then follows that all higher divided differences will be zero.) We can take advantage of this fact by not using differences beyond the column where the values are essentially constant, because this indicates that the function behaves nearly like a polynomial of that degree.

It is most important to also observe that the third derivative of a cubic polynomial is also a constant. (In this instance, $P^{(3)}(x) = 2 * 3! = 12.$) The relationship between divided differences and derivatives will be explored in detail in Chapter 5. For now we just state that for an nth-degree polynomial, $P_n(x)$, whose highest power term has the coefficient a_n, the nth divided differences will always be equal to a_n. Because the nth derivative of this polynomial is equal to $a_n * n!$, the relationship between derivatives and divided differences seems to involve $n!$. We exploit this next.

Error of Interpolation

The error term for an interpolating polynomial derived from a divided-difference table is identical to that for the equivalent Lagrangian polynomial because, as we have already observed, all polynomials of degree n that match at $n + 1$ points are identical. That means that the error term associated with the nth-degree polynomial $P_n(x)$ of Eq. (3.5) is simply Eq. (3.2), which we repeat here:

$$E(x) = (x - x_0)(x - x_1) \cdots (x - x_n)\frac{f^{(n+1)}(\xi)}{(n + 1)!}.$$

It is still not convenient to use this error expression, because the derivative of f that appears is unknown. However, if $f(x)$ is almost the same as some polynomial of degree n (and we will know that this is true because the nth divided differences will be almost constant), interpolating with an nth-degree polynomial should be nearly exact. The reason is that the

$(n + 1)$st derivative of $f(x)$ will be nearly zero and the error of the nth-degree interpolating polynomial will be very small.

What if we use a lower-degree polynomial? The error should be larger. If $f(x)$ is a known function, we can use Eq. (3.2) to bound the error. Here is an example.

EXAMPLE 3.4   Here is a divided difference table for $f(x) = x^2 e^{-x/2}$:

x_i	$f(x_i)$	$f_i^{[1]}$	$f_i^{[2]}$	$f_i^{[3]}$	$f_i^{[4]}$
1.10	0.6981	0.8593	−0.1755	0.0032	0.0027
2.00	1.4715	0.4381	−0.1631	0.0191	
3.50	2.1287	−0.0511	−0.0657		
5.00	2.0521	−0.2877			
7.10	1.4480				

Find the error of the interpolates for $f(1.75)$ using polynomials of degrees one, two, and three.

The results are shown in Table 3.3, for which Eq. (3.6) was used to do the interpolations. (MATLAB helped in finding the derivatives and evaluating the maximum and minimum values within the intervals.) The error formula does bracket the actual errors, as expected. In this case, observe that the use of a cubic polynomial does not improve the accuracy. In part, this is because we do not have the x-value well centered within the tabulated values; also, the value of the derivative is not decreasing.*   ▲

Error Estimation When $f(x)$ Is Unknown— The Next-Term Rule

Occasionally, almost always when dealing with experimental data, the function is unknown. Still, there is a way to estimate the error of the interpolation. This is because the nth-order divided difference is itself an approximation for $f^{(n)}(x)/n!$, as will be demonstrated in Chapter 5. What this means is that the error of the interpolation is given approximately by the value of the next term that would be added!

Table 3.3   Errors of interpolation for $f(1.75)$

Degree	Interpolated value	Actual error	$f^{(n+1)}$ maximum	$f^{(n+1)}$ minimum	Upper bound	Lower bound
1	1.25668	0.01996	−0.3679	0.0594	0.0299	−0.00483
2	1.28520	−0.00856	−0.8661	0.1249	0.0059	−0.0408
3	1.28611	−0.00947	1.1398	−0.0359	0.0014	−0.0439

---

*In the section on theory we show why the error is reduced by centering the x-value. Briefly, this makes the product of the $(x - x_i)$ terms smaller.

This most valuable rule for estimating the error of interpolation we call the *next-term rule*. It is easy to state and to use:

$$E_n(x) = \text{(approximately) the value of the next term that would be added to } P_n(x).$$

Here is how it works for the preceding example:

Degree	Exact error	Estimate from next-term rule
1	0.01996	0.02852
2	0.00856	0.00091
3	−0.00947	−0.00249

As you can see, the agreement is at least fair.

Interpolation Near the End of a Table

Thus far, we have assumed that the entries are indexed from the top to the bottom of the table. This would appear to indicate that our formulas do not work well for constructing polynomials from divided differences at the end of the table. Remember, however, that the ordering of the points is immaterial. We can just as well begin at the bottom and number the entries going upward, with no adjustment of Eq. (3.6) required. The table is really not changed at all, just the symbols that we use. We now use Eq. (3.6) with the newly indexed values.

Tables 3.4(a) and (b) compare the two different numbering schemes. The entries in the rows of Table 3.4(b) are exactly the same numbers as in the upward diagonals of Table 3.4(a).

Table 3.4(a) Conventional divided-difference table

x_0	f_0	$f_0^{[1]}$	$f_0^{[2]}$	$f_0^{[3]}$	$f_0^{[4]}$
x_1	f_1	$f_1^{[1]}$	$f_1^{[2]}$	$f_1^{[3]}$	
x_2	f_2	$f_2^{[1]}$	$f_2^{[2]}$		
x_3	f_3	$f_3^{[1]}$			
x_4	f_4				

Table 3.4(b) Divided-difference table indexed upwardly

x_4	f_4				
x_3	f_3	$f_3^{[1]}$			
x_2	f_2	$f_2^{[1]}$	$f_2^{[2]}$		
x_1	f_1	$f_1^{[1]}$	$f_1^{[2]}$	$f_1^{[3]}$	
x_0	f_0	$f_0^{[1]}$	$f_0^{[2]}$	$f_0^{[3]}$	$f_0^{[4]}$

Evenly Spaced Data

If the x-values are evenly spaced, getting an interpolating polynomial is considerably simplified. Instead of using divided differences, "ordinary differences" are used; the differences in f-values are not divided by the differences in x-values. A delta symbol is used to write them and, for a table of $N + 1$ $(x, f(x))$ pairs, differences up the Nth order can be computed.

We suppose that the table has entries indexed from 0 to N. First-order differences are then written as Δf_i and are computed as $\Delta f_i = f_{i+1} - f_i$, $i = 0, \ldots, (N-1)$. Second-order differences, $\Delta^2 f_i$, are the differences of the first-order differences: $\Delta^2 f_i = \Delta(\Delta f_{i+1} - \Delta f_i)$, which is easily shown to be $\Delta^2 f_i = f_{i+2} - 2f_{i+1} + f_i$, $i = 0, \ldots, (N-2)$. Higher-order differences are again the differences of the next lower-order differences. They can be computed from the original f-values:

$$\Delta^n f_i = f_{i+n} - nf_{i+n-1} + \frac{n(n-1)}{2!}f_{i+n-2} - \ldots \pm f_i, \qquad i = 0, \ldots, (N-n).$$

Observe that the coefficients are the familiar binomial coefficients.

An interpolating polynomial of degree n can be written in terms of these ordinary differences, with x evaluated at x_s:

$$P_n(x_s) = f_0 + s\Delta f_0 + \frac{s(s-1)}{2!}\Delta^2 f_0 + \frac{s(s-1)(s-2)}{3!}\Delta^3 f_0 + \ldots$$

$$+ \frac{s(s-1)\ldots(s-n+1)}{n!}\Delta^n f_0,$$

where $s = (x - x_0)/h$, with $h = \Delta x$, the uniform spacing in x-values. Observe again that the coefficients are the familiar binomial coefficients.

This form of the interpolating polynomial is called the Newton–Gregory forward polynomial. We will use this type of interpolating polynomial several times in later chapters. Several other forms of interpolating polynomials can be written in terms of the differences of the table. We do not pursue this topic further because the divided difference formulas apply to evenly spaced data, although earlier editions of this text go into considerable detail.

The next-term rule applies to this Newton–Gregory polynomial: The error of interpolation is approximated by the next term that would be added. Here is an example:

Given this table of x, $f(x)$ values, and the columns of differences, find $f(0.73)$ from a cubic interpolating polynomial.

x	$f(x)$	Δf	$\Delta^2 f$	$\Delta^3 f$	$\Delta^4 f$
0.0	0.000	0.203	0.017	0.024	0.020
0.2	0.203	0.220	0.041	0.044	0.052
0.4	0.423	0.261	0.085	0.096	0.211
0.6	0.684	0.346	0.181	0.307	
0.8	1.030	0.527	0.488		
1.0	1.557	1.015			
1.2	2.572				

In order to center the x-values around $x = 0.73$, we must use the four entries beginning with $x = 0.4$. That makes $x_0 = 0.4$ and $s = (0.73 - 0.4)/0.2 = 1.65$. Inserting the proper values into the expression for the Newton–Gregory polynomial, we get

$$f(0.73) = 0.423 + (1.65)(0.261) + \frac{(1.65)(0.65)}{2!}(0.085) + \frac{(1.65)(0.65)(-0.35)}{3!}(0.096)$$

$$= 0.423 + 0.4306 + 0.0456 - 0.0060 = 0.893.$$

The function is actually for $f(x) = \tan(x)$, so we know that the true value of $f(0.73)$ is 0.895; the error is 0.002. The next-term rule estimates the error as 0.004. This estimate is not very good due to compensating errors from round-off.

One nice feature of a table of ordinary differences is that an error in an entry for $f(x)$ can be readily detected. Such an error causes a disruption to the regular progression of values in the columns of differences. For example, if the entry for $x = 0.6$ has two digits reversed (0.648 rather than 0.684) and the table is recomputed, the columns for $\Delta^2 f$ and $\Delta^3 f$ lose their regularity.

Differences Versus Divided Differences

Obviously, the table of function differences that we have been discussing is closely related to the table of divided differences. Except for dividing function differences by a difference of x-values in the latter, these two tables are the same when the x-values are evenly spaced. To make this crystal clear, compare the tables for the simple case of $f(x) = 2x^3$ with $h = 0.5$, as shown in Tables 3.5(a) and (b).

Table 3.5(a) Table of function differences for $f(x) = 2x^3$, $h = 0.5$

x_i	f_i	Δf_i	$\Delta^2 f_i$	$\Delta^3 f_i$	$\Delta^4 f_i$	$\Delta^5 f_i$
0.00	0.00	0.25	1.50	1.50	0.00	0.00
0.50	0.25	1.75	3.00	1.50	0.00	0.00
1.00	2.00	4.75	4.50	1.50	0.00	
1.50	6.75	9.25	6.00	1.50		
2.00	16.00	15.25	7.50			
2.50	31.25	22.75				
3.00	54.00					

Table 3.5(b) Table of divided differences for $f(x) = 2x^3$, $h = 0.5$

x_i	f_i	$f_i^{[1]}$	$f_i^{[2]}$	$f_i^{[3]}$	$f_i^{[4]}$	$f_i^{[5]}$
0.00	0.00	0.50	3.00	2.00	0.00	0.00
0.50	0.25	3.50	6.00	2.00	0.00	0.00
1.00	2.00	9.50	9.00	2.00	0.00	
1.50	6.75	18.50	12.00	2.00		
2.00	16.00	30.50	15.00			
2.50	31.25	45.50				
3.00	54.00					

As expected, the columns of third differences are constant in both tables. For divided differences, this constant is equal to just 2, the coefficient of x^3. For the difference table, it is equal to that coefficient times $(3!)(h^3)$, or $2 * 6 * 0.5^3 = 1.5$.

For first differences, the divided differences are equal to the function differences divided by h (0.5 here). Second divided differences are equal to second function differences divided by $(h)(2h)$ (0.5 in this example). Third divided differences are equal to third function differences divided by $(h)(2h)(3h)$ (0.75 in this instance). The pattern should now be clear:

$$f_i^{[n]} = \frac{\Delta^n f_i}{n! h^n}.$$

If the values are not evenly spaced, a comparison is impossible because the table of function differences is not defined.

This difference between the two kinds of tables has a great effect on the relation between differences and derivatives, a topic that we explore in Chapter 5.

3.4 Interpolating with a Cubic Spline

There is another way to fit a polynomial to a set of data. A cubic spline fits a "smooth curve" to the points, borrowing from the idea of a device used in drafting. This device is a flexible rod, bent to conform to the points (and usually held in place by weights). This device is better than a French curve, for how the French curve is placed to draw the curve is very subjective.

The spline curve can be of varying degrees. Suppose we have the following set of $n + 1$ data points * (which need not be evenly spaced):

$$(x_i, y_i), \quad i = 0, 1, 2, \ldots, n.$$

In general, we fit a set of nth-degree polynomials between each pair of adjacent points, $g_i(x)$, from x_i to x_{i+1}. If the degree of the spline is one (just straight lines between the points), the "curve" would appear as shown in the accompanying figure. The problem with this *linear spline* is that the slope is discontinuous at the points (knots). Splines of degree greater than one do not have this problem.

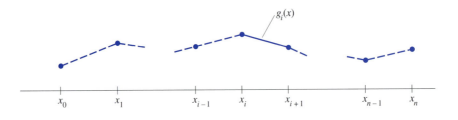

---

*The usual terminology for these data points is *knots*.

An interpolating polynomial that passes through all the points also is continuous in its slope. However, higher-degree interpolating polynomials have another problem. Here is an example of how bad the situation can be.

Let $f(x) = \cos^{10}(x)$ over the interval $[-2, 2]$. Fit polynomials of degrees 2, 4, 6, and 8 that match $f(x)$ at equispaced points on the interval. Figure 3.1 shows that none of the polynomials fit well to $f(x)$. The problem is that $f(x)$ is nearly flat except for the "bump" between about $x = -1$ and $x = 1$. The flat portions require $P_n(x)$ to have zeros outside of $[-1, 1]$, creating the oscillations. One remedy to the problem is to fit different polynomials to subregions of $f(x)$. Fitting first-degree polynomials is equivalent to a linear spline.

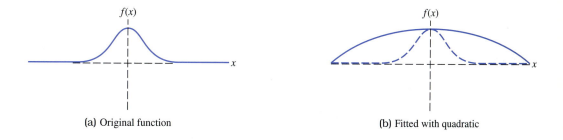

(a) Original function   (b) Fitted with quadratic

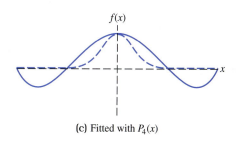

(c) Fitted with $P_4(x)$

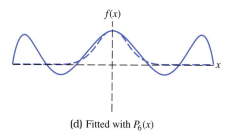

(d) Fitted with $P_6(x)$

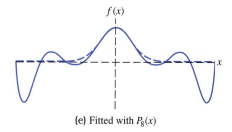

(e) Fitted with $P_8(x)$

Figure 3.1

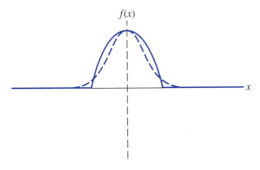

Figure 3.2

Figure 3.2 shows a much better fit using a quadratic between $x = -.65$ and $x = .65$ with $P(x) = 0$ outside this region. Although the fit is better (and we could improve it further), there are discontinuities in the slopes at the joins of the polynomials and $f(x)$ has no such breaks in its slope. Fitting a cubic or quartic polynomial to four or five points in the neighborhood of $x = 0$ would do a better job, but the slope would still be discontinuous at the ends. Spline curves of degree greater than one have a continuous slope.

As we have said, the concept of a "spline curve" derives from a device used by draftsmen to draw a smooth curve. The drafting spline bends according to the laws of beam flexure so that both the slope and curvature are continuous. Our mathematical spline curve must use polynomials of degree three (or more) to match this behavior. Although splines can be of any degree, *cubic splines* are by far the most popular; only these will be described here.

We will create a succession of cubic splines over successive intervals of the data. These polynomials will have the same slope and curvature at the points (knots) where they join. We do not require that the intervals be of the same width.

At the endpoints of the data set on which we fit $f(x)$ with spline curves, there is no "joining" polynomial. This means that the slope and curvature there are not constrained. This will be covered later in the development.

We write the equation for a cubic polynomial in the ith interval, $g_i(x_i)$, between the points (x_i, y_i) and (x_{i+1}, y_{i+1}). It looks like the solid curve shown here:

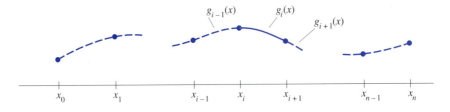

and has the equation

$$g_i(x) = a_i(x - x_i)^3 + b_i(x - x_i)^2 + c_i(x - x_i) + d_i. \qquad (3.7)$$

Thus the cubic spline function we want is of the form

$$g(x) = g_i(x) \text{ on the interval } [x_i, x_{i+1}], \qquad \text{for } i = 0, 1, \ldots, n - 1$$

and meets these conditions:

$$g_i(x_i) = y_i, \qquad i = 0, 1, \ldots, n - 1 \qquad \text{and} \qquad g_{n-1}(x_n) = y_n; \qquad \text{(3.8a)}$$

$$g_i(x_{i+1}) = g_{i+1}(x_{i+1}), \qquad i = 0, 1, \ldots, n - 2; \qquad \text{(3.8b)}$$

$$g_i'(x_{i+1}) = g_{i+1}'(x_{i+1}), \qquad i = 0, 1, \ldots, n - 2; \qquad \text{(3.8c)}$$

$$g_i''(x_{i+1}) = g_{i+1}''(x_{i+1}), \qquad i = 0, 1, \ldots, n - 2. \qquad \text{(3.8d)}$$

[Equations (3.8) say that the cubic spline fits to each of the points (3.8a), is continuous (3.8b), and is continuous in slope and curvature (3.8c) and (3.8d), throughout the region spanned by the points.]

If there are $n + 1$ points, the number of intervals and the number of $g_i(x)$'s are n. Thus there are four times n unknowns, which are the $\{a_i, b_i, c_i, d_i\}$ for $i = 0, 1, \ldots, n - 1$. Equation (3.8a) immediately gives

$$d_i = y_i, \qquad i = 0, 1, \ldots, n - 1. \qquad \text{(3.9)}$$

Equation (3.8b) then gives

$$y_{i+1} = a_i(x - x_{i+1})^3 + b_i(x - x_{i+1})^2 + c_i(x - x_{i+1}) + y_i$$
$$= a_i h_i^3 + b_i h_i^2 + c_i h_i + y_i, \qquad i = 0, 1, \ldots, n - 1. \qquad \text{(3.10)}$$

[In the last part of Eq. (3.10), we used $h_i = (x_{i+1} - x_i)$, the width of the ith interval.]

To relate the slopes and curvatures of the joining splines, we differentiate Eq. (3.7):

$$g_i'(x) = 3a_i h_i^2 + 2b_i h_i + c_i, \qquad \text{(3.11)}$$

$$g_i''(x) = 6a_i h_i + 2b_i, \qquad \text{for } i = 0, 1, \ldots, n - 1. \qquad \text{(3.12)}$$

Observe that the second derivative of a cubic is linear, so $g_i''(x)$ is linear within $[x_i, x_{i+1}]$. The development is simplified if we write the equations in terms of the second derivatives—that is, if we let $S_i = g_i''(x_i)$ for $i = 0, 1, \ldots, n - 1$ and $S_n = g_{n-1}''(x_n)$.

From Eq. (3.12), we have

$$S_i = 6a_i(x_i - x_i) + 2b_i$$
$$= 2b_i;$$
$$S_{i+1} = 6a_i(x_{i+1} - x_i) + 2b_i$$
$$= 6a_i h_i + 2b_i.$$

Hence we can write

$$b_i = \frac{S_i}{2}, \qquad \text{(3.13)}$$

$$a_i = \frac{S_{i+1} - S_i}{6h_i}. \qquad \text{(3.14)}$$

We substitute the relations for a_i, b_i, d_i given by Eqs. (3.9), (3.13), and (3.14) into Eq. (3.7) and then solve for c_i:

$$y_{i+1} = \left(\frac{S_{i+1} - S_i}{6h_i}\right)h_i^3 + \frac{S_i}{2}h_i^2 + c_i h_i + y_i;$$

$$c_i = \frac{y_{i+1} - y_i}{h_i} - \frac{2h_i S_i + h_i S_{i+1}}{6}.$$

We now invoke the condition that the slopes of the two cubics that join at (x_i, y_i) are the same. For the equation in the ith interval, Eq. (3.8c) becomes, with $x = x_i$,

$$y_i' = 3a_i(x_i - x_i)^2 + 2b_i(x_i - x_i) + c_i = c_i.$$

In the previous interval, from x_{i-1} to x_i, the slope at its right end will be

$$y_i' = 3a_{i-1}(x_i - x_{i-1})^2 + 2b_{i-1}(x_i - x_{i-1}) + c_{i-1}$$
$$= 3a_{i-1}h_{i-1}^2 + 2b_{i-1}h_{i-1} + c_{i-1}.$$

Equating these, and substituting for a, b, c, d their relationships in terms of S and y, we get

$$y_i' = \frac{y_{i+1} - y_i}{h_i} - \frac{2h_i S_i + h_i S_{i+1}}{6}$$
$$= 3\left(\frac{S_i - S_{i-1}}{6h_{i-1}}\right)h_{i-1}^2 + 2\left(\frac{S_{i-1}}{2}\right)h_{i-1} + \frac{y_i - y_{i-1}}{h_{i-1}} - \frac{2h_{i-1}S_{i-1} + h_{i-1}S_i}{6}.$$

Simplifying this equation, we get

$$h_{i-1}S_{i-1} + (2h_{i-1} + 2h_i)S_i + h_i S_{i+1} = 6\left(\frac{y_{i+1} - y_i}{h_i} - \frac{y_i - y_{i-1}}{h_{i-1}}\right) \tag{3.15}$$
$$= 6(f[x_i, x_{i+1}] - f[x_{i-1}, x_i]).$$

The last part of Eq. (3.15) involves divided differences.

Equation (3.15) applies at each internal point, from $i = 1$ to $i = n - 1$, there being $n + 1$ total points. This gives $n - 1$ equations relating the $n + 1$ values of S_i. We get two additional equations involving S_0 and S_n when we specify conditions pertaining to the end intervals of the whole curve. To some extent, these end conditions are arbitrary. Four* alternative choices are often used: Observe that the fourth end condition is not a knot condition.

---

* A fifth condition is sometimes encountered—a function is periodic and the data cover a full period. In this case, $S_0 = S_n$ and the slopes are also the same at the first and last points.

1. Take $S_0 = 0$ and $S_n = 0$. This makes the end cubics approach linearity at their extremities. This condition, called a *natural spline*, matches precisely to the drafting device. This technique is used very frequently.
2. Another often used condition is to force the slopes at each end to assume specified values. When that information is not known, the slope might be estimated from the points. If $f'(x_0) = A$ and $f'(x_n) = B$, we use these relations (note that divided differences are employed):

$$\text{At left end:} \quad 2h_0 S_0 + h_0 S_1 = 6(f[x_0, x_1] - A).$$

$$\text{At right end:} \quad h_{n-1} S_{n-1} + 2h_{n-1} S_n = 6(B - f[x_{n-1}, x_n]).$$

3. Take $S_0 = S_1$, $S_n = S_{n-1}$. This is equivalent to assuming that the end cubics approach parabolas at their extremities.
4. Take S_0 as a linear extrapolation from S_1 and S_2, and S_n as a linear extrapolation from S_{n-1} and S_{n-2}. Only this condition gives cubic spline curves that match exactly to $f(x)$ when $f(x)$ is itself a cubic. For condition 4, we use these relations:

$$\text{At left end:} \quad \frac{S_1 - S_0}{h_0} = \frac{S_2 - S_1}{h_1}, \quad S_0 = \frac{(h_0 + h_1)S_1 - h_0 S_2}{h_1}.$$

$$\text{At right end:} \quad \frac{S_n - S_{n-1}}{h_{n-1}} = \frac{S_{n-1} - S_{n-2}}{h_{n-2}}, \quad \text{(3.16)}$$

$$S_n = \frac{(h_{n-2} + h_{n-1})S_{n-1} - h_{n-1} S_{n-2}}{h_{n-2}}.$$

Relation 1, where $S_0 = 0$ and $S_n = 0$, is called a *natural spline*. It is often felt that this flattens the curve too much at the ends; in spite of this, it is frequently used. Relation 4 frequently suffers from the other extreme, giving too much curvature in the end intervals. Probably the best end condition to use is condition 2, provided reasonable estimates of the derivative are available.

If we write the equation of $S_1, S_2, \ldots, S_{n-1}$ (Eq. (3.15)) in matrix form, we get

$$
\begin{bmatrix}
h_0 & 2(h_0 + h_1) & h_1 & & & & \\
 & h_1 & 2(h_1 + h_2) & h_2 & & & \\
 & & h_2 & 2(h_2 + h_3) & h_3 & & \\
 & & & & \ddots & & \\
 & & & & h_{n-2} & 2(h_{n-2} + h_{n-1}) & h_{n-1}
\end{bmatrix}
\begin{bmatrix}
S_0 \\ S_1 \\ S_2 \\ S_3 \\ \vdots \\ S_{n-1} \\ S_n
\end{bmatrix}
$$

$$
= 6
\begin{bmatrix}
f[x_1, x_2] - f[x_0, x_1] \\
f[x_2, x_3] - f[x_1, x_2] \\
f[x_3, x_4] - f[x_2, x_3] \\
\vdots \\
f[x_{n-1}, x_n] - f[x_{n-2}, x_{n-1}]
\end{bmatrix}.
$$

In this matrix array there are only $n - 1$ equations, but $n + 1$ unknowns. We can eliminate two unknowns (S_0 and S_n) using the relations that correspond to the end-condition assumptions. In the first three cases, this reduces the S vector to $n - 1$ elements, and the coefficient matrix becomes square, of size ($n - 1 \times n - 1$). Furthermore, the matrix is always tridiagonal (even in case 4), and hence is solved speedily and can be stored economically.

For each end condition, the coefficient matrices become

Condition 1 $S_0 = 0, S_n = 0$:

$$\begin{bmatrix} 2(h_0 + h_1) & h_1 & & & \\ h_1 & 2(h_1 + h_2) & h_2 & & \\ & h_2 & 2(h_2 + h_3) & h_3 & \\ & & & \ddots & \\ & & & h_{n-2} & 2(h_{n-2} + h_{n-1}) \end{bmatrix}$$

Condition 2 $f'(x_0) = A$ and $f'(x_n) = B$:

$$\begin{bmatrix} 2h_0 & h_0 & & & \\ h_0 & 2(h_0 + h_1) & h_1 & & \\ & h_1 & 2(h_1 + h_2) & h_2 & \\ & & & \ddots & \\ & & & h_{n-1} & 2h_{n-1} \end{bmatrix}$$

Condition 3 $S_0 = S_1, S_n = S_{n-1}$:

$$\begin{bmatrix} (3h_0 + 2h_1) & h_1 & & & \\ h_1 & 2(h_1 + h_2) & h_2 & & \\ & h_2 & 2(h_2 + h_3) & h_3 & \\ & & & \ddots & \\ & & & h_{n-2} & (2h_{n-2} + 3h_{n-1}) \end{bmatrix}$$

Condition 4 S_0 and S_n are linear extrapolations:

$$\begin{bmatrix} \dfrac{(h_0 + h_1)(h_0 + 2h_1)}{h_1} & \dfrac{h_1^2 - h_0^2}{h_1} & & & \\ h_1 & 2(h_1 + h_2) & h_2 & & \\ & h_2 & 2(h_2 + h_3) & & h_3 \\ & & & \ddots & \\ & & \dfrac{h_{n-2}^2 - h_{n-1}^2}{h_{n-2}} & \dfrac{(h_{n-1} + h_{n-2})(h_{n-1} + 2h_{n-2})}{h_{n-2}} \end{bmatrix}.$$

With condition 4, after solving the set of equations, we must compute S_0 and S_n using Eq. (3.16). For conditions 1, 2, and 3, no computations are needed. For each of the first three cases, the right-hand-side vector is the same; it is given in Eq. (3.15). If the data are evenly spaced, the matrices reduce to a simple form.

After the S_i values are obtained, we get the coefficients a_i, b_i, c_i, and d_i for the cubics in each interval. From these we can compute points on the interpolating curve.

$$a_i = \frac{S_{i+1} - S_i}{6h_i};$$

$$b_i = \frac{S_i}{2};$$

$$c_i = \frac{y_{i+1} - y_i}{h_i} - \frac{2h_i S_i + h_i S_{i+1}}{6};$$

$$d_i = y_i.$$

EXAMPLE 3.5 Fit the data of Table 3.6 with a natural cubic spline curve, and evaluate the spline values $g(0.66)$ and $g(1.75)$. [The true relation is $f(x) = 2e^x - x^2$.] We see that $h_0 = 1.0$, $h_1 = 0.5$, and $h_2 = 0.75$. The divided differences that we can use to get the right-hand sides of our equations are $f[0, 1] = 2.4366$, $f[1, 1.5] = 4.5536$, and $f[1.5, 2.25] = 9.5995$.

For a natural cubic spline, we use end condition 1 and solve

$$\begin{bmatrix} 3.0 & 0.5 \\ 0.5 & 2.5 \end{bmatrix} \begin{bmatrix} S_1 \\ S_2 \end{bmatrix} = \begin{bmatrix} 12.7020 \\ 30.2754 \end{bmatrix},$$

giving $S_1 = 2.2920$ and $S_2 = 11.6518$. ($S_0 = S_3 = 0$, of course.) Using these S's, we compute the coefficients of the individual cubic splines to arrive at

i	Interval	$g_i(x)$
0	[0.0, 1.0]	$0.3820(x - 0)^3 + 0(x - 0)^2 + 2.0546(x - 0) + 2.0000$
1	[1.0, 1.5]	$3.1199(x - 1)^3 + 1.146(x - 1)^2 + 3.2005(x - 1) + 4.4366$
2	[1.5, 2.25]	$-2.5893(x - 1.5)^3 + 5.8259(x - 1.5)^2 + 6.6866(x - 1.5) + 6.7134$

Figure 3.3 shows the cubic spline curve. (You should verify that these equations satisfy all the conditions that were given for cubic spline curves.)

Table 3.6

x	$f(x)$
0.0	2.0000
1.0	4.4366
1.5	6.7134
2.25	13.9130

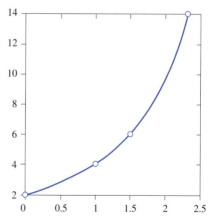

Figure 3.3

We use g_0 to find $g(0.66)$: It is 3.4659. (True = 3.4340)
We use g_2 to find $g(1.75)$: It is 8.7087. (True = 8.4467)

Some observations on this example: (a) We were given four points that define three intervals, (b) on each of the three intervals a $g(x)$ is defined, and (c) because each g has four coefficients, we must evaluate 12 unknown coefficients. However, by introducing the S's, we only had to solve two equations!   ▲

EXAMPLE 3.6   The data in the following table are from astronomical observations of a type of variable star called a *Cepheid variable* and represent variations in its apparent magnitude with time:

Time	0.0	0.2	0.3	0.4	0.5	0.6	0.7	0.8	1.0
Apparent magnitude	0.302	0.185	0.106	0.093	0.240	0.579	0.561	0.468	0.302

Use each of the four end conditions to compute cubic splines, and compare the values interpolated from each spline function at intervals of time of 0.05.

The augmented matrices whose solutions give values for $S_1, S_2, \ldots, S_7$ are shown in Table 3.7. A computer program was used to obtain the results shown in Table 3.8.

The results from the four end conditions between $x = 0.2$ and $x = 0.8$ are nearly identical; they differ by less than 0.001. Only in the end portions is there some difference.

Table 3.7

Condition 1

Matrix coefficients are

—	0.60	0.10	−1.23
0.10	0.40	0.10	3.96
0.10	0.40	0.10	9.60
0.10	0.40	0.10	11.52
0.10	0.40	0.10	−21.42
0.10	0.40	0.10	−4.50
0.10	0.60	—	0.60

Condition 2

Matrix coefficients are

—	0.40	0.10	0.00
0.20	0.60	0.10	−1.23
0.10	0.40	0.10	3.96
0.10	0.40	0.10	9.60
0.10	0.40	0.10	11.52
0.10	0.40	0.10	−21.42
0.10	0.40	0.10	−4.50
0.10	0.60	0.20	0.60
0.10	0.40	—	0.00

Table 3.7 *Continued*

	Condition 3		
Matrix coefficients are			
—	0.80	0.10	−1.23
0.10	0.40	0.10	3.96
0.10	0.40	0.10	9.60
0.10	0.40	0.10	11.52
0.10	0.40	0.10	−21.42
0.10	0.40	0.10	−4.50
0.10	0.80	—	0.60

	Condition 4		
Matrix coefficients are			
—	1.20	−0.30	−1.23
0.10	0.40	0.10	3.96
0.10	0.40	0.10	9.60
0.10	0.40	0.10	11.52
0.10	0.40	0.10	−21.42
0.10	0.40	0.10	−4.50
−0.30	1.20	—	0.60

Table 3.8

t	Values, condition 1	Values, condition 2*	Values, condition 3	Values, condition 4
0.00	**0.302**	**0.302**	**0.302**	**0.302**
0.05	0.278	0.276	0.282	0.297
0.10	0.252	0.250	0.256	0.271
0.15	0.222	0.221	0.224	0.231
0.20	**0.185**	**0.185**	**0.185**	**0.185**
0.25	0.143	0.143	0.142	0.141
0.30	**0.106**	**0.106**	**0.106**	**0.106**
0.35	0.087	0.087	0.088	0.088
0.40	**0.093**	**0.093**	**0.093**	**0.093**
0.45	0.133	0.133	0.133	0.133
0.50	**0.240**	**0.240**	**0.240**	**0.240**
0.55	0.424	0.424	0.424	0.424
0.60	**0.579**	**0.579**	**0.579**	**0.579**
0.65	0.608	0.608	0.608	0.608
0.70	**0.561**	**0.561**	**0.561**	**0.561**
0.75	0.511	0.511	0.511	0.511
0.80	**0.468**	**0.468**	**0.468**	**0.468**
0.85	0.426	0.426	0.426	0.430
0.90	0.385	0.385	0.384	0.392
0.95	0.343	0.343	0.343	0.350
1.00	**0.302**	**0.302**	**0.302**	**0.302**

*Note that in the values for condition 2 we used forward and backward differences to approximate the slope at either end of the curve; that is, $V'(0.0) = -0.585$ and $V'(1.0) = -0.830$.

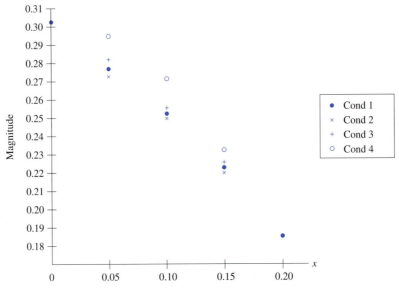

Figure 3.4

Figure 3.4 shows the points for the four conditions from $x = 0.0$ to $x = 0.2$. Condition 4 gives values that are significantly different from the others. ▲

EXAMPLE 3.7   Cubic splines that fit the function $f(x) = \cos^{10}(x)$ at $x = -2, -1, -0.5, 0, 0.5, 1,$ and 2 were constructed. (This is the same function that showed how poorly ordinary interpolating polynomials worked.) Figure 3.5 compares the cubic spline interpolates with the true function. The two plots are barely distinguishable. ▲

   The very powerful computer algebra system, *Mathematica,* can get both spline curves and the Bezier curves of the next section. MATLAB can interpolate with a spline fit to the data. We will explore these applications later in this chapter.

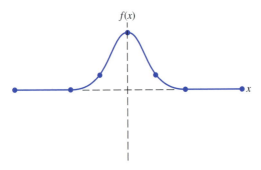

Figure 3.5

3.5 Bezier Curves and B-Spline Curves

In addition to the splines we have studied in the previous section, there are others that are important. In particular, Bezier curves and B-splines are widely used in computer graphics and computer-aided design. B-splines are often used to numerically integrate and differentiate functions that are defined only through a set of data points. These two types of curves are not really interpolating splines, because the curves do not normally pass through all of the points. In this respect they show some similarity to least-squares curves, which are discussed in a later section. However, both Bezier curves and B-splines have the important property of staying within the polygon determined by the given points. We will be more explicit about this property later. In addition, these two new spline curves have a nice geometric property in that in changing one of the points we change only one portion of the curve, a "local" effect. For the cubic spline curve of the previous section, changing just one point has a "global" effect in that the entire curve from the first to the last point is affected. Finally, for the cubic splines just studied, the points were given data points. For the two curves we study in this section the points in question are more likely "control" points that we select to determine the shape of the curve we are working on.

For simplicity, we consider mainly the cubic version of these two curves. In what follows, we will express $y = f(x)$ in parametric form. The parametric form represents a relation between x and y by two other equations, $x = F_1(u)$, $y = F_2(u)$. The independent variable u is called the *parameter*. For example, the equation for a circle can be written, with θ as the parameter, as

$$x = r \cos(\theta),$$
$$y = r \sin(\theta).$$

If we express x and y in terms of a parameter, u, the point (x, y) becomes $(x(u), y(u))$. We will use this with values of the parameter u between 0 and 1.

We discuss Bezier curves first. Bezier curves are named after the French engineer, P. Bezier of the Renault Automobile Company. He developed them in the early 1960s to fill a need for curves whose shape can be readily controlled by changing a few parameters. Bezier's application was to construct pleasing surfaces for car bodies.

Suppose we are given a set of control points, $p_i = (x_i, y_i)$, $i = 0, 1, \ldots, n$. (These points are also referred to as *Bezier points*.) Figure 3.6 is an example.

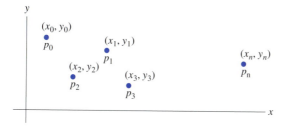

Figure 3.6

These points could be chosen on a computer screen, using a pointing device. The points do not necessarily progress from left to right. We treat the coordinates of each point as a two-component vector,

$$p_i = \begin{pmatrix} x_i \\ y_i \end{pmatrix}.$$

The set of points, in parametric form, is

$$P(u) = \begin{bmatrix} x(u) \\ y(u) \end{bmatrix}, \qquad 0 \le u \le 1.$$

The nth-degree Bezier polynomial determined by $n + 1$ points is given by

$$P(u) = \sum_{i=0}^{n} \binom{n}{i} (1 - u)^{n-i} u^i p_i,$$

where

$$\binom{n}{i} = \frac{n!}{i!(n - i)!}.$$

(The preceding formula really represents two other scalar equations, one for x_i and the other for y_i.) For $n = 2$, this would give the quadratic equation defined by three points, p_0, p_1, p_2:

$$P(u) = (1)(1 - u)^2 p_0 + 2(1 - u)(u)p_1 + (1)u^2 p_2,$$

because, for $n = 2$ and $i = 0, 1, 2$, we have $\binom{2}{0} = 1$, $\binom{2}{1} = 2$, $\binom{2}{2} = 1$. The preceding equation represents the pair of equations

$$x(u) = (1 - u)^2 x_0 + 2(1 - u)(u)x_1 + u^2 x_2,$$
$$y(u) = (1 - u)^2 y_0 + 2(1 - u)(u)y_1 + u^2 y_2.$$

Observe that, if $u = 0$, $x(0)$ is identical to x_0 and similarly for $y(0)$. If $u = 1$, the point referred to is (x_2, y_2). As u takes on values between 0 and 1, a curve is traced that goes from the first point to the third of the set. Ordinarily the curve will not pass through the central point of the three. (If the points are collinear, the curve is the straight line through them all.) In effect, the points of the second-degree Bezier curve have coordinates that are weighted sums of the coordinates of the three points that are used to define it. From another point of view, one can think of the Bezier equations as weighted sums of three polynomials in u, where the weighting factors are the coordinates of the three points.

Applying the general defining equation for $n = 3$, we get the cubic Bezier polynomial that we now consider in some detail. The properties of other Bezier polynomials are the same as for the cubic. Here is the Bezier cubic:

$$x(u) = (1 - u)^3 x_0 + 3(1 - u)^2 u x_1 + 3(1 - u)u^2 x_2 + u^3 x_3,$$
$$y(u) = (1 - u)^3 y_0 + 3(1 - u)^2 u y_1 + 3(1 - u)u^2 y_2 + u^3 y_3.$$

Observe again that $(x(0), y(0)) = p_0$ and $(x(1), y(1)) = p_3$, and that the curve will not ordinarily go through the intermediate points. As illustrated in the example curves in Fig. 3.7, changing the intermediate "control" points changes the shape of the curve. The examples are in Figs. 3.7(a) through 3.7(e). The first three of these show Bezier curves defined by one group of four points.

Figures 3.7(d) and 3.7(e) demonstrate how cubic Bezier curves can be continued beyond the first set of four points; one just subdivides seven points (p_0 to p_6) into two groups of four, with the central one (p_3) belonging to both sets. Figure 3.7(e) shows that p_2, p_3, and p_4 must be collinear to avoid a discontinuity in the slope at p_3.

It is of interest to list the properties of Bezier cubics:

1. $P(0) = p_0$, $P(1) = p_3$.
2. Because $dx/du = 3(x_1 - x_0)$ and $dy/du = 3(y_1 - y_0)$ at $u = 0$, the slope of the curve at $u = 0$ is $dy/dx = (y_1 - y_0)/(x_1 - x_0)$, which is the slope of the secant line between p_0 and p_1. Similarly, the slope at $u = 1$ is the same as the secant line between the last two points. This is indicated in the figures by dashed lines.
3. The Bezier curve is contained in the convex hull determined by the four points.

The *convex hull* of a set of points is the smallest convex set that contains the points. A set, *C*, is *convex* if and only if the line segment between any two points in the set lies entirely in set *C*. The following sketches show examples of the convex hull of four points.

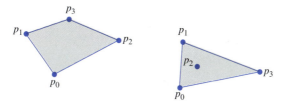

It is often convenient to represent the Bezier curve in matrix form. For Bezier cubics, this is

$$P(u) = [u^3, u^2, u, 1] \begin{bmatrix} -1 & 3 & -3 & 1 \\ 3 & -6 & 3 & 0 \\ -3 & 3 & 0 & 0 \\ 1 & 0 & 0 & 0 \end{bmatrix} \begin{bmatrix} p_0 \\ p_1 \\ p_2 \\ p_3 \end{bmatrix}$$

$$= u^T M_2 p.$$

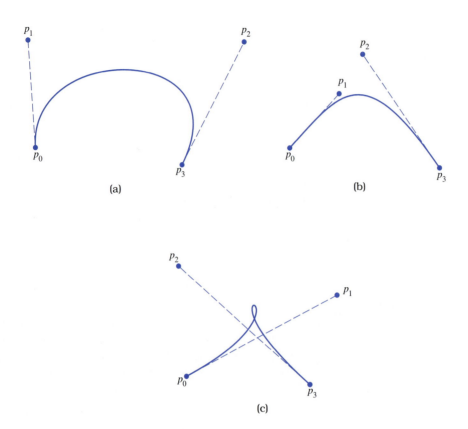

(a)

(b)

(c)

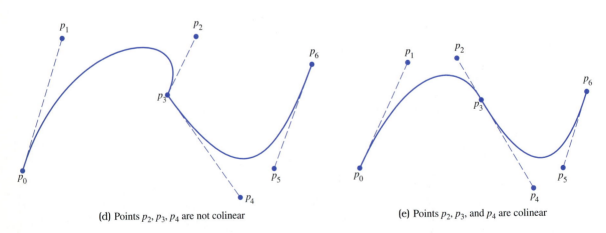

(d) Points p_2, p_3, p_4 are not colinear

(e) Points p_2, p_3, and p_4 are colinear

Figure 3.7

An algorithm for drawing a piece of a Bezier curve is

Given four points, (x_i, y_i), $i = 0, \ldots, 3$:

DO FOR $u = 0$ TO 1 STEP 0.01:
    Compute
$$x = (1 - u)^3 x_0 + 3(1 - u)^2 u x_1 + 3(1 - u)u^2 x_2 + u^3 x_3,$$
$$y = (1 - u)^3 y_0 + 3(1 - u)^2 u y_1 + 3(1 - u)u^2 y_2 + u^3 y_3.$$
    Plot (x, y).
ENDDO.

To continue the curve, repeat this process for the next set of four points, beginning with point p_3.

B-Spline Curves

We now discuss B-splines. These curves are like Bezier curves in that they do not ordinarily pass through the given data points. (The least-squares curves that are described in Section 3.7 are similar in this respect.) They can be of any degree, but we will concentrate on the cubic form. Cubic B-splines resemble the ordinary cubic splines of the previous section in that a separate cubic is derived for each pair of points in the set. However, the B-spline need not pass through any points of the set that are used in its definition.

We begin the description by stating the formula for a cubic B-spline in terms of parametric equations whose parameter is u.

Given the points $p_i = (x_i, y_i)$, $i = 0, 1, \ldots, n$, the cubic B-spline for the interval (p_i, p_{i+1}), $i = 1, 2, \ldots, n - 1$, is

$$B_i(u) = \sum_{k=-1}^{2} b_k p_{i+k}, \quad \text{where}$$

$$b_{-1} = \frac{(1 - u)^3}{6},$$

$$b_0 = \frac{u^3}{2} - u^2 + \frac{2}{3}, \tag{3.17}$$

$$b_1 = -\frac{u^3}{2} + \frac{u^2}{2} + \frac{u}{2} + \frac{1}{6},$$

$$b_2 = \frac{u^3}{6}, \quad 0 \le u \le 1.$$

As before, p_i refers to the point (x_i, y_i); it is a two-component vector. The coefficients, the b_k's, serve as a basis and do not change as we move from one set of points to the next. Observe that they can be considered weighting factors applied to the coordinates of a set of four points. The weighted sum, as u varies from 0 to 1, generates the B-spline curve.

If we write out the equations for x and y from Eq. (3.17), we get

$$x_i(u) = \frac{1}{6}(1-u)^3 x_{i-1} + \frac{1}{6}(3u^3 - 6u^2 + 4)x_i$$

$$+ \frac{1}{6}(-3u^3 + 3u^2 + 3u + 1)x_{i+1} + \frac{1}{6}u^3 x_{i+2};$$

$$y_i(u) = \frac{1}{6}(1-u)^3 y_{i-1} + \frac{1}{6}(3u^3 - 6u^2 + 4)y_i$$

$$+ \frac{1}{6}(-3u^3 + 3u^2 + 3u + 1)y_{i+1} + \frac{1}{6}u^3 y_{i+2}.$$

Note the notation here: $x_i(u)$ and $y_i(u)$ are functions (of u) and x_i, y_i are components of the point p. (The end portions are a special situation that we discuss later.)

As we have said, the u-cubics act as weighting factors on the coordinates of the four successive points to generate the curve. For example, at $u = 0$, the weights applied are 1/6, 2/3, 1/6, and 0. At $u = 1$, they are 0, 1/6, 2/3, and 1/6. These values vary throughout the interval from $u = 0$ to $u = 1$. As an exercise, you are asked to graph these factors. This will give you a visual impression of how the weights change with u.

Let us now examine two B-splines determined from a set of exactly four points. Figures 3.8(a) and 3.8(b) show the effect of varying just one of the points. As you would expect, when p_2 is moved upward and to the left, the curve tends to follow; in fact, it is pulled to the opposite side of p_1. You may be surprised to see that the curve is never very close to the two intermediate points, though it begins and ends at positions somewhat adjacent. It will be helpful to think of the curve generated from the defining equation for B_1 as associated with a curve that goes from near p_1 to p_2. It is also helpful to remember that points p_0, p_1, p_2, and p_3 are used to get B_1.

Because a set of four points is required to generate only a portion of the B-spline, that associated with the two inner points, we must consider how to get the B-spline for more than four points as well as how to extend the curve into the region outside of the middle pair. We use a method analogous to the cubic splines of Section 3.5 marching along one point at a time, forming new sets of four. We abandon the first of the old set when we add the new one.

(a)   (b)

Figure 3.8

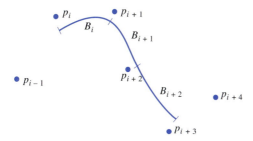

Figure 3.9 Successive B-splines joined together

The conditions that we want to impose on the B-spline are exactly the same as for ordinary splines: continuity of the curve and its first and second derivatives. It turns out that the equations for the weighting factors (the u-polynomials, the b_k) are such that these requirements are met. Figure 3.9 shows how three successive parts of a B-spline might look.

We can summarize the properties of B-splines as follows:

1. Like the cubic splines of Section 3.4, B-splines are pieced together so they agree at their joints in three ways:

a. $B_i(1) = B_{i+1}(0) = \dfrac{p_i + 4p_{i+1} + p_{i+2}}{6}$,

b. $B'_i(1) = B'_{i+1}(0) = \dfrac{-p_i + p_{i+2}}{2}$,

c. $B''_i(1) = B''_{i+1}(0) = p_i - 2p_{i+1} + p_{i+2}$.

The subscripts here refer to the portions of the curve and the points in Fig. 3.9.

2. The portion of the curve determined by each group of four points is within the convex hull of these points.

Now we consider how to generate the ends of the joined B-spline. If we have points from p_0 to p_n, we already can construct B-splines B_1 through B_{n-2}. We need B_0 and B_{n-1}. Our problem is that, using the procedure already defined, we would need additional points outside the domain of the given points. We probably also want to tie down the curve in some way—having it start and end at the extreme points of the given set seems like a good idea. How can we do this?

First, we can add more points without creating artificiality by making the added points coincide with the given extreme points. If we add not just a single fictitious point at each end of the set, but two at each end, we will find that the new curves not only join properly with the portions already made, but start and end at the extreme points as we wanted. (It looks like we have added two extra portions, but reflection shows these are degenerate, giving only a single point.)

In summary: We add fictitious points p_{-2}, p_{-1}, p_{n+1}, and p_{n+2}, with the first two identical with p_0 and the last two identical with p_n. (There are other methods to handle the starting and ending segments of B-splines that we do not cover.)

The matrix formulation for cubic B-splines is helpful. Here it is:

$$B_i(u) = \frac{1}{6}[u^3, u^2, u, 1]\begin{bmatrix} -1 & 3 & -3 & 1 \\ 3 & -6 & 3 & 0 \\ -3 & 0 & 3 & 0 \\ 1 & 4 & 1 & 0 \end{bmatrix}\begin{bmatrix} p_{i-1} \\ p_i \\ p_{i+1} \\ p_{i+2} \end{bmatrix}$$

$$= \frac{u^T M_b p}{6}.$$

(3.18)

This applies on the interval [0, 1] and for the points (p_i, p_{i+1}).

B-splines differ from Bezier curves in three ways:

1. For a B-spline, the curve does not begin and end at the extreme points.
2. The slopes of the B-splines do not have any simple relationship to lines drawn between the points.
3. The endpoints of the B-splines are in the vicinity of the two intermediate given points, but neither the x- nor the y-coordinates of these endpoints normally equal the coordinates of the intermediate points.

An algorithm for drawing a B-spline curve is as follows:

Given $n + 1$ points, $p_i = (x_i, y_i)$, $i = 0, \ldots, n$:
    Set $p_{-1} = p_{-2} = p_0$.
    Set $p_{n+1} = p_{n+2} = p_n$.
    DO FOR $i = 0$ to $n - 1$:
        DO FOR $u = 0$ TO 1 STEP 0.01:
        Compute
        $x = (1 - u)^3/6x_{i-1} + (3u^3 - 6u^2 + 4)/6x_i$
            $+ (-3u^3 + 3u^2 + 3u + 1)/6x_{i+1} + u^3/6x_{i+2}$,
        $y = (1 - u)^3/6y_{i-1} + (3u^3 - 6u^2 + 4)/6y_i$
            $+ (-3u^3 + 3u^2 + 3u + 1)/6y_{i+1} + u^3/6y_{i+2}$.
        Plot (x, y).
        ENDDO (u).
    ENDDO (i).

We conclude this section by looking at several examples of B-splines. The five parts of Fig. 3.10 show B-splines that are defined by the same sets of points as the Bezier curves in Fig. 3.7. (Fictitious points have been added to complete the end portions of these B-splines.) There are significant differences.

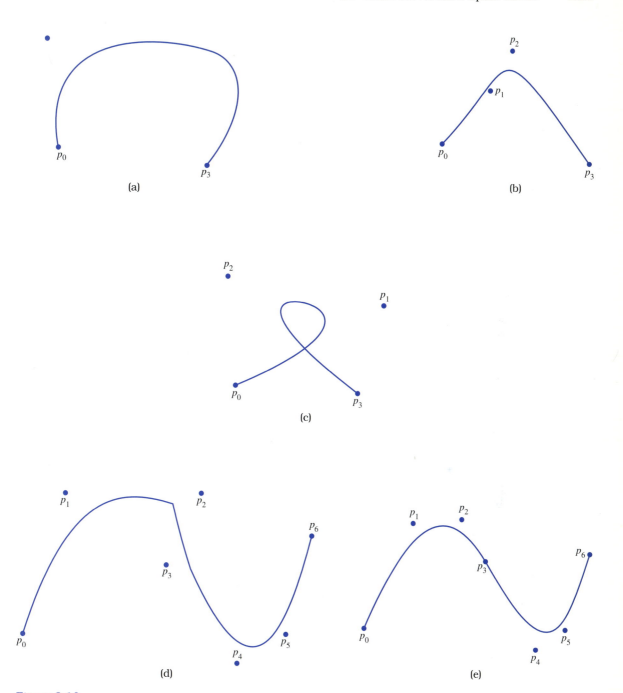

Figure 3.10

3.6 Polynomial Approximation of Surfaces

When a function z is a polynomial function* of two variables x and y—say, of degree 3 in x and of degree 2 in y—we would have

$$z = f(x, y) = a_0 + a_1x + a_2y + a_3x^2 + a_4xy + a_5y^2 + a_6x^3$$
$$+ a_7x^2y + a_8xy^2 + a_9x^3y + a_{10}x^2y^2 + a_{11}x^3y^2.$$

(3.19)

Such a function describes a surface; (x, y, z) is a point on it. The functional relation is seen to involve many terms. If we are concerned with four independent variables (three space dimensions plus time, say), even low-degree polynomials would be quite intractable. Except for special purposes, such as when we need an explicit representation, perhaps to permit ready differentiation at an arbitrary point, we can avoid such complications by handling each variable separately. We will treat only this case.

Note the immediate simplification of Eq. (3.19) if we let y take on a constant value, say, $y = c$. Combining the y factors with the coefficients, we get

$$z\big|_{y=c} = b_0 + b_1x + b_2x^2 + b_3x^3.$$

This will be our attack in interpolating at the point (a, b) in a table of two variables—hold one variable constant—say, $y = y_1$—and the table becomes a single-variable problem. The preceding methods then apply to give $f(a, y_1)$. If we repeat this at various values of y, $y = y_2, y_3, \ldots, y_n$, we will get a table with x constant at the value $x = a$ and with y varying. We then interpolate at $y = b$.

EXAMPLE 3.8   Estimate $f(1.6, 0.33)$ from the values in Table 3.9. Use quadratic interpolation in the x-direction and cubic interpolation for y. We select one of the variables to hold constant, say, x. (This choice is arbitrary because we would get the same result, except for differences due to round-off, if we had chosen to hold y constant.) We decide to interpolate for y within the three rows of the table at $x = 1.0$, 1.5, and 2.0, because the desired value at

Table 3.9   Tabulation of a function of two variables $z = f(x, y)$

x \ y	0.1	0.2	0.3	0.4	0.5	0.6
0.5	0.165	0.428	0.687	0.942	1.190	1.431
1.0	0.271	0.640	1.003	1.359	1.703	2.035
1.5	0.447	0.990	1.524	2.045	2.549	3.031
2.0	0.738	1.568	2.384	3.177	3.943	4.672
2.5	1.216	2.520	3.800	5.044	6.241	7.379
3.0	2.005	4.090	6.136	8.122	10.030	11.841
3.5	3.306	6.679	9.986	13.196	16.277	19.198

---

*We approximate a nonpolynomial function by a polynomial that agrees with the function, just as we have done with a function of one variable.

Table 3.10

	y	z	Δz	$\Delta^2 z$	$\Delta^3 z$
$x = 1.0$	0.2	0.640	0.363	−0.007	−0.005
	0.3	1.003	0.356	−0.012	
	0.4	1.359	0.344		
	0.5	1.703			
$x = 1.5$	0.2	0.990	0.534	−0.013	−0.004
	0.3	1.524	0.521	−0.017	
	0.4	2.045	0.504		
	0.5	2.549			
$x = 2.0$	0.2	1.568	0.816	−0.023	−0.004
	0.3	2.384	0.793	−0.027	
	0.4	3.177	0.766		
	0.5	3.943			

Table 3.11

	x	z	Δz	$\Delta^2 z$
$y = 0.33$	1.0	1.1108	0.5710	0.3717
	1.5	1.6818	0.9427	
	2.0	2.6245		

$x = 1.6$ is most nearly centered within this set. We choose y-values of 0.2, 0.3, 0.4, and 0.5 so that $y = 0.33$ is centralized. The shading in Table 3.9 shows the region of fit for our polynomials.

Because the x-values are evenly spaced, we elect to use Newton–Gregory forward polynomials. Table 3.10 shows the ordinary differences that we need.

We need the subtables from $y = 0.2$ to $y = 0.5$, because, for a cubic interpolation, four points are required. Using any convenient formula (remember that all cubics that agree at four points are identical), we get Table 3.11. In the last tabulation we carry one extra decimal to guard against round-off errors. Interpolating again, we get $z = 1.8406$, which we report as $z = 1.841$.   ▲

The function tabulated in Table 3.9 is $f(x, y) = e^x \sin y + y - 0.1$, so the true value is $f(1.6, 0.33) = 1.8350$. Our error of 0.0056 occurs because quadratic interpolation for x is inadequate in view of the large second difference. In retrospect, it would have been better to use quadratic interpolation for y, because the third differences of the y-subtables are small, and let x take on a third-degree relationship. (You may want to verify that this reduces the error to 0.0022.)

It is instructive to observe which of the values in Table 3.9 entered into our computation. The shaded rectangle covers these values. This is the "region of fit" for the interpolating polynomial that we have used. The principle of choosing values so that the point at which the interpolating polynomial is used is centered in the region of fit obviously applies here in exact analogy to the one-way table situation. It also applies to tables of three and four variables in the same way. Of course, the labor of interpolating in such multidimensional cases soon becomes burdensome.

A rectangular region of fit is not the only possibility. We may change the degree of interpolation as we subtabulate the different rows or columns. Intuitively, it would seem best to use higher-degree polynomials for the rows near the interpolating point, decreasing the degree as we get farther away. The coefficient of the error term, when this is done, will be found to be minimized thereby, though for multidimensional interpolating polynomials the error term is quite complex. The region of fit will be diamond-shaped when such tapered degree functions are used.

We may adapt the Lagrangian form of interpolating polynomial to the multidimensional case also. It is perhaps easiest to employ a process similar to the preceding example. Holding one variable constant, we write a series of Lagrangian polynomials for interpolation at the given value of the other variable, and then combine these values in a final Lagrange form. The net result is a Lagrangian polynomial in which the function factors are replaced by Lagrangian polynomials. The resulting expression for the previous example would be

$$
\begin{aligned}
&\frac{(y - 0.3)(y - 0.4)(y - 0.5)}{(0.2 - 0.3)(0.2 - 0.4)(0.2 - 0.5)} \\
&\times \left[\frac{(x - 1.5)(x - 2.0)}{(1.0 - 1.5)(1.0 - 2.0)}(0.640) + \frac{(x - 1.0)(x - 2.0)}{(1.5 - 1.0)(1.5 - 2.0)}(0.990) + \frac{(x - 1.0)(x - 1.5)}{(2.0 - 1.0)(2.0 - 1.5)}(1.568) \right] \\
&+ \frac{(y - 0.2)(y - 0.4)(y - 0.5)}{(0.3 - 0.2)(0.3 - 0.4)(0.3 - 0.5)} \\
&\times \left[\frac{(x - 1.5)(x - 2.0)}{(1.0 - 1.5)(1.0 - 2.0)}(1.003) + \frac{(x - 1.0)(x - 2.0)}{(1.5 - 1.0)(1.5 - 2.0)}(1.534) + \frac{(x - 1.0)(x - 1.5)}{(2.0 - 1.0)(2.0 - 1.5)}(2.384) \right] \\
&+ \frac{(y - 0.2)(y - 0.3)(y - 0.5)}{(0.4 - 0.2)(0.4 - 0.3)(0.4 - 0.5)} \\
&\times \left[\frac{(x - 1.5)(x - 2.0)}{(1.0 - 1.5)(1.0 - 2.0)}(1.359) + \frac{(x - 1.0)(x - 2.0)}{(1.5 - 1.0)(1.5 - 2.0)}(2.045) + \frac{(x - 1.0)(x - 1.5)}{(2.0 - 1.0)(2.0 - 1.5)}(3.177) \right] \\
&+ \frac{(y - 0.2)(y - 0.3)(y - 0.4)}{(0.5 - 0.2)(0.5 - 0.3)(0.5 - 0.4)} \\
&\times \left[\frac{(x - 1.5)(x - 2.0)}{(1.0 - 1.5)(1.0 - 2.0)}(1.703) + \frac{(x - 1.0)(x - 2.0)}{(1.5 - 1.0)(1.5 - 2.0)}(2.549) + \frac{(x - 1.0)(x - 1.5)}{(2.0 - 1.0)(2.0 - 1.5)}(3.943) \right] .
\end{aligned}
\tag{3.20}
$$

The equation is easy to write, but its evaluation by hand is laborious. If one is writing a computer program for interpolation in such multivariate situations, the Lagrangian form is recommended. There is a special advantage in that equal spacing in the table is not required. The Lagrangian form is also perhaps the most straightforward way to write out the polynomial as an explicit function.

When the given points are not evenly spaced, Lagrangian polynomials or the method of divided differences should be used for interpolation. With the latter, exactly the same principle is involved: Hold one variable constant while subtables of divided differences are constructed, then combine the interpolated values from these subtables into a new table.

Parallel processing can save many time steps in the preceding computations. Each value in the column of differences of Tables 3.10 and 3.11 can be computed at the same time. (We must wait for the interpolations from Table 3.10 to be completed before we do Table 3.11, of course.) Every factor of Eq. (3.20) can be evaluated in parallel.

Using Cubic Splines, Bezier Surfaces, and B-Spline Surfaces

Another alternative is to use cubic splines for interpolation in multivariate cases. Here again it is perhaps best to hold one variable constant while constructing one-way splines, then combine the results from these in the second phase. The computational effort would be significant, however.

Interpolating for values of functions of two independent variables can also be thought of as constructing a surface that is defined by the given points. Rather than finding values on a surface that contains the given points, we can construct surfaces that are analogous to Bezier curves and B-spline curves where the surface does not normally contain the given points.

So far we have been able to interpolate on simple surfaces where we are given z as a function of x and y. Suppose now we are given a set of points, $p_i = \{(x_i, y_i, z_i), i = 0, \ldots, n\}$, and we wish to fit a surface to those points. This would be the case if we were trying to draw a mountain, an airplane, or a teapot. But first we consider the representation of more general surfaces. Let $p = (x, y, z)$ be any point on the surface. Then the coordinates of each point are represented as the equations

$$x = x(u, v),$$

$$y = y(u, v),$$

$$z = z(u, v),$$

where u, v are the independent variables that range over a given set of values and x, y, z are the dependent variables. This is a slight change of notation from the first part of this section.

An example of this would be the equations of a sphere of radius r about the origin: $(0, 0, 0)$. Here any point on the surface of the sphere is given by

$$x = r \cos(u)\sin(v),$$

$$y = r \sin(u)\sin(v),$$

$$z = r \cos(v),$$

where u ranges in value from 0 to 2π and v ranges from 0 to π. Figure 3.11 illustrates this.

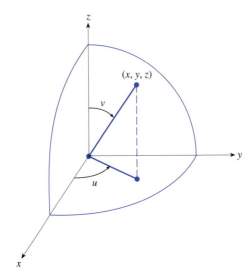

Figure 3.11

We will only describe constructing a B-spline surface. (A most interesting and informative description of Bezier surfaces can be found in Crow, 1987. See also Pokorny and Gerald, 1989.)

From the previous section, we know that a cubic B-spline curve segment starting near the point p_i to near the point p_{i+1} is determined by the four points

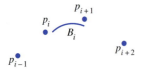

where $p_i(u) = (x_i(u), y_i(u))$ in two dimensions, or $p_i(u) = (x_i(u), y_i(u), z_i(u))$ if we had been working in three dimensions. The segment was then extended by introducing p_{i+3}, deleting p_{i-1}, and generating the curve for $0 \le u \le 1$.

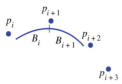

The process is continued until we have B_{n-2}. Finally, the first and last segments are generated by starting with p_0, p_0, p_0, p_1 and ending with p_{n-1}, p_n, p_n, p_n.

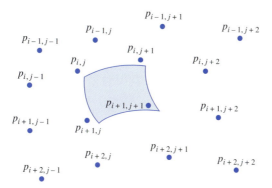

Figure 3.12

In an analogous manner the interpolating B-spline surface patch depends on 16 points, as Fig. 3.12 shows. Here $p_{i,j} = (x_{i,j}, y_{i,j}, z_{i,j})$, a point in E^3. This patch is generated by computing the points $p_{i,j}(u, v)$, for $0 \leq u \leq 1$ and $0 \leq v \leq 1$. Here we have changed the subscripts on the points $p_{i,j}$ so as to fit into matrix notation.

For simplicity, we will consider only the x-coordinate in detail. Comparable formulations hold for the y- and z-coordinates. The simplest formulation for $x_{ij}(u, v)$ is based on the matrix formulation of Eq. (3.18) and is given by

$$x_{ij}(u, v) = \frac{1}{36} [u^3, u^2, u, 1] M_b X_{i,j} M_b^T \begin{bmatrix} v^3 \\ v^2 \\ v \\ 1 \end{bmatrix}, \tag{3.21}$$

where $X_{i,j}$ is the 4 × 4 matrix

$$\begin{bmatrix} x_{i-1,j-1} & x_{i-1,j} & x_{i-1,j+1} & x_{i-1,j+2} \\ x_{i,j-1} & x_{i,j} & x_{i,j+1} & x_{i,j+2} \\ x_{i+1,j-1} & x_{i+1,j} & x_{i+1,j+1} & x_{i+1,j+2} \\ x_{i+2,j-1} & x_{i+2,j} & x_{i+2,j+1} & x_{i+2,j+2} \end{bmatrix},$$

which are just the x-coordinates of the 16 points of Fig. 3.12. The matrix M_b is the matrix we saw before in Eq. (3.18):

$$M_b = \begin{bmatrix} -1 & 3 & -3 & 1 \\ 3 & -6 & 3 & 0 \\ -3 & 0 & 3 & 0 \\ 1 & 4 & 1 & 0 \end{bmatrix}.$$

The y and z equations are then obtained merely by substituting the corresponding matrices $Y_{i,j}$ and $Z_{i,j}$, which are formed from the y and z components of the 16 points. Because each of these equations is cubic in u and v, they are referred to as *bicubic equations*. The coordinates of the points on a patch are given by

$$x(u, v) = \frac{1}{36} [u^3, u^2, u, 1] M_b X_{i,j} M_b^T [v^3, v^2, v, 1]^T,$$

$$y(u, v) = \frac{1}{36} [u^3, u^2, u, 1] M_b Y_{i,j} M_b^T [v^3, v^2, v, 1]^T,$$

$$z(u, v) = \frac{1}{36} [u^3, u^2, u, 1] M_b Z_{i,j} M_b^T [v^3, v^2, v, 1]^T,$$

as u and v range between 0 and 1. It is easily verified that the weights applied to each of the 16 points are

$$\begin{bmatrix} 1 & 4 & 1 & 0 \\ 4 & 16 & 4 & 0 \\ 1 & 4 & 1 & 0 \\ 0 & 0 & 0 & 0 \end{bmatrix} \quad \text{At } p_{i,j}(u, v) \text{ (for } u = 0, v = 0\text{), and}$$

$$\begin{bmatrix} 0 & 0 & 0 & 0 \\ 0 & 1 & 4 & 1 \\ 0 & 4 & 16 & 4 \\ 0 & 1 & 4 & 1 \end{bmatrix} \quad \text{At } p_{i,j}(u, v) \text{ (for } u = 1, v = 1\text{)}$$

where each (i, j)th element is the coefficient for the corresponding point in Fig. 3.12. In effect, these matrices are templates that overlay the points shown in Fig. 3.12.

The surface patch is extended by adding another row or column of points and deleting a corresponding row or column of points. One should verify that the current and previous patches are connected smoothly along the edge where they join. An initial or final patch can be obtained by repeating a corner, as was suggested for the B-spline curve. This will ensure that the patch actually starts or ends at a point. For the surface we would repeat a point nine times, instead of three times as was done for the curve.

For a more detailed and informative discussion of interpolating curves and surfaces, the reader should consult Pokorny and Gerald (1989).

3.7 Least-Squares Approximations

Suppose we wish to fit a curve to a set of approximate data, such as from the determination of the effects of temperature on a resistance by students in their physics laboratory. They have recorded the temperature and resistance measurements as shown in Fig. 3.13, where the graph suggests a linear relationship. We want to suitably determine the constants a and b in the equation relating resistance R and temperature T,

$$R = aT + b, \tag{3.22}$$

so that in subsequent use the resistance can be predicted at any temperature. The line as sketched by eye represents the data fairly well, but if we replotted the data and asked some-

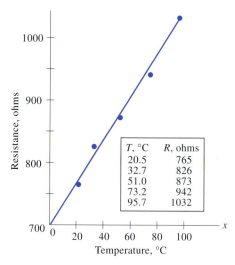

T, °C	R, ohms
20.5	765
32.7	826
51.0	873
73.2	942
95.7	1032

Figure 3.13

one else to draw a line, rarely would exactly the same line be obtained. One of our require-ments for fitting a curve to data is that the process be *unambiguous.* We would also like, in some sense, to minimize the deviations of the points from the line. The deviations are measured by the distances from the points to the line—how these distances are measured depends on whether both variables are subject to error. We will assume that the error of reading the temperatures in Fig. 3.13 is negligible, so that all the errors are in the resistance measurements, and we will use vertical distances. (If both were subject to error, we might use perpendicular distances and would modify the following. In this way the problem also becomes considerably more complicated. We will treat only the simpler case.)

We might first suppose we could minimize the deviations by making their sum a mini-mum, but this is not an adequate criterion. Consider the case of only two points (Fig. 3.14). Obviously, the best line passes through each point, but any line that passes through the midpoint of the segment connecting them has a sum of errors equal to zero.

Then what about making the sum of the magnitudes of the errors a minimum? This also is inadequate, as the case of three points shows (Fig. 3.15). Assume that two of the points are at the same x-value (which is not an abnormal situation as frequently experiments are duplicated). The best line will obviously pass through the average of the duplicated tests.

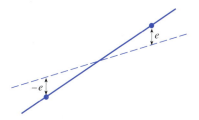

Figure 3.14

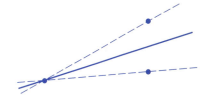

Figure 3.15

However, any line that falls between the dotted lines shown will have the same sum of the magnitudes of the vertical distances. We wish an unambiguous result, so we cannot use this as a basis for our work.

We might accept the criterion that we make the magnitude of the maximum error a minimum (the so-called *minimax* criterion), but for the problem at hand this is rarely done.* This criterion is awkward because the absolute-value function has no derivative at the origin, and it also is felt to give undue importance to a single large error. The usual criterion is to minimize the sum of the *squares* of the errors, the "least-squares" principle.†

In addition to giving a unique result for a given set of data, the least-squares method is also in accord with the *maximum-likelihood* principle of statistics. If the measurement errors have a so-called normal distribution and if the standard deviation is constant for all the data, the line determined by minimizing the sum of squares can be shown to have values of slope and intercept that have maximum likelihood of occurrence.

Let Y_i represent an experimental value, and let y_i be a value from the equation

$$y_i = ax_i + b,$$

where x_i is a particular value of the variable assumed to be free of error. We wish to determine the best values for a and b so that the y's predict the function values that correspond to x-values. Let $e_i = Y_i - y_i$. The least-squares criterion requires that

$$S = e_1^2 + e_2^2 + \cdots + e_N^2$$
$$= \sum_{i=1}^{N} e_i^2$$
$$= \sum_{i=1}^{N} (Y_i - ax_i - b)^2$$

be a minimum. N is the number of (x, Y)-pairs. We reach the minimum by proper choice of the parameters a and b, so they are the "variables" of the problem. At a minimum for S, the two partial derivatives $\partial S/\partial a$ and $\partial S/\partial b$ will both be zero. Hence, remembering that the x_i and Y_i are data points unaffected by our choice of values for a and b, we have

$$\frac{\partial S}{\partial a} = 0 = \sum_{i=1}^{N} 2(Y_i - ax_i - b)(-x_i),$$
$$\frac{\partial S}{\partial b} = 0 = \sum_{i=1}^{N} 2(Y_i - ax_i - b)(-1).$$

Dividing each of these equations by -2 and expanding the summation, we get the so-called *normal equations*

---

*We will use this criterion in the next chapter.
†The various criteria for a "best fit" can be described by minimizing a norm of the error vector. Relate each criterion to its corresponding vector norm to review the definition of such norms.

$$a \sum x_i^2 + b \sum x_i = \sum x_i Y_i,$$
$$a \sum x_i + bN = \sum Y_i.$$

(3.23)

All the summations in Eq. (3.23) are from $i = 1$ to $i = N$. Solving these equations simultaneously gives the values for slope and intercept a and b.

For the data in Fig. 3.13 we find that

$$N = 5, \qquad \sum T_i = 273.1, \qquad \sum T_i^2 = 18{,}607.27, \qquad \sum R_i = 4438,$$
$$\sum T_i R_i = 254{,}932.5.$$

Our normal equations are then

$$18{,}607.27a + 273.1b = 254{,}932.5,$$
$$273.1a + 5b = 4438.$$

From these we find $a = 3.395$, $b = 702.2$, and hence write Eq. (3.22) as

$$R = 702 + 3.39T.$$

Nonlinear Data

In many cases, of course, data from experimental tests are not linear, so we need to fit to them some function other than a first-degree polynomial. Popular forms that are tried are the exponential forms

$$y = ax^b$$

or

$$y = ae^{bx}.$$

We can develop normal equations for these analogously to the preceding development for a least-squares line by setting the partial derivatives equal to zero. Such nonlinear simultaneous equations are much more difficult to solve* than linear equations. Thus the exponential forms are usually linearized by taking logarithms before determining the parameters:

$$\ln y = \ln a + b \ln x$$

or

$$\ln y = \ln a + bx.$$

We now fit the new variable $z = \ln y$ as a linear function of $\ln x$ or x as described earlier. Here we do not minimize the sum of squares of the deviations of Y from the curve, but

---

*They were treated in Chapter 2.

rather the deviations of ln Y. In effect, this amounts to minimizing the squares of the percentage errors, which itself may be a desirable feature. An added advantage of the linearized forms is that plots of the data on either log-log or semilog graph paper show at a glance whether these forms are suitable by whether a straight line represents the data when so plotted.

In cases when such linearization of the function is not desirable, or when no method of linearization can be discovered, graphical methods are frequently used; one merely plots the experimental values and sketches in a curve that seems to fit well. Special forms of graph paper, in addition to log-log and semilog, may be useful (probability, log-probability, and so on). Transformation of the variables to give near linearity, such as by plotting against $1/x$, $1/(ax + b)$, $1/x^2$, and other polynomial forms of the argument may give curves with gentle enough changes in slope to allow a smooth curve to be drawn. S-shaped curves are not easy to linearize; the Gompertz relation

$$y = ab^{c^x}$$

is sometimes employed. The constants a, b, and c are determined by special procedures. Another relation that fits data to an S-shaped curve is

$$\frac{1}{y} = a + be^{-x}.$$

In awkward cases, subdividing the region of interest into subregions with a piecewise fit in the subregions can be used.

The objection to the graphical technique is its *lack of uniqueness*. Two individuals will usually not draw the same curve through the points. One's judgment is frequently distorted by one or two points that deviate widely from the remaining data. Often one tends to pay too much attention to the extremities in comparison to the points in the central parts of the region of interest.

Further problems are caused if we wish to integrate or differentiate the function. Our discussion of least-squares polynomials is one solution to these difficulties.

Least-Squares Polynomials

Because polynomials can be readily manipulated, fitting such functions to data that do not plot linearly is common. We now consider this case. It will turn out that the normal equations are linear for this situation, which is an added advantage. In the development, we use n as the degree of the polynomial and N as the number of data pairs. Obviously, if $N = n + 1$, the polynomial passes exactly through each point and the methods discussed earlier in this chapter apply, so we will always have $N > n + 1$ in the following.

We assume the functional relationship

$$y = a_0 + a_1x + a_2x^2 + \cdots + a_nx^n, \tag{3.24}$$

with errors defined by

$$e_i = Y_i - y_i = Y_i - a_0 - a_1x_i - a_2x_i^2 - \cdots - a_nx_i^n.$$

We again use Y_i to represent the observed or experimental value corresponding to x_i, with x_i free of error. We minimize the sum of squares,

$$S = \sum_{i=1}^{N} e_i^2 = \sum_{i=1}^{N} (y_i - a_0 - a_1 x_1 - a_2 x_i^2 - \cdots - a_n x_i^n)^2.$$

At the minimum, all the partial derivatives $\partial S/\partial a_0$, $\partial S/\partial a_1$, $\ldots$, $\partial S/\partial a_n$ vanish. Writing the equations for these gives $n + 1$ equations:

$$\frac{\partial S}{\partial a_0} = 0 = \sum_{i=1}^{N} 2(Y_i - a_0 - a_1 x_i - \cdots - a_i x_i^n)(-1),$$

$$\frac{\partial S}{\partial a_1} = 0 = \sum_{i=1}^{N} 2(Y_i - a_0 - a_1 x_i - \cdots - a_i x_i^n)(-x_i),$$

$$\vdots \qquad \vdots$$

$$\frac{\partial S}{\partial a_n} = 0 = \sum_{i=1}^{N} 2(Y_i - a_0 - a_1 x_i - \cdots - a_n x_i^n)(-x_i^n).$$

Dividing each by -2 and rearranging gives the $n + 1$ normal equations to be solved simultaneously:

$$
\begin{aligned}
a_0 N + a_1 \sum x_i + a_2 \sum x_i^2 + \cdots + a_n \sum x_i^n &= \sum Y_i, \\
a_0 \sum x_i + a_1 \sum x_i^2 + a_2 \sum x_i^3 + \cdots + a_n \sum x_i^{n+1} &= \sum x_i Y_i, \\
a_0 \sum x_i^2 + a_1 \sum x_i^3 + a_2 \sum x_i^4 + \cdots + a_n \sum x_i^{n+2} &= \sum x_i^2 Y_i, \quad \textbf{(3.25)} \\
\vdots \qquad\qquad & \qquad \vdots \\
a_0 \sum x_i^n + a_1 \sum x_i^{n+1} + a_2 \sum x_i^{n+2} + \cdots + a_n \sum x_i^{2n} &= \sum x_i^n Y_i.
\end{aligned}
$$

Putting these equations in matrix form shows an interesting pattern in the coefficient matrix.

$$
\begin{bmatrix}
N & \sum x_i & \sum x_i^2 & \sum x_i^3 & \cdots & \sum x_i^n \\
\sum x_i & \sum x_i^2 & \sum x_i^3 & \sum x_i^4 & \cdots & \sum x_i^{n+1} \\
\sum x_i^2 & \sum x_i^3 & \sum x_i^4 & \sum x_i^5 & \cdots & \sum x_i^{n+2} \\
& & & \vdots & & \\
\sum x_i^n & \sum x_i^{n+1} & \sum x_i^{n+2} & \sum x_i^{n+3} & \cdots & \sum x_i^{2n}
\end{bmatrix}
a =
\begin{bmatrix}
\sum Y_i \\
\sum x_i Y_i \\
\sum x_i^2 Y_i \\
\vdots \\
\sum x_i^n Y_i
\end{bmatrix}. \quad \textbf{(3.26)}
$$

All the summations in Eqs. (3.25) and (3.26) run from 1 to N.

Solving large sets of linear equations is not a simple task. Methods for this are the subject of Chapter 2. These particular equations have an added difficulty in that they have the undesirable property known as *ill-conditioning*. Its result is that round-off errors in solving them cause unusually large errors in the solutions, which of course are the desired values of the coefficients a_i in Eq. (3.24). Up to $n = 4$ or 5, the problem is not too great (that is, double-precision arithmetic in computer solutions is only desirable, not essential), but beyond this point special methods are needed. Such special methods use orthogonal polynomials in an equivalent form of Eq. (3.24). We will not pursue this matter further,* although we will treat one form of orthogonal polynomials in a later chapter in connection with representation of functions. From the point of view of the experimentalist, functions more complex than fourth-degree polynomials are rarely needed, and when they are, the problem can often be handled by fitting a series of polynomials to subsets of the data.

The matrix of Eq. (3.26) is called the *normal matrix* for the least-squares problem. There is another matrix that corresponds to this, called the *design matrix*. It is of the form

$$
A = \begin{bmatrix}
1 & 1 & 1 & \cdots & 1 \\
x_1 & x_2 & x_3 & \cdots & x_N \\
x_1^2 & x_2^2 & x_3^2 & \cdots & x_N^2 \\
\vdots & & & & \vdots \\
x_1^n & x_2^n & x_3^n & \cdots & x_N^n
\end{bmatrix}.
$$

It is easy to show that AA^T is just the coefficient matrix of Eq. (3.26). It is also easy to see that Ay, where y is the column vector of Y-values, gives the right-hand side of Eq. (3.26). (You ought to try this for, say, a 3×3 case to reassure yourself.) This means that we can rewrite Eq. (3.26) in matrix form, as

$$
AA^Ta = Ba = Ay.
$$

We can use Gaussian elimination to solve the system (but only for low-degree polynomials). However, because B has special properties, another method can be used that avoids the problem of ill-conditioning.

1. The matrix $B = AA^T$ is symmetric and positive definite. An $n \times n$ matrix, M, is said to be positive semidefinite if, for every n-component vector, $x^TMx \geq 0$. If we add the condition that $x^TMx = 0$ only if x is the zero vector, M is said to be positive definite. (You should show that B is positive definite and symmetric.)

2. In linear algebra, it is shown that B can be diagonalized by an orthogonal matrix P:

$$
PBP^T = PAA^TP^T = D,
$$

where the diagonal elements of D are the eigenvalues of B. Note that orthogonality implies that $PP^T = I$, the identity matrix.

3. B is positive definite, so all of its eigenvalues are nonnegative. This means that we can define a matrix S as

$$
S = \sqrt{D}, \quad \text{or} \quad S^2 = D.
$$

The diagonal elements of S are called the singular values of A.

4. We can rewrite Eq. (3.26) and its solution as follows:

$$
AA^Ta = P^TDPa = (SP)^T(SP)a = Ay,
$$
$$
a = P^TD^{-1}PAy.
$$

---

*Ralston (1965) is a good source of further information. The ill-conditioning problem, though very real, is often academic, because seldom is a degree above 4 or 5 needed to give a curve that fits the data with adequate precision.

This last eliminates having to multiply out AA^T and, by extending this approach, leads to an important method for solving Eq. (3.26) called *singular-value decomposition.* (See Press, *Numerical Recipes,* 1992, on this topic.)

MATLAB has a command: $[\text{U, S, V}] = \text{svd(A)}$ that computes the singular value decomposition of matrix A. The combination $\text{U}*\text{S}*\text{V}'$ is equal to A and the singular values of A are on the diagonal of S. *Mathematica* can do the same. (When A is symmetric and semidefinite, the singular values are the eigenvalues.) We do not pursue this idea further.

We illustrate the use of Eqs. (3.25) to fit a quadratic to the data of Table 3.12. Figure 3.16 shows a plot of the data. (The data are actually a perturbation of the relation $y = 1 - x + 0.2x^2$. It will be of interest to see how well we approximate this function.) To

Table 3.12   Data to illustrate curve-fitting

x_i	0.05	0.11	0.15	0.31	0.46	0.52	0.70	0.74	0.82	0.98	1.17
Y_i	0.956	0.890	0.832	0.717	0.571	0.539	0.378	0.370	0.306	0.242	0.104

$$\sum x_i = 6.01 \qquad\qquad N = 11$$
$$\sum x_i^2 = 4.6545 \qquad\qquad \sum Y_i = 5.905$$
$$\sum x_i^3 = 4.1150 \qquad\qquad \sum x_i Y_i = 2.1839$$
$$\sum x_i^4 = 3.9161 \qquad\qquad \sum x_i^2 Y_i = 1.3357$$

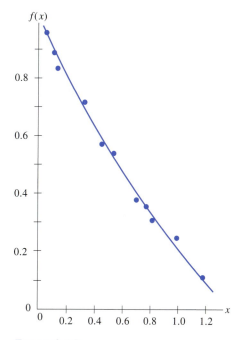

Figure 3.16

set up the normal equations, we need the sums tabulated in Table 3.12. A calculator can give these directly as accumulated totals. We need to solve the set of equations

$$11a_0 + 6.01a_1 + 4.6545a_2 = 5.905,$$
$$6.01a_0 + 4.6545a_1 + 4.1150a_2 = 2.1839,$$
$$4.6545a_0 + 4.1150a_1 + 3.9161a_2 = 1.3357.$$

The result is $a_0 = 0.998$, $a_1 = -1.018$, $a_2 = 0.225$, so the least-squares method gives

$$y = 0.998 - 1.018x + 0.225x^2.$$

Compare this to $y = 1 - x + 0.2x^2$. We do not expect to reproduce the coefficients exactly because of the errors in the data.

What Degree Polynomial Should Be Used?

In the general case, we may wonder what degree of polynomial should be used. As we use higher-degree polynomials, we of course will reduce the deviations of the points from the curve until, when the degree of the polynomial, n, equals $N - 1$, there is an exact match (assuming no duplicate data at the same x-value) and we have an interpolating polynomial. The answer to this problem is found in statistics. One increases the degree of approximating polynomial as long as there is a statistically significant decrease in the variance, σ^2, which is computed by

$$\sigma^2 = \frac{\sum e_i^2}{N - n - 1}. \tag{3.27}$$

For the preceding example, when the degree of the polynomial made to fit the points is varied from 1 to 7, we obtain the results shown in Table 3.13.

The criterion of Eq. (3.27) chooses the optimum degree as 2. This is no surprise, in view of how the data were constructed. It is important to realize that the numerator of Eq. (3.27), the *sum of the deviations squared* of the points from the curve, should continually decrease as the degree of the polynomial is raised. It is the denominator of Eq. (3.27) that makes σ^2

Table 3.13

Degree	Equation	σ^2 (Eq. 3.27)	$\sum e^2$
1	$y = 0.952 - 0.760x$	0.0010	0.0092
2	$y = 0.998 - 1.018x + 0.225x^2$	0.0002	0.0018
3	$y = 1.004 - 1.079x + 0.351x^2 - 0.069x^3$	0.0003	0.0018
4	$y = 0.998 - 0.838x - 0.522x^2 + 1.040x^3 - 0.454x^4$	0.0003	0.0016
5	$y = 1.031 - 1.704x + 4.278x^2 - 9.477x^3 + 9.394x^4 - 3.290x^5$	0.0001	0.0007
6	$y = 1.038 - 1.910x + 5.952x^2 - 15.078x^3 + 18.277x^4 - 9.835x^5 + 1.836x^6$	0.0002	0.0007
7	$y = 1.032 - 1.742x + 4.694x^2 - 11.898x^3 + 16.645x^4 - 14.346x^5 + 8.141x^6 - 2.293x^7$	0.0002	0.0007

increase as we go above the optimum degree. In this example, this behavior is observed for $n = 3$. Above $n = 3$, a second effect sets in. Due to ill-conditioning, the coefficients of the least-squares polynomials are determined with poor precision. This modifies the expected increases of the values of σ^2.

Before leaving this section, we illustrate how to apply these methods to a more complicated function.

EXAMPLE 3.9 The results of a wind tunnel experiment on the flow of air on the wing tip of an airplane provide the following data:

R/C: 0.73, 0.78, 0.81, 0.86, 0.875, 0.89, 0.95, 1.02, 1.03, 1.055, 1.135, 1.14, 1.245, 1.32, 1.385, 1.43, 1.445, 1.535, 1.57, 1.63, 1.755;

V_θ/V_∞:   0.0788, 0.0788, 0.064, 0.0788, 0.0681, 0.0703, 0.0703, 0.0681, 0.0681, 0.079, 0.0575, 0.0681, 0.0575, 0.0511, 0.0575, 0.049, 0.0532, 0.0511, 0.049, 0.0532, 0.0426;

where R is the distance from the vortex core, C is the aircraft wing chord, V_θ is the vortex tangential velocity, and V_∞ is the aircraft free-stream velocity. Let $x = R/C$ and $y = V_\theta/V_\infty$. We would like our curve to be of the form

$$g(x) = \frac{A}{x}(1 - e^{-\lambda x^2}),$$

and our least-squares equation becomes

$$S = \sum_{i=1}^{21} (Y_i - g(x_i))^2$$

$$= \sum_{i=1}^{21} \left(Y_i - \frac{A}{x_i}(1 - e^{-\lambda x_i^2}) \right)^2.$$

Setting $S_A = S_\lambda = 0$ gives the following equations:

$$\sum_{i=1}^{21} \left(\frac{1}{x_i}\right)(1 - e^{-\lambda x_i^2})\left(Y_i - \frac{A}{x_i}(1 - e^{-\lambda x_i^2}) \right) = 0,$$

$$\sum_{i=1}^{21} x_i(e^{-\lambda x_i^2})\left(Y_i - \frac{A}{x_i}(1 - e^{-\lambda x_i^2}) \right) = 0.$$

When this system of nonlinear equations is solved, we get

$$g(x) = \frac{0.07618}{x}(1 - e^{-2.30574x^2}).$$

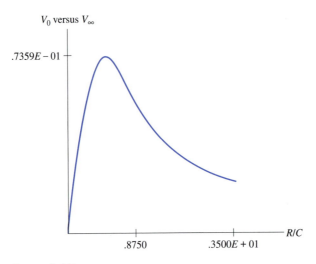

V_0 versus V_∞

.7359E – 01

.8750   .3500E + 01   R/C

Figure 3.17

For these values of A and λ, $S = 0.000016$. The graph of this function is presented in Fig. 3.17. ▲

Here is an algorithm for obtaining a least-squares polynomial:

> Given N data pairs, (x_i, Y_i), $i = 1, \ldots, N$, obtain an nth degree least-squares polynomial by the following:
> Form the coefficient matrix, M, with $n + 1$ rows (r) and $n + 1$ columns (c), by:
>
> $$\text{Set } M_{rc} = \sum_{i=1}^{N} x_i^{r+c-2}.$$
>
> Form the right-hand-side vector b, with $n + 1$ rows (r), by:
>
> $$\text{Set } b_r = \sum_{i=1}^{N} x_i^{r-1} Y_i,$$
>
> Solve the linear system $Ma = b$ to get the coefficients in
>
> $$y = a_0 + a_1 x + a_2 x^2 + \cdots + a_n x^n,$$
>
> which is the desired polynomial that fits the data.

3.8 Theoretical Matters

In this section we assemble several items of theory whose conclusions were referred to earlier in this chapter.

Why Polynomials?

Throughout this chapter the emphasis has been on fitting polynomials to functions. Why choose polynomials over other functions? The answer lies in the Weierstrass approximation theorem. This states the following:

> If $f(x)$ is continuous on a finite interval $[a, b]$, there exists a polynomial $P_n(x)$ of degree n such that
>
> $$|f(x) - P_n(x)| < \text{ERROR},$$
>
> throughout the interval $[a, b]$, for any given $\text{ERROR} > 0$. (The degree required of $P_n(x)$ is a function of ERROR.)

What this means is that we can achieve what is called *uniform approximation*—the maximum error in $[a, b]$ is bounded rather than some average of the errors.

A proof of this theorem can be found in Ralston (1965).

Observe, however, that polynomials cannot fit to a function that has discontinuities because every polynomial is continuous and has continuous derivatives. Obviously, they can be used if we fit only portions of the function between its points of discontinuity.

Identical Polynomials

We have written interpolating polynomials in several forms: with Lagrangian polynomials, using divided differences, using terms from a difference table (if the data are equispaced), or in the standard form:

$$P_n(x) = a_0 + a_1x + a_2x^2 + \cdots + a_nx^n.$$

We now show that every polynomial of degree n that has the same value at $n + 1$ distinct points is identical.

First, the conclusion seems intuitively true because the $n + 1$ data pairs are exactly enough to determine the $n + 1$ coefficients of the polynomial, and, because every expression of the polynomial can be reduced to the standard form, they must be identical.

A more formal and compelling proof is by contradiction:

> Suppose there are two different polynomials of degree n that agree at $n + 1$ distinct points. Call these $P_n(x)$ and $Q_n(x)$, and write their difference:
>
> $$D(x) = P_n(x) - Q_n(x),$$
>
> where $D(x)$ is a polynomial of at most degree n. But because P and Q match at the $n + 1$ points, their difference $D(x)$ is equal to zero at all $n + 1$ of these x-values; that is, $D(x)$ is a polynomial of degree n at most but has $n + 1$ distinct zeros. However, this is impossible unless $D(x)$ is identically zero. Hence $P_n(x)$ and $Q_n(x)$ are not different—they must be the same polynomial.

A most important consequence of this uniqueness property of interpolating polynomials is that their error terms are also identical (though we may want to express the error term in different forms). We only have to derive the error term for one form of interpolating polynomial to have the error term for all forms of interpolating polynomials.

The Error Term for Interpolation

We begin the development of an expression for the error of $P_n(x)$, an nth-degree interpolating polynomial, by writing the error function in a form that has the known property that it is zero at the $n + 1$ points, from x_0 through x_n, where $P_n(x)$ and $f(x)$ are the same. We call this function $E(x)$:

$$E(x) = f(x) - P_n(x) = (x - x_0)(x - x_1) \cdots (x - x_n)g(x).$$

The $n + 1$ linear factors give $E(x)$ the zeros we know it must have, and $g(x)$ accounts for its behavior at values other than at $x_0, x_1, \ldots, x_n$. Obviously, $f(x) - P_n(x) - E(x) = 0$, so

$$f(x) - P_n(x) - (x - x_0)(x - x_1) \cdots (x - x_n)g(x) = 0. \tag{3.28}$$

To determine $g(x)$, we now use the interesting mathematical device of constructing an auxiliary function (the reason for its special form becomes apparent as the development proceeds). We call this auxiliary function $W(t)$, and define it as

$$W(t) = f(t) - P_n(t) - (t - x_0)(t - x_1) \cdots (t - x_n)g(x).$$

Note in particular that x has *not* been replaced by t in the $g(x)$ portion. (W is really a function of both t and x, but we are only interested in variations of t.) We now examine the zeros of $W(t)$.

Certainly at $t = x_0, x_1, \ldots, x_n$, the W function is zero ($n + 1$ times), but it is also zero if $t = x$ by virtue of Eq. (3.28). There are then a total of $n + 2$ values of t that make $W(t) = 0$. We now impose the necessary requirements on $W(t)$ for the *law of mean value* to hold. $W(t)$ must be continuous and differentiable. If this is so, there is a zero to its derivative $W'(t)$ between each of the $n + 2$ zeros of $W(t)$, a total of $n + 1$ zeros. If $W''(t)$ exists, and we suppose it does, there will be n zeros of $W''(t)$, and likewise $n - 1$ zeros of $W'''(t)$, and so on, until we reach $W^{(n+1)}(t)$, which must have at least one zero in the interval that has x_0, x_n, or x as endpoints. Call this value of $t = \xi$. We then have

$$W^{(n+1)}(\xi) = 0 = \frac{d^{n+1}}{dt^{n+1}}[f(t) - P_n(t) - (t - x_0) \cdots (t - x_n)g(x)]_{t=\xi} \tag{3.29}$$

$$= f^{(n+1)}(\xi) - 0 - (n + 1)!g(x).$$

The right-hand side of Eq. (3.29) occurs because of the following arguments. The $(n + 1)$st derivative of $f(t)$, evaluated at $t = \xi$, is obvious. The $(n + 1)$st derivative of $P_n(t)$ is zero because every time any polynomial is differentiated its degree is reduced by one, so that the nth derivative is of degree zero (a constant) and its $(n + 1)$st derivative is zero. We apply the same argument to the $(n + 1)$st-degree polynomial in t that occurs in the last term—its $(n + 1)$st derivative is a constant that results from the t^{n+1} term and is $(n + 1)!$.

Of course $g(x)$ is independent of t and goes through the differentiations unchanged. The form of $g(x)$ is now apparent:

$$g(x) = \frac{f^{(n+1)}(\xi)}{(n + 1)!}, \quad \xi \text{ between } (x_0, x_n, x)$$

The conditions on $W(t)$ that are required for this development (continuous and differentiable $n + 1$ times) will be met if $f(x)$ has these same properties, because $P_n(x)$ is continuous and differentiable. We now have our error term:

$$E(x) = (x - x_0)(x - x_1) \cdots (x - x_n)\frac{f^{(n+1)}(\xi)}{(n + 1)!}, \tag{3.30}$$

with ξ on the smallest interval that contains $\{x, x_0, x_1, \ldots, x_n\}$.

If we modify this, expressing it in terms of $s = (x - x_0)/h$, it becomes more compatible with the Newton–Gregory forward polynomials. Remembering that

$$x_1 = x_0 + h, \quad x_2 = x_0 + 2h, \quad \ldots,$$

so that

$$(x - x_0) = sh, \quad (x - x_1) = sh - h = (s - 1)h,$$
$$(x - x_2) = sh - 2h = (s - 2)h, \quad \ldots,$$

we find that Eq. (3.30) becomes

$$E(x_s) = \frac{(s)(s - 1)(s - 2) \cdots (s - n)}{(n + 1)!} h^{n+1} f^{(n+1)}(\xi)$$

$$= \binom{s}{n + 1} h^{n+1} f^{(n+1)}(\xi), \quad \xi \text{ between } (x_0, x_n, x_s) \tag{3.31}$$

If we look at the formula for a Newton–Gregory interpolating polynomial, we find that the next term after the last one that is included is $\binom{s}{n+1}\Delta^{n+1}f_0$. We get the error term of Eq. (3.31) by substituting $h^{n+1}f^{(n+1)}(\xi)$ for the $(n + 1)$st difference. This is true for all interpolating polynomials, even those based on divided differences, because of the relationship between divided differences and ordinary differences.

This is the justification for our *next-term rule*.

Centering the x-Value

We have remarked several times that centering the x-value within the range of the x_i used to construct the interpolating polynomial will give better results. This is easy to see. Consider a set of data pairs with these x-values:

i	x_i
0	0.2
1	0.4
2	0.5
3	0.7
4	0.8
5	1.0

Suppose that $x = 0.63$ and that we want to develop a second-degree polynomial (that will utilize three data pairs). It is obvious that the $(x - x_i)$ factors in the error term will have the smallest product (in magnitude) if we choose x_i-values of 0.5, 0.7, and 0.8, because these three x_i's are nearest to $x = 0.63$. (There may be unusual situations when the derivative portion of the error term contradicts this rule, but such an event is so rare that we usually ignore it, particularly when we do not know the size of the derivative term.)

Recognizing which data pairs are nearest to x is easier when the data are equispaced. It is particularly hard when the data pairs are not arranged in order of the x_i.

We can further demonstrate this rule by observing the errors for second-degree interpolating polynomials based on the data for $f(x) = \sin(x)$ when we vary the range of x_i's:

i-values	x_i-values	Estimate of $f(0.8)$	Actual error
0, 1, 2	0.1, 0.5, 0.9	0.71445	2.904E-3
1, 2, 3	0.5, 0.9, 1.3	0.71895	−1.593E-3
2, 3, 4	0.9, 1.3, 1.7	0.71450	2.854E-3
3, 4, 5	1.3, 1.7, 2.1	0.75083	−3.347E-2

The results agree with our rule: $x = 0.8$ is best centered when we take values at $i = 1, 2, 3$, and the error is smaller when that is done.

In Summary

To summarize, then, the error of polynomial interpolation is reduced by making its range as symmetrical as possible about the point of interpolation and by choosing a higher degree of polynomial, up to the point where round-off or the effect of local irregularities causes offsetting errors. But another factor has the greatest importance of all—the step size h. With a given set of tabulated data, we may not be able to do much about the step size. However, if we are designing a new set of tables, the step size may be open to our selection; it is then advantageous to make it small. (This makes our tables bulkier and adds proportionately to the computational effort, of course.)

This effect of h is so important that a special notation is often used to focus attention on it. We write

$$\text{Error} = O(h^n)$$

if there is a constant K such that if h is "small enough," and

$$|\text{Error}| \leq Kh^n$$

where $h > 0$ and where K is some constant not equal to zero. The expression $O(h^n)$ is read "order of h to the nth power." For example, the error of a quadratic interpolating polynomial whose range is (x_0, x_2) would be of order h *cubed* because

$$\text{Error} = \frac{(s)(s-1)(s-2)}{6} h^3 f'''(\xi), \quad \text{or} \quad \text{Error} = O(h^3).$$

As h gets small, $f'''(\xi) \to f'''(x_0)$ because ξ is squeezed between x_0 and x_2, and hence approaches a constant value.

A Contradiction?

As we have pointed out, one way to improve the accuracy of a polynomial that approximates a known function is to use closely spaced points of agreement. But that is not always true! We saw in Section 3.4 that some functions, those that are smooth everywhere except for a local "bump," are difficult to approximate with ordinary interpolating polynomials.

Here is another example:

$$\text{Let} \quad f(x) = \frac{1}{1 + 25x^2} \quad \text{on} \quad [-1, 1].$$

Now suppose we find interpolating polynomials that fit to $f(x)$ at uniformly spaced points on $[-1, 1]$. It turns out that the maximum error does not decrease as we increase the degree of the interpolating polynomials! The following table gives, for polynomials of odd degree, the approximate value of the maximum error.

Degree	Maximum error	Degree	Maximum error
3	0.7070	13	1.0700
5	0.4327	15	2.1041
7	0.2572	17	4.2190
9	0.3001	19	8.8526
11	0.5567		

From the data, it is clear that the maximum errors are increasing after degree $= 7$. (This phenomenon occurs after degree $= 2$ for polynomials of even degree.)

Why such behavior, particularly in view of the Weierstrass theorem? Because the theorem does not apply if we constrain the matching with $f(x)$ to occur at evenly spaced points! In Chapter 4, we will find that fitting the function at special points does cause the maximum error to decrease with the degree.

3.9   Using MATLAB and *Mathematica*

MATLAB can easily interpolate with a polynomial that is defined from a set of points. If we want the cubic polynomial that passes through the first four points of the table in Section 3.2, we do this: (1) define the x-values in an array, (2) do the same for the y-values, (3) get the polynomial of degree n (where the number of points is $n + 1$) with the command: `polyfit(x, y, n)`. Here is what MATLAB shows:

```
x = [1.0 2.7 3.2 4.8]
x =
     1.0000   2.7000   3.2000   4.8000
y = [14.2 17.8 22.0 38.3]
y =
     14.2000   17.8000   22.0000   38.3000
f = polyfit(x,y,3)
f =
     -0.5275   6.4952   -16.1177   24.3499
```

which are the coefficients of the interpolating polynomial. We find the y-value that corresponds to $x = 3.0$ with

```
xx = polyval(f, 3.0)
xx =
     20.2120
```

and the result matches that of the Lagrangian technique, as expected. (In this we rearranged the x-y pairs in order of increasing x, but that is not necessary.)

This same `polyfit` command gets the least-squares polynomial fit if the number of points is greater than the degree plus 1. As an example, let us fit the data of Table 3.12 with least-squares polynomials of degrees 1 and 2. As before, we first define the x- and y-values as arrays. Then we do: `polyfit(x, y, 2)`, and see the same polynomial coefficients as in Table 3.13.

If we redo the fit operation with other degrees for the polynomial, we get the other coefficients of Table 3.13 but we get the coefficients of higher-degree polynomials with better precision because MATLAB uses 16-digit arithmetic.

We can interpolate within a table of data pairs without explicitly getting the interpolating polynomial with the command: `yi = interp1(x, y, xi, 'linear')`. In using this, x and y are arrays of the x- and y-values and xi is the point for which the interpolated y-value is desired. The data must be arranged with the x-value increasing monotonically. This command does a linear interpolation using the appropriate values from the table. (The last parameter can be omitted; a linear fit is the default.)

A better value is obtained from fitting a cubic interpolating polynomial to the appropriate set of four points by using 'cubic' as the last parameter but for this, the *x*-values must be evenly spaced. If the fourth parameter is 'spline', the interpolated value is from a cubic spline. With all of these commands, if xi is an array, we get an array of interpolated values. We get all the values of the last column of Table 3.8 with the command: yi = spline(x,y,xi), where x and y are arrays of the data points and xi is the *x*-values of the table. MATLAB computes spline interpolates using condition 4.

Interpolating on a surface is also provided. The command is:

```
interp2(x, y, z, xi, yi, 'method')
```

where the "method" can be 'linear' (the default) or 'cubic'. If 'cubic' is specified, the *x*- and *y*-values must be evenly spaced. Both options require that the *x*- and *y*-values be monotonic.

Difference Tables

Because we get interpolating polynomials so easily with the polyfit command in MATLAB, it is perhaps redundant to have it get difference tables. However, this is also very easy to do. The differences of a function are obtained by defining $f(x)$ as an array of function values, then applying the command: diff(fx, n), where n is the order of difference. The array can be either a row or a column vector. There is no command to get divided differences directly; a sequence of commands could do it. We leave this as a challenge to the student.

Plotting

The powerful plotting commands of MATLAB are useful if we want to visualize how interpolating or least-squares polynomials compare to the data points. Here are the commands to plot the cubic polynomial through the first four points of the first table in Section 3.2 and superimpose the points:

```
x = [1.0 2.7 3.2 4.8]; y = [14.2 17.8 22.0 38.3];
   f = polyfit(x,y,3); fs = poly2sym(f);
      fplot(fs,[0 5]); grid; hold; plot(x,y,'o')
```

and we see Figure 3.18.

We can do a similar sequence of commands to plot least-squares polynomials and the data to which they fit. It is also possible to plot a spline curve and the points from which it is derived. For example, this sequence of commands in MATLAB produced Figure 3.3:

```
x = [0.0 1.0 1.5 2.25];
y = [2.0 4.4366 6.7134 13.9130];
h = 0.0 : 0.05 : 2.25;
out = interp1(x, y, h, 'spline');
plot (x, y, 'o', h, out)
```

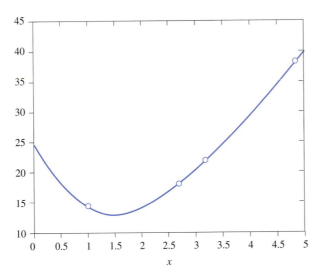

Figure 3.18

Using *Mathematica*

The *Mathematica* program is perhaps the most extensive of all current computer algebra systems. It is a very large program (maybe the correct word is "huge") with 843 built-in functions, called "objects" in the *Mathematica* documentation. In addition to the built-in objects, *Mathematica* comes with almost 150 other "standard packages"—functions that are created using the powerful programming capabilities that are incorporated in the system.

Mathematica is available for several computer platforms—Macintosh, PCs (DOS and Windows versions), Unix-based systems, among others. The computational parts of these different versions are identical, but there are different "front ends" (user interfaces) for the various versions; these interfaces take advantage of the special capabilities of the particular computer system.

Because of its size, *Mathematica* requires memory and hard disk capacities that minimum computer systems do not provide. It also takes several seconds to load, even on a fast personal computer. Because of its size and complexity, *Mathematica* is not an easy program to learn, although doing the simplest and most common things is not hard.

Mathematica can carry out many of the numerical procedures of this chapter. Functions that create interpolating polynomials, plot spline and Bezier curves, and obtain least-squares fits to data points are available. We describe how each of these procedures is done in this system through examples. These examples show that *Mathematica* displays the results of computations in a dialog: $In[n]$ identifies the user input and $Out[n]$ labels the response, with the value of n increasing for successive requests.

Most of these illustrative examples can also be done in Maple.

Difference Tables

Mathematica has no built-in procedure for computing differences of a set of function values (this function is not necessary, because the interpolating polynomial can be obtained without them). Still, it is easy to get the differences and instructive to do so. We start by defining a list of function values, in [1], and then we use this list in a **Table** operation to construct the differences in [2]. **Table** creates a new list that, in this case, contains the differences of the original list of function values.) Request [3] asks for the display to be a column of values:

```
In[1] :=
   f = {.203, .423, .684, 1.030, 1.557, 2.572}
Out[1] =
   {0.203, 0.423, 0.684, 1.03, 1.557, 2.572}
In[2] :=
   dif = Table [ %[[i+i]] - %[[i]], {i, Length[f] - 1} ]
Out[2] =
   {0.22, 0.261, 0.346, 0.527, 1.015}
In[3] :=
   ColumnForm [ dif ]
Out[3] =
   0.22
   0.261
   0.346
   0.527
   1.015
```

Observe that *Mathematica* uses braces to enclose a list of values, which are then treated as a single entity. We name the list "f" and then obtain the differences of "f" through the **Table** operation. Its parameters need some explanation.

The % symbol refers to the previous result. (%% would refer to the result before the previous one, %%% to the one before that, and %n refers to the result of request [n].) We obtain the ith element of the list through %[[i]] and the $(i + 1)$st element through %[[i + 1]]. The last parameter, {i, Length[f] - 1}, causes i to run from 1 to the length of list f, minus 1.

Interpolation

There are at least two ways to develop an interpolating polynomial that passes through each of a set of points. The most obvious is the function **InterpolatingPolynomial**. (*Mathematica* almost always uses very long names for its operations to avoid ambiguity.) We begin with a set of data points and then ask for the polynomial:

```
In[4] :=
   data = {{1, 1}, {2, 4}, {4, 3}, {5, 4}}
```

```
Out[4] =
  {{1, 1}, {2, 4}, {4, 3}, {5, 4}}
In[5] :=
  poly = InterpolatingPolynomial [ data, x ]
Out[5] =
```

$$1 + (3 + (-(\frac{7}{6}) + \frac{5(-4 + x)}{12}) (-2 + x)) (-1 + x)$$

The result from [5] can be put into more familiar form with a request to expand (to multiply out the terms):

```
In[6] :=
  Expand [ % ]
Out[6] =
```

$$-(\frac{23}{3}) + \frac{37x}{3} - \frac{49x^2}{12} + \frac{5x^3}{12}$$

We can evaluate the polynomial as follows:

```
In[7] :=
  poly /. x→3.5
Out[7] =
  3.34375
```

In the last operation, the symbol /. means to evaluate "poly," and we specify that this evaluation is for $x = 3.5$. The result, 3.34375, is the interpolated value (from a cubic interpolating polynomial, as there are four data points) at $x = 3.5$.

A second way to get an interpolating polynomial is through **Interpolation.** This develops a so-called **InterpolatingFunction:**

```
In[8] :=
  interp = Interpolation [ data ]
Out[8] =
  InterpolatingFunction[{1, 5}, <>]
```

Observe that the interpolating function is not displayed explicitly. We do learn that its range is from $x = 1$ to $x = 5$. We also evaluate it in a different way, as shown in request [9] below. The result is identical to the previous interpolating polynomial because `Interpolation` creates cubic interpolating polynomials by default. We can change this default (see [10]):

```
In[9] :=
  interp [ 3.5 ]
Out[9] =
  3.34375
In[10] :=
  interp2 = Interpolation [ data, InterpolationOrder→2 ]
Out[10] =
  InterpolatingFunction[{1, 5}, <>]
```

```
In[11] :=
   interp2[ 3.5 ]
Out[11] =
   4.125
In[12] :=
   interp2[ 4.5 ]
Out[12] =
   3.375
```

This last interpolating function created quadratics through successive triples of the data points. You should verify the truth of this statement.

Interpolating on a Surface

If we give **Interpolation** a set of values defined on a grid of (x, y) points, we can interpolate on the surface defined by the points (which are in three-dimensional space). Here is an example where the points are actually values of $f(x, y) = 3x - y^2$:

```
In[13] :=
   tab15 =
   {{2, 1, 5}, {2, 1.5, 3.75}, {2, 2., 2.}, {2, 2.5, -0.25},
    {3, 1, 8}, {3, 1.5, 6.75}, {3, 2., 5.}, {3, 2.5, 2.75},
    {4, 1, 11}, {4, 1.5, 9.75}, {4, 2., 8.}, {4, 2.5, 5.75},
    {5, 1, 14}, {5, 1.5, 12.75}, {5, 2., 11.}, {5, 2.5, 8.75}};
In[14] :=
   interp2 = Interpolation [ tab15 ]
Out[14] =
   InterpolatingFunction[{{2, 5}, {1, 2.5}}, <>]
In[15] :=
   interp2 [3, 2]
Out[15] =
   5.
In[16] :=
   interp2 [3.2, 1.7]
Out[16] =
   6.71
In[17] :=
   3*3.2 - 1.7^2
Out[17] =
   6.71
```

Some explanations are in order. When we terminated the input in [13] with a semicolon, the output was suppressed. In [15] we evaluated the interpolating function at $x = 3, y = 2$, getting the point-value at $(3, 2)$, where $f(x, y) = 5$. In [16], we asked for $f(3.2, 1.7)$ and got 6.71. This was verified in [17].

Least-Squares Fitting of Data

Mathematica has a built-in function to do least-squares fitting of data. The desired relation does not have to be linear, and more than one independent variable is permitted. We work with a list of data points and supply a pattern for the equation to be fitted. How this is done can be understood from some examples.

EXAMPLE 3.10   Fit a least-squares line to (20.5, 765), (32.7, 826), (51.0, 873), (73.2, 942), (95.7, 1032). (These are the data for the first example in Section 3.7 of this chapter.) Here is how we do this:

```
In[18] :=
  d1 = {{20.5, 765}, {32.7, 826}, {51.0, 873}, {73.2, 942},
       {95.7, 1032}
Out[18] =
  {{20.5, 765}, {32.7, 826}, {51., 873}, {73.2, 942},
   {95.7, 1032}}
In[19] :=
  Fit[d1, {1, t}, t]
Out[19] =
  702.172 + 3.39487t
```

The result in [19] agrees with the result in Section 3.7.   ▲

EXAMPLE 3.11   We reproduce the second example in Section 3.7 with

```
In[20] :=
  d2 = {{.05, .956}, {.11, .890}, {.15, .832}, {.31, .717},
   {.46, .571}, {.52, .539}, {.70, .378}, {.74, .370}, {.82, .306},
   {.98, .242}, {1.17, .104}}
Out[20] =
  {{0.05, 0.956}, {0.11, 0.89}, {0.15, 0.832}, {0.31, 0.717},
   {0.46, 0.571}, {0.52, 0.539}, {0.7, 0.378}, {0.74, 0.37},
   {0.82, 0.306}, {0.98, 0.242}, {1.17, 0.104}}
In[21] :=
  Fit[d2, (1, x, x^2}, x]
Out[21] =
  0.997968 - 1.01804 + 0.224682x²
```
▲

EXAMPLE 3.12   In this example we fit to the values of a known function: $f(x, y) = 2x^2 + 3y - 1$. We fit a function of two variables in this fashion:

```
In[22] :=
  d3 = {{2, 3, 16}, {3, 2, 23}, {4, 1, 34}}
Out[22] =
  {{2, 3, 16}, {3, 2, 23}, {4, 1, 34}}
```

```
In[23] :=
   Fit[ d3, {1, x^2, y}, {x, y}]
Out[23] =
   -1. + 2.x² + 3.y
```

In this example we have exactly enough data to compute the coefficients so the result, $2x^2 + 3y - 1$, is the exact answer. (We then see that this is still another way to get an interpolating polynomial: supply exactly enough data to compute the coefficients through a least-squares fit.) If we supply more data points than parameters in the function being fit, a least-squares evaluation of the coefficients results. ▲

Plotting

Mathematica is notable for its plotting capabilities. Both two- and three-dimensional plots are easy to create and can be displayed with or without grid lines and with or without labels on the axes and for the plot as a whole. (There are still more options.)

We might want to plot an interpolating polynomial to see how well the polynomial matches the function (that is, if we know the function!). This next compares an interpolating polynomial to the function itself. We take $f(x) = xe^{-x/2}$, and begin with a table of data pairs:

x	$f(x)$
-0.5	-0.6420
2.3	0.7283
5.7	0.3297
7.2	0.1967

We will fit a cubic to these four points and draw its graph, then draw the graph for $f(x)$, and, finally, superimpose the two graphs. Here is how we do this in *Mathematica:*

```
In[24] :=
   f[x_] :=x*E^(-x/2)
In[25] :=
   x_f = {{-.5, f[-.5]}, (2.3, f[2.3]}, {5.7, f[5.7]}, {7.2, f[7.2]}}
Out[25] =
   {{-0.5, -0.642013}, {2.3, 0.728265}, {5.7, 0.329713},
    {7.2, 0.196731}}
In[26] :=
   InterpolatingPolynomial [%, x]
Out[26] =
   -0.642013 + (0.489385 + (-0.0978397 + 0.0134636(-5.7 + x))
      (-2.3 + x))(0.5 + x)
In[27] :=
   pltpoly = Plot [ %, {x, -1.8}]
Out[27] =
-Graphics-
```

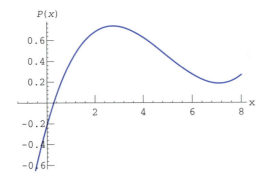

```
In[28] :=
   pltfx = Plot [ x*E^(-x/2), {x, -1, 8}]
Out[28] =
   -Graphics-
```

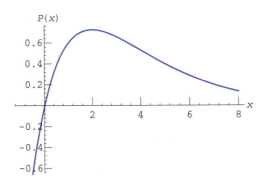

```
In[29] :=
   Show [ pltpoly, pltfx ]
Out[29] =
   -Graphics-
```

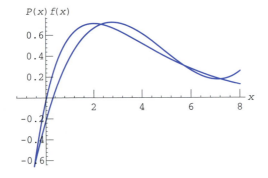

Notice that the two curves cross at the data points.

Spline Curves

We can plot cubic spline curves and Bezier curves by using one of *Mathematica*'s standard packages. Before we use the package, we must load it; we do this in [31]. We create the cubic spline in [32] and plot it in [33]. The straight lines connect the data points.

This same package also allows us to create Bezier curves. In [34], we combine the creation and plotting of the Bezier curve defined by our four points:

```
In[30] :=
  spdata = {{1, 1}, {2, 4}, {3, 3}, {4, 4}}
Out[30] =
  {{1, 1}, {2, 4}, {3, 3}, {4, 4}}
In[31] :=
  <<Graphics'Spline'
In[32] :=
  splin = Spline[spdata, Cubic]
Out[32] =
  Spline[{{1, 1}, {2, 4}, {3, 3}, {4, 4}}, Cubic, <>]
In[33] :=
  Show[Graphics[{Line[spdata], splin}]]
Out[33] =
  -Graphics-
```

```
In[34] :=
  Show[Graphics[{Line[spdata], Spline[spdata, Bezier]}]]
Out[34] =
  -Graphics-
```

Chapter Summary

If you fully understand this chapter, you are now able to

1. Construct interpolating polynomials for any set of data by using the Lagrangian formulation or by using divided differences. You can explain why the divided-difference technique is preferred, and you can use either method to interpolate from a given set of data pairs.

2. Understand Neville's method for interpolation, and explain how it only needs to do a succession of linear interpolations to accomplish the task.

3. Write and use an expression for the error of the interpolate. You know the difficulty and limitations of such error estimates. You can explain and use the next-term rule.

4. Build a difference table from a set of evenly spaced data, and construct an interpolating polynomial from the table.

5. Show a situation where the error of interpolation increases with the degree of polynomial, and tell why this occurs. You know about methods to overcome the problem.

6. Set up the equations for a cubic spline and explain how the end portions can be established. You know why a cubic spline is an improved method for interpolation.

7. Construct Bezier curves and B-spline curves. You can explain how these curves differ from the interpolating polynomials and some important applications for them.

8. Apply the techniques of this chapter to three-dimensional problems.

9. Fit a line or curve to experimental data using the least-squares method. You know why the use of an interpolating polynomial is inappropriate.

10. Explain the elements of some theory behind the methods of this chapter, including why polynomials are so often preferred over other functions, that the nth-degree polynomial through $n + 1$ points is unique, how the error term for interpolation can be derived, why the data pairs should be chosen to make the x-value centered within them, why the next-term rule can estimate the error, and what the "$O(\ )$" expression means with respect to errors.

11. Use *Mathematica* and/or MATLAB to get an interpolating polynomial and to fit a set of data with a least-squares curve.

Computer Programs

The two programs of Chapter 3 are (1) a Fortran 90 program that gets least-squares polynomials that fit a set of data points, and (2) a C program that uses divided differences to pass a polynomial through a set of data points and then interpolates with the polynomial to create a table of (x, y)-values.

Program 3.1 Least-Squares Polynomials

Figure 3.19 is the listing of the program. Program output follows.

```
Program Least_Squares
!
!!!!!!!!!!!!!!!!!!!!!!!!!!!!!!!!!!!!!!!!!!!!!!!!!!!!!!!!!!!!!!!!!!!!!!!!!!
!                             CHAPTER 3                                 !
!                                                                       !
!  Gerald/Wheatley, APPLIED NUMERICAL ANALYSIS(Sixth Edition)           !
!                  Addison Wesley Longman, 1999                         !
!                                                                       !
!!!!!!!!!!!!!!!!!!!!!!!!!!!!!!!!!!!!!!!!!!!!!!!!!!!!!!!!!!!!!!!!!!!!!!!!!!
!                                                                       !
!  This program is used in fitting a polynomial to a set of data.       !
!  The program reads in n pairs of X- and Y-values and computes the     !
!  coefficients of the normal equations for the least-squares           !
!  method.                                                              !
!                                                                       !
!!!!!!!!!!!!!!!!!!!!!!!!!!!!!!!!!!!!!!!!!!!!!!!!!!!!!!!!!!!!!!!!!!!!!!!!!!
!                                                                       !
!     Parameters are :                                                  !
!                                                                       !
!     X, Y  - array of X- and Y-values                                  !
!     N     - number of data pairs                                      !
!     MS,MF - the range of degree of polynomials to be computed         !
!             the maximum degree is 9.                                  !
!     A     - Augmented array of the coefficients of the normal         !
!             equations.                                                !
!     C     - array of coefficients of the least-squares                !
!             polynomials.                                              !
!                                                                       !
!!!!!!!!!!!!!!!!!!!!!!!!!!!!!!!!!!!!!!!!!!!!!!!!!!!!!!!!!!!!!!!!!!!!!!!!!!
!
      Implicit None
      Real(Kind = 8), Dimension(100) :: X,Y,C,Xn
      Real(Kind = 8), Dimension(10,11) :: A
      Real(Kind = 8) :: Sum,Beta
      Integer, Parameter :: n = 11
      Integer :: ms = 1, MF = 7, msPLUS1, mfPLUS1,mfPLUS2,i,j, &
                 iMINUS1,ipt,icoef,jcoef
!
!  -------------------------------------------------------------------
!
!
      X(1:n) = (/0.05,0.11,0.15,0.31,0.46,0.52,0.7,0.74,0.82, &
                0.98,1.17/)
      Y(1:n) = (/0.956, 0.89, 0.832, 0.717, 0.571, 0.539, 0.378, &
                0.37, 0.306, 0.242, 0.104/)
!
! This program will find coefficients for each
! polynomial from degree MS to degree MF.
!  -------------------------------------------------------------------
!
! Compute matrix of coefficients and R.H.S. For mf degree.
! However, first check to see If max degree requested is too
```

Figure 3.19

```
!  large. it cannot exceed N-1. If it does, reduce to equal n-1
!  and write message.
!
      If ( MF > (n-1) ) Then
        MF = n - 1
        Write(*,200) MF
        End If
!
      mfPLUS1 = MF + 1
      mfPLUS2 = MF + 2
!
!  -----------------------------------------------------------------
!
!  Put ones into a new array. This will hold the powers of
!  the X-values as we proceed.
!
      Do i = 1,n
         Xn(i) = 1.0
         End Do
!
!  -----------------------------------------------------------------
!
!  Compute first column and N+1st column of A. I moves down the
!  rows, J sums over the N-values.
!
      Do i = 1,mfPLUS1
        A(i,1) = 0.0
        A(I,mfPLUS2) = 0.0
        Do j = 1,n
          A(i,1) = A(i,1) + Xn(j)
          A(I,mfPLUS2) = A(I,mfPLUS2) + Y(j)*Xn(j)
          Xn(j) = Xn(j) * X(j)
          End Do
      End Do
!
!  -----------------------------------------------------------------
!
!  Compute the last row of A. I moves across the columns,
!  and j sums over the n-values.
!
      Do i = 2,mfPLUS1
        A(mfPLUS1,I) = 0.0
        Do j = 1,n
          A(mfPLUS1,i) = A(mfPLUS1,i) + Xn(j)
          Xn(j) = Xn(j) * X(j)
          End Do
      End Do
!
!  -----------------------------------------------------------------
!
!  Now fill in the rest of the A matrix. I moves down the rows,
!  J moves across the columns.
!
```

Figure 3.19 *Continued*

```
      Do j = 2,mfPLUS1
        Do i = 1,MF
           A(i,j) = A(i+1,j-1)
           End Do
        End Do
!
! --------------------------------------------------------------
!
! Write out the matrix of normal equations.
!
      Write(*,*) ' *************************************************'
      Write(*,'(/)')
      Write(*, 204)
      Write(*,*) ' The normal equations are: '
      Write(*, '(/)')
      Write(*, 201) ((A(i,j), j=1,mfPLUS2), i=1,mfPLUS1)
      Write(*, '(//)')
!
! Now call a subroutine to solve the system. Do this for each
! degree from MS to MF. Get the LU decomposition of A.
!
      Call Crout_Reduction(A,mfPLUS1)
!
! --------------------------------------------------------------
!
! Reset the R.H.S. into C. we need to do this for each degree.
!
      msPLUS1 = MS + 1
      Do i = msPLUS1,mfPLUS1
        Do j = 1,I
           C(j) = A(J,mfPLUS2)
           End Do
        Call Solve_LU(A,C,I)
        iMINUS1 = I - 1
!
! --------------------------------------------------------------
!
! Now write out the coefficients of the least-squares polynomial.
!
      Write(*,202) iMINUS1, ( C(j), j=1,I)
!
! --------------------------------------------------------------
!
! Compute and print the value of
!       Beta = Sum of Deviations squared/(N-M-1).
!
      Beta = 0.0
      Do ipt = 1,n
        Sum = 0.0
        Do icoef = 2,i
           jcoef = i - icoef + 2
           Sum = ( Sum + C(jcoef) ) * X(ipt)
           End Do
```

Figure 3.19 *Continued*

```
                Sum = Sum + C(1)
                Beta = Beta + ( Y(ipt) - Sum )**2
                End Do
            Beta = Beta / (N - I)
            Write(*,203) Beta
        End Do
!
! -----------------------------------------------------------------
!
    200 Format(//' Degree of polynomial cannot exceed n-1.',/ &
                   ' requested maximum degree too large - ', &
                   'reduced to ',I3)
    201 Format(T2,9F8.2)
    202 Format(/' For degree of ',I2,' coefficients are'//  &
                   ' ',' ',11F9.3)
    203 Format(T10,' Beta is ',F10.5//)
    204 Format(T20, ' Output for Leasqr.f90 '/)
        Stop
        Contains
        Subroutine Crout_Reduction(a,n)
!
! -----------------------------------------------------------------
!
!       Subroutine Crout_Reduction :
!       Using the Crout reduction method transforms the matrix A
!       into the product of two matrices, L and U, where here the
!       upper-triangular matrix U has ones on its diagonal. The
!       L and U matrices are Returned in the matrix A.
!
!       This algorithm was present in Chapter 2.5.
!
! -----------------------------------------------------------------
!
        Implicit None
        Real(Kind = 8), Dimension(:,:), Intent(In Out) :: A
        Real(Kind = 8) :: Sum
        Integer, Intent(In) :: n
        Integer :: i,j,jM1, iMINUS1, k
!
! -----------------------------------------------------------------
!
        Do  i = 1,n
        Do  j = 2,N
          Sum = 0.0
          If ( J <= I ) Then
            JM1 = J - 1
            Do  K = 1,JM1
                Sum = Sum + A(I,K)*A(K,J)
                End Do
            A(i,j) = A(i,j) - Sum
          Else
            iMINUS1 = I - 1
```

Figure 3.19 *Continued*

```
                 If ( iMINUS1 /= 0 ) Then
                    Do   K = 1,iMINUS1
                         Sum = Sum + A(I,K)*A(K,J)
                         End Do
                    End If
!
!   ------------------------------------------------------------------
!
!   Test for small values on the diagonal
!
                 If ( Abs(A(I,I)) < 1.0E-10 ) Then
                    Write(*,100) I
                    Return
                 Else
                    A(i,j) = ( A(i,j) - Sum ) / A(i,i)
                    End If
               End If
        End Do
        End Do
           Return
!
      100 Format(' Reduction not completed because small value', &
                 ' found for divisor in row ',I3)
           End Subroutine Crout_Reduction
           Subroutine Solve_LU(A,B,n)
!
!   ------------------------------------------------------------------
!
!       Subroutine Solve_LU :
!                             This subroutine finds the solution to a set
!       of N linear equations that corresponds to the right-hand-side
!       vector B. the a matrix is the LU decomposition equivalent to
!       the coefficient matrix of the original equations, as produced
!       BY Crout_Reduction. the solution vector is Returned in the
!       B vector.
!
!   ------------------------------------------------------------------
!
        Implicit None
        Real(Kind = 8), Dimension(:,:), Intent(In) :: A
        Real(Kind = 8), Dimension(:), Intent(Out) :: B
        Real(Kind = 8) ::Sum
        Integer, Intent(In) :: n
        Integer :: I, j,k,iMINUS1
!
!   ------------------------------------------------------------------
!
!   Forward Substitution
!
        B(1) = B(1) / A(1,1)
        Do   i = 2,N
             iMINUS1 = i - 1
```

Figure 3.19 *Continued*

```
                   Sum = 0.0
                   Do   K = 1,iMINUS1
                         Sum = Sum + A(I,K)*B(K)
                         End Do
                B(i) = ( B(i) - Sum ) / A(i,i)
                   End Do
!
!    ------------------------------------------------------------------
!
!    Now we are ready for back-substitution. Recall that the elements
!    of U on the diagonal are all ones.
!
      Do   j = (n-1),1,-1
            sum = B(j)
            Do   k = (j+1),n
                  sum = sum - A(j,k)*B(k)
                  End Do
            B(j) = sum
            End Do
!
      Return
      End Subroutine Solve_LU
   End Program Least_Squares
```

```
***************Output for leasqr.f90 ********************

  The normal equations are:

    11.00    6.01    4.65    4.11    3.92    3.92    4.07    4.34    5.90
     6.01    4.65    4.11    3.92    3.92    4.07    4.34    4.72    2.18
     4.65    4.11    3.92    3.92    4.07    4.34    4.72    5.22    1.34
     4.11    3.92    3.92    4.07    4.34    4.72    5.22    5.84    1.00
     3.92    3.92    4.07    4.34    4.72    5.22    5.84    6.59    0.83
     3.92    4.07    4.34    4.72    5.22    5.84    6.59    7.50    0.74
     4.07    4.34    4.72    5.22    5.84    6.59    7.50    8.57    0.70
     4.34    4.72    5.22    5.84    6.59    7.50    8.57    9.84    0.68


  For degree of  1 coefficients are

            0.952  -0.760
            Beta is   0.00102


  For degree of  2 coefficients are

            0.998  -1.018    0.225
            Beta is   0.00023


  For degree of  3 coefficients are
            1.004  -1.079    0.351  -0.069
            Beta is   0.00026
```

Figure 3.19 *Continued*

```
    For degree of   4 coefficients are
            0.988  -0.837   -0.527   1.046   -0.456
            Beta is   0.00027


    For degree of   5 coefficients are
            1.037  -1.824   4.895  -10.753   10.537   -3.659
            Beta is   0.00013


    For degree of   6 coefficients are
            1.041  -1.946    5.886  -14.081   15.818   -7.599   1.112
            Beta is   0.00017


    For degree of   7 coefficients are
            1.060   -2.562   12.442  -44.819   88.980  -99.704   59.448
                -14.607
            Beta is   0.00021
```

Figure 3.19 *Continued*

Program 3.2 Interpolating in a Table of Data

Figure 3.20 is the listing. Output from the program follows the listing.

```
/*  *********************************************************************
 *                                                                     *
 *   Chapter 3      Interpolation Using Divided Differences            *
 *                                                                     *
 *   Gerald/Wheatley, APPLIED NUMERICAL ANALYSIS (sixth edition)       *
 *                 Addison Wesley Longman, 1999                        *
 *                                                                     *
 *********************************************************************
 *                                                                     *
 *   This program uses the divided differences to get the             *
 *   interpolating polynomial that goes through a given set of        *
 *   data points.                                                      *
 *                                                                     *
 *   PROCEDURE ddcoef finds the coefficents of the interpolating      *
 *   polynomial.                                                       *
 *                                                                     *
 *   FUNCTION ddvalue uses the coefficients of the previous           *
 *   procedure to evaluate the polynomial at a given value, u.        *
 *                                                                     *
 *                                                                     *
 *********************************************************************  */
```

Figure 3.20

```
#include <stdio.h>
#include <math.h>

float x[10], y[10],       /* the given data points */
      dd[10];             /* the vector of coefficients for p(x) */
float u;
int   i, j, n;


/*
  Procedure ddcoef begins here.
  INPUTS: x,y -  the given data points.
          n    - the number of data points
  OUTPUT: dd -  the coefficients of p(x)
*/

ddcoef(x, y, dd, n)
float x[], y[], dd[];
int    n;

{
  int i, j, k;
  float temp1, temp2;

  for (i = 1; i <= n; ++i)
    dd[i] = y[i];
  for (j = 2; j <= n; ++j)
  {
    temp1 = dd[j-1];
    printf("\n");
    for (k = j; k <= n; ++k)
    {
      temp2 = dd[k];
      dd[k] = (dd[k] - temp1)/(x[k] - x[k-j+1]);
      temp1 = temp2;
    }                       /* end of for k loop */
  }                         /* end of for j loop */
}                           /* end of ddcoef */

/* ****************************************************************
 *                                                               *
 *     Function ddvalue begins now                               *
 *     INPUT:  u - the x-value                                   *
 *     OUTPUT: ddvalue - the corresponding y-value, i.e. p(u)    *
 *                                                               *
 **************************************************************** */

float ddvalue(u)
float u;

{
  float sum;
  int i;
```

Figure 3.20 *Continued*

```
   sum = 0.0;

/*
    Compute value by nested multiplication from highest term
*/
    for ( i = n; i >= 2; --i )
       sum = (sum + dd[i]) * (u - x[i-1]);
    sum = sum + dd[1];
    return(sum);
}                               /* end of function ddvalue */


void main()                    /* start of main program */
{

/*
   Set up four data points
*/
   n = 5;
   x[1] = 3.2; x[2] = 2.7; x[3] = 1.0;
   x[4] = 4.8; x[5] = 5.6;
   y[1] = 22.0; y[2] = 17.8; y[3] = 14.2;
   y[4] = 38.3; y[5] = 51.7;

/*
    Generate the coefficients for the polynomial
*/
   ddcoef(x, y, &dd, n);       /* compute coefficients */
   clrscr();
   printf("\n");
   printf("The coefficients for the polynomial are:\n\n");
   for ( i = 1; i <= n; ++i )
      printf("%.5f ," dd[i]);
   printf("\n\n");
   printf("\t\t\t*********\n\n");
   getch();
   printf("\t\t  U\t\t   P(U)\t\t\n\n");
/*
    Set up a table of values from 1 to 7
*/
   u = 1.0;
   do
   {
     printf("\t\t%.3f\t\t %.5f\n," u, ddvalue(u));
     u = u + 0.2;
   }
   while ( u < 5.601 );

   printf("\n\n");
   printf("\t\t\t*********\n\n");
   getch();
}                                     /* end of main */
```

Figure 3.20 *Continued*

```
**************************************************************

                    OUTPUT FOR DIVDIF.C

The coefficients for the polynomial are:

  22.00000    8.40000    2.85562    -0.52748    0.25584


                    **********


              U                          P(U)

            1.000                      14.20000
            1.200                      12.89775
            1.400                      12.25424
            1.600                      12.16433
            1.800                      12.53272
            2.000                      13.27390
            2.200                      14.31221
            2.400                      15.58181
            2.600                      17.02667
            2.800                      18.60060
            3.000                      20.26723
            3.200                      22.00000
            3.400                      23.78221
            3.600                      25.60695
            3.800                      27.47714
            4.000                      29.40553
            4.200                      31.41471
            4.400                      33.53706
            4.600                      35.81480
            4.800                      38.30000
            5.000                      41.05450
            5.200                      44.15002
            5.400                      47.66806
            5.600                      51.69998
                    *********
```

Figure 3.20 *Continued*

Exercises

Section 3.2

1. Write the Lagrangian interpolating polynomial that passes through each point:

x	-2.3	0.5	3.1
y	2.1	-1.3	4.2

Plot the points, and sketch the parabola that passes through them.

▶ **2.** Given the four points, $(1, 2)$, $(3, 4)$, $(5, 3)$, $(9, 8)$, write the cubic in Lagrangian form that passes through them. Multiply out each term to express in the standard form, as $ax^3 + bx^2 + cx + d$.

3. Given that $\ln(2) = 0.69315$, $\ln(3) = 1.0986$, and $\ln(6) = 1.7918$, interpolate with a Lagrangian polynomial for the natural logarithm of each integer from 1 to 10. Tabulate these together with the error of each point.

4. If $e^{0.2}$ is approximated by Lagrangian interpolation among the values of $e^0 = 1$, $e^{0.1} = 1.1052$, and $e^{0.3} = 1.3499$, find the maximum and minimum estimates of the error. Compare to the actual error.

5. Repeat Exercise 4, but this time extrapolate to get $e^{0.4}$.

▶ **6.** Compute the Neville table from the following data and, from it, get interpolates for $f(4)$, using polynomials of degrees 2, 3, and 4.

x	$f(x)$
3	20.718
5	130.09
7	470.41
−2	−1.9817
−3	−17.993

7. Construct the Neville table for Exercise 4 to approximate $e^{0.2}$. Does the estimate from a polynomial of degree 2 agree with that in Exercise 4? What if we were to use linear interpolation? Would this be the same?

8. Repeat Exercise 5, this time using Neville's method.

9. Show that the entries in the top line of the Neville table of Exercise 7 do in fact represent the results of interpolating for $e^{0.2}$ with polynomials of increasing degrees.

10. Suppose a Neville table for the following n data points is computed with parallel processors. How many fewer time steps are required compared to doing it with a single CPU?

a. For $n = 8$
b. For $n = 16$
c. For $n = 11$

Section 3.3

11. Construct a divided-difference table from:

x	$f(x)$
0.5	−1.1518
−0.2	0.7028
0.7	−1.4845
0.1	−0.14943
0.0	0.13534

12. Repeat Exercise 2, except this time use divided differences. Compare the polynomial in standard form with that obtained in Exercise 2.

13. Repeat Exercise 4, but now use divided differences.

14. Use the divided difference table of Exercise 11 to estimate $f(0.15)$, using

a. a polynomial of degree 2 through the first three points.
b. a polynomial of degree 2 through the last three points.
c. a polynomial of degree 3 through the first four points.
d. a polynomial of degree 3 through the last four points.
e. a polynomial of degree 4.
f. Why are the results different?

▶ **15.** In Exercise 14, which three points are best to use for constructing the quadratic if we want

a. $f(0.15)$?
b. $f(−0.1)$?
c. $f(1.2)$?

16. Repeat Exercise 5, this time using divided differences.

17. The function in Exercise 11 is unknown, but that does not hinder our use of the table for interpolation. Interpolate with a cubic polynomial that passes through the first four points to get $f(0.2)$. Estimate the error from the next-term rule.

▶ **18.** In Exercise 13, you estimated a value for $e^{0.2}$. How does the actual error compare to the error bounds from Eq. (3.2)?

19. Complete the difference table for the following data:

x	1.20	1.25	1.30	1.35	1.40	1.45	1.50
$f(x)$	0.1823	0.2231	0.2624	0.3001	0.3365	0.3716	0.4055

20. In Exercise 19, what degree of polynomial is required to exactly fit to all seven data pairs? What lesser-degree polynomial will nearly fit the data? Justify your answer. (*Hint:* Look at the table for $f(x)$, a cubic polynomial.)

21. Form a difference table for $f(x) = 2x^3 - 4x^2 - 2x + 3$ for the interval $[-1, 1]$ with a spacing of 0.2. (If you do this by hand, remember that nested multiplication is more efficient.) Are the third differences a constant as expected? Is the value of this constant equal to $a_0 n! h^n$?

▶ **22.** Using the data in Exercise 19, compute the value of $\Delta^3 f_0$, if x_0 is the second entry in the table ($x_0 = 1.25$), directly from the f-values, not from the table.

23. Without computing the divided-difference table for the data in Exercise 19, what is $f[x_0, x_1, x_2, x_3]$ if x_0 is 1.25? Compute this in two ways, first from the table values, then from the answer to Exercise 22.

24. Use a Newton–Gregory interpolating polynomial of degree 3 to estimate the value of $f(1.37)$ from the data of Exercise 19. Select the best point to call x_0. Estimate the error by the next-term rule.

25. Repeat Exercise 24, but now get $f(0.77)$. Is the estimated error larger than in Exercise 24? If so, explain.

▶ **26.** The following table is already computed. Use a Newton–Gregory interpolating polynomial of degree 2 to estimate $f(0.203)$, taking $x_0 = 0.125$. Then add one term to get $f(0.203)$ from a third-degree polynomial. Estimate the errors of each from the next-term rule.

x	$f(x)$	Δf	$\Delta^2 f$	$\Delta^3 f$	$\Delta^4 f$
0.125	0.79168				
		−0.01834			
0.250	0.77334		−0.01129		
		−0.02963		0.00134	
0.375	0.74371		−0.00995		0.00038
		−0.03958		0.00172	
0.500	0.70413		−0.00823		0.00028
		−0.04781		0.00200	
0.625	0.65632		−0.00623		
		−0.05404			
0.750	0.60228				

27. Repeat Exercise 26, but now get $f(0.612)$, taking $x_0 = 0.375$.

28. What would be the answers to Exercise 27 if we took $x_0 = 0.125$?

29. Use the data of Table 14 to find a value for $y(0.54)$ using a cubic that fits at $x = 0.3, 0.5, 0.7,$ and 0.9.

▶ **30.** What is the minimum degree of polynomial that will exactly fit all seven pairs of data in Exercise 29? (Answer is *not* sixth-degree.)

31. Construct a divided-difference table for the data in Exercise 29. How do the values compare to those in the given table?

32. The precision of $f(x)$ data has a considerable effect on a table of differences. Demonstrate this fact by recomputing the table of Exercise 26 after rounding to three decimal places. Repeat this calculation, but chop after three places.

x	y	Δy	$\Delta^2 y$	$\Delta^3 y$
0.1	0.003			
		0.064		
0.3	0.067		0.017	
		0.081		0.002
0.5	0.148		0.019	
		0.100		0.003
0.7	0.248		0.022	
		0.122		0.004
0.9	0.370		0.026	
		0.148		0.005
1.1	0.518		0.031	
		0.179		
1.3	0.697			

Table 14

Section 3.4

33. For $f(x)$ as defined here, find polynomials of degrees 2, 3, 4, and 5 that fit $f(x)$ at equally spaced points in $[-1, 1]$. Plot these values and observe that the fit is poor.

$$f(x) = \begin{cases} 0, & -1 < x < -0.25 \\ 1 - |4x|, & -0.25 < x < 0.25 \\ 0, & 0.25 < x < 1.0 \end{cases}$$

▶ **34.** Find the coefficient matrix and the right-hand side for fitting a cubic spline curve to the following data. Use linearity end conditions (condition 1).

x	$f(x)$
0.15	0.1680
0.27	0.2974
0.76	0.7175
0.89	0.7918
1.07	0.8698
2.11	0.9972

35. Solve the set of equations in Exercise 34, and then determine the coefficients of the various cubics. Plot the cubic spline curve. Compare the interpolates at $x = 0.33$, $x = 0.92$, and $x = 2.05$ with the tabulated values for $ERF(x)$ (the so-called error function).

36. Repeat Exercises 34 and 35, this time for each of the other end conditions. How different are the interpolates, and which end condition gives the least average error?

37. Fit a natural cubic spline to the function in Exercise 33, matching to the function at five equally spaced points between -1 and 1. Plot the cubic spline curve and compare it to the plots of the polynomials of Exercise 33, particularly with the fourth-degree polynomial.

38. Repeat Exercise 37, but use end conditions 3 and 4. Plot these points and compare to the plot from end condition 1.

▶ **39.** Repeat Exercise 38, but now force the end slopes to be zero.

40. If the data given are periodic and cover one period, the first and last points will have identical f-values and the beginning and ending slopes will be the same. Develop the relations that give a cubic spline curve for such periodic data.

41. The data in Example 3.6 are from the kind of periodic data referred to in Exercise 40. Use the relation developed in Exercise 40 to get the periodic cubic spline curve. Which of the results of Example 3.6 are closest to this spline?

Section 3.5

42. Show that the matrix forms of the equations for Bezier and B-spline curves are equivalent to the algebraic equations given in Section 3.5.

▶ **43.** Write the matrix form of the equations for Bezier curve of order 4.

44. Prove that the convex hull does enclose all the points for both Bezier and B-spline curves. (*Hint:* Use the fact that any point p in the convex hull for the set of points $\{p_0, p_1, \ldots, p_n\}$ can be written as $p = \Sigma \, a_i p_i$ where all the a's are nonnegative and their sum is one.)

▶ **45.** The slopes at the ends of the cubic B-spline curve seem to be the same as the slopes between adjacent points. Is this true?

46. Suppose that we have constructed a connected B-spline curve and then one of the points is changed. What part(s) of the connected curve is (are) affected? Are Bezier curves similar in this respect? What about a cubic spline? Do the terms *local control* and *global control* apply to the phenomena you observe?

47. A higher-degree B-spline curve is a natural extension of our cubic B-splines. What about reducing the degree

to give a quadratic B-spline curve? What assumptions are reasonable for such quadratics?

48. Compute and then graph the cubic Bezier curve defined by this set of points.

Point	x	$f(x)$
1	100	100
2	50	150
3	200	150
4	50	50
5	100	200
6	50	100
7	200	100
8	50	50
9	100	200
10	100	100

49. By letting u vary from 0 to 1, plot the weighting factors that produce a Bezier curve.

50. Repeat Exercise 49, this time for a B-spline curve.

Section 3.6

51. In Section 3.6 it is asserted that the order in which the interpolation is done does not matter. Verify that this is true by interpolating within the data of Table 3.9 to find values at $y = 0.33$ (for x constant at 1.0, 1.5, and 2.0), using cubic interpolation with y-values of 0.2, 0.3, 0.4, and 0.5. Then interpolate from these to get $f(1.6, 0.33)$, and compare to the value 1.841 obtained in the text.

▶ **52.** In Example 3.8 after the computations were completed, it was observed that a cubic in x and a quadratic in y would be preferred. Do this to obtain an estimate of $f(1.6, 0.33)$ and compare it to the true value, 1.8350. Use the best "region of fit."

53. Example 3.8 used a rectangular region of fit when a more nearly circular region should be advantageous. Interpolate from the data of Table 3.9 to evaluate $f(1.62, 0.31)$ by a set of polynomials that fit at $x = 1.5$ to 2.0 when y is 0.2 or 0.4, and fits at $x = 0.5$ to 2.5 when y is 0.3. Do this by forming a set of difference tables. This is awkward to do if we begin with x held constant, but there is no problem if we begin with y held constant.

▶ **54.** Interpolate for $f(3.32, 0.71)$ from the following data. Use cubics in each direction and with the best region of fit. Because the x- and y-values are not evenly spaced, you will need to use Lagrangian polynomials or divided differences.

x \ y	0.1	0.4	0.6	0.9	1.2
1.1	1.100	0.864	0.756	0.637	0.550
3.0	8.182	6.429	5.625	4.737	4.091
3.7	12.445	9.779	8.556	7.205	6.223
5.2	24.582	19.314	16.900	14.232	12.291
6.5	38.409	30.179	26.406	22.237	19.205

55. Find a value at (3.7, 0.6) on the B-spline surface constructed from the 16 points in the upper-left corner of the data of Exercise 54.

56. A Bezier surface can be constructed in a manner similar to that for a B-spline surface. Repeat Exercise 55, but for a Bezier surface.

Section 3.7

57. The "least-squares" line for the data in Fig. 3.13 has the equation

$$R = 702.2 + 3.395T.$$

A line drawn by eye was $R = 700 + 3.5T$. Compute the deviations of the actual data from each of these lines. Then compare the sum of the squares of these deviations. Observe that, even though the maximum deviation is about the same for each, the sum of squares differs significantly. How do the average errors compare?

58. Show that the point (X, Y), where X is the mean of all x-values and Y is the mean of all y-values, falls on the least-squares line. Often a change of variable is made to put the point at the origin, thus reducing the magnitude of the numbers worked with, which is an advantage if the least-squares line is computed by hand.

▶ **59.** Find the least-squares line that fits the following data, assuming that the x-values are free from error. (The data are actually $y = 3x + 2$ plus a random variation.)

x	y
1	5.04
2	8.12
3	10.64
4	13.18
5	16.20
6	20.04

60. Repeat Exercise 59, but now consider all the errors to be in the x-values with y free of error. Modify the nor-

mal equations to get the least-squares line $x = ay + b$. Observe that this is not the same line obtained in Exercise 59.

▶ **61.** Multivariate analysis finds a function of more than one independent variable. Suppose that z is a function of both x and y. Find the normal equations to fit

$$z = ax + by + c$$

and then use the following data to fit the least-squares plane to them.

x	0	1.2	2.1	3.4	4.0	4.2	5.6	5.8	6.9
y	0	0.5	6.0	0.5	5.1	3.2	1.3	7.4	10.2
z	1.2	3.4	−4.6	9.9	2.4	7.2	14.3	3.5	1.3

62. Draw the straight line between (2, 3) and (6, 5). Write its equation in the form $y = ax + b$. How much does this line shift to give the least-squares line that fits to these two points and a third point, if the third point is

a. (4, 5)?
b. (4, 2)?
c. (5, 5)?
d. (3, 2)?

63. In Section 3.7 the statement is made that $A * A^T$, where A is the design matrix given in Section 3.7, is equal to the coefficient matrix of Eq. (3.26). Show that this is true, and show that $A * y$, where y is the column vector of y-values, is the same as the right-hand side of Eq. (3.26).

64. Show that $A * A^T$, where A is the design matrix, is symmetric and positive definite.

65. Observe that the following data seem to be fit by a curve $y = ae^{bx}$ by plotting on semilog paper and noting that the points appear to fall on a straight line. (The data are the solubilities of n-butane in anhydrous hydrofluoric acid at high pressures and were used in the design of petroleum refineries.) Find values of a and b from the plot.

Temperature, °F	Solubility, wt. %
77	2.4
100	3.4
185	7.0
239	11.1
285	19.6

66. Plot the data of Exercise 65 on ordinary graph paper, and observe that the data are nonlinear.

▶ **67.** Find the least-squares values for the parameters of $y = ae^{bx}$ by fitting the data of Exercise 65 to the relation $\ln(y) = \ln(a) + bx$.

68. It is suspected (from theoretical considerations) that the rate of flow from a fire hose is proportional to some power of the pressure at the nozzle. Do the following data confirm the speculation? What is the least-squares value of the exponent? (Assume the pressure data are more accurate.)

Flow, gallons per minute	94	118	147	180	230
Pressure, psi	10	16	25	40	60

69. Plot the data of Exercise 68 on log-log paper, and observe that they nearly fall on a line of slope 2. That means that a quadratic would be a good function for fitting them. Find the least-squares values for the constants in

$$\text{Flow} = aP^2 + bP + c, \quad \text{where } P = \text{pressure.}$$

70. The data in Exercise 59, although actually perturbations from a straight line, plot better along a curve because of the accidental occurrence of three negative deviations in succession. Fit a quadratic to the data. How do the sums of squares of the deviations compare between the line and the curve?

71. If the degree of the least-squares polynomial is exactly one less than the number of points, the polynomial passes through all of the points; that is, it is an interpolating polynomial. Fit a fifth-degree polynomial to the six points of Exercise 59. Now plot this polynomial and compare it to plots of the least-squares line of Exercise 59 and the least-squares quadratic of Exercise 70. Compare the computed y-values at $x = 1.5, 2.5, 3.5, 4.5,$ and 5.5 from the three relations.

72. How do the maximum and minimum values of the slopes of the three curves of Exercise 71 compare to the true slope of 3? What does this mean with respect to getting the slope of experimental data?

▶ **73.** The following data seem to fit to a cubic equation but determine by least squares the optimum degree.

x	$f(x)$	x	$f(x)$
0.1	1.9	9.4	-3.1
1.1	7.9	11.1	-13.0
1.6	24.9	11.4	-28.7
2.4	24.9	12.2	-39.5

x	$f(x)$	x	$f(x)$
2.5	34.9	13.2	-48.6
4.1	42.7	14.1	-40.2
5.2	29.7	15.6	-51.6
6.1	49.8	16.1	-30.5
6.6	36.1	17.6	-34.6
7.1	23.7	17.9	-16.4
8.2	13.0	19.1	-13.4
9.1	20.5	20.0	-1.1

74. Repeat Exercise 73, but this time use every other point $(x = 0.1, 1.6, 2.5, \ldots)$. Do you get the same results? Repeat again but with the other half of the points.

75. The data of Exercise 73 suggest a function of the form $y = A + B * \sin(Cx)$. How can least squares be used to determine the coefficients? What difficulties will there be in solving the normal equations? Suppose that it were known that $C = \pi/10$. Would this make it easier to get the values of A and B?

Section 3.8

76. Write $x^3 - 2x^2 + x - 7$ in at least two different ways that do not look identical.

77. Find bounds to the errors of each of the results of Exercise 3, and compare them to the actual errors.

▶ **78.** Find bounds to the errors when each of the polynomials of Exercise 33 is used to estimate $f(0.1)$. Compare to the actual errors.

79. Make a table for $f(x) = e^{-x}(x^2)$ for $x = 0.2, 0.3, 0.6, 0.9,$ and 1.0. Construct quadratic interpolating polynomials using three successive points beginning at $x = 0.2, x = 0.3,$ and $x = 0.6$. What are the errors if we use each of these to estimate $f(0.5)$? Compare to the bounds for the errors.

80. Fit $f(x) = 2x - 1 + \exp(-10x^2)$ between $x = -1$ and $x = 1$ with polynomials of degrees 2, 3, 4, 5, and 6 that match to $f(x)$ at equally spaced points. Compare the graphs of these to the plot of $f(x)$.

Section 3.9

Use *Mathematica* or another symbolic algebra program to solve Exercises 81–88.

81. Construct the divided-difference table for the data of Exercise 11.

82. This is more challenging: Construct the Neville table for the data of Exercise 6.

83. Obtain the cubic interpolating polynomial that fits to the data of Exercise 2.

▶ **84.** Solve Exercise 51 with a computer algebra program.

85. Solve Exercise 61 with a computer algebra program.

▶ **86.** Solve Exercise 67 with a computer algebra program.

87. Plot the cubic spline curve that passes through the points of Exercise 34.

88. Plot the Bezier curve that is defined by the data of Exercise 48.

Applied Problems and Projects

89. S. H. P. Chen and S. C. Saxena report experimental data for the emittance of tungsten as a function of temperature [*Ind. Eng. Chem. Fund.* 12, 220 (1973)]. Their data follow. They found that the equation

$$e(T) = 0.02424 \left(\frac{T}{303.16} \right)^{1.27591}$$

correlated the data for all temperatures accurately to three digits. What degree of interpolating polynomial is required to match to their correlation at points midway between the tabulated temperatures? Discuss the pros and cons of polynomial interpolation in comparison to using their correlation.

$T, °K$	300	400	500	600	700	800	900	1000	1100
e	0.024	0.035	0.046	0.058	0.067	0.083	0.097	0.111	0.125

$T, °K$	1200	1300	1400	1500	1600	1700	1800	1900	2000
e	0.140	0.155	0.170	0.186	0.202	0.219	0.235	0.252	0.269

90. In studies of radiation-induced polymerization, a source of gamma rays was employed to give measured doses of radiation. However, the dosage varied with position in the apparatus, with these figures being recorded:

Position, in. from base point	0	0.5	1.0	1.5	2.0	3.0	3.5	4.0
Dosage, 10^5 rads/hr	1.90	2.39	2.71	2.98	3.20	3.20	2.98	2.74

For some reason, the reading at 2.5 in. was not reported, but the value of radiation there is needed. Fit interpolating polynomials of various degrees to the data to supply the missing information. What do you think is the best estimate for the dosage level at 2.5 in.?

91. Studies of the kinetics of elution of copper compounds from ion-exchange resins gave the following data. The normality of the leaching liquid was the most important factor in determining the diffusivity. The data were obtained at convenient values of normality; we desire a table of D for integer values of normality ($N = 0.0, 1.0, 2.0, 3.0, 4.0, 5.0$). Use the data to construct such a table.

N	$D \times 10^6$, cm²/sec	N	$D \times 10^6$, cm²/sectr
0.0521	1.65	0.9863	3.12
0.1028	2.10	1.9739	3.06
0.2036	2.27	2.443	2.92
0.4946	2.76	5.06	2.07

92. Experiment with the placing of intermediate points in Exercise 37 to see whether you can reduce the average error of the cubic spline curve.

93. When the steady-state heat-flow equation is solved numerically, temperatures $u(x, y)$ are calculated at the nodes of a gridwork constructed in the domain of interest. (This is the content of Chapter 7.) When a certain problem was solved, the values given in the following table were obtained. This procedure does not give the temperatures at points other than the nodes of the grid; if they are desired, one can interpolate to find them. Use the data to estimate the values of the temperature at the points (0.7, 1.2), (1.6, 2.4), and (0.65, 0.82).

x \ y	0.0	0.5	1.0	1.5	2.0	2.5
0.0	0.0	5.00	10.00	15.00	20.00	25.00
0.5	5.00	7.51	10.05	12.70	15.67	20.00
1.0	10.00	10.00	10.00	10.00	10.00	10.00
1.5	15.00	12.51	9.95	7.32	4.33	0.0
2.0	20.00	15.00	10.00	5.00	0.00	−5.00

94. Star S in the Big Dipper (Ursa Major) has a regular variation in its apparent magnitude. Leon Campbell and Laizi Jacchia give data for the mean light curve of this star in their book *The Story of Variable Stars* (Blakeston, 1941). A portion of these data is given here.

Phase		−110	−80	−40	−10	30	80	110
Magnitude		7.98	8.95	10.71	11.70	10.01	8.23	7.86

The data are periodic in that the magnitude for phase $= -120$ is the same as for phase $= +120$. The spline functions discussed in Section 3.4 do not allow for periodic behavior. For a periodic function, the slope and second derivatives are the same at the two endpoints. Taking this into account, develop a spline that interpolates the preceding data.

   Other data given by Campbell and Jacchia for the same star are

Phase		−100	−60	−20	20	60	100
Magnitude		8.37	9.40	11.39	10.84	8.53	7.89

How well do interpolants based on your spline function agree with this second set of observations?

95. Develop the matrices to make Eq. (3.21) generate points on a Bezier surface. Show that this passes through the 12 points on the borders of the group of 16 in Fig. 3.12. How could a Bezier surface be created that passes through the innermost set of four points?

96. A fictitious chemical experiment produces seven data points:

t		−1	−0.96	−0.86	−0.79	0.22	0.5	0.930
y		−1	−0.151	0.894	0.986	0.895	0.5	−0.306

   a. Plot the points and interpolate a smooth curve by intuition.
   b. Plot the unique sixth-degree polynomial that interpolates these points.
   c. Use a spline program to evaluate enough points to plot this curve.
   d. Compare your results with the graph in Fig. 3.21.

97. There are several different ways to construct interpolating polynomials from a table of evenly spaced data. Earlier editions of this book discussed this in detail. (A device called a "lozenge diagram" can be helpful.) Using the data of the table in Exercise 29, construct these polynomials, each of degree four:

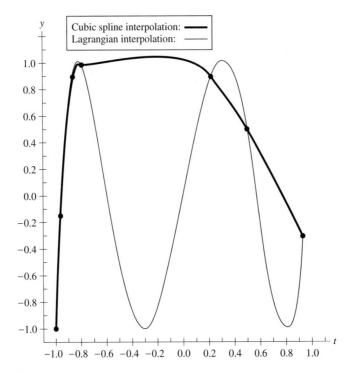

Figure 3.21

a. Newton–Gregory backward, $x_0 = 1.1$.
b. Stirling, $x_0 = 0.7$.
c. Gauss forward, $x_0 = 0.7$.

Do all of these give the same interpolate for $y(0.54)$?

98. Suppose that the entry for $y(0.4)$ in the table of Exercise 29 were erroneously entered as 0.284 (two digits reversed). How does this affect the difference table? Is there a relation that tells how such an isolated error is propagated through the table of differences?

4

Approximation of Functions

Contents of This Chapter

In this chapter we consider two ways to approximate the values of a known function. Both are of great importance. We first discuss how to represent a function by a polynomial or a ratio of polynomials in the most efficient way. "Efficient way" refers to obtaining values, at any point in an interval, with the smallest error for a given number of arithmetic operations. This topic is essential when developing computer procedures that compute sines, cosines, logarithms, and other nonalgebraic functions. We limit our search of approximating functions to polynomials and ratios of these because such computation is the only kind a computer can do. (Modern math-coprocessor chips can do more.)

The second topic of the chapter is the technique of approximating a function with a series containing only sines and cosines. Such series are called Fourier series. The coefficients of the Fourier series representation of a function are found by evaluating the definite integrals of a product of the function and certain cosine and sine terms. This evaluation can be facilitated by computer algebra systems. (The coefficients can also be obtained by numerical procedures; we consider this technique in the next chapter.)

To some extent this chapter considers a problem that is in contrast to that of the previous chapter. There we tried to find a function (usually a polynomial) that matches a set of data points. Here we try to find a function (again, usually involving polynomials but also other forms) that matches a function of some other type.

4.1 Chebyshev Polynomials

Develops the theory of a class of orthogonal polynomials that are the basis for fitting nonalgebraic functions with polynomials of maximum efficiency.

4.2 Economized Power Series

Shows how the Chebyshev polynomials can be used to create polynomial approximations that are significantly more efficient than a Maclaurin series.

4.3 Approximation with Rational Functions

Applies the previous techniques to generate the coefficients of rational functions (ratios of polynomials) that are even more efficient. The section also discusses continued fractions, which offer a means of evaluating a rational function with fewer operations.

4.4 Fourier Series

Discusses these approximating series composed of sine and cosine terms. The classical method of determining the coefficients involves many integrations. However, for almost all functions, a Fourier series can approximate a function throughout an interval quite closely. Fourier series are important in many areas, particularly in the analytical solution of partial-differential equations.

4.5 Theoretical Matters

Presents a small part of the large theory associated with this chapter's topics.

4.6 Using Computer Algebra Systems

Describes those elements of the symbolic operations of the three computer algebra systems that can help in the computations of this chapter.

Chapter Summary

Reviews the important topics of this chapter.

Computer Program

Illustrates how some of the procedures in the chapter can be implemented on a computer.

Parallel Processing

The numerical procedures of this chapter offer many possibilities for applying parallel processing, but all have been discussed before:

Sections 4.2 and 4.3 evaluate polynomials.

Section 4.3 solves systems of equations.

Section 4.4 evaluates a sum.

4.1 Chebyshev Polynomials

We turn now to the problem of representing a function with a minimum error. This is a central problem in the software development of digital computers because it is more economical to compute the values of the common functions using an efficient approximation than to store a table of values and employ interpolation techniques. Because digital computers are essentially only arithmetic devices, the most elaborate function they can compute is a rational function, a ratio of polynomials. We will hence restrict our discussion to representation of functions by polynomials or rational functions.

One way to approximate a function by a polynomial is to use a truncated Taylor series. This is not the best way, in most cases. To study better ways, we first need to introduce the Chebyshev polynomials.

The familiar Taylor-series expansion represents the function with very small error near the point of the expansion, but the error increases rapidly (proportional to a power) as we employ it at points farther away. In a digital computer, we have no control over where in an interval the approximation will be used, so the Taylor series is not usually appropriate. We would prefer to trade some of its excessive precision at the center of the interval to reduce the errors at the ends.

We can do this while still expressing functions as polynomials by the use of Chebyshev polynomials. The first few of these are*

$$
\begin{aligned}
T_0(x) &= 1, \\
T_1(x) &= x, \\
T_2(x) &= 2x^2 - 1, \\
T_3(x) &= 4x^3 - 3x, \\
T_4(x) &= 8x^4 - 8x^2 + 1, \\
T_5(x) &= 16x^5 - 20x^3 + 5x, \\
T_6(x) &= 32x^6 - 48x^4 + 18x^2 - 1, \\
T_7(x) &= 64x^7 - 112x^5 + 56x^3 - 7x, \\
T_8(x) &= 128x^8 - 256x^6 + 160x^4 - 32x^2 + 1, \\
T_9(x) &= 256x^9 - 576x^7 + 432x^5 - 120x^3 + 9x, \\
T_{10}(x) &= 512x^{10} - 1280x^8 + 1120x^6 - 400x^4 + 50x^2 - 1.
\end{aligned}
\tag{4.1}
$$

The members of this series of polynomials can be generated from the two-term recursion formula

$$
T_{n+1}(x) = 2xT_n(x) - T_{n-1}(x), \qquad T_0(x) = 1, \qquad T_1(x) = x. \tag{4.2}
$$

---

*The commonly accepted symbol $T(x)$ comes from the older spelling, Tschebycheff.

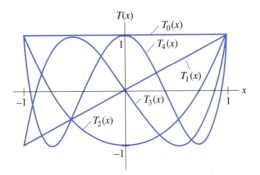

Figure 4.1

Note that the coefficient of x^n in $T_n(x)$ is always 2^{n-1}. In Fig. 4.1 we plot the first four polynomials of Eq. (4.1).

These polynomials have some unusual properties. They form an orthogonal set, in that

$$\int_{-1}^{1} \frac{T_n(x)T_m(x)}{\sqrt{1-x^2}}\,dx = \begin{cases} 0, & n \neq m, \\ \pi, & n = m = 0, \\ \dfrac{\pi}{2}, & n = m \neq 0. \end{cases} \tag{4.3}$$

The orthogonality of these functions will not be of immediate concern to us.

The Chebyshev polynomials are also terms of a Fourier series,* because

$$T_n(x) = \cos n\theta, \tag{4.4}$$

where $\theta = \arccos x$. Observe that $\cos 0 = 1$, $\cos \theta = \cos(\arccos x) = x$.

To demonstrate the equivalence of Eq. (4.4) to Eqs. (4.1) and (4.2), we recall some trigonometric identities, such as

$$\cos 2\theta = 2\cos^2 \theta - 1,$$
$$T_2(x) = 2x^2 - 1;$$
$$\cos 3\theta = 4\cos^3 \theta - 3\cos \theta,$$
$$T_3(x) = 4x^3 - 3x;$$
$$\cos(n+1)\theta + \cos(n-1)\theta = 2\cos \theta \cos n\theta,$$
$$T_{n+1}(x) + T_{n-1}(x) = 2xT_n(x).$$

---

* We discuss Fourier series later in this chapter.

Because of the relation $T_n(x) = \cos n\theta$, it is apparent that the Chebyshev polynomials have a succession of maximums and minimums of alternating signs, each of magnitude one. Further, because $|\cos n\theta| = 1$ for $n\theta = 0, \pi, 2\pi, \ldots$, and because θ varies from 0 to π as x varies from 1 to -1, $T_n(x)$ assumes its maximum magnitude of unity $n + 1$ times on the interval $[-1, 1]$. Figure 4.1 illustrates this for T_n, $n = 0, 1, 2, 3, 4$.

Most important for our present application of these polynomials is the fact that, of all polynomials of degree n where the coefficient of x^n is unity, the polynomial

$$\frac{1}{2^{n-1}} T_n(x)$$

has a smaller upper bound to its magnitude in the interval $[-1, 1]$ than any other. We prove this in a later section. Because the maximum magnitude of $T_n(x)$ is one, the upper bound referred to is $1/2^{n-1}$. This is important because we will be able to write power-series representations of functions whose maximum errors are given in terms of this upper bound.

4.2   Economized Power Series

We are now ready to use Chebyshev polynomials to "economize" a power series. Consider the Maclaurin series for e^x:

$$e^x = 1 + x + \frac{x^2}{2} + \frac{x^3}{6} + \frac{x^4}{24} + \frac{x^5}{120} + \frac{x^6}{720} + \cdots .$$

If we would like to use a truncated series to approximate e^x on the interval $[0, 1]$ with a precision of 0.001, we will have to retain terms through that in x^6, because the error after the term in x^5 will be more than 1/720. Suppose we subtract

$$\left(\frac{1}{720}\right)\left(\frac{T_6}{32}\right)$$

from the truncated series. We note from Eq. (4.1) that this will exactly cancel the x^6 term and at the same time make adjustments in other coefficients of the Maclaurin series. Because the maximum value of T_6 on the interval $[0, 1]$ is unity, this will change the sum of the truncated series by only

$$\frac{1}{720} \cdot \frac{1}{32} < 0.00005,$$

which is small with respect to our required precision of 0.001. Performing the calculations, we have

$$e^x \doteq 1 + x + \frac{x^2}{2} + \frac{x^3}{6} + \frac{x^4}{24} + \frac{x^5}{120} + \frac{x^6}{720}$$

$$- \frac{1}{720}\left(\frac{1}{32}\right)(32x^6 - 48x^4 + 18x^2 - 1), \tag{4.5}$$

$$e^x \doteq 1.000043 + x + 0.499219x^2 + \frac{x^3}{6} + 0.043750x^4 + \frac{x^5}{120}.$$

This gives a fifth-degree polynomial that approximates e^x on [0, 1] almost as well as the sixth-degree one derived from the Maclaurin series. (The actual maximum error of the fifth-degree expression is 0.000270; for the sixth-degree expression it is 0.000226.) We hence have "economized" the power series in that we get nearly the same precision with fewer terms.

By subtracting $\frac{1}{120}(T_5/16)$ we can economize further, getting a fourth-degree polynomial that is almost as good as the economized fifth-degree one. It is left as an exercise to do this and to show that the maximum error is now 0.000781, so that we have found a fourth-degree power series that meets an error criterion that requires us to use two additional terms of the original Maclaurin series. Because of the relative ease with which they can be developed, such economized power series are frequently used for approximations to functions and are much more efficient than power series of the same degree obtained by merely truncating a Taylor or Maclaurin series. Table 4.1 compares the errors of these power series.

The maximum error in the economized fifth-degree polynomial is only slightly greater than in the sixth-degree Maclaurin series. The economized fourth-degree polynomial incurs a maximum error about three and one-half times as much, but still within the 0.001 limit that was initially imposed, and will require significantly reduced computational effort. In addition, there is a proportionately reduced memory-space requirement to store the con-

Table 4.1 Comparison of errors of economized power series and a Maclaurin series for e^x

x	e^x	Maclaurin, sixth-degree	Economized, fifth-degree	Economized, fourth-degree	Maclaurin, fourth-degree
0	1.00000	1.00000	1.00004	1.00004	1.00000
0.2	1.22140	1.22140	1.22142	1.22098	1.22140
0.4	1.49182	1.49182	1.49179	1.49133	1.49173
0.6	1.82212	1.82211	1.82208	1.82212	1.82140
0.8	2.22554	2.22549	2.22553	2.22605	2.22240
1.0	2.71828	2.71806	2.71801	2.71749	2.70833
Maximum error		0.00023	0.00027	0.00078	0.00995

stants of the polynomial. In contrast, a fourth-degree Maclaurin series has an error nearly ten times greater than the 0.001 tolerance, and its error is over twelve times that of the fourth-degree economized form.

Chebyshev Series

By rearranging the Chebyshev polynomials, we can express powers of x in terms of them:

$$1 = T_0,$$

$$x = T_1,$$

$$x^2 = \frac{1}{2}(T_0 + T_2),$$

$$x^3 = \frac{1}{4}(3T_1 + T_3),$$

$$x^4 = \frac{1}{8}(3T_0 + 4T_2 + T_4),$$

$$x^5 = \frac{1}{16}(10T_1 + 5T_3 + T_5),$$ 　(4.6)

$$x^6 = \frac{1}{32}(10T_0 + 15T_2 + 6T_4 + T_6),$$

$$x^7 = \frac{1}{64}(35T_1 + 21T_3 + 7T_5 + T_7),$$

$$x^8 = \frac{1}{128}(35T_0 + 56T_2 + 28T_4 + 8T_6 + T_8),$$

$$x^9 = \frac{1}{256}(126T_1 + 84T_3 + 36T_5 + 9T_7 + T_9).$$

By substituting these identities into an infinite Taylor series and collecting terms in $T_i(x)$, we create a Chebyshev series. For example, we can get the first four terms of a Chebyshev series by starting with the Maclaurin expansion for e^x. Such a series converges more rapidly than does a Taylor series on $[-1, 1]$:

$$e^x = 1 + x + \frac{x^2}{2} + \frac{x^3}{6} + \frac{x^4}{24} + \cdots.$$

Replacing terms by Eq. (4.6), but omitting polynomials beyond $T_3(x)$ because we want only four terms,* we have

---

*The number of terms that are employed determines the accuracy of the computed values, of course.

Table 4.2   Comparison of Chebyshev series for e^x with Maclaurin series:
$e^x = 0.9946 + 0.9973x + 0.5430x^2 + 0.1772x^3$;
$e^x = 1 + x + 0.5x^2 + 0.1667x^3$

x	e^x	Chebyshev	Error	Maclaurin	Error
−1.0	0.3679	0.3631	0.0048	0.3333	0.0346
−0.8	0.4493	0.4536	−0.0042	0.4346	0.0147
−0.6	0.5488	0.5534	−0.0046	0.5440	0.0048
−0.4	0.6703	0.6712	−0.0009	0.6693	0.0010
−0.2	0.8187	0.8154	0.0033	0.8187	0.0001
0	1.0000	0.9946	0.0054	1.0000	0.0000
0.2	1.2214	1.2172	0.0042	1.2213	0.0001
0.4	1.4918	1.4917	0.0001	1.4907	0.0012
0.6	1.8221	1.8267	−0.0046	1.8160	0.0061
0.8	2.2255	2.2307	−0.0051	2.2054	0.0202
1.0	2.7183	2.7121	0.0062	2.6667	0.0516

$$e^x = T_0 + T_1 + \frac{1}{4}(T_0 + T_2) + \frac{1}{24}(3T_1 + T_3) + \frac{1}{192}(3T_0 + 4T_2 + \cdots)$$

$$+ \frac{1}{1920}(10T_1 + 5T_3 + \cdots) + \frac{1}{23{,}040}(10T_0 + 15T_2 + \cdots) + \cdots$$

$$= 1.2661T_0 + 1.1302T_1 + 0.2715T_2 + 0.0443T_3 + \cdots.$$

To compare the Chebyshev expansion with the Maclaurin series, we convert back to powers of x, using Eq. (4.1):

$$e^x = 1.2661 + 1.1302(x) + 0.2715(2x^2 - 1) + 0.0443(4x^3 - 3x) + \cdots. \tag{4.7}$$
$$e^x = 0.9946 + 0.9973x + 0.5430x^2 + 0.1772x^3 + \cdots.$$

Table 4.2 and Fig. 4.2 compare the error of the Chebyshev expansion, Eq. (4.7), with the Maclaurin series, using terms through x^3 in each case. The figure shows how the Chebyshev expansion attains a smaller maximum error by permitting the error at the origin to increase. The errors can be considered to be distributed more or less uniformly throughout the interval. In contrast to this, the Maclaurin expansion, which gives very small errors near the origin, allows the error to bunch up at the ends of the interval.

If the function is to be expressed directly as an expansion in Chebyshev polynomials, the coefficients can be obtained by integration. Based on the orthogonality property, the coefficients are computed from *

$$a_i = \frac{2}{\pi} \int_{-1}^{1} \frac{f(x)T_i(x)}{\sqrt{1 - x^2}} dx,$$

---

*The integration is not easy because the integrand is infinite at the endpoints.

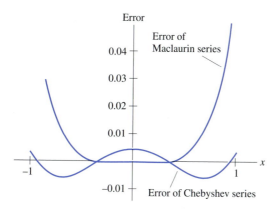

Figure 4.2

and the series is expressed as

$$f(x) = \frac{a_0}{2} + \sum_{i=1}^{\infty} a_i T_i(x).$$

A change of variable will be required if the desired interval is other than $[-1, 1]$. In some cases, the definite integral that defines the coefficients can be profitably evaluated by the numerical procedures that we discuss in the next chapter.

Because the coefficients of the terms of a Chebyshev expansion usually decrease even more rapidly than the terms of a Maclaurin expansion, we can get an estimate of the magnitude of the error from the next nonzero term after those that were retained. For the truncated Chebyshev series given by Eq. (4.7), the $T_4(x)$ term would be

$$\frac{1}{192}(T_4) + \frac{1}{23{,}040}(6T_4) + \cdots = 0.00525 T_4.$$

Because the maximum value of $T_4(x)$ on $(-1, 1)$ is 1.0, we estimate the maximum errors of Eq. (4.7) to be 0.00525. The maximum error in Table 4.2 is 0.0062. This good agreement is caused by the very rapid decrease in coefficients in this example.

The computational economy to be gained by economizing a Maclaurin series, or by using a Chebyshev series, is even more dramatic when the Maclaurin series is slowly convergent. The previous example for $f(x) = e^x$ is a case in which the Maclaurin series converges rapidly. The power of the methods of this section is better demonstrated in the following example.

EXAMPLE 4.1   A Maclaurin series for $(1 + x)^{-1}$ is

$$(1 + x)^{-1} = 1 - x + x^2 - x^3 + x^4 - \cdots \qquad (-1 < x < 1).$$

Table 4.3 compares the accuracy of truncated Maclaurin series with the economized series derived from them.

In Table 4.3, we see that the error of the Maclaurin series is small for $x = 0.2$, and this also would be true for other values near $x = 0$, whereas the economized polynomial has

Table 4.3 Comparison of errors: Maclaurin, Chebyshev, and economized series for $1/(1 + x)$

	Degree	Maclaurin Value	Maclaurin Error	Chebyshev Value	Chebyshev Error	Economized* Value	Economized* Error
At $x = 0.2$	2	0.840000	0.006667	0.841035	0.007702	0.758600	−0.074733
	4	0.833600	0.000267	0.833316	−0.000017	0.764594	−0.068739
	6	0.833344	$11 * 10^{-6}$	0.833359	0.000026	0.803646	−0.029687
	8	0.833334	$1 * 10^{-6}$	0.833365	0.000032	0.822786	−0.010547
	10	0.833333	0	0.833364	0.000031		
At $x = 0.8$	2	0.840000	0.284445	0.549866	−0.005899	0.812600	0.257045
	4	0.737600	0.182045	0.555518	−0.000038	0.738314	0.122759
	6	0.672064	0.116509	0.555561	0.000006	0.628558**	0.073003
	8	0.630121	0.074566	0.555568	0.000012	0.602106***	0.046551
	10	0.603277	0.047722	0.555568	0.000012		

* Economized series were derived from Maclaurin series whose degree is greater by 2.
** A series of degree equal to 4 has a value of 0.658246 with an error of 0.102691.
*** A series of degree equal to 6 has a value of 0.598199 with an error of 0.042644.

less accuracy. At $x = 0.8$, the situation is reversed, however. Economized polynomials of degrees 8 and 6, derived from truncated Maclaurin series of degrees 10 and 8, actually have smaller errors than their precursors. Further economization, giving polynomials of degrees 6 and 4, have lesser or only slightly greater errors than their precursors, at significant savings of computational effort and with smaller storage requirements in a computer's memory for the coefficients of the polynomials. ▲

4.3 Approximation with Rational Functions

We have seen that expansion of a function in terms of Chebyshev polynomials gives a power-series expansion that is much more efficient on the interval $(−1, 1)$ than the Maclaurin expansion, in that it has a smaller maximum error with a given number of terms. These are not the best approximations for use in most digital computers, however. In this application, we measure efficiency by the computer time required to evaluate the function, plus some consideration of storage requirements for the constants. Because the arithmetic operations of a computer can directly evaluate only polynomials, we limit our discussion of more efficient approximations to rational functions, which are the ratios of two polynomials.

Our discussion of methods for finding efficient rational approximations will be elementary and introductory only. Obtaining truly best approximations is a difficult subject. In its present stage of development it is as much art as science, and requires successive approximations from a "suitably close" initial approximation. Our study will serve to introduce the student to some of the ideas and procedures used. The topic is of great importance, however, because the saving of just 1 msec of time in the generation of a frequently used elementary function may save many dollars' worth of machine time each year.

Padé Approximations

We start with a discussion of Padé approximations. Suppose we wish to represent a function as the quotient of two polynomials:

$$f(x) \doteq R_N(x) = \frac{a_0 + a_1 x + a_2 x^2 + \cdots + a_n x^n}{1 + b_1 x + b_2 x^2 + \cdots + b_m x^m}, \qquad N = n + m.$$

The constant term in the denominator can be taken as unity without loss of generality, because we can always convert to this form by dividing numerator and denominator by b_0. The constant b_0 will generally not be zero, for, in that case, the fraction would be undefined at $x = 0$. The most useful of the Padé approximations are those with the degree of the numerator equal to, or one more than, the degree of the denominator. Note that the number of constants in $R_N(x)$ is $N + 1 = n + m + 1$.

The Padé approximations are related to Maclaurin expansions in that the coefficients are determined in a similar fashion to make $f(x)$ and $R_N(x)$ agree at $x = 0$ and also to make the first N derivatives agree at $x = 0$.*

We begin with the Maclaurin series for $f(x)$ (we use only terms through x^N) and write

$$
\begin{aligned}
f(x) - R_N(x) &\doteq (c_0 + c_1 x + c_2 x^2 + \cdots + c_N x^N) - \frac{a_0 + a_1 x + \cdots + a_n x^n}{1 + b_1 x + \cdots + b_m x^m} \\
&= \frac{(c_0 + c_1 x + \cdots + c_N x^N)(1 + b_1 x + \cdots + b_m x^m) - (a_0 + a_1 x + \cdots + a_n x^n)}{1 + b_1 x + \cdots + b_m x^m}.
\end{aligned}
$$

(4.8)

The coefficients c_i are $f^{(i)}(0)/(i!)$ of the Maclaurin expansion. Now if $f(x) = R_N(x)$ at $x = 0$, the numerator of Eq. (4.8) must have no constant term. Hence

$$c_0 - a_0 = 0.$$

For the first N derivatives of $f(x)$ and $R_N(x)$ to be equal at $x = 0$, the coefficients of the powers of x up to and including x^N in the numerator must all be zero also. This gives N additional equations for the a's and b's. The first n of these involve a's, the rest only b's and c's:

$$b_1 c_0 + c_1 - a_1 = 0,$$
$$b_2 c_0 + b_1 c_1 + c_2 - a_2 = 0,$$
$$b_3 c_0 + b_2 c_1 + b_1 c_2 + c_3 - a_3 = 0,$$
$$\vdots$$
$$b_m c_{n-m} + b_{m-1} c_{n-m+1} + \cdots + c_n - a_n = 0,$$
$$b_m c_{n-m+1} + b_{m-1} c_{n-m+2} + \cdots + c_{n+1} = 0,$$
$$b_m c_{n-m+2} + b_{m-1} c_{n-m+3} + \cdots + c_{n+2} = 0,$$
$$\vdots$$
$$b_m c_{N-m} + b_{m-1} c_{N-m+1} + \cdots + c_N = 0.$$

(4.9)

---

*A similar development can be derived for the expansion about a nonzero value of x, but the manipulations are not as easy. By a change of variable we can always make the region of interest contain the origin.

Note that, in each equation, the sum of the subscripts on the factors of each product is the same, and is equal to the exponent of the x-term in the numerator. The $N + 1$ equations of Eqs. (4.8) and (4.9) give the required coefficients of the Padé approximation. We illustrate by an example.

EXAMPLE 4.2

Find arctan $x \doteq R_9(x)$. Use in the numerator a polynomial of degree 5.

The Maclaurin series through x^9 is

$$\arctan x \doteq x - \frac{1}{3}x^3 + \frac{1}{5}x^5 - \frac{1}{7}x^7 + \frac{1}{9}x^9. \tag{4.10}$$

We form, analogously to Eq. (4.8),

$$
\begin{aligned}
&f(x) - R_9(x) \\
&= \frac{\left(x - \frac{1}{3}x^3 + \frac{1}{5}x^5 - \frac{1}{7}x^7 + \frac{1}{9}x^9\right)(1 + b_1 x + b_2 x^2 + b_3 x^3 + b_4 x^4) - (a_0 + a_1 x + \cdots + a_5 x^5)}{(1 + b_1 x + b_2 x^2 + b_3 x^3 + b_4 x^4)}
\end{aligned}
\tag{4.11}
$$

Making coefficients through that of x^9 in the numerator equal to zero, we get

$$a_0 = 0,$$
$$a_1 = 1,$$
$$a_2 = b_1,$$
$$a_3 = -\frac{1}{3} + b_2,$$
$$a_4 = -\frac{1}{3}b_1 + b_3,$$
$$a_5 = \frac{1}{5} - \frac{1}{3}b_2 + b_4,$$
$$\frac{1}{5}b_1 - \frac{1}{3}b_3 = 0,$$
$$-\frac{1}{7} + \frac{1}{5}b_2 - \frac{1}{3}b_4 = 0,$$
$$-\frac{1}{7}b_1 + \frac{1}{5}b_3 = 0,$$
$$\frac{1}{9} - \frac{1}{7}b_2 + \frac{1}{5}b_4 = 0.$$

We first solve for the b's from the linear system of the last four equations, then substitute to get the a's.

$$b_1 = 0, \quad b_2 = \frac{10}{9}, \quad b_3 = 0, \quad b_4 = \frac{5}{21},$$
$$a_0 = 0, \quad a_1 = 1, \quad a_2 = 0, \quad a_3 = \frac{7}{9}, \quad a_4 = 0, \quad a_5 = \frac{64}{945}.$$

A rational function that approximates arctan x is then

$$\arctan x \doteq \frac{x + \dfrac{7}{9}x^3 + \dfrac{64}{945}x^5}{1 + \dfrac{10}{9}x^2 + \dfrac{5}{21}x^4}. \tag{4.12}$$

In Table 4.4 we compare the errors for this Padé approximation (Eq. 4.12) to the Maclaurin series expansion (Eq. 4.10). Enough terms are available in the Maclaurin series to give five-decimal precision at $x = 0.2$ and 0.4, but at $x = 1$ (the limit for convergence of the series) the error is sizable. Even though we used no more information in establishing it, the Padé formula is surprisingly accurate, having an error only 1/275 as large at $x = 1$. It is then particularly astonishing to realize that the Padé approximation is still not the best one of its form, for it violates the minimax principle. If the extreme precision near $x = 0$ is relaxed, we can make the maximum error smaller in the interval.

Figure 4.3 shows how closely the Padé approximation matches arctan(x), especially on $[-1, 1]$.

Table 4.4 Comparison of Padé approximation to Maclaurin series for arctan x

x	True value	Padé (Eq. 4.12)	Error	Maclaurin (Eq. 4.10)	Error
0.2	0.19740	0.19740	0.00000	0.19740	0.00000
0.4	0.38051	0.38051	0.00000	0.38051	0.00000
0.6	0.54042	0.54042	0.00000	0.54067	−0.00025
0.8	0.67474	0.67477	−0.00003	0.67982	−0.00508
1.0	0.78540	0.78558	−0.00018	0.83492	−0.04952

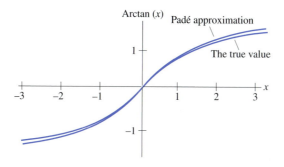

Figure 4.3

The error of a Padé approximation can often be roughly estimated by computing the next nonzero term in the numerator of Eq. (4.12). For Example 4.2, the coefficient of x^{10} is zero, and the next term is

$$\left(-\frac{1}{7}b_4 + \frac{1}{9}b_2 - \frac{1}{11} \right)x^{11} = \left[-\frac{1}{7}\left(\frac{5}{21}\right) + \frac{1}{9}\left(\frac{10}{9}\right) - \frac{1}{11} \right]x^{11}$$

$$= -0.0014x^{11}.$$

Dividing by the denominator, we have

$$\text{Error} \doteq \frac{-0.0014x^{11}}{1 + 1.1111x^2 + 0.2381x^4}.$$

At $x = 1$ this estimate gives -0.00060, which is about three times too large, but still of the correct order of magnitude. It is not unusual that such estimates are rough; analogous estimates of error by using the next term in a Maclaurin series behave similarly. The validity of the rule of thumb that "next term approximates the error" is poor when the coefficients do not decrease rapidly.

The preference for Padé approximations with the degree of the numerator the same as or one more than the degree of the denominator rests on the empirical fact that the errors are usually less for these. There are, however, even more efficient rational functions.

Continued Fractions

Before we discuss such better approximations in the form of rational functions, remarks on the amount of effort required for the computation using Eq. (4.12) are in order. If we were to implement the equation in a computer as it stands, we would, of course, use the constants in decimal form, and we would evaluate the polynomials in nested form:

$$\text{Numerator} = [(0.0677x^2 + 0.7778)x^2 + 1]x,$$

$$\text{Denominator} = (0.2381x^2 + 1.1111)x^2 + 1.$$

Because additions and subtractions are generally much faster than multiplications or divisions, we generally neglect them in a count of operations. We have then three multiplications for the numerator, two for the denominator, plus one to get x^2, and one division, for a total of seven operations. The Maclaurin series is evaluated with six multiplications, using the nested form. If division and multiplication consume about the same time, there is about a standoff in effort, but greater precision for Eq. (4.12).*

Because small differences in effort accumulate for a frequently used function, it is

---

*On some computers, division is slower than multiplication. This will modify the conclusion reached here. On recent models of computers, multiply/divide is not much slower than add/subtract, so we might want to include the adds/subtracts in the operation count. For both the rational function and the Maclaurin series, there are four of these.

of interest to see whether we can further decrease the number of operations to evaluate Eq. (4.12). By means of a succession of divisions, we can re-express it in continued-fraction form: *

$$\frac{0.0677x^5 + 0.7778x^3 + x}{0.2381x^4 + 1.1111x^2 + 1} = \frac{0.2844x^5 + 3.2667x^3 + 4.2x}{x^4 + 4.6667x^2 + 4.2}$$

$$= \frac{0.2844x(x^4 + 11.4846x^2 + 14.7659)}{x^4 + 4.6667x^2 + 4.2}$$

$$= \frac{0.2844x}{(x^4 + 4.6667x^2 + 4.2)/(x^4 + 11.4846x^2 + 14.7659)}$$

$$= \frac{0.2844x}{1 - (6.8179x^2 + 10.5659)/(x^4 + 11.4846x^2 + 14.7659)}$$

$$= \frac{0.2844x}{1 - 6.8179(x^2 + 1.5497)/(x^4 + 11.4846x^2 + 14.7659)}$$

$$= \frac{0.2844x}{1 - 6.8179/[(x^4 + 11.4846x^2 + 14.7659)/(x^2 + 1.5497)]}$$

$$= \frac{0.2844x}{1 - 6.8179/[x^2 + 9.9348 - 0.6304/(x^2 + 1.5497)]}.$$

In this last form, we see that three divisions and two multiplications are needed (one multiplication by x and one to get x^2), for a total of five operations. We have saved two steps. In most cases there is an even greater advantage to the continued-fraction form; in this example the missing powers of x favored the evaluation as polynomials.

Better Rational Function Approximations

One can get somewhat improved rational function approximations by starting with the Chebyshev expansion and operating analogously to the method for Padé approximations. We illustrate with an approximation for e^x. The Chebyshev series was derived in Section 4.2, Eq. (4.7):

$$e^x = 1.2661T_0 + 1.1302T_1 + 0.2715T_2 + 0.0443T_3.$$

Using this approximation, we form the difference

$$f(x) - \frac{P_n(x)}{Q_m(x)}$$

$$= \frac{(1.2661 + 1.302T_1 + 0.2715T_2 + 0.0443T_3)(1 + b_1T_1) - (a_0 + a_1T_1 + a_2T_2)}{1 + b_1T_1}.$$

---

*Acton (1970) is an excellent reference.

Here we have chosen the numerator as a second-degree Chebyshev polynomial and the denominator as first degree. We again make the first $N = n + m$ powers of x in the numerator vanish. Expanding the numerator, we get

$$\text{Numerator} = 1.2661 + 1.1302T_1 + 0.2715T_2 + 0.0443T_3 + 1.2661b_1T_1 + 1.1302b_1T_1^2$$
$$+ 0.2715b_1T_1T_2 + 0.0443b_1T_1T_3 - a_0 - a_1T_1 - a_2T_2.$$

Before we can equate coefficients to zero, we need to resolve the products of Chebyshev polynomials that occur. Recalling that $T_n(x) = \cos n\theta$, we can use the trigonometric identity

$$\cos n\theta \cos m\theta = \frac{1}{2}[\cos(n + m)\theta + \cos(n - m)\theta],$$

$$T_n(x)T_m(x) = \frac{1}{2}[T_{n+m}(x) + T_{|n-m|}(x)].$$

The absolute value of the difference $n - m$ occurs because $\cos(z) = \cos(-z)$. Using this relation we can write the equations

$$a_0 = 1.2661 + \frac{1.1302}{2}b_1,$$

$$a_1 = 1.1302 + \left(\frac{0.2715}{2} + 1.2661\right)b_1,$$

$$a_2 = 0.2715 + \left(\frac{1.1302}{2} + \frac{0.0443}{2}\right)b_1,$$

$$0 = 0.0443 + \frac{0.2715}{2}b_1.$$

Solving, we first get $b_1 = -0.3263$, then we get $a_0 = 1.0817, a_1 = 0.6727, a_2 = 0.0799$, and

$$e^x \doteq \frac{1.0817 + 0.6727T_1 + 0.0799T_2}{1 - 0.3263T_1},$$

$$e^x \doteq \frac{1.0018 + 0.6727x + 0.1598x^2}{1 - 0.3263x}. \tag{4.13}$$

The last expression results when the Chebyshev polynomials are written in terms of powers of x. In Table 4.5 the error of this rational approximation is compared to the Chebyshev expansion. We see that the maximum error is reduced by 22%. Note that we do not, nevertheless, yet have a "best approximation." The error should reach equal maximums at five points in the interval—instead the error is large near $x = 1$ and too small elsewhere.

The basis for this last statement is the minimax theorem. Based on a theorem due to Chebyshev, we may state a principle whereby we may determine whether the approximation represented by a given polynomial or rational function is optimum, in the sense that it gives the least maximum error of any rational function of the same degree of numerator

Table 4.5   Comparison of rational approximations (Eq. 4.13) with
Chebyshev series for e^x

x	e^x	Chebyshev	Error	Rational function	Error
−1.0	0.3679	0.3631	0.0048	0.3686	−0.0007
−0.8	0.4493	0.4536	−0.0042	0.4488	0.0006
−0.6	0.5488	0.5534	−0.0046	0.5484	0.0005
−0.4	0.6703	0.6712	−0.0009	0.6707	−0.0004
−0.2	0.8187	0.8154	0.0033	0.8201	−0.0014
0	1.0000	0.9946	0.0054	1.0018	−0.0018
0.2	1.2214	1.2172	0.0042	1.2225	−0.0011
0.4	1.4918	1.4917	0.0001	1.4911	0.0008
0.6	1.8221	1.8267	−0.0046	1.8191	0.0030
0.8	2.2255	2.2307	−0.0051	2.2224	0.0032
1.0	2.7183	2.7121	0.0062	2.7227	−0.0044

and denominator on a given interval. An expression is "minimax" if and only if there are at least $N + 2$ maxima in the deviations, and these are all equal in magnitude and of alternating sign, on the interval of approximation. (Here N is the sum of the degrees of numerator and denominator of the rational function.) In the discussion that follows, we shall be referring to the magnitude of the errors and will not be concerned about their sign.

A second important consequence of this principle is that we can put bounds on the error of the minimax expression from the range of the errors of a function that is not minimax. Suppose we have an expression like Eq. (4.13). It has five maxima of alternating sign on the interval $[-1, 1]$, as shown by Table 4.5. This is the correct number, but we know the function isn't minimax because the maxima aren't equal in magnitude. Although some of the maxima are not given precisely by the table, we can see that the smallest is 0.0006 (at $x = -0.8$) and the largest is 0.0044 (at $x = 1$). From this range of maximum error values, we can bound the maximum error of the minimax rational function [of degree (2, 1)]: The minimax expression will have a maximum error on $[-1, 1]$ no less than 0.0006 and no greater than 0.0044.

Similarly, the truncated Chebyshev series of degree 3 whose errors are tabulated in Table 4.2 is not truly minimax; it has five alternating maxima to its errors but they are not quite equal in magnitude. From an examination of Table 4.2 we can say, however, that the minimax polynomial of degree 3 will have a maximum error bounded by 0.0046 and 0.0062. (A tighter bound might result from a more careful computation of the errors of Eq. (4.7).) Such a prediction about the amount of improvement that will be provided by a minimax expression can help us decide whether the additional effort to find it is worthwhile.

To obtain the optimum rational function that approximates the function with equal-magnitude errors distributed through the interval is beyond the scope of this text. The approach that is used is to improve an initial estimate of a function, such as Eq. (4.13), by successive trials, often modifying the constants on the basis of experience until eventually one has a satisfactory formula. Systematic methods of determining the constants in such minimax rational approximations have also been determined. They are iteration methods beginning from an initial "sufficiently good" approximation. They are expensive to compute because the iterations involve solving a set of nonlinear equations. Ralston (1965) describes one such method. Prenter (1975) discusses the approximation of functions of several variables.

An Algorithm to Construct a Padé Approximation, $R_N(x)$

Given a power series $P_N(x)$ (a Maclaurin series if the expansion is about $x = 0$) that includes powers of x to x^N:

$$\text{SET } R_N(x) = \sum_{j=0}^{n} a_j x^j \bigg/ \left(1 + \sum_{i=1}^{m} b_i x^i\right), \ (n = m \text{ or } n = m + 1, N = n + m).$$

Subtract $R_N(x)$ from $P_N(x)$ to form a fraction.
Multiply out the numerator and combine like powers of x.
Set the coefficient of each power of x to zero to give a set of equations in the a's and b's.
Solve the equations for the coefficients of $R_n(x)$.

Note: An improved rational function approximation is obtained if the initial power series is a Chebyshev series.

4.4 Fourier Series

Polynomials are not the only functions that can be used to approximate known functions. Another means for representing known functions are approximations that use sines and cosines, called *Fourier series* after the French mathematician who first proposed, in the early 1800s, that "any function can be represented by an infinite sum of sine and cosine terms."

Fourier used these series in his studies of heat conduction. His belief that any function can be represented in the form of a sum of sine and cosine terms with the proper coefficients, possibly with an infinite number of terms, was disputed by other mathematicians because he did not adequately develop the theory. Actually, the belief is false, for there are functions (mostly esoteric) that do not have a representation as a Fourier series. However, most functions can be so represented; we will discuss the conditions for this in another section.

Representing a function as a trigonometric series is important in solving some partial-differential equations analytically. In this section we will see how to determine the coefficients of a Fourier series.

Because a Fourier series is a sum of sine and/or cosine terms, it will obviously always be a periodic function.

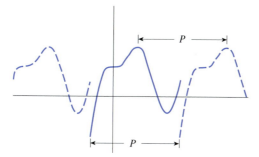

Figure 4.4 Plot of a periodic function of period P

Any function, $f(x)$, is periodic of period P if it has the same value for any two x-values that differ by P, or

$$f(x) = f(x + P) = f(x + 2P) = \cdots = f(x - P) = f(x - 2P) = \cdots.$$

Figure 4.4 shows such a periodic function. Additional occurrences are shown as dashed on the plot. Observe that the period can be started at any point on the x-axis. $\text{Sin}(x)$ and $\cos(x)$ are periodic of period 2π; $\sin(2x)$ and $\cos(2x)$ are periodic of period π; $\sin(nx)$ and $\cos(nx)$ are periodic of period $2\pi/n$.

We now discuss how to find the A's and B's in a Fourier series of the form

$$f(x) \approx \frac{A_0}{2} + \sum_{n=1}^{\infty} [A_n \cos(nx) + B_n \sin(nx)]. \tag{4.14}$$

(We read the symbol "$\approx$" in Eq. (4.14) as "is represented by.") The determination of the coefficients of a Fourier series (when a given function, $f(x)$, can be so represented) is based on the *property of orthogonality* for sines and cosines. For integer values of n, m:

$$\int_{-\pi}^{\pi} \sin(nx)\, dx = 0; \tag{4.15}$$

$$\int_{-\pi}^{\pi} \cos(nx)\, dx = \begin{cases} 0, & n \neq 0, \\ 2\pi, & n = 0; \end{cases} \tag{4.16}$$

$$\int_{-\pi}^{\pi} \sin(nx)\cos(mx)\, dx = 0; \tag{4.17}$$

$$\int_{-\pi}^{\pi} \sin(nx)\sin(mx)\, dx = \begin{cases} 0, & n \neq m, \\ \pi, & n = m; \end{cases} \tag{4.18}$$

$$\int_{-\pi}^{\pi} \cos(nx)\cos(mx)\, dx = \begin{cases} 0, & n \neq m, \\ \pi, & n = m. \end{cases} \tag{4.19}$$

Although the term *orthogonal* should not be interpreted geometrically, it is related to the same term used for orthogonal (perpendicular) vectors whose dot product is zero. Many functions, besides sines and cosines, are orthogonal, such as the Chebyshev polynomials that were discussed previously.

To begin, we assume that $f(x)$ is periodic of period 2π and can be represented as in Eq. (4.14). We find the values of A_n and B_n in Eq. (4.14) in the following way.

1. Multiply both sides of Eq. (4.14) by $\cos(0x) = 1$, and integrate term by term between the limits of $-\pi$ and π. (We assume that this is a proper operation; you will find that it works.)

$$\int_{-\pi}^{\pi} f(x)\, dx = \int_{-\pi}^{\pi} \frac{A_0}{2}\, dx + \sum_{n=1}^{\infty} \int_{-\pi}^{\pi} A_n \cos(nx)\, dx + \sum_{n=1}^{\infty} \int_{-\pi}^{\pi} B_n \sin(nx)\, dx \quad \text{(4.20)}$$

Because of Eqs. (4.15) and (4.16), every term on the right vanishes except the first, giving

$$\int_{-\pi}^{\pi} f(x)\, dx = \frac{A_0}{2}(2\pi), \quad \text{or} \quad A_0 = \frac{1}{\pi}\int_{-\pi}^{\pi} f(x)\, dx. \quad \text{(4.21)}$$

Hence A_0 is found and it is equal to twice the average value of $f(x)$ over one period.

2. Multiply both sides of Eq. (4.14) by $\cos(mx)$, where m is any positive integer, and integrate:

$$\int_{-\pi}^{\pi} \cos(mx)f(x)\, dx = \int_{-\pi}^{\pi} \frac{A_0}{2} \cos(mx)\, dx + \sum_{n=1}^{\infty} \int_{-\pi}^{\pi} A_n \cos(mx)\cos(nx)\, dx$$

$$+ \sum_{n=1}^{\infty} \int_{-\pi}^{\pi} B_n \cos(mx)\sin(nx)\, dx. \quad \text{(4.22)}$$

Because of Eqs. (4.16), (4.17), and (4.19), the only nonzero term on the right is when $m = n$ in the first summation, so we get a formula for the A's:

$$A_n = \frac{1}{\pi} \int_{-\pi}^{\pi} f(x)\cos(nx)\, dx, \quad n = 1, 2, 3, \ldots. \quad \text{(4.23)}$$

3. Multiply both sides of Eq. (4.14) by $\sin(mx)$, where m is any positive integer, and integrate:

$$\int_{-\pi}^{\pi} \sin(mx)f(x)\, dx = \int_{-\pi}^{\pi} \frac{A_0}{2} \sin(mx)\, dx + \sum_{n=1}^{\infty} \int_{-\pi}^{\pi} A_n \sin(mx)\cos(nx)\, dx$$

$$+ \sum_{n=1}^{\infty} \int_{-\pi}^{\pi} B_n \sin(mx)\sin(nx)\, dx. \quad \text{(4.24)}$$

Because of Eqs. (4.15), (4.17), and (4.18), the only nonzero term on the right is when $m = n$ in the second summation, so we get a formula for the B's:

$$B_n = \frac{1}{\pi} \int_{-\pi}^{\pi} f(x)\sin(nx)\, dx, \quad n = 1, 2, 3, \ldots. \quad \text{(4.25)}$$

By comparing Eqs. (4.21) and (4.23), you now see why Eq. (4.14) had $A_0/2$ as its first term. That makes the formula for all of the A's the same:

$$A_n = \frac{1}{\pi} \int_{-\pi}^{\pi} f(x)\cos(nx)\, dx, \qquad n = 0, 1, 2, \ldots . \tag{4.26}$$

It is obvious that getting the coefficients of Fourier series involves many integrations. We observe that this can be facilitated by a computer algebra system.

Fourier Series for Periods Other Than 2π

What if the period of $f(x)$ is not 2π? No problem—we just make a change of variable. If $f(x)$ is periodic of period P, the function can be considered to have one period between $-P/2$ and $P/2$. The functions $\sin(2\pi x/P)$ and $\cos(2\pi x/P)$ are periodic between $-P/2$ and $P/2$. (When $x = -P/2$, the angle becomes $-\pi$; when $x = P/2$, it is π.) We can repeat the preceding developments for sums of $\cos(2n\pi x/P)$ and $\sin(2n\pi x/P)$, or rescale the preceding results. In any event, the formulas become, for $f(x)$ periodic of period P:

$$A_n = \frac{2}{P} \int_{-P/2}^{P/2} f(x)\cos\left(\frac{n\pi x}{P/2}\right) dx, \qquad n = 0, 1, 2, \ldots , \tag{4.27}$$

$$B_n = \frac{2}{P} \int_{-P/2}^{P/2} f(x)\sin\left(\frac{n\pi x}{P/2}\right) dx, \qquad n = 1, 2, 3, \ldots . \tag{4.28}$$

Because a function that is periodic with period P between $-P/2$ and $P/2$ is also periodic with period P between A and $A + P$, the limits of integration in Eqs. (4.27) and (4.28) can be from 0 to P.

EXAMPLE 4.3   Let $f(x) = x$ be periodic between $-\pi$ and π. (See Fig. 4.5.) Find the A's and B's of its Fourier expansion.

For A_0:

$$A_0 = \frac{1}{\pi} \int_{-\pi}^{\pi} f(x)\, dx = \frac{1}{\pi} \int_{-\pi}^{\pi} x\, dx = \left. \frac{x^2}{2\pi} \right]_{-\pi}^{\pi} = 0. \tag{4.29}$$

For the other A's:

$$A_n = \frac{1}{\pi} \int_{-\pi}^{\pi} x \cos(nx)\, dx = \frac{1}{\pi} \left(\frac{\cos(nx)}{n^2} + \frac{x \sin(nx)}{n} \right) \Bigg]_{-\pi}^{\pi} = 0. \tag{4.30}$$

For the B's:

$$B_n = \frac{1}{\pi} \int_{-\pi}^{\pi} x \sin(nx)\, dx = \frac{1}{\pi} \left(\frac{\sin(nx)}{n^2} - \frac{x \cos(nx)}{n} \right) \Bigg]_{-\pi}^{\pi}$$

$$= \frac{2(-1)^{n+1}}{n}, \qquad n = 1, 2, 3, \ldots . \tag{4.31}$$

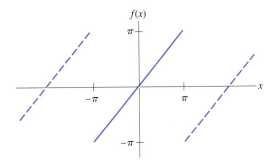

Figure 4.5 Plot of $f(x) = x$, periodic of period 2π

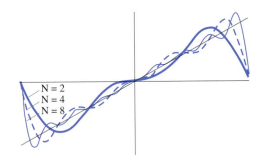

Figure 4.6 Plot of Eq. (4.32) for N = 2, 4, 8

We then have

$$x \approx 2 \sum_{n=1}^{\infty} \frac{(-1)^{n+1}}{n} \sin(nx), \qquad -\pi < x < \pi. \tag{4.32}$$

Figure 4.6 shows how the series approximates to the function when only two, four, or eight terms are used.   ▲

EXAMPLE 4.4   Find the Fourier coefficients for $f(x) = |x|$ on $-\pi$ to π:

$$A_0 = \frac{1}{\pi} \int_{-\pi}^{0} (-x)\, dx + \frac{1}{\pi} \int_{0}^{\pi} x\, dx = \frac{2}{\pi} \int_{0}^{\pi} x\, dx = \pi; \tag{4.33}$$

$$A_n = \frac{1}{\pi} \int_{-\pi}^{0} (-x)\cos(nx)\, dx + \frac{1}{\pi} \int_{0}^{\pi} x \cos(nx)\, dx$$

$$= \frac{2}{\pi} \left(\frac{\cos(nx)}{n^2} + \frac{x \sin(nx)}{n} \right) \Bigg]_{0}^{\pi}$$

$$= \begin{cases} 0, & n = 2, 4, 6, \ldots, \\ \dfrac{-4}{n^2 \pi}, & n = 1, 3, 5, \ldots; \end{cases} \tag{4.34}$$

$$B_n = \frac{1}{\pi} \int_{-\pi}^{0} (-x)\sin(nx)\, dx + \frac{1}{\pi} \int_{0}^{\pi} x \sin(nx)\, dx = 0. \tag{4.35}$$

Because the definite integrals in Eq. (4.34) are nonzero only for odd values of n, it simplifies to change the index of the summation. The Fourier series is then

$$|x| \approx \frac{\pi}{2} - \frac{4}{\pi} \sum_{n=1}^{\infty} \frac{\cos((2n-1)x)}{(2n-1)^2}. \tag{4.36}$$

Figure 4.7 shows how the series approximates the function when two, four, or eight terms are used.

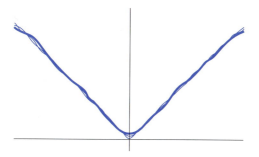

Figure 4.7 Plot of Eq. (4.36) for N = 2, 4, 8   ▲

When you compare Eqs. (4.32) and (4.36) and their plots in Figs. 4.6 and 4.7, you will notice several differences:

1. The first contains only sine terms, the second only cosines.
2. Equation (4.32) gives a value at both endpoints that is the average of the end values for $f(x)$, where there is a discontinuity.
3. Equation (4.36) gives a closer approximation when only a few terms are used.

Example 4.5 will further examine these points.

EXAMPLE 4.5   Find the Fourier coefficients for $f(x) = x(2 - x) = 2x - x^2$ over the interval $(-2, 2)$ if it is periodic of period 4. Equations (4.27) and (4.28) apply.

$$A_0 = \frac{2}{4} \int_{-2}^{2} (2x - x^2)\, dx = \frac{-8}{3} \tag{4.37}$$

$$A_n = \frac{2}{4} \int_{-2}^{2} (2x - x^2)\cos\left(\frac{n\pi x}{2}\right) dx = \frac{16(-1)^{n+1}}{n^2 \pi^2}, \qquad n = 1, 2, 3, \ldots \tag{4.38}$$

$$B_n = \frac{2}{4} \int_{-2}^{2} (2x - x^2)\sin\left(\frac{n\pi x}{2}\right) dx = \frac{8(-1)^{n+1}}{n\pi}, \qquad n = 1, 2, 3, \ldots \tag{4.39}$$

$$x(2 - x) \approx \frac{-4}{3} + \frac{16}{\pi^2} \sum_{n=1}^{\infty} \frac{(-1)^{n+1}}{n^2} \cos\left(\frac{n\pi x}{2}\right) + \frac{8}{\pi} \sum_{n=1}^{\infty} \frac{(-1)^{n+1}}{n} \sin\left(\frac{n\pi x}{2}\right) \tag{4.40}$$

You will notice that both sine and cosine terms occur in the Fourier series and that the discontinuity at the endpoints shows itself in forcing the Fourier series to reach the average value. It should also be clear that the series is the sum of separate series for $2x$ and $-x^2$. Figure 4.8 shows how the series of Eq. (4.40) approximates to the function when 40 terms are used. It is obvious that many more terms are needed to reduce the error to negligible proportions because of the extreme oscillation near the discontinuities, called the *Gibbs phenomenon*. The conclusion is that a Fourier series often involves a lot of computation as well as awkward integrations to give the formula for the coefficients.

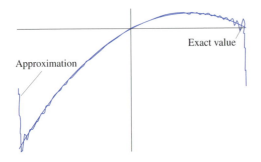

Figure 4.8 Plot of Eq. (4.40) for N = 40   ▲

Fourier Series for Nonperiodic Functions and Half-Range Expansions

The development until now has been for a periodic function. What if $f(x)$ is not periodic? Can we approximate it by a trigonometric series? We assume that we are interested in approximating the function only over a limited interval and we do not care whether the approximation holds outside of that interval. This situation frequently occurs when we want to solve partial-differential equations analytically.

Suppose we have a function defined for all x-values, but we are only interested in representing it over $(0, L)$.* Figure 4.9 is typical. Because we will ignore the behavior of the function outside of $(0, L)$, we can redefine the behavior outside that interval as we wish. Figures 4.10 and 4.11 show two possible redefinitions.

In the first redefinition, we have reflected the portion of $f(x)$ about the y-axis and have extended it as a periodic function of period $2L$. This creates an *even* periodic function. If we reflect it about the origin and extend it periodically, we create an *odd* periodic function of period $2L$. More formally, we define even and odd functions through these relations:

$$f(x) \text{ is even if } f(-x) = f(x), \tag{4.41}$$

$$f(x) \text{ is odd if } f(-x) = -f(x). \tag{4.42}$$

It is easy to see that $\cos(Cx)$ is an even function and that $\sin(Cx)$ is an odd function for any real value of C.

There are two important relationships for integrals of even and odd functions. (If you think of the integrals in a geometric interpretation, these relationships are obvious.)

$$\text{If } f(x) \text{ is even, } \int_{-L}^{L} f(x)\, dx = 2 \int_{0}^{L} f(x)\, dx. \tag{4.43}$$

$$\text{If } f(x) \text{ is odd, } \int_{-L}^{L} f(x)\, dx = 0. \tag{4.44}$$

---

*If the range of interest is $[a, b]$, a simple change of variable can make this $[0, L]$.

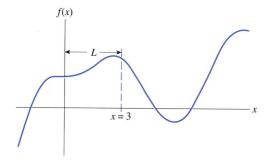

Figure 4.9 A function, $f(x)$, of interest on [0, 3]

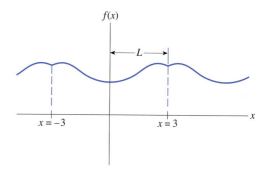

Figure 4.10 Plot of a function reflected about the y-axis, an even function

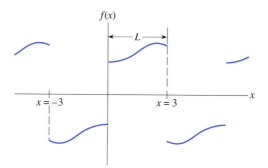

Figure 4.11 Plot of a function reflected about the origin, an odd function

It is also easy to show that the product of two even functions is even, that the product of two odd functions is even, and that the product of an even and an odd function is odd. This means that, if $f(x)$ is even, $f(x)\cos(nx)$ is even and $f(x)\sin(nx)$ is odd. Further, if $f(x)$ is odd, $f(x)\cos(nx)$ is odd and $f(x)\sin(nx)$ is even. Because of Eq. (4.41), the Fourier series expansion of an even function will contain only cosine terms (all the B-coefficients are zero). Also, if $f(x)$ is odd, its Fourier expansion will contain only sine terms (all the A-coefficients are zero). These facts are important when we develop the "half-range" expansion of a function.

Therefore, if we want to represent $f(x)$ between 0 and L as a Fourier series and are interested only in approximating it on the interval $(0, L)$, we can redefine f within the interval $(-L, L)$ in two importantly different ways: (1) We can redefine the portion from $-L$ to 0 by reflecting about the y-axis. We then generate an even function. (2) We can reflect the portion between 0 and L about the origin to generate an odd function. Figures 4.10 and 4.11 showed these two possibilities.

Thus two different Fourier series expansions of $f(x)$ on $(0, L)$ are possible, one that has only cosine terms or one that has only sine terms. We get the A's for the even extension of $f(x)$ on $(0, L)$ from

$$A_n = \frac{2}{L} \int_0^L f(x)\cos\left(\frac{n\pi x}{L}\right) dx, \qquad n = 0, 1, 2, \ldots. \tag{4.45}$$

We get the B's for the odd extension of $f(x)$ on $(0, L)$ from

$$B_n = \frac{2}{L} \int_0^L f(x)\sin\left(\frac{n\pi x}{L}\right) dx, \qquad n = 1, 2, 3, \ldots. \tag{4.46}$$

EXAMPLE 4.6   Find the Fourier cosine series expansion of $f(x)$, given that

$$f(x) = \begin{cases} 0, & 0 < x < 1, \\ 1, & 1 < x < 2. \end{cases} \tag{4.47}$$

Figure 4.12 shows the even extension of the function.

Because we are dealing with an even function on $(-2, 2)$, we know that the Fourier series will have only cosine terms. We get the A's with

$$A_0 = \frac{2}{2} \int_1^2 (1) \, dx = 1; \tag{4.48}$$

$$A_n = \frac{2}{2} \int_1^2 (1)\cos\left(\frac{n\pi x}{2}\right) dx = \begin{cases} 0, & n \text{ even}, \\ \dfrac{2(-1)^{(n+1)/2}}{n\pi}, & n \text{ odd}. \end{cases}$$

Then the Fourier cosine series is

$$f(x) \approx \frac{1}{2} + \frac{2}{\pi} \sum_{n=1}^{\infty} \frac{(-1)^n \cos((2n-1)(\pi x/2))}{(2n-1)}. \tag{4.49}$$

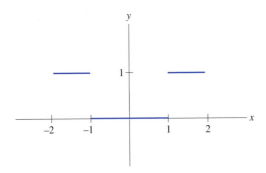

Figure 4.12 Plot of Eq. (4.47) reflected about the y-axis

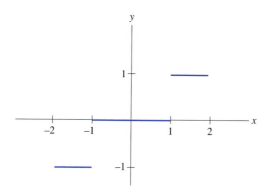

Figure 4.13 Plot of Eq. (4.47) reflected about the origin

EXAMPLE 4.7 Find the Fourier sine series expansion for the same function as in Example 4.6. Figure 4.13 shows the odd extension of the function.

We know that all of the A-coefficients will be zero, so we need to compute only the B's:

$$
\begin{aligned}
B_n &= \frac{2}{2} \int_1^2 (1)\sin\left(\frac{n\pi x}{2}\right) dx \\
&= \frac{2}{n\pi}\left[-\cos(n\pi) + \cos\left(\frac{n\pi}{2}\right)\right], \qquad n = 1, 2, 3, \ldots
\end{aligned}
\tag{4.50}
$$

The term in brackets gives the sequence $1, -2, 1, 0, 1, -2, 1, 0, \ldots$. Because this sequence is awkward to reduce (except by use of the mod function), we simply write

$$
f(x) = \frac{2}{\pi} \sum_{n=1}^{\infty} \frac{[\cos(n\pi/2) - \cos(n\pi)]}{n} \sin\left(\frac{n\pi x}{2}\right).
\tag{4.51}
$$
▲

Summary of Formulas for Computation of Fourier Coefficients

A function that is periodic of period P and meets certain criteria (see below) can be represented by Eq. (4.52):

$$
f(x) = \frac{A_0}{2} + \sum_{n=1}^{\infty} A_n \cos\left(\frac{n\pi x}{P/2}\right) + \sum_{n=1}^{\infty} B_n \sin\left(\frac{n\pi x}{P/2}\right).
\tag{4.52}
$$

The coefficients can be computed with

$$
A_n = \frac{2}{P} \int_{-P/2}^{P/2} f(x)\cos\left(\frac{n\pi x}{P/2}\right) dx, \qquad n = 0, 1, 2, \ldots,
\tag{4.53}
$$

$$
B_n = \frac{2}{P} \int_{-P/2}^{P/2} f(x)\sin\left(\frac{n\pi x}{P/2}\right) dx, \qquad n = 1, 2, 3, \ldots
\tag{4.54}
$$

(The limits of the integrals can be from a to $a + P$.)

If $f(x)$ is an even function, only the A's will be nonzero. Similarly, if $f(x)$ is odd, only the B's will be nonzero. If $f(x)$ is neither even nor odd, its Fourier series will contain both cosine and sine terms.

Even if $f(x)$ is not periodic, it can be represented on just the interval $(0, L)$ by redefining the function over $(-L, 0)$ by reflecting $f(x)$ about the y-axis or, alternatively, about the origin. The first creates an even function, the second an odd function. The Fourier series of the redefined function will actually represent a periodic function of period $2L$ that is defined for $(-L, L)$.

When L is the half-period, the Fourier series of an even function contains only cosine terms and is called a Fourier cosine series. The A's can be computed by

$$A_n = \frac{2}{L} \int_0^L f(x)\cos\left(\frac{n\pi x}{L}\right) dx, \qquad n = 0, 1, 2, \ldots . \tag{4.55}$$

The Fourier series of an odd function contains only sine terms and is called a Fourier sine series. The B's can be computed by

$$B_n = \frac{2}{L} \int_0^L f(x)\sin\left(\frac{n\pi x}{L}\right) dx, \qquad n = 1, 2, 3, \ldots . \tag{4.56}$$

If $f(x)$ (or its redefined extension) has a finite discontinuity, the Fourier series will converge to the average of the two limiting values at the discontinuity. The Fourier series converges slowly at a point of discontinuity and exhibits more pronounced oscillations (the Gibbs phenomenon) near that point. If $f(x)$ (or the redefined function) has a discontinuity in its first derivative, convergence will be slower at that point.

4.5 Theoretical Matters

There is much theory on functional approximation and trigonometric interpolations. (A Fourier series can be thought of as an interpolation formula.) We only scratch the surface, however.

Error Bounds for Chebyshev Polynomials

We have asserted that, of all polynomials of degree n whose highest power of x has a coefficient of one, $T_n(x)/2^{n-1}$ has the smallest error bounds on $[-1, 1]$. The proof is by contradiction.

Let $P_n(x)$ be a polynomial whose leading term* is x^n and suppose that its maximum magnitude on $[-1, 1]$ is less than that of $T_n(x)/2^{n-1}$. Write

$$\frac{T_n(x)}{2^{n-1}} - P_n(x) = P_{n-1}(x),$$

*We restrict the polynomials to those whose leading term is x^n so that all are scaled alike.

where $P_{n-1}(x)$ is a polynomial of degree $n - 1$ or less, as the x^n terms cancel. The polynomial $T_n(x)$ has $n + 1$ extremes (counting endpoints), each of magnitude one, so $T_n(x)/2^{n-1}$ has $n + 1$ extremes each of magnitude $1/2^{n-1}$, and these successive extremes alternate in sign. By our supposition about $P_n(x)$, at each of these maximums or minimums, the magnitude of $P_n(x)$ is less than $1/2^{n-1}$; hence $P_{n-1}(x)$ must change its sign at least for every extreme of $T_n(x)$, which is then at least $n + 1$ times. Hence $P_{n-1}(x)$ crosses the axis at least n times and would have n zeros. However, this is impossible if $P_{n-1}(x)$ is only of degree $n - 1$, unless it is identically zero. The premise must then be false and $P_n(x)$ has a larger magnitude than the polynomial we are testing or, alternatively, $P_n(x)$ is exactly the same polynomial.

Orthogonal Polynomials

The Chebyshev polynomials are only one of many that are orthogonal. We will discuss the Legendre polynomials in Chapter 5 in connection with Gaussian quadrature. They are orthogonal because

$$\int_{-1}^{1} L_n L_m \, dx = \begin{cases} 0, & n \neq m, \\ >0, & n = m. \end{cases}$$

Other orthogonal polynomials include the Hermite, Laguerre, Jacobi, Weber, and Gegenbauer. (Observe that these are named for famous mathematicians. Each has special applications.)

A better definition of orthogonality is to say that a set of functions Q_i are orthogonal with respect to the weight function $W(x)$ if

$$\int_{-1}^{1} Q_n(x)W(x)Q_m(x) \, dx = \begin{cases} 0, & n \neq m, \\ >0, & n = m. \end{cases}$$

If the value is unity when $n = m$, the set is said to be *orthonormal*.

Theory of Fourier Series

We will only summarize the important theorems concerning Fourier series. Proofs can be found in Conte and de Boor (1980), Fike (1968), Brigham (1974), Ramirez (1985), and Ralston (1965). In the following theorems, $f(x)$ refers to the periodic function being represented or to the periodic extension given by a redefinition. It is essential that $f(x)$ be integrable if we are to compute the coefficients of a Fourier series by the formulas given previously.

1. $f(x)$ is said to be piecewise continuous on $(0, L)$ if it is continuous on $(0, L)$ except for a finite number of finite discontinuities. If $f(x)$ and/or $f'(x)$ is piecewise continuous on $(0, L)$, $f(x)$ is said to be piecewise smooth. An infinite series is said to converge pointwise to $f(x)$ if the sum of n terms of the series converges to $f(x)$ at the point in $(0, L)$ as $n \to \infty$. An infinite series is said to converge uniformly if it converges pointwise to $f(x)$ at all points in $(0, L)$.

2. If $f(x)$ is continuous and piecewise smooth, its Fourier series converges uniformly to $f(x)$. If $f(x)$ is piecewise smooth, the series converges pointwise to $f(x)$ at all points where $f(x)$ is continuous and converges to the average value where $f(x)$ has a finite discontinuity.

3. If $f(x)$ is piecewise continuous, its Fourier series can be integrated term by term to yield a series that converges pointwise to the integral of $f(x)$.

4. If $f(x)$ is continuous and $f'(x)$ is piecewise smooth, then the Fourier series of $f(x)$ can be differentiated term by term to give a series that converges pointwise to $f'(x)$ wherever $f''(x)$ exists.

The theory of Fourier series is a major topic in mathematics. Most mathematical texts that cover Fourier series at least outline proofs of the preceding theorems.

4.6 Using Computer Algebra Systems

In this chapter we provide a side-by-side comparison of each of three computer algebra systems: Maple, *Mathematica,* and MATLAB. Each of these will be applied to topics of this chapter. The topics are covered in this order:

Chebyshev polynomials

Taylor series

Economized power series

Chebyshev series

Padé approximation

Continued fractions

Chebyshev/Padé approximation

Minimax approximation

Fourier series

Chebyshev Polynomials

Suppose we want the Chebyshev polynomial of degree 5. Maple provides a command to get Chebyshev polynomials in its "orthopoly package." One can access all of the orthogonal polynomials defined in it by first issuing the command:

```
with(orthopoly)
```

but there is an alternative "long" way to call for the function:

```
orthopoly[T](5, x)
```

which then displays the polynomial in "pretty print":

```
16x⁵ - 20x³ + 5x
```

Mathematica does the same with a built-in function:

```
ChebyshevT[5, x]
```

but the display shows the terms in reverse order:

```
5x - 20x³ + 16x⁵
```

MATLAB has no command (in the Student Edition, version 4) but this M-file* will compute them:

```
function T = Tch(n)
if n == 0
   disp('1')
elseif n == 1
   disp('x')
else
   t0 = '1';
   t1 = 'x';
   for i = 2:n
      T = symop('2*x','*',t1,'-',t0);
      t0 = t1;
      t1 = T;
   end        %for
end            %if
```

After storing this with the file name `tch.m`, entering the commands

```
Tch(5)
collect
pretty
```

gives

```
16x⁵ - 20x³ + 5x
```

Taylor Series

We show how to get the Taylor series of degree 6 for e^x about $x = 0$.

Maple produces a Taylor series about $x = 0$ (a Maclaurin series) with this command:

```
taylor(exp(x), x = 0);
```

but this will be of the default Order = 6 (giving a fifth-degree polynomial).

---

*We described how M-files can be used to store the definition of a function in Chapter 1.

If we desire the sixth-degree polynomial, we must issue the command:

```
Order := 7;
```

After doing this, doing

```
p := taylor(exp(x),x = 0);
```

gives the correct sixth-degree polynomial:

```
p := 1 + x + 1/2x² + 1/6x³ + 1/24x⁴ + 1/120x⁵ + 1/720x⁶ + O(x⁷)
```

plus the error term, $O(x^7)$, at the end. It is essential to notice that this result is a symbolic quantity and, if we want to compute with it, it must be converted to an ordinary polynomial with the command:

```
p := convert(",polynom);
```

which also drops the error term:

```
p := 1 + x + 1/2x² + 1/6x³ + 1/24x⁴ + 1/120x⁵ + 1/720x⁶
```

Mathematica does exactly the same as Maple (except the degree is specified in the command):

```
Series[Exp[x], {x, 0, 6}]
```

with the output automatically in pretty print. The error term is written as $O[x^7]$.

Mathematica differs from Maple in that the output does not have to be converted before being used in other computations.

MATLAB behaves similarly in getting a Taylor series. To get a sixth-degree Maclaurin series, we issue this command:

```
taylor('exp(x)', 7)
```

which produces, if we "pretty" it, the same polynomial but with built-up fractions. It also includes the error term $O(x^7)$, which must be removed before using the series in other computations.

Economized Polynomials

Because all of the computer algebra systems can get both a Maclaurin series and the Chebyshev polynomials, subtracting the proper multiple of $T_n(x)$ from the series of degree n will give the first economized polynomial.

Maple's `taylor(exp(x), x = 0)`, with Order set at 7, gives a symbolic expression that includes the error term. We must remove this error term in order to subtract the proper multiple of the Chebyshev series from it. As shown above, this can be done with

```
p := convert(", polynom)
```

if we have just obtained the Maclaurin series (the quotation mark is a reference to the last answer). Now, doing

```
p - orthopoly[T](6, x)/6!/2^5
```

produces Eq. (4.5) but with the coefficients expressed as ratios of integers:

$$\frac{23041}{23040} + x + \frac{639}{1280}x^2 + 1/6x^3 + 7/160x^4 + 1/120x^5$$

Now, if we do `evalf(")`, we can get the coefficients in floating point form:

$1.000043403 + x + .4992187500x^2 + .1666666667x^3 + 0.437500000x^4 + .008333333333x^5$

Mathematica is similar to Maple but we do not have to remove the error term. We can just do

```
Series[Exp[x], {x, 0, 6}] - ChebyshevT[6, x]/6!/2^5
```

to obtain Eq. (4.5) but with the coefficients as ratios of integers. The result is exactly as just given for Maple except the error term is carried forward:

$$\frac{23041}{23040} + x + \frac{639x^2}{1280} + \frac{x^3}{6} + \frac{7x^4}{160} + \frac{x^5}{120} + O[x^7]$$

With MATLAB, we must first remove the error term from the Maclaurin series, just as with Maple. Also, because the result of `taylor('exp(x)',7)` not only contains the error term but is symbolic, we must remove the error term with symbolic operations:

```
ts = taylor('exp(x)',7)
ts = symsub(ts,'0(xn7))
```

and then

```
cs = Tch(6)
cs = collect(cs)
pretty
```

which displays the Chebyshev series in the usual form:

```
32*x^6-48*x^4+18*x^2-1
```

Now, to get the multiple of the Chebyshev series, we must do some multiplies, but, because `cs` is also symbolic, we do it symbolically (`symop` can take up to 16 arguments):

```
cs2 = symop(cs,'/','6!','/','2^5')
```

which does it (and we can get the usual formulation with "`pretty`"). Now we are ready to subtract:

```
es = symsub(ts,cs2)
```

and get Eq. (4.5), but with the coefficients as ratios of integers:

```
23041/23040+x+639/1280*x^2+1/6*x^3+7/160*x^4+1/120*x^5
```

We can get the coefficients as floating point values with

```
vpa(es)
```

which gives them with the default precision of 16 digits. If we prefer only 7 digits, we do

```
vpa(es,7)
```

and see

```
1.000043+x+.4992188*x^2+.1666667*x^3+4.375000e-2*x^4
+8.333333e-3*x^5
```

Chebyshev Series

Only Maple has a built-in command to get a Chebyshev series:

```
chebyshev(exp(x),x = - 1..1);
```

which produces a series with terms up to $T_8(x)$:

```
  1.266065878 T(0, x) + 1.130318208 T(1, x) + .2714953396 T(2, x)
+ .04433684985 T(3, x) + .005474240443 T(4, x)
+ .0005429263119 T(5, x) + .00004497732296 T(6, x)
+ .3198436462 10⁻⁵ T(7, x) + .1992124806 10⁻⁶ T(8, x)
```

Padé Approximations

Here is how we can get a Padé approximation to arctan(x) with fifth-degree polynomials in both numerator and denominator.

Maple can do this with:

```
numapprox[pade](arctan(x),x=0,[5,5]);
```

to give a result very much like Eq. (4.12):

$$\frac{\frac{64}{945}x^5 + 7/9x^3 + x}{1 + 10/9x^2 + 5/21x^4}$$

Mathematica can do it too, if we first load a package called

```
<<Calculus'Pade';
```

and then do

```
Pade[ArcTan[x],{x,0,5,5}]
```

which gives a result exactly like Eq. (4.12):

$$\frac{x + \dfrac{7x^3}{9} + \dfrac{64x^5}{945}}{1 + \dfrac{10x^2}{9} + \dfrac{5x^4}{21}}$$

MATLAB does not have a command to get a Padé approximation.

Continued Fractions

Only Maple can get a continued fraction from a ratio of polynomials. If we have obtained the Padé approximation to arctan(x) as above, with: `numapprox[pade](arctan(x),` `x=0,[5,5])`; and then do

`numapprox[confracform](",x);`

we see

$$\frac{64}{225}x + \frac{1309}{675}\cfrac{1}{x + \cfrac{8743}{2805}\cfrac{1}{x + \cfrac{236196}{1167815}\cfrac{1}{x + \cfrac{1683}{1249}1/x}}}$$

which is not in the same form as that of Section 4.3.

Chebyshev/Padé Approximation

Maple is again the only program that can get a Padé-like approximation beginning with a Chebyshev series. It does this with

`numapprox[chebpade](exp(x),x,[2,1]);`

to produce

```
(1.086272879 T(0, x) + .6843619105 T(1, x) + .08464994161 T(2, x))/
(T(0, x) - .3181281121 T(1, x))
```

which is a more accurate result than in Eq. (4.13).

Minimax Approximation

Both Maple and *Mathematica* can get a minimax approximation from built-in commands, but MATLAB does not have this capability. To get a minimax rational function for e^x between $x = 0$ and $x = 1$, of degrees 2 and 1, the Maple command:

`numapprox[minimax](exp(x),x=0..1, [2,1]);`

gives

$$\frac{1.164275859 + (.8302764291 + .2779023279 \text{ x}) \text{ x}}{1.164066084 - .3281321682 \text{ x}}$$

Mathematica's command:

```
<<NumericalMath'Approximations'
```

first gives a warning message, which requires us to do

```
Clear[MiniMaxApproximation]; Remove[MiniMaxApproximation]
```

and now, with

```
MiniMaxApproximation[Exp[x],{x,{0,1},2,1}]
```

we obtain

```
{(1.000108385918444 + 0.711987749200353 x + 0.2334396361907942 x²)/
  (1 - 0.2843549535172196 x)}
```

which is almost identical to that from Maple, if we normalize that result to give a constant term of unity in the denominator. Both of these programs use the Remez algorithm.

Fourier Series

Mathematica has a built-in command to get the Fourier series for a function; the others can get the coefficients by integration, of course. (Maple's `fourier` command gets the Fourier transforms, not the series.)

   With *Mathematica,* we must first load a package:

```
<<Calculus'FourierTransform'
```

and, because there is a warning message, do

```
Clear[FourierTrigSeries];Remove[FourierTrigSeries]
```

and then

```
FourierTrigSeries[x*(2-x),{x,-2,2},4]
```

which gives the series. However, the sine terms are hard to interpret until we do

```
Collect[%,Pi]
```

which gives a correct result (see Eq. (4.40)), but all the cosine terms are grouped together with a denominator of `Pi²` and the sine terms are similarly grouped with a denominator of `Pi`.

Integration for Fourier Series Coefficients

All of the computer algebra systems can do the required integrations to get the coefficients of a Fourier series. If $f(x) = x(2 - x)$ over $[-2, 2]$ and we want A_3, here are the commands that are used:

In Maple, we do

```
2/4*int(x*(2-x)*cos(3*Pi*x/2), x= -2..2);
```

Mathematica's command is

```
2/4*Integrate[x*(2-x)*Cos[3*Pi*x/2],{x,-2,2}]
```

With MATLAB, two commands are needed because the first result is symbolic and the integration operation does not permit a multiplier (although the 2/4 could be included in the integrand):

```
a3 = int('x*(2-x)*cos(3*pi*x/2)', -2, 2)
symmul(a3,'2/4')
```

In all cases, the correct result is obtained:

```
16/π²
```

Chapter Summary

Here's what you are able to do if you understand the topics of the chapter:

1. Explain how to get a Chebyshev polynomial of degree n, what is meant by a set of orthogonal functions, and what is the most important property of the Chebyshev polynomials.

2. Use Chebyshev polynomials to economize a power series, and explain why such approximations are preferred in computers.

3. Obtain the coefficients of a Padé approximation, and describe how one can get a further improvement in efficiency. You can explain the minimax principle.

4. Tell how the coefficients of a trigonometric series can be evaluated.

5. Use a computer program like the one of this chapter.

Computer Program

This chapter has just one program, a Fortran 90 program that gets the Padé approximation to a known function (which is a ratio of polynomials that approximates the function better than a Maclaurin series).

Figure 4.14 shows the listing of the program. It computes the coefficients of the polynomials in the numerator and denominator for the Padé approximation from the coefficients of a Maclaurin series. Output from the program for an approximation to the arctan function is given after the listing of the program. It includes a table of computed values and the true values for arctan(x).

```
    Program Pade_approximation
!
!
!!!!!!!!!!!!!!!!!!!!!!!!!!!!!!!!!!!!!!!!!!!!!!!!!!!!!!!!!!!!!!!!!!!!!
!                                                                 !
!                          Chapter 4                              !
!     Gerald/Wheatley, APPLIED NUMERICAL ANALYSIS, (Sixth edition) !
!                   Addison Wesley Longman, 1999                  !
!                                                                 !
!!!!!!!!!!!!!!!!!!!!!!!!!!!!!!!!!!!!!!!!!!!!!!!!!!!!!!!!!!!!!!!!!!!!!
!                                                                 !
!     This program creates a Pade approximation from the coefficients !
!     of a Maclaurin series. It also prints a table of values     !
!     computed from the Pade approximation.                       !
!                                                                 !
!!!!!!!!!!!!!!!!!!!!!!!!!!!!!!!!!!!!!!!!!!!!!!!!!!!!!!!!!!!!!!!!!!!!!
!                                                                 !
!     Variables used:                                             !
!                                                                 !
!       N        Degree of the Maclaurin series                  !
!       Nn,ND    Degree of numerator, denominator of Pade (NN+ND=N) !
!       a,b,c    Arrays of coefficients For numerator, denominator, !
!                   and of the Maclaurin series                   !
!       AA       Augmented matrix of equations that are solved to !
!                   get the Pade coefficients                     !
!       i,j,k    Loop indices                                     !
!       v        Holds values computed from the approximation     !
!                                                                 !
!                                                                 !
!!!!!!!!!!!!!!!!!!!!!!!!!!!!!!!!!!!!!!!!!!!!!!!!!!!!!!!!!!!!!!!!!!!!!
!                                                                 !
!        This version computes the Maclaurin coefficients. This   !
!        example is for the function arctan(x).                   !
!                                                                 !
!!!!!!!!!!!!!!!!!!!!!!!!!!!!!!!!!!!!!!!!!!!!!!!!!!!!!!!!!!!!!!!!!!!!!
!                                                                 !
      Implicit None
!     Define some variables, set up arrays
      Integer, Parameter :: n = 9, numerator_deg = 5, denominator_deg = 4
      Real, Dimension(denominator_deg, denominator_deg+1) :: AA
      Real, Dimension(0:n) :: a,b,c
      Real :: num, ratio, sum, v, x
      Integer :: i, j, k
!
!   Compute the Maclaurin coefficients in vector c()
```

Figure 4.14

```
!
      Do i = 0,n
          c(i) = 0.0
          End Do
      Do i = 1, n, 2
          c(i) = 1.0 / Real(i)
          End Do
      Do i = 3,n,4
          c(i) = -c(i)
          End Do
!
!
!
      Write(*,*)  "Here are the coefficients of the Maclaurin series"
      Write(*,*)
      Write(*,*)  "power of x     Coefficient"
      Do i = 0,n
          Write(*,100) i, c(i)
          End Do
100   Format(T3, I2,T14,F10.4)
      Write(*,*)
!
!
!     Set up the equations for the B's (denominator coefficients)
!
      Do i = 1,denominator_deg
          AA(i, i) = c(numerator_deg)
          End Do
      Do i = 1,denominator_deg-1
          Do j = i + 1, denominator_deg
              AA(i, j) = c(numerator_deg-j + i)
              End Do ! j
          End Do ! i
      Do i = 2 , denominator_deg
          Do j = 1,i + 1
              AA(i, j) = c(numerator_deg + i-j)
              End Do ! j
          End Do ! i
!
!  Now do right hand side
!
      Do i = 1,denominator_deg
          AA(i, denominator_deg + 1) = -c(numerator_deg + i)
          End Do
!
! Now solve for the B's. First reduce.
!
    Do i = 1,denominator_deg-1
      Do j = i + 1,denominator_deg
          ratio = AA(j, i) / AA(i, i)
          AA(j,i+1:denominator_deg+1) = AA(j,i+1:denominator_deg+1) - &
                                        ratio*AA(i,i+1:
```

Figure 4.14 *Continued*

```
denominator_deg+1)
        End Do ! j
    End Do ! i
!
! Now back substitute
!
    b(denominator_deg) = AA(denominator_deg, denominator_deg + 1) / &
AA(denominator_deg, denominator_deg)
    Do j = denominator_deg-1,1,-1
        sum = 0
        Do k = denominator_deg, j + 1, -1
            sum = sum + AA(j, k) * b(k)
            End Do
        b(j) = (AA(j, denominator_deg + 1)-sum) / AA(j, j)
        End Do ! j
        b(0) = 1
!
! Now get the A's
!
    Do i = 1,numerator_deg
        sum = 0
        Do j = 1,i
            IF (j > denominator_deg) Exit
            sum = sum + b(j) * c(i-j)
            End Do ! j
        A(i) = sum + c(i)
        End Do ! i
    a(0) = c(0)
!
!    Print the coefficients of the Pade approximation
!
    Write(*,*) "The coefficients of the Pade approximation are"
    Write(*,*)
    Write(*,*)  " exp on x    coef in numerator"
    Write(*,*)
    Write(*,200) (i, a(i), i = 0, numerator_deg)
200 Format(T5,I2, T16,F9.4)
    Write(*,*)
    Write(*,*)
    Write(*,*)  " exp on x    coef in denominator"
    Write(*,*)
    Write(*,300) (i, b(i), i = 0,denominator_deg)
300 Format(T5,I2,T13,F12.4)
    Write(*,*)
    Write(*,*)
!
! Now do a table of approximations using Pade
!
    Write(*,*)
    Write(*,*) "Here are values computed from the Pade approximation"
    Write(*,*)
```

Figure 4.14 *Continued*

```
      Write(*,*) "   x ", "           Value          arctan"
      Write(*,*)
      x = 0.0
      Do
          If (x > 1.001) Exit
          v = a(numerator_deg)
          Do i = numerator_deg,1,-1
              v = v*x + a(i-1)
              End Do
      num = v
          v = b(denominator_deg)
          Do i = denominator_deg,1,-1
             v = v*x + b(i-1)
             End Do
      v = num / v
      Write(*,400)  x, v, atan(x)
      x = x + 0.2
      End Do
 400 Format(T3, F3.1, T14, F10.7,T30, F10.7)
      Stop
  End Program Pade_Approximation


 *********************Output for Pade Program*******************

  Here are the coefficients of the Maclaurin series

  power of x      Coefficient
     0               0.0000
     1               1.0000
     2               0.0000
     3              -0.3333
     4               0.0000
     5               0.2000
     6               0.0000
     7              -0.1429
     8               0.0000
     9               0.1111

  The coefficients of the Pade approximation are

   exp on x     coef in numerator

      0             0.0000
      1             1.0000
      2             0.0000
      3             0.7778
      4             0.0000
      5             0.0677
```

Figure 4.14 *Continued*

```
exp on x      coef in denominator

   0                1.0000
   1                0.0000
   2                1.1111
   3                0.0000
   4                0.2381



Here are values computed from the Pade approximation

    x              Value              arctan

  0.0            0.0000000          0.0000000
  0.2            0.1973956          0.1973956
  0.4            0.3805064          0.3805064
  0.6            0.5404218          0.5404196
  0.8            0.6747710          0.6747410
  1.0            0.7855856          0.7853982
```

Figure 4.14 *Continued*

Exercises

Section 4.1

1. Compute $T_{11}(x)$ and $T_{12}(x)$.

2. Show that Eq. (4.3) is satisfied for these values of (m, n): (0, 1), (1, 1), (1, 2). (Don't do this numerically!)

3. Extend the graphs of $T_1(x)$, $T_2(x)$, $T_3(x)$ to the interval $[-2, 2]$. Observe that the maximum magnitude for the Chebyshev polynomials is not equal to one outside of $[-1, 1]$.

▶ 4. Graph $T_5(x)$ for $x = -1$ to $x = 1$. Locate the zeros more accurately than can be read from the graph; use any method of Chapter 1.

5. Expand $\cos(6x)$ in terms of $\cos(x)$, and show that this is equivalent to $T_6(x)$. (You may have to use a formula for the sum of two angles to do the expansion.)

Section 4.2

6. Reduce the fifth-degree polynomial of Eq. (4.5) to a fourth-degree polynomial, and confirm that its maximum error is 0.000791.

7. Graph the error of the economized fourth-degree polynomial of Exercise 6 over $[-1, 1]$, and compare it to

the graph of the errors of the fourth-degree Maclaurin series for e^x.

▶ 8. The function arctan(x) can be represented by this power series:

$$\text{arctan}(x) = x - \frac{x^3}{3} + \frac{x^5}{5} - \frac{x^7}{7} + \frac{x^9}{9} - \cdots .$$

Economize this three times to give a third-degree polynomial. Graph the errors, and compare this graph to the errors of the ninth-degree expansion.

9. Find the first few terms of the Chebyshev series for $\sin(x)$ by rewriting the Maclaurin series in terms of the $T(x)$'s and collecting terms. Then express this series as a power series. Compare the errors of both series after truncating to give third-degree polynomials.

10. A series expansion for $(1 + x/5)^{1/2}$ is

$$1 + \frac{x}{10} - \frac{x^2}{200} + \frac{x^3}{2000} - \frac{x^4}{16,000} + \frac{7x^5}{800,000} - \cdots .$$

Convert this to a Chebyshev series, including terms to T_2. What is the maximum error of the truncated Chebyshev series? Compare this to the error of the power series when it is truncated to second degree.

▶ **11.** To get the smaller error of a Chebyshev series or an economized power series requires that the approximation be for the interval $[-1, 1]$. Show what change of variable will change $f(x)$ on $[a, b]$ to $f(y)$ on $[-1, 1]$.

Section 4.3

▶ **12.** Find Padé approximations for the following functions, with numerators and denominators each of degree three:
a. $\sin^2(x)$, b. $\cos(x^3)$, c. e^x.

13. Compare the errors on $[-1, 1]$ for the Padé approximations of Exercise 12 with the errors of the corresponding Maclaurin series that has terms up to x^6.

14. Express the following rational fractions in continued-fraction form. In each part, compare the number of multiplication and division operations with that resulting from evaluating the polynomials by Horner's method (in nested form).

a. $\dfrac{x^2 - 2x + 2}{x^2 + 2x - 2}$

b. $\dfrac{2x^3 + x^2 + x + 3}{x^2 - x - 4}$

c. $\dfrac{2x^4 + 45x^3 + 381x^2 + 1353x + 1511}{x^3 + 21x^2 + 157x + 409}$

▶ **15.** Express each of the Padé approximations of Exercise 12 as continued fractions.

16. Estimate the errors of each of the Padé approximations in Exercise 12 by computing the coefficient of the next nonzero of the numerator. Compare these to the actual errors at $x = 1$ and at $x = -1$.

17. A Chebyshev series for $\cos(\pi x/4)$ is
$$0.851632 - 0.146437T_2 + 0.00192145T_4$$
$$- 9.965 * 10^{-6}T_6.$$
Use this series to develop a Padé-like rational function by the method of Section 4.3, where the function is $R_{4,2}$.

18. Fike (1968) gives this example of a rational fraction approximation to $\Gamma(1 + x)$ on $[0, 1]$:
$$R_{3,4}(x) =$$
$$\frac{0.999999 + 0.601781x + 0.186145x^2 + 0.0687440x^3}{1 + 1.17899x - 0.122321x^2 - 0.0260996x^3 + 0.060992x^4}.$$
Is this a minimax approximation? If not, what are the bounds of the errors of the $R_{3,4}$ minimax approximation?

Section 4.4

19. Which of these functions is periodic? What is the period if it is periodic?

a. $\sin(2x) + 2\cos(x)$ c. $\sin^3(x)$
b. $e^{-10x}\cos(x)$ d. e^{2ix}, where $i = \sqrt{-1}$

20. Example 4.4 plots $f(x) = |x|$ between $x = -\pi$ and $x = \pi$. $f(x)$ is also periodic. Extend the plot of Fig. 4.7 for the interval $[-10, 10]$.

21. Find the Fourier coefficients for $f(x) = x^3$ if $f(x)$ is periodic and one period extends from $x = 0$ to $x = 2$.

▶ **22.** Find the Fourier coefficients for $g(x) = x^2 - 1$ if it is periodic and one period extends from $x = 0$ to $x = 2$.

23. Show that the Fourier series of
$$x^3 + x^2 - 1$$
is just the sum of those for $f(x)$ (Exercise 21) and $g(x)$ (Exercise 22).

24. Is the Fourier series for $[f(x)] * [g(x)]$ (Exercises 21 and 22) the product of the two series?

25. Suppose we are interested in $f(x) = e^{-x}\sin(2x - 1)$ only in the interval $[0, 2]$. Sketch the half-range extensions that give

a. an even function.
b. an odd function.

26. Repeat Exercise 25, but for the range of interest $[-1, 3]$.

▶ **27.** Find the Fourier coefficients for the periodic functions of parts (a) and (b) of Exercise 25.

28. Repeat Exercise 27, this time for the function of Exercise 26.

Section 4.5

▶ **29.** How do the maxima and minima of these polynomials compare to those of $T_4(x)/8$ within the interval $x = -1$, $x = 1$?

a. $(x^2 - 0.8)(x^2 - 0.15)$.
b. $(x^2 - 0.9)(x^2 - 0.14)$.

30. The Legendre polynomials (which we will discuss in the next chapter) resemble the Chebyshev polynomials in that they have the same number of zeros within $[-1, 1]$ and the same number of maxima and minima. These Legendre polynomials can be obtained through this recursion formula:
$$L_0(x) = 1, \qquad L_1(x) = x,$$
$$(n + 1)L_{n+1}(x) = (2n + 1)xL_n(x) - nL_{n-1}(x).$$
Compare the graphs of several of these polynomials with Chebyshev polynomials of the same degree. Why are the Legendre polynomials less suited to economizing a power series?

31. Verify Eq. (4.3) after making the substitution from Eq. (4.4) for x. (Do this analytically.)

32. $\sin(nx)$ is orthogonal over $[-\pi, \pi]$ (see Eq. 4.18). Make a change of variable so it is orthogonal over $[-1, 1]$. Then compare its graph with that of $T_4(x)$ when n is such that there are exactly five maxima/minima on the interval.

33. Repeat Exercise 31 but for $\cos(nx)$.

Section 4.6

34. Use Maple to find T_5, T_6, T_7, T_8.

▶ 35. Compute Legendre polynomials (see Exercise 30) with *Mathematica*. Display all of these through degree 6.

36. Use MATLAB to get the economized polynomials of degree 5 from the Maclaurin series of degree 6 for:

     a. $f(x) = e^x \sin(\pi x/2)$.
     b. $f(x) = \cos^2(\pi x)/e^x$.
     c. $f(x) = |x|, -2 < x < 2$, and periodic of period 4.

37. Reduce each of the results of Exercise 36 to degree 4.

▶ 38. Compare the errors of each of the economized series of Exercise 37 with the original Maclaurin series on $[-1, 1]$.

39. DERIVE is a computer algebra system that we have not described previously. It is perhaps the easiest to use because it is menu-driven. Use it to get six terms of the Fourier series for each of the functions defined in Exercise 36.

40. Use DERIVE to compare the graphs of each series of Exercise 39 with the plots of the original functions. Do this for x-values between -5 and 5.

Applied Problems and Projects

41. In Section 3 of this chapter, the Padé rational functions were developed to approximate $f(x)$ on the interval $[-1, 1]$. If we want to approximate $f(x)$ on a different interval, say, $[a, b]$, a simple linear transformation can change the interval to $[-1, 1]$. What if we want to approximate a function on an interval with one or both endpoints infinite? Devise a transformation for such cases.

42. Investigate, for some computer system available to you, how some or all of the following transcendental functions are approximated in Fortran 90. Classify these into Taylor series formulas, Chebyshev polynomials, rational functions, or other types. Which of these are minimax?

     a. $\sin(x)$
     b. $\cos(x)$
     c. $\tan(x)$
     d. $\operatorname{atan}(x)$
     e. $\exp(x)$
     f. $\ln(x)$

43. Repeat Exercise 42 for other computer languages: BASIC, Pascal, and C.

44. As illustrated by Fig. 4.8, the sum of n terms of the Fourier series for a function that has a jump discontinuity has larger oscillations near the discontinuity—the Gibbs phenomenon. For $f(x)$ equal to a square wave, investigate whether the departure of the sum of the series from $f(x)$ for the last "hump" in the curve (the "ear"), decreases when n is increased. What can you conclude about the size of the ear?

45. One way to eliminate the Gibbs phenomenon (see Exercise 44) is to abolish the jump discontinuity by subtracting a linear function from $f(x)$. Suppose the linear function is $L(x)$. For $f(x)$ equal to square wave, find the $L(x)$ for which $f(x) - L(x)$ has no jump discontinuities. Compare the accuracy of the sum of 10 terms of the Fourier series for $f(x)$ with that for the sum of 10 terms of the Fourier series for $g(x) = f(x) - L(x)$.

46. Another way to ameliorate the problem of the Gibbs phenomenon is to use the so-called Lanczos's factors. Search the literature to find out more about this method. Apply it to obtain an improved approximation to a square wave.

47. Chapter 3 described the fitting of functions with polynomials and this chapter describes fitting them with sinusoids (Fourier series). Another possibility is to fit with exponentials, $y(x) = \Sigma c_i \exp(a_i x)$. Is it possible to do this? Under what conditions is it possible? How can the values of c_i and a_i be determined? Specifically, fit a four-term sum to these data and compare to the exact solution, $y = \sin(\pi x/6)$:

$$
\begin{array}{cccccc}
\textbf{x:} & 1 & 2 & 3 & 4 \\
\textbf{y:} & \dfrac{1}{2} & \sqrt{\dfrac{3}{2}} & 1 & \sqrt{\dfrac{3}{2}}
\end{array}
$$

5

Numerical Differentiation and Numerical Integration

5.4 Extrapolation Techniques

Allows you to get an improved estimate of the value of a derivative that is equivalent to using a formula based on a higher-degree polynomial. These techniques are often used in computer programs. The Richardson extrapolation procedure has wide application beyond its use for derivatives.

5.5 Newton–Cotes Integration Formulas

Derives formulas by integrating an interpolation formula analogous to the development of differentiation formulas. The three most important of these formulas for integration are based on polynomials of degrees 1, 2, and 3, with the integration interval the same as the region of fit for the polynomial.

5.6 The Trapezoidal Rule—A Composite Formula

Describes the first of the Newton–Cotes formulas when applied to a succession of evenly spaced data points. It is very widely used in computer programs. When the trapezoidal rule is used for a known function, its results may be extrapolated, a technique known as Romberg integration.

5.7 Simpson's Rules

Consists of the composite rules obtained when the Newton–Cotes formulas based on polynomials of degrees 2 and 3 are applied to a succession of evenly spaced intervals. These rules are more accurate than the trapezoidal rule, particularly so for Simpson's $\frac{1}{3}$ rule, based on a quadratic interpolating polynomial. Extrapolation may be used here also.

5.8 Other Ways to Derive Integration Formulas

Explains the application of symbolic methods to obtain integration formulas and a second method, the method of undetermined coefficients. This last technique has many applications.

5.9 Gaussian Quadrature

Details a procedure for getting the integral of a known function that uses fewer function evaluations for a required accuracy than do the previous methods. Because the cost of computer time is usually related directly to the number of function evaluations, this procedure is widely used in modern computer programs. You are also introduced to Legendre polynomials, a representative of the very important *orthogonal polynomials*.

5.10 Adaptive Integration

Describes a technique that can reduce the number of function evaluations when Simpson's $\frac{1}{3}$ rule is used, by properly selecting the size of intervals within subregions of the integration region. A kind of binary search is used. An interesting bookkeeping problem is involved.

5.11 Multiple Integrals

Explains how multiple integrals can also be evaluated numerically by extending the methods for a single integral.

5.1 Getting Derivatives and Integrals Numerically

Rita and Ed were again at lunch in the cafeteria of Ruscon Engineering.

"How did you make out on the interpolation problem?" Ed asked.

"It went all right," Rita replied. "In fact, we have computer programs now to interpolate either with polynomials or with cubic splines. My boss now wants me to work on some extensions."

"What do you mean, extensions?" Ed asked.

"Well, we need the time derivative of the position. That's the velocity, of course. And some calculations that lead to the fuel consumption require us to integrate a function when we know the values only at discrete times."

"I should think that would follow directly from what you did before," Ed said. "If you have a polynomial that gives you the position as a function of time, can't you just differentiate or integrate that polynomial?"

"Very true," Rita replied. "The catch is that we never really develop the polynomial; we just work with the position values and their differences. My boss warns me also that estimating the errors of derivatives and integrals determined from function values that are known only at discrete points in time is pretty tricky. I suspect that there are ways to tackle these problems that are similar to the methods for interpolation."

Rita is right; there are methods for getting derivatives and integrals from a table of function values, and these methods resemble those for interpolation. In this chapter we explore these methods. These same methods are also useful to estimate the value of the derivative or integral when $f(x)$ is known.

Our computer algebra systems can get derivatives and integrals using these same methods but they can also get them symbolically, obtaining the analytical result.

5.2   Derivatives from Difference Tables

We can use a divided-difference table to estimate values for derivatives. Recall that the interpolating polynomial of degree n that fits at points $p_0, p_1, p_2, \ldots, p_n$ is, in terms of divided differences,

$$
\begin{aligned}
f(x) = P_n(x) &+ \text{error} \\
= f[x_0] &+ f[x_0, x_1](x - x_0) \\
&+ f[x_0, x_1, x_2](x - x_0)(x - x_1) \\
&+ \cdots + f[x_0, x_1, \ldots, x_n] \prod_{i=0}^{n-1} (x - x_i) \\
&+ \text{error}.
\end{aligned}
\tag{5.1}
$$

(We may sometimes use an alternate notation for the divided difference: $f_0^{[n]}$.)

If $P_n(x)$ matches well to $f(x)$, we should get a polynomial that approximates the derivative, $f'(x)$, by differentiating it. Recall that the derivative of a product of n terms is a sum of n of these product terms with one member of each term in the sum replaced by its derivative. For example,

$$
\frac{d}{dx}(u * v * w) = u' * v * w + u * v' * w + u * v * w'
$$

or, for the products in $P_n(x)$ of Eq. (5.1),

$$
\frac{d}{dx} \prod_{i=0}^{n-1} (x - x_i) = \sum_{i=0}^{n-1} \frac{(x - x_0)(x - x_1) \cdots (x - x_{n-1})}{(x - x_i)}
$$

$$
= \sum_{i=0}^{n-1} \prod_{\substack{j=0 \\ j \neq i}}^{n-1} (x - x_j).
$$

Carrying this out, we get this approximation for $f'(x)$:

$$P_n'(x) = f[x_0, x_1] + f[x_0, x_1, x_2][(x - x_1) + (x - x_0)] + \cdots$$

$$+ f[x_0, x_1, \ldots, x_n] \sum_{i=0}^{n-1} \frac{(x - x_0)(x - x_1) \cdots (x - x_{n-1})}{(x - x_i)}. \qquad (5.2)$$

To get the error term for Eq. (5.2), we have to differentiate the error term for $P_n(x)$:

$$\text{Error} = (x - x_0)(x - x_1) \cdots (x - x_n) \frac{f^{(n+1)}(\xi)}{(n + 1)!}. \qquad (5.3)$$

When this error term is differentiated, we will find a sum that has in one of its terms

$$\frac{d}{dx}[f^{(n+1)}(\xi)],$$

which is impossible to evaluate because ξ depends on x in an unknown way. However, if we take $x = x_i$ (where x_i is one of the tabulated points), the difficult term drops out and we get this expression for the error:

Error of the approximation to $f'(x)$, when $x = x_i$, is

$$\text{Error} = \left[\prod_{\substack{j=0 \\ j \neq i}}^{n} (x_i - x_j) \right] \frac{f^{(n+1)}(\xi)}{(n + 1)!}, \qquad \xi \text{ in } [x, x_0, x_n]. \qquad (5.4)$$

Observe that the error is not zero even when x is a tabulated value, although the interpolating polynomial agrees with $f(x)$ at this point. In fact, the error of the derivative is less at some x-values between the points.*

It is not surprising that the next-term rule applies here as it did for interpolating polynomials. We use a known polynomial function in the following example to show that the procedure works correctly.

EXAMPLE 5.1   Let $f(x) = x^2 - x + 1$, and tabulate for $x = 0, 2, 3, 5, 6$ (five points). Here is the divided difference table:

i	x_i	f_i	$f_i^{[1]}$	$f_i^{[2]}$	$f_i^{[3]}$	$f_i^{[4]}$
0	0	1	1	1	0	0
1	2	3	4	1	0	
2	3	7	7	1		
3	5	21	10			
4	6	31				

---

* We discuss errors of numerical differentiation in some detail in Section 5.15.

(In this table, we have used our nonstandard notation for the divided differences, thus simplifying the writing of equations that follow.) Observe that the second differences are constant and the third are zero, as we expect for a quadratic polynomial.

What is the estimated value for the derivative if x is 4.1 and we use a cubic interpolating polynomial that starts at $i = 1$? Here is the arithmetic:

$$f'(x) = f_1^{[1]} + f_1^{[2]}[(x - x_2) + (x - x_1)]$$
$$+ f_1^{[3]}[(x - x_2)(x - x_3) + (x - x_1)(x - x_3) + (x - x_1)(x - x_2)]. \tag{5.5}$$

Substituting values gives

$$f'(4.1) = 4 + 1 * [(4.1 - 3) + (4.1 - 2)]$$
$$+ 0 * [(4.1 - 3)(4.1 - 5) + (4.1 - 2)(4.1 - 5) + (4.1 - 2)(4.1 - 3)]$$
$$= 4 + 1 * (1.1 + 2.1)$$
$$+ 0 * [(1.1)(-0.9) + (2.1)(-0.9) + (2.1)(1.1)]$$
$$= 7.2,$$

which is the exact answer. This is as we expect, because $f(x)$ is a quadratic and the cubic between $x = 2$ and $x = 6$ is actually a quadratic. In fact, if we expand Eq. (5.5), we get $f'(x) = 2x - 1$, the analytical expression for the derivative. (If the tabulated function is not a polynomial, we would not get exact answers, of course.)

If we were to use an interpolating polynomial of degree 1 (linear interpolation) to get the derivative polynomial, starting at point $x = x_i$, we would have just

$$f_1'(x) = f[x_i, x_{i+1}] + \text{error} = f_i^{[1]} + \text{error}.$$

For $x = 4.1$, if we take $i = 2$ (the best value, why?), we estimate $f'(4.1)$ as 7. The true value is 7.2, so the error is 0.2. Let us use the next-term rule to estimate the error. The next term is

$$f_2^{[2]}[(x - x_3) + (x - x_2)] = 1 * [(4.1 - 5) + (4.1 - 3)] = 0.2,$$

which is precisely the actual error!   ▲

As we saw in Chapter 3, if we want to estimate the derivative for an x-value near the end of the table, we appear to be severely limited in the degree of interpolating polynomial. We can overcome this limitation by reordering i-values, putting them in reverse order. Our formulas still work correctly, but we must remember to go diagonally upward to get the values for a given value of i.

Because an interpolating polynomial fits better to the function if the x-values used in its construction are such that the x-value for getting the derivative is centered within them, we should choose the starting point (the i-value) to make this true. (If the x-values are in order, our task is easier.)

An Algorithm to Obtain an Estimate of the Derivative from a Divided-Difference Table

Given $n + 1$ data pairs, (x_i, f_i), $i = 0, \ldots, n$:

(Do the table)
DO FOR $i = 0$ TO n STEP 1
    SET $f(i, 0) = f(i)$,
ENDDO (FOR i).
DO FOR $j = 1$ TO n STEP 1
    DO FOR $i = 0$ TO $n - j$ STEP 1:
        $f(i, j) = [f(i + 1, j - 1) - f(i, j - 1)]/[x(i + j) - x(i)]$,
    ENDDO (FOR i).
ENDDO (FOR j).
(Get user inputs)
INPUT:
    $x = x$-value,
    $i = i$-value to start,
    ND = degree of polynomial.
(Compute derivative)
SET SUM = 0.
DO FOR $j =$ ND TO 2 STEP -1
    SET SUMP = 0.
    DO FOR $k = 0$ TO $j - 1$ STEP 1
        SET $p = 1$.
        DO FOR $l = 0$ TO $j - 1$ STEP 1
            IF $l = k$ THEN
                $p = p * [x - x(i + 1)]$
            ENDIF.
        ENDDO (FOR l).
        SET SUMP = SUMP + p.
    ENDDO (FOR k).
    SET SUM = SUM + $f(i, 1)$.
ENDDO (FOR j).
(Display result)
DISPLAY SUM as derivative value at x.

(The computations of the derivative value can be repeated with new user inputs.)

Evenly Spaced Data

When the data are evenly spaced, we can use a table of function differences to construct the Newton–Gregory polynomial. We write this in terms of $s = (x - x_i)/h$:

$$P_n(s) = f_i + s\Delta f_i + \frac{s(s-1)}{2!}\Delta^2 f_i + \frac{s(s-1)(s-2)}{3!}\Delta^3 f_i$$

$$+ \cdots + \prod_{j=0}^{n-1}(s-j)\frac{\Delta^n f_i}{n!} + \text{error};$$

$$\text{Error} = \left[\prod_{j=0}^{n}(s-j)\right]\frac{f^{(n+1)}(\xi)}{(n+1)!}, \qquad \xi \text{ in } [x, x_1, \ldots, x_n].$$

(In this formula, i is the index value where we enter the difference table.)

The derivative of $P_n(s)$ should approximate $f'(x)$. We do exactly the same as we did for the polynomial constructed from a divided-difference table, getting

$$\frac{d}{dx}P_n(s) = \frac{d}{ds}P_n(s)\frac{ds}{dx}$$

$$= \frac{1}{h}\left[\Delta f_i + \sum_{j=2}^{n}\left\{\sum_{k=0}^{j-1}\prod_{\substack{\ell=0\\ \ell \neq k}}^{j-1}(s-\ell)\right\}\frac{\Delta^j f_i}{j!}\right]. \tag{5.6}$$

(The $1/h$ factor comes from $ds/dx = d/dx\,(x-x_i)/h = 1/h$.)

Again, the error term involves an unknown quantity unless x is one of the tabulated values. When $x = x_i$, $s = 0$. In this case, we get this analog of Eq. (5.4) when an interpolating polynomial of degree n is used:

$$\begin{array}{c}\text{Error}\\ \text{(when } x = x_i\text{)}\end{array} = \frac{(-1)^n h^n}{n+1}f^{(n+1)}(\xi), \qquad \xi \text{ in } [x_1, \ldots, x_n]. \tag{5.7}$$

Equation (5.6) is a formula for estimating derivatives from a table of differences that we enter at index value i. Here is an example (that again uses a known function, $f(x) = e^x$).

EXAMPLE 5.2   Estimate the value of $f'(3.3)$ with a cubic polynomial that is created if we enter the table at $i = 2$, given this difference table:

i	x_i	f_i	Δf_i	$\Delta^2 f_i$	$\Delta^3 f_i$	$\Delta^4 f_i$	$\Delta^5 f_i$
0	1.30	3.669	3.017	2.479	2.041	1.672	1.386
1	1.90	6.686	5.496	4.520	3.713	3.058	2.504
2	2.50	12.182	10.016	8.233	6.771	5.562	
3	3.10	22.198	18.249	15.004	12.333		
4	3.70	40.447	33.253	27.337			
5	4.30	73.700	60.590				
6	4.90	134.290					

For our example, $h = 0.6$ and we take $x_i = 2.5$ (the best choice). With $x = 3.3$, we have $s = (3.3 - 2.5)/0.6 = 4/3$. Here is the equation that approximates the derivative (it will be of degree 2, of course):

$$P_2(x) = \frac{1}{h}\left\{ \Delta f_2 + \frac{1}{2!}\left[\sum_{k=0}^{1}\prod_{\substack{\ell=0\\\ell\neq k}}^{1}(s-\ell)\right]\Delta^2 f_2 + \frac{1}{3!}\left[\sum_{k=0}^{2}\prod_{\substack{\ell=0\\\ell\neq k}}^{2}(s-\ell)\right]\Delta^3 f_2\right\}$$

$$= \frac{1}{0.6}\left\{ 10.016 + \frac{1}{2}\left[\left(\frac{4}{3}-1\right)+\left(\frac{4}{3}-0\right)\right] * 8.233\right.$$

$$+ \frac{1}{6}\left[\left(\frac{4}{3}-1\right)\left(\frac{4}{3}-2\right)+\left(\frac{4}{3}-0\right)\left(\frac{4}{3}-2\right)\right.$$

$$\left.\left. + \left(\frac{4}{3}-0\right)\left(\frac{4}{3}-1\right)\right] * 6.771\right\}$$

$$= 26.875 \qquad \text{(versus 27.113, the exact value of } f'(3.3)\text{).}$$

If we use the next-term rule to estimate the error of the derivative based on a cubic inter-polating polynomial, we get a value of 0.315 compared to the actual error of 0.238. This is not an unreasonable estimate. ▲

Figure 5.1 plots the polynomial that estimates the derivative (the curve that is nearly straight), the true derivative curve, and the points where the interpolating polynomial fits the function. (In this special case, these points fall on the derivative curve because, for $f(x) = e^x$, the function and its derivative are the same.) Observe that the line for the de-rivative polynomial does not agree with the exact derivative curve at the points where the interpolating polynomial agrees with the function but that there are three places where the derivative polynomial does agree with the exact value for $f'(x)$.

A higher-degree polynomial would estimate values for the derivative more exactly.

Simpler Formulas

Equation (5.6) is awkward to use when we do hand computations (but a computer program, though using several nested loops, has no problem). If we stipulate that the x-value must be in the difference table, the computation is simplified considerably. Thus, for an estimate of $f'(x_i)$, we get

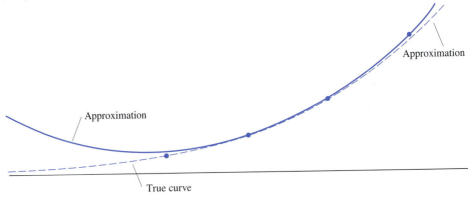

Figure 5.1

$$f'(x) = \frac{1}{h}\left[\Delta f_i - \frac{1}{2}\Delta^2 f_i + \frac{1}{3}\Delta^3 f_i - \cdots + (-1)^{n-1}\frac{\Delta^n f_i}{n}\right]_{x=x_i}, \qquad (5.8)$$

because, at $x = x_i$, $s = 0$ and

$$\frac{1}{n!}\prod_{j=0}^{n-1}(s-j) = \frac{(-1)^{n-1}}{n}.$$

Equation (5.8) is easy to use for hand computations because the multipliers of the differences are so simple. There are other consequences as well. Suppose we write the error term as an "order of" expression. If we use just one term of Eq. (5.8) (this means that we are linearly interpolating, using a polynomial of degree 1), we have

$$f'(x_i) = \frac{1}{h}[\Delta f_i] - \frac{1}{2}hf''(\xi), \qquad \text{error is } O(h).$$

With two terms,

$$f'(x_i) = \frac{1}{h}\left[\Delta f_i - \frac{1}{2}\Delta^2 f_i\right] + \frac{1}{3}h^2 f^{(3)}(\xi), \qquad \text{error is } O(h^2).$$

In general, with n terms, the error is $O(h^n)$. As already pointed out, we also can estimate the size of the error from the next-term rule.

The derivative formula of Eq. (5.8) is called a *forward-difference approximation* because the differences all involve f-values that lie forward in the table from f_i. Suppose we use a second-degree polynomial that matches the table at x_i, x_{i+1}, and x_{i+2} but evaluate it for $f'(x_{i+1})$, using $s = 1$. Eq. (5.6) then becomes

$$f'(x_{i+1}) = \frac{1}{h}\left[\Delta f_i + \frac{1}{2}\Delta^2 f\right] + O(h^2).$$

Rewriting in terms of the f-values, we have

$$f'(x_{i+1}) = \frac{1}{h}\left[(f_{i+1} - f_i) + \frac{1}{2}(f_{i+2} - 2f_{i+1} + f_i)\right] + \text{error}$$

$$= \frac{1}{h}\frac{f_{i+2} - f_i}{2} + \text{error}, \qquad (5.9)$$

$$\text{error} = -\frac{1}{6}h^2 f^{(3)}(\xi) = O(h^2).$$

Equation (5.9) is called a *central-difference formula* because the x-value is centered within the range of x-values used in its construction. It is most important to note that we get an improved order of error even though only two function values are involved. The coefficient of the error term is also smaller, an added benefit. This formula is widely used to estimate the derivative both from tabulated values and even when the function is known.

EXAMPLE 5.3   For $f(x) = xe^{-x/2}$, estimate the value of $f'(0.3)$ using $h = 0.1, 0.05$, and 0.025. Compare the results from one term of Eq. (5.8) and from Eq. (5.9). In each case, compute the actual error.

We use these two formulas to get the estimates:

From Eq. (5.8) $f'(0.3) = \dfrac{1}{h}[f(0.3 + h) - f(0.3)]$.

From Eq. (5.9) $f'(0.3) = \dfrac{1}{2h}[f(0.3 + h) - f(0.3 - h)]$.

Table 5.1 shows the results. The results from the forward-difference formula have errors much greater than those from central differences. The former are about halved as h is halved (successive errors are decreased by a factor of 1.98), whereas the errors from the preferred formula decrease in proportion to h^2 (successive errors are decreased by a factor of 3.98). These errors are also much smaller.

Table 5.1

	h	Estimate of $f'(0.3)$	Error
From Eq. (5.8)	0.1	0.69280	−0.0388
	0.05	0.71195	−0.0196
	0.025	0.72171	−0.0099
From Eq. (5.9)	0.1	0.73262	0.00102
	0.05	0.73186	0.00026
	0.025	0.73167	0.00006

▲

5.3 Higher-Order Derivatives

We can differentiate the formulas of Section 5.2 to obtain formulas for higher-order derivatives, but we prefer to use a simpler method. The *symbolic operator method* can be used to derive formulas for higher-order derivatives based on evenly spaced data. There are two interrelated operators:

Difference operator: $\Delta f(x_0) = f(x_0 + h) - f(x_0)$,

Or: $\Delta f(x_i) = \Delta f_i = f_{i+1} - f_i$.

Stepping operator: $Ef(x_0) = f(x_0 + h)$,

Or: $Ef_i = f_{i+1}$.

$E^n f_i = f_{i+n}$

Relation between E and Δ: $E = 1 + \Delta$.

We will also use the symbol D to represent the differentiation operator: $D(f) = df/dx$, $D^n(f) = d^nf/dx^n$.

Let us begin with

$$f_{i+s} = E^sf_i,$$

where $s = (x - x_i)/h$. If we take the derivative as if this were an algebraic expression, we get

$$Df_{i+s} = \frac{d}{dx}f(x_{i+s}) = \frac{d}{dx}(E^sf_i)$$

$$= \frac{1}{h}\frac{d}{ds}(E^sf_i) = \frac{1}{h}(\ln E)E^sf_i.$$

If $s = 0$, we get

$$Df_i = f'_i = \frac{1}{h}(\ln E)f_i = \frac{1}{h}\ln(1 + \Delta)f_i, \qquad (5.10)$$

which means that D and Δ are related by

$$D = \frac{1}{h}\ln(1 + \Delta).$$

Now we will use a Maclaurin expansion for $\ln(1 + \Delta)$ to transform Eq. (5.10) to

$$f'_i = \frac{1}{h}\left(\Delta f_i - \frac{1}{2}\Delta^2f_i + \frac{1}{3}\Delta^3f_i - \frac{1}{4}\Delta^4f_i + \cdots\right), \qquad (5.11)$$

which is identical to Eq. (5.8). We have derived the formula for the derivative of $f(x)$ at $x = x_i$ through symbol manipulation. Now the second derivative of $f(x)$ is $f''(x) = D^2f$, for which we can get the operator by multiplying the operator of Eq. (5.11) by itself. Further, we can get a formula for the nth derivative by multiplying n times!

Here are the results for D^2f_i if we perform the multiplication:

$$f''_i = \frac{1}{h^2}\left(\Delta^2f_i - \Delta^3f_i + \frac{11}{12}\Delta^4f_i - \frac{5}{6}\Delta^5f_i + \cdots\right). \qquad (5.12)$$

Formulas for higher derivatives we leave as an exercise. A most important consequence of such expansions is that the first term is

$$D^nf_i = \frac{1}{h^n}\Delta^nf_i, \qquad (5.13)$$

which we have used previously in the next-term rule. An even better approximation to $f^{(n)}(x_i)$ when n is even is $\Delta^nf_{i-n/2}$, which centers x_i within the range of data points.

Divided Differences

If the data are evenly spaced, a divided-difference table is really the same as a table of function differences except that the h-value is included in the entries. This relation holds

$$\frac{\Delta^n f_i}{(h^n)(n!)} = f_i^{[n]} = f[x_i, x_{i+1}, \ldots, x_{i+n}],$$

which means that the nth divided difference and the nth derivative are related by another expression:

$$f_i^{[n]} = f[x_i, x_{i+1}, \ldots, x_{i+n}] = \frac{D^n f(x_i)}{n!} \approx \frac{f^{(n)}(x_i)}{n!}. \tag{5.14}$$

EXAMPLE 5.4   Given Table 5.2, estimate the second derivative at $x = 1.3$ using one, two, and three terms of Eq. (5.12). Also estimate the errors of each.
   Using one term,

$$f''(1.3) = \frac{1}{h^2}(\Delta^2 f_1) = \frac{1}{0.04} * (-0.02185) = -0.5463.$$

Using two terms,

$$f''(1.3) = \frac{1}{h^2}(\Delta^2 f_1 - \Delta^3 f_1)$$

$$= \frac{1}{0.04} * (-0.02185 - 0.00739) = -0.7310.$$

Using three terms,

$$f''(1.3) = \frac{1}{h^2}\left(\Delta^2 f_1 - \Delta^3 f_1 + \frac{11}{12}\Delta^4 f_1\right)$$

$$= \frac{1}{0.04} * \left[-0.02185 - 0.00739 + \frac{11}{12}(0.00019)\right]$$

$$= -0.7267.$$

Table 5.2

i	x_i	f_i	Δf_i	$\Delta^2 f_i$	$\Delta^3 f_i$	$\Delta^4 f_i$	$\Delta^5 f_i$
0	1.10	3.24075	-0.27596	-0.02885	0.00700	0.00039	-0.00020
1	1.30	2.96479	-0.30481	-0.02185	0.00739	0.00019	-0.00021
2	1.50	2.65999	-0.32666	-0.01446	0.00758	-0.00002	
3	1.70	2.33333	-0.34112	-0.00688	0.00756		
4	1.90	1.99221	-0.34800	0.00067			
5	2.10	1.64421	-0.34733				
6	2.30	1.29688					

We can estimate the errors in each case from the next-term rule. Doing so gives -0.1848, 0.0044, and 0.0044 for the successive results. (We would expect the errors to decrease in the last two, but round-off errors are beginning to affect the computations.)

The function tabulated in Table 5.2 is $4 * \sin(x)/x$ for which the true value of $f''(1.3)$ is -0.72243. (A symbolic algebra program was of great help in getting this result!) We then compute the actual errors as -0.1761, 0.0086, and 0.0043. The estimates are certainly reasonable, though not exact.   ▲

Central-Difference Formulas

In the previous section, a central-difference formula for the derivative was shown to be more accurate than a forward-difference formula. Let us use the symbolic method to derive the central-difference formula.

It is convenient to define a *backward-difference operator:*

$$\nabla f_0 = f_0 - f_{-1}, \qquad \nabla f_i = f_i - f_{i-1},$$

which is related to the stepping operator, E, by

$$E^{-1} = 1 - \nabla.$$

If we use this relation in Eq. (5.10), we get

$$\frac{d}{dx}f(x)_{x=x_i} = f'_i = \frac{1}{h}(\ln E)f_i = \frac{1}{h}\left[\ln\left(\frac{1}{1-\nabla}\right)\right]f_i, \tag{5.15}$$

which shows that another symbolic relation for the derivative operator, D, is

$$D = \frac{1}{h}\ln\left(\frac{1}{1-\nabla}\right).$$

Expanding the logarithm term gives a formula for the derivative in terms of backward differences:

$$f'_i = \frac{1}{h}\left(\nabla + \frac{1}{2}\nabla^2 + \frac{1}{3}\nabla^3 + \frac{1}{4}\nabla^4 + \cdots\right)f_i. \tag{5.16}$$

Suppose we take the average of the derivative values given by Eqs. (5.11) and (5.16) by adding the two and dividing by 2:

$$f'_i = \frac{1}{2h}\left[(\Delta + \nabla) - \frac{1}{2}(\Delta^2 - \nabla^2) + \frac{1}{3}(\Delta^3 + \nabla^3) - \frac{1}{4}(\Delta^4 - \nabla^4) + \cdots\right]f_i. \tag{5.17}$$

The first term of this formula, rewritten in terms of function values, is

$$f'_i = \frac{(f_{i+1} - f_i) + (f_i - f_{i-1})}{2h} = \frac{1}{h}\frac{f_{i+1} - f_{i-1}}{2},$$

which is our central-difference formula. We can get central-difference formulas for the derivative based on higher-degree interpolating polynomials by using more terms of Eq. (5.17).

As before, we can get the central-difference operators for higher derivatives by multiplying the operator of Eq. (5.17) by itself. It turns out that this technique gives very complicated expressions. Because only the first term of the expression is commonly used, we can take a simpler approach.

We observed that a quadratic polynomial that fits from points $i - 1$ to $i + 1$ has its central point at i [which we can call point (i)]. We used this fact to get a central-difference formula for the first derivative (Eq. 5.9) from the forward-difference formula of Eq. (5.8).

We can take a similar approach for the second derivative. We write Eq. (5.13) for point $(i - 1)$ but interpret it as the second derivative at point (i):

$$f''(x_i) \approx \frac{1}{h^2}(\Delta^2 f_{i-1}) = \frac{f_{i+1} - 2f_i + f_{i-1}}{h^2}. \tag{5.18}$$

Equation (5.18) gives an approximation of error $O(h^2)$.

EXAMPLE 5.5   Apply Eq. (5.18) to estimate $f''(1.3)$ (that is, at $i = 1$) from Table 5.2.
We have

$$f''(1.3) = \frac{1}{h^2}\Delta^2 f_0 = \frac{-0.02885}{0.2^2} = -0.7213,$$

which is much closer to the true value of -0.72243 than even with three terms of the forward-difference formula used in Example 5.4.   ▲

There are other ways to obtain formulas for the various derivatives from a difference table. Appendix B explains the method of undetermined coefficients. Earlier editions of this book described the use of a "lozenge diagram," a device to determine the coefficients in formulas such as Eq. (5.17) but with more terms. These diagrams can be built up from one known formula such as Eq. (5.8) or (5.12) to give a device that works for any desired formula, including some that we have not discussed here.

An Algorithm to Compute First and Second Derivatives from Central-Difference Formulas

Given a function $f(x)$:

(Get user inputs)
INPUT
     x = x-value,
     h = delta-x value.
(Compute derivatives)
SET $d_1 = [f(x + h) - f(x - h)]/(2h)$.
SET $d_2 = [f(x + h) - 2f(x) + f(x - h)]/h^2$.
(Display results)
DISPLAY d_1 and d_2 as estimates of $f'(x), f''(x)$.

(The computations of the derivative values can be repeated with new user inputs.)

5.4 Extrapolation Techniques

There is another way to improve the accuracy of the estimates of derivatives from a table of evenly spaced values. This technique is equivalent to using formulas based on higher-degree polynomials without explicitly finding the formula. This method is best explained through an example.

EXAMPLE 5.6   Estimate the value of $f'(2.4)$ and $f''(2.4)$ from Table 5.3, for which the x's are evenly spaced.

We will use central-difference formulas, Eqs. (5.9) and (5.18), both of which have an error of $O(h^2)$.

We begin with computations for $f'(2.4)$. Equation (5.9) gives

$$f'(2.4) \approx \frac{f(2.5) - f(2.3)}{2(0.1)} = \frac{0.0491256 - 0.0747636}{0.2},$$ (5.19)

$$f'(2.4)[\text{exact}] = -0.12819 + C_1(0.1)^2,$$

where we show the exact value as the estimate plus the error, a quantity that is proportional to h^2. (C_1 is the proportionality constant.)

Suppose we repeat the calculation, this time using $h = 0.2$. (We can't make h smaller using only a table of values. It may seem that we are losing ground, but be patient.) This computation gives

$$f'(2.4) \approx \frac{f(2.6) - f(2.2)}{2(0.2)} = \frac{0.038288 - 0.089584}{0.4},$$ (5.20)

$$f'(2.4)[\text{exact}] = -0.12824 + C_2(0.2)^2.$$

The values of C_1 and C_2 in Eqs. (5.19) and (5.20) are not usually identical, but we will assume they are the same. (Each of the C's actually involves the value of $f^{(3)}(\xi)$, where the ξ's are not identical but should not differ by much. It can be shown that we make an error

Table 5.3

i	x_i	f_i
0	2.0	0.123060
1	2.1	0.105706
2	2.2	0.089584
3	2.3	0.074764
4	2.4	0.061277
5	2.5	0.049126
6	2.6	0.038288
7	2.7	0.028722
8	2.8	0.020371
9	2.9	0.013164
10	3.0	0.007026

of $O(h^4)$ in taking the two values as equal.) Based on this assumption that the C's are equal, we can solve the two equations to eliminate the C's:

$$f'(2.5)[\text{exact}] = -0.12819 + \frac{1}{(0.2/0.1)^2} - 1[-0.12819 - (-0.12824)]$$

$$= -0.12817 + O(h^4).$$

The computation is an instance of a general rule: Given two estimates of a value that have errors of $O(h^n)$, where the h's are in the ratio of 2 to 1, we can extrapolate to a better estimate of the exact value as follows:

$$\text{Better estimate} = \text{more accurate} + \frac{1}{2^n - 1}(\text{more accurate} - \text{less accurate}). \qquad \textbf{(5.21)}$$

(The more accurate value is the one computed with the smaller value for h.)

We can repeat the scheme using h-values of 0.2 and 0.4. Table 5.4 shows the result as "first-order extrapolations." The first-order extrapolations have errors of $O(h^4)$ so the second-order extrapolation comes from

$$f'(2.4)[\text{exact}] = -0.12817 + \frac{1}{2^4 - 1}[-0.12817 - (-0.12820)]$$

$$= -0.12817 + O(h^6).$$

Because there was no change in the fifth decimal place from the first extrapolation, we are tempted to say that the improved estimate is good to that many places. This conclusion is correct, except that round-off in the original data can affect the results and invalidate our conclusion. (The data were constructed from $f(x) = e^{-x}\sin(x)$, and $f'(2.4)$ is -0.128171.)

Table 5.4

h	Initial estimate	First-order extrapolation	Second-order extrapolation
0.1	−0.12819		
		−0.12817	
0.2	−0.12824		−0.12817
		−0.12820	
0.4	−0.12836		

▲

Second-Derivative Computations

We proceed in exactly the same manner to get an improved estimate of $f''(2.4)$. The initial estimates are as follows:

For $h = 0.1$,

$$f''(2.4) = \frac{0.049126 - 2(0.061277) + 0.074764}{(0.1)^2} = 0.13360.$$

For $h = 0.2$,

$$f''(2.4) = \frac{0.038288 - 2(0.061277) + 0.089584}{(0.2)^2} = 0.13295.$$

For $h = 0.4$,

$$f''(2.4) = \frac{0.020371 - 2(0.061277) + 0.123060}{(0.4)^2} = 0.13048.$$

Table 5.5 shows these values in the initial estimates column. The first- and second-order improvements are obtained from these initial estimates in exactly the same way as for the first derivatives in Table 5.4.

Again we are tempted to say that $f''(2.4)$ equals 0.13382 to five decimal places, but in this instance round-off errors have made the conclusion not true. [The exact value of $f''(2.4)$ is 0.13379.]

Additional extrapolations could be made, using the fact that the error of the second-order extrapolation is $O(h^6)$, but it is not recommended to go that far because of round-off effects.

Table 5.5

h	Initial estimate	First-order extrapolation	Second-order extrapolation
0.1	0.13360		
		0.13382	
0.2	0.13295		0.13382
		0.13377	
0.4	0.13048		

Richardson Extrapolations

We can apply this same technique when we want to differentiate a known function numerically. In this application, known as Richardson's extrapolation method, we can make the h-values smaller rather than use larger values as is required when the function is known only as a table.

We begin at some arbitrarily selected value of h and compute $f'(x)$ from

$$f'(x) = \frac{f(x + h) - f(x - h)}{2h}.$$

We then compute a second value for $f'(x)$ with h half as large. From these two computations, we extrapolate using Eq. (5.21). This improved value has an error of $O(h^4)$.

Normally, one builds a table by continuing with higher-order extrapolations with the h-value halved at each stage.

EXAMPLE 5.7   Build a Richardson table for $f(x) = x^2 \cos(x)$ to evaluate $f'(1)$. Start with $h = 0.1$. Repeat for $f'(2)$.

Tables 5.6 and 5.7 show the results. The exact answers are $f'(1) = 0.2391336$ and $f'(2) = -5.3017770$. The Richardson technique indicates convergence when two successive values on the same line are the same. Because single precision was used in computing these tables, convergence is to values slightly different from the exact values, to $f'(1) = 0.239132$ and $f'(2) = -5.301808$.

Table 5.6   Richardson table starting with $h = 0.1$ for derivative at $x = 1$

h	f'	First extrapolation	Second extrapolation	Third extrapolation
0.1	0.226736			
0.05	0.236031	0.239129		
0.025	0.238358	0.239133	0.239134	
0.0125	0.238938	0.239132	0.239132	0.239132

Table 5.7   Richardson table starting with $h = 0.1$ for derivative at $x = 2$

h	f'	First extrapolation	Second extrapolation	Third extrapolation
0.1	−5.296478			
0.05	−5.300449	−5.301773		
0.025	−5.301454	−5.301789	−5.301790	
0.0125	−5.301719	−5.301807	−5.301808	−5.301808

An Algorithm to Compute a Table of Richardson Extrapolations for Computing the Derivative

Given a function, $f(x)$:

(Get user inputs)
INPUT
    $x = x$-value,
    $h = h$-value to start,
    MAXST = maximum number of stages,
    TOL = tolerance value for termination.
    $d(0, 1) = 0$

(Compute lines of the table)
DO FOR ST = 0 TO MAXST STEP 1
   SET $d(ST, 0) = [f(x + h) - f(x - h)]/(2h)$.
   DO FOR j = 1 TO ST STEP 1
      SET $d(ST, j) = d(ST, j - 1) + [d(ST, j - 1)$
        $-d(ST - 1, j - 1)]/(2^{2j} - 1)$,
   ENDDO (FOR j).
   IF $|d(ST, ST) - d(ST, ST - 1)|$ < TOL THEN
     EXIT
   ENDIF.
   SET $h = h/2$.
ENDDO (FOR ST).

On termination, the last computed value is the extrapolated estimate of the derivative.

For convenience, here we collect formulas for computing derivatives.

Formulas for Computing Derivatives

Formulas for the first derivative:

$$f'(x_0) = \frac{f_1 - f_0}{h} + O(h)$$

$$f'(x_0) = \frac{f_1 - f_{-1}}{2h} + O(h^2) \qquad \text{Central difference}$$

$$f'(x_0) = \frac{-f_2 + 4f_1 - 3f_0}{2h} + O(h^2)$$

$$f'(x_0) = \frac{-f_2 + 8f_1 - 8f_{-1} + f_{-2}}{12h} + O(h^4) \qquad \text{Central difference}$$

Formulas for the second derivative:

$$f''(x_0) = \frac{f_2 - 2f_1 + f_0}{h^2} + O(h)$$

$$f''(x_0) = \frac{f_1 - 2f_0 + f_{-1}}{h^2} + O(h^2)$$

$$f''(x_0) = \frac{-f_3 + 4f_2 - 5f_1 + 2f_0}{h^2} + O(h^2)$$

$$f''(x_0) = \frac{-f_2 + 16f_1 - 30f_0 + 16f_{-1} -}{12h^2}$$

Formulas for the third derivative:

$$f'''(x_0) = \frac{f_3 - 3f_2 + 3f_1 - f_0}{h^3} + O(h)$$

$$f'''(x_0) = \frac{f_2 - 2f_1 + 2f_{-1} - f_{-2}}{2h^3} + O(h^2) \quad \text{Averaged difference}$$

Formulas for the fourth derivative:

$$f^{iv}(x_0) = \frac{f_4 - 4f_3 + 6f_2 - 4f_1 + f_0}{h^4} + O(h)$$

$$f^{iv}(x_0) = \frac{f_2 - 4f_1 + 6f_0 - 4f_{-1} + f_{-2}}{h^4} + O(h^2) \quad \text{Central difference}$$

5.5 Newton–Cotes Integration Formulas

The usual strategy in developing formulas for numerical integration is similar to that for numerical differentiation. We pass a polynomial through points defined by the function, and then integrate this polynomial approximation of the function. This permits us to integrate a function known only as a table of values. When the values are equispaced, our familiar Newton–Gregory forward polynomial is a convenient starting point, so

$$\int_a^b f(x)\ dx \doteq \int_a^b P_n(x_s)\ dx. \tag{5.22}$$

The formula we get from Eq. (5.22) will not be exact because the polynomial is not identical with $f(x)$. We get an expression for the error by integrating the error term of $P_n(x_s)$.*

$$\text{Error} = \int_a^b \binom{s}{n+1} h^{n+1} f^{(n+1)}(\xi)\ dx.$$

---

*This error term comes from applying the next-term rule to the Newton–Gregory interpolating polynomial. $\binom{s}{n+1}$ is the number of combinations of s things taken $(n+1)$ at a time, which is an abbreviation for

$$\frac{s(s-1)(s-2)\ \dots\ (s-n+1)}{n!}.$$

Section 5.15 discusses errors in more detail.

There are various ways that we can employ Eq. (5.22). The interval of integration (a, b) can match the range of fit of the polynomial, (x_0, x_n). In this case, we get the Newton–Cotes formulas; these are a set of integration rules corresponding to the varying degrees of the interpolating polynomial. The first three, with the degree of the polynomial 1, 2, or 3, are particularly important, and we discuss them at length in the following sections.

If the degree of the polynomial is too high, errors due to round-off and local irregularities can cause a problem. This explains why it is only the lower-degree Newton–Cotes formulas that are often used.

The range of the polynomial and the interval of integration do not have to be the same. If the interval of integration extends outside the range of fit, however, we are extrapolating, and this incurs larger errors. If we only desire to get the integral of a known function, we will normally avoid extrapolation. As we will see in the next chapter, however, integrating the polynomial outside its range of fit leads to some important methods for solving differential equations. Using an interval of integration that is a subset of the points at which the polynomial agrees with the function also has special application in solving differential equations numerically.

The utility of numerical integration extends beyond the need to integrate a function known only as a table of values. Most computer programs for integration of functions whose form is known use these numerical techniques rather than the analytical methods of the calculus. Although programs like MATLAB and *Mathematica* do use analytical procedures, in a large number of cases no closed form for the integral exists. Numerical integration applies regardless of the complexity of the integrand or the existence of a closed form for the integral.

Let us now develop our three important Newton–Cotes formulas. During the integration, we will need to change the variable of integration from x to s, because the Newton–Gregory polynomials are expressed in terms of $s = (x - x_0)/h$. Observe that $dx = h\,ds$. For $n = 1$,

$$\int_{x_0}^{x_1} f(x)\, dx \doteq \int_{x_0}^{x_1} (f_0 + s\Delta f_0)\, dx = h \int_{s=0}^{s=1} (f_0 + s\Delta f_0)\, ds$$

$$= hf_0 s \Big]_0^1 + h\Delta f_0 \frac{s^2}{2} \Big]_0^1 = h\left(f_0 + \frac{1}{2}\Delta f_0\right) \qquad (5.23)$$

$$= \frac{h}{2}[(2f_0 + (f_1 - f_0)] = \frac{h}{2}(f_0 + f_1).$$

$$\text{Error} = \int_{x_0}^{x_1} \frac{s(s-1)}{2} h^2 f''(\xi)\, dx = h^3 f''(\xi_1) \int_0^1 \frac{s^2 - s}{2}\, ds$$

$$= h^3 f''(\xi_1) \left(\frac{s^3}{6} - \frac{s^2}{4}\right)\Big]_0^1 = -\frac{1}{12} h^3 f''(\xi_1), \qquad x_0 < \xi_1 < x_1. \qquad (5.24)$$

The details of getting the error term here and below are spelled out in a later section.

For $n = 2$,

$$\int_{x_0}^{x_2} f(x)\, dx \doteq \int_{x_0}^{x_2} \left(f_0 + s\Delta f_0 + \frac{s(s-1)}{2}\Delta^2 f_0 \right) dx$$

$$= h \int_0^2 \left(f_0 + s\Delta f_0 + \frac{s(s-1)}{2}\Delta^2 f_0 \right) ds \tag{5.25}$$

$$= h f_0 s \Big]_0^2 + h\Delta f_0 \frac{s^2}{2}\Big]_0^2 + h\Delta^2 f_0 \left(\frac{s^3}{6} - \frac{s^2}{4} \right)\Big]_0^2$$

$$= h \left(2f_0 + 2\Delta f_0 + \frac{1}{3}\Delta^2 f_0 \right) = \frac{h}{3}(f_0 + 4f_1 + f_2),$$

with an error term of

$$\text{Error} = -\frac{1}{90} h^5 f^{(4)}(\xi), \qquad x_0 < \xi < x_2. \tag{5.26}$$

Similarly, for $n = 3$, we find

$$\int_{x_0}^{x_3} f(x)\, dx = \int_{x_0}^{x_3} P_3(x_s)\, dx = \frac{3h}{8}(f_0 + 3f_1 + 3f_2 + f_3). \tag{5.27}$$

$$\text{Error} = -\frac{3}{80} h^5 f^{(4)}(\xi_1), \qquad x_0 < \xi_1 < x_3. \tag{5.28}$$

In summary, the basic Newton–Cotes formulas are

$$\int_{x_0}^{x_1} f(x)\, dx = \frac{h}{2}(f_0 + f_1) - \frac{1}{12} h^3 f''(\xi),$$

$$\int_{x_0}^{x_2} f(x)\, dx = \frac{h}{3}(f_0 + 4f_1 + f_2) - \frac{1}{90} h^5 f^{(4)}(\xi),$$

$$\int_{x_0}^{x_3} f(x)\, dx = \frac{3h}{8}(f_0 + 3f_1 + 3f_2 + f_3) - \frac{3}{80} h^5 f^{(4)}(\xi).$$

An important item to observe is that the error terms for both $n = 2$ and $n = 3$ are $O(h^5)$. This means that the error of integration using a quadratic is similar to the integral using a cubic; it is a consequence of finding the integral of the next term equal to zero in deriving the error term of Eq. (5.26). Note also that the coefficient in Eq. (5.26), $-\frac{1}{90}$, is smaller than that in Eq. (5.28), $-\frac{3}{80}$. The formula based on a quadratic is unexpectedly more accurate.

This phenomenon is true of all the even-order Newton–Cotes formulas; each has an order of h in its error term the same as for the formula of next higher order. This fact suggests that the even-order rules are especially useful.

5.6 The Trapezoidal Rule—A Composite Formula

The first of the Newton–Cotes formulas, based on approximating $f(x)$ on (x_0, x_1) by a straight line, is also called the *trapezoidal rule*. We have derived it by integrating $P_1(x_s)$, but the familiar and simple trapezoidal rule can also be considered to be an adaptation of the definition of the definite integral as a sum. To evaluate $\int_a^b f(x)\,dx$, we subdivide the interval from a to b into n subintervals, as in Fig. 5.2. The area under the curve in each subinterval is approximated by the trapezoid formed by replacing the curve by its secant line drawn between the endpoints of the curve. The integral is then approximated by the sum of all the trapezoidal areas. (If we found the limiting value of this sum as the widths of the intervals approach zero, we would have the exact value of the integral, but in numerical integration the widths are finite.) There is no necessity to make the subintervals equal in width, but our formula is simpler if this is done. Let h be the constant Δx. Because the area of a trapezoid is its average height times the base, for each subinterval,

$$\int_{x_i}^{x_{i+1}} f(x)\,dx \doteq \frac{f(x_i) + f(x_{i+1})}{2}(\Delta x) = \frac{h}{2}(f_i + f_{i+1}), \qquad (5.29)$$

and for $[a, b]$ subdivided into n subintervals of size h,

$$\int_a^b f(x)\,dx \doteq \sum_{i=0}^{n-1} \frac{h}{2}(f_i + f_{i+1}) = \frac{h}{2}(f_0 + f_1 + f_1 + f_2 + \cdots + f_{n-1} + f_n);$$

$$\int_a^b f(x)\,dx = \frac{h}{2}(f_0 + 2f_1 + 2f_2 + \cdots + 2f_{n-1} + f_n). \qquad (5.30)$$

Equation (5.29) is identical to Eq. (5.23). Equation (5.30) is called the *composite trapezoidal rule:* It lets us apply the formula over an extended region where $f(x)$ is far from linear by applying the procedure to subintervals in which it can be approximated by linear segments. The formula is beautifully simple, and its applicability to unequally spaced values is useful in finding the integral of an experimentally determined function. It is obvious from Fig. 5.2 that the method is subject to large errors unless the subintervals are small, for replacing a curve by a straight line is hardly accurate.

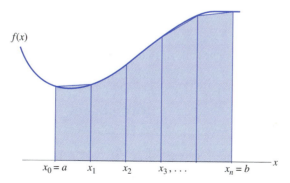

Figure 5.2

Table 5.8

x	$f(x)$	x	$f(x)$
1.6	4.953	2.8	16.445
1.8	6.050	3.0	20.086
2.0	7.389	3.2	24.533
2.2	9.025	3.4	29.964
2.4	11.023	3.6	36.598
2.6	13.464	3.8	44.701

EXAMPLE 5.8   Suppose we wish to integrate the function tabulated in Table 5.8 over the interval from $x = 1.8$ to $x = 3.4$. The composite trapezoidal rule gives

$$\int_{1.8}^{3.4} f(x)\, dx = \frac{0.2}{2}[6.050 + 2(7.389) + 2(9.025) + 2(11.023) + 2(13.464)$$

$$+ 2(16.445) + 2(20.086) + 2(24.533) + 29.964] = 23.9944.$$

The data in Table 5.8 are for $f(x) = e^x$, so the true value of the integral is $e^{3.4} - e^{1.8} = 23.9144$. We are off in the second decimal.

We have previously shown the error of the trapezoidal rule as Eq. (5.24). We repeat it here:

$$\text{Local error of trapezoidal rule} = -\frac{1}{12}h^3 f''(\xi_1), \qquad x_0 < \xi_1 < x_1.$$

This error, it should be emphasized, is the error of only a single step, and is hence called the *local error*. We normally apply the trapezoidal formula to a series of subintervals to get the integral over a large interval from $x = a$ to $x = b$. We are interested in the total error, which is called the *global error*.

To develop the formula for global error of the trapezoidal rule, we note that it is the sum of the local errors:

$$\text{Global error} = -\frac{1}{12}h^3[f''(\xi_1) + f''(\xi_2) + \cdots + f''(\xi_n)]. \tag{5.31}$$

In Eq. (5.31) each of the values of ξ_i is found in the n successive subintervals. If we assume that $f''(x)$ is continuous on (a, b), there is some value of x in (a, b)—say, $x = \xi$—at which the value of the sum in Eq. (5.31) is equal to $n \cdot f''(\xi)$. As $nh = b - a$, the global error becomes

$$\text{Global error of trapezoidal rule} = -\frac{1}{12}h^3 n f''(\xi) = \frac{-(b-a)}{12}h^2 f''(\xi) = O(h^2). \tag{5.32}$$

The fact that the global error is $O(h^2)$ while the local error is $O(h^3)$ is reasonable, because, for example, if h is halved the number of subintervals is doubled, so we add together twice as many errors.

When the function $f(x)$ is known, Eq. (5.32) permits us to estimate the error of numerical integration by the trapezoidal rule. In applying this equation we bracket the error by calculating with the maximum and the minimum values of $f''(x)$ on the interval $[a, b]$.

For this example, our error expression gives these estimates:

$$\text{Error} = -\frac{1}{12}h^3 n f''(\xi), \qquad 1.8 \le \xi \le 3.4,$$

$$= -\frac{1}{12}(0.2)^3(8)\begin{Bmatrix} e^{1.8} & \text{(min)} \\ e^{3.4} & \text{(max)} \end{Bmatrix} = \begin{Bmatrix} -0.0323 & \text{(min)} \\ -0.1598 & \text{(max)} \end{Bmatrix}.$$

Alternatively,

$$\text{Error} = -\frac{1}{12}(0.2)^2(3.4 - 1.8)\begin{Bmatrix} e^{1.8} & \text{(min)} \\ e^{3.4} & \text{(max)} \end{Bmatrix} = \begin{Bmatrix} -0.0323 \\ -0.1598 \end{Bmatrix}.$$

The actual error was -0.080. ▲

If we had not known the function for which we have tabulated values, we would have estimated $h^2 f''(\xi)$ from the second differences.

An Algorithm for Composite Trapezoidal Rule Integration

Given a function $f(x)$:

(Get user inputs)
INPUT
    a, b = endpoints of interval,
    n = number of intervals.
(Do the integration)
SET $h = (b - a)/n$.
SET SUM = 0.
DO FOR $i = 1$ TO $n - 1$ STEP 1
    SET $x = a + h * i$,
    SET SUM = SUM + 2 $* f(x)$
ENDDO (FOR i).
SET SUM = SUM + $f(a) + f(b)$.
SET ANS = SUM $* h/2$.

The value of the integral is given by ANS.

Romberg Integration

We can improve the accuracy of the trapezoidal rule integral by a technique that is similar to Richardson extrapolation. This technique is known as *Romberg integration.*

Because the integral determined with the trapezoidal method has an error of $O(h^2)$, we can combine two estimates of the integral that have h-values in a 2:1 ratio by Eq. (5.21), which we repeat here:

$$\text{Better estimate} = \text{more accurate} + \frac{1}{2^n - 1}(\text{more accurate} - \text{less accurate}). \quad \textbf{(5.33)}$$

When we apply this equation to get the integral of a known function, we begin with an arbitrary value for h in Eq. (5.30). A second estimate is then made with the value of h halved. From these two estimates we extrapolate to get an improved estimate using Eq. (5.33). This has an error of $O(h^4)$.

Obviously, this can be extended to produce a table of successively better estimates. When we find that the values converge, we have the best estimate that we can make in the light of round-off error. As shown in Section 5.15, each new extrapolation has error orders that increase: $O(h^4), O(h^6), O(h^8), \ldots$.

We can reduce the number of computations because, when h is halved, all of the old points at which the function was evaluated to get Eq. (5.30) appear in the new computation and we thus can avoid repeating the evaluations. Figure 5.3 illustrates this point.

This next example shows how the *Romberg table* appears for the function $f(x) = e^{-x^2}$ integrated between the limits of 0.2 and 1.5. This integral has no closed form solution. It is closely related to the *error function,* a quantity that is so important in statistics and other branches of applied mathematics that values have been tabulated.

EXAMPLE 5.9   Use Romberg integration to find the integral of e^{-x^2} between the limits of $a = 0.2$ and $b = 1.5$. Take the initial subinterval size as $h = (b - a)/2 = 0.65$.

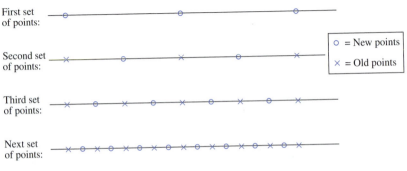

Figure 5.3

Our first estimate is

$$\text{Integral} = \frac{h}{2}[f(a) + 2f(a + h) + f(b)]$$

$$= \frac{0.65}{2}[e^{-0.2^2} + 2e^{-0.85^2} + e^{-1.5^2}]$$

$$= 0.66211.$$

The next estimate uses $h = 0.65/2 = 0.325$:

$$\text{Integral} = \frac{h}{2}[f(a) + 2f(a + h) + 2f(a + 2h) + 2f(a + 3h) + f(b)]$$

$$= \frac{0.325}{2}[e^{-0.2^2} + 2e^{-0.525^2} + 2e^{-0.85^2} + 2e^{-1.175^2} + e^{-1.5^2}]$$

$$= 0.65947.$$

Observe that only two new function evaluations appear in the second estimate. We now extrapolate:

$$\text{Improved} = 0.65947 + \frac{1}{3}[0.65947 - 0.66211]$$

$$= 0.65859.$$

Table 5.9 exhibits the calculations when we repeat the estimations, halving the h-value each time.

Table 5.9   Romberg table of integrals over interval from 0.2 to 1.5 with an initial h of 0.65

0.66211			
0.65947	0.65859		
0.65898	0.65881	0.65882	
0.65886	0.65882	0.65882	0.65882

The Romberg Method for a Tabulated Function

We can apply the Romberg method to integrate a function known only as a table of evenly spaced function values, but now we cannot make h smaller. Instead, we use estimates of the integral with h doubled each time, just as we did to improve the estimates of derivatives in Section 5.4. Here is an example.

EXAMPLE 5.10   Use the data of Table 5.8 to get the integral between the limits of $x = 1.8$ and $x = 3.4$. Begin with $h = 0.4$.

Our first estimate is

$$\text{Integral} = \frac{0.4}{2}[6.050 + 2(9.025) + 2(13.464) + 2(20.086) + 29.964]$$

$$= 24.2328.$$

We now use $h = 0.8$ to get the next estimate:

$$\text{Integral} = \frac{0.8}{2}[6.050 + 2(13.464) + 29.964]$$

$$= 25.1768.$$

We then extrapolate with Eq. (5.32):

$$\text{Improved} = 24.2328 + \frac{1}{3}[24.2328 - 25.1768]$$

$$= 23.9181.$$

If we had started with $h = 0.2$, we would have had the results shown in Table 5.10, where an additional extrapolation is made to give the value 23.9147. (The exact value of the integral is 23.9144. Round-off errors make it impossible to reach the exact value from these tabulated function values.)

Table 5.10 Romberg table of integrals over interval from 1.8 to 3.4 from tabulated values

$h = 0.2$	23.9944	23.9149	23.9147
$h = 0.4$	24.2328	23.9181	
$h = 0.8$	25.1768		

The Romberg method is applicable to a wide class of functions—to any Riemann-integrable function, in fact. Smoothness and continuity are not required. However, when $f(x)$ is discontinuous, we should make the evenly spaced points fall on the discontinuities. This can be done if we break the interval into subintervals that are bounded by the discontinuities.

An Algorithm for Romberg Integration

Given a function, $f(x)$:

(Get user inputs)
INPUT:
    a, b = endpoints of interval,
    NST = number of stages.

(Do the integration)
SET $h = (b - a)/2$.
(Do the lines of the table)
SET SUM $= f(a) + 2 * f(a + h) + f(b)$.
SET NTGRL$(0, 0)$ = SUM $* h/2$. (First value)
SET $d = 2 * h$. (d is distance between added points)
DO FOR ST $= 1$ TO NST STEP 1
   SET $h = h/2$.
   SET $d = d/2$.
   DO FOR $i = 1$ TO 2ST STEP 1
      SET $x = a - h + d * i$.
      SET SUM $=$ SUM $+ 2 * f(x)$.
   ENDDO (FOR i).
   SET NTGRL$(\text{ST}, 0) =$ SUM $* h/2$.

(Now extrapolate)
   DO FOR $j = 1$ TO ST STEP 1
      SET NTGRL(ST, j) = NTGRL$(\text{ST}, j - 1)$ +
         [NTGRL$(\text{ST}, j - 1)$ − NTGRL$(\text{ST} - 1, j - 1)]/(2^j - 1)$.
   ENDDO (FOR j).
ENDDO (FOR ST).

The last computed value is the estimate of the integral.

An alternative stopping criterion is when two successive computations in a line differ by less than some tolerance value.

5.7 Simpson's Rules

The composite Newton–Cotes formulas based on quadratic and cubic interpolating polynomials are known as *Simpson's rules*. The first, based on a quadratic, is known as *Simpson's $\frac{1}{3}$ rule* and that based on a cubic is known as *Simpson's $\frac{3}{8}$ rule*. These names come from the coefficients of the formulas.

Simpson's $\frac{1}{3}$ Rule

The second-degree Newton–Cotes formula integrates a quadratic over two intervals of equal width. (We shall call these intervals *panels.*) We build the composite rule from Eq. (5.25), which we repeat here:

$$f(x) \; dx = \frac{h}{3}[f_0 + 4f_1 + f_2]. \tag{5.34}$$

This very popular formula has a local error of $O(h^5)$:

$$\text{Error} = -\frac{h^5}{90}f^{(4)}(\xi).$$

We build upon this formula to get a composite rule that is applied to a subdivision of the interval of integration into n panels (n must be even):

$$\int f(x)\, dx = \frac{h}{3}[f(a) + 4f_1 + 2f_2 + 4f_3 + 2f_4 + \cdots + 4f_{n-1} + f(b)] \quad \textbf{(5.35)}$$

with an error term of

$$\text{Error} = -\frac{(b - a)}{180}h^4 f^{(4)}(\xi).$$

The order of the global error changes to $O(h^4)$. The denominator in the error term changes to 180 because we integrate over pairs of panels (meaning that the local rule is applied $n/2$ times). The fact that the error is $O(h^4)$ is of special importance.

EXAMPLE 5.11   Use Simpson's $\frac{1}{3}$ rule to evaluate the integral of e^{-x^2} over the interval 0.2 to 1.5, using 2, 4, 6, . . . subdivisions until the values converge to five decimal places.

    Table 5.11 shows the results. The exact answer to five decimal places is 0.65882.

Table 5.11

n	Estimate of integral
2	0.65181
4	0.65860
6	0.65878
8	0.65881
10	0.65882

▲

    If we compare these values to those in the Romberg table in Table 5.9, we will see that the result with four panels agrees with the first Romberg extrapolation and that the result with eight panels agrees with the second Romberg extrapolation. We then see that doing a Romberg extrapolation of results from the trapezoidal rule gives an identical result to the Simpson $\frac{1}{3}$ rule integral with the same number of panels. For comparison, the trapezoidal rule without extrapolation requires about 50 subintervals to get five-decimal-place accuracy.

    We can apply a Romberg-type extrapolation to the Simpson's rule values of Table 5.11. Because the errors are of $O(h^4)$, we compute from the results with $n = 4$ and 8:

$$\text{Improved} = \text{more accurate} + \frac{1}{15}[\text{more} - \text{less}]$$

$$= 0.65881 + \frac{1}{15}[0.65881 - 0.65860] = 0.65882,$$

which is the correct value to five decimal places. This extrapolated result has an error of $O(h^6)$, so the next extrapolation would have a multiplier in the correction term of 1/63.

Algorithm for Simpson's $\frac{1}{3}$ Rule Integration

Given a function, $f(x)$:

(Get user inputs)
INPUT:
    a, b = endpoints of interval.
    n = number of intervals (n must be even).
(Do the integration)
SET $h = (b - a)/n$.
SET SUM = 0.
DO FOR i = 1 TO $n/2$ STEP 1
    SET $x = a - h + 2 * h * i$,
    SET SUM = SUM + 4 * $f(x)$,
    IF $i \neq n/2$ THEN:
        SET SUM = SUM + 2 * $f(x + h)$
    ENDIF.
ENDDO (FOR i).
SET SUM = SUM + $f(a)$ + $f(b)$.
SET ANS = SUM * $h/3$.

The value of the integral is given by ANS.

Using Tabulated Values

We can apply Simpson's $\frac{1}{3}$ rule to a table of evenly spaced function values in an obvious way. If the number of intervals from the table is not even, we can do a subinterval at one end or the other with the trapezoidal rule and the rest with Simpson's rule. (We should select the end subinterval for applying the trapezoidal rule where the function is more nearly linear.)

EXAMPLE 5.12   Apply Simpson's $\frac{1}{3}$ rule to the data of Table 5.12. What difference results if we apply the trapezoidal rule at the left end rather than the right end?

    If we use the values from $x = 0.7$ to $x = 1.9$ (six panels) in Simpson's $\frac{1}{3}$ rule, we get 1.51938 for this integral. We add to this the trapezoidal integral from $x = 1.9$ to $x = 2.1$, which is 0.29740, to get 1.81678 for the integral from $x = 0.7$ to $x = 2.1$.

    Conversely, if we select the first interval to integrate by the trapezoidal rule and the remainder by Simpson's $\frac{1}{3}$ rule, we add 0.15620 to 1.66142 to get 1.81762 for the integral over the entire range.

    The exact value for the integral over the entire range is 1.81759; the second choice is the better one. Our criterion seems to work for choosing the end for applying the trapezoidal rule. (Looking at the first differences shows that the function is more nearly linear at the start.)   ▲

Table 5.12

x	$f(x)$
0.7	0.64835
0.9	0.91360
1.1	1.16092
1.3	1.36178
1.5	1.49500
1.7	1.55007
1.9	1.52882
2.1	1.44513

Simpson's $\frac{3}{8}$ Rule

The composite rule based on fitting four points with a cubic leads to Simpson's $\frac{3}{8}$ rule. We start with Eq. (5.27), which integrates over three panels:

$$\int_{x_0}^{x_3} f(x)\ dx = \frac{3h}{8}[f_0 + 3f_1 + 3f_2 + f_3] - \frac{3}{80}h^5 f^{(4)}(\xi). \qquad (5.36)$$

(We remark again that the local order of error, $O(h^5)$, is the same as for Simpson's $\frac{1}{3}$ rule and that the coefficient is larger.)

We apply this to sets of triple panels to get the composite rule:

$$\int_a^b f(x)\ dx = \frac{3h}{8}[f(a) + 3f_1 + 3f_2 + 2f_3 + 3f_4 + 3f_5 + 2f_6 \qquad (5.37)$$
$$+ \cdots + 2f_{n-3} + 3f_{n-2} + 3f_{n-1} + f(b)]$$

with an error term of

$$\text{Error} = -\frac{(b-a)}{80}h^4 f^{(4)}(\xi).$$

We apply this just as we did with the $\frac{1}{3}$ rule. If the function is known, we divide the interval of integration into n panels where n is divisible by 3.

EXAMPLE 5.13   Apply Simpson's $\frac{3}{8}$ rule to the same function as in Example 5.11 [$f(x) = e^{-x^2}$], integrating from $x = 0.2$ to $x = 1.5$. Compare the results with 3, 6, 9, ... panels.

Table 5.13 shows the results. Comparing these results with those in Table 5.11, we see

Table 5.13

n	Estimate of integral
3	0.65593
6	0.65872
9	0.65881
12	0.65882

that the correct result was not achieved until $n = 12$, whereas with Simpson's $\frac{1}{3}$ rule, it took only 8 subdivisions.   ▲

One might wonder why Simpson's $\frac{3}{8}$ rule is ever used if it is less efficient than the $\frac{1}{3}$ rule. One useful application is the calculation of a tabulated function with an odd number of panels by doing the first (or the last) three with the $\frac{3}{8}$ rule and the rest with the $\frac{1}{3}$ rule. As an example, we apply this to the values of Table 5.12.

If we use the $\frac{3}{8}$ rule over points 0 through 3 and then use the $\frac{1}{3}$ rule, the values are $0.61753 + 1.20015 = 1.81768$. With the other choice, we add 0.91326 to 0.90445 to get 1.81771. There is little difference, although the former is slightly better.

We summarize the rules for integration:

Formulas for Integration (Uniform spacing, $\Delta x = h$)

Trapezoidal rule:

$$\int_a^b f(x)\ dx = \frac{h}{2}(f_1 + 2f_2 + 2f_3 + \cdots + 2f_n + f_{n+1})$$

$$- \frac{(b-a)}{12} h^2 f''(\xi), \qquad a \le \xi \le b$$

Simpson's $\frac{1}{3}$ rule:

$$\int_a^b f(x)\ dx = \frac{h}{3}(f_1 + 4f_2 + 2f_3 + 4f_4 + 2f_5 + \cdots + 4f_n + f_{n+1})$$

$$- \frac{(b-a)}{180} h^4 f^{(4)}(\xi), \qquad a \le \xi \le b$$

(requires an even number of panels)

Simpson's $\frac{3}{8}$ rule:

$$\int_a^b f(x)\ dx = \frac{3h}{8}(f_1 + 3f_2 + 3f_3 + 2f_4 + 3f_5 + 3f_6 + \cdots + 3f_n + f_{n+1})$$

$$- \frac{(b-a)}{80} h^4 f^{(4)}(\xi), \qquad a \le \xi \le b$$

(requires a number of panels divisible by 3)

5.8   Other Ways to Derive Integration Formulas

It is interesting to examine other ways of deriving integration formulas than by integrating the interpolating polynomial. One way is to use symbolic methods, similar to those of Section 5.3.

In terms of the stepping operator E,

$$f(x_s) = f_s = E^s f_0.$$

Multiplying by $dx = h\,ds$ and integrating from x_0 to x_1 ($s = 0$ to $s = 1$), we obtain

$$\int_{x_0}^{x_1} f(x)\,dx = h \int_0^1 E^s f_0\,ds = \left[\frac{hE^s}{\ln E}f_0\right]_0^1 = \frac{h(E-1)}{\ln E}f_0.$$

Let $E = 1 + \Delta$ and expand $\ln(1 + \Delta)$ as a power series:

$$\ln(1 + \Delta) = \Delta - \frac{1}{2}\Delta^2 + \frac{1}{3}\Delta^3 - \frac{1}{4}\Delta^4 + \cdots.$$

On dividing Δ by this series, we get

$$\int_{x_0}^{x_1} f(x)\,dx = \frac{h\Delta}{\Delta - \dfrac{1}{2}\Delta^2 + \dfrac{1}{3}\Delta^3 - \dfrac{1}{4}\Delta^4 + \cdots}f_0 \tag{5.38}$$

$$= h\left(f_0 + \frac{1}{2}\Delta f_0 - \frac{1}{12}\Delta^2 f_0 + \frac{1}{24}\Delta^3 f_0 - \cdots\right).$$

The coefficients are considerably easier to get by this technique than by the term-by-term integration of Section 5.5.

Equation (5.38) is not a Newton–Cotes formula unless we use only the first two terms, making the interval of integration and the range of fit of the polynomial agree (the trapezoidal rule). When n terms are used, the formula represents a polynomial of degree n, fitting from x_0 to x_n, but integrated only from x_0 to x_1. This is especially useful in connection with differential-equation methods.

We can develop the formula for an nth-degree interpolating polynomial integrated over two panels, from x_0 to x_2, in a similar fashion:

$$\int_{x_0}^{x_2} f(x)\,dx = \left[\frac{hE^s}{\ln E}f_0\right]_0^2 = \frac{h(E^2-1)}{\ln E}f_0 = \frac{h(E+1)(E-1)}{\ln E}f_0.$$

Again letting $E = 1 + \Delta$ so $E - 1 = \Delta$ and $E + 1 = 2 + \Delta$, and dividing Δ by the series for $\ln(1 + \Delta)$, we get

$$\int_{x_0}^{x_2} f(x)\,dx = h(2 + \Delta)\left(f_0 + \frac{1}{2}\Delta f_0 - \frac{1}{12}\Delta^2 f_0 + \frac{1}{24}\Delta^3 f_0 - \cdots\right)$$

$$= h\left(2f_0 + \Delta f_0 - \frac{1}{6}\Delta^2 f_0 + \frac{1}{12}\Delta^3 f_0 - \cdots + \Delta f_0 + \frac{1}{2}\Delta^2 f_0 - \frac{1}{12}\Delta^3 f_0 + \cdots\right) \tag{5.39}$$

$$= h\left(2f_0 + 2\Delta f_0 + \frac{1}{3}\Delta^2 f_0 + 0 - \cdots\right).$$

Obviously the method can be extended. One reason that we might desire formulas such as Eqs. (5.38) and (5.39) is for constructing formulas that permit integration over m panels based on polynomials of degree n.

We now present still another interesting method of deriving formulas that can be applied to a variety of situations including the development of integration formulas. It may be called the method of *undetermined coefficients.*[*] We express the formula as a sum of $n + 1$ terms with unknown coefficients, and then evaluate the coefficients by requiring that the formula be exact for all polynomials of degree n or less. We illustrate it here by finding Simpson's $\frac{1}{3}$ rule by this other technique. Express the integral as a weighted sum of three equispaced function values:

$$\int_{-1}^{1} f(x) \, dx = af(-1) + bf(0) + cf(+1). \qquad (5.40)$$

The symmetrical interval of integration simplifies the arithmetic. We stipulate that the function is to be evaluated at three equally spaced points, the two end values and the midpoint. The formula contains three terms, so we can require it to be correct for all polynomials of degree 2 or less. If that is true, it certainly must be true for the three special cases of $f(x) = x^2$, $f(x) = x$, and $f(x) = 1$. We rewrite Eq. (5.40) three times, applying each definition of $f(x)$ in turn:

$$f(x) = 1: \qquad \int_{-1}^{1} dx = 2 = a(1) + b(1) + c(1) = a + b + c;$$

$$f(x) = x: \qquad \int_{-1}^{1} x \, dx = 0 = a(-1) + b(0) + c(1) = -a + c;$$

$$f(x) = x^2: \qquad \int_{-1}^{1} x^2 \, dx = \frac{2}{3} = a(1) + b(0) + c(1) = a + c.$$

Solving the three equations simultaneously gives $a = \frac{1}{3}$, $b = \frac{4}{3}$, $c = \frac{1}{3}$. Here the spacing between points was unity; obviously the integral is proportional to $\Delta x = h$. We then get Simpson's $\frac{1}{3}$ rule:

$$\int_{-h}^{h} f(x) \, dx = h\left[\frac{1}{3}f(-h) + \frac{4}{3}f(0) + \frac{1}{3}f(h)\right].$$

5.9 Gaussian Quadrature

Our previous formulas for numerical integration were all predicated on evenly spaced x-values; this means the x-values were predetermined. With a formula of three terms, then, there were three parameters, the coefficients (weighting factors) applied to each of the functional values. A formula with three parameters corresponds to a polynomial of the second degree, one less than the number of parameters. Gauss observed that if we remove the requirement that the function be evaluated at predetermined x-values, a three-term formula will contain six parameters (the three x-values are now unknowns, plus the three weights) and should correspond to an interpolating polynomial of degree 5. Formulas based on this principle are called *Gaussian quadrature formulas.* They can be applied only when $f(x)$ is known explicitly, so that it can be evaluated at any desired value of x.

---

[*] Appendix B describes this method in more detail, illustrating it with several examples.

We will determine the parameters in the simple case of a two-term formula containing four unknown parameters:

$$\int_{-1}^{1} f(t) \doteq af(t_1) + bf(t_2).$$

The method is the same as that illustrated in the previous section, by determining unknown parameters. We use a symmetrical interval of integration to simplify the arithmetic, and call our variable t. (This notation agrees with that of most authors. As the variable of integration is only a dummy variable, its name is unimportant.) Our formula is to be valid for any polynomial of degree 3; hence it will hold if $f(t) = t^3, f(t) = t^2, f(t) = t$, and $f(t) = 1$:

$$
\begin{aligned}
f(t) = t^3: \quad & \int_{-1}^{1} t^3 \, dt = 0 = at_1^3 + bt_2^3; \\
f(t) = t^2: \quad & \int_{-1}^{1} t^2 \, dt = \frac{2}{3} = at_1^2 + bt_2^2; \\
f(t) = t: \quad & \int_{-1}^{1} t \, dt = 0 = at_1 + bt_2; \\
f(t) = 1: \quad & \int_{-1}^{1} dt = 2 = a + b.
\end{aligned}
\tag{5.41}
$$

Multiplying the third equation by t_1^2, and subtracting from the first, we have

$$0 = 0 + b[t_2^3 - t_2 t_1^2] = b(t_2)(t_2 - t_1)(t_2 + t_1). \tag{5.42}$$

We can satisfy Eq. (5.42) by either $b = 0$, $t_2 = 0$, $t_1 = t_2$, or $t_1 = -t_2$. Only the last of these possibilities is satisfactory, the others being invalid, or else reduce our formula to only a single term, so we choose $t_1 = -t_2$. We then find that

$$a = b = 1,$$

$$t_2 = -t_1 = \sqrt{\frac{1}{3}} = 0.5773,$$

$$\int_{-1}^{1} f(t) \, dt \doteq f(-0.5773) + f(0.5773).$$

It is remarkable that adding these two values of the function gives the exact value for the integral of any cubic polynomial over the interval from -1 to 1.

Suppose our limits of integration are from a to b, and not -1 to 1 for which we derived this formula. To use the tabulated Gaussian quadrature parameters, we must change the interval of integration to $(-1, 1)$ by a change of variable. We replace the given variable by another to which it is linearly related according to the following scheme:

If we let

$$x = \frac{(b - a)t + b + a}{2} \qquad \text{so that } dx = \left(\frac{b - a}{2}\right) dt,$$

then

$$\int_{a}^{b} f(x) \, dx = \frac{b - a}{2} \int_{-1}^{1} f\left(\frac{(b - a)t + b + a}{2}\right) dt.$$

EXAMPLE 5.14   Evaluate $I = \int_0^{\pi/2} \sin x \, dx$. (Obviously, $I = 1.0$, so we can readily see the error of our estimate.)

To use the two-term Gaussian formula, we must change the variable of integration to make the limits of integration from -1 to 1.

Let

$$x = \frac{(\pi/2)t + \pi/2}{2}, \qquad \text{so } dx = \frac{\pi}{4} \, dt.$$

Observe that when $t = -1$, $x = 0$; when $t = 1$, $x = \pi/2$. Then

$$I = \frac{\pi}{4} \int_{-1}^{1} \sin\left(\frac{\pi t + \pi}{4}\right) dt.$$

The Gaussian formula calculates the value of the new integral as a weighted sum of two values of the integrand, at $t = -0.5773$ and at $t = 0.5773$. Hence,

$$I = \frac{\pi}{4}[(1.0)(\sin(0.10566\pi)) + (1.0)(\sin(0.39434\pi))]$$

$$= 0.99847.$$

The error is 1.53×10^{-3}.   ▲

The power of the Gaussian method derives from the fact that we need only two functional evaluations. If we had used the trapezoidal rule, which also requires only two evaluations, our estimate would have been $(\pi/4)(0.0 + 1.0) = 0.7854$, an answer quite far from the mark. Simpson's $\frac{1}{3}$ rule requires three functional evaluations and gives $I = 1.0023$, with an error of -2.3×10^{-3}, somewhat greater than for Gaussian quadrature.

Gaussian quadrature can be extended beyond two terms. The formula is then given by

$$\int_{-1}^{1} f(t) \, dt \doteq \sum_{i=1}^{n} w_i f(t_i), \qquad \text{for } n \text{ points.} \tag{5.43}$$

This formula is *exact* for functions $f(t)$ that are polynomials of degree $2n - 1$ or less! Moreover, by extending the method we used previously for the 2-point formula, for each n we obtain a system of $2n$ equations:

$$w_1 t_1^k + \cdots + w_n t_n^k = \begin{cases} 0, & \text{for } k = 1, 3, 5, \ldots, 2n - 1; \\ \dfrac{2}{k + 1}, & \text{for } k = 0, 2, 4, \ldots, 2n - 2. \end{cases}$$

This approach is obvious. However, this set of equations, obtained by writing $f(t)$ as a succession of polynomials, is not easily solved. We wish to indicate an approach that is easier than the methods for a nonlinear system that we used in Chapter 2.

It turns out that the t_i's for a given n are the roots of the nth-degree Legendre polynomial. The Legendre polynomials are defined by recursion:

$$(n + 1)L_{n+1}(x) - (2n + 1)xL_n(x) + nL_{n-1}(x) = 0,$$
$$\text{with } L_0(x) = 1, \qquad L_1(x) = x.$$

Then $L_2(x)$ is

$$L_2(x) = \frac{3xL_1(x) - (1)L_0(x)}{2} = \frac{3}{2}x^2 - \frac{1}{2},$$

whose zeros are $\pm\sqrt{\frac{1}{3}} = \pm 0.5773$, precisely the t-values for the two-term formula.

By using the recursion relation, we find

$$L_3(x) = \frac{5x^3 - 3x}{2},$$

$$L_4(x) = \frac{35x^4 - 30x^2 + 3}{8}, \qquad \text{and so on.}$$

The methods of Chapter 1 allow us to find the roots of these polynomials. After they have been determined, the set of equations analogous to Eqs. (5.41) can easily be solved for the weighting factors because the equations are linear with respect to these unknowns.

Table 5.14 lists the zeros of Legendre polynomials up to degree 5, giving values that we need for Gaussian quadrature where the equivalent polynomial is up to degree 9. For example, $L_3(x)$ has zeros at $x = 0, +0.77459667$, and -0.77459667.

Table 5.14   Values for Gaussian quadrature

Number of terms	Values of t	Weighting factor	Valid up to degree
2	−0.57735027	1.0	3
	0.57735027	1.0	
3	−0.77459667	0.55555555	5
	0.0	0.88888889	
	−0.77459667	0.55555555	
4	−0.86113631	0.34785485	7
	−0.33998104	0.65214515	
	0.33998104	0.65214515	
	0.86113631	0.34785485	
5	−0.90617975	0.23692689	9
	−0.53846931	0.47862867	
	0.0	0.56888889	
	0.53846931	0.47862867	
	0.90617975	0.23692689	

Before continuing with an example of the use of Gaussian quadrature, it is of interest to summarize the properties of Legendre polynomials.

1. The Legendre polynomials are *orthogonal* over the interval $[-1, 1]$. That is,

$$\int_{-1}^{1} L_n(x)L_m(x)\, dx \begin{cases} = 0 \text{ if } n \neq m; \\ > 0 \text{ if } n = m. \end{cases}$$

This is a property of several other important functions, such as $\{\cos(nx), n = 0, 1, \dots \}$. Here we have

$$\int_{0}^{2\pi} \cos(mx)\cos(nx)\, dx \begin{cases} = 0 \text{ if } n \neq m; \\ > 0 \text{ if } n = m. \end{cases}$$

In this case we say that these functions are orthogonal over the interval $[0, 2\pi]$.

2. Any polynomial of degree n can be written as a sum of the Legendre polynomials:

$$P_n(x) = \sum_{i=0}^{n} c_i L_i(x).$$

3. The n roots of $L_n(x) = 0$ lie in the interval $[-1, 1]$.

Using these properties, we are able to show that Eq. (4.43) is exact for polynomials of degree $2n - 1$ or less.

The weighting factors and t-values for Gaussian quadrature have been tabulated. (Love, 1966, gives values for up to 200-term formulas.) We are content to give a few of the values in Table 5.14.

We illustrate the three-term formula with an example.

EXAMPLE 5.15 Evaluate $I = \int_{0.2}^{1.5} e^{-x^2}\, dx$ using the three-term Gaussian formula,

$$x = \frac{(1.5 - 0.2)t + 1.5 + 0.2}{2} = 0.65t + 0.85.$$

Then

$$I = \frac{1.5 - 0.2}{2} \int_{-1}^{1} e^{-(0.65t+0.85)^2}\, dt$$

$$= 0.65[0.555\ldots e^{-[0.65(-0.774\ldots)+0.85]^2} + 0.888\ldots e^{-[0.65(0.0)+0.85]^2}$$

$$+ 0.555\ldots e^{-[0.65(0.774\ldots)+0.85]^2}]$$

$$= 0.65860. \quad \text{(compare to exact value 0.65882)}$$

A four-term Gaussian quadrature gives 0.65883, off from the correct value only in the last digit. Compare this result, which required only four function evaluations, to the result of Example 5.11, where Simpson's $\frac{1}{3}$ rule required eight evaluations to achieve the same accuracy. ▲

5.10 Adaptive Integration

The trapezoidal rule and Simpson's $\frac{1}{3}$ rule are often used to find the integral of $f(x)$ over a fixed interval $[a, b]$ using a uniform value for Δx. When $f(x)$ is a known function, we can choose the value for $\Delta x = h$ arbitrarily. The problem is that we do not know a priori what value to choose for h to attain a desired accuracy. Romberg-type integration is a way to find the necessary h. We start with two panels, $h = h_1 = (b - a)/2$, and apply one of the formulas. Then we let $h_2 = h_1/2$ and apply the formula again, now with four panels, and compare the results. If the new value is sufficiently close, we terminate and use a Richardson extrapolation to further reduce the error. If the second result is not close enough to the first, we again halve h and repeat the procedure. We continue in this way until the last result is close enough to its predecessor.

We illustrate this obvious procedure with an example.

EXAMPLE 5.16 Integrate $f(x) = 1/x^2$ over the interval $[0.2, 1]$ using Simpson's $\frac{1}{3}$ rule. Use a tolerance value of 0.02 to terminate the halving of $h = \Delta x$. From calculus, we know that the exact answer is 4.0.

We introduce a special notation that will be used throughout this section.

$$S_n[a, b] = \text{the computed value using Simpson's } \frac{1}{3} \text{ rule with } \Delta x = h_n \text{ over } [a, b].$$

If we use this notation, the composite Simpson rule becomes

$$I(f) = S_n[a, b] - \frac{(b - a)}{180} h_n^4 f^4(\xi), \qquad a < \xi < b.$$

Using this with $h_1 = (1.0 - 0.2)/2 = 0.4$, we compute $S_1[0.2, 1.0]$. We continue halving h, $h_{n+1} = h_n/2$, computing its corresponding $S_{n+1}[a, b]$ until $|S_{n+1} - S_n| < 0.02$, the tolerance value. The following table shows the results.

n	h_n	S_n	$|S_{n+1} - S_n|$
1	0.4	4.948148	
			0.761111
2	0.2	4.187037	
			0.162819
3	0.1	4.024218	
			0.022054
4	0.05	4.002164	
			0.002010
5	0.025	4.000154	

From the table we see that, at $n = 5$, we have met the tolerance criterion, because $|S_5 - S_4| < 0.02$. A Romberg extrapolation gives

$$RS[a, b] = S_5 + \frac{S_5 - S_4}{15} = 4.00002.$$

(We use $RS[a, b]$ to represent the Romberg extrapolation from Simpson's rule.) ▲

The Adaptive Scheme

The disadvantage of this technique is that the value of h is the same over the entire interval of integration, whereas the behavior of $f(x)$ may not require such uniformity. Consider Fig. 5.4. It is obvious that, in the subinterval $[c, b]$, h can be much larger than in subinterval $[a, c]$, where the curve is much less smooth. We could subdivide the entire interval $[a, b]$ nonuniformly by personal intervention after examining the graph of $f(x)$. We prefer to avoid such intervention.

Adaptive integration automatically allows for different h's on different subintervals of $[a, b]$, choosing values adequate for a specified accuracy. We do not specify where the size change for h occurs; this can occur anywhere within it. We use something like a binary search to locate the point where we should change the size of h. Actually, the total interval $[a, b]$ may be broken into several subintervals, with different values for h within each of them. This depends on the tolerance value, TOL, and the nature of $f(x)$.

To describe this strategy, we repeat the preceding example to find the integral of $f(x) = 1/x^2$ between $x = 0.2$ and $x = 1$. We choose a value for TOL of 0.02, and do the computations in double precision to minimize the effects of round-off.

We begin as before by specifying just two subintervals in $[a, b]$. The first computation is a Simpson integration over $[0.2, 1]$ with $h_1 = 0.4$. The result, which we call $S_1[0.2, 1]$, is 4.94814815. The next step is to integrate over each half of $[0.2, 1]$ but with h half as large, $h_2 = 0.2$. We get

$$S_2[0.2, 0.6] = 3.51851852 \qquad \text{and} \qquad S_2[0.6, 1] = 0.66851852.$$

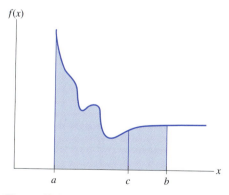

Figure 5.4

We now test the accuracy of our initial computations by seeing whether the difference between $S_1[0.2, 1]$ and the sum of $S_2[0.2, 0.6]$ and $S_2[0.6, 1]$ is greater than TOL. (Actually we compare the magnitude of this difference.)

$$S_1[0.2, 1] - (S_2[0.2, 0.6] + S_2[0.6, 1]) = 0.7611111.$$

Because this result is greater than TOL = 0.02, we must use a smaller value for h.

We continue by applying the strategy to one-half of the original interval. We arbitrarily choose the right half and compute $S_2[0.6, 1]$ with $h = h_2 = (1 - 0.6)/2 = 0.2$, comparing it to $S_3[0.6, 0.8] + S_3[0.8, 1]$ (both of these use $h_3 = h_2/2 = 0.1$). We also halve the value for TOL, getting

$$S_2[0.6, 1] - (S_3[0.6, 0.8] + S_3[0.8, 1]) = 0.66851852 - (0.41678477 + 0.25002572)$$
$$= 0.66851852 - 0.66681049$$
$$= 0.001708 \qquad \text{versus TOL} = 0.01.$$

This passes the test, so we take advantage of the results that we have available and do a Richardson extrapolation to get

$$RS[0.6, 1] = 0.66681049 + \frac{1}{15}(0.66681049 - 0.66851852)$$
$$= 0.66669662.$$

We now move to the next adjacent subinterval, [0.2, 0.6], and repeat the procedure. We compute

$$S_2[0.2, 0.6] = 3.51851852, \qquad \text{with } h_2 = 0.2;$$
$$S_3[0.2, 0.4] = 2.52314815; \qquad S_3[0.4, 0.6] = 0.83425926;$$
$$S_2[0.2, 0.6] - (S_3[0.2, 0.4] + S_3[0.4, 0.6]) = 0.161111 \qquad \text{versus TOL} = 0.01,$$

which fails, so we proceed to another level with the right half:

$$S_3[0.4, 0.6] = 0.83425926, \qquad \text{with } h_3 = 0.1;$$
$$S_4[0.4, 0.5] = 0.50005144; \qquad S_4[0.5, 0.6] = 0.33334864;$$
$$S_3[0.4, 0.6] - (S_4[0.4, 0.5] + S_4[0.5, 0.6]) = 0.000859 \qquad \text{versus TOL} = 0.005,$$

which passes. We extrapolate:

$$RS[0.4, 0.6] = 0.8333428.$$

The next adjacent interval is [0.2, 0.4]. For this we use TOL = 0.005. We find that this does not meet the criterion, so we next do [0.3, 0.4]. We do meet the TOL level of 0.0025:

$$S_4[0.3, 0.4] = 0.83356954, \qquad \text{with } h_4 = 0.05;$$
$$S_5[0.3, 0.35] = 0.47620166; \qquad S_5[0.35, 0.4] = 0.35714758;$$
$$S_4[0.3, 0.4] - (S_5[0.3, 0.35] + S_5[0.35, 0.4]) = 0.000220 \qquad \text{versus TOL} = 0.0025,$$

which passes, so

$$RS[0.3, 0.4] = 0.83333492.$$

Our last subinterval is [0.2, 0.3]. We find that we again meet the test. We give only the extrapolated result

$$RS[0.2, 0.3] = 1.666686.$$

Adding all of the *RS*-values gives the final answer:

$$\text{Integral over } [0.2, 1] = 4.00005957.$$

By employing adaptive integration, we reduced the number of function evaluations from 33 to 17.

Bookkeeping and the Avoidance of Repeating Function Evaluations

It should be obvious that we were recomputing many of the values of $f(x)$ in the previous integration. We can avoid these recalculations if we store these computations in such a way as to retrieve them appropriately. We also need to keep track of the current subinterval, the previous subintervals that we return to, and the appropriate value for h and TOL for each subinterval. The mechanism for storing these quantities is a *stack,* a data structure that is a last-in first-out device that resembles a stack of dishes in a restaurant. Actually, we use just a two-dimensional array of seven columns and as many rows as levels that we wish to accommodate. (Often a large number of levels is provided—say, 200—even though we hardly ever need so many.)

After an initial calculation to get $h_1 = (b - a)/2$, $c = a + h_1$, $f(a), f(c), f(b)$, and $S_1[a, b]$, we store a set of seven values: $a, f(a), f(c), f(b), h$, TOL, $S[a, b]$. We retrieve these values into variables that represent these quantities and continue with the first stage of the computations.

Whenever the test fails after computing for the current subinterval, we store two sets of values in two rows of the seven columns:

First row: $a, f(a), f(d), f(c), h_n$, TOL, $S[a, c]$,

Next row: $c, f(c), f(e), f(b), h_n$, TOL, $S[c, b] \leftarrow$ TOP,

where the letters a, d, c, e, b refer to points in the last subinterval that are evenly spaced from left to right in that order. We also use a pointer to the last row stored. It is named TOP to indicate it is the "top" of the stack (even though it points to the last row stored as we normally view an array). Whenever we store a set of values, we add one to TOP; whenever we retrieve a set of values, we subtract one so that TOP always points to the row that is next available for retrieval.

We begin each iteration by retrieving the row of quantities pointed to by TOP (the one labeled "Next row" above). In this way, we can reuse the previously computed function values to get values for computing the rightmost remaining subinterval. (Observe that the next subinterval begins at the c-value for the last subinterval.)

The following algorithm implements the adaptive integration scheme that we have described.

Algorithm for Computing $I(f) = \int_a^b f(x)\, dx$

SET VALUE = 0.0,
Evaluate: $h_1 = (b - a)/2,\ c = a + h_1,\ Fa = f(a)$,
        $Fc = f(c),\ Fb = f(b),\ Sab = S_1(a, b)$.
STORE($a, Fa, Fc, Fb, h_1,$ TOL, Sab).
SET TOP = 1.
REPEAT
    RETRIEVE($a, Fa, Fc, Fb, h_1,$ TOL, Sab).
    SET TOP = TOP − 1.
    Evaluate: $h_2 = h_1/2,\ d = a + h_2,\ e = a + 3h_2,\ Fd = f(d)$,
        $Fe = f(e)$,
        $Sac = S_2(a, c),\ Scb = S_2(c, b),\ S_2(a, b) = Sac + Scb$.
    IF $|S_2(a, b) - S_1(a, b)| <$ TOL THEN
        Compute $RS(a, b)$,
        VALUE = VALUE + $RS(a, b)$,
    ELSE
        $h_1 = h_2,$ TOL = TOL/2,
        SET TOP = TOP + 1,
        STORE($a, Fa, Fd, Fc, h_1,$ TOL, Sac),
        SET TOP = TOP + 1,
        STORE($c, Fc, Fe, Fb, h_1,$ TOL, Scb)
    UNTIL TOP = 0.
$I(f)$ = VALUE.

A Fortran program listed later in this chapter shows how recursion simplifies this algorithm.

5.11   Multiple Integrals

We consider first the case when the limits of integration are constants. In the calculus we learned that a double integral may be evaluated as an iterated integral; in other words, we may write

$$\iint\limits_A f(x, y)\, dA = \int_a^b \left(\int_c^d f(x, y)\, dy \right) dx = \int_c^d \left(\int_a^b f(x, y)\, dx \right) dy. \quad \textbf{(5.44)}$$

In Eq. (5.44) the rectangular region A is bounded by the lines

$$x = a, \qquad x = b, \qquad y = c, \qquad y = d.$$

In computing the iterated integrals, we hold x constant while integrating with respect to y (vice versa in the second case).

We can easily adapt the previous integration formulas to get a multiple integral. Recall that any one of the integration formulas is just a linear combination of values of the function, evaluated at varying values of the independent variable. In other words, a quadrature formula is just a weighted sum of certain functional values. The inner integral is written then as a weighted sum of function values with one variable held constant. We then add together a weighted sum of these sums. If the function is known only at the nodes of a rectangular grid through the region, we are constrained to use these values. The Newton–Cotes formulas are a convenient set to employ. There is no reason why the same formula must be used in each direction, although it is often particularly convenient to do so.

EXAMPLE 5.17   We illustrate this technique by evaluating the integral of the function of Table 5.15 over the rectangular region bounded by

$$x = 1.5, \quad x = 3.0, \quad y = 0.2, \quad y = 0.6.$$

Let us use the trapezoidal rule in the x-direction and Simpson's $\frac{1}{3}$ rule in the y-direction. (Because the number of panels in the x-direction is not even, Simpson's $\frac{1}{3}$ rule does not apply readily.) It is immaterial which integral we evaluate first. Suppose we start with y constant:

$$y = 0.2: \quad \int_{1.5}^{3.0} f(x, y)\, dx = \int_{1.5}^{3.0} f(x, 0.2)\, dx = \frac{h}{2}(f_1 + 2f_2 + 2f_3 + f_4)$$

$$= \frac{0.5}{2}[0.990 + 2(1.568) + 2(2.520) + 4.090]$$

$$= 3.3140;$$

$$y = 0.3: \quad \int_{1.5}^{3.0} f(x, 0.3)\, dx = \frac{0.5}{2}[1.524 + 2(2.384) + 2(3.800) + 6.136]$$

$$= 5.0070.$$

Table 5.15   Tabulation of a function of two variables, $u = f(x, y)$

x \ y	0.1	0.2	0.3	0.4	0.5	0.6
0.5	0.165	0.428	0.687	0.942	1.190	1.431
1.0	0.271	0.640	1.003	1.359	1.703	2.035
1.5	0.447	0.990	1.524	2.045	2.549	3.031
2.0	0.738	1.568	2.384	3.177	3.943	4.672
2.5	1.216	2.520	3.800	5.044	6.241	7.379
3.0	2.005	4.090	6.136	8.122	10.030	11.841
3.5	3.306	6.679	9.986	13.196	16.277	19.198

Similarly, at

$$y = 0.4, \quad I = 6.6522;$$
$$y = 0.5, \quad I = 8.2368;$$
$$y = 0.6, \quad I = 9.7435.$$

We now sum these in the y-direction according to Simpson's rule:

$$f(x, y) \, dx = \frac{0.1}{3} [3.3140 + 4(5.0070) + 2(6.6522) + 4(8.2368) + 9.7435]$$

$$= 2.6446$$ ▲

(In this example our answer does not check well with the analytical value of 2.5944 because the x-intervals are large. We could improve our estimate somewhat by fitting a higher-degree polynomial than the first to provide the integration formula. We can even use values outside the range of integration for this, using the techniques of Section 5.8 to derive the formulas.)

The previous example shows that double integration by numerical means reduces to a double summation of weighted function values. The calculations we have just made could be written in the form

$$\int f(x, y) \, dx \, dy = \sum_{j=1}^{m} v_j \sum_{i=1}^{n} w_i f_{ij}$$

$$= \frac{\Delta y \Delta x}{3 \ 2} [(f_{1,1} + 2f_{2,1} + 2f_{3,1} + f_{4,1})$$

$$+ 4(f_{1,2} + 2f_{2,2} + 2f_{3,2} + f_{4,2}) + \cdots$$

$$+ (f_{1,5} + 2f_{2,5} + 2f_{3,5} + f_{4,5})].$$

It is convenient to write this in pictorial operator form, in which the weighting factors are displayed in an array that is a map to the location of the functional values to which they are applied.

$$\int f(x, y) \, dx \, dy = \frac{\Delta y}{3} \frac{\Delta x}{2} \begin{Bmatrix} 1 & 4 & 2 & 4 & 1 \\ 2 & 8 & 4 & 8 & 2 \\ 2 & 8 & 4 & 8 & 2 \\ 1 & 4 & 2 & 4 & 1 \end{Bmatrix} f_{i,j}. \quad (5.45)$$

We interpret the numbers in the array of Eq. (5.45) in this manner: We use the values 1, 4, 2, 4, and 1 as weighting factors for functional values in the top row of the portion of Table 5.15 that we integrate over (values were $x = 1.5$ and y varies from 0.2 to 0.6). Similarly, the second column of the array in Eq. (5.45) represents weighting factors that are applied to a column of function values where $y = 0.4$ and x varies from 1.5 to 3.0. Observe that the values in the pictorial operator of Eq. (5.45) follow immediately from the Newton–Cotes coefficients for single-variable integration.

Other combinations of Newton–Cotes formulas give similar results. It is probably easiest for hand calculation to use these pictorial integration operators. Pictorial integration is readily adapted to any desired combination of integration formulas. Except for the difficulty of representation beyond two dimensions, this operator technique also applies to triple and quadruple integrals.

There is an alternative representation to such pictorial operators that is easier to translate into a computer program. We also derive it somewhat differently. Consider the numerical integration formula for one variable

$$\int_{-1}^{1} f(x) \, dx \doteq \sum_{i=1}^{n} a_i f(x_i). \tag{5.46}$$

We have seen in Section 5.9 that such formulas can be made exact if $f(x)$ is any polynomial of a certain degree. Assume that Eq. (5.46) holds for polynomials up to degree s.*

We now consider the multiple integral formula

$$\int_{-1}^{1} \int_{-1}^{1} \int_{-1}^{1} f(x, y, z) \, dx \, dy \, dz \doteq \sum_{i=1}^{n} \sum_{j=1}^{n} \sum_{k=1}^{n} a_i a_j a_k f(x_i, y_j, z_k). \tag{5.47}$$

We wish to show that Eq. (5.47) is exact for all polynomials in x, y, and z up to degree s. Such a polynomial is a linear combination of terms of the form $x^\alpha y^\beta z^\gamma$, where α, β, and γ are nonnegative integers whose sum is equal to s or less. If we can prove that Eq. (5.47) holds for the general term of this form, it will then hold for the polynomial.

To do this we assume that

$$f(x, y, z) = x^\alpha y^\beta z^\gamma.$$

Then, because the limits are constants and the integrand is factorable,

$$I = \int_{-1}^{1} \int_{-1}^{1} \int_{-1}^{1} x^\alpha y^\beta z^\gamma \, dx \, dy \, dz$$
$$= \left(\int_{-1}^{1} x^\alpha \, dx \right) \left(\int_{-1}^{1} y^\beta \, dy \right) \left(\int_{-1}^{1} z^\gamma \, dz \right).$$

Replacing each term according to Eq. (5.46), we get,

$$I = \left(\sum_{i=1}^{n} a_i x_i^\alpha \right) \left(\sum_{j=1}^{n} a_j y_j^\beta \right) \left(\sum_{k=1}^{n} a_k z_k^\gamma \right) = \sum_{i=1}^{n} a_i x_i^\alpha \sum_{j=1}^{n} a_j y_j^\beta \sum_{k=1}^{n} a_k z_k^\gamma. \tag{5.48}$$

We need now an elementary rule about the product of summations. We illustrate it for a simple case. We assert that

$$\left(\sum_{i=1}^{3} u_i \right) \left(\sum_{j=1}^{2} v_j \right) = \sum_{i=1}^{3} \left(\sum_{j=1}^{2} u_i v_j \right)$$
$$= \sum_{i=1}^{3} \sum_{j=1}^{2} u_i v_j.$$

---

*For Newton–Cotes formulas, $s = n - 1$ for n even and $s = n$ for n odd. For Gaussian quadrature formulas, $s = 2n - 1$, and the x_i will be unevenly spaced.

The last equality is purely notational. We prove the first by expanding both sides:

$$\left(\sum_{i=1}^{3} u_i \right) \left(\sum_{j=1}^{2} v_j \right) = \sum_{i=1}^{3} u_i \sum_{j=1}^{2} v_j$$

$$= (u_1 + u_2 + u_3)(v_1 + v_2)$$

$$= u_1 v_1 + u_1 v_2 + u_2 v_1 + u_2 v_2 + u_3 v_1 + u_3 v_2;$$

$$\sum_{i=1}^{3} \sum_{j=1}^{2} u_i v_j = (u_1 v_1 + u_1 v_2) + (u_2 v_1 + u_2 v_2) + (u_3 v_1 + u_3 v_2).$$

On removing parentheses, we see the two sides are the same. Using this principle, we can write Eq. (5.48) in the form

$$I = \sum_{i=1}^{n} \sum_{j=1}^{n} \sum_{k=1}^{n} a_i a_j a_k x_i^\alpha y_j^\beta z_k^\gamma, \tag{5.49}$$

which shows that the questioned equality of Eq. (5.47) is valid, and we can write a program for a triple integral by three nested DO loops. The coefficients a_i are chosen from any numerical integration formula. If the three one-variable formulas corresponding to Eq. (5.47) are not identical, an obvious modification of Eq. (5.49) applies. In some cases a change of variable is needed to correspond to Eq. (5.46).

If we are evaluating a multiple integral numerically where the integrand is a known function, our choice of the form of Eq. (5.46) is wider. Of higher efficiency than the Newton–Cotes formulas is Gaussian quadrature. Because it also fits the pattern of Eq. (5.46), the formula of Eq. (5.49) applies. We illustrate this with a simple example.

EXAMPLE 5.18   Evaluate

$$I = \int_0^1 \int_{-1}^0 \int_{-1}^1 yz e^x \, dx \, dy \, dz$$

by Gaussian quadrature using a three-term formula for x and two-term formulas for y and z. We first make the changes of variables to adjust the limits for y and z to $(-1, 1)$:

$$y = \frac{1}{2}(u - 1), \qquad dy = \frac{1}{2} \, du;$$

$$z = \frac{1}{2}(v + 1), \qquad dz = \frac{1}{2} \, dv.$$

Our integral becomes

$$I = \frac{1}{16} \int_{-1}^1 \int_{-1}^1 \int_{-1}^1 (u - 1)(v + 1)e^x \, dx \, du \, dv.$$

The two- and three-point Gaussian formulas are, from Section 5.9:

$$\int_{-1}^{1} f(x)\ dx = (1)f(-0.5774) + (1)f(0.5774),$$

$$\int_{-1}^{1} f(x)\ dx = \left(\frac{8}{9}\right)f(-0.7746) + \left(\frac{8}{9}\right)f(0) + \left(\frac{5}{9}\right)f(0.7746).$$

The integral is then

$$I = \frac{1}{16} \sum_{i=1}^{2} \sum_{j=1}^{2} \sum_{k=1}^{3} a_i a_j b_k (u_i + 1)(v_j - 1)e^{x_k},$$

$$a_1 = 1, \qquad a_2 = 1,$$

$$b_1 = \frac{5}{9}, \qquad b_2 = \frac{8}{9}, \qquad b_3 = \frac{5}{9},$$

and values of u, v, and x as given.

A few representative terms of the sum are

$$I = \frac{1}{16}\left[(1)(1)\left(\frac{5}{9}\right)(-0.5774 + 1)(-0.5774 - 1)e^{-0.7446}\right.$$

$$+ (1)(1)\left(\frac{8}{9}\right)(-0.5774 + 1)(-0.5774 - 1)e^{0}$$

$$+ (1)(1)\left(\frac{5}{9}\right)(-0.5774 + 1)(-0.5774 - 1)e^{0.7746}$$

$$+ (1)(1)\left(\frac{5}{9}\right)(0.5774 + 1)(-0.5774 - 1)e^{-0.7746}$$

$$\left. + \cdots \right].$$

On evaluating, we get $I = -0.58758$. The analytical value is

$$-\frac{1}{4}(e - e^{-1}) = -0.58760.$$

▲

5.12 Multiple Integration with Variable Limits

If the limits of integration are not constant, so that the region in the x, y-plane upon which the integrand $f(x, y)$ is to be summed is not rectangular, we must modify the procedure of Section 5.11. We consider a simple example to illustrate. Evaluate

$$\int\int f(x,\ y)\ dy\ dx$$

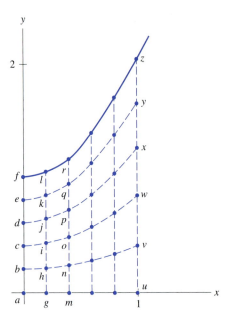

Figure 5.5

over the region bounded by the lines $x = 0$, $x = 1$, $y = 0$, and the curve $y = x^2 + 1$. The region is sketched in Fig. 5.5. If we draw vertical lines spaced at $\Delta x = 0.2$ apart, shown as dashed lines in Fig. 5.5, it is obvious that we can approximate the inner integral at constant x-values along any one of the vertical lines (including $x = 0$ and $x = 1$). If we use the trapezoidal rule with five panels for each of these, we get the series of sums

$$S_1 = \frac{h_1}{2}(f_a + 2f_b + 2f_c + 2f_d + 2f_e + f_f),$$

$$S_2 = \frac{h_2}{2}(f_g + 2f_h + 2f_i + 2f_j + 2f_k + f_l),$$

$$S_3 = \frac{h_3}{2}(f_m + 2f_n + \cdots),$$

$$\vdots$$

$$S_6 = \frac{h_6}{2}(f_u + 2f_v + 2f_w + 2f_x + 2f_y + f_z).$$

The subscripts here indicate the values of the function at the points so labeled in Fig. 5.5. The values of the h_i are not equal in each of the equations, but in each they are the vertical distances divided by 5. The combination of these sums to give an estimate of the double integral will then be

$$\text{Integral} = \frac{0.2}{2}(S_1 + 2S_2 + 2S_3 + 2S_4 + 2S_5 + S_6).$$

To be even more specific, suppose that $f(x, y) = xy$. Then

$$S_1 = \frac{1.0/5}{2}(0 + 0 + 0 + 0 + 0 + 0) = 0,$$

$$S_2 = \frac{1.04/5}{2}(0 + 0.0832 + 0.1664 + 0.2496 + 0.3328 + 0.208) = 0.1082,$$

$$S_3 = \frac{1.16/5}{2}(0 + 0.1856 + 0.3712 + 0.5568 + 0.7424 + 0.464) = 0.2691,$$

$$S_4 = \frac{1.36/5}{2}(0 + 0.3264 + 0.6528 + 0.9792 + 1.3056 + 0.816) = 0.5549,$$

$$S_5 = \frac{1.64/5}{2}(0 + 0.5248 + 1.0496 + 1.5744 + 2.0992 + 1.312) = 1.0758,$$

$$S_6 = \frac{2.0/5}{2}(0 + 0.8 + 1.6 + 2.4 + 3.2 + 2.0) = 2.0;$$

$$\text{Integral} = \frac{0.2}{2}(0 + 0.2164 + 0.5382 + 1.1098 + 2.1516 + 2.0)$$

$$= 0.6016 \qquad \text{versus analytical value of } 0.583333.$$

The extension of this to more complicated regions and the adaptation to the use of Simpson's rule should be obvious. If the functions that define the region are not single-valued, we must divide the region into subregions to avoid the problem, but we must also do this when we integrate analytically.

The previous calculations were not very accurate because the trapezoidal rule has relatively large errors. Gaussian quadrature should be an improvement, even using fewer points within the region. Let us use 3-point quadrature in the x-direction and 4-point quadrature in the y-direction. As in Section 5.9, we must change the limits of integration:

$$\int_0^1 \int_0^{x^2+1} xy \, dy \, dx$$

to

$$\frac{1}{4} \int_{-1}^1 \int_{-1}^1 \frac{s+1}{2} \left[\frac{(x^2(s) + 1)t + (x^2(s) + 1)}{2} \right] dt \, ds$$

in which we make the following substitutions:

$$x = \frac{s+1}{2} \qquad y = \frac{(x^2(s) + 1)t + (x^2(s) + 1)}{2}.$$

The integral is approximated by the sum

$$\sum_{i=1}^3 \sum_{j=1}^4 w_i W_j f(s_i, t_j),$$

where the w_i's, W_j's, s_i's, and t_j's are the values taken from Table 5.14. Using that table, we set $w_1 = 0.55555555$, $w_3 = w_1$, and $w_2 = 0.88888889$; we set $s_1 = -0.77459667$, $s_3 = -s_1$, and $s_2 = 0.0$. The values for the W_j's and t_j's are obtained in the same way. For each fixed i, $i = 1, 2, 3$, let S_i be the corresponding value obtained using Gaussian quadrature for a fixed s_i, where $S_i = \sum_{j=1}^{4} W_j f(s_i, t_j)$.

The following intermediate values are easily verified:

$$S_1 = (0.00279158 + 0.02487506 + 0.05050174 + 0.03741447) = 0.11558285,$$

$$S_2 = (0.01886891 + 0.16813600 + 0.34135240 + 0.25289269) = 0.78125000,$$

$$S_3 = (0.06845742 + 0.61000649 + 1.23844492 + 0.91750833) = 2.83441716.$$

We sum these values as follows:

$$\frac{w_1 S_1 + w_2 S_2 + w_3 S_3}{4} = 0.58333334,$$

which agrees with the exact answer to seven places. In this case we used only 12 evaluations of the function (exceptionally simple to do here, but usually more costly) compared to the 36 used with the trapezoidal rule.

To keep track of the intermediate computations, it is convenient to use a template such as

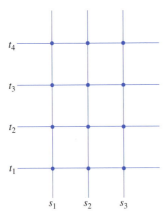

and to compute the S_i's along the verticals. The points (s_i, t_i) within the region are often called *Gauss points*.

5.13 Applications of Cubic Splines

In addition to their obvious use for interpolation, splines (Chapter 3) can be used for finding derivatives and integrals of functions, even when the function is known only as a table of values. The smoothness of splines can give improved accuracy in some cases, because of the requirement that each portion have the same first and second derivatives as its neighbor where they join.

For the cubic spline that approximates $f(x)$, we can write, for the interval $x_i \le x \le x_{i+1}$,

$$f(x) = a_i(x - x_i)^3 + b_i(x - x_i)^2 + c_i(x - x_i) + d_i,$$

where the coefficients are determined as in Section 3.4. The method outlined in that section computes S_i and S_{i+1}, the values of the second derivative at each end of the subinterval. From these S-values and the values of $f(x)$, we compute the coefficients of the cubic:

$$a_i = \frac{S_{i+1} - S_i}{6(x_{i+1} - x_i)},$$

$$b_i = \frac{S_i}{2},$$

$$c_i = \frac{f(x_{i+1}) - f(x_i)}{x_{i+1} - x_i} - \frac{2(x_{i+1} - x_i)S_i + (x_{i+1} - x_i)S_{i+1}}{6},$$

$$d_i = f(x_i).$$

Approximating the first and second derivatives is straightforward; we estimate these as the values of the derivatives of the cubic:

$$f'(x) = 3a_i(x - x_i)^2 + 2b_i(x - x_i) + c_i, \tag{5.50}$$

$$f''(x) = 6a_i(x - x_i) + 2b_i. \tag{5.51}$$

At the $n + 1$ points x_i where the function is known and the spline matches $f(x)$, these formulas are particularly simple:

$$f'(x_i) \doteq c_i,$$

$$f''(x_i) \doteq 2b_i.$$

(We note that a cubic spline is not useful for approximating derivatives of order higher than second. A higher degree of spline function would be required for these values.)

Approximating the integral of $f(x)$ over the n intervals where $f(x)$ is approximated by the spline is similarly straightforward:

$$\int_{x_1}^{x_{n+1}} f(x)\, dx = \int_{x_1}^{x_{n+1}} P_3(x)\, dx$$

$$= \sum_{i=1}^{n} \left[\frac{a_i}{4}(x - x_i)^4 + \frac{b_i}{3}(x - x_i)^3 + \frac{c_i}{2}(x - x_i)^2 + d_i(x - x_i) \right]_{x_i}^{x_{i+1}}$$

$$= \sum_{i=1}^{n} \left[\frac{a_i}{4}(x_{i+1} - x_i)^4 + \frac{b_i}{3}(x_{i+1} - x_i)^3 \right.$$

$$\left. + \frac{c_i}{2}(x_{i+1} - x_i)^2 + d_i(x_{i+1} - x_i) \right].$$

If the intervals are all the same size, $(h = x_{i+1} - x_i)$, this equation becomes

$$\int_{x_1}^{x_{n+1}} f(x)\, dx = \frac{h^4}{4} \sum_{i=1}^{n} a_i + \frac{h^3}{3} \sum_{i=1}^{n} b_i + \frac{h^2}{2} \sum_{i=1}^{n} c_i + h \sum_{i=1}^{n} d_i.$$

We illustrate the use of splines to compute derivatives and integrals by a simple example.

EXAMPLE 5.19   Compute the integral and derivatives of $f(x) = \sin \pi x$ over the interval $0 \le x \le 1$ from the spline that fits at $x = 0, 0.25, 0.5, 0.75$, and 1.0. (See Table 5.16.) We use end condition 1: $S_1 = 0, S_5 = 0$. Solving for the coefficients of the cubic spline, we get the results shown in Table 5.17.

The estimated values for $f'(x)$ and $f''(x)$ computed with Eqs. (5.50) and (5.51) are shown in Table 5.18. The errors of these estimates from the exact values ($f'(x) = \pi \cos(\pi x)$ and $f''(x) = -\pi^2 \sin(\pi x)$) are shown in the last two columns.

In general, the cubic spline gives good estimates of the derivatives, the maximum error being 2.5% for the first derivative and 5.0% for the second.

It is of interest to compare these values with estimates of the derivatives from a fourth-degree interpolating polynomial that fits $f(x)$ at the same five points. Table 5.19 exhibits these estimates. For the first derivative, the spline curve gives better results near the ends of the range for $f(x)$; the polynomial gives better results near the midpoint. Both are very good in this example.

Comparison of estimates for the second derivative shows a similar relationship, except for the fourth-degree polynomial, which is very bad at the endpoints.

We readily compute the integral from the cubic spline:

$$\int_0^1 f(x)\, dx = \frac{(0.25)^4}{4}(0) + \frac{(0.25)^3}{3}(-12.5376) + \frac{(0.25)^2}{2}(3.1340)$$

$$+ \; 0.25(2.4142)$$

$$= 0.6362 \qquad (\text{exact} = 0.6366; \text{ error} = +0.0004).$$

Table 5.16

i, point number	x	$f(x)$
1	0	0
2	0.25	0.7071
3	0.50	1.0000
4	0.75	0.7071
5	1.0	0

Table 5.17

i	x	S_i	a_i	b_i	c_i	d_i
1	0	0	-4.8960	0	3.1340	0
2	0.25	-7.3440	-2.0288	-3.6720	2.2164	0.7071
3	0.50	-10.3872	2.0288	-5.1936	0	1.000
4	0.75	-7.3440	4.8960	-3.6720	-2.2164	0.7071

Table 5.18 Estimates of $f'(x)$ and $f''(x)$ from a cubic spline

x	$f'(x)$	$f''(x)$	Error in $f'(x)$	Error in $f''(x)$
0.00	3.1344	0.0000	0.007146	0.000000
0.05	3.0977	−1.4689	0.005191	−0.075053
0.10	2.9876	−2.9378	0.000275	−0.112090
0.15	2.8039	−4.4067	−0.004766	−0.074028
0.20	2.5469	−5.8756	−0.005287	0.074363
0.25	2.2164	−7.3445	0.005053	0.365600
0.30	1.8340	−7.9529	0.012627	−0.031778
0.35	1.4211	−8.5613	0.005155	−0.232546
0.40	0.9778	−9.1698	−0.007015	−0.216781
0.45	0.5041	−9.7782	−0.012668	0.030114
0.50	−0.0000	−10.3866	0.000000	0.517038
0.55	−0.5041	−9.7782	0.012668	0.030113
0.60	−0.9778	−9.1698	0.007015	−0.216781
0.65	−1.4211	−8.5613	−0.005155	−0.232547
0.70	−1.8340	−7.9529	−0.012628	−0.031778
0.75	−2.2164	−7.3445	−0.005053	0.365598
0.80	−2.5469	−5.8756	0.005287	0.074362
0.85	−2.8039	−4.4067	0.004766	−0.074028
0.90	−2.9876	−2.9378	−0.000275	−0.112088
0.95	−3.0977	−1.4689	−0.005190	−0.075051
1.00	−3.1344	0.0000	−0.007146	0.000003

Table 5.19 Estimates of $f'(x)$ and $f''(x)$ from a cubic spline

x	$f'(x)$	$f''(x)$	Error in $f'(x)$	Error in $f''(x)$
0.00	3.0849	1.1505	0.056643	−1.150496
0.05	3.0894	−0.9358	0.013513	−0.608130
0.10	2.9950	−2.8025	−0.007196	−0.247359
0.15	2.8128	−4.4496	−0.013630	−0.031101
0.20	2.5537	−5.8771	−0.012126	0.075874
0.25	2.2288	−7.0849	−0.007321	0.106083
0.30	1.8489	−8.0732	−0.002311	0.088523
0.35	1.4251	−8.8418	0.001151	0.047960
0.40	0.9684	−9.3909	0.002436	0.004319
0.45	0.4897	−9.7203	0.001778	−0.027803
0.50	−0.0000	−9.8301	−0.000000	−0.039509
0.55	−0.4897	−9.7203	−0.001779	−0.027803
0.60	−0.9684	−9.3909	−0.002437	0.004321
0.65	−1.4251	−8.8418	−0.001152	0.047961
0.70	−1.8489	−8.0732	0.002311	0.088524
0.75	−2.2288	−7.0849	0.007320	0.106084
0.80	−2.5537	−5.8771	0.012125	0.075876
0.85	−2.8128	−4.4496	0.013630	−0.031100
0.90	−2.9950	−2.8025	0.007196	−0.247358
0.95	−3.0894	−0.9358	−0.013513	−0.608130
1.00	−3.0849	1.1505	−0.056643	−1.150496

For this example, the accuracy of the integral is significantly better than that obtained with Simpson's $\frac{1}{3}$ rule, which gives 0.6381, error $= -0.0015$.

This example illustrates the fact that getting derivatives numerically is more difficult than getting integrals by numerical methods.*   ▲

5.14 An Application of Numerical Integration—Fourier Transforms

In Chapter 4 we saw that the coefficients of a Fourier series are obtained by integration. We can get these numerically as an alternative to analytical integrations. Example 5.20 compares the accuracy of doing this with the trapezoidal rule and with Simpson's rule.

EXAMPLE 5.20   Evaluate the coefficients for the half-range expansions for the function $f(x) = x$ on $[0, 2]$ by numerical integration.

For the even extension (Fourier cosine series), we get the A's from

$$A_n = \left(\frac{2}{2}\right) \int_0^2 x \cos\left(\frac{n\pi x}{2}\right) dx, \qquad n = 0, 1, 2, \ldots.$$

For the odd extension (Fourier sine series), we get the B's from

$$B_n = \left(\frac{2}{2}\right) \int_0^2 x \sin\left(\frac{n\pi x}{2}\right) dx, \qquad n = 1, 2, 3, \ldots.$$

Table 5.20 shows the results from numerical integrations with 20 subdivisions of the interval $[0, 2]$ using both the trapezoidal rule and Simpson's rule and compares the results with analytical integrations. Table 5.21 does the same but for 200 intervals. It is clear (as expected) that we can use numerical integrations. The accuracy with Simpson's rule is good—essentially perfect agreement is seen in Table 5.21. The errors do increase as n increases (this is particularly apparent in Table 5.20).   ▲

The implication of Example 5.20 is that we can compute Fourier coefficients on a table of data, say from measurements for a periodic phenomenon. Doing so performs what is termed a *discrete Fourier transform*. Other names for this procedure are *harmonic analysis* and *finite Fourier transform*. The data, which originally are most often functions of time, can now be interpreted in terms of the angles that appear in a Fourier series, more often expressed as the frequencies. It is a transform because we change a function of t (time) to an equivalent function of frequencies.

---

*The section on theory explains why this is true.

Table 5.20   Comparison of numerical integration with analytical results: 20 subdivisions of [0, 2]

	Trapezoidal rule		Simpson's rule		Analytical integration	
n	A_n	B_n	A_n	B_n	A_n	B_n
0	2		2		2	
1	−0.81224	1.27062	−0.81056	1.27324	−0.81057	1.27323
2	0	−0.63138	0	−0.63665	0	−0.63662
3	−0.09175	0.41653	−0.08999	0.42453	−0.09006	0.42441
4	0	−0.30777	0	−0.31860	0	−0.31831
5	−0.03414	0.24142	−0.03219	0.25523	−0.03242	0.25465

Table 5.21   Comparison of numerical integration with analytical results: 200 subdivisions of [0, 2]

	Trapezoidal rule		Simpson's rule		Analytical integration	
n	A_n	B_n	A_n	B_n	A_n	B_n
0	2		2		2	
1	−0.81059	1.27321	−0.81057	1.27324	−0.81057	1.27323
2	0	−0.63657	0	−0.63662	0	−0.63662
3	−0.09008	0.42433	−0.09006	0.42441	−0.09006	0.42441
4	0	−0.31821	0	−0.31831	0	−0.31831
5	−0.03244	0.25452	−0.03242	0.25465	−0.03242	0.25465

Why should we want to so transform a set of experimental data? Because knowing which frequencies of a Fourier series are most significant (have the largest coefficients) gives information on the fundamental frequencies of the system. This knowledge is important because an applied periodic external force that includes components of the same frequency as one of these fundamental frequencies causes extremely large disturbances. (Such a periodic force may come from vibrations from rotating machinery, from wind, or from earthquakes.) We normally want to avoid such extreme responses for fear that the system will be damaged. Here is an example.

EXAMPLE 5.21   An experiment (actually, these are contrived data) showed the displacements given in Table 5.22 when the system was caused to vibrate in its natural modes. The values represent a periodic function on the interval for t of [2, 10] because they repeat themselves after $t = 10$.

We will use trapezoidal integration to find the Fourier series coefficients for the data. Doing so gives these values for the A's and B's:

Table 5.22   Measurements of displacements versus time

t	Displacement	t	Displacement
2.000	3.804	6.250	3.746
2.250	6.503	6.500	5.115
2.500	7.496	6.750	4.156
2.750	6.094	7.000	1.593
3.000	3.003	7.250	−0.941
3.250	−0.105	7.500	−1.821
3.500	−1.598	7.750	−0.329
3.750	−0.721	8.000	2.799
4.000	1.806	8.250	5.907
4.250	4.350	8.500	7.338
4.500	5.255	8.750	6.380
4.750	3.878	9.000	3.709
5.000	0.893	9.250	0.992
5.250	−2.048	9.500	−0.116
5.500	−3.280	9.750	1.047
5.750	−2.088	10.000	3.802
6.000	0.807		

n	A	B
0	4.6015	
1	1.5004	−0.5006
2	−0.0009	0.0016
3	−0.0017	0.0016
4	0.0008	4.0011
5	−0.0017	0.0000
6	−0.0009	0.0022
7	−0.0005	−0.0023
8	−0.0008	0.0009

This shows that only A_0, A_1, B_1, and B_4 are important. There would be no amplification of motion from forces that do not include the frequencies corresponding to these.
   (Table 5.22 was constructed from

$$f(t) = 2.3 + 1.5 \cos(t) - 0.5 \sin(t) + 4 \sin(4t),$$

plus a small random variation whose values ranged from −0.01 to +0.01. It is the random variations that cause nonzero values for the insignificant A's and B's.)   ▲

The Fast Fourier Transform

If we need to do a finite Fourier transform on lots of data, the effort used in carrying out the computations is exorbitant. In the preceding examples, where we reevaluated cosines and sines numerous times, we should have recognized that many of these values are the same. When we evaluate the integrals for a finite Fourier transform, we compute sines and cosines for angles around the origin, as indicated in Fig. 5.6.

When we need to find $\cos(nx)$ and $\sin(nx)$, we move around the circle; when $n = 1$, we use each value in turn. For other values of n, we use every nth value, but it is easy to see that these repeat previous values. The *fast Fourier transform* (often written as *FFT*) takes advantage of this fact to avoid the recomputations.

In developing the FFT algorithm, the preferred method is to use an alternative form of the Fourier series. Instead of

$$f(x) \approx \frac{A_0}{2} + \sum_{n=1}^{\infty} [A_n \cos(nx) + B_n \sin(nx)], \qquad (\text{period} = 2\pi), \qquad \textbf{(5.52)}$$

we will use an equivalent form in terms of complex exponentials. Utilizing Euler's identity (using i as $\sqrt{-1}$),

$$e^{ijx} = \cos(jx) + i \sin(jx),$$

we can write Eq. (5.52) as

$$f(x) = \sum_{j=0}^{\infty} (c_j e^{ijx} + c_{-j} e^{-ijx})$$

$$= 2c_0 + \sum_{j=1}^{\infty} [(c_j + c_{-j})\cos(jx) + i(c_j - c_{-j})\sin(jx)] \qquad \textbf{(5.53)}$$

$$= \sum_{j=-\infty}^{\infty} c_j e^{ijx}.$$

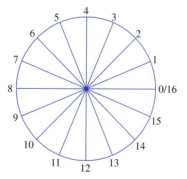

Figure 5.6 Angles used in computing for 16 points

We can match up the A's and B's of Eq. (5.52) to the c's of (5.53):

$$A_j = c_j + c_{-j}, \qquad B_j = i(c_j - c_{-j}), \tag{5.54}$$

$$c_j = \frac{A_j - iB_j}{2}, \qquad c_{-j} = \frac{A_j + iB_j}{2}.$$

For $f(x)$ real, it is easy to show that $c_0 = \bar{c}_0$ and $c_j = \bar{c}_{-j}$, where the bars represent complex conjugates.

For integers j and k, it is true that

$$\int_0^{2\pi} (e^{ikx})(e^{ijx})\, dx = \int_0^{2\pi} e^{i(k+j)x}\, dx = \begin{cases} 0 & \text{for } k \neq -j, \\ 2\pi & \text{for } k = -j. \end{cases}$$

(You can verify the first of these through Euler's identity.) This allows us to evaluate the c's of Eq. (5.53) by the following.

For each fixed k, we get

$$f(x)e^{-ikx} = \sum_{j=-\infty}^{\infty} c_k e^{i(j-k)x},$$

$$\int_0^{2\pi} f(x)e^{-ikx}\, dx = 2\pi c_k, \qquad \text{or} \tag{5.55}$$

$$c_k = \frac{1}{2\pi} \int_0^{2\pi} f(x)e^{-ikx}\, dx, \qquad k = 0, \pm 1, \pm 2, \dots.$$

EXAMPLE 5.22 (You should verify each of these.)

1. Let $f(x) = x$; then

$$c_k = \frac{1}{2\pi} \int_0^{2\pi} xe^{ikx}\, dx = -\frac{i}{k}, \qquad k \neq 0.$$

2. Let $f(x) = x(2\pi - x)$; then

$$c_k = \frac{1}{2\pi} \int_0^{2\pi} x(2\pi - x)e^{-ikx}\, dx = -\frac{2}{k^2}, \qquad k \neq 0.$$

3. Let $f(x) = \cos(x)$; then

$$c_k = \frac{1}{2\pi} \int_0^{2\pi} \cos(x)e^{-ikx}\, dx = \begin{cases} \dfrac{1}{2} & \text{for } k = 1 \text{ or } -1, \\ 0 & \text{for all other } k. \end{cases}$$

Note that for Eq. (5.52) this makes $A_1 = 1$ and all the other A_j's $= 0$. ▲

Thus for a given $f(x)$ that satisfies continuity conditions, we have

$$c_j = \frac{1}{2\pi} \int_0^{2\pi} f(x)e^{-ijx}\, dx, \qquad j = 0, \pm 1, \pm 2, \dots.$$

The magnitudes of the Fourier series coefficients $|c_j|$ are the *power spectrum of f;* these show the frequencies that are represented in $f(x)$. If we know $f(x)$ in the time domain, we can identify f by computing the c_j's. In getting the Fourier series, we have transformed from the time domain to the frequency domain, an important aspect of wave analysis.

Suppose we have N values for $f(x)$ on the interval $[0, 2\pi]$ at equispaced points, $x_k = 2\pi k/N$, $k = 0, 1, \ldots, N - 1$. Because $f(x)$ is periodic, $f_N = f_0$, $f_{N+1} = f_1$, and so on. Instead of formal analytical integration, we would use a numerical integration method to get the coefficients. Even if $f(x)$ is known at all points in $[0, 2\pi]$, we might prefer to use numerical integration. This would use only certain values of $f(x)$, often those evaluated at uniform intervals. It is also often true that we do not know $f(x)$ everywhere, because we have sampled a continuous signal. In that case, however, it is better to use the discrete Fourier transform, which can be defined as

$$X(n) = \sum_{k=0}^{N-1} x_0(k)e^{-i2\pi nk/N}, \qquad n = 0, 1, 2, \ldots, N - 1. \qquad (5.56)$$

In Eq. (5.56), we have changed notation to conform more closely to the literature on FFT. $X(n)$ corresponds to the coefficients of N frequency terms, and the $x_0(k)$ are the N values of the signal samples in the time domain. You can think of n as indexing the X-terms and k as indexing the x_0-terms. Equation (5.56) corresponds to a set of N linear equations that we can solve for the unknown $X(n)$. Because the unknowns appear on the left-hand side of (5.56), this requires only the multiplication of an N-component vector by an $N \times N$ matrix.

It will simplify the notation if we let $W = e^{-i2\pi/N}$, making the right-hand-side terms of Eq. (5.56) become $x_0(k)W^{nk}$. To develop the FFT algorithm, suppose that $N = 4$. We write the four equations for this case:

$$X(0) = W^0 x_0(0) + W^0 x_0(1) + W^0 x_0(2) + W^0 x_0(3),$$
$$X(1) = W^0 x_0(0) + W^1 x_0(1) + W^2 x_0(2) + W^3 x_0(3),$$
$$X(2) = W^0 x_0(0) + W^2 x_0(1) + W^4 x_0(2) + W^6 x_0(3),$$
$$X(3) = W^0 x_0(0) + W^3 x_0(1) + W^6 x_0(2) + W^9 x_0(3).$$

In matrix form:

$$\begin{bmatrix} X(0) \\ X(1) \\ X(2) \\ X(3) \end{bmatrix} = \begin{bmatrix} W^0 & W^0 & W^0 & W^0 \\ W^0 & W^1 & W^2 & W^3 \\ W^0 & W^2 & W^4 & W^6 \\ W^0 & W^3 & W^6 & W^9 \end{bmatrix} x_0. \qquad (5.57)$$

In solving the set of N equations in the form of Eq. (5.57), we will have to make N^2 complex multiplications plus $N(N - 1)$ complex additions. Using the FFT, however, greatly reduces the number of such operations. Although there are several variations on the algorithm, we will concentrate on the Cooley–Tukey formulation.

The matrix of Eq. (5.57) can be factored to give an equivalent form for the set of equations. At the same time we will use the fact that $W^0 = 1$ and $W^k = W^{k \bmod(N)}$:

$$
\begin{bmatrix} X(0) \\ X(2) \\ X(1) \\ X(3) \end{bmatrix} = \begin{bmatrix} 1 & W^0 & 0 & 0 \\ 1 & W^2 & 0 & 0 \\ 0 & 0 & 1 & W^1 \\ 0 & 0 & 1 & W^3 \end{bmatrix} \begin{bmatrix} 1 & 0 & W^0 & 0 \\ 0 & 1 & 0 & W^0 \\ 1 & 0 & W^2 & 0 \\ 0 & 1 & 0 & W^2 \end{bmatrix} x_0.
\tag{5.58}
$$

You should verify that the factored form (Eq. 5.58) is exactly equivalent to Eq. (5.57) by multiplying out. Note carefully that the elements of the X-vector are scrambled. (The development can be done formally and more generally by representing n and k as binary values, but it will suffice to show the basis for the FFT algorithm by expanding on this simple $N = 4$ case.)

By using the factored form, we now get the values of $X(n)$ by two steps (stages), in each of which we multiply a matrix times a vector. In the first stage, we transform x_0 into x_1 by multiplying the right matrix of Eq. (5.58) and x_0. In the second stage, we multiply the left matrix and x_1, getting x_2. We get X by unscrambling the components of x_2. By doing the operation in stages, the number of complex multiplications is reduced to $N(\log_2 N)$. For $N = 4$, this is a reduction by one-half, but for large N it is very significant; if $N = 1024$, there are 10 stages and the reduction in complex multiplies is a hundredfold!

It is convenient to represent the sequence of multiplications of the factored form (Eq. 5.58 or its equivalent for larger N) by flow diagrams. Figure 5.7 is for $N = 4$ and Fig. 5.8 is for $N = 16$. Each column holds values of x_{ST}, where the subscript tells which stage is being computed; ST ranges from 1 to 2 for $N = 4$ and from 1 to 4 for $N = 16$. [The number of stages, for N a power of 2, is $\log_2(N)$.] In each stage, we get x-values of the next stage from those of the present stage. Every new x-value is the sum of the two x-values from the previous stage that connect to it, with one of these multiplied by a power of W. The diagram tells which x_{ST} terms are combined to give an $x_{\mathrm{ST}+1}$ term, and the numbers shown within the lines are the powers of W that are used. For example, looking at Fig. 5.8 we see that

$$x_2(6) = x_1(2) + W^8 x_1(6),$$
$$x_3(13) = x_2(13) + W^6 x_2(15),$$
$$x_4(9) = x_3(8) + W^9 x_3(9), \qquad \text{and so on.}$$

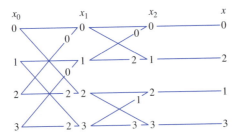

Figure 5.7

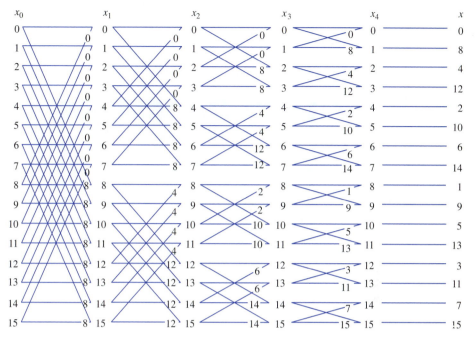

Figure 5.8

The last columns in Figs. 5.7 and 5.8 indicate how the final x-values are unscrambled to give the X-values. This relationship can be found by expressing the index k of x in the last stage as a binary number and reversing the bits; this gives n in $X(n)$. For example, in Fig. 5.8 we see that $x_4(3) = X(12)$ and $x_4(11) = X(13)$. From the bit-reversing rule, we get

$$3 = 0011_2 \rightarrow 1100_2 = 12, \qquad 11 = 1011_2 \rightarrow 1101_2 = 13.$$

Observe also that the bit-reversing rule can give the powers of W that are involved in computing the next stage. For the last stage, the powers are identical to the numbers obtained by bit reversal. At each previous stage, however, only the first half of the powers are employed, but each power is used twice as often. It is of interest to see how we can generate these values. Computer languages that facilitate bit manipulations make this an easy job, but there is a good alternative. Observe how the powers in Fig. 5.7 differ from those in Fig. 5.8 and how they progress from stage to stage. The following table pinpoints this.

Stage	N = 4				N = 16														
1:	0	0	2	2	0	0	0	0	0	0	0	0	8	8	8	8	8	8	8	8
2:	0	2	1	3	0	0	0	0	8	8	8	8	4	4	4	4	12	12	12	12
3:					0	0	8	8	4	4	12	12	2	2	10	10	6	6	14	14
4:					0	8	4	12	2	10	6	14	1	9	5	13	3	11	7	15

Can you see what a similar table for $N = 2$ would look like? Its single row would be 0   1. Now we see that the row of powers for the last stage can be divided into two halves with the numbers in the second half always 1 greater than the corresponding entry in the first half. The row above is the left half of the current row with each value repeated. This observation leads to the following algorithm.

Algorithm to Generate Powers of *W* in FFT

For N a power of 2, let $Q = \log_2(N)$.

Initialize an array P of length N to all zeros.
SET st = 1.
REPEAT
    Double the values of $P(K)$ for $K = 1 .. 2^{st-1}$,
    Let each $P(K + 2^{st-1}) = P(K) + 1$ for $K = 0 .. 2^{st-1} - 1$,
    Increment st,
UNTIL st > Q.

The successive new values for powers of W are now in array P.

EXAMPLE 5.23   Use the algorithm to generate the powers of W for $N = 8$:

$$Q = \log_2(8) = 3.$$

K	0	1	2	3	4	5	6	7
Initial P array:	0	0	0	0	0	0	0	0
ST = 1, doubled:	0	0	0	0	0	0	0	0
add 1:	0	1	0	0	0	0	0	0
ST = 2, doubled:	0	2	0	0	0	0	0	0
add 1:	0	2	1	3	0	0	0	0
ST = 3, doubled:	0	4	2	6	0	0	0	0
add 1:	0	4	2	6	1	5	3	7

The last row of values corresponds to the bits of the binary numbers 000 to 111, after reversal. ▲

Our discussion has assumed that N is a power of 2; for this case the economy of the FFT is a maximum. When N is not a power of 2 but can be factored, there are adaptations of the general idea that reduce the number of operations, but they are more than $N \log_2(N)$. See Brigham (1974) for a discussion of this as well as a fuller treatment of the theory behind FFT.

More recently there has been interest in another transform, called the discrete Hartley transform. A discussion of this transform would parallel our discussion of the Fourier transform. Moreover, it has been shown that this transform can be converted into a fast Hartley transform (FHT) that reduces to $N \log_2(N)$ computations. For a full coverage of the FHT, one should consult Bracewell (1986). The advantages of the FHT over the Fourier transform are its faster and easier computation. Moreover, it is easy to compute the FFT from the Hartley transform. However, the main power of the FHT is that all the computations are done in real arithmetic, so that we can use a language like Pascal that does not have a complex data type. An interesting and easy introduction into the FHT is found in O'Neill (1988).

An Algorithm to Perform a Fast Fourier Transform

Given n data points, (x_i, f_i), $i = 0, \ldots, n - 1$ (with n a power of 2) and x on $[0, 2\pi]$:

SET $yr_i = f_i$, $i = 0, \ldots, n - 1$.

SET $yi_i = 0$, $i = 0, \ldots, n - 1$.

SET $c_i = \cos(2i\pi/n)$, $i = 0, \ldots, n - 1$.   (These are trigonometric

SET $s_i = \sin(2i\pi/n)$, $i = 0, \ldots, n - 1$.   values that are used.)

SET nst $= \log(n)/\log(2)$. (nst = number of stages)

SET $p_i = 0$, $i = 0, \ldots, n - 1$.   (Use the previous algorithm

DO FOR st $= 1$ TO nst: to get

    SET $p_i = 2p_i$, $i = 1, \ldots, 2^{st-1}$.   "bit reversal"

    SET $p_{i+2} = p_i + 1$, $i = 0, \ldots, 2^{st-1}$.   values)

ENDDO (FOR st).

SET stage $= 1$.   (These values

SET nsets $= 1$.   are for the

SET del $= n/2$.   first stage—

SET $k = 0$. k indexes y-value being computed.)

REPEAT

    DO FOR set $= 1$ TO nsets:

        DO FOR $i = 0$ TO $n/\text{nsets} - 1$:

            SET $j = i$ MOD del $+$ (set $- 1$) $*$ del $* 2$.   (Indexes old y-values)

            SET $\ell = p_{\text{INT}(k/\text{del})}$.   (Indexes c_i, s_i values)

            SET $yyr_k = yr_j + c_\ell * yr_{j+\text{del}} - s_\ell * yi_{j+\text{del}}$.

            SET $yyi_k = yi_j + c_\ell * yi_{j+\text{del}} - s_\ell * yr_{j+\text{del}}$.

            SET $k = k + 1$.

        ENDDO (FOR i).

    ENDDO (FOR set).

    SET $yr_i = yyr_i$, $i = 0 .. n - 1$.   (Reset

    SET $yi_i = yyi_i$, $i = 0 .. n - 1$.   values

    SET stage $=$ stage $+ 1$.   for

    SET nsets $=$ nsets $* 2$.   next

    SET del $=$ del$/2$.   stage.)

    SET $k = 0$.

UNTIL stage $>$ nst.

When terminated, the A's and B's of the Fourier series are contained in the yr and yi arrays. These must be divided by $n/2$ and should be unscrambled using the p-array values as indices.

Note: The algorithm can be used when the f_i are complex numbers; set the imaginary parts into yi_i in this case.)

EXAMPLE 5.24   Use the FFT algorithm to obtain the finite Fourier series coefficients for the same data as in Table 5.22 (These are perturbed values from

$$f(t) = 2.3 + 1.5 \cos(t) - 0.5 \sin(t) + 4 \sin(t)).$$

A computer program that implements the algorithm gave these results:

n	A_n	B_n
0	**4.6017**	
1	**1.4993**	**− 0.4994**
2	0.0017	−0.0010
3	0.0003	−0.0005
4	0.0015	**3.9990**
5	0.0019	0.0009
6	−0.0004	−0.0009
7	−0.0003	−0.0019
8	0.0017	−0.0008
9	−0.0023	0.0019
10	−0.0024	−0.0011
11	0.0003	0.0020
12	0.0008	−0.0033
13	−0.0004	0.0011
14	0.0025	0.0003
15	−0.0005	0.0013
16	−0.0010	

The results are essentially the same as those of Example 5.21, which were computed by the trapezoidal rule.

Observe that we compute exactly as many A's and B's as there are data points. This is not only reasonable (we cannot "manufacture" information) but is in accord with *information theory*. (Some mention of this theory is made in the next section.)   ▲

5.15 Theoretical Matters

This section assembles several theoretical topics that are important for numerical differentiation and integration.

Error Terms for Derivatives

In Section 5.2 the error terms for derivative formulas were computed by differentiating the error term for the corresponding interpolation polynomial. It is of interest to get expressions for the errors in a different way.

Consider this Taylor series:

$$f(x_i + h) = f(x_i) + f'(x_i)(h) + f''(\xi)\frac{h^2}{2!}, \qquad \xi \text{ in } [x_i, x_i + h],$$

where the last term is the error of the series. Solve for $f'(x_i)$:

$$f'(x_i) = \frac{f(x_i + h) - f(x_i)}{h} + \text{error},$$

$$\text{error} = -f''(\xi)\frac{h}{2} = O(h), \qquad \xi \text{ in } [x_i, x_i + h],$$

which is the forward-difference formula for the first derivative, which we have written with an alternative notation:

$$f'_i = \frac{f_{i+1} - f_i}{h} - f''(\xi)\frac{h}{2}.$$

Similarly, from the two Taylor series for $f(x_i + h)$ and $f(x_i - h)$, we can solve for the central-difference formula:

$$f(x_i + h) = f(x_i) + f'(x_i)(h) + f''(x_i)\frac{h^2}{2} + f'''(\xi_1)\frac{h^3}{3!},$$

$$f(x_i - h) = f(x_i) - f'(x_i)(h) + f''(x_i)\frac{h^2}{2} - f'''(\xi_2)\frac{h^3}{3!}, \qquad \text{(5.59)}$$

$$f'_i = \frac{f_{i+1} - f_{i-1}}{2h} + \text{error},$$

$$\text{error} = -f'''(\xi)\frac{h^2}{6} = O(h^2), \qquad \xi \text{ in } [x_{i-1}, x_{i+1}].$$

In this we have used the intermediate value theorem to rewrite the error in terms of ξ. This formula is sometimes called a *three-point formula* because three points, at x_i, x_{i+1}, and x_{i-1}, are used (even though the central one does not appear).

Proceeding in a similar fashion, we can obtain additional formulas for derivatives from

$$f(x_i \pm nh) = f_i \pm f_i'(nh) + f_i'' \frac{(nh)^2}{2!} \pm f_i''' \frac{(nh)^3}{3!} + \cdots, \qquad n = 1, 2, 3, \ldots.$$

Three-point forward-difference formula for f':

$$f_i' = \frac{-f_{i+2} + 4f_{i+1} - 3f_i}{2h} + \text{error},$$

$$\text{error} = f'''(\xi)\frac{h^2}{3} = O(h^2).$$

Five-point forward-difference formula for f':

$$f_i' = \frac{-3f_{i+4} + 16f_{i+3} - 36f_{i+2} + 48f_{i+1} - 25f_i}{12h} + \text{error},$$

$$\text{error} = f^{(5)}(\xi)\frac{h^4}{5} = O(h^4).$$

Five-point central difference formula for f':

$$f_i' = \frac{-f_{i+2} + 8f_{i+1} - 8f_{i-1} + f_{i-2}}{12h} + \text{error},$$

$$\text{error} = f^{(5)}(\xi)\frac{h^4}{30} = O(h^4).$$

Three-point forward-difference formula for f'':

$$f_i'' = \frac{f_{i+2} - 2f_{i+1} + f_i}{h^2} = \text{error},$$

$$\text{error} = -f'''(\xi)(h) = O(h).$$

Three-point central-difference formula for f'':

$$f_i'' = \frac{f_{i+1} - 2f_i + f_{i-1}}{h^2} + \text{error},$$

$$\text{error} = -f^{(4)}(\xi)\frac{h^2}{12} = O(h^2).$$

Five-point forward-difference formula for f'':

$$f_i'' = \frac{11f_{i+4} - 56f_{i+3} + 114f_{i+2} - 104f_{i+1} + 35f_i}{12h^2} + \text{error},$$

$$\text{error} = -f^{(5)}(\xi)\frac{5h^3}{6} = O(h^3).$$

Five-point central-difference formula for f'':

$$f''_i = \frac{-f_{i+2} + 16f_{i+1} - 30f_i + 16f_{i-1} - f_{i-2}}{12h^2} + \text{error},$$

$$\text{error} = f^{(6)}(\xi)\frac{h^4}{90} = O(h^4).$$

Error Terms in Richardson Extrapolations

In Section 5.4 improved estimates of $f'(x)$ were obtained from central-difference estimates, with the value of h differing by a factor of two. The rationale was based on the assumption that the errors were proportional to h^2 (when the error was just $O(h^2)$). We can establish the validity of the extrapolation by beginning with more terms of the Taylor series that gave Eq. (5.59):

$$f'_i = \frac{f_{i+1} - f_{i-1}}{2h} + a_1 h^2 + a_2 h^4 + a_3 h^6 + \cdots, \qquad (5.60)$$

where the added terms represent the error. If we let $F(h)$ be the first term on the right and rearrange, we have

$$F(h) = f'_i - a_1 h^2 - a_2 h^4 - a_3 h^6 - \cdots. \qquad (5.61)$$

Now, if h is halved, we have

$$F\left(\frac{h}{2}\right) = f'_i - a_1\left(\frac{h}{2}\right)^2 - a_2\left(\frac{h}{2}\right)^4 - a_3\left(\frac{h}{2}\right)^6 - \cdots. \qquad (5.62)$$

The Richardson extrapolation adds $\frac{1}{3}$ of the difference between Eq. (5.62) and Eq. (5.61) to Eq. (5.62) to give

$$G(h) = f'_i + \frac{1}{4}a_2 h^4 + \frac{5}{16}a_3 h^6 + \cdots,$$

and we have improved our estimate to one with an error of $O(h^4)$.

Obviously, we can repeat this step, but now we need a multiplier of 1/15 to eliminate the first added term. Then we can extrapolate again, using multipliers of $1/(2^n - 1)$, where n is the exponent on h in the term being eliminated.

This scheme is commonly used for any computation with an error of the form of Eq. (5.60). In particular, it is valid for trapezoidal rule integrals—Romberg integration.

Error Terms for Newton–Cotes Formulas

The Newton–Cotes integration formulas can be obtained by integrating a Newton–Gregory forward-interpolating polynomial over one, two, or three equal subdivisions. The corresponding error term comes from integrating the error term of the polynomial. In Section 5.5

the error term for the trapezoidal rule was shown to be

$$\text{Error of trapezoidal rule} = \frac{-h^3 f''(\xi)}{12}.$$

It is instructive to carry more terms in the error expression. We begin with $P(x)$:

$$P(x) = f_i + s\Delta f_i + s(s-1)\frac{\Delta^2 f_i}{2!} + s(s-1)(s-2)\frac{\Delta^3 f_i}{3!} + \cdots, \qquad \textbf{(5.63)}$$

where $s = (x - x_i)/h$.

The integral of the first two terms from x_i to $x_i + h$ gives the trapezoidal rule, as we have seen. An expression for the error is obtained by integrating the remaining terms:

$$\text{Error} = -h^3 \frac{\Delta^2 f_i}{12} + h^4 \frac{\Delta^3 f_i}{24} - 19h^5 \frac{\Delta^4 f_i}{720} + 3h^6 \frac{\Delta^5 f_i}{160} - \cdots.$$

(A symbolic algebra program was very helpful in evaluating these integrals.) We observe that using only the first term gives the normal expression for the error:

$$\text{Error} = -\frac{h^3 f''(\xi)}{12} = O(h^3).$$

Let us do the same for Simpson's $\frac{1}{3}$ rule. The rule is given by the integral of the first three terms of Eq. (5.63) between the limits of x_i to $x_i + 2h$ and the error expression by integrating the rest of the polynomial between the same limits. Carrying out the integrations gives

$$\text{Error} = 0 - h^5 \frac{\Delta^4 f_i}{90} + h^6 \frac{\Delta^5 f_i}{90} - 37h^7 \frac{\Delta^6 f_i}{3780} + \cdots.$$

Because the first term of this error expression is zero, Simpson's $\frac{1}{3}$ rule has an unexpectedly small error. If we replace $\Delta^4 f_i$ with its equivalent in terms of derivatives, we get the normal expression for the error:

$$\text{Error} = -\frac{h^5 f^{(4)}(\xi)}{90} = O(h^5).$$

Simpson's $\frac{3}{8}$ rule does not have a zero for the first term in the error expression, so its error term is the same, $O(h^5)$.

Error Terms for the Composite Rules

When we apply the Newton–Cotes formula to a succession of intervals, the error is the sum of the errors of the individual error terms. We refer to the error of the composite rule as a *global error,* and to the error of the single application as the *local error.* For example, suppose we use the trapezoidal rule to integrate from $x = a$ to $x = b$, subdividing the interval of integration into n equal subintervals of size h. We then have

$$\text{Global error} = \frac{-h^3 [f''(\xi_1) + f''(\xi_2) + \cdots + f''(\xi_n)]}{12}.$$

If $f''(x)$ is continuous on $[a, b]$, there is a value, ξ, for which the sum of the n terms can be written as

$$\text{Global error} = \frac{-nh^3f''(\xi)}{12} = \frac{-(b-a)h^2f''(\xi)}{12} = O(h^2)$$

because $(b-a) = nh$.

The reduction in order of error occurs for any of the composite Newton–Cotes rules. Both of Simpson's rules then have global errors of $O(h^4)$.

The Next-Term Rule

We have made much of the fact that the error term for differentiation, integration, and even interpolation can be easily written using the *next-term rule*. By this we mean that the term after the last one used in the formula can be converted into the error term by either of these replacements:

$$\Delta^n f_i = h^n f^{(n)}(x_i) \qquad \text{or} \qquad f_i^{[n]} = \frac{f^{(n)}(x_i)}{n!}.$$

The basis for this is the development of Section 5.3.

Round-Off and Accuracy of Derivatives

It is important to realize the effect of round-off error on the accuracy of derivatives computed by finite-difference formulas. As we have seen, decreasing the value of h or increasing the degree of the interpolating polynomial increases the accuracy of our derivative formulas. (These procedures reduce the truncation error.)

As h is reduced, however, we are required to subtract function values that are almost equal, and this, we have seen, incurs a large error due to round-off. The increase in round-off error as h gets smaller causes the best accuracy to occur at some intermediate point. The data in Table 5.23 demonstrate this phenomenon. The test was made with both single-precision arithmetic (giving 6–7 significant figure accuracy) and double-precision (14–15 figures). The function was $f(x) = e^x$, at $x = 0$. The true value of both $f'(x)$ and $f''(x)$ is 1.0, of course. Central-difference approximations were employed.

In the single-precision test, the optimum value of h was only 0.01 for the first derivative, and the best accuracy of the second derivative was at $h = 0.1$, a surprisingly large value. When the round-off errors were reduced by using double-precision, essentially exact results were obtained for the first derivative throughout the range of h values. For the second derivative, $h = 0.1 \times 10^{-2}$ or $h = 0.1 \times 10^{-3}$ gives the best accuracy. Note how large the error grows; it is especially marked for second derivatives when h becomes much smaller than the optimum. Higher derivatives than the second will be expected to show an even more exaggerated influence of round-off.

Computer models vary widely in the number of digits of accuracy they provide, due to the different number of bits provided for the storage of floating-point values and also, to a

Table 5.23   Derivatives computed by central-difference formulas $f(x) = e^x$, at $x = 0$

h	$f'(x)$	$f''(x)$
Results with single precision:		
0.1E 00	0.1001663E 01	0.1000761E 01
0.1E−01	0.1000001E 01	0.9959934E 00
0.1E−02	0.9999569E 00	0.8940693E 00
0.1E−03	0.9959930E 00	−0.8344640E 02
0.1E−04	0.9775158E 00	−0.4768363E 04
0.1E−05	0.9834763E 00	−0.5960462E 05
0.1E−06	0.5960463E 00	−0.1192093E 08
0.1E−07	0.2980230E 01	−0.5960458E 09
0.1E−08	0.2980229E 02	−0.5960459E 11
0.1E−09	0.2980229E 03	−0.5960456E 13
Results with double precision:		
0.1E 00	0.1001667E 01	0.1000834E 01
0.1E−01	0.1000016E 01	0.1000008E 01
0.1E−02	0.1000000E 01	0.1000000E 01
0.1E−03	0.1000000E 01	0.9999999E 00
0.1E−04	0.1000000E 01	0.9999989E 00
0.1E−05	0.1000000E 01	0.1000031E 01
0.1E−06	0.9999999E 01	0.9894834E 00
0.1E−07	0.9999999E 00	−0.4163322E 00
0.1E−08	0.9999999E 00	−0.4163319E 02
0.1E−09	0.1000000E 01	0.2775545E 04

lesser degree, due to the format of their floating-point representations. Another factor that affects the accuracy of functional values is the algorithm used in their evaluation. Chapter 4 gave some insight into this. Different software packages that are used with computer operating systems employ different algorithms, which are of varying accuracy. Further, as Chapter 4 showed, the error of approximating the function differs with the value of the argument.

No analytical way of predicting the effects of round-off error seems to exist. The best we can do is to warn budding numerical analysts so they will be properly wary when computing derivatives numerically. In the words of many authorities, *the process of differentiating is basically an unstable one*—meaning that small errors made during the process cause greatly magnified errors in the final result.

Differentiation of "noisy" data encounters a similar problem. If the data being differentiated are from experimental tests, or are observations subject to errors of measurement, the errors so influence the derivative values calculated by numerical procedures that they may be meaningless. The usual recommendation is to smooth the data first, using methods that are discussed in Chapter 3. Passing a cubic spline through the points and then getting the derivative of this approximation to the data has become quite popular. A least-squares curve may also be used. The strategy involved is straightforward—we don't try to represent the function by one that fits exactly to the data points, because this fits to the errors as well as to the trend of the information. Rather, we approximate with a smoother curve that we hope is closer to the truth than the data themselves. The problem, of course, is how much smoothing should be done. One can go too far and "smooth" beyond the point where only errors are eliminated.

A final situation should be mentioned. Some functions, or data from a series of tests, are inherently "rough." By this we mean that the function values change rapidly; a graph would show sharp local variations. When the derivative values of the function incur rapid changes, a sampling of the information may not reflect them. In this instance, the data indicate a smoother function than actually exists. Unless enough data are at hand to show the local variations, valid values of the derivatives just cannot be obtained. The only solution is more data, especially near the "rough" spots. And then we are beset by problems of accuracy of the data!

Fortunately this problem does not occur with numerical integration. As you have seen, all the integration formulas add function values together. Because the errors can be positive or negative and the probability for each is the same, errors tend to cancel out. That means that integration is a smoothing process. We assess integration as *inherently stable.* This is generally true of computations that are *global,* in contrast to those that are *local* in nature such as differentiation.

Errors in Multiple Integration and Extrapolations

The error term of a one-variable quadrature formula is an additive one just like the other terms in the linear combination (although of special form). It would seem reasonable that it would go through the multiple summations in a similar fashion, so we should expect error terms for multiple integration that are analogous to the one-dimensional case. We illustrate that this is true for double integration using the trapezoidal rule in both directions, with uniform spacings, choosing n intervals in the x-direction and m in the y-direction.

From Section 5.6 we have

$$\text{Error of } \int_a^b f(x)\ dx = -\frac{b-a}{12}h^2 f''(\xi) = O(h^2),$$

$$h = \Delta x = \frac{b-a}{n}.$$

In developing Romberg integration, we found that the error term could be written as

$$\text{Error} = O(h^2) = Ah^2 + O(h^4) \doteq Ah^2 + Bh^4,$$

where A is a constant and the value of B depends on a fourth derivative of the function. Appending this error term to the trapezoidal rule, we get (equivalent to Eq. (5.61)):

$$\int_a^b f(x,\ y)\ dx \ \Bigg|_{y=y_j} = \frac{h}{2}(f_{0,j} + 2f_{1,j} + 2f_{2,j} + \cdots + f_{n,j}] + A_j h^2 + B_j h^4.$$

Summing these in the y-direction and retaining only the error terms, we have

$$\int_c^d \int_a^b f(x,\ y)\ dx\ dy = \frac{k}{2}\frac{h}{2} \sum_{i=0}^n \sum_{j=0}^m a_i a_j f_{i,j} + \frac{k}{2}(A_0 + 2A_1 + 2A_2 + \cdots + A_m)h^2$$

$$+ \frac{k}{2}(B_0 + 2B_1 + 2B_2 + \cdots + B_m)h^4 + \bar{A}k^2 + \bar{B}k^4, \qquad \text{(5.64)}$$

$$k = \Delta y = \frac{d - c}{m}.$$

In Eq. (5.64), $\bar{A}$ and $\bar{B}$ are the coefficients of the error term for y. The coefficients A and B for the error terms in the x-direction may be different for each of the $(m + 1)$ y-values, but each of the sums in parentheses in Eq. (5.64) is $2n$ times some average value of A or B, so the error terms become

$$\text{Error} = \frac{k}{2}(nA_{av})h^2 + \frac{k}{2}(nB_{av})h^4 + \bar{A}k^2 + \bar{B}k^4. \qquad \text{(5.65)}$$

Because both Δx and Δy are constant, we may take $\Delta y = k = \alpha \Delta x = \alpha h$, where $\alpha = \Delta y/\Delta x$, and Eq. (5.65) can be written, with $nh = (b - a)$,

$$\text{Error} = \left(\frac{b - a}{2}A_{av}\alpha\right) h^2 + \left(\frac{b - a}{2}B_{av}\alpha\right) h^4 + A\alpha^2 h^2 + B\alpha^4 h^4$$

$$= K_1 h^2 + K_2 h^4.$$

Here K_2 will depend on fourth-order partial derivatives. This confirms our expectation that the error term of double integration by numerical means is of the same form as for single integration.

Because this is true, a Romberg integration may be applied to multiple integration, whereby we extrapolate to an $O(h^4)$ estimate from two trapezoidal computations at a 2-to-1 interval ratio. From two such $O(h^4)$ computations we may extrapolate to one of $O(h^6)$ error.

Information Theory—The Sampling Theorem

In performing a discrete Fourier transform, we work with samples of some function of t, $f(t)$. We normally have data taken at evenly spaced intervals of time. If the interval between samples is D sec, its reciprocal, $1/D$, is called the *sampling rate* (the number of samples per second).

Corresponding to the sampling interval, D, is a critical frequency, called the *Nyquist critical frequency*, f_c, where

$$f_c = \frac{1}{2}D.$$

The reason this is a critical frequency is seen from the following argument. Suppose we sample a sine wave whose frequency is f_c and get a value corresponding to its positive peak amplitude. The next sample will be at the negative peak, the next beyond that at the positive peak, and so on—that is, critical sampling is at a rate of two samples per cycle. We can construct the magnitude of the sine wave from these two samples. If the frequency is less than f_c, we will have more than two samples per cycle and again we can construct the wave correctly. On the other hand, if the frequency is greater than f_c, we have fewer than two samples per cycle and we have inadequate information to determine $f(t)$.

The significance of this theorem is that if the phenomenon described by $f(t)$ has no frequencies greater than f_c, then $f(t)$ is completely determined from samples at the rate $1/D$. Unfortunately, this also means that if there are frequencies in $f(t)$ greater than f_c, all these frequencies are spuriously crowded into the range $[0, f_c]$, causing a distortion of the power spectrum. This distortion is called *aliasing*.

All of this is very clear if we think of the results of an FFT on the samples. If we have N samples of the phenomenon, we certainly cannot determine more than a total of exactly N of the Fourier coefficients, the A's and B's. The last of these will be $A_{n/2}$ (assuming an even number of samples). We see that this corresponds to the Nyquist frequency.

5.16 Using MATLAB

In this section we present examples that show how MATLAB handles differentiation and integration. There are three examples relating to derivatives and eight that involve integrations.

A. Get the derivative of $x * e^{-x/2}$ at $x = 3.0$ from a central difference. Use $h = 0.025$.   We can do this in MATLAB as follows:

```
x = [.3 − .025 .3 + .025];   (creates a two element array)
y = x .* exp(x ./(−2));   (creates another array, using 'dot' operations)
dydx = diff(y) ./diff(x)   giving   0.7317
```

B. For $f(x) = e^x/x$, find the derivative analytically and evaluate this at $x = 0.3$; also get the eighth derivative.   MATLAB can do this by:

```
f = 'exp(x)/x';   (it must be symbolic)
df = diff(f, 'x')   which gives:   exp(x)/x − exp(x)/x^2
numeric (subs (df, 3, 'x')   and we see:   4.4635
df8 = diff (f, 8)   which gives the correct answer, some terms are:
exp(x)/x − 8 exp(x)/x^2 + . . . − 40320 exp(x)/x^8 + 40320 exp(x)/x^9
```

C. Fit a fourth-degree polynomial to five evenly spaced points for $f(x) = \sin(\pi x)$ between 0 and 1, then approximate $P'(x)$ and $P''(x)$ for $x = 0.63$. Compare the results with the analytical values.   MATLAB can do the example with:

```
x = [0 .25 .5 .75 1]   (creates x-array), displayed as:
   0 0.2500 0.5000 0.7500 1.0000
```

```
y = sin(pi * x)    (creates y-array):
   0 0.7071 1.0000 0.7071 0.0000
p = polyfit (x, y, 4);    (gets the polynomial coefficients; suppress the display)
ps = poly2sym(p);    (converts to symbolic, required in order to differentiate)
vpa (ps, 5)    (to see it in normal form):
         3.6602*x^4 - 7.3204*x^3 + .57528*x^2 + 3.0849*x + 1.0112e-16
d1 = diff (ps);    (gets the derivative in symbolic form, not displayed)
vpa(d1, 5)    (now we display it):   14.641*x^3 - 21.961*x^2 + 1.1506*x + 3.0849
subs(ans, .63)    (evaluates at x = 0.63):   -1.245604773000000
```

We repeat to get the second derivative:

```
d2 = diff (d1);
vpa (subs(d2, .63))    (which results in):   -9.08785130000000
```

D. First get the indefinite integral of $e^{-x}(x^2 + 1)$, then get the definite integral between the limits of 0.3 and 2.7.   MATLAB solves this example in this way:

```
I = int('exp(-x)*(x^2 + 1)')    and we see:
        -exp(-x)*x^2 - 2*x*exp(-x) - 3*exp(-x)
```

which looks better if we do:

```
factor (I):
        -exp(-x)*(x^2 + 2*x + 3)
```

To get the definite integral, we just add two parameters:

```
Iab = int('exp(-x)*(x^2 + 1)', .3, 2.7)    giving:
        -1569/100*exp(-27/10) + 369/100*exp(3/10)
```

which we can evaluate with:

```
    vpa (Iab, 5)    to see:
    1.6791
```

E. The integral of e^{-x^2} cannot be expressed in terms of "ordinary functions," but it is well known in terms of the "error function" and values of the definite integral are tabulated. See how MATLAB reacts when asked to integrate this and to find its value between $x = 0.2$ and $x = 1.5$.   MATLAB recognizes that the integral gives the error function:

```
ff = 'exp(-x^2)';
iffx = int(ffx)    which gives
        1/2*pi^(1/2)*erf(x)
```

and numeric (int(ffx, 0.2, 1.5)) shows the answer:

```
    0.6588.
```

F. Numerically integrate $x * \sin(x)$ between $x = 0$ and $x = 1$. If possible, compare the trapezoidal rule results with those from Simpson's rule. In MATLAB, numerical integration can be accomplished with two commands: `'quad'` and `'quad8'`. The first of these uses an adaptive Simpson's rule. The second command uses an eight-panel Newton–Cotes technique. The result from `quad ('x.*sin', 0, 1)` is 0.3011669, which agrees with the analytical answer to five digits. The command `'trap(x, y)'`, where x and y are arrays of x- and y-values, does a trapezoidal integration. If the interval of integration is subdivided into 100 panels, this command gives the result as 0.3012, accurate to four or more digits.

G. The function $f(x) = e^x/\sin(x)$ has a severe singularity at $x = 0$. Investigate what happens when this is integrated between $x = 0$ and $x = 1$. MATLAB cannot handle the singularity. The output is just an echo of the input.

H. For $f(x) = (x^2 + \sin(x))/(x + 1)$, fit a fourth-degree polynomial to five evenly spaced points between $x = 2$ and $x = 6$, then integrate this; compare to the analytical answer. MATLAB can do this example with:

```
x = [2 3 4 5 6];      (creating the x-array)
f = (x.^2 + sin(x)) ./(x + 1);    (creating the f(x) array)
p = polyfit (x, f, 4);    (gets the coefficients of the polynomial as an array)
pp = poly2sym (p);    (converts to symbolic form)
numeric (int (pp, 2, 6))    gives
   12.6714
numeric (int ('(x^2 + sin(x))/(x + 1)', 'x', 2, 6))
   12.6705.    produces the analytical answer:
```

I. How can Gaussian quadrature be performed? MATLAB does not have Gaussian quadrature built in. A program could be written to get the Legendre polynomials and their zeros. These then could be used to carry out Gaussian quadrature. (*Mathematica* actually uses an improved Gaussian quadrature technique in the `'Nintegrate'` function. The Package NumericMath `'GaussianQuadrature'` has functions that list the weights and points used in integrating to a specified precision.)

J. Evaluate the integral of $y * z * e^x$ for $x = [-1 .. 1]$, $y = [-1 .. 0]$, and $z = [0 .. 1]$. We can nest the integrations in MATLAB:

```
int (int (int('y*z*exp(x)', 'x', -1, 1), 'y', -1, 0), 'z', 0, 1)
   -1/4 exp(1) + 1/4 exp(-1)    and see: which the 'numeric' command translates to
      -0.5876
```

K. Integrate $f(x, y) = x * y$ over the region bounded by $x = [0 .. 1]$ and the curve $y = x^2 + 1$. MATLAB gives the correct answer with:

```
int(int('x * y', 'y', 0, 'x^2 + 1'), 'x', 0, 1)
```

producing:

    7/12

and `numeric(ans)` gives:

    0.5833

5.17   Parallel Processing

Our discussion in this section will concern only those cases when parallel processing can speed the computations of numerical differentiation and integration. As pointed out in Chapter 3, parallel processing can obtain the columns of difference tables individually and speed up the creation of the table thereby.

In formulas like Eq. (5.9), the product terms and their sum can be speeded up by parallel processing, but as we normally use only a few factors and terms, the overhead makes parallel processing a minor advantage. This is also true for Eqs. (5.12) and (5.18). Each column of a Richardson table (Section 5.4) can be computed in parallel, but usually we do not do them in this way. Rather we alternate between column and row computations to determine when to stop the extrapolation.

The Newton–Cotes formulas of Section 5.5 also have too few terms to be good candidates for parallel processing.

When we integrate using the composite trapezoidal and Simpson's rules, parallel processing can be advantageous. Each of the function evaluations can be done at the same time (and there may be parallelization possibilities within these), and the summations can be speeded up as well. This opportunity applies to Romberg integration.

Gaussian quadrature is similar but, as we generally use only a few terms, the speedup is marginal.

Because adaptive integration employs Simpson's rule, the preceding remarks apply to this method. The same is true for multiple integration because this technique is also just successive applications of a trapezoidal or Simpson rule.

Getting integrals and derivatives from cubic splines cannot profit much from parallel processing because there are not enough terms to justify its use. Of course, parallel processors can be effective in computing the elements of the system matrix and in solving the system (as explained in Chapter 2) to give the coefficients of the individual cubics.

In summary, except for doing integrations with many subintervals, the advantages of parallel processing are not significant for the operations of Chapter 5.

Chapter Summary

If you are not able to do the following, we advise you to restudy portions of this chapter. You should now be able to

1.   Develop formulas for differentiation and integration from an interpolating polynomial.

2. Get an expression for error of the formula you have developed.

3. Derive formulas by the symbolic method and from a Taylor series.

4. Apply these formulas to obtain derivatives and integrals and estimate the error in the result.

5. Use extrapolation techniques to get improved estimates of derivatives and integrals. You should be able to explain why this extrapolation works.

6. Evaluate an integral using Gaussian quadrature. You know why this method obtains the integral with fewer function evaluations than with the other integration techniques. You can develop a Legendre polynomial from the recursion formula and tell what is meant by an orthogonal polynomial.

7. Explain why adaptive integration is a more efficient way to obtain the integrals. You can use this technique.

8. Apply the various integration methods to evaluate improper integrals and multiple integrals.

9. Use a cubic spline to obtain derivatives and integrals. You know the limitations of the cubic spline for getting higher derivatives.

10. Discuss the principles behind the error terms for the various formulas of the chapter.

11. Explain why numerical differentiation is called an unstable process and why numerical integration is not an unstable process.

12. Use MATLAB to do integration and differentiation.

13. Employ computer programs (perhaps even write your own) to carry out differentiation and integration.

Computer Programs

There are two programs in this section of Chapter 5. A C program that computes the derivatives from a table of function values is the first one. The second is a Fortran 90 program that does adaptive integration.

Program 5.1

Figure 5.9 is the listing of the C program; its output follows the listing. The program reads in the table of evenly spaced data pairs from a file and computes a difference table. Then the program requests an x-value from the user and computes both first and second derivatives from the table, using central differences where possible. (The x-value must be one that is in the table.) Another x-value is then asked for and the derivatives computed, or, if the user prefers, the program terminates. The output shown at the end of the listing is for a table of values for $f(x) = e^x$; the exact derivative values are equal to the function values.

```
/* **********************************************************
 *                                                          *
 *   Chapter 5        Derivatives from a Table              *
 *                                                          *
 *   Gerald/Wheatley, APPLIED NUMERICAL ANALYSIS (sixth edition) *
 *                Addison Wesley Longman, 1999              *
 *                                                          *
 * **********************************************************
 *                                                          *
 *   This program computes the first and second derivatives *
 *   from a table of evenly spaced values. These data are read *
 *   from a file. Once the data have been obtained, a difference *
 *   table is computed and displayed.                       *
 *                                                          *
 *   A value for which the derivatives are to be estimated is *
 *   obtained from the user (this must match one of the tabulated *
 *   x values) and the derivatives are computed. Fourth degree *
 *   polynomials are used whenever possible. The operation is *
 *   repeated until terminated by the user.                 *
 *                                                          *
 * ********************************************************** /

#include <stdio.h>

void main()
  {
    float x[10], y[10], tbl[10][10], xval, h, d1, d2;
    int i, n, j, index;
    char inbuf[20];
    FILE *file;

/* Get the data */
    clrscr();
    file = fopen("b:\xydata.txt", "r");
    fscanf (file, "%d", &n);
    for ( i = 1; i <= n; i++)
      fscanf(file, "%f %f", &x[i], &y[i]);
    fclose(file);

/* Compute the table */
    for ( i = 1; i <= n; i++)
      tbl[i][0] = y[i];
    for (j = 1; j <= 4; j++)
      for (i=1; 1<=n-j; i++)
        tbl[i][j] = tbl[i+1][j-1] - tbl[i][j-1];

/* Display the table /*
    for (i = 1; i <= n; i++)
    {
      printf("%10.3f", x[i]);
      if (i <= n-4)
        for (j = 0; j <= 4; j++)
          printf("%10.3f", tbl[i][j]);
```

Figure 5.9

```
              else
                for (j = 0; j <= n-i; j++)
                  printf("%10.3f", tbl[i][j]);
              printf("\n");
          }                                  /* end of for */

  /* Repeat the operation until user presses ^C */
    do
    {

  /* Get x-value from user */
      do
      {
        printf("\n\n Enter your x-value  ");
        printf("\n (To quit, press ^C) : ");
        gets(inbuf);
        sscanf(infut, "%f", &xval);

  /* Make sure x-value is in the table */
        index = 0;
        for (i = 1; i <= n; i++)
          if (xval == x[i])
          index = i;
          if (index == 0)
            printf ("\n Your value not in the table, try again");
        }
        while (index == 0);               /* end of inner do/while loop */

  /* Find the derivatives, use a 4th degree polynomial, centered if
  possible */
        h = x[2] - x[1];
        if (index == 1)
          {
            d1 = (tbl[1][1] - tbl[1][2]/2 + tbl[1][3]/3 - tbl[1][4]/4)/h;
            d2 = (tbl[1][2] - tbl[1][3] + 11*tbl[1][4]/12)/h/h;
            printf("At x = %7.3f, f'(x) = %7.3f", x[index], d1);
            printf(" and f''(x) is %7.3f.", d2);
          }                      /* end of first if */
        if (index == 2)
          {
            d1 = (tbl[1][1] + tbl[2][1])/2 - tbl[1][3]/6 + tbl[1][4]/
            12)/h;
            d2 = (tbl[1][2] - tbl[1][4]/12)/h/h;
            printf("At x = %7.3f, f'(x) = %7.3f", x[index], d1);
            printf(" and f''(x) is %7.3f.", d2);
          }                        /* end of second if */
        if (index == n-1)
{
          {
            d1 = ((tbl[n-1][1] + tbl[n-2][1])/2 - tbl[n-3][3]/6
                  - tbl[n-4][4]/12)/h;
            d2 = (tbl[n-2][2] - tbl[n-4][4]/12)/h/h;
            printf("At x = %7.3f, f'(x) = %7.3f", x[index], d1);
            printf(" and f''(x) is %7.3f.", d2);
          }                          /* end of third if */
```

Figure 5.9 *Continued*

```
      if (index == n)
        {
          d1 = (tbl[n-1][1] + tbl[n-2][2]/2 + tbl[n-3][3]/3
          + tbl[n-4][4]/4)/h;
          d2 = (tbl[n-2][2] + tbl[n-3][3] + 11*tbl[n-4][4]/12)/h/h;
          printf("At x = %7.3f, f'(x) = %7.3f", x[index], d1);
          printf(" and f''(x) is %7.3f.", d2);
        }                         /* end of fourth if */
      if (index > 2 && index < (n-1))
        {
          d1 = ((tbl[index][1]+tbl[index-1][1])/2 -
                (tbl[index-1][3]+tbl[index-1][3])/12)/h;
          d2 = (tbl[index-1][2] - tbl[index-2][4]/12)/h/h;
          printf("At x = %7.3f, f'(x) = %7.3f", x[index], d1);
          printf(" and f''(x) is %7.3f.", d2);
        }                         /* end of last if */
      }
    while (1 == 1);                 /* end of do/while (forever!) */

  }                                 /* end of main */

        ****************** Output ******************

          1.300    3.669    0.813    0.179    0.041    0.007
          1.500    4.482    0.992    0.220    0.048    0.012
          1.700    5.474    1.212    0.268    0.060    0.012
          1.900    6.686    1.480    0.328    0.072
          2.100    8.166    1.808    0.400
          2.300    9.974    2.208
          2.500   12.182

 Enter your x-value
 (To quit, press ^C) : 1.5
At x =   1.500, f'(x) =    4.481 and f''(x) is    4.460.

 Enter your x-value
 (To quit, press ^C) : 2.5
At x =   2.500, f'(x) =   12.175 and f''(x) is   12.075.

 Enter your x-value
 (To quit, press ^C) : 1.8

 Your value not in the table, try again

 Enter your x-value
 (To quit, press ^C) : 1.7
At x =   1.700, f'(x) =    5.470 and f''(x) is    5.485.

 Enter your x-value
 (To quit, press ^C) : ^C
```

Figure 5.9 *Continued*

Program 5.2

The second program is listed in Figure 5.10. It is a Fortran program that does adaptive integration. In this implementation, the function $1/e^{x*x}$ is integrated between $x = 0.2$ and $x = 1.5$.

```
Program R_Adaptive
!
!!!!!!!!!!!!!!!!!!!!!!!!!!!!!!!!!!!!!!!!!!!!!!!!!!!!!!!!!!!!!!!!
!                       CHAPTER 5                             !
!                                                             !
!                                                             !
! Gerald/Wheatley, APPLIED NUMERICAL ANALYSIS (Sixth Edition) !
!                 Addison Wesley Longman, 1999                !
!                                                             !
!!!!!!!!!!!!!!!!!!!!!!!!!!!!!!!!!!!!!!!!!!!!!!!!!!!!!!!!!!!!!!!!
!                                                             !
!    This program implements the adaptive Simpson Integration !
!    presented in Chapter 5. This program solves the problem  !
!    using recursion that is available in Fortran 90.         !
!                                                             !
!!!!!!!!!!!!!!!!!!!!!!!!!!!!!!!!!!!!!!!!!!!!!!!!!!!!!!!!!!!!!!!!
!                                                             !
        Implicit None
        Real(Kind = 8) :: a_left = 0.2, b_right = 1.5, &
                          initial_tolerance = 0.01
        Integer :: function_count = 0, &
                   count = 0   ! actual number of Function calls

        Write(*, '(//)')
        Write(*,100) adapt(a_left, b_right,initial_tolerance)
        Write(*, '(//)')
        Write(*,200) function_count
100     Format(' The integral value is: ',F15.5)
200     Format(' The number of function calls is : ', I5)
        Stop

        Contains
!
!       This Function uses the adaptive Simpson method as
!       presented in Section 5.10. The Variables used here are
!       defined the same as given in the algorithm.
!
! --------------------------------------------------------------
!
        Recursive Function adapt(a, b, tol ) Result(answer)
        Implicit None
```

Figure 5.10

```
              Real(Kind = 8), Intent(In) :: a,b,tol
              Real(kind = 8)  :: h1, h2,b1,  &
                               c, d, e, &
                               fa, fb, fc, fd, fe, &
                               S1ab, S2ab, S2ac, S2cb, &
                               answer
  !
  !
  !



  !<-------left_half_info-------><-------right_half_info---------->
  !a            d               c               e              b
  !
  ! |_____|_____|_____|_____|
  ! <-----h2------>
  ! <-----------h1--------------->
  !
              count = count + 1
              If (count > 512) Then
                 Write(*,*) ' Maximum number of subdivisions exceeded '
                 Stop
                 End If
              h1 = (b-a)/2.0
              h2 = h1/2.0
              d = a + h2
              e = a + 3.0*h2
              c = a + h1
              b1  = a + 2.0*h1
              fa  = f(a)
              fb  = f(b)
              fc  = f(c)
              fd  = f(d)
              fe  = f(e)
              S1ab = h1*(fa + 4.0*fc +  fb)/3.0
              S2ac - h2*(fa + 4.0*fd +  fc)/3.0
              S2cb - h2*(fc + 4.0*fe +  fb)/3.0
              S2ab = S2ac + S2cb
              If (ABS(S2ab - S1ab) < tol*h1 ) Then
                 answer =  S2ab + (S2ab - S1ab)/15.0
              Else
                 answer =  adapt(a, c, Tol) + adapt(c,b1,tol)
              End If
              Return
              End Function Adapt


  !
  ! ----------------------------------------------------------------
  !                    The integrand is defined
  ! ----------------------------------------------------------------
  !
```

Figure 5.10 *Continued*

```
Function f(x)
Implicit None
Real(Kind = 8) :: f
Real(Kind = 8), Intent(In) :: x
Function_count = Function_count + 1
f = 1.0/exp((x*x))
Return
End Function f

End Program R_Adaptive

************ Output for Adaptive Simpson Program ************

The integral value is:        0.65882

The number of function calls is  :    15
```

Figure 5.10 *Continued*

Exercises

Section 5.2

1. For the following divided-difference table, compute a value for $f'(0.242)$ from a quadratic polynomial that fits the table at $i = 1, 2$, and 3 (the best choice?).

i	x_i	f_i	$f_i^{[1]}$	$f_i^{[2]}$	$f_i^{[3]}$
0	0.15	0.1761	2.4355	-5.7505	15.3476
1	0.21	0.3222	1.9754	-3.9088	8.7492
2	0.23	0.3617	1.7409	-2.9464	5.9642
3	0.27	0.4314	1.4757	-2.2307	
4	0.32	0.5051	1.2973		
5	0.35	0.5441			

2. Estimate the error in the answer to Exercise 1 from the next-term rule. The function actually is $1 + \log_{10}(x)$. Compare the estimate to the actual error.

3. The differences in the table of Exercise 1 are actually the divided differences of $f(x)$ accurate to six decimal places even though the function values are shown to only four decimals. Recompute the differences using the tabulated function values and repeat Exercise 2. How much does the rounding affect the errors? Is rounding more important than truncation?

▶ **4.** Repeat Exercise 1 but this time for $f'(x)$ at $x = 0.21$, 0.22, 0.23, 0.24, 0.25, 0.26, and 0.27. Plot the estimates and compare to a graph of the true values. Make another plot of the errors versus x. At what point is the error smallest?

5. As described in Exercise 3, the differences tabulated in Exercise 1 are based on more accurate function values. Recompute the divided difference table using the tabulated function values, then repeat Exercise 4. How does rounding change the errors you found in Exercise 4?

6. Use Eq. (5.4) to find bounds for the errors at $x = 0.21$, 0.23, and 0.27 in Exercise 4. Do these bounds bracket the errors found in Exercise 4?

7. Use the next-term rule to estimate the error in Exercise 4. Compare these errors with the actual errors. Are the estimates always larger?

8. Repeat Exercise 7 but with the recomputed table done in Exercises 3 and 5.

9. Using the table of Exercise 1, get $f'(0.242)$ from quadratics that begin with i-values of

a. $i = 0$.
b. $i = 2$.
c. $i = 3$.

Do these results confirm that using $i = 1$ is the best starting value to use?

10. What degree of polynomial gives the most accurate value of $f'(0.242)$ from the table of Exercise 1?

▶ 11. The following difference table is for $f(x) = x + \sin(x)/3$. Use it to find

   a. $f'(0.72)$ from a cubic polynomial.
   b. $f'(1.33)$ from a quadratic.
   c. $f'(0.50)$ from a fourth-degree polynomial.

In each part, choose the best starting i-value.

i	x_i	f_i	Δf_i	$\Delta^2 f_i$	$\Delta^3 f_i$	$\Delta^4 f_i$
0	0.30	0.3985	0.2613	−0.0064	−0.0022	0.0003
1	0.50	0.6598	0.2549	−0.0086	−0.0018	0.0004
2	0.70	0.9147	0.2464	−0.0104	−0.0014	0.0005
3	0.90	1.1611	0.2360	−0.0118	−0.0010	
4	1.10	1.3971	0.2241	−0.0128		
5	1.30	1.6212	0.2113			
6	1.50	1.8325				

12. What are the errors in Exercise 11? For part (c), use Eq. (5.7) to find bounds on the error. Do these bounds bracket the actual error?

13. Use the next-term rule to estimate the errors for each part of Exercise 11. Compare these errors with the actual errors. Is the estimate always larger?

▶ 14. Use the central-difference formula, Eq. (5.9), to compute $f'(0.50)$ from the data of Exercise 11. Find the maximum and minimum estimates of the error, and compare them to the actual error.

15. Repeat Exercise 14, this time using a forward-difference formula obtained by using only the first term of Eq. (5.8). Also compute the error by the next-term rule.

16. Repeat Exercise 15, but now use two terms of Eq. (5.8).

17. The differences in Exercise 11 are based on function values accurate to six decimal places even though the table shows only four. Recompute the differences using the tabulated values, then repeat Exercises 11, 12, and 13. How do these new results compare to the previous ones?

18. Use a central-difference formula to estimate $f'(0.66)$ if $f(x) = e^x/(x - 2)$. Also compute with one term of Eq. (5.8), the forward-difference formula. Compare the errors of each for

   a. $h = 0.1$.
   b. $h = 0.01$.
   c. $h = 0.001$.

19. Repeat Exercise 18, but now round all calculated values to three decimal places to see the effect of round-off.

Section 5.3

20. Multiply the operator of Eq. (5.11) by itself to get formulas for the third and fourth derivatives of $f(x_i)$. (A symbolic algebra program can help in doing the multiplications!) Confirm that the first term of these formulas matches Eq. (5.13).

21. Use the formulas of Exercise 20 to compute the third and fourth derivatives of $f(x) = x^2 * e^x$ at $x = 0.3$ using $h = 0.05$.

▶ 22. Use one, two, and three terms of Eq. (5.12) to estimate the second derivative of $f(x) = \sin(x)/(x + 2)$ at $x = 0.5$ with $h = 0.05$. Estimate the errors by the next-term rule, and compare to the actual errors.

23. Derive a central-difference approximation to $f^{(4)}(x)$ in a way similar to that of Eq. (5.17). What order of error will it have? Can you do the same for the third derivative?

24. Use the formula of Exercise 23 to estimate the fourth derivative of $f(x) = \sin(x)/(x + 2)$ at $x = 0.5$ with $h = 0.05$. Repeat with $h = 0.025$. Do the errors of these two estimates agree with the order of the error?

25. Another way to derive formulas for derivatives is from a Taylor-series expansion of $f(x)$ about the point x_0. For example, if we now subtract the series for $x = x_0 - h$ from that where $x = x_0 + h$ and solve for $f'(x_0)$, we get the central-difference formula as well as its error term and order of error. Carry out this operation to confirm this.

▶ 26. Repeat Exercise 25 to get the forward-difference formula for $f'(x_0)$. Repeat again to get a backward-difference formula in terms of x_0 and $x_0 - h$.

27. Use the Taylor-series technique of Exercise 25 to derive both central- and forward-difference formulas for the second derivative together with their error terms.

Section 5.4

28. Use the extrapolation technique of Section 5.4 to get $f'(0.90)$ from the table of Exercise 11 to an $O(h^6)$ accuracy.

29. Show that the first extrapolation for $f'(x_0)$ with h-values differing by 2 to 1 is the same as the formula

$$f_0' = \frac{1}{H}\left(\frac{\Delta f_{-1} + \Delta f_0}{2} - \frac{1}{6}\frac{\Delta^3 f_{-2} + \Delta^3 f_{-1}}{2}\right),$$

where H is the smaller of the h's.

x	$f(x)$
1.0	1.543
1.1	1.669
1.2	1.811
1.3	1.971
1.4	2.151
1.5	2.352
1.6	2.577
1.7	2.828
1.8	3.107

30. Consider whether extrapolations similar to that of Section 5.4 can be used for unevenly spaced data and a divided-difference table. (Taylor-series expansions as described in Exercise 25 may be helpful.) If you succeed in getting a formula for $f'(x)$, use it to estimate $f'(0.27)$ for the data in Exercise 1. What order of error results?

▶ 31. Apply Richardson extrapolation to get $f'(0.32)$ accurate to five significant figures for $f(x) = \sin^2(x/2)$, starting with $h = 0.1$ and using central differences. When the extrapolations agree to five significant figures, are they that accurate?

32. Repeat Exercise 31, but now for $f''(0.32)$.

▶ 33. Can Richardson extrapolation be applied to forward-difference approximations? If you think they can, repeat Exercise 31 but this time with forward differences.

34. Write a computer program to implement the algorithm of Section 5.4 to create a Richardson table.

Section 5.5

35. Use the Newton–Cotes formula for $n = 1$ applied twice to estimate the integral of $f(x) = e^{-x} * \sin(x)$ over $[1, 3]$ with each application having $h = 1$. Compare the actual error with the error bounds computed from Eq. (5.24) applied to each subinterval.

36. Repeat Exercise 35, this time using the Newton–Cotes formula for $n = 2$ applied once to the whole interval ($h = 1$). Use Eq. (5.26) to get the error bounds.

37. Repeat Exercise 35, but now use the Newton–Cotes formula for $n = 3$ applied once to the whole interval ($h = 2/3$). Use Eq. (5.28) to get the error bounds.

▶ 38. Derive the Newton–Cotes formulas for $n = 4$ and $n = 5$.

39. Beginning with the divided-difference form of the interpolating polynomial, rederive the Newton–Cotes formulas for $n = 1, 2$, and 3. (Assume, of course, that the points are evenly spaced.)

Section 5.6

40. The following table has values for $f(x)$. Integrate between $x = 1.0$ and $x = 1.8$ using the trapezoidal rule (Eq. 5.30) with

a. $h = 0.1$.
b. $h = 0.2$.
c. $h = 0.4$.

▶ 41. The function tabulated in Exercise 40 is $\cosh(x)$. What are the errors in parts (a), (b), and (c)? How closely are these proportional to h^2? What other errors are present besides the truncation error?

42. In Example 5.8, the integral of e^x between $x = 1.8$ and $x = 3.4$ was computed with the trapezoidal rule using $h = 0.2$ and has an error of -0.08. It is shown that this falls within the bounds given by Eq. (5.32), but if we don't know what $f(x)$ is, we cannot use that equation. Estimate the error using the second differences of the tabulated data.

43. If we want the integral of Example 5.8 correct to five decimal places (error < 0.000005), how small must h be?

44. Find the integral of $f(x)$ between $x = 0$ and $x = 2$ for

x	$f(x)$
0.00	1.0000
0.12	0.8869
0.53	0.5886
0.87	0.4190
1.08	0.3396
1.43	0.2393
2.00	0.1353

▶ 45. Use Romberg integration to evaluate the integral of $f(x) = 1/x$ between $x = 1$ and $x = 2$. Carry six decimals, and continue until there is no change in the fifth place. Compare to the analytical value, $\ln(2)$.

46. Apply Romberg extrapolation to the results of Exercise 40.

47. Extrapolate to the limit to get the integral of the following data between 0 and 1.00:

x	$f(x)$
0.00	0.3989
0.25	0.3867
0.50	0.3521
0.75	0.3011
1.00	0.2420

Section 5.7

48. Repeat Exercise 40, but use Simpson's $\frac{1}{3}$ rule.

49. Use the error expression for Eq. (5.35) to find bounds on the error of Exercise 48. Does the actual error fall within these bounds? [The function tabulated in Exercise 40 is $\cosh(x)$].

▶ **50.** Evaluate the integral of e^x between $x = 0$ and $x = 1$ with a value of h small enough to guarantee five-decimal-place accuracy. What is the maximum size for h?

51. Show that one extrapolation of the trapezoidal rule is identical to Simpson's $\frac{1}{3}$ rule with a comparable value for h.

52. Compute the integral of $f(x) = \sin(x)/x$ between $x = 0$ and $x = 1$ using Simpson's $\frac{1}{3}$ rule with $h = 0.5$ and then with $h = 0.25$. (Remember that the limit of $\sin(x)/x$ at $x = 0$ is 1.) From these two results, extrapolate to get a better result. What is the order of the error after the extrapolation? Compare your answer with the true answer.

53. Repeat Exercise 50 but use Simpson's $\frac{3}{8}$ rule.

54. The function tabulated here is given at points that divide seven panels. That means that Simpson's $\frac{1}{3}$ rule does not fit, but we can use a combination of the $\frac{1}{3}$ and $\frac{3}{8}$ rules. Compare the results when the $\frac{3}{8}$ rule is applied at different subintervals within [3.0, 6.5]. Which of these gives the smallest error? [The function of the table is $f(x) = 3 + \cos(x)/(x - 1.5)$.]

x	$f(x)$
3.0	2.34000
3.5	2.53177
4.0	2.73854
4.5	2.92973
5.0	3.08105
5.5	3.17717
6.0	3.21337
6.5	3.19532

▶ **55.** Simpson's $\frac{1}{3}$ rule, although based on passing a quadratic through three evenly spaced points, actually gives the exact answer if $f(x)$ is a cubic. The implication is that the area under any cubic between $x = a$ and $x = b$ is identical to the area of a parabola that matches the cubic at $x = a$, $x = b$, and $x = (a + b)/2$. Prove this.

Section 5.8

56. Use the symbolic method to derive a formula for three-panel integration, using a polynomial of degree m.

57. Integration from x_0 to x_1, as in Eq. (5.38), is an arbitrary choice taken for convenience only. If the limits were taken from x_{-1} to x_0, a similar formula for one-panel integration would result, but different coefficients would be obtained. Carry out the integrations.

58. Perform computations similar to those in Exercise 57, but get a two-panel integration formula from x_{-1} to x_1.

▶ **59.** Use the method of undetermined coefficients to derive the trapezoidal rule.

60. Repeat Exercise 59 but for Simpson's $\frac{3}{8}$ rule.

61. Use the method of undetermined coefficients to derive the central-difference formulas, Eqs. (5.9) and (5.17).

Section 5.9

▶ **62.** Evaluate the integral of e^x between $x = 0$ and $x = 1$ using three-term Gaussian quadrature. Compare the accuracy of your answer with that of Exercise 50. If you get five-decimal-place accuracy, how many fewer function evaluations are required with Gaussian quadrature?

63. Evaluate the integral of $\sin(x)/x$ between $x = 0$ and $x = 1$ with a four-term Gaussian formula. What is the error? How small must h be with Simpson's $\frac{1}{3}$ rule to match this accuracy?

▶ **64.** By using Gaussian formulas of increasing complexity, determine how many terms are needed to evaluate the integral of $x^3 * \sin(x^2)e^{x-3}$ over the interval $[-1.5, 2.7]$ to get accuracy to six significant figures.

65. An n-term Gaussian formula assumes that a polynomial of degree $2n - 1$ is used to fit the function between $x = a$ and $x = b$. Does this mean that the error is the same as for a Newton–Cotes integration formula based on a polynomial of degree $2n - 1$?

66. The t-values given in Table 5.14 are the zeros of Legendre polynomials that can be constructed as shown in Section 5.9. Using any method from Chapter 1, confirm the values for $n = 3, 4,$ and 5.

Section 5.10

67. Use a calculator to verify the computations in Example 5.16.

68. Redo Example 5.16 using an adaptive trapezoidal rule method and obtaining the same accuracy. How many more function evaluations would be required if the adaptive scheme were not used?

▶ **69.** Evaluate the integral of $x^3 (\sin(x) + \cos(x))$ between $x = 0$ and $x = 2$ using a tolerance value, TOL, sufficiently small as to get an answer within 0.1% of the true answer, 3.657126.

70. Most programs for adaptive integration will compute the appropriate step size if they use the procedure of Section 5.10. However, in some cases this leads to significant errors. For instance, the integral of $\sin^2(16x)$ between $x = 0$ and $x = \pi/2$ is $\pi/4$, but it is easy to see that the values of $S_1[0, \pi/2]$ and $S_2[0, \pi/2]$ both equal zero, where $h_1 = \pi/4$ and $h_2 = \pi/8$.

How can we solve this problem correctly with the adaptive method of Section 5.10? (It is interesting to know that the HP-15C calculator avoids this error.)

71. Implement the algorithm for adaptive integration in a computer program using the C language.

Section 5.11

72. In Example 5.17 the statement is made, "It is immaterial which integral we evaluate first." Confirm this statement by repeating Example 5.17, but this time integrating with respect to y first and holding x constant.

73. Write pictorial operators similar to that of Eq. (5.45) for

▶ a. Simpson's $\frac{1}{3}$ rule in the x-direction and the trapezoidal rule in the y-direction.
 b. Simpson's $\frac{1}{3}$ rule in both directions.
 c. Simpson's $\frac{3}{8}$ rule in both directions.
▶ d. What conditions are placed on the number of panels in both directions by parts (a), (b), and (c)?

74. Because Simpson's $\frac{1}{3}$ rule is exact when $f(x)$ is a cubic, evaluation of the following triple integral should be exact. Confirm by evaluating both numerically and analytically. Use Eq. (5.49) adapted for this integral.

$$\int_0^1 \int_0^2 \int_{-1}^0 x^3 yz^2 \, dx \, dy \, dz$$

75. Draw a pictorial operator that represents the formula used in Exercise 74. You may want to do this with three widely separated planes such as:

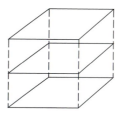

76. Evaluate the following integral, and compare your answers to the analytical solution. Use $h = 0.1$ in both directions in parts (a) and (b),

 a. using the trapezoidal rule in both directions.
 b. using Simpson's $\frac{1}{3}$ rule in both directions.
 c. using Gaussian quadrature, three-term formulas in both directions.

$$\int_{-0.2}^{1.4} \int_{0.4}^{2.6} e^x \cos(y) \, dy \, dx$$

77. Solve Exercise 76 by performing the trapezoidal rule integrations first with $h = 0.2$ (in both directions), then with $h = 0.1$, and extrapolate. The answer should match parts (b) of the exercise. Does it?

▶ **78.** Integrate with varying values of Δx and Δy using the trapezoidal rule in both directions, and show that the error decreases about in proportion to h^2:

$$\int_0^1 \int_0^1 (x^2 + y^2) \, dx \, dy.$$

Section 5.12

79. Integrate $\iint \cos(y) * \sin(x) \, dx \, dy$ over the region defined by the portion of a unit circle that lies in the first quadrant. Integrate first with respect to x holding y constant, using $h = 0.25$. Subdivide the horizontal lines into four panels.

 a. Use the trapezoidal rule.
 b. Use Simpson's $\frac{1}{3}$ rule.

80. The order of integration in multiple integrations can usually be changed. Repeat Exercise 79, but now integrate first with respect to y. Compare the answer with that of Exercise 79.

81. Integrate $\exp(-x^2 y^2)$ over the region bounded by the two parabolas $y = x^2$ and $y = 2x^2 - 1$. Observe that, because the integrand is an even function and the region is symmetrical about the y-axis, the integration may be over half the area and then doubled. Choose reasonable values for the number of panels in each direction and use any formula you prefer.

82. Use Gaussian quadrature to solve Exercise 79, with

▶ a. three-term formulas.
  b. four-term formulas.

83. In performing finite element analysis (described in Chapter 9), integrations over a rectangular region with sides parallel to the axes are required. Gaussian quadrature is a preferred method. If two-term formulas are employed, we need to evaluate the integrand at only four points within the rectangle. These points are called "Gauss points." Where are the Gauss points for the rectangle with opposite corners at (2.1, 3.7) and (5.2, 7.0)?

Section 5.13

84. The following table is for $f(x) = 1/(x + 2)$. Find values for $f'(x)$ and $f''(x)$ at $x = 1.5$, 2.0, and 2.5 from cubic spline functions that approximate $f(x)$. Compare to the true values to determine the errors. Also compare to derivative values computed from central-difference formulas.

  a. Use end condition 1.
  b. Use end condition 3.
  c. Use end condition 4.

x	1.0	1.5	2.0	2.5	3.0
$f(x)$	0.333	0.286	0.250	0.222	0.200

85. Plot the values of $f'(x)$ and $f''(x)$ from the cubic splines of Exercise 84 on [1.0, 3.0], and compare to plots of the true values.

86. The comparisons in Exercise 84 may favor the cubic spline because they are based on cubic polynomials, whereas the central-difference formulas are based on quadratics. Repeat Exercise 84, but now use interpolating polynomials of degrees 3 and 4.

▶ **87.** Repeat Exercise 84 but this time use cubic splines that have the correct slopes at the ends, condition 2.

▶ **88.** Integrate sech(x) over [0, 2] by integrating the natural cubic spline curve (end condition 1) that fits at five evenly spaced points on [0, 2]. Compare the result to the analytical value. Also compare to the integral from Simpson's $\frac{1}{3}$ rule.

89. Repeat Exercise 87 using end conditions 2, 3, and 4. For condition 2, use the analytical values for $f'(x)$.

90. Repeat Exercise 88, but now force the values of $f''(x)$ at the ends to the analytical values of the second derivative of sech(x).

Section 5.14

91. Use trapezoidal rule integration with 20 panels to get the Fourier coefficients for these functions. Compare the answers to those from analytical integration.

  a. $f(x) = x^3$ on [0, 2].
  b. $f(x) = x^2 - 1$ on [−1, 3].
  c. $f(x) = e^{-x} * \sin(2x - 1)$ on [1, 4].

▶ **92.** Repeat Exercise 91, but now use Simpson's rule.

93. How many panels would be required in Exercise 91 if we are to match the analytical values to five decimal places?

94. How many panels would be required in Exercise 92 if we are to match the analytical values to five decimal places?

95. Verify that Eqs. (5.57) and (5.58) are really identical.

96. Make a flow diagram like that in Fig. 5.8 for $n = 8$.

▶ **97.** Use the algorithm of Section 5.14 to generate the powers of W for use in the FFT for $n = 16$. Do these agree with those in Fig. 5.8?

98. Repeat Exercise 97 but now use the bit-reversing rule.

Section 5.15

99. Use the Taylor series method to derive expressions for $f'(x)$ and $f''(x)$ and their error terms using f-values that precede f_0. (These are called backward-difference formulas.)

100. Repeat Exercise 99, but now use the symbolic operator method.

101. Repeat Exercise 99, but this time use the method of undetermined coefficients. (See Appendix B.)

102. Develop the expanded form for the error terms of the Newton–Cotes integration formulas for $n = 4$ and $n = 5$.

103. Prove that *all* Newton–Cotes integration formulas of even order have error terms of order 1 greater than expected.

▶ **104.** Demonstrate by varying the size of h that the local error for Simpson's $\frac{1}{3}$ rule is $O(h^5)$ and that the global error is $O(h^4)$. Repeat using several choices for $f(x)$.

105. Demonstrate with several different polynomials that the two-term Gaussian quadrature formula is exact for all polynomials of degree 3 or less.

106. Assume that $f(x) = \sin(x^2/2)$. Make a table similar to

Table 5.20 to show that the error of $f'(x)$ and $f''(x)$ does not continually decrease as h is made smaller.

107. Repeat Exercise 106, but for forward-difference formulas. Are the best values for h the same?

108. Repeat Exercise 106, but now use the trapezoidal rule to integrate between $x = 0$ and $x = 0.4$. How does the optimum h-value compare with that for differentiation?

109. Repeat Exercise 108, but now use Simpson's $\frac{1}{3}$ rule.

Section 5.16

110. Use MATLAB to get the analytical values of the derivatives for the functions in Exercises 2, 11, 18, and 31.

111. Use MATLAB to get the analytical values for the integrals in Exercises 35, 41, 50, and 52.

112. Use MATLAB to do the multiplications in Exercise 20.

113. Use MATLAB to get the zeros in Exercise 66.

Applied Problems and Projects

114. Differential thermal analysis is a specialized technique that can be used to determine transition temperatures and the thermodynamics of chemical reactions. It has special application in the study of minerals and clays. Vold [*Anal. Chem.* 21, 683 (1949)] describes the technique. In this method, the temperature of a sample of the material being studied is compared to the temperature of an inert reference material when both are heated simultaneously under identical conditions. The furnace housing the two materials is normally heated so that its temperature T_f increases (approximately) linearly with time (t), and the difference in temperatures (ΔT) between the sample and the reference is recorded. Some typical data are

t, min	0	1	2	3	4	5	6	7
ΔT, °F	0.00	0.34	1.86	4.32	8.07	13.12	16.80	18.95
T_f, °F	86.2	87.8	89.4	91.0	92.7	94.3	95.9	97.5

t	8	9	10	11	12	13	14	15	16
ΔT	18.07	16.69	15.26	13.86	12.58	11.40	10.33	8.95	6.46
T_f	99.2	100.8	102.3	103.9	105.5	107.1	108.6	110.2	111.8

t	17	18	19	20	21	22	23	24	25
ΔT	4.65	3.37	2.40	1.76	1.26	0.88	0.63	0.42	0.30
T_f	113.5	115.1	116.8	118.4	120.0	121.6	123.2	124.9	126.5

The ΔT values increase to a maximum, then decrease, due to the heat evolved in an exothermic reaction. One item of interest is the time (and furnace temperature) when the reaction is complete. Vold shows that the logarithm of ΔT should decrease linearly after the reaction is over; while the chemical reaction is occurring, the data depart from this linear relation. Vold used a graphical method to find this point. Perform numerical computations to find, from the preceding data, the time and the furnace temperature when the reaction terminates. Compare the merits of doing it graphically or numerically.

115. The temperature difference data in Problem 114 can be used to compute the heat of reaction. To do this, the integral of the values of ΔT is required, from the point where the reaction begins

(which is at the point where ΔT becomes nonzero) to the time when the reaction ceases, as found in Exercise 106. Determine the value of the required integral. Which of the methods of this chapter should give the best value for the integral?

116. *Fugacity* is a term used by engineers to describe the available work from an isothermal process. For an ideal gas, the fugacity f is equal to its pressure P, but for real gases,

$$\ln \frac{f}{P} = \int_0^P \frac{C-1}{P}\, dp,$$

where C is the experimentally determined *compressibility factor*. For methane, values of C are

P (atm)	C	P (atm)	C
1	0.9940	80	0.3429
10	0.9370	120	0.4259
20	0.8683	160	0.5252
40	0.7043	250	0.7468
60	0.4515	400	1.0980

Write a program that reads in the P and C values and uses them to compute and print f corresponding to each pressure given in the table. Assume that the value of C varies linearly between the tabulated values (a more precise assumption would fit a polynomial to the tabulated C values). The value of C approaches 1.0 as P approaches 0.

117. The stress developed in a rectangular bar when it is twisted can be computed if one knows the values of a torsion function U that satisfies a certain partial-differential equation. Chapter 7 describes a numerical method that can determine values of U. To compute the stress, it is necessary to integrate $\int \int U \, dx \, dy$ over the rectangular region for which the data given here apply. Determine the stress. (You may be able to simplify the integration because of the symmetry in the data.)

x \ y	0.0	0.2	0.4	0.6	0.8	1.0	1.2
0.0	0	0	0	0	0	0	0
0.2	0	2.043	3.048	3.354	3.048	2.043	0
0.4	0	3.123	4.794	5.319	4.794	3.123	0
0.6	0	3.657	5.686	6.335	5.686	3.657	0
0.8	0	3.818	5.960	6.647	5.960	3.818	0
1.0	0	3.657	5.686	6.336	5.686	3.657	0
1.2	0	3.123	4.794	5.319	4.794	3.123	0
1.4	0	2.043	3.048	3.354	3.048	2.043	0
1.6	0	0	0	0	0	0	0

118. Make a critical comparison of the accuracy of Newton–Cotes integration formulas compared to Gaussian quadrature. Test the formulas for a variety of functions for which you can calculate the integrals analytically. Select some functions that are smooth, some that have sharp changes in value, and some with periodic behavior.

119. Write a general-purpose subroutine that performs Gaussian quadrature. You should have the subroutine change the limits of the integration appropriately and call a function subprogram to compute function values, with the name of the function subprogram passed as an argument. It should also receive as an argument the degree of the formula to be employed.

120. The data in Exercise 84 exhibit round-off in three of the entries. By recalculating the spline function, with more and less accurate values of these entries, determine how the values of S are affected by the precision of the data. Also calculate how the precision affects the estimates of $f'(x)$ and $f''(x)$. Compare the effects of the precision of the original data, using a cubic spline, with the effects of precision on the values of $f'(x)$ and $f''(x)$ when central-difference formulas are used.

121. Employ the technique that resulted in Eqs. (5.38) and (5.39) to get integration rules for the integral over three and four panels. Repeat this by the method of undetermined coefficients. Repeat again by expanding $f(x)$ as a Taylor series, this time finding the error terms.

6

Numerical Solution of Ordinary Differential Equations

Contents of This Chapter

The subject of ordinary differential equations is not only one of the most beautiful parts of mathematics, but it is also an essential tool for modeling many physical situations: spring–mass systems, resistor–capacitor–inductance circuits, bending of beams, chemical reactions, pendulums, the motion of a rotating mass around another body, and so forth. These equations have also demonstrated their usefulness in ecology and economics. The predator–prey problem has become a classic example of differential equations.

The prominence of ordinary differential equations in applied mathematics is due to the fact that most scientific laws are more readily expressed in terms of rates of change. For example,

$$\frac{du}{dt} = -0.27(u - 60)^{5/4}$$

is an equation describing (approximately) the rate of change of temperature u of a body losing heat by natural convection with constant-temperature surroundings. This is termed a first-order differential equation because the highest-order derivative is the first.

If the equation contains derivatives of nth order, it is said to be an nth-order differential equation. For example, a second-order equation describing the oscillation of a weight acted upon by a spring, with resistance to motion proportional to the square of the velocity, might be

$$\frac{d^2x}{dt^2} + 4\left(\frac{dx}{dt}\right)^2 + 0.6x = 0,$$

where x is the displacement and t is time.

The solution to a differential equation is the function that satisfies the differential equation and that also satisfies certain initial conditions on the function. In solving a differential equation analytically, we usually find a general solution containing arbitrary constants and then evaluate the arbitrary constants so that the expression agrees with the initial conditions. For an nth-order equation, n independent initial conditions must usually be known.* The analytical methods are limited to certain special forms of equations; elementary courses normally treat only linear equations with constant coefficients when the degree of the equation is higher than first. Neither of these examples is linear.

Numerical methods have no such limitations to only standard forms. We obtain the solution as a tabulation of the values of the function at various values of the independent variable, however, and not as a functional relationship. We must also pay a price for our ability to solve practically any equation, in that we must recompute the entire table if the initial conditions are changed.

Our procedure will be to explore several methods of solving first-order equations, and then to show how these same methods can be applied to systems of simultaneous first-order equations and to higher-order differential equations. We will use for our typical first-order equation the form

$$\frac{dy}{dx} = f(x, \, y),$$

$$y(x_0) = y_0.$$

6.1 The Spring–Mass Problem—A Variation

Presents a variation of the usual spring–mass problem in that two second-order differential equations must be solved.

6.2 The Taylor-Series Method

Gives a straightforward adaptation of classic calculus to develop the solution as an infinite series. The catch is that a computer usually cannot be programmed to construct the terms and one doesn't know how many terms should be used.

6.3 Euler and Modified Euler Methods

Describes methods that are simple to use but subject to error unless the step size Δx is made very small.

6.4 Runge–Kutta Methods

Presents methods that are very popular because of their good efficiency; they are used in most computer programs for differential equations. They are single-step methods, as are the Euler methods. In this section we compare the methods presented so far.

---

*In later chapters we will study differential equations that are subject to boundary conditions as well as initial conditions.

6.5 Multistep Methods

Covers methods that are even more efficient than the previous methods but cannot be used at the beginning of the interval of integration; they can be employed after several steps with a single-step method. These methods have a built-in ability to monitor the error in the solution.

6.6 Milne's Method

Examines a multistep method that appears very attractive until one finds that it may be unstable.

6.7 The Adams–Moulton Method

Describes a multistep method that does not suffer from the fault of instability. This section compares the methods of Adams–Moulton and Milne for accuracy and stability through an example.

6.8 Convergence Criteria

Details criteria that impose additional limitations on the step size beyond the requirement for accuracy; here you are exposed to some theory.

6.9 Systems of Equations and Higher-Order Equations

Includes the usual situations in applied problems; fortunately our methods are readily extended to cover them. The various methods are again applied to the same example to show how they compare.

6.10 Comparison of Methods/Stiff Equations

Summarizes the various methods and makes a critical comparison of their strengths and weaknesses. The somewhat esoteric notion of "stiff" equations is examined.

6.11 Theoretical Matters

Looks briefly at the requirements for the existence and uniqueness of the solution to a differential equation. It also examines the concept of the stability of differential equations.

6.12 Using Maple and MATLAB

Considers some of the features of these computer algebra systems that can solve differential equations, both numerically and analytically.

Chapter Summary

Lists the important topics of this chapter.

Computer Programs

Gives the listing and description of two implementations of the algorithms of this chapter.

Parallel Processing

As you will see when you study the algorithms of this chapter, numerical methods for solving ordinary differential equations (posed as an initial value problem) do not lend themselves to parallel processing. We quote from a conference of 1991 on parallel processing (*SIAM News*, November 1991):

> Attempts to apply parallelism to ordinary differential equations are met with an inherent difficulty—the initial value problem propagates information forward in time and therefore has an essentially serial character. If a large system is to be integrated, it is possible to divide the equations among processors and use parallelism across space, but issues of synchronization and communication can become very important. Because these issues are to some extent problem-dependent, little progress has been made in developing a general theory. . . .
>
> The progress that has been made is in low-degree parallelism, in which a few processors are applied to different parts of the same step using Runge–Kutta-like methods.

The report pointed out that research in this area is continuing and that there are several ways in which parallelism can be profitable.

6.1   The Spring–Mass Problem—A Variation

Every student of differential equations has studied the ordinary spring–mass problem. In this, the motion of a weight that is suspended from a linear spring is determined. We shall consider a variation of this: the motion of two masses that move on a frictionless surface.

Figure 6.1 shows our system. Mass 1 is a block that rolls along a horizontal surface and whose motion is controlled by the linear spring whose spring constant is k_1. The second mass, m_2, is a wheel of radius r_2 that rolls on the top of mass 1 and is attached to another spring whose spring constant is k_2. The equations of motion for this system are:

$$(m_1 + 0.5m_2)\frac{d^2x_1}{dt^2} - 0.5m_2\frac{d^2x_2}{dt^2} + k_1x_1 = 0,$$

$$-0.5m_2\frac{d^2x_2}{dt^2} + 1.5m_2\frac{d^2x_1}{dt^2} + k_2x_2 = 0.$$

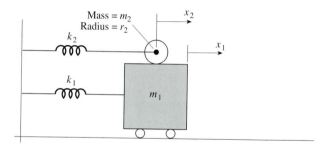

Figure 6.1

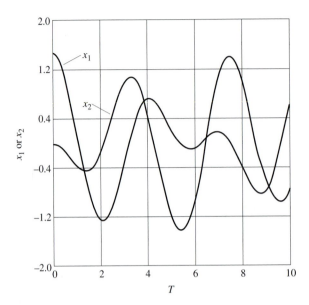

Figure 6.2

Assume that $m_1 = m_2 = 5$ Kg and that $k_1 = k_2 = 15$ N/m (Newtons per meter). If the initial conditions are $x_1(0) = 1.5$ m, $x_2(0) = 0$ (both measured from the equilibrium positions of the masses), and if the initial velocities (dx/dt) of both masses are 0, the graph of the solution is given in Figure 6.2. The two curves show how x_1 and x_2 vary with time. Section 6.12 will show how Maple can be used to get the solution and to draw the curves.

6.2 The Taylor-Series Method

The first method we discuss is not strictly a numerical method, but it is sometimes used in conjunction with the numerical schemes, is of general applicability, and serves as an introduction to the other techniques we will study. Consider the example problem

$$\frac{dy}{dx} = -2x - y, \qquad y(0) = -1. \qquad (6.1)$$

(This particularly simple example is chosen to illustrate the method so that you can readily check the computational work. The analytical solution,

$$y(x) = -3e^{-x} - 2x + 2$$

is obtained immediately by application of standard methods and will be compared with our numerical results to show the error at any step.)

We develop the relation between y and x by finding the coefficients of the Taylor series in which we expand y about the point $x = x_0$:

$$y(x) = y(x_0) + y'(x_0)(x - x_0) + \frac{y''(x_0)}{2!}(x - x_0)^2 + \frac{y'''(x_0)}{3!}(x - x_0)^3 + \cdots .$$

If we let $x - x_0 = h$, we can write the series as

$$y(x) = y(x_0) + y'(x_0)h + \frac{y''(x_0)}{2}h^2 + \frac{y'''(x_0)}{6}h^3 + \cdots . \tag{6.2}$$

Because $y(x_0)$ is our *initial condition,* the first term is known from the initial condition $y(0) = -1$. (Because the expansion is about the point $x = 0$, our Taylor series is actually the Maclaurin series in this example.)

We get the coefficient of the second term by substituting $x = 0$, $y = -1$ in the equation for the first derivative, Eq. (6.1):

$$y'(x_0) = y'(0) = -2(0) - (-1) = 1.$$

We get the second- and higher-order derivatives by successively differentiating the equation for the first derivative. Each of these derivatives is evaluated corresponding to $x = 0$ to get the various coefficients:

$$y''(x) = -2 - y', \qquad y''(0) = -2 - 1 = -3,$$
$$y'''(x) = -y'', \qquad y'''(0) = 3,$$
$$y^{(4)}(x) = -y''', \qquad y^{(4)}(0) = -3.$$

We then write our series solution for y, letting $x = h$ be the value at which we wish to determine y:

$$y(h) = -1 + 1.0h - 1.5h^2 + 0.5h^3 - 0.125h^4 + \text{error}.$$

The solution to the differential equation, $dy/dx = -2x - y$, where $y(0) = -1$, is then given by Table 6.1. As we computed the last two entries, we were doubtful about their accuracy without having used more terms of the Taylor series, because the successive terms were decreasing less and less rapidly. Thus we need more terms than we have calculated to get four-decimal-place accuracy.

Table 6.1

x	y	y, analytical
0.0	-1.00000	-1.00000
0.1	-0.91451	-0.91451
0.2	-0.85620	-0.85619
0.3	-0.82251	-0.82245
0.4	-0.81120	-0.81096
0.5	-0.82031	-0.81959

When a convergent Taylor series is truncated, the error is simple to express. We merely take the next term and evaluate the derivative at the point $x = \xi$, $0 < \xi < h$, instead of at the point $x = x_0$. This is exactly what we did to write error terms in the previous chapters. The error term of the Taylor series after the h^4 term is

$$\text{Error} = \frac{y^{(5)}(\xi)}{5!} h^5, \qquad 0 < \xi < h.$$

However, this cannot be computed, because evaluating the derivative at $x = \xi$ is impossible with ξ unknown, and even bounding it in the interval $[0, h]$ is impossible because the derivatives are known only at $x = 0$ and not at $x = h$.

Numerical analysis is sometimes termed an art instead of a science because, in situations like this, the number of Taylor-series terms to be included is a matter of judgment and experience. We normally truncate the Taylor series when the contribution of the last term is negligible to the number of decimal places to which we are working. However, this is correct only when the succeeding terms become small rapidly enough—in some cases the sum of the many neglected small terms is significant.

The Taylor series is easily applied to a higher-order equation. For example, if we are given

$$y'' = 3 + x - y^2, \qquad y(0) = 1, \qquad y'(0) = -2,$$

we can find the derivative terms in the Taylor series as follows:

$y(0)$, and $y'(0)$ are given by the initial conditions.

$y''(0)$ comes from substitution into the differential equation from $y(0)$ and $y'(0)$.

$y'''(0)$ and higher derivatives are found by differentiating the equation for the previous order of derivative and substituting previously computed values.

(In Section 6.12, we show how Maple can readily get the Taylor-series solutions).

Automatic Differentiation

In the preceding example, getting the derivatives was very easy, even when doing them by hand. Suppose, however, that we want to do hand computations to get the Taylor-series solution of

$$y' = f(x, y) = \frac{x}{(y - x^2)}, \qquad \text{with } y(0) = 1.$$

In this instance, the derivatives quickly become messy. However, software exists that can develop the terms of the series. These procedures perform *automatic differentiation* as opposed to *symbolic differentiation*. In automatic differentiation, machine code is produced that finds values of derivatives of functions when dy/dx is defined through a *code list*.

We will not give a thorough explanation, only an example, but L. R. Rall (1981) and Corliss and Chang (1982) are good sources for more information. Here is our example:

Solve $y' = f(x, y) = \dfrac{x}{(y - x^2)}$ using automatic differentiation with $y(0) = 1$.

To do this, we first create a code list, which is just a name for a sequence of statements that define dy/dx, with only a single operation on each line:

```
T1 = x*x
T2 = y - T1
dy/dx = x/T2    (which is f(x, y)).
```

We will use a simplified notation for the terms of the Taylor series:

$$(y)_k = \left(\frac{1}{k!}\right)\left[\frac{d^k y}{dx^k}\right], \qquad k = 0, \ldots, n.$$

And we will use $(x)_0 = x_0$. We then have $(y)_0 = y(x_0)$.

The software for automatic differentiation includes the standard rules for differentiation in recursive form, such as the derivatives of $(u + v)_k$, $(u - v)_k$, $(u * v)_k$, and $(u/v)_k$, plus the elementary functions, including sin, cos, ln, exp, and so on.

In our example, we have $(x)_0 = 0$, $(x)_1 = 1$ (because $dx/dx = 1$), so that $(x)_k = 0$ for all higher derivatives of x. From the initial condition, $(y)_0 = 1$ and from the expression for $y'(x)$, $(y)_1 = 0$. It is not hard to determine that $(y)_2 = 0.5$. The automatic differentiation software develops a recursion formula for the additional coefficients of the Taylor series. This formula is something like this:

$$(y)_k = \alpha_k \sum_{i=1}^{k-1} i(y)_i(y)_{k-1},$$

where the multiplier, α_k, is a complicated function of k.

Similar recursion formulas will be derived by the software for any differential equation that can be compiled into a code list, and these can have any initial condition.

For our example, all the odd-order terms are zero; the even-order terms are:

Order	4	6	8
Coefficient	$\frac{1}{8}$	$\frac{1}{48}$	$\frac{-1}{384}$

Using this in the Taylor series produces $y(0.1) = 1.0050125$, $y(0.2) = 1.0202013$.

The authors are especially grateful to Professor Ramon E. Moore of The Ohio State University for calling our attention to this method for solving ordinary differential equations.

6.3 Euler and Modified Euler Methods

As we saw in the previous section, the Taylor-series method may be awkward to apply if the derivatives become complicated and in this case the error is difficult to determine. Of course, many computer packages, including Maple and MATLAB, can use symbolic methods to differentiate a function. In fact, even a calculator like the HP-48S can differentiate a function, but the analytic derivatives often grow very complicated.

We know that the error in a Taylor series will be small if the step size h is small. In fact, if we make h small enough, we may only need a few terms of the Taylor-series expansion for good accuracy. The Euler method follows this idea to the extreme for first-order differential equations—it uses only the first two terms of the Taylor series! Suppose that we have chosen h small enough that we may truncate after the first-derivative term. Then

$$y(x_0 + h) = y(x_0) + hy'(x_0) + \frac{y''(\xi)h^2}{2}, \qquad x_0 < \xi < x_0 + h,$$

where we have written the usual form of the error term for the truncated Taylor series.

In using this equation, the value of $y(x_0)$ is given by the initial condition and $y'(x_0)$ is evaluated from $f(x_0, y_0)$ which is given by the differential equation, $dy/dx = f(x, y)$. It is necessary to use this method iteratively, advancing the solution to $x = x_0 + 2h$ after $y(x_0 + h)$ has been computed, and then on to $x = x_0 + 3h$, and so on. Adopting a subscript notation for the successive y-values and representing the error by the order relation, we may write the algorithm for the Euler method in the form*

$$y_{n+1} = y_n + hy_n' + O(h^2). \tag{6.3}$$

As an example, we apply this to the previous equation

$$\frac{dy}{dx} = -2x - y, \qquad y(0) = -1,$$

where the computation can be done rather simply. It is convenient to arrange the work as in Table 6.2. Here we take $h = 0.1$.

Each of the y_n values is computed using Eq. (6.3), adding hy_n' and y_n of the previous line. Comparing the last result to the analytical answer $y(0.40) = -0.81096$, we see that there is only one-decimal-place accuracy, even though we have advanced the solution only four steps! To gain four-decimal-place accuracy, we must reduce the error by more than 400-fold. Because the global error is about proportional to h, we will need to reduce the step size about 426-fold, to <0.00024.

The trouble with this most simple method is its lack of accuracy, requiring an extremely small step size. Figure 6.3 suggests how we might improve this method with just a little additional effort.

In the simple Euler method, we use the slope at the beginning of the interval, y_n', to determine the increment to the function. This technique would be correct only if the function were linear. What we need instead is the correct average slope within the interval. This can be approximated by the mean of the slopes at both ends of the interval.

Suppose we use the arithmetic average of the slopes at the beginning and end of the interval to compute y_{n+1}:

$$y_{n+1} = y_n + h\frac{y_n' + y_{n+1}'}{2}. \tag{6.4}$$

---

*This error is just the local error. Over many steps, the global error becomes $O(h)$.

Table 6.2

x_n	y_n	y'_n	hy'_n
0.0	−1.00000	1.00000	0.10000
0.1	−0.90000	0.70000	0.07000
0.2	−0.83000	0.43000	0.04300
0.3	−0.78700	0.18700	0.01870
0.4	−0.76830	−0.03170	

(Analytical answer is −0.81096, error is −0.04266.)

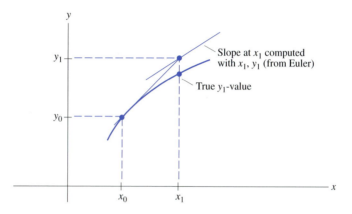

Figure 6.3

This should give us an improved estimate for y at x_{n+1}. However, we are unable to employ Eq. (6.4) directly, because the derivative is a function of both x and y and we cannot evaluate y'_{n+1} with the true value of y_{n+1} unknown. The modified Euler method works around this problem by estimating or "predicting" a value of y_{n+1} by the simple Euler relation, Eq. (6.3). It then uses this value to compute y'_{n+1}, giving an improved estimate (a "corrected" value) for y_{n+1}. Because the value of y'_{n+1} was computed using the predicted value, of less than perfect accuracy, we might want to recorrect the y_{n+1} value as many times as will make a significant difference. (However, if more than one or two recorrections are required, it is more efficient to reduce the step size or use a different method.)

We will illustrate the modified Euler method, which we also call the Euler predictor–corrector method, on the problem we previously treated. Table 6.3 shows all the steps in going from $x = 0.0$ to $x = 0.5$. To show the method as simply as possible, we did not recorrect at any step in the table.

However, in computing from $x = 0.0$ to $x = 0.2$, we will get improved results if we recorrect. The next display shows the results of recorrecting. (Lines 2 and 4 are where we recorrect.)

Table 6.3

x_n	y_n	hy'_n	$y_{n+1,p}$	$hy'_{n+1,p}$	hy'_{av}	$y_{n+1,c}$
0.0	−1.0000	0.1000	−0.9000	0.0700	0.0850	−0.9150
0.1	−0.9150	0.0715	−0.8435	0.0444	0.0579	−0.8571
0.2	−0.8571	0.0457	−0.8114	0.0211	0.0334	−0.8237
0.3	−0.8237	0.0224	−0.8013	0.0001	0.0112	−0.8124
0.4	−0.8124	0.0012	−0.8112	−0.0189	−0.0088	−0.8212
0.5	−0.8212					

$[y(0.5) = -0.81959$, the analytical value$]$

x_n	y_n	hy'_n	$y_{n+1,p}$	$hy'_{n+1,p}$	hy'_{av}	$y_{n+1,c}$
0.0	−1.0000	0.1000	−0.9000	0.0700	0.0850	−0.9150
			−0.9150	0.0715	0.0858	−0.9142
0.1	−0.9142	0.0714	−0.8428	0.0443	0.0578	−0.8564
			−0.8564	0.0456	0.0585	−0.8557
0.2	−0.8557					

We work as before, but recorrect when there is a significant difference between $y_{n+1,p}$ and $y_{n+1,c}$, that is, the predicted and corrected values for the next y_{n+1}. It is obvious that this recorrected value, −0.9142, is closer to the exact value, $y(0.1) = -0.9145$. The corrected value at $x = 0.2$, −0.8557, is also closer to the exact value. However, it would be far better to use a more efficient method such as the Runge–Kutta methods of Section 6.4 rather than using many recorrections in the modified Euler method. It is true, however, that this method is somewhat more efficient than the Euler method of Eq. (6.3).

We can find the error of the modified Euler method by comparing it with the Taylor series:

$$y_{n+1} = y_n + y'_n h + \frac{1}{2} y''_n h^2 + \frac{y'''(\xi)}{6} h^3, \qquad x_n < \xi < x_n + h.$$

Replace the second derivative by the forward-difference approximation for y'', $(y'_{n+1} - y'_n)/h$, which has error of $O(h)$, and write the error term as $O(h^3)$:

$$y_{n+1} = y_n + h \left(y'_n + \frac{1}{2} \left[\frac{y'_{n+1} - y'_n}{h} + O(h) \right] h^2 \right) + O(h^3),$$

$$y_{n+1} = y_n + h \left(y'_n + \frac{1}{2} y'_{n+1} - \frac{1}{2} y'_n \right) + O(h^3),$$

$$y_{n+1} = y_n + h \left(\frac{y'_n + y'_{n+1}}{2} \right) + O(h^3).$$

This shows that the error of one step of the modified Euler method is $O(h^3)$. This is the local error. There is an accumulation of errors from step to step, so that the error over the whole range of application, the so-called global error, is $O(h^2)$. This seems intuitively reasonable, because the number of steps into which the interval is subdivided is proportional to $1/h$; hence the order of error is reduced to $O(h^2)$ on continuing the technique. We shall treat the accumulation of errors for the simple Euler method more fully in Section 6.11.

[We take note of the fact that Maple and MATLAB can solve our example problem analytically (using symbolic methods) as well as numerically.]

6.4 Runge–Kutta Methods

The two numerical methods of the last section, though not very impressive, serve as a good introduction to our next procedures. Although we can improve the accuracy of these two methods by taking smaller step sizes, much greater accuracy can be obtained more efficiently by a group of methods named after two German mathematicians, Runge and Kutta. They developed algorithms that solve a differential equation efficiently and yet are the equivalent of approximating the exact solution by matching the first n terms of the Taylor-series expansion. We will consider only the fourth- and fifth-order Runge–Kutta methods, even though there are higher-order methods. Actually, the modified Euler method of the last section is a second-order Runge–Kutta method.

To impart some idea of how the Runge–Kutta methods are developed, we will show the derivation of a simple second-order method. Here, the increment to y is a weighted average of two estimates of the increment which we call k_1 and k_2. Thus for the equation $dy/dx = f(x, y)$,

$$
\begin{aligned}
y_{n+1} &= y_n + ak_1 + bk_2, \\
k_1 &= hf(x_n, y_n), \\
k_2 &= hf(x_n + \alpha h, y_n + \beta k_1).
\end{aligned}
\tag{6.5}
$$

We can think of the values k_1 and k_2 as estimates of the change in y when x advances by h, because they are the product of the change in x and a value for the slope of the curve, dy/dx. The Runge–Kutta methods always use the simple Euler estimate as the first estimate of Δy; the other estimate is taken with x and y stepped up by the fractions α and β of h and of the earlier estimate of Δy, k_1. Our problem is to devise a scheme of choosing the four parameters, a, b, α, β. We do so by making Eq. (6.5) agree as well as possible with the Taylor-series expansion, in which the y-derivatives are written in terms of f, from $dy/dx = f(x, y)$,

$$
y_{n+1} = y_n + hf(x_n, y_n) + \frac{h^2}{2}f'(x_n, y_n) + \cdots .
$$

An equivalent form, because $df/dx = f_x + f_y \, dy/dx = f_x + f_y f$, is

$$
y_{n+1} = y_n + hf_n + h^2 \left(\frac{1}{2}f_x + \frac{1}{2}f_y f \right)_n.
\tag{6.6}
$$

[All the derivatives in Eq. (6.6) are calculated at the point (x_n, y_n).] We now rewrite Eq. (6.6) by substituting the definitions of k_1 and k_2:

$$
y_{n+1} = y_n + ahf(x_n, y_n) + bhf[x_n + \alpha h, y_n + \beta hf(x_n, y_n)].
\tag{6.7}
$$

To make the last term of Eq. (6.7) comparable to Eq. (6.6), we expand $f(x, y)$ in a Taylor series in terms of x_n, y_n, remembering that f is a function of two variables,* retaining only first derivative terms:

$$f[x_n + \alpha h, y_n + \beta h f(x_n, y_n)] \doteq (f + f_x \alpha h + f_y \beta h f)_n. \tag{6.8}$$

On the right side of both Eqs. (6.6) and (6.8) f and its partial derivatives are all to be evaluated at (x_n, y_n).

Substituting from Eq. (6.8) into Eq. (6.7), we have

$$y_{n+1} = y_n + a h f_n + b h (f + f_x \alpha h + f_y \beta h f)_n,$$

or, rearranging,

$$y_{n+1} = y_n + (a + b) h f_n + h^2 (\alpha b f_x + \beta b f_y f)_n. \tag{6.9}$$

Equation (6.9) will be identical to Eq. (6.6) if

$$a + b = 1, \qquad \alpha b = \frac{1}{2}, \qquad \beta b = \frac{1}{2}.$$

Note that only three equations need to be satisfied by the four unknowns. We can choose one value arbitrarily (with minor restrictions); hence we have a set of second-order methods. For example, taking $a = \frac{2}{3}$, we have $b = \frac{1}{3}, \alpha = \frac{3}{2}, \beta = \frac{3}{2}$. Other choices give other sets of parameters that agree with the Taylor-series expansion. If we take $a = \frac{1}{2}$, the other variables are $b = \frac{1}{2}, \alpha = 1, \beta = 1$. This last set of parameters gives the modified Euler algorithm that we have previously discussed; the modified Euler method is a special case of a second-order Runge–Kutta method.

Fourth-order Runge–Kutta methods are most widely used and are derived in similar fashion. Greater complexity results from having to compare terms through h^4, and this gives a set of 11 equations in 13 unknowns. The set of 11 equations can be solved with 2 unknowns being chosen arbitrarily. The most commonly used set of values leads to the algorithm

$$y_{n+1} = y_n + \frac{1}{6}(k_1 + 2k_2 + 2k_3 + k_4),$$

$$k_1 = hf(x_n, y_n),$$

$$k_2 = hf\left(x_n + \frac{1}{2}h, y_n + \frac{1}{2}k_1\right),$$

$$k_3 = hf\left(x_n + \frac{1}{2}h, y_n + \frac{1}{2}k_2\right),$$

$$k_4 = hf(x_n + h, y_n + k_3).$$

$$\tag{6.10}$$

---

*Appendix A will remind readers of this expansion.

Using Eqs. (6.10) to apply the Runge–Kutta fourth-order to our problem, $dy/dx = -2x - y$, $y(0) = -1$ with $h = 0.1$, we obtain the results shown in Table 6.4. The results here are very impressive compared to those given in Table 6.1, where we computed the values using the terms of the Taylor series up to the h^4 term. Table 6.4 agrees to five decimals with the analytical result—illustrating a further gain in accuracy with less effort than with the Taylor-series method of Section 6.2—and it certainly is better than the Euler or modified Euler methods.

Figure 6.4 illustrates the four slope values that are combined in the four k's of the Runge–Kutta method.

The local error term for the fourth-order Runge–Kutta method is $O(h^5)$; the global error would be $O(h^4)$. It is computationally more efficient than the modified Euler method

Table 6.4

x_n	y_n	k_1	k_2	k_3	k_4	Average
0.0	−1.0000	0.1000	0.0850	0.0858	0.0714	0.0855
0.1	−0.9145	0.0715	0.0579	0.0586	0.0456	0.0583
0.2	−0.8562	0.0456	0.0333	0.0340	0.0222	0.0337
0.3	−0.8225	0.0222	0.0111	0.0117	0.0011	0.0115
0.4	−0.8110	0.0011	−0.0090	−0.0085	−0.0181	−0.0086
0.5	−0.81959					

$[y(0.5) = -0.81959$, the analytical value]

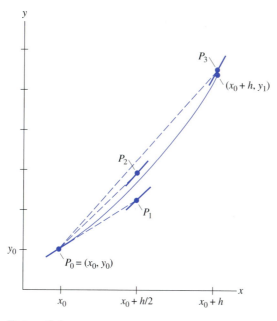

Figure 6.4

because, although four evaluations of the function are required per step rather than two, the steps can be manyfold larger for the same accuracy. The Runge–Kutta techniques have been very popular, especially the fourth-order method just presented. Because going from second to fourth order was so beneficial, we may wonder whether we should use a still higher order of formula. Higher-order (fifth, sixth, and so on) Runge–Kutta formulas have been developed and can be used to advantage in determining a suitable size for h, as we will see.

One way to determine whether the Runge–Kutta values are sufficiently accurate is to recompute the value at the end of each interval with the step size halved. If only a slight change in the value of y_{n+1} occurs, the results are accepted; if not, the step must be halved again until the results are satisfactory. This procedure is very expensive, however. For instance, to implement Eq. (6.10) this way, we would need an additional seven function evaluations to determine the accuracy of our y_{n+1}. The best case then would require $4 + 7 = 11$ function evaluations to go from (x_n, y_n) to (x_{n+1}, y_{n+1}).

A different approach uses two Runge–Kutta methods of different orders. For instance, we could use one fourth-order and one fifth-order method to move from (x_n, y_n) to (x_{n+1}, y_{n+1}). We would then compare our results at y_{n+1}. The Runge–Kutta–Fehlberg method, now one of the most popular of these methods, does just this. Only six functional evaluations (versus eleven) are required, and we also have an estimate of the error (the difference of the two y's at $x = x_{n+1}$):

An Algorithm for the Runge–Kutta–Fehlberg Method

$$k_1 = h \cdot f(x_n, y_n),$$

$$k_2 = h \cdot f\left(x_n + \frac{h}{4}, y_n + \frac{k_1}{4}\right),$$

$$k_3 = h \cdot f\left(x_n + \frac{3h}{8}, y_n + \frac{3k_1}{32} + \frac{9k_2}{32}\right),$$

$$k_4 = h \cdot f\left(x_n + \frac{12h}{13}, y_n + \frac{1932k_1}{2197} - \frac{7200k_2}{2197} + \frac{7296k_3}{2197}\right),$$

$$k_5 = h \cdot f\left(x_n + h, y_n + \frac{439k_1}{216} - 8k_2 + \frac{3680k_3}{513} - \frac{845k_4}{4104}\right), \tag{6.11}$$

$$k_6 = h \cdot f\left(x_n + \frac{h}{2}, y_n - \frac{8k_1}{27} + 2k_2 - \frac{3544k_3}{2565} + \frac{1859k_4}{4104} - \frac{11k_5}{40}\right);$$

$$\hat{y}_{n+1} = y_n + \left(\frac{25k_1}{216} + \frac{1408k_3}{2565} + \frac{2197k_4}{4104} - \frac{k_5}{5}\right), \text{ with global error } O(h^4),$$

$$y_{n+1} = y_n + \left(\frac{16k_1}{135} + \frac{6656k_3}{12825} + \frac{28561k_4}{56430} - \frac{9k_5}{50} + \frac{2k_6}{55}\right),$$

with global error $O(h^5)$;

$$\text{Error}, E \doteq \frac{k_1}{360} - \frac{128k_3}{4275} - \frac{2197k_4}{75240} + \frac{k_5}{50} + \frac{2k_6}{55}.$$

The basis for the Runge–Kutta–Fehlberg scheme is to compute two Runge–Kutta estimates for the new value of $\hat{y}_{n+1}$ but of different orders of errors. Thus, instead of comparing estimates of y_{n+1} for h and $h/2$, we compare the estimates $\hat{y}_{n+1}$ and y_{n+1} using fourth- and fifth-order (global) Runge–Kutta formulas. Moreover, both equations make use of the same k's, so only six function evaluations are needed versus the previous eleven. In addition, one can increase or decrease h depending on the value of the estimated error. As our estimate for the new y_{n+1}, we use the fifth-order (global) estimate.

As an example, we once more solve $dy/dx = -2x - y$, $y(0) = -1$ with $h = 0.1$, using the Runge–Kutta–Fehlberg method:

$$k_1 = 0.1,$$
$$k_2 = 0.0925000,$$
$$k_3 = 0.0889609,$$
$$k_4 = 0.0735157,$$
$$k_5 = 0.0713736,$$
$$k_6 = 0.0853872,$$

$$\hat{y}_1 = -0.914512212, \qquad y_1 = -0.914512251, \qquad \text{Error, } E = -0.000000040.$$

The exact value is $y(0.1) = -0.914512254$. Thus on the first step, y_1 agrees with the exact answer to 8 decimal places with only two additional function evaluations. Moreover, we have the value E to adjust our step size for the next iteration. Of course, we would use the more accurate y_{n+1} for the next step. This algorithm is well documented and implemented in the FORTRAN program, RKF45, of Forsythe, Malcolm, and Moler (1977). MATLAB has two numerical procedures, ode45 and ode23. Maple has rkf45 in its arsenal to get the numerical solution to differential equations.

A summary and comparison of the four numerical methods we have studied for solving $y' = f(x, y)$ is presented in Table 6.5.

To see empirically that the global errors of Table 6.5 hold, again consider the example $dy/dx = -2x - y$, $y(0) = -1$. Table 6.6 shows how the errors of $y(0.4)$ decrease as h is halved. The table shows the ratios of errors of successive calculations.

Table 6.5

Method	Estimate of slope over x-interval	Global error	Local error	Evaluations of $f(x, y)$ per step
Euler	Initial value	$O(h)$	$O(h^2)$	1
Modified Euler	Arithmetic average of initial and final predicted slope	$O(h^2)$	$O(h^3)$	2
Runge–Kutta (fourth-order)	Weighted average of four values	$O(h^4)$	$O(h^5)$	4
Runge–Kutta–Fehlberg	Weighted average of six values	$O(h^5)$	$O(h^6)$	6

Table 6.6

	Error in value computed at $x = 0.4$			Ratios of successive errors		
h	Euler	Modified Euler	Runge–Kutta 4th	Euler	Modified Euler	Runge–Kutta 4th
0.4000	2.11E-01	2.90E-02	2.40E-04			
0.2000	9.10E-02	6.42E-03	1.27E-05	3.3	4.5	18.9
0.1000	4.27E-02	1.44E-03	7.29E-07	2.1	4.5	17.4
0.0500	2.07E-02	3.48E-04	4.37E-08	2.1	4.1	16.7
0.0250	1.02E-02	8.54E-05	2.76E-09	2.0	4.1	15.8
0.0125	5.06E-03	2.11E-05	1.65E-10	2.0	4.0	16.7

In Table 6.6, we obtain the second row in this way: For a step size of $h = 0.2$, we compute the errors in the values for y at $x = 0.4$ using the three methods indicated at the top of columns two through four. We write down the values of the differences between the computed value and the analytical value. The last three columns represent the ratio between the previous error (larger step size h) and the current. For instance, the 3.3 in the second row is the ratio of 2.11E-01/9.10E-2 for the errors from Euler's method for $h = 0.4$ and $h = 0.2$. We do the same for the modified Euler method and the Runge–Kutta fourth-order method in columns six and seven. We see that as h gets smaller, the last three columns approach the ratios of 2.0, 4.0, and 16.0. This is what we expect, because these three methods are, respectively, $O(h)$, $O(h^2)$, and $O(h^4)$ and because at each stage the step size is halved.

We end this section by showing the Runge–Kutta–Merson method, another fourth-order method even though five different k's must be computed. It can be seen from the formula that the order is given, not by the number of k's, but by the global error.

$$k_1 = h \cdot f(x_n, y_n),$$

$$k_2 = h \cdot f\left(x_n + \frac{h}{3}, y_n + \frac{k_1}{3}\right),$$

$$k_3 = h \cdot f\left(x_n + \frac{h}{3}, y_n + \frac{k_1}{6} + \frac{k_2}{6}\right),$$

$$k_4 = h \cdot f\left(x_n + \frac{h}{2}, y_n + \frac{k_1}{8} + \frac{3k_3}{8}\right), \tag{6.12}$$

$$k_5 = h \cdot f\left(x_n + h, y_n + \frac{k_1}{2} - \frac{3k_3}{2} + 2k_4\right);$$

$$y_{n+1} = y_n + \frac{(k_1 + 4k_4 + k_5)}{6} + O(h^5);$$

$$\text{Error, } E \doteq \frac{1}{30}(2k_1 - 9k_3 + 8k_4 - k_5).$$

As we have already indicated, there are methods that use Runge–Kutta formulas of orders 5, 6, and higher. In fact, the IMSL routine DVERK uses formulas or orders 5 and 6 that were developed by J. H. Verner. In this case, the method uses eight function evaluations. Maple has an option in its procedure for solving differential equations that is called `'dverk78'`.

Although the Runge–Kutta method has been very popular in the past, it has its limitations in solving certain types of differential equations. However, for a large class of problems the methods presented in this section produce some very stunning results. Also the technique introduced by Fehlberg in comparing two different orders rather than halving step sizes increases the efficiency of the Runge–Kutta methods.

The methods so far discussed are called single-step methods. They use only the information at (x_n, y_n) to get to (x_{n+1}, y_{n+1}). In the next sections we examine methods that utilize past information from previous points to get (x_{n+1}, y_{n+1}).

6.5 Multistep Methods

Runge–Kutta-type methods (which include Euler and modified Euler as special cases) are called single-step methods because they use only the information from the last step computed. In this they have the ability to perform the next step with a different step size and are ideal for beginning the solution where only the initial conditions are available. After the solution has begun, however, there is additional information available about the function (and its derivative) if we are wise enough to retain it in the memory of the computer. A multistep method is one that takes advantage of this fact.

The principle behind a multistep method is to utilize the past values of y and/or y' to construct a polynomial that approximates the derivative function, and extrapolate this into the next interval. Most methods use equispaced past values to make the construction of the polynomial easy. The Adams method is typical. The number of past points that are used sets the degree of the polynomial and is therefore responsible for the truncation error. The order of the method is equal to the power of h in the global error term of the formula, which is also equal to one more than the degree of the polynomial.

To derive the relations for the Adams method, we write the differential equation $dy/dx = f(x, y)$ in the form

$$dy = f(x, y) \, dx,$$

and we integrate between x_n and x_{n+1}:

$$\int_{x_n}^{x_{n+1}} dy = y_{n+1} - y_n = \int_{x_n}^{x_{n+1}} f(x, y) \, dx.$$

To integrate the term on the right, we approximate $f(x, y)$ as a polynomial in x, deriving this by making it fit at several past points. If we use three past points, the approximating polynomial will be a quadratic. If we use four points, it will be a cubic. The more points we use, the better accuracy (until round-off interferes, of course).

Suppose that we fit a second-degree polynomial through the last three points: (x_n, f_n), (x_{n-1}, f_{n-1}), (x_{n-2}, f_{n-2}). We could use an interpolating polynomial formula, but, for variety, we will use *Mathematica* and the `InterpolatingPolynomial` function (see Section 3.9). When we perform the operation, we get a quadratic approximation to the derivative function (but displayed differently):

$$f(x, y) = \frac{1}{2}h^2(f_n - 2f_{n-1} + f_{n-2})x^2 + \frac{1}{2}h(3f_n - 4f_{n-1} + f_{n-2})x + f_n.$$

Now we again use *Mathematica* to integrate between the limits of $x = x_n$ and $x = x_{n+1}$. The result is a formula for the increment in y (again displayed differently):

$$y_{n+1} - y_n = \frac{h}{12}(23f_n - 16f_{n-1} + 5f_{n-2}),$$

and we have the formula to advance y:

$$y_{n+1} = y_n + \frac{h}{12}[23f_n - 16f_{n-1} + 5f_{n-2}] + O(h^4). \qquad \textbf{(6.13)}$$

Observe that Eq. (6.13) resembles the single-step formulas of the previous sections in that the increment to y is a weighted sum of the derivatives times the step size, but differs in that past values are used rather than estimates in the forward direction.

EXAMPLE 6.1   We illustrate the use of Eq. (6.13) to calculate $y(0.6)$ for $dy/dx = -2x - y$, $y(0) = -1$. We compute good values for $y(0.2)$ and $y(0.4)$ using a single-step method. In this case we obtain these values using the Runge–Kutta–Fehlberg method with $h = 0.2$. These values are given in Table 6.7.

Then, from Eq. (6.13), we have

$$y(0.6) = -0.81096 + \frac{0.2}{12}[23(0.01096) - 16(0.45619) + 5(1.0)]$$

$$= -0.84508.$$

Comparing our result with the exact solution (-0.84643), we find that the computed value has an error of 0.00135. We can reduce the size of the error by doing the calculations with a smaller step size of 0.1. We use the fifth-order values of the Runge–Kutta–Fehlberg method once again to obtain the values in Table 6.8.

Table 6.7

x	y	y, analytical	$f(x, y)$
0.0	−1.0000000	−1.0000000	1.0000000
0.2	−0.8561921	−0.8561923	0.4561921
0.4	−0.8109599	−0.8109601	0.0109599

Table 6.8

x	y	y, analytical	$f(x, y)$
0.0	−1.00000	−1.00000	1.00000
0.1	−0.91451	−0.91451	0.71451
0.2	−0.85619	−0.85619	0.45619
0.3	−0.82245	−0.82245	0.22245
0.4	−0.81096	−0.81096	0.01096
0.5	−0.81959	−0.81959	−0.18041

Using Eq. (6.13) again with the values for $f(x, y)$ at $x = 0.3$, $x = 0.4$, $x = 0.5$, from Table 6.8 we recompute $y(0.6)$:

$$y(0.6) = -0.81959 + \frac{0.1}{12}[23(-0.18041) - 16(0.01096) + 5(0.22245)]$$

$$= -0.84636,$$

which has an error of 0.00007.   ▲

In Section 6.7 we will see that we can reduce the error by using more past points for fitting a polynomial. In fact, when the derivation is done for four points to get a cubic approximation to $f(x, y)$, the following is obtained:

$$y_{n+1} = y_n + \frac{h}{24}[55f_n - 59f_{n-1} + 37f_{n-2} - 9f_{n-3}] + O(h^5). \quad \textbf{(6.14)}$$

We repeat Example 6.1 with $h = 0.1$ to compute $y(0.6)$, now using values from Table 6.8 for $x = 0.2$, $x = 0.3$, $x = 0.4$, and $x = 0.5$, and computing

$$y(0.6) = -0.81959 + \frac{0.1}{24}[55(-0.18041) - 59(0.01096)$$

$$+ 37(0.22245) - 9(0.45619)]$$

$$= -0.84644.$$

The error of this computation has been reduced to 0.00001. We summarize the results of these two formulas in Table 6.9.

Table 6.9

Number of points used	Estimate of $y(0.6)$	Error ($h = 0.1$)
3	−0.8463626	0.000072
4	−0.8464420	0.000007

6.6 Milne's Method

The method of Milne is a multistep approach that first predicts a value for y_{n+1} from three past values of the derivative. It differs from the Adams method in that it integrates over more than one interval. The past values that we require may have been computed by a Runge–Kutta method, or possibly by the Taylor-series method. In the Milne method, we suppose that four equispaced starting values of y are known, at the points x_n, x_{n-1}, x_{n-2}, and x_{n-3}. We can employ quadrature formulas to integrate as follows:

$$\frac{dy}{dx} = f(x, y),$$

$$\int_{x_{n-3}}^{x_{n+1}} \left(\frac{dy}{dx}\right) dx = \int_{x_{n-3}}^{x_{n+1}} f(x, y)\ dx \approx \int_{x_{n-3}}^{x_{n+1}} P_2(x)\ dx;$$

therefore

$$y_{n+1} - y_{n-3} =$$

$$\frac{4h}{3}(2f_n - f_{n-1} + 2f_{n-2}) + \frac{28}{90}h^5 y^{(5)}(\xi_1), \qquad x_{n-3} < \xi_1 < x_{n+1}. \tag{6.15}$$

We have integrated the function $f(x, y)$ by replacing it with a quadratic interpolating polynomial that fits at the three points, where $x = x_n$, x_{n-1}, and x_{n-2}, and integrating according to the methods of Chapter 5.* Note that we extrapolate in the integration by one panel both to the left and to the right of the region of fit. Hence the error is larger, because of the extrapolation, than it would be with only interpolation.

We can calculate f_{n+1} reasonably accurately with the value of y_{n+1}. In Milne's method, we use Eq. (6.15) as a predictor formula and then correct with

$$\int_{x_{n-1}}^{x_{n+1}} \left(\frac{dy}{dx}\right) dx = \int_{x_{n-1}}^{x_{n+1}} f(x, y)\ dx \approx \int_{x_{n-1}}^{x_{n+1}} P_2(x)\ dx,$$

$$y_{n+1,c} - y_{n-1} = \frac{h}{3}(f_{n+1} + 4f_n + f_{n-1}) - \frac{h^5}{90}y^{(5)}(\xi_2), \qquad x_{n-1} < \xi_2 < x_{n+1}. \tag{6.16}$$

In Eq. (6.16), the polynomial P_2 is not identical to that in Eq. (6.15) because they do not fit the function at the same three points. In Eq. (6.16) the polynomial fits at x_{n+1}, x_n, and x_{n-1}. Note the changed range of integration and the smaller coefficient in the error term of Eq. (6.16) because the polynomial is not extrapolated; f_{n+1} is calculated using y_{n+1} from the predictor formula. The integration formula is the familiar Simpson's $\frac{1}{3}$ rule, because we

---

*The formula can also be derived by the method of undetermined coefficients, as shown in Appendix B.

Table 6.10

x	y	$f(x, y)$
0.0	-1.0000000	1.0000000
0.1	-0.9145122	0.7145123
0.2	-0.8561923	0.4561923
0.3	-0.8224547	0.2224547
0.4	(-0.8109678) predicted	
	(-0.8109596) corrected	$(-0.8109601$ analytical)
0.5	(-0.8195969) predicted	
	(-0.8195906) corrected	$(-0.8195920$ analytical)

integrate a quadratic over two panels within the region of fit. In this method we do not try to recorrect y_{n+1}.

We illustrate once again with our familiar simple problem, $dy/dx = -2x - y$, $y(0) = -1$, where the first four values are calculated using the Runge–Kutta fifth-order formula. See Table 6.10.

Normally, the values of y_{n+1} from the predictor and the corrector do not agree. Consideration of the error terms of Eqs. (6.15) and (6.16) suggests that the true value should usually lie between the two values and closer to the corrector value. Although ξ_1 and ξ_2 are not necessarily the same value, they lie in similar intervals. If we assume that the values of $y^{(5)}(\xi_1)$ and $y^{(5)}(\xi_2)$ are equal, the error in the corrector formula is 1/28 times the error in the predictor formula. Hence the difference between the predictor and corrector formula is about 29 times the error in the corrected value. This is frequently used as a criterion of accuracy for Milne's method.* The ease with which we can monitor the error is a particular advantage of predictor–corrector methods. We are able to know immediately whether the step size is too big to give the desired degree of accuracy. This is in strong contrast to the Runge–Kutta method of the previous section (but not of Runge–Kutta–Fehlberg).

Milne's method is simple and has a good local error, $O(h^5)$. However, it is subject to instability in certain cases—the errors do not tend to zero as h is made smaller. Because of this instability, another method, a modification of Adams method, is more widely used than Milne's. In the next section, a numerical example demonstrates the instability of Milne's method. Even though such a numerical example is less conclusive than the analytical approach that we present in Section 6.11, the example is much easier to appreciate.

6.7 The Adams–Moulton Method

The Adams–Moulton method is a method that does not have the same instability problem as the Milne method, but is about as efficient. It also assumes a set of starting values already calculated by some other technique. Here we take a cubic through four points, from x_{n-3}

---

*A further criterion as to whether the single correction normally used in Milne's method is adequate is discussed in Section 6.8.

to x_n, and integrate over one step, from x_n to x_{n+1}. This is the same as the method described in Section 6.5. We assume that $dy/dx = f(x, y)$.

Thus getting our formula for the cubic interpolating polynomial by the methods explained in Chapter 3 (or through a computer algebra system) and then integrating from x_n to x_{n+1}, we get the Adams–Moulton *predictor formula:*

> Predictor: $$(6.17)$$
> $$y_{n+1} = y_n + \frac{h}{24}(55f_n - 59f_{n-1} + 37f_{n-2} - 9f_{n-3}) + \frac{251}{720}h^5 y^{(5)}(\xi_1).$$

(The last part of Eq. (6.17) is the error.) After we have computed a tentative value for f at x_{n+1}, we use it together with the f-values at x_n, x_{n-1}, and x_{n-2} to construct another cubic polynomial that we integrate from x_n to x_{n+1}. This gives the Adams–Moulton *corrector formula* (again with the last term being the error):

> Corrector: $$(6.18)$$
> $$y_{n+1} = y_n + \frac{h}{24}(9f_{n+1} + 19f_n - 5f_{n-1} + f_{n-2}) - \frac{19}{720}h^5 y^{(5)}(\xi_2).$$

(These formulas can also be developed with the method of undetermined coefficients, as shown in Appendix B.)

We illustrate the Adams–Moulton method using our earlier example, $dy/dx = -2x - y$, $y(0) = -1$. Using Eqs. (6.17) and (6.18), we construct Table 6.11. Here is how the entries in the table were obtained. By the predictor formula of (6.17), we get

$$y(0.4) = -0.8224547 + \frac{0.1}{24}[55(0.2224547) - 59(0.4561923)$$

$$+ 37(0.7145123) - 9(1.0)]$$

$$= -0.8109687.$$

Table 6.11

x	y	$f(x, y)$	
0.0	-1.0000000	1.0000000	
0.1	-0.9145122	0.7145123	
0.2	-0.8561923	0.4561923	
0.3	-0.8224547	0.2224547	
0.4	(-0.8109687) predicted		
	(-0.8109652) corrected		$(-0.8109601$ analytical)
0.5	(-0.8195978) predicted		
	(-0.8195905) corrected		$(-0.8195920$ analytical)

Then $f(0.4, -0.8109687)$ is computed, to get -0.0109688, and we use the corrector formula of Eq. (6.18) to get

$$y(0.4) = -0.8224547 + \frac{0.1}{24}[9(-0.0109688) + 19(0.2224547)$$

$$- 5(0.4561923) + 0.7145123]$$

$$= -0.8109652.$$

The computations are continued in the same manner to get $y(0.5)$. The corrected value almost agrees to five decimals with the predicted value. Comparing error terms of Eqs. (6.17) and (6.18) and assuming that the two fifth-derivative values are equal, we see that the true value should lie between the predicted and corrected values, with the error in the corrected value being about

$$\frac{19}{251 + 19} \quad \text{or} \quad \frac{1}{14.2}$$

times the difference between the predicted and corrected values. A frequently used criterion for accuracy of the Adams–Moulton method with four starting values is that the corrected value is not in error by more than 1 in the last place if the difference between predicted and corrected values is less than 14 in the last decimal place.* If this degree of accuracy is not met, we know that h is too large.

Changing the Step Size

When the predicted and corrected values agree to as many decimals as the desired accuracy, we can save computational effort by increasing the step size. We can conveniently double the step size, after we have seven equispaced values, by omitting every second one. When the difference between predicted and corrected values reaches or exceeds the accuracy criterion, we should decrease step size. If we interpolate two additional y-values with a fourth-degree polynomial, where the error will be $O(h^5)$, consistent with the rest of our work, we can readily halve the step size.† Convenient formulas for this are

$$y_{n-1/2} = \frac{1}{128}[35y_n + 140y_{n-1} - 70y_{n-2} + 28y_{n-3} - 5y_{n-4}], \quad \textbf{(6.19)}$$

$$\left(\frac{1}{128}\right)[-5y_n + 60y_{n-1} + 90y_{n-2} - 20y_{n-3} + 3y_{n-4}].$$

Use of these values with y_n, y_{n-1} gives four values of the function at intervals of $\Delta x = h/2$.

The efficiency of both the Adams–Moulton and Milne methods is about twice that of the Runge–Kutta–Fehlberg and Runge–Kutta methods. Only two function evaluations are

---

* The convergence criterion of Section 6.8 should also be met.
† An alternative but computationally more expensive way to get intermediate points would be to compute them with Runge–Kutta formulas.

needed per step for the former methods, whereas six or four are required with the single-step alternatives. All have similar error terms. Change of step size with the multistep methods is considerably more awkward, however.

Stability Considerations

We have stressed that the advantage of the Adams–Moulton method over that of Milne is that it is stable rather than unstable like Milne's method. An analytical proof of the stability of Adams–Moulton by examining a particular equation is not entirely satisfying because we cannot prove its stability for all cases in that way. (Proving that an assertion is *not* true is always easier because we need find only one counterexample.)

It is much clearer to compare the stability of Adams–Moulton with Milne through a numerical example. The table in Fig. 6.5 presents results from a computer program that

Values with $h = 0.04$	Solution using Milne method		Solution using Adams–Moulton method	
x	y	Error	y	Error
0.000000E 00	0.100000E 01	0.000000E 00	0.100000E 01	0.000000E 00
0.400000E−01	0.670320E 00	0.000000E 00	0.670320E 00	0.000000E 00
0.800000E−01	0.449329E 00	0.000000E 00	0.449329E 00	0.000000E 00
0.120000E 00	0.301195E 00	0.000000E 00	0.301195E 00	0.000000E 00
0.160000E 00	0.201667E 00	0.229836E−03	0.201552E 00	0.344634E−03
0.200000E 00	0.135271E 00	0.640750E−04	0.134873E 00	0.462234E−03
0.240000E 00	0.904572E−01	0.260949E−03	0.902753E−01	0.442863E−03
0.280000E 00	0.607594E−01	0.507981E−04	0.604129E−01	0.397373E−03
0.320000E 00	0.405497E−01	0.212613E−03	0.404273E−01	0.335068E−03
0.360000E 00	0.273069E−01	0.169165E−04	0.270547E−01	0.269119E−03
0.400000E 00	0.181599E−01	0.155795E−03	0.181052E−01	0.210475E−03
0.440000E 00	0.122874E−01	−0.100434E−04	0.121160E−01	0.161357E−03
0.480000E 00	0.811849E−02	0.111297E−03	0.810813E−02	0.121657E−03
0.520000E 00	0.554193E−02	−0.253431E−04	0.542601E−02	0.905804E−04
0.560000E 00	0.361742E−02	0.804588E−04	0.363111E−02	0.667726E−04
0.599999E 00	0.251041E−02	−0.316396E−04	0.242996E−02	0.488090E−04
0.639999E 00	0.160174E−02	0.598307E−04	0.162614E−02	0.354277E−04
0.679999E 00	0.114626E−02	−0.324817E−04	0.108822E−02	0.255615E−04
0.719999E 00	0.700667E−03	0.459237E−04	0.728243E−03	0.183482E−04
0.759999E 00	0.530960E−03	−0.305050E−04	0.487344E−03	0.131114E−04
0.799999E 00	0.299215E−03	0.362506E−04	0.326133E−03	0.933232E−05
0.839999E 00	0.252197E−03	−0.273281E−04	0.218250E−03	0.661935E−05
0.879999E 00	0.121497E−03	0.292376E−04	0.146054E−03	0.468052E−05
0.919999E 00	0.124879E−03	−0.238383E−04	0.977399E−04	0.330036E−05
0.959999E 00	0.437888E−04	0.239405E−04	0.654081E−04	0.232129E−05
0.999999E 00	0.658748E−04	−0.204745E−04	0.437714E−04	0.162897E−05
0.104000E 01	0.106330E−04	0.197998E−04	0.292920E−04	0.114074E−05
0.108000E 01	0.378257E−04	−0.174260E−04	0.196024E−04	0.797314E−06
0.112000E 01	−0.280455E−05	0.164789E−04	0.131180E−04	0.556327E−06
0.116000E 01	0.239183E−04	−0.147521E−04	0.877864E−05	0.387549E−06

Figure 6.5 Comparison of relative error in results obtained by the Milne and Adams–Moulton methods

	Solution using Milne method		Solution using Adams–Moulton method	
x	y	Error	y	Error

Values with $h = 0.04$

x	y	Error	y	Error
0.120000E 01	−0.762395E−05	0.137682E−04	0.587471E−05	0.269573E−06
0.124000E 01	0.165678E−04	−0.124492E−04	0.393138E−05	0.187252E−06
0.128000E 01	−0.876950E−05	0.115303E−04	0.263090E−05	0.129902E−06
0.132000E 01	0.123371E−04	−0.104865E−04	0.176061E−05	0.900100E−07
0.136000E 01	−0.842895E−05	0.966946E−05	0.117821E−05	0.622986E−07
0.140000E 01	0.965523E−05	−0.882369E−05	0.788466E−06	0.430742E−07
0.144000E 01	−0.755810E−05	0.811550E−05	0.527645E−06	0.297529E−07
0.148000E 01	0.779361E−05	−0.741997E−05	0.353103E−06	0.205325E−07
0.152000E 01	−0.656401E−05	0.681446E−05	0.236298E−06	0.141571E−07
0.156000E 01	0.640522E−05	−0.623733E−05	0.158132E−06	0.975331E−08
0.160000E 01	−0.561102E−05	0.572355E−05	0.105823E−06	0.671417E−08
0.164000E 01	0.531755E−05	−0.524211E−05	0.708171E−07	0.461910E−08
0.168000E 01	−0.475746E−05	0.480803E−05	0.473911E−07	0.317556E−08
0.172000E 01	0.443906E−05	−0.440517E−05	0.317144E−07	0.218119E−08
0.176000E 01	−0.401659E−05	0.403931E−05	0.212234E−07	0.149759E−08
0.180000E 01	0.371683E−05	−0.370160E−05	0.142028E−07	0.102763E−08
0.184000E 01	−0.338345E−05	0.339366E−05	0.950460E−08	0.704578E−09
0.188000E 01	0.311713E−05	−0.311028E−05	0.636053E−08	0.482931E−09
0.192000E 01	−0.284671E−05	0.285129E−05	0.425650E−08	0.330768E−09
0.196000E 01	0.261645E−05	−0.261337E−05	0.284847E−08	0.226487E−09
0.200000E 01	−0.239358E−05	0.239564E−05	0.190621E−08	0.155008E−09

Values near $x = 2.0$ with $h = 0.004$

x	y	Error	y	Error
0.191192E 01	−0.734619E−05	0.735117E−05	0.496895E−08	0.425615E−11
0.191592E 01	0.744385E−05	−0.743908E−05	0.477411E−08	0.405009E−11
0.191992E 01	−0.752346E−05	0.752805E−05	0.458691E−08	0.392220E−11
0.192392E 01	0.762249E−05	−0.761808E−05	0.440706E−08	0.379785E−11
0.192792E 01	−0.770495E−05	0.770919E−05	0.423425E−08	0.368061E−11
0.193192E 01	0.780545E−05	−0.780138E−05	0.406822E−08	0.350298E−11
0.193592E 01	−0.789078E−05	0.789469E−05	0.390870E−08	0.339284E−11
0.193992E 01	0.799286E−05	−0.798910E−05	0.375544E−08	0.328626E−11
0.194392E 01	−0.808104E−05	0.808465E−05	0.360819E−08	0.318456E−11
0.194792E 01	0.818480E−05	−0.818133E−05	0.346671E−08	0.303002E−11
0.195192E 01	−0.827585E−05	0.827918E−05	0.333078E−08	0.293365E−11
0.195592E 01	0.838139E−05	−0.837819E−05	0.320017E−08	0.284017E−11
0.195992E 01	−0.847532E−05	0.847839E−05	0.307469E−08	0.274958E−11
0.196392E 01	0.858274E−05	−0.857979E−05	0.295413E−08	0.261657E−11
0.196792E 01	−0.867956E−05	0.868240E−05	0.283830E−08	0.253331E−11
0.197192E 01	0.878896E−05	−0.878623E−05	0.272701E−08	0.245226E−11
0.197592E 01	−0.888869E−05	0.889132E−05	0.262008E−08	0.237388E−11
0.197992E 01	0.900017E−05	−0.899765E−05	0.251735E−08	0.225930E−11
0.198392E 01	−0.910284E−05	0.910526E−05	0.241864E−08	0.218714E−11
0.198792E 01	0.921648E−05	−0.921415E−05	0.232380E−08	0.211720E−11
0.199192E 01	−0.932212E−05	0.932435E−05	0.223268E−08	0.204925E−11
0.199592E 01	0.943801E−05	−0.943586E−05	0.214514E−08	0.195066E−11
0.199992E 01	−0.954665E−05	0.954871E−05	0.206103E−08	0.188805E−11

Figure 6.5 *Continued*

solved $y' = -10y$, $y(0) = 1$ (for which the analytical solution is $y = e^{-10x}$) over the interval $x = 0$ to $x = 2$. The first part of the table is computed with $h = 0.04$. In the second part, for which only partial output is given, $h = 0.004$.

Observe that the results with the Milne method grow to have very large relative errors. In fact, near $x = 2.0$, the relative error is practically 100%. The oscillatory behavior of the solution is also characteristic of instability. Even with the smaller h, the Milne method blows up. The errors are even greater near $x = 2.0$ in spite of the smaller step size. (When x is small the Milne solution is considerably more accurate when $h = 0.004$. Results are not shown for this.)

The results by the Adams–Moulton method do not show this anomalous behavior. The relative error, although growing to some degree, still stays manageable (8% at $x = 2.0$ with $h = 0.04$, <0.1% at $x = 2.0$ with $h = 0.004$). The expected decrease of error with decrease in step size is realized. The oscillation of values we notice in the results by Milne is absent. In sum, we conclude that the Adams–Moulton method gives good results, particularly at the smaller value of h, whereas the results of the Milne method are hopeless.

It is worth remarking that we usually don't have analytical answers to compare with our numerical results. Observing oscillatory behavior in itself does not mean instability, because the correct solution may *be* oscillatory. For this reason, the usual practice in numerical analysis is to entirely avoid methods that are sometimes unstable, even though they might be more accurate in some instances.

6.8 Convergence Criteria

In Section 6.3, we recorrected in the modified Euler method until no further change in y_{n+1} resulted. Usually this requires one more calculation than would otherwise be needed if we could predict whether the recorrection would make a significant change. In the methods of Milne and Adams–Moulton, we usually do not recorrect, but use a value of h small enough that this is unnecessary. We now look for a criterion to show how small h should be in the Adams–Moulton method, for $dy/dx = f(x, y)$, so that recorrections are not necessary. Let

$$y_p = \text{value of } y_{n+1} \text{ from predictor formula,}$$
$$y_c = \text{value of } y_{n+1} \text{ from corrector formula,}$$
$$y_{cc}, y_{ccc}, \text{ etc.} = \text{values of } y_{n+1} \text{ if successive recorrections are made,}$$
$$y_\infty = \text{value to which successive recorrections converge,}$$
$$D = y_c - y_p.$$

The change of y_c by recorrecting would be

$$y_{cc} - y_c = \left(y_n + \frac{h}{24}(9y_c' + 19y_n' - 5y_{n-1}' + y_{n-2}')\right)$$
$$- \left(y_n + \frac{h}{24}(9y_p' + 19y_n' - 5y_{n-1}' + y_{n-2}')\right) \tag{6.20}$$
$$= \frac{9h}{24}(y_c' - y_p').$$

In Eq. (6.20) we have used the subscript p or c to denote which y-value is used in evaluating the derivative at $x = x_{n+1}$. We now manipulate the difference $(y'_c - y'_p)$:

$$y'_c - y'_p = f(x_{n+1}, y_c) - f(x_{n+1}, y_p) = \frac{f(x_{n+1}, y_c) - f(x_{n+1}, y_p)}{(y_c - y_p)}(y_c - y_p)$$

$$= f_y(\xi_1)D, \qquad \xi_1 \text{ between } y_c \text{ and } y_p, \text{ with } D = y_c - y_p.$$

Hence

$$y_{cc} - y_c = \frac{9hD}{24}f_y(\xi_1)$$

is the difference on recorrecting. If recorrected again, the result is

$$y_{ccc} - y_{cc} = \frac{9h}{24}(y'_{cc} - y'_c)$$

$$= \frac{9h}{24}f_y(\xi_2) \cdot (y_{cc} - y_c)$$

$$= \frac{9h}{24}f_y(\xi_2)\left[\frac{9hD}{24}f_y(\xi_1)\right] \qquad \textbf{(6.21)}$$

$$= \left(\frac{9h}{24}\right)^2 [f_y(\xi)]^2 D, \qquad \xi \text{ between } y_c \text{ and } y_{cc}.$$

In Eq. (6.21), we need to impose the restrictions that f be continuous and $f_y(\xi_2)$ have the same sign; ξ lies between the extremes of y_p, y_c, and y_{cc}.

On further recorrections we will have a similar relation. We get y_∞ by adding all the corrections of y_p together:

$$y_\infty = y_p + (y_c - y_p) + (y_{cc} - y_c) + (y_{ccc} - y_{cc}) + \cdots$$

$$= y_p + D + \frac{9hf_y(\xi)}{24}D + \left(\frac{9hf_y(\xi)}{24}\right)^2 D + \left(\frac{9hf_y(\xi)}{24}\right)^3 D + \cdots .$$

The increment to y_p is a geometric series; so, if the ratio is less than unity, which is necessary if a geometric series is to have a sum,

$$y_\infty = y_p + \frac{D}{1 - r}, \qquad r = \frac{9hf_y(\xi)}{24}, \qquad \xi \text{ between } y_p \text{ and } y_\infty.$$

Hence, unless

$$|r| = \frac{h|f_y(\xi)|}{24/9} \doteq \frac{h|f_y(x_n, y_n)|}{24/9} < 1,$$

the successive recorrections diverge. Our first convergence criterion is

$$h < \frac{24/9}{|f_y(x_n, y_n)|}. \qquad \textbf{(6.22)}$$

If we wish to have y_c and y_∞ the same to within one in the Nth decimal place, then

$$y_\infty - y_c = \left(y_p + \frac{D}{1 - r}\right) - (y_p + D) = \frac{rD}{1 - r} < 10^{-N}.$$

If $r \ll 1$, the fraction

$$\frac{r}{1 - r} \doteq r;$$

and a second convergence criterion, which ensures that the first corrected value is adequate (that is, it will not be changed in the Nth decimal place by further corrections), is

$$D \cdot 10^N < \left| \frac{1}{r} \right| \doteq \frac{24/9}{h|f_y(x_n, y_n)|}. \qquad \textbf{(6.23)}$$

For the Adams–Moulton method we have the following three criteria. If all are met, the corrected value should be good to N decimals.

Convergence criteria: $\begin{cases} h < \dfrac{24/9}{|f_y|}, \\ D \cdot 10^N < \dfrac{24/9}{h|f_n|}; \end{cases}$

Accuracy criterion: $D \cdot 10^N < 14.2.$

Similar criteria for the Milne method are derived in the same way. They are

Convergence criteria: $\begin{cases} h < \dfrac{3}{|f_y|}, \\ D \cdot 10^N < \dfrac{3}{h|f_y|}; \end{cases}$ **(6.24)**

Accuracy criterion: $D \cdot 10^N < 29.$

These criteria are for a single first-order equation only. A similar analysis for a system is much more complicated.

We illustrate the use of these criteria with an example. Given the equation $dy/dx = \sin x - 3y$, $y(0) = 1$. In the neighborhood of the point $(1, 0.3)$, what maximum value of h is permitted if we wish to compute by (1) Milne's method, (2) the Adams–Moulton method, and get accuracy to five decimals? How close must the predictor and corrector values be?

1. For Milne:

$$f_y = -3, \quad \text{so} \quad h < \frac{3}{|-3|} = 1$$

to ensure convergence if we were to recorrect y_c in the Milne method. This requirement is not severe—we certainly would choose Δx smaller than this to give information throughout the range of x-values from 0 to 1. Suppose h were taken as 0.2. Then

$$D < \frac{3}{h|f_y|10^N} = \frac{3}{0.2|-3|10^5} = 5.0 \times 10^{-5}.$$

The difference between y_p and y_c cannot exceed this value; if it does, recorrections will be needed. This is more severe than $D < 29 \times 10^{-5}$ required for accuracy to one in the fifth decimal for y_c. We should monitor a Milne program to be sure that the difference between y_p and y_c does not exceed 5×10^{-5}. If it should exceed this value, we would need to reduce h so that this criterion could be met.

2. For Adams–Moulton, we calculate

$$h < \frac{24/9}{|-3|} = 0.89.$$

If $h = 0.2$,

$$D < \frac{24/9}{0.2|-3|10^5} = 4.4 \times 10^{-5}.$$

Again, this difference in y_p and y_c is more severe than $D < 14.2 \times 10^{-5}$ and should be used to control the value of h.

6.9 Systems of Equations and Higher-Order Equations

We have so far treated only the case of a first-order differential equation. Most differential equations that are the mathematical model for a physical problem are of higher order, or even a *set* of simultaneous higher-order differential equations. For example,

$$\frac{w}{g}\frac{d^2x}{dt^2} + b\frac{dx}{dt} + kx = f(x, t)$$

represents a vibrating system in which a linear spring with spring constant k restores a displaced mass of weight w against a resisting force whose resistance is b times the velocity. The function $f(x, t)$ is an external forcing function acting on the mass.

An analogous second-order equation describes the flow of electricity in a circuit containing inductance, capacitance, and resistance. The external forcing function in this case represents the applied electromotive force. Compound spring–mass systems and electrical networks can be simulated by a system of such second-order equations.

We first show how a higher-order differential equation can be reduced to a system of simultaneous first-order equations. We then show that these can be solved by an application of the methods previously studied. We treat here initial-value problems only, for which n values of the functions or derivatives (with n equal to the order of the system) are all specified at the same (initial) value of the independent variable. When some of the conditions are specified at one value of the independent variable and others at a second value, we call it a *boundary-value problem.* Methods of solving these are discussed in the next chapter.

By solving for the second derivative, we can normally express a second-order equation as

$$\frac{d^2x}{dt^2} = f\left(t,\ x,\ \frac{dx}{dt}\right), \qquad x(t_0) = x_0, \qquad x'(t_0) = x_0'. \tag{6.25}$$

The initial value of the function x and its derivative are generally specified. We convert this to a pair of first-order equations by the simple expedient of defining the derivative as a second function. Then, because $d^2x/dt^2 = (d/dt)(dx/dt)$,

$$\frac{dx}{dt} = y, \qquad x(t_0) = x_0,$$

$$\frac{dy}{dt} = f(t,\ x,\ y), \qquad y(t_0) = x_0'.$$

This pair of first-order equations is equivalent to the original Eq. (6.25). For even higher orders, each of the lower derivatives is defined as a new function, giving a set of n first-order equations that correspond to the nth-order differential equation. For a system of higher-order equations, each is similarly converted, so that a larger set of first-order equations results. Thus the nth-order differential equation

$$y^{(n)} = f(x,\ y,\ y',\ \ldots,\ y^{(n-1)}),$$

$$y(x_0) = A_1, \qquad y'(x_0) = A_2, \qquad \ldots, \qquad y^{(n-1)}(x_0) = A_n$$

is converted into a system of n first-order differential equations by letting $y_1 = y$ and

$$y_1' = y_2,$$
$$y_2' = y_3,$$
$$\vdots$$
$$y_{n-1}' = y_n,$$
$$y_n' = f(x,\ y_1,\ y_2,\ \ldots,\ y_n);$$

with initial conditions

$$y_1(x_0) = A_1, \qquad y_2(x_0) = A_2, \qquad \ldots, \qquad y_n(x_0) = A_n.$$

We now illustrate the application of the various methods to the pair of first-order equations:

$$\frac{dx}{dt} = xy + t, \qquad x(0) = 1,$$

$$\frac{dy}{dt} = ty + x, \qquad y(0) = -1. \tag{6.26}$$

Taylor-Series Method

We need the various derivatives $x', x'', x''', \ldots, y', y'', y''', \ldots$, all evaluated at $t = 0$:

$$x' = xy + t, \qquad\qquad x'(0) = (1)(-1) + 0 = -1$$
$$y' = ty + x, \qquad\qquad y'(0) = (0)(-1) + 1 = 1,$$
$$x'' = xy' + x'y + 1, \qquad x''(0) = (1)(1) + (-1)(-1) + 1 = 3,$$
$$y'' = y + ty' + x', \qquad y''(0) = -1 + (0)(1) - 1 = -2,$$
$$x''' = x'y' + xy'' + x''y + x'y', \qquad x'''(0) = -7,$$
$$y''' = y' + y' + ty'' + x'', \qquad y'''(0) = 5,$$

and so on;   and so on;

$$x(t) = 1 - t + \frac{3}{2}t^2 - \frac{7}{6}t^3 + \frac{27}{24}t^4 - \frac{124}{120}t^5 + \cdots,$$

$$y(t) = -1 + t - t^2 + \frac{5}{6}t^3 - \frac{13}{24}t^4 + \frac{47}{120}t^5 + \cdots.$$

(6.27)

At $t = 0.1$, $x = 0.9139$ and $y = -0.9092$.

Equations (6.27) are the solution to the set (6.26). Note that we need to alternate between the functions in getting the derivatives; for example, we cannot get $x''(0)$ until $y'(0)$ is known; we cannot get $y'''(0)$ until $x''(0)$ is known. After we have obtained the coefficients of the Taylor-series expansions in Eq. (6.27), we can evaluate x and y at any value of t, but the error will depend on how many terms we employ.

Euler Predictor–Corrector

We apply the predictor to each equation; then the corrector can be used. Again note that we work alternately with the two functions.

Take $h = 0.1$. Let p and c subscripts indicate predicted and corrected values, respectively:

$$x_p(0.1) = 1 + 0.1[(1)(-1) + 0] = 0.9,$$
$$y_p(0.1) = -1 + 0.1[(0)(-1) + 1] = -0.9,$$
$$x_c(0.1) = 1 + 0.1\left(\frac{-1 + [(0.9)(-0.9) + 0.1]}{2}\right) = 0.9145,$$
$$y_c(0.1) = -1 + 0.1\left(\frac{1 + [(0.1)(-0.9) + 0.9145]}{2}\right) = -0.9088.$$

In computing $x_c(0.1)$, we used the x_p and y_p. In computing $y_c(0.1)$ after $x_c(0.1)$ is known, we have a choice between x_p and x_c. There is an intuitive feel that one should use x_c, with the idea that one should always use the best available values. This does not always expedite convergence, probably due to compensating errors. Here we have used the best values to date. Recorrecting in the obvious manner gives

$$x(0.1) = 0.9135,$$
$$y(0.1) = -0.9089.$$

We can now advance the solution another step if desired, by using the computed values at $t = 0.1$ as the starting values. From this point we can advance one more step, and so on for any value of t. The errors will be the combination of local truncation error at each step plus the propagated error resulting from the use of inexact starting values.

Runge–Kutta–Fehlberg

Again there is an alternation between the x and y calculations. In applying this method, one always uses the previous k-value in incrementing the function values and the value of h to increment the independent variable. As in the previous calculations, we alternate between computations for x and for y; for example, we do $k_{1,x}$, then $k_{1,y}$, before doing $k_{2,x}$, and so on.

Keeping in mind that the equations are

$$\frac{dx}{dt} = f(t,\ x,\ y) = xy + t, \qquad x(0) = 1,$$

$$\frac{dy}{dt} = g(t,\ x,\ y) = ty + x, \qquad y(0) = -1,$$

the k-values for x and y are

for x:

$$
\begin{aligned}
k_{1,x} &= hf(0,\ 1,\ -1) \\
&= 0.1[(1)(-1) + 0] \\
&= -0.1;
\end{aligned}
$$

$$
\begin{aligned}
k_{2,x} &= hf(0.025,\ 0.975,\ -0.975) \\
&= 0.1[(0.975)(-0.975) + 0.025] \\
&= -0.092562;
\end{aligned}
$$

$$
\begin{aligned}
k_{3,x} &= hf(0.038,\ 0.965,\ -0.964) \\
&= 0.1[(0.965)(-0.964) + 0.038] \\
&= -0.089226;
\end{aligned}
$$

$$
\begin{aligned}
k_{4,x} &= hf(0.092,\ 0.919,\ -0.915) \\
&= 0.1[(0.919)(-0.915) + 0.092] \\
&= -0.074892;
\end{aligned}
$$

$$
\begin{aligned}
k_{5,x} &= hf(0.1,\ 0.913,\ -0.908) \\
&= 0.1[(0.913)(-0.908) + 0.1] \\
&= -0.072904;
\end{aligned}
$$

$$
\begin{aligned}
k_{6,x} &= hf(0.05,\ 0.954,\ -0.953) \\
&= 0.1[(0.954)(-0.953) + 0.05] \\
&= -0.085868.
\end{aligned}
$$

for y:

$$
\begin{aligned}
k_{1,y} &= hg(0,\ 1,\ -1) \\
&= 0.1[(0)(-1) + 1] \\
&= 0.1;
\end{aligned}
$$

$$
\begin{aligned}
k_{2,y} &= hg(0.025,\ 0.975,\ -0.975) \\
&= 0.1[(0.025)(-0.975) + 0.975] \\
&= 0.095062;
\end{aligned}
$$

$$
\begin{aligned}
k_{3,y} &= hg(0.038,\ 0.965,\ -0.964) \\
&= 0.1[(0.038)(-0.964) + 0.095] \\
&= 0.092845;
\end{aligned}
$$

$$
\begin{aligned}
k_{4,y} &= hg(0.092,\ 0.919,\ -0.915) \\
&= 0.1[(0.092)(-0.915) + 0.919] \\
&= 0.083461;
\end{aligned}
$$

$$
\begin{aligned}
k_{5,y} &= hg(0.1,\ 0.913,\ -0.908) \\
&= 0.1[(0.1)(-0.908) + 0.913] \\
&= 0.082178;
\end{aligned}
$$

$$
\begin{aligned}
k_{6,y} &= hg(0.05,\ 0.954,\ -0.953) \\
&= 0.1[(0.05)(-0.953) + 0.954] \\
&= 0.090628.
\end{aligned}
$$

Then using the fifth-order formula, we get

$$x(0.1) = 1 + (-0.01185 - 0.046307 - 0.037905 + 0.013123 - 0.003122)$$
$$= 0.913936;$$
$$y(0.1) = -1 + (0.01185 + 0.048185 + 0.042242 - 0.014792 + 0.003296)$$
$$= -0.909217.$$

Extending the Taylor-series solution even further shows that the Runge–Kutta–Fehlberg values are correct to more than five decimals, whereas the modified Euler values are correct to only three, so $h = 0.1$ may be too large for that method.

Advancing the solution by the Runge–Kutta–Fehlberg method will again involve using the computed values of x and y as the initial values for another step. The errors here will be much less than those for the Euler predictor–corrector method.

Adams–Moulton

After getting four starting values, we proceed with the algorithm of Eqs. (6.17) and (6.18), again alternately computing x and then y (see Table 6.12.)

In the computations we first get predicted values of x and y:

$$x(0.1) = 0.9330 + \frac{0.025}{24}[55(-0.7929) - 59(-0.8582) + 37(-0.9271) - 9(-1.0)]$$
$$= 0.913937;$$
$$y(0.1) = -0.9303 + \frac{0.025}{24}[55(0.8632) - 59(0.9060) + 37(0.9515) - 9(1.0)]$$
$$= -0.909217.$$

Table 6.12

	t	x	x'	t	y	y'
Starting values	0.000	1.0	−1.0	0.00	−1.0	1.0
	0.025	0.9759	−0.9271	0.025	−0.9756	0.9515
	0.050	0.9536	−0.8582	0.050	−0.9524	0.9060
	0.075	0.9330	−0.7929	0.075	−0.9303	0.8632
Predicted	0.10	(0.9139)	(−0.7310)	0.10	(−0.9092)	(0.8230)
Corrected		0.9139			−0.9092	

After getting x' and y' at $t = 0.1$, using $x(0.1)$ and $y(0.1)$, we then correct:

$$x(0.1) = 0.9330 + \frac{0.025}{24}[9(-0.7310) + 19(-0.7929) - 5(-0.8582) + (-0.9271)]$$

$$= 0.913936;$$

$$y(0.1) = -0.9303 + \frac{0.025}{24}[9(0.8230) + 19(0.8632) - 5(0.9060) + (0.9515)]$$

$$= -0.909217.$$

The close agreement of predicted and corrected values indicates six-decimal-place accuracy.

In this method, as we advance the solution to larger values of t, the comparison between predictor and corrector values tells us whether the step size needs to be changed.

6.10   Comparison of Methods/Stiff Equations

Comparison of Methods

It is appropriate that we summarize the various methods that have been discussed in this chapter and compare them. Table 6.13 compares the accuracy, effort required, stability, and other features of the methods.

The data in Table 6.13 lead us to draw the usual conclusion about the best scheme for solving a differential equation of higher order, or a system of N first-order equations: We begin with a fourth- or fifth-order Runge–Kutta to get a total of four values for each of the functions (this also allows us to compute four values for each of the derivatives), and

Table 6.13   Comparison of methods for differential equations

Method	Type	Local error	Global error	Function evalua-tions/ step	Stability	Ease of changing step size	Recom-mended?
Modified Euler	Single-step	$O(h^3)$	$O(h^2)$	2	Good	Good	No
Fourth-order Runge–Kutta	Single-step	$O(h^5)$	$O(h^4)$	4	Good	Good	Yes
Runge–Kutta–Fehlberg	Single-step	$O(h^6)$	$O(h^5)$	6	Good	Good	Yes
Milne	Multistep	$O(h^5)$	$O(h^4)$	2	Poor	Poor	No
Adams–Moulton	Multistep	$O(h^5)$	$O(h^4)$	2	Good	Poor	Yes

then advance the solution with Adams–Moulton. At each step after employing Adams–Moulton, we check* the accuracy and adjust the step size when appropriate.

There is still a problem during the starting phase when Runge–Kutta is being used. For instance, when a Runge–Kutta–Fehlberg method (Section 6.4) is used to start the solution for a multistep method that needs equispaced function values, an additional restriction is imposed: the step size must be uniform. This may mean that closer spaced values may need to be computed than are required to meet the accuracy criterion alone.

A method due to Hamming has been widely accepted. It is available through subroutines in some FORTRAN libraries. It begins the solution with a fourth-order Runge–Kutta, and then continues with a predictor–corrector. The equations employed are

$$y_{i+1,p} = y_{i-3} + \frac{4h}{3}(2f_i - f_{i-1} + 2f_{i-2}),$$

$$y_{i+1,m} = y_{i+1,p} - \frac{112}{121}(y_{i,p} - y_{i,c}),$$

$$y_{i+1,c} = \frac{1}{8}[9y_i - y_{i-2} + 3h(f_{i+1,m} + 2f_i - f_{i-1})], \tag{6.28}$$

$$y_{i+1} = y_{i+1,c} + \frac{9}{121}(y_{i+1,p} - y_{i+1,c}).$$

The predictor equation is the Milne predictor. Before correcting, the estimate of y_{i+1} is modified using the difference between the predicted and corrected values in the *previous* interval (this is omitted in the first interval, as these are not available). A corrector formula is used that depends on two previous y-values, though heavily weighted in favor of the last one. Finally, an adjustment is made based on the error estimate computed from the difference between predicted and corrected values. Note that, although two additional equations are employed in each step compared to the predictor–corrector methods previously described, only two evaluations of the derivative function are needed, the same as before. For many applications, it is the evaluation of the derivative function that is costly in computer time—the two extra algebraic steps don't count for much.

The special merits of Hamming's method are stability combined with good accuracy. Like Milne's and Adams's, the method has a local error of $O(h^5)$ and a global error of $O(h^4)$.

Gear (1967) has proposed a predictor–corrector method that has an $O(h^6)$ local error but uses only three previous steps rather than the four previous steps employed by Adams–Moulton and Milne. It obtains its high order of error by using recorrected values of the function and derivative values. The formulas are

---

*Ordinarily a weighted average of the N errors is monitored. Alternatively, the maximum error is controlled.

$$y_{n+1,p} = -18y_n + 9y_{n-1,c_1} + 10y_{n-2,c_2} + 9hy'_n + 18hy'_{n-1} + 3hy'_{n-2},$$

$$hy'_{n+1,p} = -57y_n + 24y_{n-1,c_1} + 33y_{n-2,c_2} + 24hy'_n + 57hy'_{n-1} + 10hy'_{n-2},$$

$$F = hy'_{n+1,p} - hf(x_{n+1}, y_{n+1,p}),$$

$$y_{n+1} = y_{n+1,p} - \frac{95}{288}F,$$

$$y_{n,c_1} = y_n + \frac{3}{160}F,$$

$$y_{n-1,c_2} = y_{n-1,c_1} - \frac{11}{1440}F,$$

$$hy'_{n+1} = hy'_{n+1,p} - F.$$

This method is stable and is applicable to systems of first-order differential equations. Gear (1971) gives a listing and a complete description of a subroutine named DIFFSUB, which includes both the Adams predictor–corrector method and Gear's stiff methods. Gear's method is also included in the Maple V package for solving differential equations.

Stiff Equations

A stiff equation results from phenomena with widely differing time scales. For example, the general solution of a differential equation may involve sums or differences of terms of the form ae^{ct}, be^{dt}, where both c and d are negative but c is much smaller than d. In such cases, using a small value for the step size can introduce enough round-off errors to cause instability.

An example is the following:

$$x' = 1195x - 1995y, \qquad x(0) = 2,$$

$$y' = 1197x - 1997y, \qquad y(0) = -2.$$

(6.29)

The analytical solution of Eq. (6.29) is

$$x(t) = 10e^{-2t} - 8e^{-800t}, \qquad y(t) = 6e^{-2t} - 8e^{-800t}.$$

Observe that the exponents are all negative and of very different magnitude, qualifying this as a stiff equation. Suppose we solve Eq. (6.29) by the simple Euler method with $h = 0.1$, applying just one step. The iterations are

$$x_{i+1} = x_i + hf(x_i, y_i) = x_i + 0.1(1195x_i - 1995y_i),$$

$$y_{i+1} = y_i + hg(x_i, y_i) = y_i + 0.1(1197x_i - 1997y_i).$$

This gives $x(0.1) = 640$, $y(0.1) = 636$, while the analytical values are $x(0.1) = 8.187$ and $y(0.1) = 4.912$. Such a result is typical (though here exaggerated) for stiff equations.

One solution to this problem is to use an implicit method rather than an explicit one. All the methods so far discussed have been explicit, meaning that the new values, x_{i+1} and y_{i+1}, are computed in terms of the previous ones, x_i and y_i. The implicit form of Euler's method is

$$x_{i+1} = x_i + hf(x_{i+1}, y_{i+1}),$$
$$y_{i+1} = y_i + h(x_{i+1}, y_{i+1}).$$

(6.30)

If the derivative functions $f(x, y)$ and $g(x, y)$ are nonlinear, this is difficult to solve. However, in Eq. (6.29) they are linear. Solving Eq. (6.29) by use of Eq. (6.30) we have

$$x_{i+1} = x_i + 0.1(1195x_{i+1} - 1995y_{i+1}),$$
$$y_{i+1} = y_i + 0.1(1197x_{i+1} - 1997y_{i+1}).$$

The system is linear, so we can write

$$\begin{bmatrix} x_{i+1} \\ y_{i+1} \end{bmatrix} = \begin{bmatrix} (1 - 1195(0.1)) & 1995(0.1) \\ -1197 & (1 + 1997(0.1)) \end{bmatrix}^{-1} \begin{bmatrix} x_i \\ y_i \end{bmatrix}$$

which has the solution $x(0.1) = 8.23$, $y(0.1) = 4.90$, reasonably close to the analytical values.

In summary, our results for the solution of Eq. (6.29) are

	$x(0.1)$	$y(0.1)$
Analytical	8.19	4.91
Euler		
Explicit	640	636
Implicit	8.23	4.90

If the step size is very small, we can get good results from the simpler Euler after the first step. With $h = 0.0001$, the table of results becomes

	$x(0.0001)$	$y(0.0001)$
Analytical	2.61	-1.39
Euler		
Explicit	2.64	-1.36
Implicit	2.60	-1.41

but this would require 1000 steps to reach $t = 0.1$, and round-off errors would be large.

If we anticipate some material from the next chapter, we can give a better description of stiffness as well as indicate the derivation of the general solution to Eq. (6.29). We rewrite Eq. (6.29) in matrix form:

$$\begin{bmatrix} x \\ y \end{bmatrix}' = A \begin{bmatrix} x \\ y \end{bmatrix}, \quad \text{where } A = \begin{bmatrix} 1195 & -1995 \\ 1197 & -1997 \end{bmatrix}.$$

The general solution, in matrix form, is

$$\begin{bmatrix} x \\ y \end{bmatrix} = ae^{-2t}v_1 + ce^{-800t}v_2,$$

where

$$v_1 = \begin{bmatrix} 5 \\ 3 \end{bmatrix} \quad \text{and} \quad v_2 = \begin{bmatrix} 1 \\ 1 \end{bmatrix}.$$

You can easily verify that $Av_1 = -2v_1$ and $Av_2 = -800v_2$. In Chapter 7 we will see that this means that v_1 is an eigenvector of A and that -2 is the corresponding eigenvalue. Similarly, v_2 is an eigenvector of A with the corresponding eigenvalue of -800. (In Chapter 7 you will learn additional methods to find the eigenvectors and eigenvalues of a matrix.)

A stiff equation can be defined in terms of the eigenvalues of the matrix A that represents the right-hand sides of the system of differential equations. When the eigenvalues of A have real parts that are negative and differ widely in magnitude as in this example, the system is stiff. In the case of a nonlinear system

$$\begin{bmatrix} x_1 \\ x_2 \\ \vdots \\ x_n \end{bmatrix}' = \begin{bmatrix} f_1(x_1, x_2, \ldots, x_n) \\ f_2(x_1, x_2, \ldots, x_n) \\ \vdots \\ f_n(x_1, x_2, \ldots, x_n) \end{bmatrix},$$

one must consider the Jacobian matrix whose terms are $\partial f_i / \partial x_j$. See Gear (1971) for more information. Both Maple and MATLAB have facilities for handling stiff equations.

6.11   Theoretical Matters

We have already mentioned several items of theory in previous sections of this chapter. There are several things that we should examine. The first is the Lipschitz condition. Following that, we will discuss instability in Milne's method, error propagation and global errors when solving differential equations numerically, and the derivation of the formulas for the multistep methods of Sections 6.5, 6.6, and 6.7.

The Lipschitz Condition

In this chapter we have studied a number of numerical techniques to solve the differential equation

$$\frac{dy}{dx} = f(x, y), \qquad \text{with } y(x_0) = y_0. \tag{6.31}$$

We solved Eq. (6.31) by generating values for $y(x)$ at discrete points along the x-axis (or the t-axis if t is the independent variable). In a mathematics course we study analytical methods of solving the problem. Unfortunately, it often happens that the analytical solution is so complicated that it is more efficient to solve it numerically.

In any case, we should examine the condition under which Eq. (6.31) really has a solution and whether the solution is unique. The *Lipschitz condition* is involved in the answer to these questions. A definition of this condition is:

Let $f(x, y)$ be defined and continuous on a region R that contains the point (x_0, y_0). We assume that the region is a closed and bounded rectangle. Then $f(x, y)$ is said to satisfy the Lipschitz condition if:

There is an $L > 0$ so that for all x, y_1, y_2 in R, we have

$$|f(x, y_1) - f(x, y_2)| < L|y_1 - y_2|.$$

If $f(x, y)$ satisfies this condition, it can be shown that there is a solution to Eq. (6.31) and that this solution is unique.

For most of the problems encountered in practice, $f(x, y)$ is continuous over region R; when this is true, we are guaranteed the existence of a solution. We can write this solution in the form

$$y(x) = y_0 + \int_{x_0}^{x} f(t, y(t)) \, dt.$$

If, in addition, $f(x, y)$ satisfies the Lipschitz condition, we then are guaranteed that the solution is unique. In the actual examples that we have considered, $f(x, y)$ is not only continuous but also $f_y(x, y)$ is bounded and continuous in the region of interest.

The example that we have used many times,

$$y' = -2x - y,$$

satisfies the Lipschitz condition, because

$$|f(x, y) - f(x, z)| = L|y - z|, \qquad L = 1 > 0.$$

From this we can infer that the solution we found was the unique one.

Instability with Milne's Method

We pointed out that a problem with Milne's method is that it can be unstable. We illustrated this instability through an example in Section 6.7 where it was compared with Adams–Moulton. A more theoretical analysis gives a better idea of the problem than a single numerical example, so we now do that.

Consider the differential equation

$$dy/dx = Ay,$$

where A is a constant. The general solution is $y = ce^{Ax}$. Suppose now that $y(x_0) = y_0$ is the initial condition; it then follows that the value of c must be $c = y_0 e^{-Ax_0}$. Hence, letting y_n be the value of the function when $x = x_n$, the analytical solution is

$$y_n = y_0 e^{A(x_n - x_0)}. \tag{6.32}$$

If we solve the differential equation by the method of Milne, we have, from the corrector formula,

$$y_{n+1} = y_{n-1} + \frac{h}{3}(y'_{n+1} + 4y'_n + y'_{n-1}).$$

Letting $y'_n = Ay_n$, from the original differential equation, and rearranging, we get

$$y_{n+1} = y_{n-1} + \frac{h}{3}(Ay_{n+1} + 4Ay_n + Ay_{n-1}),$$

(6.33)

$$\left(1 - \frac{hA}{3}\right)y_{n+1} - \frac{4hA}{3}y_n - \left(1 + \frac{hA}{3}\right)y_{n-1} = 0.$$

We would like to solve this equation for y_n in terms of y_0 to compare to Eq. (6.32). Equation (6.33) is a second-order linear difference equation that can be solved in a manner analogous to that for differential equations. The solution is

$$y_n = C_1 Z_1^n + C_2 Z_2^n,$$

(6.34)

where Z_1, Z_2 are the roots of the quadratic

$$\left(1 - \frac{hA}{3}\right)Z^2 - \frac{4hA}{3}Z - \left(1 + \frac{hA}{3}\right) = 0.$$

(6.35)

[The reader should check that Eq. (6.34) is a solution of Eq. (6.33) by direct substitution.] For simplification, let $hA/3 = r$; the roots of Eq. (6.35) are then

$$Z_1 = \frac{2r + \sqrt{3r^2 + 1}}{1 - r},$$

(6.36)

$$Z_2 = \frac{2r - \sqrt{3r^2 + 1}}{1 - r}.$$

We are interested in comparing the behavior of Eqs. (6.34) and (6.32) as the step size h becomes small. As $h \to 0$, $r \to 0$, and $r^2 \to 0$ even faster. Neglecting the $3r^2$ terms in comparison to the constant 1 under the radical in Eq. (6.36) gives

$$Z_1 \doteq \frac{2r + 1}{1 - r} = 1 + 3r + O(r^2) = 1 + Ah + O(h^2),$$

(6.37)

$$Z_2 \doteq \frac{2r - 1}{1 - r} = -1 + r + O(r^2) = -\left(1 - \frac{Ah}{3}\right) + O(h^2).$$

The last results are obtained by dividing the fractions. We now compare Eq. (6.37) with the Maclaurin series of the exponential function,

$$e^{hA} = 1 + hA + O(h^2),$$

$$e^{-hA/3} = 1 - \frac{hA}{3} + O(h^2).$$

We see that, for $h \to 0$,

$$Z_1 \doteq e^{hA}, \qquad Z_2 \doteq -e^{-hA/3}.$$

Hence the Milne solution is represented by

$$y_n = C_1(e^{hA})^n + C_2(e^{-hA/3})^n = C_1 e^{A(x_n - x_0)} + C_2 e^{-A(x_n - x_0)/3}. \qquad \textbf{(6.38)}$$

In Eq. (6.38), we have used $x_n - x_0 = nh$. The solution consists of two parts. The first term obviously agrees with the analytical solution, Eq. (6.32). The second term, called a *parasitic term,* will die out as x_n increases if A is a positive constant, but if A is negative, it will grow exponentially with x_n. Note that we get this peculiar behavior independent of h; smaller step size is of no benefit in eliminating the error.

Errors and Error Propagation

Our previous error analyses have examined the error of a single step only, the so-called *local truncation error* of the methods. Because all practical applications of numerical methods to differential equations involve many steps, the accumulation of these errors, termed the *global truncation error,* is important. We remember that there are several sources of error in a numerical calculation in addition to the truncation error.

Original Data Errors

If the initial conditions are not known exactly (or must be expressed inexactly as a terminated decimal number), the solution will be affected to a greater or lesser degree, depending on the sensitivity of the equation. Highly sensitive equations are said to be subject to *inherent instability.*

Round-Off Errors

Because we can carry only a finite number of decimal places, our computations are subject to inaccuracy from this source, no matter whether we round or whether we chop off. Carrying more decimal places in the intermediate calculations than we require in the final answer is the normal practice to minimize this, but in lengthy calculations this is a source of error that is extremely difficult to analyze and control. Furthermore, in a computer program, if we use double precision, we require a longer execution time and also more storage to hold the more precise values. If these values are for a large array, the memory space needed may exceed that available to the program. This type of error is especially acute when two nearly equal quantities are subtracted.

Truncation Errors of the Method

These are the types of error we have been discussing, because we use truncated series for approximation in our work, when an infinite series is needed for exactness. The choice of method is our best control here, with suitable selection of h.

In addition to these three types of error, when we solve differential equations numerically we must be concerned about the propagation of previous errors through the subsequent steps. Because we use the end values at each step as the starting values for the next one, it is as if incorrect original data were distorting the later values. (Round-off would almost always produce error even if our method were exact.) This effect we now examine, but only for the very simple case of the Euler method. This will show how such error studies are made, as well as suggest how difficult the analysis of more practical methods is.

We consider the first-order equation $dy/dx = f(x, y)$, $y(x_0) = y_0$. Let

$$Y_n = \text{calculated value at } x_n,$$

$$y_n = \text{true value at } x_n,$$

$$e_n = y_n - Y_n = \text{error in } Y_n; \; y_n = Y_n + e_n.$$

By the Euler algorithm,

$$Y_{n+1} = Y_n + hf(x_n, Y_n).$$

By Taylor series,

$$y_{n+1} = y_n + hf(x_n, y_n) + \frac{h^2}{2}y''(\xi_n), \qquad x_n < \xi_n < x_n + h,$$

$$e_{n+1} = y_{n+1} - Y_{n+1} = y_n - Y_n + h[f(x_n, y_n) - f(x_n, Y_n)] + \frac{h^2}{2}y''(\xi_n) \qquad (6.39)$$

$$= e_n + h\frac{f(x_n, y_n) - f(x_n, Y_n)}{y_n - Y_n}(y_n - Y_n) + \frac{h^2}{2}y''(\xi_n)$$

$$= e_n + hf_y(x_n, \eta_n)e_n + \frac{h^2}{2}y''(\xi_n), \qquad \eta_n \text{ between } y_n, Y_n.$$

In Eq. (6.39), we have used the mean-value theorem, imposing continuity and existence conditions on $f(x, y)$ and f_y. We suppose, in addition, that the magnitude of f_y is bounded by the positive constant K in the region of x, y-space in which we are interested.* Hence,

$$e_{n+1} \leq (1 + hK)e_n + \frac{1}{2}h^2y''(\xi_n). \qquad (6.40)$$

Here $y(x_0) = y_0$ is our initial condition, which we assume free of error. Because $Y_0 = y_0$, $e_0 = 0$:

$$e_1 \leq (1 + hK)e_0 + \frac{1}{2}h^2y''(\xi_0) = \frac{1}{2}h^2y''(\xi_0),$$

$$e_2 \leq (1 + hK)\left[\frac{1}{2}h^2y''(\xi_0)\right] + \frac{1}{2}h^2y''(\xi_1) = \frac{1}{2}h^2[(1 + hK)y''(\xi_0) + y''(\xi_1)].$$

---

*This is essentially the same as the Lipschitz condition, which will guarantee existence and uniqueness of a solution.

Similarly,

$$e_3 \leq \frac{1}{2}h^2[(1 + hK)^2 y''(\xi_0) + (1 + hK)y''(\xi_1) + y''(\xi_2)],$$

$$e_n \leq \frac{1}{2}h^2[(1 + hK)^{n-1}y''(\xi_0) + (1 + hK)^{n-2}y''(\xi_1) + \cdots + y''(\xi_{n-1})].$$

If $f_y \leq K$ is positive, the truncation error at every step is propagated to every later step after being amplified by the factor $(1 + hf_y)$ each time. Note that as $h \to 0$, the error at any point is just the sum of all the previous errors. If the f_y are negative and of magnitude such that $|hf_y| < 2$, the errors are propagated with diminishing effect.

We now show that the accumulated error after n steps is $O(h)$; that is, the global error of the simple Euler method is $O(h)$. We assume, in addition, that y'' is bounded, $|y''(x)| < M$, $M > 0$. After taking absolute values, Eq. (6.40) becomes

$$|e_{n+1}| \leq (1 + hK)|e_n| + \frac{1}{2}h^2 M.$$

Now we compare to the first-order difference equation:

$$Z_{n+1} = (1 + hK)Z_n + \frac{1}{2}h^2 M,$$

$$Z_0 = 0.$$

(6.41)

Obviously the values of Z_n are at least equal to the magnitudes of $|e_n|$. The solution to Eq. (6.41) is (check by direct substitution)

$$Z_n = \frac{hM}{2K}(1 + hK)^n - \frac{hM}{2K}.$$

The Maclaurin expansion of e^{hK} is

$$e^{hK} = 1 + hK + \frac{(hK)^2}{2} + \frac{(hK)^3}{6} + \cdots,$$

so that

$$1 + hK < e^{hK} \qquad (K > 0),$$

$$Z_n < \frac{hM}{2k}(e^{hK})^n - \frac{hM}{2K} = \frac{hM}{2K}(e^{nhK} - 1)$$

$$= \frac{hM}{2K}(e^{(x_n - x_0)K} - 1) = O(h).$$

(6.42)

It follows that the global error e_n is $O(h)$. (This result can be derived without difference equations.)

Derivation of the Multistep Formulas

The simplest way to derive the formulas for the Adams–Moulton method (Eqs. (6.17) and (6.18)) is by the method of undetermined coefficients in Appendix B. However, we would like to know the error term of these formulas. We can find this by deriving the formulas through fitting a third-degree polynomial, $p_3(x)$, through four points. The points are

$$[(x_{n-3}, f_{n-3}), \qquad (x_{n-2}, f_{n-2}), \qquad (x_{n-1}, f_{n-1}), \qquad (x_n, f_n)].$$

We will use $p_3(x)$ in this equation:

$$y_{n+1} = y_n + \int_{x_n}^{x_{n+1}} p_3(x)\,dx + \int_{x_n}^{x_{n+1}} E(x)\,dx.$$

Using divided differences as explained in Chapter 3, we find that

$$
\begin{aligned}
p_3(x) = f_{n-3} &+ f[x_{n-3}, x_{n-2}](x - x_{n-3}) \\
&+ f[x_{n-3}, x_{n-2}, x_{n-1}](x - x_{n-3})(x - x_{n-2}) \\
&+ f[x_{n-3}, x_{n-2}, x_{n-1}, x_n](x - x_{n-3})(x - x_{n-2})(x - x_{n-1}).
\end{aligned}
\tag{6.43}
$$

As an example of evaluating the terms of Eq. (6.43) we will do the last term. First, we expand the divided difference:

$$f[x_{n-3}, x_{n-2}, x_{n-1}, x_n] = \frac{f_n - 3f_{n-1} + 3f_{n-2} - f_{n-3}}{6h^3}.$$

We need to integrate:

$$\int_{x_n}^{x_{n+1}} (x - x_{n-3})(x - x_{n-2})(x - x_{n-1})\,dx.$$

A change of variable is helpful: let $x = x_0 + t * h$, with t varying from zero to 1. The integral then is

$$h^4 \int_0^1 (t + 3)(t + 2)(t + 1)\,dt = \frac{55h^4}{4}.$$

The other terms are evaluated similarly. We add to get the increment to y_n of Eq. (6.17):

$$\frac{h}{24}(55f_n - 59f_{n-1} + 37f_{n-2} - 9f_{n-3}).$$

To obtain the error term of Eq. (6.17) we use the expression for the error of an interpolating polynomial of Section 3.2, Eq. (3.2):

$$\int_{x_n}^{x_{n+1}} E(x)\,dx = \int_{x_n}^{x_{n+1}} \left[\frac{f^{(4)}(\xi_x)}{24}\right](x - x_{n-3})(x - x_{n-2})(x - x_{n-1})(x - x_n)\,dx$$

If we use the mean value theorem of integral calculus we can take the first term out of the integrand to get

$$\int_{x_n}^{x_{n+1}} E(x)\,dx = \frac{y^{(5)}(\xi_x)}{24} \int_{x_n}^{x_{n+1}} (x - x_{n-3})(x - x_{n-2})(x - x_{n-1})(x - x_n)\,dx \tag{6.44}$$

It is again convenient to change the variable of integration: $w = x - x_n$. This changes Eq. (6.44) to

$$\int_{x_n}^{x_{n+1}} E(x) \, dx = \frac{y^{(5)}(\xi_x)}{24} \int_{x_n}^{x_{n+1}} (w + 3h)(w + 2h)(w + h) \, dw$$

$$= \frac{251}{720} h^5 y^{(5)}(\xi_1).$$

(6.45)

(The change of variable made $x - x_{n-3} = x - (x_n - 3h)$ become $w + 3h$, and so on. The numerical part of the value for the integral in Eq. (6.45) is readily obtained with any of our computer algebra systems.)

A similar approach will obtain the corrector formulas of Eq. (6.18) and its error term.

6.12 Using Maple and MATLAB

Maple has a built-in function, `dsolve`, that provides very powerful ways to solve a differential equation and also a system of equations, both numerically and analytically. There are several options, so it is quite flexible. The list of options for a numerical solution include: `dverk78`, `gear`, `lsode`, `mgear`, `rkf45`, and `Taylor series` (these are invoked by including `method = 'option name'` in the command.)* MATLAB is similar in providing several options for solving both stiff and nonstiff equations.

In several sections of this chapter, we solved a first-order equation by several methods and compared the numerical results to the analytical. That equation is

$$dy/dx = -2x - y, \qquad y(0) = -1.$$

Maple can generate the analytical solution by symbolic methods. To do this, we first define a variable that stores the equation (we do not always show the output from Maple):

```
deq := diff(y(x),x) + y(x) + 2*x = 0;
dsolve({deq, y(0) = -1}, y(x));
```

The program then comes back with the answer:

```
y(x) = -2x + 2 - 3 exp(-x)
```

Maple has excellent plot routines for two and three dimensions, and we illustrate them with this first example. The plot programs are called by

```
with(plots);
G := dsolve({deq, y(0) = -1}, y(x));
plot(rhs(G), x = 0 .. 0.4);
```

The last command calls for a plot of the right-hand side of

```
G := y(x) = -2x + 2 - 3 exp(-x)
```

from $x = 0$ to $x = 0.4$.

---

*To see the complete list, you should look at the help file for `dsolve`.

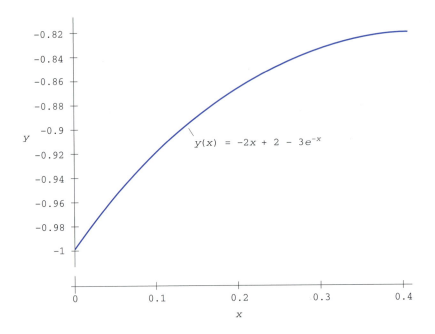

The graph gives the solution to $y' = -2x - y$, $y(0) = -1$.

MATLAB also can find the analytical solution to this example as well as many other differential equations. Its command for solving an equation analytically is also named `'dsolve'`. (Numerical solutions are obtained with different commands; we discuss this next). To solve the present example, which we repeat:

$$\frac{dy}{dx} = -2x - y, \qquad y(0) = -1,$$

we enter this:

```
dsolve('Dy = -2*x - y', 'y(0) = -1')
```

and see

```
ans =
-2*x + 2 - 3*exp(-x)
```

which is identical to the result from Maple.

If we want to find the value of this at $x = 0.5$, we enter:

```
x = 0.5; eval (ans)
```

to get

```
ans =
-0.81959197913790
```

which is the value of $y(-0.5)$ with a numerical format of long.

The Taylor-Series Method

MATLAB appears to have no built-in function to get the Taylor-series solution to a differential equation, but Maple does.

Section 6.2 described how a Taylor series can be generated for the solution of a differential equation. Getting the coefficients of the successive terms by hand is often very awkward. Maple can get these coefficients quite readily, even in cases where the successive derivatives are complicated. Here is how Maple can get the series for the first easy example of Section 6.2. The equation is:

$$\frac{dy}{dx} = -2x - y, \qquad y(0) = -1.$$

If this is a continuation of the Maple session just described, we have already defined the equation as deq, so we just call dsolve with the series option:

```
dsolve({deq, y(0) = -1}, y(x), series);
```

This resulting fifth-degree polynomial is obtained:

```
y(x) = -1 + x - 3/2 x² + 1/2 x³ - 1/8 x⁴ + 1/40 x⁵
```

which is precisely the result of Section 6.2.

For the second example of Section 6.2, a second-order equation, we can get the series solution in Maple. Here is the equation:

$$y'' = 3 + x - y^2, \qquad y(0) = 1, \qquad y'(0) = -2.$$

Maple gets the series by:

```
deq := diff(y(x), x$2) = 3 + x - y(x)^2;
dsolve({deq, y(0) = 1, D(y)(0) = -2}, y(x), series);
```

which produces

```
y(x) = 1 - 2x + x² + 5/6 x³ - 1/2 x⁴ + 7/60 x⁵ + O(x⁶).
```

In the discussion of automatic differentiation in Section 6.2, we pointed out that getting the series by hand is difficult in a case like

$$y' = \frac{x}{(y - x^2)}, \qquad y(0) = 1,$$

but Maple has no trouble. If we want the coefficients up to the 20th power, we must first change the value of the Order variable from its default value to a higher value, then make the call to dsolve:

```
Order := 22;
deq := diff(y(x), x) = x/(y(x) - x^2);
dsolve({deq, y(0) = 1}, y(x), series);
```

and we then see:

```
y(x) = 1 + 1/2 x² + 1/8 x⁴ + 1/48 x⁶ - 1/384 x⁸
     - 13/3840 x¹⁰ - 47/46080 x¹² + 73/645120 x¹⁴
     + 2247/10321920 x¹⁶ + 1681/185794560 x¹⁸
     - 15551/3715891200 x²⁰ + O(x²²).
```

Using Runge–Kutta Methods

In Maple the `dsolve` function contains options that get the solution numerically; one of these uses the RKF45 technique. Here is how we can solve the system of equations for the spring–mass problem of Section 6.1:

$$7.5\frac{d^2x}{dt^2} - 2.5\frac{d^2y}{dt^2} + 15x = 0,$$

$$-2.5\frac{d^2x}{dt^2} + 7.5\frac{d^2y}{dt^2} + 15y = 0.$$

```
deqs := {7.5*(D@@2)(x)(t) - 2.5*(D@@2)(y)(t) = -15*x(t),
         - 2.5*(D@@2)(x)(t) + 7.5*(D@@2)(y)(t) = -15*y(t)};
init2 := {x(0) = 1.5, D(x)(0) = 0, y(0) = 0, D(y)(0) = 0};
f := dsolve(deqs union init2, {x(t), y(t)}, numeric,
            value = array([0,1,2,3,4,5,6,7,8,9,10]));
```

$$
f := \begin{bmatrix}
\left[t,\ x(t),\ \dfrac{\partial}{\partial t}x(t),\ y(t),\ \dfrac{\partial}{\partial t}y(t) \right] \\
\begin{array}{rrrrr}
0 & 1.5 & 0 & 0 & 0 \\
1 & .1339720871 & -2.146291121 & -.3748068965 & -.4180793505 \\
2 & -1.288761709 & -.1744587699 & -.1339031115 & .9979116686 \\
3 & -.2972658043 & 1.616428776 & .9949368956 & .6835202582 \\
4 & .7384490188 & .1216164109 & .4601845211 & -1.683613031 \\
5 & .1992007643 & -.7533314938 & -1.281768774 & -1.045040044 \\
6 & -.06231153858 & .2660673637 & -.7886952775 & 1.873409903 \\
7 & .1838131416 & -.1353636271 & 1.172024309 & 1.246721058 \\
8 & -.4905644047 & -.9132868490 & .9061941005 & -1.583061867 \\
9 & -.7243061352 & .7633470306 & -.7649954720 & -1.073094265 \\
10 & .7434827667 & 1.585913977 & -.6808782480 & 1.009898677 \\
\end{array}
\end{bmatrix}
$$

```
with(plots);
odeplot (f,[[t,x(t)], [t,y(t)]], 0 .. 10, numpoints = 100);
```

An explanation is in order. The first command to Maple was to define the two equations. The second defined the initial conditions. We then defined *f* to be the result from a RKF45 computation of the solution, with the answer displayed as a table of values. That gave the table with columns of values for *t, x, dx/dt, y,* and *dy/dt.*

After seeing the table, we requested a plot of the solution, and got Fig. 6.6. However, this is hardly what we expected; the plot ought to be a curve, not a pair of broken lines.

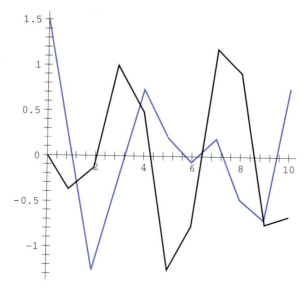

Figure 6.6

Why? Oh, yes, we actually asked for a plot of the values in the table of values for t, x, x', y, and y', not the solution! Maple connected the points with straight lines.

We can get the smooth curves (like that given in Section 6.1) if we ask for a plot from the procedure. For variety, let us use a different method, `dverk78`. So we do as follows:

```
ff := dsolve(deqs union init2, {x(t), y(t)},
        type = numeric, method = dverk78);
```

```
odeplot(ff,[[t,x(t)],[t,y(t)]], 0 . . 10, numpoints = 100);
```

Sure enough, we now have the plot that we anticipated. (See Fig. 6.7.)

Although MATLAB does not have a facility to get the Taylor-series solution, it can solve an ordinary differential equation numerically, using a Runge–Kutta method with automatic step size adjustment to meet a specified tolerance. There are two versions of this, `'ode23'` and `'ode45'`. The former uses second- and third-order formulas and the latter uses fourth- and fifth-order formulas. In some cases, the second command uses larger steps in obtaining the solution because the more accurate formulas can meet the tolerance criterion more easily.

The first step in using `ode23` is to create an M-file for the equation. So we open a new M-file in the FILE menu of MATLAB and enter:

```
function dy = de1(x,y)
dy = -2*x - y;
```

We save this short file under the name `''de1.m''`, where the m-extension is mandatory and the name must be that shown on the first line of the file, here `'de1'`. Now we enter this command:

```
[x,y] = ode23('de1', 0, .5, −1}
```

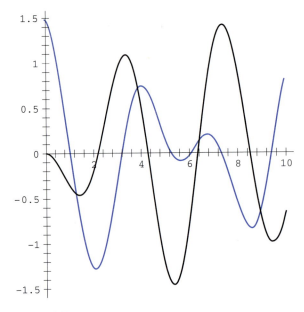

Figure 6.7

which asks to solve the equation defined in the M-file between an initial value for x of zero and a final value of 0.5, using -1 for the initial value for y. The MATLAB screen then displays a sequence of x and y values:

```
x =
            0
 0.00390625000000
 0.03515625000000
 0.06640625000000
 0.09765625000000
 0.12890625000000
 0.16015625000000
 0.19140625000000
 0.22265625000000
 0.25390625000000
 0.28515625000000
 0.31640625000000
 0.34765625000000
 0.37890625000000
 0.41015625000000
 0.44140625000000
 0.47265625000000
 0.50000000000000
```

```
y =
 -1.00000000000000
 -0.99611660838127
 -0.96667603855910
 -0.94006418632142
 -0.91619402106590
 -0.89498118984373
 -0.87634393497687
 -0.86020301420978
 -0.84648162331798
 -0.83510532109757
 -0.82600195666266
 -0.81910159897959
 -0.81433646856920
 -0.81164087131031
 -0.81095113427999
 -0.81220554356772
 -0.81534428400279
 -0.81959082376784
```

which are the successive computations that were made to achieve the default tolerance of 1.0E-3. (This tolerance can be changed by adding an additional parameter to the command. The default tolerance in `ode45` is 1.0E-6). We observe that there were seventeen steps used to compute the results between $x = 0$ and $x = 0.5$ and that the answer matches the analytical to six decimal places. If we now ask for a plot of y versus x, with the command:

```
plot(x, y, x, y, 'o'),
```

we see Figure 6.8. The circles represent the individual x-y points.

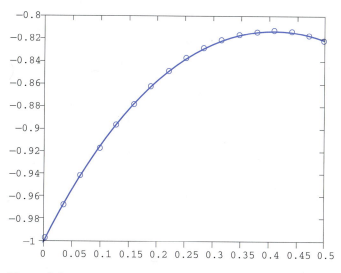

Figure 6.8

If we repeat this but use `ode45`, the answer obtained matches the analytical to ten decimals; in this example the same number of steps was used.

We can solve a system of differential equations with `ode23` (or with `ode45`). We first create an M-file in a similar fashion to the previous example. If we are to solve the system of Eqs. (6.26), which is

$$\frac{dx}{dt} = xy + t, \qquad \frac{dy}{dt} = ty + x, \qquad x(0) = 1, \qquad y(0) = -1,$$

to find x and y at $t = 0.1$, the M-file that we create is:

```
function xyd = de2(t,x)
xyd = [x(1)*x(2) + t; t*x(2) + x(1)]
```

which we store as `'de2.m'`. Observe that we have redefined the variables; x becomes $x(1)$ and y becomes $x(2)$. Now we ask MATLAB to solve it:

```
[t,x] = ode23('de2, 0, 1, [1,-1])
```

and see first a list of t-values used in the computation, then a list of the x-matrix, which has two columns: the first for $x(1) = x$, the second for $x(2) = y$. The answer for $t = 0.1$ is $x = 0.913936$ and $y = -0.909217$, exactly the same as for the RKF method of Section 6.9. Again, seventeen steps were used; this appears to be the standard number in `ode23` and `ode45` if the tolerance is easy to meet.

Chapter Summary

If you understand the material in this chapter, you should be able to

1. Use the Taylor-series procedure to solve first- or second-order differential equations. You can explain why this method is not generally used.

2. Solve first-order equations and systems of first-order equations by any of these methods:
    a. Simple Euler and modified Euler
    b. Fourth-order Runge–Kutta and Runge–Kutta–Fehlberg
    c. Multistep methods, such as Adams–Moulton

3. Explain the difference between a single-step method and a multistep method.

4. Rewrite a higher-order differential equation or system of equations in terms of an equivalent system of first-order equations.

5. Tell why the global error is different from the local error.

6. Explain the terms efficiency, accuracy, stability, and convergence as applied to the procedures of this chapter. In particular, you know why Milne's method is rejected.

7. Understand what is meant by the term *stiff differential equation* and give examples. You can solve this type of equation by an implicit method.

8. State the condition a differential equation must meet to have a unique solution.
9. Use either MATLAB or Maple to solve differential equations both analytically and numerically.

Computer Programs

There are two programs in this section of Chapter 6. Both are written in the C language. The first uses the Runge–Kutta–Fehlberg method to solve a system of ordinary differential equations. The second uses the Adams–Moulton method to advance the solution of a pair of differential equations.

Program 6.1

The RKF program is listed in Figure 6.9. As written, it can handle as many as 11 coupled first-order equations. The independent variable is *t,* and the dependent variables are held in arrays `xo` and `xend`. The program carries out the computations through calls to several procedures: `initialize` sets up the initial values, `getDERIVS` evaluates the derivative functions, `TheLargerOf` returns the larger of two values, `computeERROR` estimates the error so that the step size may be adjusted appropriately, `ComputeNewX` gets the next set of *x*-values, and `rkf` (which actually solves the system of equations by calls to the other procedures). The output from the program is given at the end of the listing.

```
/* * * * * * * * * * * * * * * * * * * * * * * * * * * * * * * * * * * * * * * * * * * * * *
 *                                                                     *
 *    Chapter 6          RKF Method for a System                       *
 *                                                                     *
 *    Gerald/Wheatley, APPLIED NUMERICAL ANALYSIS (sixth edition)      *
 *                Addison Wesley Longman, 1999                         *
 *                                                                     *
 * * * * * * * * * * * * * * * * * * * * * * * * * * * * * * * * * * * * * * * * * * * * * *
 *                                                                     *
 *    This Program solves a system of N first order differential       *
 *    equations by the RUNGE-KUTTA-FEHLBERG method. The equations      *
 *    are of the form:                                                 *
 *        DX1/DT = F1(T,X)                                             *
 *        DX2/DT = F2(T,X), etc.                                       *
 *                                                                     *
 *    The particular problem solved here is the second order           *
 *    differential equation:                                           *
 *                                                                     *
 *       Y'' = 2 - 4Y*Y/SQR(SIN(x)), 1 <= x <= 2.                      *
 *          where Y(1) = sin(1)*sin(1)                                 *
 *                Y'(1) = 2*sin(1)*cos(1)                              *
 *                                                                     *
```

Figure 6.9

```
*   The Analytical solution to this equation is:                       *
*         Y(x) = sin(x)*sin(x).                                        *
*                                                                     *
*   The computed and the analytical solution is printed out at        *
*   each time step.                                                   *
*                                                                     *
* * * * * * * * * * * * * * * * * * * * * * * * * * * * * * * * * * * * * * * * * * * * * * * * * * /

#include <stdio.h>
#include <math.h>

/* Define global variables */

#define maxN        11       /* handles up to 10 first order equations */
#define hSTART      0.1      /* starting value for delta t */
#define tol         0.00005
#define MaxCount    10
#define tINIT       1.0      /* initial time */
#define tFINAL      2.0      /* final time at last interval */

float x0[maxN],    /* state vector at beginning of interval */
      xend[maxN],  /* state vector at end of interval */
      f[maxN],     /* array that holds the values of the derivatives*/
      h,           /* adjustable step size */
      t0,          /* initial time value at beginning of interval */
      finalTime;   /* value equal to: tFINAL - hSTART/2.0 */

float workarray[7][maxN], /* a two-dimensional array to store the
k values */
      tEND, error;        /* tEND is t[i+1] = t[i] + hSTART */

int   i_rkf, count_rkf;
int   numberOFequations,  /* actual number of differential equations*/
      i, count;

/*
   This Procedure sets up all the initial values.
*/

void initialize()
{
   numberOFequations = 2;

   x0[1] = sin(1.0)*sin(1.0);
   x0[2] = 2.0*sin(1.0)*cos(1.0);
   t0 = tINIT;
   h = hSTART;
}                  /* end of initialize */

/*
  This procedure evaluates the first-order equations for
  the given time and state values.
```

Figure 6.9 *Continued*

```
      For this: dx1/dt = x2; dx2/dt = 2 - 4*x1*x1/(sin(t))^2

*/
void getDERIVS(x0, f, t0)
  float x0[], f[], t0;
{
  f[1] = x0[2];
  f[2] = 2.0 - 4.0*(x0[1]*x0[1])/(sin(t0)*sin(t0));
}                       /* end of getDERIVS */


/*
  This function returns the larger of two real numbers.
*/
float TheLargerOf(a, b)
  float a, b;
{
  if (a > b)
    return(a);
  else
    return(b);
}                       /* end of TheLargerOf */


/*
  This procedure computes the estimated error from the new
  ki's (these are in workarray) in getting the next state vector.
*/

void computeERROR ()
{
  float temp;

  error= 0.0;
  for (i = 1; i <= numberOFequations; ++i)
    {
      temp = fabs (workarray[1] [i]/360.0
                   - 128.0*workarray[3] [i]/4275.0
                   - 2197.0*workarray[4] [i]/75240.0
                   + workarray[5][i] / 50.0
                   + 2.0*workarray[6][i]/55.0);
      error = TheLargerOf (error, temp);
    }                       /* end of the for i loop */
}                               /* end of computeERROR */

/*
  This procedure updates the state vector from time[i] to time[i+1].
*/
void ComputeNewX()
{

  for (i=1; i <= numberOFequations; ++i)
    xend [i] = x0[i] + 16.0*workarray[1][i]/135.0
               + 6656.0*workarray[3][i]/12825.0
               + 28561.0*workarray[4][i]/56430.0
```

Figure 6.9 *Continued*

```
                       - 9.0*workarray[5][i]/50.0
                       + 2.0*workarray[6][i]/55.0;
}                               /* end of ComputeNewX */
/*
  This Procedure solves system of differential equations using
  the 4th/5th order method of Runge-Kutta Fehlberg from t[i] to
  t[i+i]. Use is made of the errror estimate to adjust the stepsize,
  h-value, for the next subinterval
*/

void rkf(t0, h, x0, xend)
  float t0, h, x0[], xend[];
{
    tEND = t0 + hSTART;
    count_rkf = 0;

/*
  Here begins a very long while loop
*/

  while ((t0 < tEND) && (count_rkf < MaxCount))
    {
      count_rkf = count_rkf + 1;

/*
  Get derivatives and the increments to t to use in the next step.
  (xend holds the k1's.)
*/

    getDERIVS(x0,f,t0);

    for (i_rkf = 1; i_rkf <= numberOFequations; ++i_rkf)
      {
      workarray[1][i_rkf] = h*f[i_rkf];
      xend[i_rkf] = x0[i_rkf] + workarray[1][i_rkf]/4.0;
      }                         /* end of first estimate */

/*
  Repeat this for second set of delta t's (the k2's)
*/
    getDERIVS(xend,f,t0+h/4.0);

    for (i_rkf = 1; i_rkf <= numberOFequations; ++i_rkf)
      {
        workarray[2][i_rkf] = h*f[i_rkf];
        xend[i_rkf] = x0[i_rkf] + (workarray[1][i_rkf]*3.0
                   + workarray[2][i_rkf]*9.0)/32.0;
      }                             /* end of second estimate */

/*
  Repeat this for third set (the k3's)
```

Figure 6.9 *Continued*

```
*/
    getDERIVS(xend,f,t0+3.0*h/8.0);

    for (i_rkf = 1; i_rkf <= numberOFequations; ++i_rkf)
       {
       workarray[3][i_rkf] = h*f[i_rkf];
       xend[i_rkf] = x0[i_rkf] + (workarray[1][i_rkf]*1932.0
                          - workarray[2][i_rkf]*7200.0
                          + workarray[3][i_rkf]*7296.0)/2197.0;
       }                           /* end of second estimate */

/*
  Repeat this for fourth set (the k4's)
*/
  getDERIVS(xend,f,t0+12.0*h/13.0);

  for (i_rkf = 1; i_rkf <= numberOFequations; ++i_rkf)
     {
     workarray[4][i_rkf] = h*f[i_rkf];
     xend[i_rkf] = x0[i_rkf] + 439.0*workarray[1][i_rkf]/216.0
                   - 8.0*workarray[2][i_rkf]
                   + 3680.0*workarray[3][i_rkf]/513.0
                   - 845.0*workarray[4][i_rkf]/4104.0;
     }                              /* end of fourth estimate */

/*
  Now for the fifth set (the k5's)
*/
    getDERIVS(xend,f,t0+h);

    for (i_rkf = 1; i_rkf <= numberOFequations; ++i_rkf)
       {
       workarray[5][i_rkf] = h*f[i_rkf];
       xend[i_rkf] = x0[i_rkf] - 8.0*workarray[1][i_rkf]/27.0
                     + 2.0*workarray[2][i_rkf]
                     - 3544.0*workarray[3][i_rkf]/2565.0
                     + 1859.0*workarray[4][i_rkf]/4104.0
                     - 11.0*workarray[5][i_rkf]/40.0;
       }                             /* end of fifth estimate */

/*
  Now the sixth and final values
*/
    getDERIVS(xend,f,t0+h/2.0);

    for (i_rkf = 1; i_rkf <= numberOFequations; ++i_rkf)
       workarray[6][i_rkf] = h*f[i_rkf];

/*
  Estimate the error and adjust h if necessary
```

Figure 6.9 *Continued*

```
*/
   error = 0;
   computeERROR();
   if (error < tol)
     {
     t0 = t0 + h;
     ComputeNewX();
     for (i_rkf = 1; i_rkf <= numberOFequations; ++i_rkf)
       x0[i_rkf] = xend[i_rkf];
     }                                /* end of first if */
   if (error > tol)
     {
     printf ("\n\n");
     printf ("FOR A STEPSIZE OF %.4f\t",h);
     printf ("THE ERROR ESTIMATE IS %.6f," error);
     printf("\n");
     h = h/2.0;
     }                          /* end of second if */
   if (error < h*tol/10.0)
     h = 2.0*h;
   if (t0 + h >tEND)
     h = tEND - t0;
   }                      /* end of while loop */
}                         /* end of rkf procedure *

/*
  main procedure begins here
*/

void main()
{
  initialize();
  count = 0;
  clrscr();
  printf ("AT TIME T = %.2f\n," t0);
  for (i_rkf = 1; i_rkf <= numberOFequations; ++i_rkf)
    printf("\t\t X[ %d] = %.6f\n," i_rkf, x0[i_rkf]);
  finalTime = tFINAL - hSTART/2.0;
  while ((t0 < finalTime) && (count < MaxCount))
    {
    rkf(t0, h, x0, xend);
    t0 = t0 + hSTART;
    printf ("\n AT TIME T = %.2f\n", t0);
    for (i = 1; i <= numberOFequations; ++i)
      printf ("\t\t   X[ %d] = %.6f\n",i, xend[i]);
    printf ("\t\t\t   EXACT IS : %.6f FOR X[1]\n",
              sin(t0)*sin(t0));
    getch();
    }                              /* end of while */
}                                  /* end of main */

* * * * * * * * * * * * * * * * * * * * * * * * * * * * * * * * * * * * * * * * * * * * * * * * * * * * * *
```

Figure 6.9 *Continued*

```
                    OUTPUT FOR RKFSYST.C


    AT TIME T =  1.00

                X[ 1 ] = 0.708073

                X[ 2 ] = 0.909297

    AT TIME T =  1.10

                X[ 1 ] = 0.794251

                X[ 2 ] = 0.808497

                    EXACT IS : 0.794251  FOR X[1]
    AT TIME T =  1.20

                X[ 1 ] = 0.868697

                X[ 2 ] = 0.675464

                    EXACT IS : 0.868697  FOR X[1]

    AT TIME T =  1.30

                X[ 1 ] = 0.928445

                X[ 2 ] = 0.515502

                    EXACT IS : 0.928444  FOR X[1]

    AT TIME T =  1.40

                X[ 1 ] = 0.971112

                X[ 2 ] = 0.334989

                    EXACT IS : 0.971111  FOR X[1]

    AT TIME T =  1.50

                X[ 1 ] = 0.994997

                X[ 2 ] = 0.141120

                    EXACT IS : 0.994996  FOR X[1]

    AT TIME T =  1.60

                X[ 1 ] = 0.999148
```

Figure 6.9 *Continued*

```
                                  X[ 2 ] = -0.058374

                                        EXACT IS : 0.999147 FOR X[1]

          AT TIME T =  1.70

                          X[ 1 ] = 0.983400

                          X[ 2 ] = -0.255542

                                    EXACT IS : 0.983399 FOR X[1]

          AT TIME T =  1.80

                          X[ 1 ] = 0.948380

                          X[ 2 ] = -0.442522

                                    EXACT IS : 0.948379 FOR X[1]

          AT TIME T =  1.90

                          X[ 1 ] = 0.895484

                          X[ 2 ] = -0.611860

                                    EXACT IS : 0.895484 FOR X[1]

          AT TIME T =  2.00

                          X[ 1 ] = 0.826822

                          X[ 2 ] = -0.756805

                                    EXACT IS : 0.826822 FOR X[1]
```

Figure 6.9 *Continued*

Program 6.2

Figure 6.10 shows the C program that solves a pair of first-order differential equations by the Adams–Moulton method. The program reads in the data necessary to begin the solution (the starting value for t, the step size for t, and four past values for x and y). These data are used to advance the solution for five Δt-steps, which are displayed. At the user's request, additional sets of five more computations are made until the user terminates the program.

In contrast to the first C program of this chapter, the code in this program does not call on procedures (except to compute the derivatives); rather, operations are done "in line." The reader may want to consider whether this unstructured style is clearer.

The particular pair of equations that are solved is from Section 6.9, Eq. (6.26).

```
/* * * * * * * * * * * * * * * * * * * * * * * * * * * * * * * * * * * * * * * * * * * * * * * * * *
 *                                                                                  *
 *   Chapter 6          Adams-Moulton Method                                        *
 *                                                                                  *
 *   Gerald/Wheatley, APPLIED NUMERICAL ANALYSIS (sixth edition)                    *
 *                    Addison Wesley Longman, 1999                                  *
 *                                                                                  *
 * * * * * * * * * * * * * * * * * * * * * * * * * * * * * * * * * * * * * * * * * * * * * * * * * *
 *                                                                                  *
 *   This program solves a pair of ordinary differential-equations                 *
 *   by the Adams-Moulton technique. The past four evenly spaced                    *
 *   values for the solution (obtained by some single-step method)                  *
 *   are read from a file as well as t0 and h, the value of                         *
 *   delta t. The two derivatives, dx/dt and dy/dt, are computed                    *
 *   through procedures. After printing the starting data for a                     *
 *   check, the past derivative values are obtained. Using this                     *
 *   information, the program produces additional solutions at                      *
 *   evenly spaced times until the user terminates the program.                     *
 *                                                                                  *
 * * * * * * * * * * * * * * * * * * * * * * * * * * * * * * * * * * * * * * * * * * * * * * * * * */

#include <stdio.h>
#include <math.h>

/*
   Define variables as global
*/

FILE *file;
float x[6], y[6], t, h, fx[6], fy[6], ffx, ffy;
int i, j, quit;
char inbuf[80];


/*
   These two procedures compute the derivative values
*/
float DERIVX (x, y, t)
float x, y, t;
{
  return (x*y + t);
}

float DERIVY (x, y, t)
float x, y, t;
{
  return (y*t + x);
}

/*
   main begins here
*/
```

Figure 6.10

```
void main ()
{

/*
   Read in data values. First is t0, then h. These are
   followed by the data for the variables.
   These are in the order of x's then the y's
*/
  clrscr();
  file = fopen ("b:admlt.dta," "r");
  fscanf (file, "%f", &t);
  fscanf (file, "%f", &h);

  for (i = 1; i <= 4; i++)
    fscanf (file, "%f", &x[i]);
  for (i = 1; i <= 4; i++)
    fscanf (file, "%f", &y[i]);

/* end of getting data */

/*
   Print data from file to check
*/
  printf ("Data from file is: \n");
  printf ("t0 and h are: %f %f\n", t, h);
  for (i = 1; i <= 4; i++)
    printf ("t = %.2f, x, y: %f %f\n", t+(i-1)*h, x[i], y[i]);

  printf ("\nPress RETURN to continue\n");
  getch();

/*
   Compute derivative values for past x and y values
*/
  for (i = 1; i<=4; i++)
  {
    fx[i] = DERIVX (x[i], y[i], t + (i-1)*h);
    fy[i] = DERIVY (x[i], y[i], t + (i-1)*h);
  }

/*
   Print to check
*/
  for (i = 1; i <= 4; i++)
    printf ("t = %.2f, dx and dy are: %f %f \n",t + (i-1)*h, fx[i],
fy[i]);

  printf ("\nPress RETURN to continue\n");
  getch();

  t = t + 4*h;
```

Figure 6.10 *Continued*

```
/*
    Begin a DO/While loop that repeats the calculations 5 times,
    displaying each result. Do this again until the user want to stop.
*/
  do
  {
    for (i = 1; i<= 5; i++)
    {
/*
      Get predictor values, then derivatives at the new point
*/
x[5] = x[4] + h*(55.0*fx[4]-59.0*fx[3]+37.0*fx[2]-9.0*fx[1])/24.0;
y[5] = y[4] + h*(55.0*fy[4]-59.0*fy[3]+37.0*fy[2]-9.0*fy[1])/24.0;
printf ("Predicted values at t = %.2f are: %f %f \n", t, x[5], y[5]);

      fx[5] = DERIVX (x[5], y[5], t);
      fy[5] = DERIVY (x[5], y[5], t);

/*
       Get corrected values
*/
    x[5] = x[4] + h*(9.0*fx[5]+19.0*fx[4]-5.0*fx[3]+fx[2])/24.0;
    y[5] = y[4] + h*(9.0*fy[5]+19.0*fy[4]-5.0*fy[3]+fy[2])/24.0;

    printf ("\tCorrected values at t = %.2f are:
    %f %f \n", t, x[5], y[5]);

/*
Recompute derivatives, reset the values, then advance t to repeat
*/
    fx[5] = DERIVX (x[5], y[5], t);
    fy[5] = DERIVY (x[5], y[5], t);

    for (j = 2; j <= 5; j++)
    {
      x[j-1] = x[j];
      y[j-1] = y[j];
      fx[j-1] = fx[j];
      fy[j-1] = fy[j];
    }                /* end of for j */

    t = t + h;

  }    /* end of for i */

/*
    See if user wants five more
*/
    printf ("\n Enter 0 to quit or enter 1 to do five more\n\n ");
    gets (inbuf);
    sscanf (inbuf, "%d," &quit);
```

Figure 6.10 *Continued*

```
  }
  while (quit != 0);      /* end of do/while */
}    /* end of main */

Data from file is:
t0 and h are: 0.000000  0.100000
t = 0.00, x, y:  1.000000    -1.000000
t = 0.10, x, y:  0.913936    -0.909217
t = 0.20, x, y:  0.852186    -0.834089
t = 0.30, x, y:  0.810633    -0.771093

Press RETURN to continue
t = 0.00, dx and dy are: -1.000000   1.000000
t = 0.10, dx and dy are: -0.730966   0.823014
t = 0.20, dx and dy are: -0.510799   0.685368
t = 0.30, dx and dy are: -0.325073   0.579305

Press RETURN to continue
Predicted values at t = 0.40 are: 0.786518  -0.717441
    Corrected values at t = 0.40 are: 0.786334  -0.717348
Predicted values at t = 0.50 are: 0.777310  -0.670518
    Corrected values at t = 0.50 are: 0.777193  -0.670449
Predicted values at t = 0.60 are: 0.781740  -0.628324
    Corrected values at t = 0.60 are: 0.781670  -0.628272
Predicted values at t = 0.70 are: 0.798700  -0.588911
    Corrected values at t = 0.70 are: 0.798657  -0.588867
Predicted values at t = 0.80 are: 0.827430  -0.550386
    Corrected values at t = 0.80 are: 0.827408  -0.550346

  Enter 0 to quit or enter 1 to do five more

 1
Predicted values at t = 0.90 are: 0.867499  -0.510812
    Corrected values at t = 0.90 are: 0.867493  -0.510771
Predicted values at t = 1.00 are: 0.918805  -0.468086
    Corrected values at t = 1.00 are: 0.918814  -0.468040
Predicted values at t = 1.10 are: 0.981624  -0.419798
    Corrected values at t = 1.10 are: 0.981649  -0.419743
Predicted values at t = 1.20 are: 1.056708  -0.363055
    Corrected values at t = 1.20 are: 1.056756  -0.362984
Predicted values at t = 1.30 are: 1.145462  -0.294247
    Corrected values at t = 1.30 are: 1.145542  -0.294151

  Enter 0 to quit or enter 1 to do five more

 0
```

Figure 6.10 *Continued*

Exercises

Section 6.2

1. ▶a. Solve the differential equation

$$\frac{dy}{dx} = x + y + xy, \qquad y(0) = 1$$

by Taylor-series expansion to get the value of y at $x = 0.1$ and at $x = 0.5$. Use terms through x^5.

b. Do the same for

$$\frac{dy}{dx} = x + y, \qquad y(0) = 1.$$

(The analytical solution is $y(x) = 2e^x - x - 1$.)

c. Do the same for

$$\frac{dy}{dx} = \frac{2x}{y} - xy, \qquad y(0) = 1.$$

(The analytical solution is $y(x) = \sqrt{2 - e^{-x^2}}$.)

d. Do the same for

$$dy/dx = (y/2)(1 - y/5), \qquad y(0) = 1.$$

(The analytical solution is $y(x) = 5/(1 + 4e^{-x/2})$.)

2. The general solution to a differential equation normally defines a family of curves. For the differential equation

$$\frac{dy}{dx} = x^2 y^2,$$

using the Taylor-series method, find the particular curve that passes through $(1, 0)$. Also find the curve through $(0, 1)$. Compare to the analytical solutions.

▶ **3.** Use the Taylor-series method to get y at $x = 0.2(0.2)0.6$,* given that

$$y'' = xy, \qquad y(0) = 1, \qquad y'(0) = 1.$$

4. A spring system has resistance to motion proportional to the square of the velocity, and its motion is described by

$$\frac{d^2x}{dt^2} + 0.1 \left(\frac{dx}{dt}\right)^2 + 0.6x = 0.$$

If the spring is released from a point that is a unit distance above its equilibrium point, $x(0) = 1$, $x'(0) = 0$, use the Taylor-series method to write a series expres-

sion for the displacement as a function of time, including terms up to t^6.

Section 6.3

▶ **5.** Repeat Exercise 1, but use the simple Euler method to get $y(0.1)$. Compare your results to those of Exercise 1 and determine how small h must be to obtain four-decimal accuracy.

6. Solve the differential equation

$$\frac{dy}{dx} = \frac{x}{y}, \qquad y(0) = 1,$$

by the simple Euler method with $h = 0.1$, to get $y(1)$. Then repeat with $h = 0.2$ to get another estimate of $y(1)$. Extrapolate these results, assuming errors are proportional to step size, and then compare them to the analytical result. (The analytical result is $y^2 = 1 + x^2$.)

▶ **7.** Repeat Exercise 5, but now with the modified Euler method with $h = 0.025$ so that the solution is obtained after four steps. Comparing with the results of Exercise 5, about how much less effort is required to solve these problems to obtain four-decimal accuracy?

8. Find the solution to

$$\frac{dy}{dt} = y^2 + t^2, \qquad y(1) = 0, \qquad \text{at } t = 2,$$

by the modified Euler method, using $h = 0.1$. Repeat with $h = 0.05$. From the two results, estimate the accuracy of the second computation.

▶ **9.** Solve $y' = \sin x + y$, $y(0) = 2$, by the modified Euler method to get y at $x = 0.1(0.1)0.5$.

10. A sky diver jumps from a plane, and during the time before the parachute opens, the air resistance is proportional to the $\frac{3}{2}$ power of the diver's velocity. If it is known that the maximum rate of fall under these conditions is 80 mph, determine the diver's velocity during the first 2 sec of fall using the modified Euler method with $\Delta t = 0.2$. Neglect horizontal drift and assume an initial velocity of zero.

---

*This notation means for $x = 0.2$ through $x = 0.6$ with increments of 0.2.

Section 6.4

11. Solve Exercise 5 by the Runge–Kutta method but with $h = 0.1$ so that the solution is obtained in only one step. Carry five decimals, and compare the accuracy and amount of work required with this method against the simple and modified Euler techniques in Exercises 5 and 7.

12. Solve Exercise 8 by the Runge–Kutta method, using $h = 0.2, 0.1$, and 0.05.

▶**13.** Determine y at $x = 0.2(0.2)0.6$ by the Runge–Kutta technique, given that

$$\frac{dy}{dx} = \frac{1}{x + y}, \qquad y(0) = 2.$$

14. Using the conditions of Exercise 10, determine how long it takes for the jumper to reach 90% of his or her maximum velocity, by integrating the equation using the Runge–Kutta technique with $\Delta t = 0.5$ until the velocity exceeds this value, and then interpolating. Then use numerical integration on the velocity values to determine the distance the diver falls in attaining $0.9v_{max}$.

15. It is not easy to know the accuracy with which the function has been determined by either the Euler methods or the Runge–Kutta method. A possible way to measure accuracy is to repeat the problem with a smaller step size, and compare results. If the two computations agree to n decimal places, one then assumes the values are correct to that many places. Repeat Exercise 14 with $\Delta t = 0.3$, which should give a global error about one-eighth as large, and by comparing results, determine the accuracy in Exercise 14. (Why do we expect to reduce the error eightfold by this change in Δt?)

16. Repeat Exercises 11, 12, and 13 using the fifth-order equation of (6.11) of the Runge–Kutta–Fehlberg method.

17. Solve Exercises 1, 6, and 9 using Runge–Kutta–Fehlberg.

▶**18.** Solve $y' = 2x^2 - y$, $y(0) = -1$ by the Runge–Kutta–Fehlberg method to $x = 2.0$.

19. Solve the equation in Exercise 10 by Runge–Kutta–Fehlberg.

Section 6.5

▶**20.** The equation

$$\frac{dy}{dx} = f(x, y) = 2x(y - 1), \qquad y(0) = 0,$$

has initial values as follows:

x	y	f
0	0	0
0.1	−0.01005	−0.20201
0.2	−0.04081	−0.41632
0.3	−0.09417	−0.65650
0.4	−0.17351	−0.93881

a. Using the Adams procedure described in Section 6.5, compute $y(0.5)$ by fitting a quadratic through the last three values of $f(x, y)$. Compare to the exact answer: -0.28403.

b. Repeat, using the last four points to fit a cubic.

c. Repeat again, but use five points to fit a quartic.

21. For the differential equation

$$\frac{dy}{dt} = y - t^2, \qquad y(0) = 1,$$

starting values are known:

$$y(0.2) = 1.2186, \qquad y(0.4) = 1.4682,$$
$$y(0.6) = 1.7379.$$

Use the Adams method, fitting cubics with the last four (y, t) values and advance the solution to $t = 1.2$. Compare to the analytical solution.

22. For the equation

$$\frac{dy}{dt} = t^2 - t, \qquad y(1) = 0,$$

the analytical solution is easy to find:

$$y = \frac{t^3}{3} - \frac{t^2}{2} + \frac{1}{6}.$$

If we use three points in the Adams method, what error would we expect in the numerical solution? Confirm your expectation by performing the computations.

Section 6.6

23. For the differential equation

$$\frac{dy}{dx} = y - x^2, \qquad y(0) = 1,$$

starting values are known:

$$y(0.2) = 1.2186, \qquad y(0.4) = 1.4682,$$
$$y(0.6) = 1.7379.$$

Use the Milne method to advance the solution to $x = 1.2$. Carry four decimals and compare to the analytical solution.

▶ **24.** For the differential equation $dy/dx = x/y$, the following values are given:

x	y
0	$\sqrt{1}$
1	$\sqrt{2}$
2	$\sqrt{5}$
3	$\sqrt{10}$

To how many decimal places will Milne's method give the value at $x = 4$? How many decimal places must be carried in the starting values of y to ensure this accuracy?

25. For the equation $y' = y \sin \pi x$, $y(0) = 1$, get starting values by the Runge–Kutta–Fehlberg method for $x = 0.2(0.2)0.6$, and advance the solution to $x = 1.0$ by Milne's method.

▶ **26.** Continue the results of Exercise 13 to $x = 2.0$ by the method of Milne. If you find that the corrector formula reproduces the predictor values, double the value of h after sufficient values are available.

Section 6.7

27. Solve Exercise 21 using the Adams–Moulton method.

28. Repeat Exercise 25 using the Adams–Moulton method.

▶ **29.** For the equation

$$\frac{dy}{dx} = x^3 + y^2, \qquad y(0) = 0,$$

using $h = 0.2$, compute three new values by the Runge–Kutta method (four decimals). Then advance to $x = 1.4$ using the Adams–Moulton method. If you find the accuracy criterion is not met, use Eqs. (6.19) to interpolate additional values so that four-place accuracy is maintained.

30. Derive the interpolation formulas given in Eqs. (6.19).

Section 6.8

▶ **31.** Given is the linear differential equation $dy/dx = y \sin x$.

a. What is the maximum value of h that ensures convergence of the Adams–Moulton method when continuing applications of the corrector formula are made?

b. If an h one-tenth of this maximum value is used, how close must the predictor and corrector values be so that recorrections are not required?

c. In terms of the maximum h in part (a), what size

of h is implied in the accuracy criterion, $D \cdot 10^n < 14.2$?

32. Repeat Exercise 31 for the differential equation

$$\frac{dy}{dx} = x^3 + y^2, \qquad y(0) = 0,$$

in the neighborhood of the point $(1.0, 0.15)$.

33. Repeat Exercise 31, using the Milne method. For part (c), the accuracy criterion is $D \cdot 10^n < 29$.

34. Derive Eq. (6.24).

35. Derive convergence criteria similar to Eqs. (6.22) and (6.23) for the Euler predictor–corrector method. Why can one not derive an accuracy criterion similar to those for the methods of Milne and Adams–Moulton?

Section 6.9

36. The mathematical model of an electrical circuit is given by the equation

$$0.5 \frac{d^2Q}{dt^2} + 6 \frac{dQ}{dt} + 50Q = 24 \sin 10t,$$

with $Q = 0$ and $I = dQ/dt = 0$ at $t = 0$. Express as a pair of first-order equations.

▶ **37.** In the theory of beams it is shown that the radius of curvature at any point is proportional to the bending moment:

$$EI \frac{y''}{[1 + (y')^2]^{3/2}} = M(x),$$

where y is the deflection of the neutral axis. In the usual approach, $(y')^2$ is neglected in comparison to unity, but if the beam has appreciable curvature, this is invalid. For the cantilever beam for which $y(0) = y'(0) = 0$, express the equation as a pair of simultaneous first-order equations.

38. The motion of the compound spring system as sketched in Fig. 6.11 is given by the solution of the pair of simultaneous equations

$$m_1 \frac{d^2y_1}{dt^2} = -k_1y_1 - k_2(y_1 - y_2),$$

$$m_2 \frac{d^2y_2}{dt^2} = k_2(y_1 - y_2),$$

where y_1 and y_2 are the displacements of the two masses from their equilibrium positions. The initial conditions are

$$y_1(0) = A, \qquad y_1'(0) = B, \qquad y_2(0) = C, \qquad y_2'(0) = D.$$

Express as a set of first-order equations.

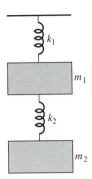

Figure 6.11

▶ **39.** Solve the pair of simultaneous equations

$$\frac{dx}{dt} = xy + t, \qquad x(0) = 0,$$

$$\frac{dy}{dt} = x - t, \qquad y(0) = 1,$$

by the modified Euler method for $t = 0.2(0.2)0.6$. (Carry three decimals rounded.) Recorrect until reproduced to three decimals.

40. Advance the solution of Exercise 39 to $x = 1.0$ ($h = 0.2$) by the Adams–Moulton method.

41. Repeat Exercise 40 but use Milne's method.

▶ **42.** Find y at $x = 0.6$, given that

$$y'' = yy', \qquad y(0) = 1, \qquad y'(0) = -1.$$

Begin the solution by the Taylor-series method, getting

$$y(0.1), \qquad y(0.2), \qquad y(0.3).$$

Then advance to $x = 0.6$ employing the Adams–Moulton technique with $h = 0.1$ on the equivalent set of first-order equations.

43. Express the third-order equation

$$y''' + ty'' - ty' - 2y = t,$$
$$y(0) = y''(0) = 0, \qquad y'(0) = 1,$$

as a set of first-order equations and solve at $t = 0.2, 0.4, 0.6$ by the Runge–Kutta method ($h = 0.2$).

44. Using the Adams–Moulton method with $h = 0.2$, advance the solution of Exercise 43 to $t = 1.0$. Estimate the accuracy of the value of y at $t = 1.0$.

45. If some simplifying assumptions are made, the equations of motion of a satellite around a central body are:

$$\frac{d^2x}{dt^2} = \frac{-x}{r^3}, \qquad \frac{d^2y}{dt^2} = \frac{-y}{r^3},$$

where

$$r = \sqrt{(x2 + y2)}, \qquad x(0) = 0.4,$$
$$y(0) = \dot{x}(0) = 0, \qquad \dot{y}(0) = 2.$$

a. Evaluate $x(t)$ and $y(t)$ for $t = 0.5(0.5)10.0$.
b. Plot the graph of the curve over this time interval.
c. Estimate the period of the orbit.

Section 6.10

▶ **46.** For a resonant spring system with a periodic forcing function, the differential equation is

$$\frac{d^2x}{dt^2} + 64x = 16 \cos 8t, \qquad x(0) = x'(0) = 0.$$

Determine the displacement at $t = 0.1(0.1)0.8$ by the method of Eq. (6.28), getting the starting values by any other method of this chapter. Compare to the analytical solution $t \sin 8t$.

47. For the first-order equation

$$\frac{dy}{dt} - 12y = \frac{3}{2}t^2 - 12t + 6t^3 - 1, \qquad y(0) = a,$$

it is not difficult to verify that the solution is

$$y = ae^{-12t} + \frac{t^3}{2} - t.$$

If $a = 0$ (so $y(0) = 0$), the solution reduces to a simple cubic in t. Still, accumulated round-off errors act as though the initial value of y is not exactly zero, causing the exponential term to appear when it should not. This makes the problem a stiff equation. Assume that $y(0) = 0$ and demonstrate that you do not get the analytical answer when you integrate from $t = 0$ to $t = 2$ using the Runge–Kutta fourth-order method with a step size of 0.005.

Section 6.11

48. Determine the Lipschitz constants for:

a. $F(x, y) = \dfrac{2y}{x}, \qquad$ for $x \geq 1$.

b. $F(x, y) = \arctan(y)$.
c. $F(x, y) = x - y^2, \qquad$ for $|y| \leq 10$.

49. Which of these satisfy the Lipschitz conditions?

▶ a. $y' = x^2 - y^2$ on the unit square
▶ b. $y' = x^2/y$ on the rectangle with corners at $(-2, 4)$, $(4, -1)$
c. $x' = tx$ on the rectangle with corners at $(1, 5)$, $(5, 1)$

50. Given $y = 1/(x - 1)$ so that $y' = -1/(x - 1)^2$. Over what x-intervals do the Lipschitz conditions hold?

▶ **51.** The Lipschitz conditions are weaker than the requirement that $\partial f/\partial y$ is continuous and bounded. Find an example where the Lipschitz conditions hold but the stronger condition does not.

▶ **52.** By making a slight change in the initial condition, determine the "amplification factor," the ratio of the change in $y(2)$ to the change of $y(0)$ for these equations. Which of them would you call unstable?

  a. $y' = 1 - x + 4y$, $y(0) = 5$
  b. $y' = 1 - x$, $y(0) = 5$
  c. $y' = -4y$, $y(0) = 5$
  d. $y' = 4y$, $y(0) = 5$

53. Estimate the propagated error at each step when the equation

$$\frac{dy}{dx} = x + y, \quad y(0) = 1,$$

is solved by the simple Euler method with $h = 0.02$, for $x = 0(0.02)0.1$. Compare to the actual errors.

▶ **54.** Follow the propagated error between $x = 1$ and $x = 1.6$ when the simple Euler method is used to solve

$$\frac{dy}{dx} = xy^2, \quad y(1) = 1.$$

Take $h = 0.1$. Compare to the actual errors at each step. The analytical solution is $y = 2/(3 - x^2)$.

55. We can derive the global error (Eq. (6.42)) for Euler's method without making use of the first-order difference equation. With the same assumptions about M and K and using the fact that the series

$$1 + s + s^2 + \cdots + s^n = \frac{s^{n+1} - 1}{s - 1},$$

show that

$$e_n \leq \frac{hM}{2K}(e^{(x_n - x_0)K} - 1).$$

(*Hint:* Let $s = 1 + hK$.)

56. Show that in the derivation of Eq. (6.17), we obtain the following values for these terms in Eq. (6.43):

  a. $f[x_{n-3}, x_{n-2}, x_{n-1}] = \dfrac{f_{n-1} - 2f_{n-2} + f_{n-3}}{2h^2}.$

  b. $f[x_{n-3}, x_{n-2}] = \dfrac{f_{n-2} - f_{n-3}}{h}.$

  c. $\displaystyle\int_{x_n}^{x_{n+1}} (x - x_{n-3})(x - x_{n-2})\, dx$

$$= h^3 \int_0^1 (t + 3)(t + 2)\, dt = \frac{55h^3}{6}.$$

Section 6.12

57. Use Maple or MATLAB to solve for the analytical solutions:

  a. $y' = x^2 + xy$, $y(1) = 2$.
  b. $x' = \sin(t)$, $x(0) = 1$.
  c. $y' + x + y = 2$, $y(x_0) = y_0$.

58. Use MATLAB and/or Maple to plot the solutions to Exercise 57, parts a and b.

▶ **59.** $y' = xy^2 + xy - y^2 - y$ is what is called a "separable" equation. Use MATLAB to get the general solution.

60. Use Maple to generate the Taylor-series solution to the equations in Exercise 57 and evaluate these at $x = 2$. In part c, use $y(2) = -3$.

61. Repeat Exercise 60 but use the RKF method. Also do this with MATLAB. Compare the solutions to those of Exercise 60 and to the analytical answers.

▶ **62.** Use Maple to get the fifth-order Taylor-series solution and evaluate for $x = 2$, for

$$y'' + 2y' = x^2 + y, \quad y(0) = 1, \quad y'(0) = -1.$$

63. Repeat Exercise 62 but now use RKF with $h = 0.05$. Solve with both Maple and MATLAB.

64. Use both Maple and MATLAB to plot the solution to Exercise 62 from $x = 0$ to $x = 2$.

Applied Problems and Projects

65. The mass in Fig. 6.12 moves horizontally on the frictionless bar. It is connected by a spring to a support located centrally below the bar. The unstretched length of the spring is $L = \sqrt{(10)} = 3.1623$ m (meters); the spring constant is $k = 100$ N/m (newtons per meter); the mass of the block is 3 kg. Let $x(t)$ be the distance from the center of the bar to the location of the block at

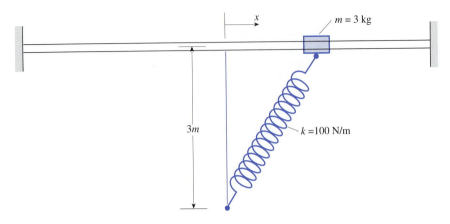

Figure 6.12

time t. Clearly the equilibrium position of the block is at $x = 1.0$ m (or $x = -1.0$ m). Let $y_0 = 10$ m (the unstretched length of the spring). This second-order differential equation describes the motion:

$$\frac{d^2x}{dt^2} = -\left(\frac{k}{m}\right) x \left(1 - \frac{y_0}{\sqrt{(x^2 + 9)}}\right).$$

a. Using both single-step and multistep methods, find the position of the block between $t = 0$ and $t = 10$ sec if $x_0 = 1.4$ and the initial velocity is zero.

b. Repeat part (a), but now with the spring stretched more at the start, $x_0 = 2.5$.

c. Use Maple and/or MATLAB to graph the motion for both parts (a) and (b). Compare your graphs to Fig. 6.13.

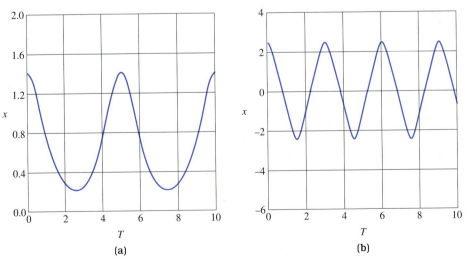

Figure 6.13

66. In finding the deflection of a beam, the y' term is often neglected (see Exercise 37). (The analytical method is easy if y' is neglected in the equation.) What is the difference in the calculated values of the maximum deflection when the nonlinear relationship (Exercise 37) is used in comparison to the simpler linear equation? For light loads, the error is negligible, of course; only for heavier ones is there a need to use the nonlinear equation. At what value of the load is the error equal to 1% of the true value?

67. Write a program that solves a first-order initial-value problem by the Adams–Moulton method, calling the RKF subroutine of the text to obtain starting values. As each new point is computed, monitor the accuracy by comparing the error estimate to a tolerance parameter, printing out a message if the tolerance is not met, but continuing the solution anyway. This might best be done by incorporating the Adams–Moulton method in a subroutine that advances the solution one step.

68. The equation $y' = 1 + y^2$, $y(0)$, has the analytical solution $y = \tan x$. The tangent function is infinite at $x = \pi/2$. Use the program you wrote in Problem 67 to solve this equation between $x = 0$ and $x = 1.6$ with a step size of 0.1. Compare the results of the program with the analytical solution.

69. What is the behavior of the Runge–Kutta fourth-order method when used over $x = 0$ to $x = 1.6$ on the equation of Problem 68? How does it compare with the multistep methods? How does its behavior change with step size?

70. Enlarge on Problem 67 by modifying your program so that the step size is halved if the error estimate exceeds the tolerance value. Use Eq. (6.19) to interpolate for the new values needed when this is done. If the error estimate is very small, say 1/20 times the tolerance, your program should double the step size. (Some programs that provide for doubling the step size keep track of how many uniformly spaced values are available and defer the doubling until there are seven. You may wish to do this, but it is a tricky bit of programming. Alternatively, one could always compute two additional values after discovering that the step size can be increased; this guarantees that there are at least seven equispaced values. Another method would be to extrapolate to obtain a value at t_{n-5} when we find, at t_{n+1}, that the error is 1/20 of the tolerance.)

71. In an electrical circuit (Fig. 6.14) that contains resistance, inductance, and capacitance (and every circuit does), the voltage drop across the resistance is iR (i is current in amperes, R is resistance in ohms), across the inductance it is L (di/dt) (L is inductance in henries), and across the capacitance it is q/C (q is charge in the capacitor in coulombs, C is capacitance in farads). We then can write, for the voltage difference between points A and B,

$$V_{AB} = L\frac{di}{dt} + Ri + \frac{q}{C}.$$

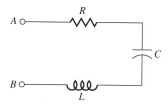

Figure 6.14

Differentiating with respect to t and remembering that $dq/dt = i$, we have a second-order differential equation:

$$L\frac{d^2i}{dt^2} + R\frac{di}{dt} + \frac{1}{C}i = \frac{dV}{dt}.$$

If the voltage V_{AB} (which has previously been 0 V) is suddenly brought to 15 V (let us say, by connecting a battery across the terminals) and maintained steadily at 15 V (so $dV/dt = 0$), current will flow through the circuit. Use an appropriate numerical method to determine how the current varies with time between 0 and 0.1 sec if $C = 1000$ μf, $L = 50$ mH, and $R = 4.7$ ohms; use Δt of 0.002 sec. Also determine how the voltage builds up across the capacitor during this time. You may want to compare the computations with the analytical solution.

72. Repeat Problem 71, but let the voltage source be a 60-Hz sinusoidal input:

$$V_{AB} = 15\sin(120\pi t).$$

How closely does the voltage across the capacitor resemble a sine wave during the last full cycle of voltage variation?

73. After the voltages have stabilized in Problem 71 (15 V across the capacitor), the battery is shorted so that the capacitor discharges through the resistance and inductor. Follow the current and the capacitor voltages for 0.1 sec, again with $\Delta t = 0.002$ sec. The oscillations of decreasing amplitude are called *damped oscillations*. If the calculations are repeated but with the resistance value increased, the oscillations will be damped out more quickly; at $R = 14.14$ ohms the oscillations should disappear; this is called *critical damping*. Perform numerical computations with values of R increasing from 4.7 to 22 ohms to confirm that critical damping occurs at 14.14 ohms.

74. In Chapter 2, Problem 84, electrical networks were discussed, but these were simple ones with resistance only. A more realistic situation is to have RLC circuits combined into a network. Using either the current or voltage laws of Kirchhoff, we can set up a set of simultaneous equations, but now these are simultaneous differential equations. For example, for the circuit in Fig. 6.15, which has two voltage sources, we have

$$(L_1 + L_3)\frac{d^2i_1}{dt^2} + (R_1 + R_3)\frac{di_1}{dt} + \left(\frac{1}{C_1} + \frac{1}{C_3}\right)i_1 - L_3\frac{d^2i_2}{dt^2} - R_3\frac{di_2}{dt} - \frac{1}{C_3}i_2 = e_1'(t),$$

$$(L_2 + L_3)\frac{d^2i_2}{dt^2} + (R_2 + R_3)\frac{di_2}{dt} + \left(\frac{1}{C_2} + \frac{1}{C_3}\right)i_2 - L_3\frac{d^2i_1}{dt^2} - R_3\frac{di_1}{dt} - \frac{1}{C_3}i_1 = e_2'(t).$$

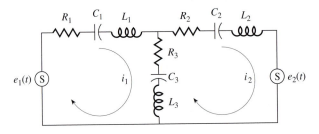

Figure 6.15

In the equations, i_1 and i_2 represent the currents in each of the loops. Solve the equations for i_1 and i_2 between $t = 0$ and $t = 0.2$ sec, if

$$e_1(t) = 100 \, \sin(120\pi t), \qquad e_2(t) = 0,$$

given that

$$\begin{aligned} R_1 &= 22 \text{ ohms}, & C_1 &= C_2 = C_3 = 10 \ \mu\text{f}, \\ R_2 &= 4.7 \text{ ohms}, & L_1 &= 2.5 \text{ mH}, \\ R_3 &= 47 \text{ ohms}, & L_2 &= L_3 = 0.5 \text{ mH}. \end{aligned}$$

75. A Foucault pendulum is one free to swing in both the x- and y-directions. It is frequently displayed in science museums to exhibit the rotation of the earth, which causes the pendulum to swing in directions that continuously vary. The equations of motion are

$$\ddot{x} - 2\omega \, \sin \psi \dot{y} + k^2 x = 0,$$
$$\ddot{y} + 2\omega \, \sin \psi \dot{x} + k^2 y = 0,$$

when damping is absent (or compensated for). In these equations the dots over the variable represent differentiation with respect to time. Here ω is the angular velocity of the earth's rotation (7.29×10^{-5} sec^{-1}), ψ is the latitude, $k^2 = g/\ell$ where ℓ is the length of the pendulum. How long will it take a 10-m-long pendulum to rotate its plane of swing by 45° at the latitude where you live? How long if located in Quebec, Canada?

76. Condon and Odishaw (1967) discuss Duffing's equation for the flux ϕ in a transformer. This nonlinear differential equation is

$$\ddot{\phi} + \omega_0^2 \phi + b\phi^3 = \frac{\omega}{N} E \cos \omega t.$$

In this equation, $E \sin \omega t$ is the sinusoidal source voltage and N is the number of turns in the primary winding, while ω_0 and b are parameters of the transformer design. Make a plot of ϕ versus t (and compare to the source voltage) if $E = 165$, $\omega = 120\pi$, $N = 600$, $\omega_0^2 = 83$, and $b = 0.14$. For approximate calculations, the nonlinear term $b\phi^3$ is sometimes neglected. Evaluate your results to determine whether this makes a significant error in the results.

77. Ethylene oxide is an important raw material for the manufacture of organic chemicals. It is produced by reacting ethylene and oxygen together over a silver catalyst. Laboratory studies gave the equation shown.

It is planned to use this process commercially by passing the gaseous mixture through tubes filled with catalyst. The reaction rate varies with pressure, temperature, and concentrations of ethylene and oxygen, according to this equation:

$$r = 1.7 \times 10^6 e^{-9716/T} \left(\frac{P}{14.7} \right) C_E^{0.328} C_O^{0.672},$$

where

$$\begin{aligned} r &= \text{reaction rate (units of ethylene oxide formed per lb of catalyst per hr)}, \\ T &= \text{temperature, } °\text{K } (°\text{C} + 273), \\ P &= \text{absolute pressure (lb/in.}^2), \\ C_E &= \text{concentration of ethylene}, \\ C_O &= \text{concentration of oxygen}. \end{aligned}$$

Under the planned conditions, the reaction will occur, as the gas flows through the tube, according to the equation

$$\frac{dx}{dL} = 6.42r,$$

where

$$x = \text{fraction of ethylene converted to ethylene oxide,}$$

$$L = \text{length of reactor tube (ft).}$$

The reaction is strongly exothermic, so that it is necessary to cool the tubular reactor to prevent overheating. (Excessively high temperatures produce undesirable side reactions.) The reactor will be cooled by surrounding the catalyst tubes with boiling coolant under pressure so that the tube walls are kept at 225°C. This will remove heat proportional to the temperature difference between the gas and the boiling water. Of course, heat is generated by the reaction. The net effect can be expressed by this equation for the temperature change per foot of tube, where B is a design parameter:

$$\frac{dT}{dL} = 24{,}302r - B(T - 225).$$

For preliminary computations, it has been agreed that we can neglect the change in pressure as the gases flow through the tubes; we will use the average pressure of $P = 22$ lb/in.^{2} absolute. We will also neglect the difference between the catalyst temperature (which should be used to find the reaction rate) and the gas temperature. You are to compute the length of tubes required for 65% conversion of ethylene if the inlet temperature is 250°C. Oxygen is consumed in proportion to the ethylene converted; material balances show that the concentrations of ethylene and oxygen vary with x, the fraction of ethylene converted, as follows:

$$C_E = \frac{1 - x}{4 - 0.375x},$$

$$C_O = \frac{1 - 1.125x}{4 - 0.375x}.$$

The design parameter B will be determined by the diameter of tubes that contain the catalyst. (The number of tubes in parallel will be chosen to accommodate the quantities of materials flowing through the reactor.) The tube size will be chosen to control the maximum temperature of the reaction, as set by the minimum allowable value of B. If the tubes are too large in diameter (for which the value of B is small), the temperatures will run wild. If the tubes are too *small* (giving a large value to B), so much heat is lost that the reaction tends to be quenched. In your studies, vary B to find the least value that will keep the maximum temperature below 300°C. Permissible values for the parameter B are from 1.0 to 10.0.

In addition to finding how long the tubes must be, we need to know how the temperature varies with x and with the distance along the tubes. To have some indication of the controllability of the process, you are also asked to determine how much the outlet temperature will change for a 1°C change in the inlet temperature, using the value of B already determined.

78. An ecologist has been studying the effects of the environment on the population of field mice. Her research shows that the number of mice born each month is proportional to the number of females in the group and that the fraction of females is normally constant in any group. This implies that the number of births per month is proportional to the total population.

She has located a test plot for further research, which is a restricted area of semiarid land. She has constructed barriers around the plot so mice cannot enter or leave. Under the conditions of the experiment, the food supply is limited, and it is found that the death rate is affected as a result, with mice dying of starvation at a rate proportional to some power of the population. (She also hypothesizes that when the mother is undernourished, the babies have less chance for survival and that starving males tend to attack one another, but these factors are only speculation.)

The net result of this scientific analysis is the following equation, with N being the number of mice at time t (with t expressed in months). The ecologist has come to you for help in solving the equation; her calculus doesn't seem to apply.

$$\frac{dN}{dt} = aN - BN^{1.7}, \qquad \text{with } B \text{ given by Table 6.14.}$$

As the season progresses, the amount of vegetation varies. The ecologist accounts for this change in the food supply by using a "constant" B that varies with the season.

If 100 mice were initially released into the test plot and if $a = 0.9$, estimate the number of mice as a function of t, for $t = 0$ to $t = 8$.

79. Redo the second program of this chapter (which implements the Adams–Moulton method for a pair of differential equations) so that it can handle up to ten simultaneous differential equations. You may want to adapt the strategy of the first program (which implements RKF) of storing values in doubly subscripted arrays.

80. A certain chemical company produces a product that is a mixture of two ingredients, A and B. In order to ensure that the product is homogeneous, A and B are fed into a well-mixed tank that holds 100 gal. The desired product must contain two parts of A to one part of B within certain specifications. The normal flows of A and B into the tank are 4 and 2 gal/min. There is no volume change when these are mixed, so the outflow is 6 gal/min. and the holding time in the tank is $100/6 = 16.66$ min. Due to an unfortunate accident, the flow of ingredient B is cut off and before this is noticed and corrected, the ratio of A to B in the tank has increased to 10 parts of A to 1 part of B. (There are still 100 gal in the tank). Set up equations that give the ratio of A to B in the tank as a function of time after the flow of B has been restored to its normal value of 2 gal/min. How long will it take until the output from the tank reaches 2 parts A to 0.99 parts B? How much product is produced (and discarded because it is not up to specification) during this time? How would you suggest that this time to reach specification be reduced?

81. The rate of decay of a radioactive material is customarily measured by its "half-life." Find out what this term means. Suppose that a nuclear storage facility receives a large shipment whose

Table 6.14

t	B	t	B
0	0.0070	5	0.0013
1	0.0036	6	0.0028
2	0.0011	7	0.0043
3	0.0001	8	0.0056
4	0.0004		

half-life is 100 years. Studies have shown that neighbors to the facility are in danger until the radioactive material decays to 10% of the original amount. How long before the people who live nearby are safe?

82. In Section 6.4, the development of a second-order Runge–Kutta method is given. Equation (6.9) can be made to match Eq. (6.6) with a variety of choices for variable a, producing equivalent values for b, α, and β. One choice, $a = 1/2$, gives the modified Euler method. Investigate how the accuracy of the solutions compare for various choices for variable a on this equation:

$$\frac{dx}{dt} = x + 2t * e^{2t}, \qquad x(1) = 3e.$$

Compute from $t = 1$ to $t = 2$ with $\Delta t = 0.1$.

7

Boundary-Value Problems

Contents of This Chapter

The previous chapter discussed the solving of differential equations where the conditions for the problem were all specified at some initial value of the independent variable. When the order of the equation is two or more, two or more such conditions are needed to get the solution. Because these conditions are specified at the start, or *initial value,* of the independent variable, such problems are called *initial-value problems.*

However, there are many times when the conditions for an equation of order two or more are not all given at the initial point but are specified at two different values for the independent variable. Because these values are usually at the endpoints (or *boundaries*) of some domain of interest, these are called *boundary-value problems.* Determining the deflection of a simply supported beam is a typical example. In this fourth-order problem, the conditions specified are normally the deflections and second derivatives of the elastic curve at the supports. Temperature distribution along a rod fall in this class when temperatures or temperature gradients are given at the ends of the rod. If the region is a flat plate with temperatures and/or gradients given along the boundary of the plate, this type of boundary-value problem has partial derivatives in the equation, and this is particularly true if the region is a three-dimensional object.

Most of this chapter is concerned with solving boundary-value problems by replacing the derivatives with finite difference approximations. When this is done, the solution is obtained by solving a set of simultaneous equations. For one-dimensional problems, a second method, called the *shooting method,* can be employed.

A special class of boundary-value problem on a one-dimensional region has a solution only for special values of some parameter of the problem. These "special values" are called the *characteristic values* for the problem and such problems are called *characteristic-value problems*. We discuss in this chapter only the most elementary forms of characteristic-value problems.

There is another method for solving boundary-value problems that has special advantages and that is widely used by engineers—the *finite-element method*. We discuss the finite element method in another chapter.

7.1   Temperature Distribution in a Rod

Derives the equation for the temperatures along a rod when heat flows along it due to a temperature gradient. If the rod is not of constant cross-section, if the thermal conductivity varies, or if heat is added from some external source (or lost to the surroundings), the equation is somewhat more complicated. In this chapter, only the steady state is considered; that is, the distribution of temperatures when these have reached their equilibrium values. The same type of equations apply to other physical situations, not just to flow of heat.

7.2   The Shooting Method

Shows how the methods of the previous chapter can be adapted to solve boundary-value problems where there is a single independent variable. A trial and error approach is employed. However, if the problem is linear, the solution to a second-order problem can be found with only two trials.

7.3   Solution Through a Set of Equations

Demonstrates that an approximation to the solution for a boundary-value problem can be obtained by replacing the derivatives with *finite-difference approximations*. This converts the problem into a system of simultaneous equations. For problems with a single independent variable, this method is somewhat less efficient than the shooting method, but no trial solutions are needed for a linear equation.

7.4   Derivative Boundary Conditions

Requires a modification of the procedures of Section 7.3; this modification is straightforward but it does necessitate an artificial extension of the region of the problem.

7.5   Characteristic-Value Problems

Describes an important special class of boundary-value problems that have a solution only for certain characteristic values of a parameter, called the *eigenvalues*. Eigenvalues and the associated *eigenvectors* are essential matrix-related quantities that have applications in many fields. Two different techniques for getting eigenvalues and eigenvectors are presented: the *power method* and the *QR method*.

7.6   Temperature Distribution in a Slab

Derives the equations for temperature distribution in a two-dimensional object. For this situation, the equation contains partial derivatives because there is more than a single independent variable. Analogous equations apply to a three-dimensional object.

7.7 Solving for the Temperatures in a Slab

Shows how two- and three-dimensional boundary-value problems can be solved. For these problems, the shooting method does not apply, but they can be solved through a set of equations obtained by replacing the derivatives with finite-difference approximations. The simplest kind of these problems satisfies *Laplace's equation;* a slightly more complicated problem is described by *Poisson's equation.* Because the set of equations can be very large but the coefficient matrix is sparse, iterative methods of solving them are frequently preferred.

7.8 The Alternating Direction Implicit Method (A.D.I.)

Introduces a technique that permits the solution to a two- or three-dimensional boundary-value problem to be through a system of equations that has a tridiagonal coefficient matrix. Such systems are particularly easy to solve. This technique is especially important for three-dimensional problems.

7.9 Irregular Regions and Nonrectangular Grids

Tackles the real-world situation in which the region is not a simple rectangle whose boundary matches to a set of uniformly spaced nodes.

7.10 Theoretical Matters

Collects a number of items of theoretical importance that underlie the problems and methods of the chapter.

7.11 Using MATLAB

Describes how this computer algebra system can assist in solving problems of this chapter.

Chapter Summary

Tests whether the procedures and topics of the chapter are well understood.

Computer Programs

Gives two programming examples that implement the methods of this chapter.

Parallel Processing

Parallel processing is possible with several of the methods of this chapter, but nothing new is introduced. In Section 7.2, an initial-value problem is solved with different values of a parameter and these could be computed in parallel. In many other sections, systems of equations are solved and for these, parallel processing can be quite effective as discussed in Chapter 2. The power method of Section 7.5 is inherently serial because each step depends on the previous one. However, within each step, a matrix multiplication is done and these computations can be speeded up with parallel processing.

7.1 Temperature Distribution in a Rod

Suppose we have a rod of length L with a uniform cross-section that has its ends held at TL degrees at the left and TR degrees at the right as shown in Figure 7.1. By concentrating our attention on an element of the rod of length dx located at a distance x from the left end, we can derive the equation that determines the temperature, u, at any point along the rod. The rod is perfectly insulated around its outer circumference so that heat flows only later- ally along the rod. It is well known that heat flows at a rate (measured in calories per second) proportional to the cross-sectional area (A), to a property of the material (k, its thermal conductivity, measured in cal/(sec $*$ cm² $*$ (°C/cm)), and to the temperature gra- dient, du/dx (measured in °C/cm), at point x. We use $u(x)$ for the temperature at point x, with x measured from the left end of the rod. Thus the rate of flow of heat into the element (at $x = x$) is

$$-kA \left(\frac{du}{dx} \right).$$

The minus sign is required because du/dx expresses how rapidly temperatures increase with x, while the heat always flows from high temperature to low.

The rate at which heat leaves the element is given by a similar equation, but now the temperature gradient must be at the point $x + dx$:

$$-kA \left[\frac{du}{dx} + \frac{d}{dx} \left(\frac{du}{dx} \right) dx \right],$$

in which the gradient term is the gradient at x plus the change in the gradient between x and $x + dx$.

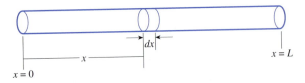

Figure 7.1

Unless heat is being added to the element (or withdrawn by some means), the rate that heat flows from the element must equal the rate that heat enters, or else the temperature of the element will vary with time. In this chapter we consider only the case of *steady-state* or *equilibrium* temperatures, so we can equate the rates of heat entering and leaving the element:

$$-kA\left(\frac{du}{dx}\right) = -kA\left[\frac{du}{dx} + \frac{d}{dx}\left(\frac{du}{dx}\right)\,dx\right].$$

When some common terms on each side of the equation are canceled, we get the very simple relation:

$$kA\frac{d}{dx}\left(\frac{du}{dx}\right)\,dx = kA\frac{d^2u}{dx^2} = 0,$$

where we have written the second derivative in its usual form. For this particularly simple example, the equation for u as a function of x is the solution to

$$\frac{d^2u}{dx^2} = 0,$$

and this is obviously just

$$u = ax + b,$$

a linear relation. This means that the temperatures vary linearly from TL to TR as x goes from 0 to L.

The rod could also lose heat from the outer surface of the element. If this is Q (cal/(sec $*$ cm^2), the rate of heat flow in must equal the rate leaving the element by conduction along the rod plus the rate at which heat is lost from the surface. This means that:

$$-kA\left(\frac{du}{dx}\right) = -kA\left[\frac{du}{dx} + \frac{d}{dx}\left(\frac{du}{dx}\right)\,dx\right] + Qp\,dx,$$

where p is the perimeter at point x. (Q might also depend on the difference in temperature within the element and the temperature of the surroundings, but we will ignore that for now).

If this equation is expanded and common terms are canceled, we get a somewhat more complicated equation whose solution is not obvious:

$$\frac{d^2u}{dx^2} = \frac{Qp}{(kA)}. \tag{7.1}$$

In Eq. (7.1), Q can be a function of x.

The situation may not be quite as simple as this. The cross-section could vary along the rod, or k could be a function of x (some kind of composite of materials, possibly). Suppose first that only the cross-section varies with x. We will have, then, for the rate of heat leaving the element:

$$-k[A + A'\,dx]\left[\frac{du}{dx} + u''\,dx\right],$$

where we have used a prime notation for derivatives with respect to x. Equating the rates in and out as before and canceling common terms results in:

$$kAu'' \, dx + kA'u' \, dx + kA'u'' \, dx^2 = Qp \, dx.$$

We can simplify this further by dropping the term with dx^2 because it goes to zero faster than the terms in dx. After also dividing out dx, this results in a second-order differential equation similar in form to some we have discussed in Chapter 6:

$$kAu'' + kA'u' = Qp. \tag{7.2}$$

The equation can be generalized even more if k also varies along the rod. We leave to the reader as an exercise to show that this results in:

$$kAu'' + (kA' + k'A)u' = Qp. \tag{7.3}$$

If the rate of heat loss from the outer surface is proportional to the difference in temperatures between that within the element and the surroundings (u_s), (and this is a common situation), we must substitute for Q:

$$Q = q(u - u_s),$$

giving:

$$kAu'' + (kA' + k'A)u' - q * pu = -q * pu_s. \tag{7.4}$$

This chapter will discuss two ways to solve equations like Eqs. (7.1–7.4).

Heat flow has been used in this section as the physical situation that is modeled, but equations of the same form apply to diffusion, certain types of fluid flow, torsion in objects subject to twisting, distribution of voltage, in fact, to any problem where the potential is proportional to the gradient.

7.2 The Shooting Method

We can rewrite Eq. (7.4) as

$$A\frac{d^2u}{dx^2} + B\frac{du}{dx} + Cu = D, \tag{7.5}$$

where the coefficients, A, B, C, and D are functions of x. (Actually, they could also be functions of both x and u, but that makes the problem more difficult to solve. In a temperature distribution problem, such nonlinearity can be caused if the thermal conductivity, k, is considered to vary with the temperature, u. That is actually true for almost all materials but, as the variation is usually small, it is often neglected and an average value is used.)

To solve Eq. (7.5), we must know two conditions on u or its derivative. If both u and u' are specified at some starting value for x, the problem is called an *initial-value problem* and the methods of Chapter 6 can be used. In this section, we consider Eq. (7.5) to have two values of u to be given and these are at two different values for x—this makes it a *boundary-value problem.* In this section we discuss how the same procedures that apply to an initial-value problem can be adapted.

The strategy is simple: Suppose we know u at $x = a$ (the beginning of a region of interest) and u at $x = b$ (the end of the region). We wish we knew u' at $x = a$; that would make it an initial-value problem. So, why not assume a value for this? Some general knowledge of the situation may indicate a reasonable guess. Or we could blindly select some value. The test of the accuracy of the guess is to see if we get the specified $u(b)$ by solving the problem over the interval $x = a$ to $x = b$. If the initial slope that we assumed is too large, we will often find that the computed value for $u(b)$ is too large. So we try again with a smaller initial slope. If the new value for $u(b)$ is too small, we have bracketed the correct initial slope. This method is called the *shooting method* because of its resemblance to the problem faced by an artillery officer who is trying to hit a distant target. The right elevation of the gun can be found if two shots are made of which one is short of the target and the other is beyond. That means that an intermediate elevation will come closer.

EXAMPLE 7.1   Solve

$$u'' - \left(1 - \frac{x}{5}\right) u = x, \qquad u(1) = 2, \qquad u(3) = -1.$$

(This is an instance of Eq. (7.5) with $A = 1$, $B = 0$, $C = -(1 - x/5)$, and $D = x$.) Assume that $u'(1) = -1.5$ (which might be a reasonable guess, because u declines between $x = 1$ and $x = 3$; this number is the average slope over the interval). If we use a program that implements the Runge–Kutta–Fehlberg method, we get the values shown in the first part of Table 7.1.

Table 7.1

x	Assume $u'(1) = -1.5$		Assume $u'(1) = -3.0$		Assume $u'(1) = -3.4950$	
	u	u'	u	u'	u	u'
1.00	2.0000	−1.5000	2.0000	−3.0000	2.0000	−3.4950
1.20	1.7614	−0.9886	1.4598	−2.5118	1.3503	−3.0145
1.40	1.6043	−0.4814	0.9921	−2.0719	0.7900	−2.5967
1.60	1.5597	0.0389	0.6192	−1.6598	0.3099	−2.2204
1.80	1.6218	0.5876	0.3275	−1.2580	−0.0997	−1.8671
2.00	1.7976	1.1783	0.1163	−0.8512	−0.4385	−1.5209
2.20	2.0967	1.8227	−0.0118	−0.4259	−0.7076	−1.1679
2.40	2.5309	2.5310	−0.0520	0.0299	−0.9043	−0.7955
2.60	3.1139	3.3116	0.0029	0.5266	−1.0237	−0.3925
2.80	3.8608	4.1706	0.1620	1.0732	−1.0586	0.0511
3.00	4.7876	5.1119	0.4360	1.6773	−1.0000	0.5439

Because the value for $u(3)$ is 4.7876 rather than the desired -1, we try again with a smaller initial slope, say $u'(1) = -3.0$, and get the middle part of Table 7.1. The resulting value for $u(3)$ is still too high: 0.4360 rather than -1. We could guess at a third trial for $u'(1)$, but let us interpolate linearly between the first two trials.[†] Doing so suggests a value for $u'(1)$ of -3.4950. Lo and behold, we get the correct answer for $u(3)$! These results are shown in the third part of Table 7.1.

Solving a boundary-value problem in this way will always require at most three trials; the interpolated initial slope for two trials is always the correct one, but *only if the problem is linear.* The proof of this statement will be given in the section on theoretical matters. Further, not only can the correct initial slope be found by interpolating from the assumed slopes of the first two trials, the intermediate values of u can also be interpolated from the trial values. Test this statement with the values in Table 7.1.

EXAMPLE 7.2 Solve

$$u'' - \left(1 - \frac{x}{5}\right)uu' = x, \qquad u(1) = 2, \qquad u(3) = -1.$$

This resembles Example 7.1, but observe that the coefficient of u' involves u, the dependent variable. This problem is nonlinear and we shall see that it is not as easy to solve. If we again use the Runge–Kutta–Fehlberg method, we get the results summarized in Table 7.2. Here the third trial, which used the interpolated value from the first two trials, does not give the correct solution. A nonlinear problem requires a kind of search operation. We could interpolate with a quadratic from the results of three trials, an adaptation of Muller's method. Table 7.3 gives the computed values for $u(x)$ between $x = 1$ and $x = 3$ with the final (good) estimate of the initial slope.

The shooting method is often quite laborious, especially with problems of fourth or higher order. With these, the necessity of assuming two or more conditions at the starting point (and matching with the same number of conditions at the end) is slow and tedious.

Table 7.2

Assumed value for $u'(1)$	Calculated value for $u(3)$
-1.5	-0.0282
-3.0	-2.0705
-2.2137^*	-1.2719
-1.9460^*	-0.8932
-2.0215^*	-1.0080
-2.0162^*	-1.0002
-2.0161^*	-1.0000

*Interpolated from two previous values

Table 7.3

x	u	u'
1.0000	2.0000	-2.0161
1.2000	1.5552	-2.4130
1.4000	1.0459	-2.6438
1.6000	0.5318	-2.6352
1.8000	0.0082	-2.3832
2.0000	-0.4272	-1.9472
2.2000	-0.7640	-1.4110
2.4000	-0.9896	-0.8441
2.6000	-1.1022	-0.2848
2.8000	-1.1047	0.2569
3.0000	-1.0000	0.7909

[†] If G = guess, and R = result: DR = desired result: G3 = G2 + (DR − R2)(G1 − G2)/(R1 − R2)

There are times when it is better to compute "backwards" from $x = b$ to $x = a$. For example, if $u(b)$ and $u'(a)$ are the known boundary values, the technique just described works best if we compute from $x = b$ to $x = a$. Another time that computing backwards would be preferred is in a fourth-order problem where three conditions are given at $x = b$ and only one at $x = a$.

▲

7.3   Solution Through a Set of Equations

There is another way to solve boundary-value problems like Example 7.1 or 7.2. We have seen in Chapter 4 that derivatives can be approximated by finite-difference quotients. If we replace the derivatives in a differential equation by such expressions, we convert it into a difference equation whose solution is an approximation to the solution of the differential equation. This method is sometimes preferred over the shooting method, but it really can be used only with linear equations. (If the differential equation is nonlinear, this technique leads to a set of nonlinear equations that are more difficult to solve. Solving such a set of nonlinear equations is best done by iteration, starting with some initial approximation to the solution vector.)

EXAMPLE 7.3   Solve the boundary-value problem of Example 7.1 but use a set of equations obtained by replacing the derivative with a central difference approximation. Divide the region into four equal subintervals and solve the equations, then divide into ten subintervals. Compare both of these solutions to the results of Example 7.1.

When the interval from $x = 1$ to $x = 3$ is subdivided into four subintervals, there are interior points (these are usually called *nodes*) at $x = 1.5$, 2.0, and 2.5. Label the nodes as x_1, x_2, and x_3. The endpoints are x_0 and x_4. We write the difference equation at the three interior nodes. The equation, $u'' - (1 - x/5)u = x$, $u(1) = 2$, $u(3) = -1$, becomes:

$$\text{At } x_1: \quad \frac{(u_0 - 2u_1 + u_2)}{h^2} - \left(\frac{1 - x_1}{5}\right)u_1 = x_1,$$

$$\text{At } x_2: \quad \frac{(u_1 - 2u_2 + u_3)}{h^2} - \left(\frac{1 - x_2}{5}\right)u_1 = x_2,$$

$$\text{At } x_3: \quad \frac{(u_2 - 2u_3 + u_4)}{h^2} - \left(\frac{1 - x_3}{5}\right)u_1 = x_3.$$

These equations are all of the form:

$$\text{At } x_i: \quad \frac{(u_{i-1} - 2u_i + u_{i+1})}{h^2} - \left(\frac{1 - x_i}{5}\right)u_i = x_i,$$

which can be rearranged into:

$$\text{At } x_i: \quad u_{i-1} - \left[2 + h^2\left(\frac{1 - x_i}{5}\right)\right]u_i + u_{i+1} = h^2 x_i.$$

Substitute $h = 0.5$, substitute the x-values at the nodes, and substitute the u-values at the endpoints and arrange in matrix form, which gives:

$$\begin{bmatrix} -2.175 & 1 & 0 \\ 1 & -2.150 & 1 \\ 0 & 1 & -2.125 \end{bmatrix} \begin{bmatrix} u_1 \\ u_2 \\ u_3 \end{bmatrix} = \begin{bmatrix} -1.625 \\ 0.5 \\ 1.625 \end{bmatrix}.$$

Observe that the system is tridiagonal and that this will always be true even when there are many more nodes, because any derivative of u involves only points to the left, to the right, and the central point.

When this system is solved, we get:

$$x_1 = 0.552, \qquad x_2 = -0.424, \qquad \text{and} \qquad x_3 = -0.964.$$

If we solve the problem again but with 10 subintervals ($h = 0.2$), we must solve a system of nine equations, because there are nine interior nodes where the value of u is unknown. The answers, together with the results from the shooting method for comparison, are:

x	Values from the finite-difference method	Values from the shooting method
1.2	1.351	1.350
1.4	0.792	0.790
1.6	0.311	0.309
1.8	−0.097	−0.100
2.0	−0.436	−0.438
2.2	−0.705	−0.708
2.4	−0.903	−0.904
2.6	−1.022	−1.024
2.8	−1.058	−1.059

There is quite close agreement. It is difficult to say from this which method is more accurate because both are subject to error. We can compare the methods and determine how making the number of subintervals greater increases the accuracy by examining the results for a problem with a known analytical answer.

EXAMPLE 7.4 Compare the accuracy of the finite-difference method with the shooting method on this second-order boundary-value problem:

$$u'' = u, \qquad u(1) = 1.17520, \qquad u(3) = 10.01787,$$

whose analytical solution is $u = \sinh(x)$.

When the problem is solved by finite-difference approximations to the derivatives, the typical equation is:

$$u_{i-1} - (2 + h^2)u_i + u_{i+1} = 0.$$

Solving with $h = 1$, $h = 0.5$, and $h = 0.25$, we get the values in Table 7.4. If we solve this with the shooting method (employing Runge–Kutta–Fehlberg), we get Table 7.5.

Table 7.4

		u-values with	
x	2 subintervals	4 subintervals	8 subintervals
1.25			1.60432
1.50		2.14670	2.13372
1.75			2.79647
2.00	3.73102	3.65488	3.63400
2.25			4.69866
2.50		7.07678	7.05698
2.75			7.79387
error at x = 2.00	0.10416	0.02802	0.00714

Table 7.5

		u-values with	
x	2 subintervals	4 subintervals	8 subintervals
1.25			1.60192
1.50		2.12931	2.12928
1.75			2.79042
2.00	3.62814	3.62692	3.62686
2.25			4.69117
2.50		7.05025	7.05020
2.75			7.78935
error at x = 2.00	0.00128	0.00006	0.00000

In both tables, the errors at $x = 2.0$ are shown. This is nearly the maximum error of any of the results.

When the results from the two methods are compared, it is clear that (1) the shooting method is much more accurate at the same number of subintervals, its errors being from 80 to over 500 times smaller; and (2) the errors for the finite-difference method decrease about four times when the number of subintervals is doubled, which is as expected.

The reader should make a similar comparison for other equations.

7.4 Derivative Boundary Conditions

The conditions at the boundary often involve the derivative of the dependent variable in addition to its value. A hot object loses heat to its surroundings proportional to the difference between the temperature at the surface of the object and the temperature of the surroundings. The proportionality constant is called the *heat-transfer coefficient* and is frequently represented by the symbol h. (This can cause confusion because we use h for the size of a subinterval. To avoid this confusion, we shall use a capital letter, H, for the heat-transfer coefficient.) The units of H are cal/sec/cm^2/°C (of temperature difference). In this section we consider a rod that loses heat to the surroundings from one or both ends. Of course, heat could be gained from the surroundings if the surroundings are hotter than the rod.

Names have been given to the various types of boundary conditions. If the value for u is specified at a boundary, it is called a *Dirichlet condition*. This is the type of problem that we have solved in Sections 7.2 and 7.3. If the condition is the value of the derivative of u, it is a *Neumann condition*. When a boundary condition involves both u and its derivative, it is called a *mixed condition*.

We now develop the relations when heat is lost from the ends of a rod that conducts heat along the rod but is insulated around its perimeter so that no heat is lost from its lateral surface. First consider the right end of the rod and assume that heat is being lost to the surroundings (implying that the surface is hotter than the surroundings). Figure 7.2 will help to visualize this. At the right end of the rod ($x = x_R$), the temperature is u_R; the temperature of the surroundings is u_{SR}. Heat then is being lost from the rod to the surroundings at a rate (measured in (cal/sec)) of

$$HA(u_R - u_{SR}),$$

where A is the area of the end of the rod. This heat must be supplied by heat flowing from inside the rod to the surface, which is at the rate of

$$-kA\frac{du}{dx},$$

where the minus sign is required because heat flows from high to low temperature. Equating these two rates and solving for du/dx (the gradient) gives (the A's cancel):

$$\frac{du}{dx} = -\left(\frac{H}{k}\right)(u_R - u_{SR}), \qquad \text{at the right end.}$$

Now consider the left end of the rod, at $x = 0$, where $u = u_L$. Assume that the temperature of the surroundings here are at some other temperature, u_{SL}. Here, heat is flowing from right to left, so we have

$$\text{Heat leaving the rod:} \qquad -HA(u_L - u_{SL}).$$

$u = u_{SR}$

$x = x_R$
$u = u_R$

Figure 7.2

For the rate at which heat flows from inside the rod we still have

$$-kA\frac{du}{dx},$$

and, after equating and solving for the gradient:

$$\frac{du}{dx} = \left(\frac{H}{k}\right)(u_L - u_{SL}), \qquad \text{at the left end.}$$

The fact that the signs in the equations for the gradients are not the same can be a source of confusion. Of course, if both ends lose heat to the surroundings, the equilibrium or steady-state temperatures of the rod will just be a linear relation between the two (possibly different) surrounding temperatures. In practical situations of heat distribution in a rod, only one end of the rod loses (or gains) heat to (from) the surroundings, the other end being held at some constant temperature.

A minor problem is presented in the cases under consideration in this section. We need to give consideration to how to approximate the gradient at the end of the rod. One could use a forward difference approximation (at the right end, a backward difference at the left), but that seems inappropriate when central differences are used to approximate the derivatives within the rod. This conflict can be resolved if we imagine that the rod is ficti- tiously extended by one subinterval at the end of the rod that is losing heat. Doing so permits us to approximate the derivative with a central difference. The "temperature" at this fictitious point is eliminated by using the equation for the gradient. The next example will clarify this.

EXAMPLE 7.5   An insulated rod is 20 cm long and is of uniform cross-section. It has its right end held at 100° while its left end loses heat to the surroundings, which are at 20°. The rod has a thermal conductivity, k, of 0.52 cal/(sec $*$ cm $*$ °C), and the heat-transfer coefficient, H, is 0.073 cal/sec/cm²/°C). Solve for the steady-state temperatures using the finite-difference method with eight subintervals.

For this example, because the boundary condition at the left end involves both the u-value at the left end and the derivative there, this example has a mixed condition at the left end, whereas it has a Dirichlet condition at the right end.

The equation that applies is Eq. (7.1) with $Q = 0$, because no heat is added at points along the rod:

$$\frac{d^2u}{dx^2} = 0.$$

The typical equation is:

$$u_{i-1} - 2u_i + u_{i+1} = 0,$$

and this applies at each node. At the left end we imagine a fictitious point at x_{-1}, and this allows us to write the equation for that node. At the left endpoint, at $x = x_0$, we write an equation for the gradient:

$$\frac{du}{dx} = \left(\frac{H}{k}\right)(u_L - u_{SL}),$$

or,

$$\frac{(u_1 - u_{-1})}{2h} = \frac{(u_1 - u_{-1})}{(2 * 2.5)}$$

$$= \left(\frac{0.073}{0.52}\right) * (u_0 - 20),$$

which we use to eliminate u_{-1}:

$$u_{-1} = u_1 - (2 * 2.5) * \left[\left(\frac{0.073}{0.52}\right)(u_0 - 20)\right]$$

$$= u_1 - 0.70192u_0 + 14.0385.$$

We will use this last for the equation written at x_0, to give, at that point:

$$u_{-1} - 2u_0 + u_1 = (u_1 - 0.70192u_0 + 14.0385) - 2u_0 + u_1 = 0,$$

or,

$$-2.70192u_0 + 2u_1 = -14.0385,$$

which is the first equation of the set.
  Here is the augmented matrix for the problem:

$$\begin{bmatrix}
-2.70192 & 2 & 0 & 0 & 0 & 0 & 0 & 0 & -14.0385 \\
1 & -2 & 1 & 0 & 0 & 0 & 0 & 0 & 0 \\
0 & 1 & -2 & 1 & 0 & 0 & 0 & 0 & 0 \\
0 & 0 & 1 & -2 & 1 & 0 & 0 & 0 & 0 \\
0 & 0 & 0 & 1 & -2 & 1 & 0 & 0 & 0 \\
0 & 0 & 0 & 0 & 1 & -2 & 1 & 0 & 0 \\
0 & 0 & 0 & 0 & 0 & 1 & -2 & 1 & 0 \\
0 & 0 & 0 & 0 & 0 & 0 & 1 & -2 & -100
\end{bmatrix}$$

for which the solution is:

i:	0	1	2	3	4	5	6	7	(8)
u_i:	41.0103	48.3840	55.7577	63.1314	70.5051	77.8789	85.2526	92.6263	(100)

Observe that the gradient all along the rod is a constant ($2.94948°\,C$/cm). ▲

  Here is another example that illustrates an important point about derivative boundary conditions.

EXAMPLE 7.6   Solve $u'' = u$, $u'(1) = 1.17520$, $u'(3) = 10.01787$, with the finite-difference method.
  This example is identical to that of Example 7.4 of Section 7.3 except that the boundary conditions are the derivatives of u rather than the values of u. (It has Neumann conditions at both ends.) For this problem, the known solution is $u = \cosh(x) + C$, and the boundary values are values of $\sinh(1)$ and $\sinh(3)$.

Because the values of u are not given at either end of the interval, we must add fictitious points at both ends; call these u_{LF} and u_{RF}. With four subintervals, ($h = 2/4 = 0.5$), we can write five equations (at each of the three interior nodes plus the two endpoints where u is unknown). We label the nodes from x_0 (at the left end) to x_4 (at the right end). Each equation is of the form:

$$u_{i-1} - 2u_i + u_{i+1} = h^2 u_i, \qquad i = 0, 1, 2, 3, 4, \qquad h^2 = 0.25,$$

where u_{-1} and u_5 are the fictitious points u_{LF} and u_{RF}.

Doing so gives this augmented matrix:

$$\begin{bmatrix} -2.25 & 1 & 0 & 0 & 0 & 0 & -u_{LF} \\ 1 & -2.25 & 1 & 0 & 0 & 0 \\ 0 & 1 & -2.25 & 1 & 0 & 0 \\ 0 & 0 & 1 & -2.25 & 1 & 0 \\ 0 & 0 & 0 & 1 & -2.25 & 0 & -u_{RF} \end{bmatrix}.$$

There are two more unknowns in this than equations: the unknown fictitious points. However, these can be eliminated by using the derivative conditions at the ends. As before, we use central difference approximation to the derivative:

$$u'(1) = 1.17520 = \frac{(u_1 - u_{LF})}{2h},$$

$$u'(3) = 10.01787 = \frac{(u_{RF} - u_3)}{2h}, \qquad (h = 0.5),$$

which we solve for the fictitious points in terms of nodal points:

$$u_{LF} = u_1 - 1.17520, \qquad u_{RF} = 10.01787 + u_3.$$

Substituting these relations for the fictitious points changes the first and last equations to

$$-2.25u_0 + 2u_1 = 1.17520,$$
$$2u_3 - 2.25u_4 = -10.01787.$$

When the five equations are solved, we get these answers:

x	Answers	cosh(x)	Error
1.0	1.55219	1.54308	-0.00911
1.5	2.33382	2.35241	0.01859
2.0	3.69870	3.76220	0.06350
2.5	5.98870	7.13229	0.14359
3.0	9.77568	10.06770	0.29202

We observe that the accuracy is much poorer than it was in Example 7.4. Take note of the fact that the numerical solution is not identical to the analytical solution; the arbitrary constant is missing (or, we may say, is equal to zero).

Using the Shooting Method

We can solve boundary-value problems where the derivative is involved at one or both end-conditions by "shooting." In fact, as this method computes both the dependent variable and its derivative, this is quite natural. Here is how Example 7.6 can be solved by the shooting method.

EXAMPLE 7.7   Solve $u'' = u$, $u'(1) = 1.17520$, $u'(3) = 10.01787$ by the shooting method.

We can begin at either end, but it seems more natural to begin from $x = 1$. To begin the solution, we must guess at a value for $u(1)$—not for the derivative as we have been doing. From this point, we compute values for u and u' by, say, RKF. If the value of $u'(3)$ is not 10.01787, we try again with a guess for $u(1)$. This will probably not give the correct value for $u'(3)$, but, because the problem is linear, we can interpolate to find the proper value to use for $u(1)$. Here are the answers when four subintervals are used:

x	$u(x)$	$u'(x)$	$\cosh(x)$
1.0	1.54319	1.17520	1.54308
1.5	2.35250	2.12932	2.35241
2.0	3.76228	3.62692	3.76220
2.5	7.13236	7.05027	7.13229
3.0	10.06767	10.01790	10.06770

▲

The results are surprisingly accurate even though the subdivision was coarse; the largest error in the $u(x)$ values is 0.00011 at $x = 1$ and the errors are less as x increases. For this example, the shooting method is much more accurate than using finite-difference approximations to the derivative.

Here is an example that has a mixed end condition.

EXAMPLE 7.8   Solve Example 7.5 by the shooting method. We restate the problem:

An insulated rod is 20-cm long and is of uniform cross-section. It has its right end held at 100° while its left end loses heat to the surroundings, which are at 20°. The rod has a thermal conductivity, k, of 0.52 cal/(sec $*$ cm $*$ °C), and the heat-transfer coefficient, H, is 0.073 cal/(sec $*$ cm^2 $*$ °C). Use the shooting method with eight subintervals.

The procedure here is similar to that used in Example 7.7 but it is necessary to begin at the right end and solve "backwards." (That is no problem; we just use a negative value for Δx.) Beginning at $x = 0$ would be very difficult because we would have to guess at both $u(0)$ and $u'(0)$.

Finding the correct value for u' at $x = 20$ is not as easy as in the previous example because we must fit to a combination of $u(0)$ and $u'(0)$. Here are the results after finding the correct value for $u'(20)$ by a trial and error technique.

i:	0	1	2	3	4	5	6	7	(8)
u_i:	41.005	48.379	55.754	63.128	70.502	77.877	85.251	92.626	(100)

(The gradient here is 2.94975 throughout.) These values match those of Example 7.5 very closely.

▲

7.5   Characteristic-Value Problems

Problems in the fields of elasticity and vibration (including applications of the wave equation of modern physics) fall into a special class of boundary-value problems known as *characteristic-value problems.* Some problems of statistics also fall into this class. We discuss only the most elementary forms of characteristic-value problems.

Consider the homogeneous* second-order equation with homogeneous boundary conditions:

$$\frac{d^2u}{dx^2} + k^2u = 0, \qquad u(0) = 0, \qquad u(1) = 0, \tag{7.5}$$

where k^2 is a parameter. (Using k^2 guarantees that the parameter is a positive number.) We first solve this equation nonnumerically to show that there is a solution only for certain particular or "characteristic" values of the parameter. These characteristic values are more often called the *eigenvalues* from the German word. The general solution is

$$u = a \, \sin(kx) + b \, \cos(kx),$$

which can easily be verified by substituting into the differential equation. The solution contains the two arbitrary constants a and b because the equation is of second order. The constants a and b are to be determined to make the general solution agree with the boundary conditions.

At $x = 0$, $u = 0 = a \sin(0) + b \cos(0) = b$. Then b must be zero. At $x = 1$, $u = 0 = a \sin(k)$; we may have either $a = 0$ or $\sin(k) = 0$ to satisfy the end condition. However, if $a = 0$, y is everywhere zero—this is called the *trivial solution,* and is usually of no interest. To get a useful solution, we must choose $\sin(k) = 0$, which is true only for certain "characteristic" values:

$$k = \pm n\pi, \qquad n = 1, 2, 3, \ldots .$$

These are the eigenvalues for the equation, and the solution to the problem is:

$$u = a \, \sin(n\pi x), \qquad n = 1, 2, 3, \ldots . \tag{7.6}$$

The constant a can have any value, so these solutions are determined only to within a multiplicative constant. Figure 7.3 sketches several of the solutions to Eq. (7.6).

These eigenvalues are the most important information for a characteristic-value problem. In a vibration problem, these give the natural frequencies of the system, which are important because, if the system is subjected to external loads applied at or very near to these frequencies, resonance causes an amplification of the motion and failure is likely.

Corresponding to each eigenvalue is an eigenfunction, $u(x)$, which determines the possible shapes of the elastic curve when the system is at equilibrium. Figure 7.3 shows such eigenfunctions. Often the smallest eigenvalue is of particular interest; at other times it is the one of largest magnitude.

We can solve Eq. (7.5) numerically, and that is what we concentrate on in this section.

---

*Homogeneous here means that all terms in the equation are functions of u or its derivatives.

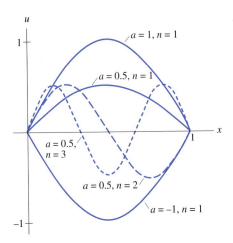

Figure 7.3

We will replace the derivatives in the differential equation with finite-difference approximations, so that we replace the differential equation with difference equations written at all nodes where the value of u is unknown (which are all the nodes of a one-dimensional system except for the endpoints).

EXAMPLE 7.9 Solve Eq. (7.5) with five subintervals. We restate the problem:

$$\frac{d^2u}{dx^2} + k^2u = 0, \qquad u(0) = 0, \qquad u(1) = 0.$$

The typical equation is

$$\frac{(u_{i-1} - 2u_i + u_{i+1})}{h^2} + k^2u_i = 0.$$

With five subintervals, $h = 0.2$, and there are four equations because there are four interior nodes. In matrix form these are

$$\begin{bmatrix} 2 - 0.04k^2 & -1 & 0 & 0 \\ -1 & 2 - 0.04k^2 & -1 & 0 \\ 0 & -1 & 2 - 0.04k^2 & -1 \\ 0 & 0 & -1 & 2 - 0.04k^2 \end{bmatrix} \begin{bmatrix} u_1 \\ u_2 \\ u_3 \\ u_4 \end{bmatrix} = \begin{bmatrix} 0 \\ 0 \\ 0 \\ 0 \end{bmatrix} \qquad (7.7)$$

where we have multiplied by -1 for convenience. Observe that this can be written as the matrix equation $(A - \lambda I)u = 0$, where I is the identity matrix and the A matrix is

$$\begin{bmatrix} 2 & -1 & 0 & 0 \\ -1 & 2 & -1 & 0 \\ 0 & -1 & 2 & -1 \\ 0 & 0 & -1 & 2 \end{bmatrix},$$

and $\lambda = 0.04k^2$.

The approximate solution to the characteristic-value problem, Eq. (7.5), is found by solving the system of Eq. (7.7). However, this system is an example of a homogeneous system (the right-hand sides are all equal to zero), and it has a nontrivial solution only if the determinant of the coefficient matrix is zero. Hence we set

$$\det(A - \lambda I) = 0.$$

Expanding the determinant will give an eighth-degree polynomial in k. (This is *not* the preferred way!). Doing so and getting the zeros of that polynomial gives these values for k:

$$k = \pm 3.09, \quad k = \pm 5.88, \quad k = \pm 8.09, \quad k = \pm 9.51.$$

The analytical values for k are

$$k = \pm 3.14 \ (\pm \pi), \quad k = \pm 7.28 \ (\pm 2\pi),$$
$$k = \pm 9.42 \ (\pm 3\pi), \quad k = \pm 12.57 \ (\pm 4\pi),$$

and we see that the estimates for k are not very good and get progressively worse. We would need a much smaller subdivision of the interval to get good values. There are other problems with this technique: expanding the determinant of a matrix of large size is computationally expensive, and solving for the roots of a polynomial of high degree is subject to large round-off errors. The system is very ill-conditioned.

We normally find the eigenvalues for a characteristic-value problem from $(A - \lambda I)u = 0$ in other ways that are not subject to the same difficulties. We describe these now. For clarity we use small matrices. ▲

The Power Method

The *power method* is an iterative technique. The basis for this is presented in the theoretical matters section. We illustrate the method through an example.

EXAMPLE 7.10 Find the eigenvalues (and the eigenvectors) of matrix A:

$$A = \begin{bmatrix} 3 & -1 & 0 \\ -2 & 4 & -3 \\ 0 & -1 & 1 \end{bmatrix}$$

(The eigenvalues of A are 5.47735, 2.44807, and 0.074577, which are found, perhaps, by expanding the determinant of $A - \lambda I$. The eigenvectors are found by solving the equations $Au = \lambda u$ for each value of λ. After normalizing, these vectors are

$$u_1 = [-0.40365, 1, -0.22335],$$
$$u_2 = [1, 0.55193, -0.38115],$$
$$u_3 = [0.31633, 0.92542, 1],$$

where the normalization has been to set the largest component equal to unity.)

We will find that both the eigenvalues and the eigenvectors are produced by the power method. We begin this by choosing a three-component vector more or less arbitrarily.

(There are some choices that don't work but usually the column vector $u = [1, 1, 1]$ is a good starting vector.) We always use a vector with as many components as rows or columns of A.

We repeat these steps:

1. Multiply $A * u$.
2. Normalize the resulting vector by dividing each component by the largest in magnitude.
3. Repeat steps (1) and (2) until the change in the normalizing factor is negligible. At that time, the normalization factor is an eigenvalue and the final vector is an eigenvector.

Step (1), with $u = [1, 1, 1]$:

$$A * u \qquad \text{gives} \qquad [2, -1, 0].$$

Step (2):

$$\text{Normalizing gives: } 2 * [1, .5, 0], \text{ and } u \text{ now is } [1, .5, 0].$$

Repeating, we get:

$$A * u = [3.5, -4, .5],$$

$$\text{normalized: } -4 * [-.875, 1, .125];$$

$$A * u = [-3.625, 7.125, -1.125],$$

$$\text{normalized: } 7.125 * [-.5918, 1, -.1837];$$

$$A * u = [-2.7755, 5.7347, -1.1837],$$

$$\text{normalized: } 5.7347 * [-.4840, 1, -.2064];$$

(after 14 iterations)

$$A * u = [-2.21113, 5.47743, -1.22333],$$

$$\text{normalized: } 5.47743 * [-.40368, 1, -.22334].$$

The fourteenth iteration shows a negligible change in the normalizing factor: we have approximated the largest eigenvalue and the corresponding eigenvector. (Twenty iterations will give even better values.) Although not very rapid, the method is extremely simple and a cinch to program. Any of the computer algebra systems can do this for us.   ▲

The Inverse Power Method

The previous example showed how the power method gets the eigenvalue of largest magnitude. What if we want the one of smallest magnitude? All we need to do to get this is to work with the inverse of A. For the matrix A of Example 7.10, its inverse is:

$$\begin{bmatrix} 1 & 1 & 3 \\ 2 & 3 & 9 \\ 2 & 3 & 10 \end{bmatrix}.$$

Applying the power method to this matrix gives a value for the normalizing factor of 13.4090 and a vector of [.3163, .9254, 1]. For the original matrix A, the eigenvalue is the reciprocal, 0.07457. The eigenvector that corresponds is the same; no change is needed.

Shifting with the Power Method

As we have seen, the power method may not converge very fast. We can accelerate the convergence as well as get eigenvalues of magnitude intermediate between the largest and smallest by *shifting*. Suppose we wish to determine the eigenvalue that is nearly equal to some number *s*. If *s* is subtracted from each of the diagonal elements of A, the resulting matrix has eigenvalues the same as for A but with *s* subtracted from them. This means that there is an eigenvalue for the shifted matrix that is nearly zero. We now use the inverse power method on this shifted matrix, and the reciprocal of this very small eigenvalue is usually very much larger in magnitude than any other. As shown in the section on theoretical matters, this causes the convergence to be rapid. Observe that if we have some knowledge of what the eigenvalues of A are, we can use this shifted power method to get the value of any of them.

How can we estimate the eigenvalues of a matrix? *Gerschgorin's theorem* can help here. This theorem is especially useful if the matrix has strong diagonal dominance. The first of Gerschgorin's theorems says that the eigenvalues lie in circles whose centers are at a_{ii} with a radius equal to the sum of the magnitudes of the other elements in row *i*. (Eigenvalues can have complex values, so the circles are in the complex plane.)

EXAMPLE 7.11   Given matrix A:

$$\begin{bmatrix} 4 & -1 & 1 \\ 1 & 1 & 1 \\ -2 & 0 & -6 \end{bmatrix}$$

find all of its eigenvalues using the shifted power method.

Gerschgorin's theorem says that there are eigenvalues within -6 ± 2, 1 ± 2, and 4 ± 2. We shift first by -6 and get an eigenvalue equal to -5.76851 (vector = [$-.11574$, -13065, 1]) using the inverse power method in four iterations; the tolerance on change in the normalization factor was 0.0001. (Getting this largest magnitude eigenvalue through the regular power method required 23 iterations.) If we repeat but shift by 1, the inverse power method gives 1.29915 as an eigenvalue (vector = [.41207, 1, $-.11291$]) in six iterations. (Using just the inverse power method to get this smallest of the eigenvalues required eight iterations.)

For this 3×3 matrix, we do not have to get the other eigenvalue; the sum of the eigenvalues equals the trace of the matrix. So, if we subtract $(-5.76851 + 1.29915)$ from -1 (giving 3.46936) we get the third eigenvalue. It is always true that the sum of the eigenvalues of matrix A equals the trace of A.

However, getting this third eigenvalue from the trace doesn't give the corresponding eigenvector. If we want that, we should use the shifted inverse power method on the original matrix.

Shifting by 4 in this example runs into a problem; a division by zero is attempted. We overcome this problem by distorting the shift amount slightly. Shifting by 3.9 and employing the inverse power method gives the eigenvalue: 3.46936, and the vector [1, .31936, −.21121] in six iterations. (If a division by zero occurs, it is advisable to distort the shift amount slightly.)

▲

The power method with its variations is fine for small matrices. However, if a matrix has two eigenvalues of equal magnitude, the method fails in that the successive normalization factors alternate between two numbers. The duplicated eigenvalue in this case is the square root of the product of the alternating normalization factors. If we want all the eigenvalues for a larger matrix, there is a better way.

The QR Method, Part 1—Similarity Transformations

If matrix A is diagonal or upper- or lower-triangular, its eigenvalues are just the elements on the diagonal. This can be proved by expanding the determinant of $(A - \lambda I)$. This suggests that, if we can transform A to upper-triangular, we have its eigenvalues! We have done such a transformation before: the Gauss elimination method does it. Unfortunately, this transformation changes the eigenvalues!!

There are other transformations that do not change the eigenvalues. These are called *similarity transformations*. For any nonsingular matrix, M, the product $M * A * M^{-1} = B$, transforms A into B, and B has the same eigenvalues as A. The trick is to find matrix M such that A is transformed into a similar upper-triangular matrix from which we can read off the eigenvalues of A from the diagonal of B. The QR technique does this. We first change one of the subdiagonal elements of A to zero; we then continue to do this for all the elements below the diagonal until A has become upper-triangular. The process is slow; many iterations are required, but the procedure does work.

Suppose that A is 4×4. Here is a matrix, Q, also 4×4, that will create a zero in position a_{42} to zero:

$$Q = \begin{bmatrix} 1 & 0 & 0 & 0 \\ 0 & c & 0 & s \\ 0 & 0 & 1 & 0 \\ 0 & -s & 0 & c \end{bmatrix},$$

where

$$d = \sqrt{(a_{22}^2 + a_{42}^2)},$$

$$c = \frac{a_{22}}{d},$$

$$s = \frac{a_{42}}{d}.$$

EXAMPLE 7.12   Given this matrix A, create a zero in position $(4, 2)$ by multiplying by the proper Q matrix.

$$A = \begin{bmatrix} 7 & 8 & 6 & 6 \\ 1 & 6 & -1 & -2 \\ 1 & -2 & 5 & -2 \\ 3 & 4 & 3 & 3 \end{bmatrix}$$

We compute:

$$d = \sqrt{(6^2 + 4^2)} = 7.21110,$$

$$c = \frac{6}{d} = 0.83205,$$

$$s = \frac{4}{d} = 0.55470.$$

The Q matrix is

$$\begin{bmatrix} 1 & 0 & 0 & 0 \\ 0 & .83205 & 0 & .55470 \\ 0 & 0 & 1 & 0 \\ 0 & -.55470 & 0 & .83205 \end{bmatrix}.$$

When we multiply Q by A, we get for $Q * A$:

$$\begin{bmatrix} 7 & 8 & 6 & 6 \\ 2.49615 & 7.21110 & .83205 & .55470 \\ 1 & -2 & 5 & -2 \\ 1.94145 & 0 & 3.05085 & 4.43760 \end{bmatrix}$$

where the element in position $(4, 2)$ is zero as we wanted. However, we do not yet have a similarity transformation. (The trace has been changed, meaning that the eigenvalues are not the same as those of A.) To get the similarity transformation that is needed, we must now postmultiply by the inverse of Q. Getting the inverse (which is Q^{-1}) is easy in this case because for any Q as defined here, its inverse is just its transpose! (When this is true for a matrix, it is called a *rotation matrix*.) If we now multiply $Q * A * Q^{-1}$, we get

$$\begin{bmatrix} 7 & 9.98460 & 6 & 0.55470 \\ 2.49615 & 7.30769 & 0.83205 & -3.53846 \\ 1 & -2.77350 & 5 & -0.55470 \\ 1.94145 & 2.46154 & 3.05085 & 3.69231 \end{bmatrix},$$

for which the trace is the same as that of the original A and whose eigenvalues are the same. However, it seems that we have not really done what we desired; the element in position $(4, 2)$ is zero no longer! There has been some improvement, though. Observe that the sum of the magnitudes of the off-diagonal elements in row four is smaller than in matrix A. This means that 3.69231 is closer to one of the eigenvalues (which will turn out to be 1) than the original value, 4. Also, the element in position $(2, 2)$ (7.30769) is closer to another eigenvalue (which is equal to 7) than the original number, 6.

This suggests that we should continue doing such similarity transformations to reduce all below-diagonal elements to zero. It takes many iterations, but, after doing 111 of these, we get

$$\begin{bmatrix} 10 & 1.5811 & -11.0680 & -3.0000 \\ 0 & 7 & -1.0000 & 0.0000 \\ 0 & 0 & 4 & -3.1623 \\ 0 & 0 & 0 & 1 \end{bmatrix},$$

where the numbers have been rounded to four decimals. (All the below-diagonal elements have a value of 0.00001 or less.) We have found the eigenvalues of A; these are 10, 7, 4, and 1. ▲

The QR Method, Part 2— Making the Matrix Upper Hessenberg

The trouble with doing such similarity transformations repeatedly is poor efficiency. We can improve the method by first doing a *Householder transformation,* which is a similarity transformation that creates zeros in matrix A for all elements below the "subdiagonal." (This means all elements below the diagonal except for those immediately below the diagonal. We might call such a matrix "almost triangular.") The name for such a matrix is *upper Hessenberg.* The Householder transformation changes matrix A into upper Hessenberg. Once an $n \times n$ matrix has been converted to upper Hessenberg, there are only $n - 1$ elements to reduce compared to $(n)(n - 1)/2$.

There is another technique that further speeds up the reduction of matrix A to upper triangular. We can employ shifting (similar to that done in the power method). The easiest way to shift is to do it with the element in the last row and last column.

Here are the steps that we will use:

1. Convert to upper Hessenberg.
2. Shift by a_{nn}, then do similarity transformations for all columns from 1 to $n - 1$.
3. Repeat step (2) until all elements to the left of a_{nn} are essentially zero. An eigenvalue then appears in position a_{nn}.
4. Ignore the last row and column, and repeat steps (2) and (3) until all elements below the diagonal of the original matrix are essentially zero. The eigenvalues then appear on the diagonal.

How do we convert matrix A to upper Hessenberg without changing the eigenvalues? This is best explained through an example.

EXAMPLE 7.13   Convert the same matrix A (as in Example 7.12) to upper Hessenberg.
We recall that A is

$$\begin{bmatrix} 7 & 8 & 6 & 6 \\ 1 & 6 & -1 & -2 \\ 1 & -2 & 5 & -2 \\ 3 & 4 & 3 & 4 \end{bmatrix}.$$

We can create zeros in the first column and rows 3 and 4 by $B * A * B^{-1}$, where

$$B = \begin{bmatrix} 1 & 0 & 0 & 0 \\ 0 & 1 & 0 & 0 \\ 0 & -b_3 & 1 & 0 \\ 0 & -b_4 & 0 & 1 \end{bmatrix}, \qquad B^{-1} = \begin{bmatrix} 1 & 0 & 0 & 0 \\ 0 & 1 & 0 & 0 \\ 0 & b_3 & 1 & 0 \\ 0 & b_4 & 0 & 1 \end{bmatrix},$$

$$b_3 = a_{31}/a_{21} = 1/1 = 1,$$
$$b_4 = a_{41}/a_{21} = 3/1 = 3.$$

Observe that the B matrix is the identity matrix with the two zeros below the diagonal in column 2 replaced with $-b_3$ and $-b_4$, where these values are the elements of column 1 of matrix A that are to be made zero divided by the subdiagonal element in column 1. The inverse of this B matrix is B with the signs changed for the new elements in its column 2.

If we now perform the multiplications $B_1 * A * B_1^{-1}$, we get

$$\begin{bmatrix} 7 & 32 & 6 & 8 \\ 1 & -1 & -1 & -2 \\ 0 & -2 & 6 & 0 \\ 0 & 22 & 6 & 10 \end{bmatrix},$$

which has zeros below the subdiagonal of column 1 and the same eigenvalues as the original matrix A.

We continue this in column 2, where now

$$B_2 = \begin{bmatrix} 1 & 0 & 0 & 0 \\ 0 & 1 & 0 & 0 \\ 0 & 0 & 1 & 0 \\ 0 & 0 & -b_4 & 1 \end{bmatrix}, \qquad \text{with} \qquad b_4 = a_{42}/a_{32} = 22/-2 = -11.$$

Here B_2^{-1} is the same as B_2 except that the sign of b_4 is changed. Now premultiplying the last matrix by B_2 and postmultiplying by B_2^{-1} gives the lower Hessenberg matrix:

$$B_2 B_1 A B_1^{-1} B_2^{-1} = \begin{bmatrix} 7 & 32 & -60 & 6 \\ 1 & -1 & 21 & -2 \\ 0 & -2 & 6 & 0 \\ 0 & 0 & -38 & 10 \end{bmatrix},$$

which is what was desired. ▲

There is a potential problem with this reduction to the Hessenberg matrix. If the divisor used to create the B matrices is zero or very small, either a division by zero occurs or the round-off error is great. We can avoid these problems by interchanging both rows and columns to put the element of largest magnitude in the subdiagonal position. It is essential to do the interchanges for both rows and columns so that the diagonal elements remain the same.

The QR Method, Part 3—The Steps Combined

If we (1) convert matrix A to upper Hessenberg, and, (2) perform QR operations on this, the final matrix that results is

$$\begin{bmatrix} 10 & 9.8315 & 4.9054 & -3.2668 \\ 0 & 1 & 1.8256 & 2.7199 \\ 0 & 0 & 4 & -1.6958 \\ 0 & 0 & 0 & 7 \end{bmatrix},$$

in which the same eigenvalues appear on the diagonal as when QR operations were done on the original A matrix. However, only seven QR iterations were required after reduction to Hessenberg compared to 111 if that step is omitted. The other elements are different because row and column interchanges were done in creating the last result.

7.6   Temperature Distribution in a Slab

Until now, the discussion in this chapter has been concerned with problems where there is a single independent variable, and, when steady-state temperatures were discussed, an insulated rod was the example. More often objects that extend in two or three dimensions are of interest. We again think of temperature distribution, but now consider a thin slab of uniform thickness and material. Figure 7.4 illustrates the slab of thickness t that has an element of size $dx \times dy$. Let u, the dependent variable, represent the temperature within the element. The location of the element is at (x, y), measured from the lower left corner of the slab. We again consider heat flowing through the element in the direction of positive x and positive y. The rate at which heat flows into the element in the x-direction is

$$-(\text{conductivity})(\text{area})(\text{gradient}) = -kA\frac{\partial u}{\partial x}$$

$$= -k(t \ dy)\frac{\partial u}{\partial x},$$

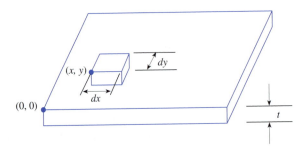

Figure 7.4

where the derivative is a partial derivative because there are two independent variables, x and y.

Similarly, the rate at which heat flows into the element in the y-direction is

$$-kA\frac{\partial u}{\partial y} = -k(t\ dx)\frac{\partial u}{\partial y}.$$

In a way that is identical to that of Section 7.1, we equate the rate at which heat enters the element (but now there are two directions to consider), to the rate of flow of heat out of the element plus the rate of heat loss from the surface of the plate (which we take as Q cal/cm^2).

Rate of flow in = rate of flow out + rate of heat lost from the surface.

The rate at which heat flows from the element by conduction is the sum of that flowing in the x-direction and in the y-direction; here the gradients are at $x + dx$ and $y + dy$:

$$\text{rate of flow out in } x\text{-direction} = -k(t\ dy)\left[\frac{\partial u}{\partial x} + \frac{\partial^2 u}{\partial x^2}\ dx\right],$$

$$\text{rate of flow out in } y\text{-direction} = -k(t\ dx)\left[\frac{\partial u}{\partial y} + \frac{\partial^2 u}{\partial y^2}\ dy\right],$$

so the total flow of heat from the element is

$$-k(t\ dy)\left[\frac{\partial u}{\partial x} + \frac{\partial^2 u}{\partial x^2}\ dx\right] - k(t\ dx)\left[\frac{\partial u}{\partial y} + \frac{\partial^2 u}{\partial y^2}\ dy\right] + Q(dx\ dy).$$

The sum of the flows into the element must equal the rate at which heat flows from the element plus the heat loss from the surface of the element if the temperature of the element is to remain constant (and we are here considering only the steady-state), so that we have, after some rearrangement:

$$kt\left(\frac{\partial^2 u}{\partial x^2} + \frac{\partial^2 u}{\partial y^2}\right)(dx\ dy) = Q(dx\ dy),$$

or

$$\left(\frac{\partial^2 u}{\partial x^2} + \frac{\partial^2 u}{\partial y^2}\right) = \frac{Q}{kt}. \tag{7.8}$$

If the object under consideration is three-dimensional, a similar development leads to

$$\left(\frac{\partial^2 u}{\partial x^2} + \frac{\partial^2 u}{\partial y^2} + \frac{\partial^2 u}{\partial z^2}\right) = \frac{Q}{k},$$

where now Q is the rate of heat loss per unit volume.

(The loss of heat in the three-dimensional case would have to be through an imbedded "heat-sink," perhaps a cooling coil. It is easier to visualize heat generation within the object, perhaps because there is an electrical current passing through it.)

Mathematicians have a special name and symbol for the sum of the second partial derivatives. This is called the *Laplacian,* and is represented by the symbol $\nabla^2 u$. Equation (7.8) is frequently seen as

$$\nabla^2 u = \frac{Q}{kt}.$$

If the thickness of the plate varies with x and y, a development that parallels that of Section 7.1 gives

$$t\nabla^2 u + \frac{\partial t}{\partial x}\left(\frac{\partial u}{\partial x}\right) + \frac{\partial t}{\partial y}\left(\frac{\partial u}{\partial y}\right) = \frac{Q}{k}. \tag{7.9}$$

If both the thickness and the thermal conductivity are variable:

$$kt\nabla^2 u + \left(k\frac{\partial t}{\partial x} + t\frac{\partial k}{\partial x}\right)\left(\frac{\partial u}{\partial x}\right) + \left(k\frac{\partial t}{\partial y} + t\frac{\partial k}{\partial y}\right)\left(\frac{\partial u}{\partial y}\right) = Q. \tag{7.10}$$

Representation as a Difference Equation

We can approximate the second derivatives in Eqs. (7.8–7.10) with central-difference approximations:

$$\frac{\partial^2 u}{\partial x^2} = \frac{(uL - 2u0 + uR)}{(\Delta x)^2},$$

where uL and uR are temperatures at nodes to the left and to the right, respectively, of a central node whose temperature is $u0$. The nodes are Δx apart. A similar formula approximates $\partial^2 u/\partial y^2$:

$$\frac{\partial^2 u}{\partial y^2} = \frac{(uA - 2u0 + uB)}{(\Delta y)^2},$$

in which uA and uB are at nodes above and below the central node. It is customary to make $\Delta x = \Delta y = h$. So, if we combine these, we get

$$\nabla^2 u = \frac{(uL + uR + uA + uB - 4u0)}{h^2}.$$

7.7 Solving for the Temperatures in a Slab

Because the equations for steady-state heat flow in a slab are partial-differential equations, the methods of Chapter 6 do not apply. However, we can use finite-difference approximations to the derivatives to set up a system of equations that will give an approximation to the temperature at points in the slab. If the region is rectangular and the boundary conditions are Dirichlet conditions (that is, the temperature, u, is specified all along the boundary), we place nodes within the slab that are spaced uniformly. (This restriction to a uniform spacing can be removed, but that introduces complications that we will avoid. Another chapter describes the finite element method, which facilitates nonuniform spacing of nodes.) With uniformly spaced nodes, the values of Δx and Δy are the same; we use h for these. It is convenient to represent the temperature (u) at a node as u_i where the subscript on u represents the nodal count from (conveniently) the upper left corner of the slab. (We begin with a rectangular slab whose dimensions are multiples of h.)

We will use central difference approximations of the second derivatives in an example that demonstrates the finite-difference method:

EXAMPLE 7.14   Solve for the steady-state temperatures in a rectangular slab that is 20 cm wide and 10 cm high. All edges are kept at $0°$ except the right edge, which is at $100°$. There is no heat gained or lost from the surface of the slab. Place nodes in the interior spaced 2.5 cm apart (giving an array of nodes in three rows and seven columns) so that there are a total of 21 internal nodes.

Figure 7.5 is a sketch of the slab with the nodes numbered in succession by rows. (We could also number them according to their row and column, with node (1, 1) at the upper left and node (3, 7) at the lower right. However, it is better to number them with a single subscript by rows when we are setting up the equations, as we have done in the figure. (In a second example and in the next section the alternative numbering system will be preferred.) Let u_i be the temperature at node (i).

The equation that governs this situation is Eq. (7.8) with $Q = 0$:

$$\left(\frac{\partial^2 u}{\partial x^2} + \frac{\partial^2 u}{\partial y^2} \right) = 0. \tag{7.11}$$

We use these approximations for the second-order derivatives at a central node, where the temperature is $u0$:

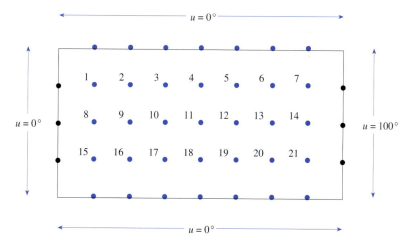

Figure 7.5

$$\frac{\partial^2 u}{\partial x^2} = \frac{(uL - 2u0 + uR)}{0.25^2};$$

$$\frac{\partial^2 u}{\partial y^2} = \frac{(uA - 2u0 + uB)}{0.25^2},$$

where uL and uR are nodes to the left and right of the central node. Similarly, nodes uA and uB are nodes above and below the central node. Substituting these into Eq. (7.11) gives:

$$\frac{(uL + uR + uA + uB - 4u0)}{.125} = 0.$$

There is a simple device we can use to remember this approximation to the Laplacian. We call it a "pictorial operator":

$$\nabla^2 u = \frac{1}{h^2} \left\{ \begin{matrix} & 1 & \\ 1 & -4 & 1 \\ & 1 & \end{matrix} \right\} u0. \tag{7.12}$$

This pictorial operator says: Add the temperatures at the four neighbors to $u0$, subtract 4 times $u0$, then divide by h^2, and you have an approximation to the Laplacian.

We can now write the 21 equations for the problem. Because in this example we set the Laplacian for every node equal to zero, we can drop the h^2 term. A node that is adjacent to a boundary will have the boundary value(s) in its equation; this will be subtracted from the right-hand side of that equation before we solve the system. Rather than write out all the equations, we will only show a few of them:

For node 1: $0 + u_2 + 0 + u_8 - 4u_1 = 0$, which, when the nodes are put in order, becomes:

$$-4u_1 + u_2 + u_8 = 0.$$

$$\text{For node 7:} \quad u_6 - 4u_7 + u_{14} = -100.$$

$$\text{For node 9:} \quad u_2 + u_8 - 4u_9 + u_{10} + u_{16} = 0.$$

$$\text{For node 14:} \quad u_7 + u_{13} - 4u_{14} + u_{21} = -100.$$

$$\text{For node 18:} \quad u_{11} + u_{17} - 4u_{18} + u_{19} = 0.$$

If we write out all 21 equations in matrix form, we have a 21×21 coefficient matrix, a right-hand-side vector of size 21, and 21 u-values for the unknown vector. When these are solved by Gaussian elimination, these are the results:

Column	Row 1	Row 2	Row 3
1	0.3530	0.4988	0.3530
2	0.9132	1.2894	0.9132
3	2.0103	2.8323	2.0103
4	4.2957	7.0193	4.2931
5	9.1531	12.6537	9.1531
6	19.6631	27.2893	19.6631
7	43.2101	53.1774	43.2101

▲

Rows 1 and 3 are the same; this is to be expected from the symmetry of boundary conditions at the top and bottom of the region. Nodes near the hot edge are warmer than those farther away.

The accuracy of the solution is improved when the nodes are closer together; they decrease about proportional to h^2, which we anticipate because the central difference approximation to the derivative is of $O(h^2)$. Another way to improve the accuracy is to use a nine-point approximation to the Laplacian. This uses the eight nodes that are adjacent to the central node and has an error of $O(h^6)$. A pictorial operator for this is

$$\nabla^2 u = \frac{1}{(6h^2)} \left\{ \begin{matrix} 1 & 4 & 1 \\ 4 & -20 & 4 \\ 1 & 4 & 1 \end{matrix} \right\}. \tag{7.13}$$

If Example 7.14 is solved using this nine-point formula and with $h = 2.5$ cm, the answers will be within ± 0.0032 of the "analytical" solution (from a series solution given by classical methods for partial differential equations).

Iterative Methods

The difficulty with getting the solution to a problem in the way that was done in the last example is that a very large matrix is needed when the nodal spacing is close. In that example, if $h = 1.25$, the number of equations increases from 21 to 105; if h were 0.625, there would be 465 equations. The coefficient matrix for 465 equations has $465^2 = 216,225$ elements! Not only is this an extravagant use of computer memory to store the values but also the solution time may be excessive. However, the matrix is *sparse,* meaning that most of the elements are zero. (Only about 1% of the elements in the last case are nonzero.)

Iterative methods that were discussed in Chapter 2 are an ideal technique for solving a sparse matrix. We do need to arrange the equations so that there is diagonal dominance (and this is readily possible for the problems of this chapter). We can write the equations in a form useful for iteration from this pictorial operator:

$$u0 = \frac{1}{4} \left\{ \begin{matrix} & 1 & \\ 1 & 0 & 1 \\ & 1 & \end{matrix} \right\}, \tag{7.14}$$

which is, when nodes are specified using row and column subscripts:

$$u_{i,j} = \frac{(u_{i-1,j} + u_{i+1,j} + u_{i,j-1} + u_{i,j+1})}{4}.$$

We can enter the Dirichlet boundary conditions into the equations by substituting these specified values for the boundary nodes that are adjacent to interior nodes.

The name given to this method of solving boundary-value problems is *Liebmann's method.* We illustrate with the same example problem as Example 7.14.

EXAMPLE 7.15 Solve Example 7.14 but now use Liebmann's method. Use $h = 2.5$ cm.

We will designate the temperatures at the nodes by $u_{i,j}$, where i and j are the row and column for the node. Row 1 is at the top; column 1 is at the left and there are three rows

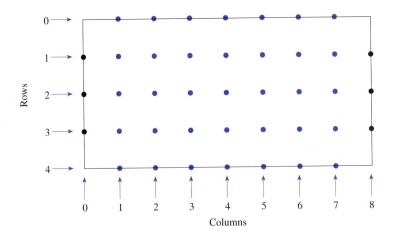

Rows

Columns

Figure 7.6

and seven columns for interior nodes. The boundary conditions will be stored in row 0 and row 4, and in column 0 and column 8.

Figure 7.6 shows how nodes are numbered for this problem—we use double subscripts to indicate the row and column.

Here is the typical equation for node (i, j):

$$u_{i,j} = \frac{(u_{i,j-1} + u_{i,j+1} + u_{i-1,j} + u_{i+1,j})}{4}, \qquad \text{with } i = 1 \ldots 3, \qquad j = 1 \ldots 7.$$

It is best to begin the iterations with approximate values for the $u_{i,j}$, but beginning with all values set to zero will also work. Another way to begin the iterations is with all interior node values set to the average of the boundary values. If this is done, 26 iterations give answers that change by less than 0.0001 and that essentially duplicate those of Example 7.14. (If the starting values are all equal to zero, it takes 30 iterations.) ▲

Accelerating Convergence in Liebmann's Method

Chapter 2 mentioned Southwell's technique for solving a set of linear equations and pointed out that "overrelaxation" can speed the convergence. The previous example is an instance where the use of an overrelaxation factor can hasten convergence. Numerical analysts give the name *successive overrelaxation* to this method, and they abbreviate this to *S.O.R.*

To use the S.O.R. techniques, the calculations are made with this formula:

$$u_{i,j} = u_{i,j} + \omega * \frac{(u_{i,j-1} + u_{i,j+1} + u_{i-1,j} + u_{i+1,j} - 4 * u_{i,j})}{4},$$

$$\text{with } i = 1 \ldots 3, \qquad j = 1 \ldots 7,$$

where the $u_{i,j}$ terms on the right are the current values of that variable and the one on the left becomes the new value. The ω-term is called the *overrelaxation factor*.

Solving Example 7.15 with various values for the overrelaxation factor gives these results:

Overrelaxation factor	Number of iterations
1.0	26
1.1	22
1.2	18
1.3	15
1.4	18
1.5	21

From this we see that overrelaxation can decrease the number of iterations required by almost 1/2.

The optimum value to use for ω, the overrelaxation factor, is not always predictable. There are methods that use the results of the first few iterations to find a good value. For a rectangular region with Dirichlet boundary conditions, there is a formula:

Optimum ω = smaller root of this quadratic equation = 0:

$$\left[\cos\left(\frac{\pi}{p}\right) + \cos\left(\frac{\pi}{q}\right)\right]\omega^2 - 16\omega + 16, \tag{7.15}$$

where p and q are the number of subdivisions of each of the sides. This formula suggests using $\omega = 1.267$ for the previous example. This is about the same as the value $\omega_{opt} = 1.3$ that was found by trial and error.

Poisson's Equation

The previous examples were for an equation known as *Laplace's equation:*

$$\nabla^2 u = 0.$$

If the right-hand-side is nonzero, we have *Poisson's equation:*

$$\nabla^2 u = R,$$

where R can be a function of position in the region, (x, y). To solve a Poisson equation, we only need to make a minor modification to the methods described for Laplace's equation.

EXAMPLE 7.16 Solve for the *torsion function*, ϕ, in a bar of rectangular cross-section, whose dimensions are 6 in. $\times$ 8 in. (The tangential stresses are proportional to the partial derivatives of the torsion function when the bar is twisted.) The equation for ϕ is

$$\nabla^2\phi = -2, \qquad \text{with } \phi = 0 \text{ on the outer boundary of the bar's cross-section.}$$

If we subdivide the cross-section of the bar into 1-in. squares, there will be 35 interior nodes at the corners of these squares ($h = 1$). If we use the iterative technique, the equation for ϕ is

$$\phi_{i,j} = \frac{(\phi_{i,j-1} + \phi_{i,j+1} + \phi_{i-1,j} + \phi_{i+1,j} + 2)}{4}, \qquad i = 1 \ldots 7, \quad j = 1 \ldots 5.$$

Table 7.6 Torsion function at interior nodes for Example 7.16

2.042	3.047	3.353	3.047	2.043
3.123	4.794	5.319	4.794	3.123
3.657	5.686	7.335	5.686	3.657
3.818	5.960	7.647	5.960	3.818
3.657	5.686	7.335	5.686	3.657
3.123	4.794	5.319	4.794	3.123
2.042	3.048	3.354	3.048	2.043

Convergence will be hastened if we employ overrelaxation. Equation (7.15) predicts ω_{opt} to be 1.383. Using overrelaxation with this value for ω converges in 13 iterations to the values in Table 7.6. ▲

If overrelaxation is not employed, it takes 25 iterations to get the values of Table 7.6. Again, overrelaxation cuts the number of iterations about in half.

Derivative Boundary Conditions

Just as we saw in Section 7.4 for a one-dimensional problem, two-dimensional problems may have derivative boundary conditions. These may be of either Neumann or mixed type. We can define a more universal type of boundary conditions by the relation:

$$Au + B = Cu', \qquad \text{where } A, B, \text{ and } C \text{ are constants.}$$

If $C = 0$, we have a Dirichlet condition: $u = -B/A$. If $A = 0$, the condition is Neumann: $u' = B/C$. If none of the constants are zero, it is mixed condition. This relation can match a boundary condition for heat loss from the surface:

$$-ku' = H(u - u_s)$$

by taking $A = H$, $B = -H * u_s$, $C = -k$.

Here is an example that shows how this universal type of boundary conditions can be handled.

EXAMPLE 7.17 Find the steady-state temperatures in a slab that is 5 cm × 9 cm and is 0.5 cm thick. Everywhere within the slab, heat is being generated at the rate of 0.6 cal/sec/cm³. The two 5-cm edges are held at 20° while heat is lost from the bottom 9-cm edge at a rate such that $\partial u / \partial y = 15$. The top edge exchanges heat with the surroundings according to $-k \, \partial u / \partial y = H * (u0 - u_s)$, where k, the thermal conductivity, is 0.16; H, the heat transfer coefficient, is 0.073; and u_s, the temperature of the surroundings, is 25°. ($u0$ in this case is the temperature of a node on the top edge.) No heat is gained or lost from the surfaces of the slab. Place nodes within the slab (and on the edges) at a distance 1 cm apart so that there are a total of 60 nodes.

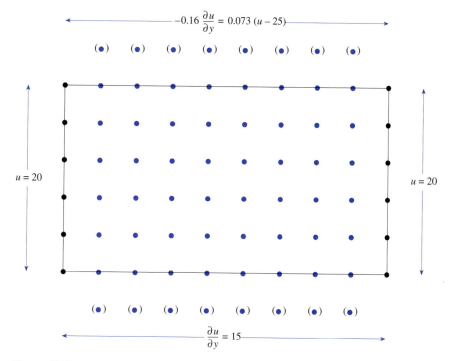

Figure 7.7

Figure 7.7 illustrates the problem. In Figure 7.7, rows of fictitious nodes are shown above and below the top and bottom nodes in the slab. These are needed because there are derivative boundary conditions on the top and bottom edges.

The Dirichlet conditions on the left and the right will be handled by initializing the entire array of nodal temperatures to 20°, and omitting these left- and right-edge nodes from the iterations that find new values for the nodal temperatures.

(These edge nodes are the uL or uR in the formula:

$$u0 = \frac{(uL + uR + uA + uB)}{4} - \frac{Q * h^2}{kt}.$$

For computations along the bottom edge (row 6), where $\partial u/\partial y = 15$, the gradient, $\partial u/\partial y$, will be computed by

$$\frac{\partial u}{\partial y} = (uA - uF) = 15, \quad uF = uA - 2 * h * 15,$$

where uA is at a node in the fifth row and uF is a fictitious node. (Take note of the fact that, if the gradient here is positive, heat flows in the negative y-direction, so heat is being lost as specified.) The equation for computing temperatures along the bottom edge is then

$$u0 = \frac{(uL + uR + uA + uB)}{4} - \frac{Q * h^2}{kt}, \quad \text{with } uB = uF.$$

Table 7.7 Temperatures after 28 iterations for Example 7.17

20.000	73.510	107.915	128.859	138.826	138.826	128.859	107.915	73.510	20.000
20.000	90.195	137.476	167.733	180.743	180.743	167.733	137.476	90.195	20.000
20.000	99.793	155.061	189.855	207.669	207.669	189.855	155.061	99.794	20.000
20.000	103.918	163.119	200.956	219.410	219.410	200.956	163.119	103.919	20.000
20.000	102.762	162.539	201.442	220.604	220.604	201.442	162.539	102.762	20.000
20.000	94.589	152.834	191.669	210.959	210.959	191.669	152.834	94.589	20.000

For computations along the top edge where the relation is

$$-k\frac{\partial u}{\partial y} = H * (u0 - us),$$

temperatures will be computed using a fictitious node above $u0$, uF, from

$$u0 = \frac{(uL + uR + uA + uB)}{4} - \frac{Q * h^2}{kt},$$

where $uA = uF$, and, because

$$\partial u/\partial y = \frac{(uF - uB)}{2h},$$

where uB is a node in the second row. So we have

$$-k\frac{\partial u}{\partial y} = \frac{-k * (uF - uB)}{2h} = H * (u0 - us),$$

which gives

$$uA = uF = uB - \left(\frac{2 * H * h}{k}\right) * u0 + \left(\frac{2 * H * h}{k}\right) * us.$$

When these replacements are included in a program and overrelaxation is employed ($\omega = 1.57$), the results after 28 iterations are as shown in Table 7.7. ▲

7.8 The Alternating Direction Implicit Method (A.D.I.)

When the partial-differential equations of this chapter are solved (using the finite-difference method), the resulting coefficient matrix is sparse. The sparseness increases as the number of nodes increases: If there are 21 nodes, 81% of the values are zeros; if there are 105 nodes, 96% are zeros; for a 30 × 30 × 30 three-dimensional system only 0.012% of the $729 * 10^6$ values are nonzero!

The coefficient matrices are not only sparse in two- and three-dimensional problems, they are also *banded,* meaning that the nonzero values fall along diagonal bands within the matrix. There are solution methods that take advantage of this banding, but, because the location of the bands depends strongly on the number of nodes in rows and columns, it is not simple to accomplish. Only for a tridiagonal coefficient matrix is getting the solution straightforward.

One way around the difficulty, as we have shown, is iteration. This is an effective way to decrease the amount of memory needed to store the nonzero coefficients and to (usually) speed up the solution process. However, as we saw in Section 7.3, the system of equations for the one-dimensional case always has a tridiagonal coefficient matrix, and, for this, neither the computational time nor the storage requirements is excessive. We ask "Is there a way to get a tridiagonal coefficient matrix when the region has two or three dimensions?" The answer to this question is yes, and the technique to achieve this is called the *alternating direction implicit method,* usually abbreviated to the *A.D.I. method.*

The trick to get a tridiagonal coefficient matrix for computing the temperatures in a slab is this: First make a traverse of the nodes across the rows and consider the values above and below each node to be constants. These "constants" go on the right-hand sides of the equations, of course. (We know that these "constant" values really do vary, but we will handle that variation in the next step). After all the nodes have been given new values with the horizontal traverse, we now make a traverse of the nodes by columns, assuming for this step that the values at nodes to the right and left are constants. There is an obvious bias in these computations, but the bias in the horizontal traverse is balanced by the opposing bias of the second step. If the object is three-dimensional, three passes are used: first in the *x*-direction, then in the *y*-direction, and finally in the *z*-direction.

A.D.I. is particularly useful in three-dimensional problems but it is easier to explain with a two-dimensional example. When we attack Laplace's equation in two dimensions, we write the equations as

$$\nabla^2 u = \frac{(uL - 2u0 + uR)}{(\Delta x)^2} + \frac{(uA - 2u0 + uB)}{(\Delta y)^2} = 0,$$

where, as before, uL, uR, uA, and uB stand for temperatures at the left, right, above, and below the central node, respectively, where it is $u0$. When, as is customary, $\Delta x = \Delta y$, these can be canceled. The row-wise equations for the $(k + 1)$ iteration are:

$$(uL - 2u0 + uR)^{(k+1)} = -(uA - 2u0 + uB)^k, \tag{7.16}$$

where the right-hand nodal values are the constants for the equations. When we work column-wise, the equations are for the $(k + 2)$ iteration

$$(uA - 2u0 + uB)^{(k+2)} = -(uL - 2u0 + uR)^{(k+1)}, \tag{7.17}$$

where, again, the right-hand nodal values are the constants.

We can speed up the convergence of the iterations by introducing an acceleration factor, ρ, to make Eq. (7.16) become

$$u0^{(k+1)} = u0^{(k)} + \rho(uL - 2u0^{(k)} + uR)^{(k)} + \rho(uA - 2u0 + uB)^{(k+1)},$$

and Eq. (7.17) becomes:

$$u0^{(k+2)} = u0^{(k+1)} + \rho(uA - 2u0 + uB)^{(k+1)} + \rho(uL - 2u0 + uR)^{(k+2)},$$

where the last terms in both use the values from the previous traverse.

Rearranging further, we get the tridiagonal systems:

$$-uL^{(k+1)} + \left(\frac{1}{\rho} + 2\right) u0^{(k+1)} - uR^{(k+1)} = \left[uA + \left(\frac{1}{\rho} - 2\right) u0 + uB\right]^{(k)}, \qquad (7.18)$$

and

$$-uA^{(k+2)} + \left(\frac{1}{\rho} + 2\right) u0^{(k+2)} - uB^{(k+2)} = \left[uL + \left(\frac{1}{\rho} - 2\right) u0 + uR\right]^{(k+1)}, \qquad (7.19)$$

for the horizontal and vertical traverses respectively.

In writing a program for the A.D.I. method, we must take note of the fact that the coefficient matrices for the two traverses are not identical because the boundary values enter differently. Here is a deliberately simple example that illustrates the procedure.

EXAMPLE 7.18 A rectangular plate is 6 in. × 8 in. The top edge (an 8-in. edge) is held at $100°$, the right edge at $50°$, and the other two edges at $0°$. Use the A.D.I. method to find the steady-state temperatures at nodes spaced 1 in. apart within the plate.

There are $5 * 7 = 35$ interior nodes, so there are 35 equations in each set (the horizontal and vertical traverses). With $\rho = 0.9$, and starting with all interior values set to $0°$, the values of Table 7.8 result after 28 iterations, which is when the maximum change in any of the values is less than 0.001. (If we begin with the interior nodes set to the average of the boundary values, these values are reached in 24 iterations with $\rho = 1.1$.)

Table 7.8 Temperatures at interior nodes for Example 7.18

48.523	67.828	74.203	78.669	79.341	77.984	71.464
27.262	44.122	53.644	58.803	61.178	61.132	57.873
17.404	28.754	37.982	42.188	45.434	47.495	48.804
9.599	17.508	23.344	27.534	30.878	34.518	40.209
4.484	8.336	11.352	13.728	17.024	19.492	27.425

▲

For this particular example, the number of nodes is small enough that Liebmann's method with overrelaxation could be used. That method is somewhat more efficient because it requires only 15 iterations to attain the same accuracy.

7.9 Irregular Regions and Nonrectangular Grids

Until now, this chapter has dealt only with regions whose boundaries coincide with the nodes of an evenly spaced grid. There are many times when such nodes do not fall on the boundary. For example, a circular region (in two dimensions) cannot use evenly spaced nodes and have nodes fall on the boundary. There are several ways that we can handle this kind of problem:

1. If the nodes are very close together, we can "distort" the boundary to line up with the nodes.
2. We can place nodes on the boundary and modify the approximations to the derivatives to allow for the unequal spacing of nodes near the boundary.
3. We can sometimes use a different coordinate system to define the nodal placement; for a circular region, this means using polar coordinates.

The first two approaches incur additional errors. For (1), the actual boundary is not the same as the distorted boundary so the "boundary" values used for nodes near the distorted boundary are incorrect. For (2) and (3), we need to derive new formulas for approximating a second derivative. We attack (2) first.

Figure 7.8 illustrates a situation where the four nodes around the central node have different spacing. As shown in the figure, the distances to points L, R, A, and B from point 0, the central node, are hL, hR, hA, and hB. These points are nodes to the left, right, above, and below the central node, respectively. The u-values at these points are uL, uR, uA, and uB. Approximate the first derivatives between points L and 0 and between points 0 and R with:

$$\left(\frac{\partial u}{\partial x}\right)_{L,0} = \frac{(u0 - uL)}{hL}, \qquad \left(\frac{\partial u}{\partial x}\right)_{0,R} = \frac{(uR - u0)}{hR}.$$

These can be interpreted as central difference approximations at points halfway between points L and 0 and halfway between 0 and R. We then approximate the second derivative with:

$$\frac{\partial^2 u}{\partial x^2} = \frac{\left[\left(\frac{\partial u}{\partial x}\right)_{0,R} - \left(\frac{\partial u}{\partial x}\right)_{L,0}\right]}{\left[\frac{(hL + hR)}{2}\right]}$$

$$= \frac{2}{(hL + hR)}\left[\frac{uL}{hL} - \frac{(hL + hR)}{(hL * hR) * u0} + \frac{uR}{hR}\right], \tag{7.20}$$

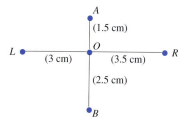

L •——— (3 cm) ——— O ——— (3.5 cm) ——— • R

A • (1.5 cm) / (2.5 cm) / • B

Figure 7.8

but this is not a central difference approximation at exactly point 0. We can use it to approximate the second derivative there but doing so incurs an error of $O(h)$. We can do the same to approximate $\partial^2 u/\partial y^2$ by using the points in a vertical line.

The Laplacian in Polar Coordinates

Regions made up of circular segments are often encountered in applied problems. The Laplacian in polar coordinates is defined as

$$\nabla^2 u = \frac{\partial^2 u}{\partial r^2} + \frac{1}{r}\frac{\partial u}{\partial r} + \frac{1}{r^2}\frac{\partial^2 u}{\partial \theta^2}.$$

If we place nodes as indicated in Figure 7.9, with the four points labeled L, R, A, and B arranged around a central node 0, we can approximate the derivatives in the formula above with

$$\nabla^2 u = \frac{(uL - 2u0 + uR)}{(\Delta r)^2} + \left(\frac{1}{r}\right) * \left[\frac{(uR - uL)}{(2\Delta r)}\right] + \frac{1}{r^2} * \left[\frac{(uA - 2u0 + uB)}{(\Delta\theta)^2}\right].$$

The Three Techniques Compared

The three techniques will be illustrated with examples.

EXAMPLE 7.19 A semicircular plate with a radius of 1 has the straight boundary (which we take as coinciding with the x-axis) held at 0° while the curved boundary is held at 100°. Place nodes at a spacing of $h = 0.2$. Figure 7.10 shows the region and nodes that are numbered from 1 to

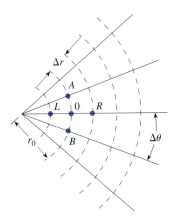

Figure 7.9

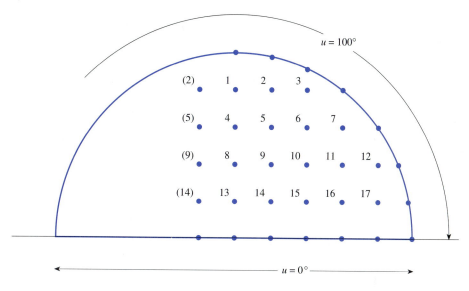

Figure 7.10

17. (We take advantage of the symmetry of the problem to reduce the number of nodes.) Compare the numerical solution with the "analytical" solution given by this infinite series:

$$u(r,\ \theta) = \frac{4c}{\pi} * \sum \left(\frac{1}{(2n-1)} * \left(\frac{r}{a} \right)^{2n-1} * \sin[(2n-1)\theta] \right),$$

with n running from $1 \ldots \infty$,

where a is the radius ($a = 1$), c is the temperature on the circumference ($c = 100$), and r and θ are the polar coordinates of a point on the plate where the origin is taken at the center of the semicircle. If we use approach (2), with some groups of nodes having unequal spacing in Figure 7.10, one arm of the "five-point pictorial operator" is shorter than the others at points 2, 3, 12, and 17.

A little analytical geometry finds that the short arms at points 2 and 17 are equal to 0.8990(0.2) and the short arms at points 3 and 12 are equal to 0.5826(0.2). We use these values in Eq. (7.20) (and the analogous one for $\partial^2 u/\partial y^2$) to set up equations for the problem. Table 7.9 compares the results from the numerical procedure with the series solution. For

Table 7.9 Computed and analytical temperatures

1: 86.053 85.906	4: 69.116 68.807	8: 48.864 48.448	13: 25.466 25.133
2: 87.548 87.417	5: 70.733 70.482	9: 50.436 50.000	14: 27.501 27.109
3: 92.124 92.094	6: 75.994 75.772	10: 55.606 55.151	15: 30.102 29.527
	7: 85.471 85.405	11: 65.891 65.593	16: 38.300 37.436
		12: 84.189 84.195	17: 57.206 57.006

each node (specified by its number), the numerical and series solutions are shown side by side.   ▲

The differences range from 0.006 (at point 12) to 0.864 (at point 16) with the greatest differences at nodes half-way between the borders.

EXAMPLE 7.20   Repeat Example 7.19 but distort the boundary to fit to evenly spaced nodes. Figure 7.11 shows how (distorted) boundary nodes are placed above points 2 and 3 and to the right of points 12 and 17. When 17 simultaneous equations are set up for this version of the problems, we get the results shown in Table 7.10. The third columns are again values from the series solution.

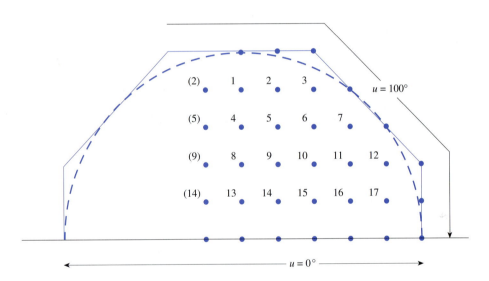

Figure 7.11

Table 7.10   Results for Example 7.20 using a distorted boundary

1: 85.152 85.906	4: 68.100 68.807	8: 47.993 48.448	13: 24.956 25.133
2: 87.254 87.417	5: 69.627 70.482	9: 49.458 50.000	14: 25.915 27.109
3: 90.237 92.094	6: 74.696 75.772	10: 54.297 55.151	15: 29.246 29.527
	7: 84.621 85.405	11: 63.788 65.593	16: 37.773 37.436
		12: 79.462 84.195	17: 54.059 57.006

▲

The computed values are not as accurate as in Example 7.19; the differences range from 0.177 (at point 13) to 4.733 (at point 12). Points near where the boundary is distorted have larger errors, as we might expect.

EXAMPLE 7.21   Repeat Example 7.19 but now arrange the nodes along radial lines from the center with $\Delta r = 0.2$ and $\Delta \theta = \pi/8$ and use the equations for the Laplacian in polar coordinates. Figure 7.12 shows how the nodes are numbered. The results are given in Table 7.11.

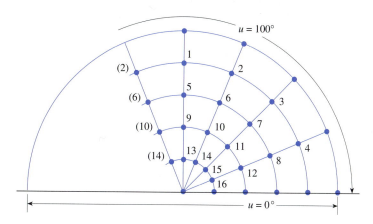

Figure 7.12

Table 7.11   Results when using the Laplacian in polar coordinates

1: 85.661 85.906	5: 68.371 68.808	9: 48.048 48.448	13: 25.003 25.133
2: 84.425 84.792	6: 67.126 67.673	10: 45.543 45.938	14: 23.301 27.393
3: 79.459 80.394	7: 58.058 58.862	11: 37.457 37.730	15: 18.245 18.241
4: 63.679 67.186	8: 39.165 39.624	12: 22.374 22.252	16: 10.147 10.066

▲

The differences between these computed values and values from the series solution range from 0.004 (at point 15) to 2.507 (at point 4). It is likely that the discontinuity of the boundary conditions near point 4 causes the largest error to occur there.

Although these examples are not likely to be representative, in this comparison the use of a rectangular set of nodes with adjusted equations at points where the boundary does not fit to the nodes gives better accuracy.

Actually, the best way to solve a boundary-value problem with regions that are not rectangular is with the finite element method, which we describe in another chapter.

7.10 Theoretical Matters

In this section we summarize many important theoretical items that underlie the solution of boundary-value problems and eigenvalues/eigenvectors. We also discuss some applications of eigenvalues in applied mathematics.

Existence and Uniqueness of the Solution to a Boundary-Value Problem

For a solution to a boundary-value problem to exist and be unique requires certain conditions to be met. For example, a linear boundary-value problem of the form

$$\frac{d^2u}{dx^2} = pu' + qu + r, \qquad \text{for } x \text{ on } [a, b],$$

with

$$u(a) = uL, \qquad u(b) = uR,$$

where p, q, and r are functions of x only, has a unique solution if two conditions are met:

$$p, q, \text{ and } r \text{ must be continuous on } [a, b],$$

and

$$q > 0 \text{ on } [a, b].$$

If the problem is nonlinear, more severe conditions apply that involve the partial derivatives of the right-hand side with respect to u and u'.

This simple example illustrates why these conditions are necessary.

EXAMPLE 7.22   (1) Obtain the general solution to $u'' = u' + u + x^2$ and evaluate the arbitrary constants if $u(1) = 1$, $u(3) = -1$. The problem should have a unique solution because, with $p = 1$, $q = 1$, and $r = x^2$, the conditions are satisfied. Using the standard techniques for linear second-order differential equations, we find that the general solution is

$$u(x) = e^{x/2}[c_1 e^{(\sqrt{5/2})x} - c_2 e^{(-\sqrt{5/2})x}] - x^2 + 2x - 4.$$

Imposing the boundary conditions lets us solve for the arbitrary constants:

$$c_1 = 0.038154 \qquad \text{and} \qquad c_2 = -7.06409.$$

(Any of the computer algebra systems can be of great help in getting these constants.)

(2) Repeat but with $q = -1$, so that the differential equation is

$$u'' = u' - u + x^2 \qquad \text{with } u(1) = 1, \qquad u(3) = -1.$$

Because q is not > 0, we do not expect a unique solution, but let us try to find it. The general solution now is

$$u(x) = e^{x/2}\left[c_1 \sin\left(\sqrt{\frac{3}{2}} * x\right) - c_2 \cos\left(\sqrt{\frac{3}{2}} * x\right)\right] + x^2 + 2x.$$

Now, when the boundary conditions are imposed, we find that we cannot solve for the constants! Still, there are special situations where a solution does exist: If $u(0) = 0$ and $u(\pi/3) = 1$, we can solve for the constants $c_1 = -1.6479$, $c_2 = 0$. ▲

The preceding conditions on a linear boundary-value problem must be interpreted to say that, if the conditions are met, a solution exists for any real values of the boundary conditions. When they are violated, the solution may not exist except in special cases.

A Linear Boundary-Value Problem Can Be Solved After Just Two Trials

It is not difficult to show that the correct solution to a second-order boundary-value problem by the shooting method is exactly (except for truncation and round-off errors) a linear combination of two trial solutions.

Suppose that $x_1(t)$ and $x_2(t)$ are two trial solutions of a boundary-value problem

$$x'' + Fx' + Gx = H, \qquad x(t_0) = A, \qquad x(t_f) = B$$

(where F, G, and H are functions of t only) and both trial solutions begin at the correct value of $x(t_0)$.

We then state that

$$y(t) = \frac{c_1 x_1 + c_2 x_2}{c_1 + c_2}$$

is also a solution. We show that this is true, because, since x_1 and x_2 are solutions, it follows that

$$x_1'' + Fx_1' + Gx_1 = H, \qquad \text{and} \qquad x_2'' + Fx_2' + Gx_2 = H.$$

If we substitute y into the original equations, with

$$y' = \frac{c_1 x_1' + c_2 x_2'}{c_1 + c_2}, \qquad \text{and} \qquad y'' = \frac{c_1 x_1'' + c_2 x_2''}{c_1 + c_2},$$

we get

$$\frac{c_1 x_1'' + c_2 x_2''}{c_1 + c_2} + \frac{c_1 x_1' + c_2 x_2'}{c_1 + c_2}F + \frac{c_1 x_1 + c_2 x_2}{c_1 + c_2}G = \frac{c_1 x_1'' + c_1 Fx_1' + c_1 Gx_1 + c_2 x_2'' + c_2 Fx_2' + c_2 Gx_2}{c_1 + c_2}$$

$$= \frac{c_1 H}{c_1 + c_2} + \frac{c_2 H}{c_1 + c_2} = H,$$

which shows that y is also a solution that begins at the correct value for $x(t_0)$. The implication of this is that, if c_1 and c_2 are chosen so that $y(t_f) = x(t_f) = B$, $y(t)$ is the correct solution to the boundary-value problem.

It also must be true that $y'(t_0)$ is the correct initial slope.

The Basis for the Power Method

The utility of the power method is that it finds the eigenvalue of largest magnitude and its corresponding eigenvector in a simple and straightforward manner. It has the disadvantage that convergence is slow if there is a second eigenvalue of nearly the same magnitude. The following discussion proves this and also shows why some starting vectors are unsuitable.

The method works because the eigenvectors are a set of *basis vectors*. A set of basis vectors is said to *span the space*, meaning that any n-component vector can be written as a unique linear combination of them. Let $v^{(0)}$ be any vector and $x_1, x_2, \ldots, x_n$ be eigenvectors. Then, for a starting vector, $v^{(0)}$,

$$v^{(0)} = c_1 x_1 + c_2 x_2 + \cdots + c_n x_n.$$

If we multiply $v^{(0)}$ by matrix A, because the x_i are eigenvectors with corresponding eigenvalues λ_i and remembering that $Ax_i = \lambda_i x$, we have,

$$v^{(1)} = Av^{(0)} = c_1 Ax_1 + c_2 Ax_2 + \cdots + c_n Ax_n \qquad (7.22)$$
$$= c_1 \lambda_1 x_1 + c_2 \lambda_2 x_2 + \cdots + c_n \lambda_n x_n.$$

Upon repeated multiplication by A, after m such multiplies, we get,

$$v^{(m)} = A^m v^{(0)} = c_1 \lambda_1^m x_1 + c_2 \lambda_2^m x_2 + \cdots + c_n \lambda_n^m x_n.$$

Now, if some one of the eigenvalues, call it λ_1, is larger than all the rest, it follows that all the coefficients in the last equation become negligibly small in comparison to λ_1^m as m gets large, so

$$A^m v^{(0)} \rightarrow c_1 \lambda_1^m x_1,$$

which is some multiple of eigenvector x_1 with the normalization factor λ_1, provided only that $c_1 \neq 0$. This is the principle behind the power method. Observe that if another of the eigenvalues is exactly of the same magnitude as λ_1, there never will be convergence to a single value. Actually, in this case, the normalization values alternate between two numbers and the eigenvalues are the square root of the product of these values. If another eigenvalue is not equal to λ_1, but is near to it, convergence will be slow. Also, if the starting vector, $v^{(0)}$, is such that the coefficient c_1 in Eq. (7.22) equals zero, the method will not work. (This last will be true if the starting vector is "perpendicular" to the eigenvector that corresponds to λ_1—that is, the dot-product equals zero). On the other hand, if the starting vector is almost "parallel" to the eigenvector of λ_1, all the other coefficients in Eq. (7.22) will be very small in comparison to c_1 and convergence will be very rapid.

The preceding discussion also shows why shifting and then using the inverse power method can often speed up convergence to the eigenvalue that is near the shift quantity.

Here we create, in the shifted matrix, an eigenvalue that is nearly zero, so that using the inverse method makes the reciprocal of this small number much larger than any other eigenvalue.

Applications of Eigenvalues—Part 1

The eigenvalues of a matrix can elucidate the rate of convergence of iterative methods for solving a set of linear equations. Think of these equations as $Ax = b$. Both the Jacobi and Gauss–Seidel methods can be expressed in the form

$$x^{(n+1)} = Gx^{(n)} = -Bx^{(n)} + b'. \qquad (7.23)$$

(Of course, both methods require that matrix A be diagonally dominant, or nearly so.) The two methods differ, and the difference can be expressed through these matrix equations, where A is written as $L + D + U$:

Jacobi: $\qquad x^{(n+1)} = -D^{-1}(L + U)x^{(n)} + D^{-1}b, \qquad (7.24)$

Gauss–Seidel: $x^{(n+1)} = -(L + D)^{-1}Ux^{(n)} + (L + D)^{-1}b. \qquad (7.25)$

As Eq. (7.23) makes clear, the rate of convergence depends on how matrix B affects the iterations.

We now discuss how matrix B operates in these two methods. If an iterative method converges, $x^{(n+1)}$ will converge to x, where this last is the solution vector. Because it is the solution, it follows that $Ax = b$. Equation (7.23) becomes, for $x^n = x$,

$$x = -Bx + b'.$$

Let $e^{(n)}$ be the error in the nth iteration

$$e^{(n)} = x^{(n)} - x.$$

When there is convergence, $e^{(n)} \to 0$, the zero vector, as n gets large. Using Eq. (7.23), it follows that

$$e^{(n+1)} = -Be^{(n)} = B^2e^{(n-1)} = -B^3e^{(n-2)} = \ldots = (-B)^{n+1}e^{(0)}.$$

Now, if $B^n \to 0$, the zero matrix, it is clear that $e^{(n)} \to 0$. To show when this occurs, we need a principle from linear algebra:

Any square matrix B can be written as UDU^{-1}. If the eigenvalues of B are distinct, then D is a diagonal matrix with the eigenvalues of B on its diagonal. (If some of the eigenvalues of B are repeated, then D may be triangular but the argument holds in either case.)

From this we write

$$B = UDU^{-1}, \qquad B^2 = UD^2U^{-1}, \qquad B^3 = UD^3U^{-1}, \ldots, \qquad B^n = UD^nU^{-1}.$$

Now, if all the eigenvalues of B (these are on the diagonal of D) have magnitudes less than one, it is clear that D^n will approach the zero matrix and that means that B^n will also. We then see that iterations converge depending on the eigenvalues of matrix B: they must all be less than one in magnitude. Further, the rate of convergence is more rapid if the largest eigenvalue is small. We also see that even if matrix A is not diagonally dominant, there may still be convergence if the eigenvalues of B are less than unity.

This example will clarify the argument.

EXAMPLE 7.23   Compare the rates of convergence for the Jacobi and Gauss–Seidel methods for $Ax = b$, where

$$A = \begin{bmatrix} 6 & -2 & 1 \\ -2 & 7 & 2 \\ 1 & 2 & -5 \end{bmatrix} \qquad b = \begin{bmatrix} 11 \\ 5 \\ -1 \end{bmatrix}.$$

For this example, we have

$$D \begin{bmatrix} 6 & 0 & 0 \\ 0 & 7 & 0 \\ 0 & 0 & -5 \end{bmatrix} \qquad L = \begin{bmatrix} 0 & 0 & 0 \\ -2 & 0 & 0 \\ 1 & 2 & 0 \end{bmatrix} \qquad U = \begin{bmatrix} 0 & -2 & 1 \\ 0 & 0 & 2 \\ 0 & 0 & 0 \end{bmatrix}$$

and

$$D^{-1} = \begin{bmatrix} 1/6 & 0 & 0 \\ 0 & 1/7 & 0 \\ 0 & 0 & -1/5 \end{bmatrix}.$$

For the Jacobi method, we need to compute the eigenvalues of this B matrix:

$$B = D^{-1}(L + U) = \begin{bmatrix} 1/6 & 0 & 0 \\ 0 & 1/7 & 0 \\ 0 & 0 & -1/5 \end{bmatrix} * \begin{bmatrix} 0 & -2 & 1 \\ -2 & 0 & 2 \\ 1 & 2 & 0 \end{bmatrix} = \begin{bmatrix} 0 & -1/3 & 1/6 \\ -2/7 & 0 & 2/7 \\ -1/5 & -2/5 & 0 \end{bmatrix},$$

for which the eigenvalues are $-0.1425 + 0.3366i$, $-0.1425 - 0.3366i$, and 0.2851. The largest in magnitude is 0.3655.

For the Gauss–Seidel method, we need the eigenvalues of this B matrix:

$$B = (L + D)^{-1}U = \begin{bmatrix} 1/6 & 0 & 0 \\ 1/21 & 1/7 & 0 \\ 11/210 & 2/35 & -1/5 \end{bmatrix} * \begin{bmatrix} 0 & -2 & 1 \\ 0 & 0 & 2 \\ 0 & 0 & 0 \end{bmatrix} = \begin{bmatrix} 0 & -1/3 & 1/6 \\ 0 & -2/21 & 1/3 \\ 0 & -11/105 & 1/6 \end{bmatrix},$$

which has these eigenvalues: 0, $0.0357 + 0.1333i$, and $0.0357 - 0.1333i$. The largest in magnitude for the Gauss–Seidel method is 0.1380. We then see that (as expected) the Gauss–Seidel method will converge faster. If we solve this example problem with both methods, starting with $[0 \quad 0 \quad 0]$ and ending the iterations when the largest change in any element of the solution is less than 0.00001, we find that Gauss–Jordan takes only seven iterations whereas the Jacobi method takes twelve. ▲

Applications of Eigenvalues—Part 2

We have used overrelaxation (the S.O.R. method) to speed the convergence of the iterations in solving a set of equations by the Gauss–Seidel technique. In view of the last discussion, this must be to reduce the eigenvalue of largest magnitude in the iteration equation. We have used S.O.R. in the following form:

$$x_i^{(n+1)} = x_i^{(n)} + \omega/a_{ii}(b_i - \Sigma a_{ij}x_j^{(n+1)} - \Sigma a_{ij}x_j^{(n)}), \qquad i = 1, 2, \ldots, N,$$

with the first summation from $j = 1$ to $j = i - 1$ and the second from $j = i$ to $j = N$. As shown before, the standard Gauss–Seidel iteration can be expressed in matrix form:

$$x^{(n+1)} = -(L + D)^{-1}Ux^{(n)} + (L + D)^{-1}b, \qquad (7.26)$$

which is more convenient for the present purpose. We want the overrelaxation equation to be in a similar form. From $A = L + D + U$, we can write

$$\omega(b - Ax) = \omega(b - (L + D + U)x) = 0.$$

Now, if we add Dx to both sides of this, we get

$$Dx - \omega Lx - \omega Dx - \omega Ux + \omega b = Dx,$$

which can be rearranged into

$$x^{(n+1)} = (D + \omega L)^{-1}[(1 - \omega)D - \omega U]x^{(n)} + \omega(D + \omega L)^{-1}b, \qquad (7.27)$$

and this is the S.O.R. form with ω equal to the overrelaxation factor. It is not easy to show in the general case that the eigenvalue of largest magnitude in Eq. (7.27) is smaller than that in Eq. (7.26), but we can do it for a simple example.

EXAMPLE 7.24 Show that overrelaxation will speed the convergence of iterations in solving

$$\begin{bmatrix} 2 & 1 \\ 1 & 3 \end{bmatrix} x = \begin{bmatrix} 6 \\ -2 \end{bmatrix}.$$

For this, the Gauss–Seidel iteration matrix is

$$-(L + D)^{-1}U = \begin{bmatrix} 0 & -1/2 \\ 0 & 1/6 \end{bmatrix},$$

whose eigenvalues are 0 and 1/6.

For the overrelaxation equation, the iteration matrix is

$$(D + \omega L)^{-1}[(1 - \omega)D - \omega U] = \begin{bmatrix} 1 - \omega & -\omega/2 \\ -\omega(1 - \omega)/3 & (\omega^2/6 - \omega + 1). \end{bmatrix}. \qquad (7.28)$$

We want the eigenvalues of this, which are, of course, functions of ω. We know that, for any matrix, the product of its eigenvalues equals its determinant (why?), so we set

$$\lambda_1 * \lambda_2 = \det(\text{iteration matrix}) = (\omega - 1)^2.$$

To get the smallest possible value for λ_1 and λ_2, we set them equal, so $\lambda_1 = \lambda_2 = (\omega - 1)$. We also know that, for any matrix, the sum of its eigenvalues equals its trace, so

$$\lambda_1 + \lambda_2 = 2(\omega - 1) = \frac{\omega^2}{6} - 2\omega + 2,$$

which has a solution $\omega = 1.045549$. Substituting this value of ω into Eq. (7.28) gives

$$\begin{bmatrix} -0.0455 & -0.5228 \\ 0.0159 & 0.1366 \end{bmatrix},$$

whose eigenvalues are 0.0455 and 0.0455, and both of these are smaller than the largest for the Gauss–Seidel matrix, which is $1/6 = 0.16667$.   ▲

Applications of Eigenvalues—Part 3

Another application of eigenvalues is in the solution of differential equations with constant coefficients. This example will show that the eigenvalues and eigenvectors of a matrix appear in the solution.

EXAMPLE 7.25   Solve this system:

$$\begin{array}{rcl} x' &=& 10x \\ y' &=& x - 3y - 7z, \\ z' &=& 2y + 6z, \end{array} \qquad \begin{array}{rcl} x(0) &=& 1, \\ y(0) &=& -1, \\ z(0) &=& 2. \end{array}$$

We can rewrite this system of equations using matrices:

$$X'(t) = \begin{bmatrix} 10 & 0 & 0 \\ 1 & -3 & 7 \\ 0 & 2 & 6 \end{bmatrix} X(t), \qquad X(0) = \begin{bmatrix} 1 \\ -1 \\ 2 \end{bmatrix}.$$

Using the standard methods for solving this system, we find that the general solution is

$$X(t) = Ae^{10t}\begin{bmatrix} 1 \\ \frac{2}{33} \\ \frac{1}{33} \end{bmatrix} + Be^{4t}\begin{bmatrix} 0 \\ -1 \\ 1 \end{bmatrix} + Ce^{-t}\begin{bmatrix} 0 \\ 1 \\ \frac{-2}{7} \end{bmatrix}, \qquad (7.29)$$

which can be expressed more clearly by

$$x(t) = Ae^{10t},$$

$$y(t) = \frac{2}{33}Ae^{10t} - Be^{4t} + Ce^{-t},$$

$$z(t) = \frac{1}{33}Ae^{10t} + Be^{4t} - \frac{2}{7}Ce^{-t}.$$

It seems somewhat remarkable that the eigenvalues of the coefficient matrix for $X'(t)$, $[10, 4, -1]$, are the exponents on t in Eq. (7.29) and its eigenvectors are the vectors in that equation.   ▲

Theory for Laplace's and Poisson's Equations

We only state a few items from the rich theory associated with partial-differential equations.

(1) If the boundary conditions for a Laplace problem are Dirichlet conditions, ($u = f(s)$, where s is a point on the boundary), the problem has a unique solution that is determined by the boundary conditions. Further, the maximum–minimum principle applies: No point in the interior of the region has a u-value greater than the maximum or less than the minimum of the u-values on the boundary.

The truth of this is apparent if we think of the problem as one of temperature distribution. Any interior point that is hotter than the maximum on the boundary must lose heat until its temperature is no hotter than that point. Likewise, any interior point that is colder than the minimum on the boundary must gain heat until it is at least equal to that temperature.

(2) A superposition principle applies for Laplace's equation. This says that, if u_1 and u_2 are both solutions to

$$u_{xx} + u_{yy} = 0,$$

then the weighted sum $c_1 u_1 + c_2 u_2$ is also a solution. This is easily demonstrated by substituting the sum into the equation. The usual nonnumerical methods for solving the equation usually depend on this principle. One first determines that there are an infinite set of solutions (ignoring the boundary conditions). Then a linear combination from this set is chosen so that the boundary conditions are met. This often leads to a Fourier series, which also has a rich body of theory.

(3) If the boundary conditions are Neumann conditions (the outward normal gradient—the *flux*—is specified at all points on the boundary) the solution is not unique in that, if u is a solution, so is $u + c$ where c is a constant. The truth of this is also apparent if we visualize the problem as one of temperature distribution. Heat is gained or lost from/to the surroundings in proportion to the flux and the temperature of the surroundings. If the flux is maintained and the temperature of the surroundings is changed, the temperatures in the object must change in order to maintain the same difference in temperatures between the object and the surroundings.

(4) The superposition principle does not apply to the Poisson equation, but a subtraction principle does. This says that, if u_1 and u_2 are both solutions to

$$u_{xx} + u_{yy} = f(x,\ y),$$

then $u_2 - u_1$ satisfies $u_{xx} + u_{yy} = 0$. As a consequence, we can find the solution to the Poisson equation by adding a particular solution for the problem to the general solution of the associated Laplace equation.

7.11 Using MATLAB

MATLAB can solve boundary-value problems analytically in some cases. Example 7.1 of Section 7.2 is not one of these, but if we modify the problem by changing it to avoid the multiplication of x and u, MATLAB can solve it. By default (which can be changed), MATLAB expects the independent variable to be t and the dependent variable to be x, so we restate the modified problem as

$$x'' - \left(1 - \frac{t}{5}\right) - x = t, \qquad x(1) = 2, \qquad x(3) = -1.$$

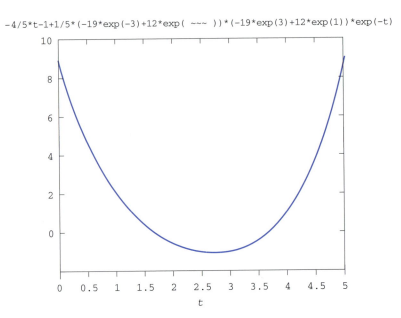

Figure 7.13

Entering this command gives a solution

```
x = dsolve('D2x = t + (1 - t/5) + x', 'x(1) = 2, x(3) = -1')
x =
       -4/5*t - 1 + 1/5*(-19*exp(-3) + 12*exp(-1))/(exp(-1)*exp(3)
         -exp(-3)*exp(1))*exp(t) - 1/5/(exp(-1)*exp(3)
           -exp(-3)*exp(1))*(-19*exp(3) + 12*exp(1))*exp(-t)
```

This seemingly complicated solution can be simplified somewhat, but the shape of the solution is more apparent from a plot. The command `'ezplot'` is an "easy" way to get this (and it also adds labels and the function that we are plotting at the top):

```
ezplot(x, [0, 5])
```

which displays Figure 7.13.

Solving with the Shooting Method

MATLAB cannot solve Example 7.1 analytically, as we have said, but it can solve it numerically with either `ode23` or `ode45` through the shooting method. (Both of these are Runge–Kutta–Fehlberg formulas with automatic step-size adjustment. The second usually uses larger steps because of its higher-order formula.) If we want to use `ode45` on the problem, we must first create an M-file to compute the derivatives:

```
function dx = shoot(t, x)
dx = [x(2); t + (1 - t/5)*x(1)];
```

in which we redefined the variables: $x(1) = dx/dt$, $x(2) = x$. Observe that we have changed the problem as given in Example 7.1 by renaming the variables, $(u \rightarrow x, x \rightarrow t)$ so the problem that we are to solve reads

```
x" = t + (1 - t/5)*x,     x(1) = 2,     x(3) = -1.
```

If we then issue a command that specifies $x(1) = 2$ and $x'(1) = -1.5$:

```
[t, x] = ode45('shoot', 1, 3, [2, -1.5])
```

for the first trial, we get a value for $x(3)$ of 4.7876. Repeating, but with the value of $x'(1) = -3$, results in $x(3) = 0.4360$. Using an initial slope equal to -3.4950 (the interpolated value) in a third trial gives the desired result: $x(3) = -1.0000$. These results are identical to Example 7.1.

It would be worthwhile to write a program to interpolate for successive starting values after doing the first two computations, especially if the problem is nonlinear.

Using a Set of Equations

There is no easy way to solve the problem through a set of equations with MATLAB—it is not solving the equations that is the difficulty, it is setting up the equations.

Eigenvalues and Eigenvectors

MATLAB can find the eigenvalues and eigenvectors of a square matrix. Here is an example:

EXAMPLE 7.27   Find the eigenvalues of

$$\begin{bmatrix} 10 & 0 & 0 \\ 1 & -3 & -7 \\ 0 & 2 & 6 \end{bmatrix}.$$

Solution:
   We define A in MATLAB:

```
A = [10, 0, 0; 1, -3, -7; 0, 2, 6]
A =
     10   0   0
      1  -3  -7
      0   2   6
```

and then do

```
e = eig(A)
e =

      4
     -1
     10
```

If we want both the eigenvalues and eigenvectors:

```
[V, D] = eig(A)
V =

        0          0     0.9977
   -0.7071     0.9615   0.0605
    0.7071    -0.2747   0.0302
D =
   4   0   0
   0  -1   0
   0   0  10
```

where the eigenvectors appear as the columns of V (they are scaled so each has a norm of one) and the eigenvalues are on the diagonal of matrix D. Observe that MATLAB gets all the eigenvectors at once.

Suppose we want to get the eigenvalues of A after its element in row 1, column 2 is changed to one. If that is what we want, we just enter:

```
A(1, 2) = 1;
eig(A)
ans =
    10.0606
    -1.1250
     4.0644
```

MATLAB uses a QR algorithm to get the eigenvalues after converting to Hessenberg form as described in Section 7.5.

Of course, the eigenvalues are the roots of the characteristic polynomial of the matrix. If we want the characteristic polynomial, we can get it:

```
charpoly(A)
ans =
    x^3 - 13*x^2 + 25*x + 46
```

which we see in symbolic form. To check that the roots are in fact the eigenvalues, we first put into algebraic form:

```
pp = [1, -13, 25, 46]
pp =
     1 -13 25 46
```

and then ask for

```
roots(pp)
ans =
    10.0606
     4.0644
    -1.1250
```

which is the same as before, as expected.   ▲

Chapter Summary

Can you do all of the following? You can if you really understand Chapter 7.

1. Solve boundary-value problems with a single independent variable either using the shooting method or a set of equations.

2. Explain why no more than two trials are needed to solve a linear boundary-value problem.

3. Understand what is meant by a characteristic-value problem.

4. Find the largest and smallest eigenvalues for a matrix by the power method. You know why shifting works and can use it to accelerate the convergence of the method.

5. Know what a similarity transformation is, how it can be used to convert a matrix to upper Hessenberg, and why this is advantageous in the QR method.

6. Derive the equations that govern the distribution of temperatures in a two- or three-dimensional object.

7. Solve for the steady-state temperatures in a slab with a variety of boundary conditions.

8. Explain why solving a system of equations by iteration is advantageous when the matrix is sparse and why overrelaxation can speed the convergence.

9. Use the A.D.I. method in two- and three-dimensional problems and tell why it is often preferred.

10. Outline the argument that relates the speed of convergence of an iterative solution of a system of equations to the eigenvalues of certain matrices.

11. Apply eigenvalues to explain why the power method and its variations works and to show that S.O.R. speeds the convergence of an iterative solution.

Computer Programs

Two of the algorithms of Chapter 7 are implemented: A Fortran 90 program that uses the power method to find the eigenvalue of largest magnitude, and a C program that carries out the A.D.I. method over a rectangular region.

Program 7.1

Figure 7.14 lists the Fortran program. Output from a test run is seen after the listing.

Program PPower

```
!!!!!!!!!!!!!!!!!!!!!!!!!!!!!!!!!!!!!!!!!!!!!!!!!!!!!!!!!!!!!
!                                                           !
!   Chapter 7                                               !
!                                                           !
!  Gerald/Wheatley, APPLIED NUMERICAL ANALYSIS(Sixth Edition) !
!                    Addison Wesley Longman, 1999           !
!                                                           !
!                                                           !
!!!!!!!!!!!!!!!!!!!!!!!!!!!!!!!!!!!!!!!!!!!!!!!!!!!!!!!!!!!!!
!                                                           !
!                                                           !
!   The Subroutine Power that is called here implements the  !
!   basic iterative method for finding the largest eigenvalue !
!   (in magnitude) and its corresponding eigenvalue.         !
!                                                           !
!!!!!!!!!!!!!!!!!!!!!!!!!!!!!!!!!!!!!!!!!!!!!!!!!!!!!!!!!!!!!
!                                                           !
        Implicit None
        Real, Dimension(10,10) :: A
        Real, Dimension(10) :: X
        Real :: Tol = 0.001
        Integer, Parameter :: n = 4, nLIMIT = 15
!
!  The 4 x 4 array, A, is initialized
!
        A(1, 1:n) = (/ 7,   8,   6,   6 /)
        A(2, 1:n) = (/ 1,   6,  -1,  -2 /)
        A(3, 1:n) = (/ 1,  -2,   5,  -2 /)
        A(4, 1:n) = (/ 3,   4,   3,   4 /)
!
!  The initial vector, X, is initialized.
!
        X(1:n) = (/ 1, 0, 0, 0 /)
!
        Call PrintA(A,n)
!
        Call Power(A,X,Tol,nLIMIT,n)
!
!
        Stop

        Contains
```

Figure 7.14

```
!
!
! Subroutine Power :
!                         This Subroutine computes the largest eigen-
! value and its corresponding eigenvector by the Power method.
!
! ------------------------------------------------------------
       Subroutine Power(A,X,Tol,nLIMIT,n)
!
! Parameters are :
!
! A      - an N X N matrix whose eigens are being determined
! X      - estimate for the eigenvector used to begin iterations.
!          If no approximation to this is known, the usual choice
!          is a vector with all components equal to unity. X also
!          returns the final eigenvector to the caller.
! c      - returns the value of the eigenvalue
! Tol    - tolerance value used to determine convergence. Iterations
!          continue until successive estimates of the eigenvalue
!          are the same within Tol in value.
! nlimit - limit to the number of iterations If not convergent
! n      - size of the matrix and the vector X
! XWRK   - vector used to store intermediate values. It must be
!          dimensioned to hold at least N elements in the
!          main Program.
!
! ------------------------------------------------------------
!
       Implicit None
       Real, Dimension(:,:), Intent(In) :: A
       Real, Dimension(:), Intent(in Out) :: X
       Real, Dimension(10) :: Xwrk
       Real :: c
       Real, Intent(In) :: Tol
       Real :: save
       Integer, Intent(In) :: n, nlimit
       Integer :: iROW, jCOL, iteration
       Logical :: convergent


! ------------------------------------------------------------
!
!Begin the iterations. Get the product of A and X, then normalize.
!The normalization factors should converge to the eigenvalue, c, on
!repeated multiplications. We store the current value of c for
!comparison with the next value to test convergence. To begin, we
!make save = 0.

       save = 0.0
       Do iteration = 1,nlimit
          Do iROW = 1,n
             Xwrk(iROW) = 0.0
```

Figure 7.14 *Continued*

```
                    Do   jCOL = 1,n
                        Xwrk(iROW) = Xwrk(iROW) +pl A(iROW,jCOL)*X(jCOL)
                        End Do   ! jCOL
                    End Do   ! iROW
!
! ------------------------------------------------------------
!
! Find the largest element of the product vector for normalizing.
!
            c = 0.0
            Do iROW = 1,n
                If ( Abs(c) < Abs(Xwrk(iROW)) ) c = Xwrk(iROW)
                End Do
!
! ------------------------------------------------------------
!
! Now normalize the product vector and put into X for next iteration.
!
            Do iROW = 1,n
!               X(iROW) = Xwrk(iROW) / C
!               End Do
!
            X(1:n) = Xwrk(1:n)/c
!
            Call Print_X_c(x, c, n)
!
! ------------------------------------------------------------
!
!See whether tolerance is met. If so, we are done. If not, we continue
!the iterations.

        save = c
        convergent = ( Abs(c - save) <= Tol )
  End do
!
! ------------------------------------------------------------
!
! If we do not meet the tolerance for convergence, we write
! a message and return last values calculated.
        If ( .not. convergent) Then
            Write(*,200) nlimit
            End If
  200   Format(/' Convergence not reached in ',I5,' iterations')
        Return
        End Subroutine Power
!
! ------------------------------------------------------------
!       This subroutine prints out the input n by n matrix, A
! ------------------------------------------------------------
!
        Subroutine PrintA(A,n)
!
```

Figure 7.14 *Continued*

```
            Implicit None
            Real, Dimension(:,:), Intent(In) :: A
            Integer, Intent(In) :: n
            Integer :: i,j
  !
            Write(*, 100)
            Do I = 1, N
               Write(*,200) (A(i,j), j = 1,n)
               End do
            Write(*,'(//)')
100       Format(//, '   the Input matrix is'/)
200       Format(10F7.2)
            Return
            End Subroutine PrintA
  !
  ! ------------------------------------------------------------
  !    This subroutine prints out the scalar value, c,
  !    and the normalized vector, X.
  ! ------------------------------------------------------------
            Subroutine Print_X_c(X, c,n)
            Implicit None
            Real, Dimension(:), Intent(In) :: X
            Real, Intent(In) :: c
            Integer, Intent(In) :: n
            Integer :: i
  !
            Write(*,100) (X(i), i = 1,n)
            Write(*,200) C
100       Format(' the normalized vector:  ', 10F8.4)
200       Format(' the value for c IS: ', F10.6/)
            Return
            End Subroutine Print_X_c
End Program PPower


    ************ Output for Program to Find Eigenvalues ************

The Input matrix is

7.00    8.00    6.00    6.00
1.00    6.00   -1.00   -2.00
1.00   -2.00    5.00   -2.00
3.00    4.00    3.00    4.00

The normalized vector:    1.0000  0.1429  0.1429  0.4286
The value for c IS:   7.000000

The normalized vector:    1.0000  0.0741  0.0494  0.4938
The value for c IS:  11.571428

The normalized vector:    1.0000  0.0375  0.0102  0.4994
The value for c IS:  10.851852
```

Figure 7.14 *Continued*

```
The normalized vector:      1.0000  0.0209 -0.0022  0.4999
The value for c IS:  10.358361

The normalized vector:      1.0000  0.0126 -0.0052  0.5000
The value for c IS:  10.153433

The normalized vector:      1.0000  0.0080 -0.0051  0.5000
The value for c IS:  10.069348

The normalized vector:      1.0000  0.0053 -0.0041  0.5000
The value for c IS:  10.033590

The normalized vector:      1.0000  0.0036 -0.0031  0.5000
The value for c IS:  10.017592

The normalized vector:      1.0000  0.0025 -0.0023  0.5000
The value for c IS:  10.009959

The normalized vector:      1.0000  0.0017 -0.0016  0.5000
The value for c IS:  10.006031

The normalized vector:      1.0000  0.0012 -0.0012  0.5000
The value for c IS:  10.003847

The normalized vector:      1.0000  0.0008 -0.0008  0.5000
The value for c IS:  10.002542

The normalized vector:      1.0000  0.0006 -0.0006  0.5000
The value for c IS:  10.001719

The normalized vector:      1.0000  0.0004 -0.0004  0.5000
The value for c IS:  10.001180

The normalized vector:      1.0000  0.0003 -0.0003  0.5000
The value for c IS:  10.000816
```

Figure 7.14 *Continued*

Program 7.2

Figure 7.15 is a listing of the C program that solves Laplace's equation on a rectangle with Dirichlet boundary conditions. Output from it will be found after the listing.

The technique of Section 7.8 is followed here. The number of rows and columns of nodes are established within the program as well as the boundary values. The interior nodes are initialized at zero and the boundary values are set into vectors for each edge of the rectangular region. The tridiagonal equations are set up with different matrices for the horizontal and vertical traverses, using a value of rho that is input by the user.

Once this preliminary work is done, the systems of equations (separate sets of equations for the two traverses), are solved through the *LU* decomposition and back-substitution. The two sets of equations, with appropriate right-hand sides, are solved alternately. After each even-order computation (which should eliminate the bias), the largest change in any of the nodal values is displayed and the user is asked if more computations should be performed. When the user is satisfied that the results are sufficiently accurate, the last set of results can be obtained.

```
/* * * * * * * * * * * * * * * * * * * * * * * * * * * * * * * * * * * * * * * * * * * * * *
 *                                                                                        *
 *    Chapter 7            Steady State Temperatures in a Rectangle                       *
 *                                                                                        *
 *    Gerald/Wheatley, APPLIED NUMERICAL ANALYSIS (sixth edition)                         *
 *                    Addison Wesley Longman, 1999                                        *
 *                                                                                        *
 * * * * * * * * * * * * * * * * * * * * * * * * * * * * * * * * * * * * * * * * * * * * * *
 *                                                                                        *
 *    This program uses the A.D.I method to find the steady state                         *
 *    temperatures within a rectangular region. This implementation                       *
 *    is for Dirichlet boundary conditions. The parameters are the                        *
 *    same as in Example 7.14 except there are more nodes.                                *
 *                                                                                        *
 *    A value for rho is obtained from the user and this is used                          *
 *    to set up the matrices for the horizontal and vertical                              *
 *    traverses. The LU decompositions are computed and two                               *
 *    iterations are done. The largest change in any of the                              *
 *    temperatures is then displayed. If this is too large, the                          *
 *    user can request additional pairs of iterations until this                         *
 *    change is sufficiently small. The result of the last                               *
 *    iteration is then displayed.                                                        *
 *                                                                                        *
 * * * * * * * * * * * * * * * * * * * * * * * * * * * * * * * * * * * * * * * * * * * * * */

#include <stdio.h>
#include <math.h>
#include <conio.h>

void main()
{

  float u[500], v[500], uMat[30][30], vMat[30][30];
  float uCoef[500][4], vCoef[500][4], bcR[30], bcL[30], bcT[30],
  bcB[30];
  float rho, rP2, rM2, larg;
  int nRu, nCu, nRv, nCv, nU_V, i, j, k, iter=0, stop;
  char inbuf[20];

/* Give values to some variables */
  nRu = nCv = 7;
  nCu = nRv = 15;
  nU_V = nRu*nCu;
```

Figure 7.15

```
/* Put initial values (for u or v) into v-vector */
  for (i=1; i<=nU_V; i++)
    v[i] = 0.0;

/* Put boundary conditions into u and v matrices */
  for (i=0; i<=nCu+1; i++)
  {
    uMat[0][i] = vMat[i][0] = 0.0;              /* top of u is the left
for v */
    uMat[nRu+1][i] = vMat[i][nCv+1] = 0.0;      /* bott of u is the
right for v */
  }                                  /* end of for */
  for (i=1; i<=nRu; i++)
  {
    uMat[i][0] = vMat[0][i] = 0.0; /* left of u is the top for v */
    uMat[i][nCu+1] = vMat[nRv+1][i] = 100.0; /* right of u is the bott
for v */
  }                                  /* end of second for */

/* Get value for rho from user and compute rP2, rM2 */
  clrscr();
  printf ("Enter a value for rho:\n");
  gets (inbuf);
  sscanf (inbuf, "%f", &rho);
  rP2 = 1.0/rho + 2;
  rM2 = 1.0/rho - 2;
/*
    Set up the two coefficient matrices, which are -1 rP2 -1
    except zero where u (v) is on boundary. Do first for u
*/

  nU_V = nRu*nCu;

  for (i=1; i<=nU_V+1; i++)
  {
    uCoef[i][1] = uCoef[i][3] = -1.0;
    uCoef[i][2] = rP2;
    if (i%nCu == 1)
      uCoef[i][1] = uCoef[i-1][3] = 0.0;
  }                                  /* end of for */

  for (i=1; i<=nU_V+1; i++)
  {
    vCoef[i][1] = vCoef[i][3] = -1.0;
    vCoef[i][2] = rP2;
    if (i%nCv == 1)
      vCoef[i][1] = vCoef[i-1][3] = 0.0;
  }

/*
    Get LU for both coefficient matrices, putting into uCoef and vCoef
```

Figure 7.15 *Contiinued*

```
*/
   for (i=2; i<=nU_V; i++)
   {
     uCoef[i-1][3] /= uCoef[i-1][2];
     vCoef[i-1][3] /= vCoef[i-1][2];
     uCoef[i][2] -= uCoef[i][1] * uCoef[i-1][3];
     vCoef[i][2] -= vCoef[i][1] * vCoef[i-1][3];
   }                              /* end of for */

  /*
     Begin loop here to solve the problem:
        (1) Put v's into vMat, used to
        (2) set up rhs for the u's, which we now
        (3) solve for the u's.
        (4) Put u's into uMat, used to
        (5) set up rhs for the v's, which we now
        (6) solve for the v's
        (7) Find maximum change in node values and display.

     After (7), ask if the user wants to continue. If not,
        display the last set of v's.
  */
   do
   {                              /* Start of do loop */

/* Put v's into vMat */
     for (i=1; i<=nRv; i++)
     {
       for (j=1; j<=nCv; j++)
       {
         k = (j-1)%nCv;
         k += (i-1)*nCv + 1;
         vMat[i][j] = v[k];
       }                                /* end of for j */
     }                                  /* end of for i */

/* Set up rhs for the u's, storing in the u-vector */
     for (i=1; i<=nU_V; i++)
     {
       j = (i-1)%nCu + 1;
       k = (i-j)/nCu;
       u[i] = vMat[j][k] + rM2*vMat[j][k+1] + vMat[j][k+2];
     }                                  /* end of for */
/* Add in boundary conditions on left and right */
     for (i=1; i<=nU_V; i+=nCu)
       u[i] += uMat[(i-1)/nCu+1][0];
     for (i=nCu; i<=nU_V; i+=nCu)
       u[i] += uMat[i/nCu][nCu+1];

  /*
     Solve the equations -- LUx=b -> Ux=y; Ly=b. Get y, then x.
     The b-vector is u[]. Use u[] for both y and x.
```

Figure 7.15 *Contiinued*

```
*/
    u[1] /= uCoef[1][2];
    for (i=2; i<=nU_V; i++)
      u[i] = (u[i] - uCoef[i][1]*u[i-1]) / uCoef[i][2];

    for (i=(nU_V-1); i>=1; --i)
      u[i] -= uCoef[i][3]*u[i+1];

/* Repeat this for the new v-vector */
/* Put u's into uMat */
    for (i=1; i<=nRu; i++)
    {
      for (j=1; j<=nCu; j++)
      {
        k = (j-1)%nCu;
        k += (i-1)*nCu + 1;
        uMat[i][j] = u[k];
      }                       /* end of for j */
    }                         /* end of for i */

/* Set up rhs for the v's, storing in the v-vector */
    for (i=1; i<=nU_V; i++)
    {
      j = (i-1)%nCv + 1;
      k = (i-j)/nCv;
      v[i] = uMat[j][k] + rM2*uMat[j][k+1] + uMat[j][k+2];
    }                         /* end of for */

/* Add in boundary conditions on top and bott */
    for (i=1; i<=nU_V; i+=nCv)
      v[i] += vMat[(i-1)/nCv+1][0];
    for (i=nCv; i<=nU_V; i+=nCv)
      v[i] += vMat[i/nCv][nCv+1];

/*
    Solve the equations—LUx=b -> Ux=y; Ly=b. Get y, then x.
    The b-vector is v[]. Use v[] for both y and x.
*/
    v[1] /= vCoef[1][2];
    for (i=2; i<=nU_V; i++)
      v[i] = (v[i]-vCoef[i][1]*v[i-1]) / vCoef[i][2];
    for (i=(nU_V-1); i>=1; --i)
      v[i] -= vCoef[i][3]*v[i+1];

/* Put the v's into the vMat */
/* Put v's into vMat */
    for (i=1; i<=nRv; i++)
    {
      for (j=1; j<=nCv; j++)
      {
        k = (j-1)%nCv;
```

Figure 7.15 *Contiinued*

```
            k += (i-1)*nCv + 1;
            vMat[i][j] = v[k];
        }                       /* end of for j */
    }                           /* end of for i */

/* Get the largest change between the u's and v's */
    iter +=2;
    larg = 0;
    for (i=1; i<=nRu; i++)
        for (j=1; j<=nCu; j++)
        {
            if (fabs(uMat[i][j] - vMat[j][i]) > larg)
                larg = fabs(uMat[i][j] - vMat[j][i]);
        }                       /* end of for j */

    printf ("At iteration %3d, max change is %f\n",iter, larg);
    getch();

/* Ask user whether to continue */
    printf ("Enter a 0 to quit and see last result,\n");
    printf ("or anything else to do two more iterations ");
    gets (inbuf);
    sscanf (inbuf, "%d," &stop);

} while (stop != 0);        /* End of do/while loop */

/*clrscr();*/

/* Put the v's into the vMat */
for (i=1; i<=nRv; i++)     /* Put v's into vMat */
{
    for (j=1; j<=nCv; j++)
    {
        k = (j-1)%nCv;
        k += (i-1)*nCv + 1;
        vMat[i][j] = v[k];
    }                           /* end of for j */
}                               /* end of for i */

/* Print the vMat omitting bc. Arrange like the u's */
printf ("\nHere is the final solution:\n");

for (i=1; i<=nCv; i++)
{
    for (j=1; j<=nRv; j++)
        printf ("%7.4f ",vMat[j][i]);
    printf("\n");
}
getch();

}    /* End of main */
```

Figure 7.15 *Continued*

```
****************** Output from ADI.C ******************

Enter a value for rho:
1.8
At iteration   2, max change is 37.821461
Enter a 0 to quit and see last result,
  or anything else to do two more iterations
At iteration   4, max change is 5.824394
Enter a 0 to quit and see last result,
  or anything else to do two more iterations
At iteration   6, max change is 2.586609
Enter a 0 to quit and see last result,
  or anything else to do two more iterations
                        .
                        .
                        .
At iteration  28, max change is 0.000694
Enter a 0 to quit and see last result,
  or anything else to do two more iterations  0

Here is the final solution:
0.0772  0.1662  0.2806  0.4377  0.6617  0.9867  1.4630  2.1649  3.2043
 4.7551  7.0990  10.7318  16.6373  27.0965  48.3359
0.1427  0.3072  0.5185  0.8089  1.2225  1.8226  2.7009  3.9925  5.8978
 8.7173  12.9094  19.1912  28.7214  43.4125  66.2473
0.1865  0.4013  0.6773  1.0566  1.5967  2.3799  3.5252  5.2062  7.6768
11.3066  16.6298  24.4022  35.6442  51.5850  73.2407
0.2018  0.4344  0.7332  1.1436  1.7281  2.5755  3.8142  5.6309  8.2968
12.2029  17.9015  26.1436  37.8685  54.0424  75.1310
0.1865  0.4013  0.6773  1.0566  1.5967  2.3799  3.5252  5.2062  7.6768
11.3066  16.6298  24.4022  35.6442  51.5850  73.2407
0.1427  0.3072  0.5185  0.8089  1.2225  1.8226  2.7009  3.9925  5.8978
 8.7173  12.9094  19.1912  28.7214  43.4125  66.2473
0.0772  0.1662  0.2806  0.4377  0.6617  0.9867  1.4630  2.1649  3.2043
 4.7551  7.0990  10.7318  16.6373  27.0965  48.3359
```

Figure 7.15 *Contiinued*

Exercises

Section 7.1

1. Show that Eq. (7.3) results when both k and A vary along a rod.

▶ **2.** Suppose that a rod of length L is made from two dissimilar materials welded together end-to-end. From $x = 0$ to $x = X$, the thermal conductivity is k_1; from $x = X$ to $x = L$, it is k_2. How will the temperatures vary along the rod if $u = 0°$ at $x = 0$ and $u = 100°$ at $x = L$? Assume that Eq. (7.3) applies with $Q = 0$ and that the cross-section is constant.

3. What if k varies with temperature: $k = a + bu + cu^2$? What is the equation that must be solved to determine the temperature distribution along a rod of constant cross-section?

4. Repeat Exercise 3 but now allow for a variable cross-section: $A = mx + n$.

Section 7.2

▶ **5.** Solve the boundary-value problem

$$y'' - xy' + 3y = 11x, \quad y(1) = 1.5, \quad y(2) = 15$$

by the shooting method. Assume two values for $y'(1)$ (which is near 5). Use $h = 0.25$, and compare results with Runge–Kutta–Fehlberg and the modified Euler methods. Why are the results different? Can you get essentially the same results with the modified Euler method as with Runge–Kutta–Fehlberg? Find the analytical solution, and compare to the computed values of $y(x)$.

6. In Exercise 5 you obtained the solution by using an initial slope interpolated from the first two trials and then repeating the integration method. Show that the values of $y(x)$ can be computed without using an integration method, by interpolating from the values obtained in the trials.

▶ **7.** In Exercise 5, the exact analytical value for $y'(1)$ is 5.5000. You probably found a different value with $h = 0.25$ and the modified Euler method. Explain the discrepancy. Can you match the analytical solution using the modified Euler method and a smaller value for h? If not, explain.

8. The shooting method can work backward as well as forward through the interval. Solve Exercise 5 by working backward from $x = 2$ (h is then negative). Do you get the same results?

9. Use the shooting method to solve

$$y'' + yy' = e^{x/2}, \quad y(-1) = 2, \quad y(1) = 3.$$

Use either the computer program of Chapter 6 or another one that you have written to solve with assumed values for the initial slope. Why is this more difficult to solve than Exercise 5?

10. Exercise 9 solves a nonlinear boundary-value problem. The solution will not be exact because there are errors in the numerical procedure, depending on the step size. If the step size is reduced, the solution should be more accurate. Vary the step size in Exercise 9 until you believe the value of $y(0)$ is correct to within $5 * 10^{-5}$. Justify your belief that you have attained this accuracy.

11. Use the shooting method to solve

$$\frac{d^3y}{dt^3} + 10\frac{dy}{dt} - 5y^3 = t^3 - ty, \quad y(0) = 0,$$

$$y(2) = -1, \quad y''(2) = 0.$$

Take $h = 0.2$.

Section 7.3

▶ **12.** Given this boundary-value problem:

$$\frac{d^2y}{d\theta^2} + \frac{y}{4} = 0, \quad y(0) = 0, \quad y(\pi) = 2,$$

which has the solution $y = 2\sin(\theta/2)$,

a. Solve, using finite difference approximations to the derivative with $h = \pi/4$ and tabulate the errors.
b. Solve again by finite differences but with a value of h small enough to reduce the maximum error to 0.5%. Can you predict from part (a) how small h should be?
c. Solve again by the shooting method. Find how large h can be to have maximum error of 0.5%.

13. Solve Exercise 5 through a set of equations. How small must h be to essentially match the shooting method with $h = 0.25$ in the Runge–Kutta procedure?

14. Solve this boundary-value problem by finite differences, first using $h = 0.2$, then with $h = 0.1$:

$$y'' + xy' - x^2y = 2x^3, \quad y(0) = 1, \quad y(1) = -1.$$

Assuming that errors are proportional to h^2, extrapolate to get an improved answer. Then, using a very small h-value in the shooting method, see if this agrees with your improved answer.

▶ **15.** Solve the equation in Exercise 5 with finite-difference approximations. What value of h is required to match the analytical answer to four significant figures?

16. Solve Exercise 9 through a set of equations. You will find that these are nonlinear. You may want to linearize by replacing y in the second term with estimated values. From a first answer, you could replace the y-values with improved estimates, repeating until the estimates and the answer agree.

Section 7.4

17. Repeat Exercise 12, except with these derivative boundary conditions:

$$y'(0) = 0, \quad y'(\pi) = 1.$$

In part (a), compare to $y = -2\cos(\theta/2)$.

18. Suppose the right boundary in Exercise 17 were $y'(2\pi) = 0$. An analytical solution is still $y = -2\cos(\theta/2)$. Can you find this solution from a set of difference equations?

▶ 19. Solve through finite differences with four subintervals:

$$\frac{d^2y}{dx^2} + y = 0, \qquad y'(0) + y(0) = 2,$$

$$y'\left(\frac{\pi}{2}\right) + y\left(\frac{\pi}{2}\right) = -1.$$

20. Write a computer program to set up the equations for Exercise 19, but this time with end conditions of

$$y'(0) + y(0) = A, \qquad y'\left(\frac{\pi}{2}\right) + y\left(\frac{\pi}{2}\right) = B.$$

21. The most general form of boundary conditions normally encountered for second-order equations is a linear combination of the function and its derivative at both ends of the interval. Set up equations, with $h = 0.2$, to solve

$$y'' - xy' + x^2y = x^3, \qquad \text{subject to}$$

$$y(0) + y'(0) + y(1) + y'(1) = 4,$$

$$y(0) - y'(0) + y(1) - y'(1) = 3.$$

22. Solve the third-order boundary-value problem with $h = 0.2$:

$$y''' - y' = e^x, \quad y(0) = 0, \quad y(1) = 1, \quad y'(1) = 0.$$

Approximate the third derivative in terms of an average of third differences:

$$y_0''' = \frac{y_2 - 2y_1 + 2y_{-1} - y_{-2}}{2h^3} + O(h^3)$$

at $x = 0.4, 0.6$, and 0.8. Using the derivative condition at $x = 1$ will eliminate the assumed function value at $x = 1.2$, but we are still short one equation. Obtain this by writing the equation at $x = 0.2$ using an unsymmetrical approximation for y''':

$$y_0''' = \frac{-y_3 + 6y_2 - 12y_1 + 10y_0 - 3y_{-1}}{2h^3} + O(h^2).$$

23. Solve the nonlinear problem

$$y'' = 2 - \frac{4y^2}{\sin^2(x)}, \qquad y'(1) = 0.9093,$$

$$y(2) = 0.8268,$$

by a set of difference equations. Compare to the analytical solution, $y = \sin^2(x)$.

Section 7.5

24. Consider the characteristic-value problem with k restricted to real values:

$$y'' - k^2y = 0, \qquad y(0) = 0, \qquad y(1) = 0.$$

a. Show analytically that there is no solution except the trivial solution $y = 0$.
b. Show, by setting up a set of difference equations corresponding to the differential equation with $h = 0.2$, that there are no real values for k for which a solution to the set exists.
c. Show, using the shooting method, that it is impossible to match $y(1) = 0$ for any real value of k [except if $y'(0) = 0$, which gives the trivial solution].

▶ 25. For the equation

$$y'' - 3y' + 2k^2y = 0, \qquad y(0) = 0, \qquad y(1) = 0,$$

find the principal eigenvalue and compare to $|k| = 2.46166$,

a. using $h = \frac{1}{2}$.
b. using $h = \frac{1}{3}$.
c. using $h = \frac{1}{4}$.
d. Assuming errors are proportional to h^2, extrapolate from parts (a) and (c) to get an improved estimate.

26. Estimate the principal eigenvalue of

$$y'' + k^2x^2y = 0, \qquad y(0) = 0, \qquad y(1) = 0.$$

27. Using the principal eigenvalue, $k = 2.46166$, in Exercise 25, find y as a function of x over $[0, 1]$. This is the corresponding eigenfunction.

28. The second eigenvalue for Exercise 26 is 12.61105. Find the corresponding eigenfunction.

▶ 29. Parallel the computations of Exercise 25 to estimate the second eigenvalue. Compare to the analytical value of 4.56773.

30. Find the dominant eigenvalue and the corresponding eigenvector by the power method:

a. $\begin{bmatrix} 3 & 1 \\ 2 & 9 \end{bmatrix}$ b. $\begin{bmatrix} 2 & 3 \\ 6 & 5 \end{bmatrix}$ c. $\begin{bmatrix} 2 & 3 \\ 3 & -2 \end{bmatrix}$

d. $\begin{bmatrix} 6 & 2 & 0 \\ 2 & 4 & 1 \\ 0 & 1 & -1 \end{bmatrix}$ e. $\begin{bmatrix} 1 & 2 & 3 \\ 0 & 1 & 3 \\ 2 & 2 & 1 \end{bmatrix}$

[In part (c), the two eigenvalues are equal but of opposite sign.]

31. For the two matrices

$$A = \begin{bmatrix} -5 & 2 & 1 \\ 1 & -9 & -1 \\ 2 & -1 & 7 \end{bmatrix},$$

$$B = \begin{bmatrix} -4 + 2i & -1 & -5i \\ -3 & 7 + i & -i \\ 2 & -1 & 4 - i \end{bmatrix},$$

   a. put bounds on the eigenvalues using Gerschgorin's theorem.

   b. can you tell from part (a) whether either of the matrices is singular?

32. Use the power method or its variations to determine all of the eigenvalues for the matrices of parts (a), (b), and (c) of Exercise 25.

33. Invert the matrices of Exercise 30, and then find the smallest eigenvalues with the power method.

▶ **34.** Find all the eigenvalues of matrix A of Exercise 31. Get these from the characteristic polynomial. Then invert A and repeat. Show that the two sets are reciprocals and that the corresponding eigenvectors are the same.

35. Repeat Exercise 34, but now use the power method to get the dominant eigenvalue. Then shift by that amount and get another. Finally use tr(A) to get the third.

36. For matrix A of Exercise 31, find three matrices each of which will convert one of the below-diagonal elements to zero.

37. Use the matrices of Exercise 36 successively to make one element below the diagonal of A equal to zero, then multiply that product and the inverse of the rotation matrix (which is easy to find because it is just its transpose). We keep the eigenvalues the same because the two multiplications are a similarity transformation.

   Repeat this process until all elements below the diagonal are less than 1.0E-4. When this is done, compare the elements now on the diagonal to the eigenvalues of A obtained by iteration. (This will take many steps. You will want to write a short computer program to carry it out.)

38. Use similarity transformations to reduce the matrix to upper Hessenberg. (Do no column or row interchanges.)

$$C = \begin{bmatrix} 3 & -1 & 2 & 7 \\ 1 & 2 & 0 & -1 \\ 4 & 2 & 1 & 1 \\ 2 & -1 & -2 & 2 \end{bmatrix}$$

39. Repeat Exercise 38, this time with row/column interchanges that maximize the magnitude of divisors.

▶ **40.** Repeat Exercise 37 but first convert to upper Hessenberg. How many fewer iterations are required?

Section 7.6

41. Show that Eq. (7.9) results if the thickness of the slab varies with position (x, y).

42. Show that Eq. (7.10) applies when both thickness and thermal conductivity vary with position in a slab.

43. Get the equivalent of Eq. (7.10) for a three-dimensional object.

44. Repeat Exercise 42, but now for k varying with temperature, such as $k = a + bu + cu^2$.

Section 7.7

45. The mixed second derivative $\partial^2 u/(\partial x\, \partial y)$ can be considered as

$$\frac{\partial}{\partial x}\left(\frac{\partial u}{\partial y}\right) = \frac{\partial^2 u}{\partial x\, \partial y} = \frac{\partial}{\partial y}\left(\frac{\partial u}{\partial x}\right).$$

If the nodes are spaced apart a distance h in both the x- and y-directions, show that this derivative can be represented by the pictorial operator

$$\frac{1}{4h^2}\begin{Bmatrix} -1 & & 1 \\ 1 & & -1 \end{Bmatrix} + O(h^2).$$

▶ **46.** If $d^2 u/dx^2$ is represented as this fourth-order central-difference formula

$$\frac{d^2 u}{dx^2} = \frac{-u_{i+2} + 16u_{i+1} - 30u_i + 16u_{i-1} - u_{i-2}}{12h^2},$$

find the fourth-order operator for the Laplacian. (This requires the function to have a continuous sixth derivative.)

47. Derive the nine-point approximation of the Laplacian in Eq. (7.13).

▶ **48.** A rectangular plate of constant thickness has heat flow only in the x- and y-directions (k is constant). If the top and bottom edges are perfectly insulated and the left edge is at 100° and the right edge at 200°, it is obvious that there is no heat flow except in the x-direction and that temperatures vary linearly with x and are constant along vertical lines.

   a. Show that such a temperature distribution satisfies Eq. (7.12) and also Eq. (7.13).

b. Show that this temperature distribution also satisfies the relation derived in Exercise 46. What about nodes adjacent to the edges?

49. Solve for the steady-state temperatures in a rectangular plate, 12 in. × 15 in., if one 15-in. edge is held at 100° and the other 15-in. edge is held at 50°; both 12-in. edges are held at 20°. The material is aluminum. Space the nodes 3 in. apart in both directions. Consider heat to flow only laterally. Sketch the approximate location of the 65° isothermal curve.

▶ 50. Solve for the steady-state temperatures in the plate of the figure when the edge temperatures are as shown. The plate is 10 cm × 8 cm, and the nodal spacing is 2 cm.

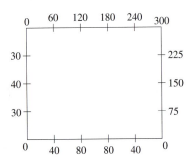

51. Repeat Exercise 50, but use the nine-point formula.

▶ 52. Suppose the differential equation for the plate is
$$3u_{xx} + 2u_{yy} = 0.$$
What will the pictorial operator of Eq. (7.12) look like?

53. The region on which we solve Laplace's equation does not have to be rectangular. We can apply the methods of Section 7.7 to any region where the nodes fall on the boundary. Solve for the steady-state temperatures at the eight interior points of this figure.

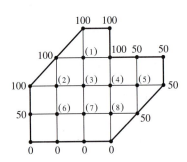

54. Solve Exercise 50 by Liebmann's method with all elements of the initial u-vector equal to zero. Then repeat with all elements equal to 300°, the upper bound to the steady-state temperatures. Repeat again with the initial values all equal to the arithmetic average of the boundary temperatures. Compare the number of iterations needed to reach a given tolerance for convergence in each case. What is the effect of the tolerance value that is used?

55. Repeat Exercise 54, but now use overrelaxation with the factor given by Eq. (7.15).

▶ 56. Find the torsion function ϕ for a 2 in. × 2 in. square bar.
   a. Subdivide the region into nine equal squares, so that there are four interior nodes. Because of symmetry, all of the nodes will have equal ϕ-values.
   b. Repeat but subdivide into 36 equal squares with 25 interior nodes. Use the results of part (a) to get starting values for iteration.

57. Solve
$$\nabla^2 u = 2 + x^2 + y^2$$
over a hollow square bar, 5 in. in outside dimension and with walls 2-in. thick (so that the inner square hole is 1 in. on a side). The origin for x and y is the center of the object. On the inner and outer surfaces, $u = 0$.

58. Solve
$$\nabla^2 u = f(x, y)$$
with $f(x, y) = xy$. Use $h = \frac{1}{3}$. The values of u on the boundary are everywhere zero. The region is a square with corners at $(0, 0)$ and $(2, 2)$.

59. Repeat part (b) of Exercise 56, but now use overrelaxation. Find the optimum overrelaxation factor experimentally. Does this match the value from Eq. (7.15)?

▶ 60. Repeat Exercise 57, but this time use overrelaxation. Vary the overrelaxation factor until you find the optimum. Compare to the value from Eq. (7.15).

61. Solve for the steady-state temperatures in the region of Exercise 53, except now the plate is insulated along the edge at each point marked zero. All other edge temperatures are held at the values as marked.

62. Solve a modification of Example 7.14, where along all edges there is an outward gradient of −15°C/cm. Is it possible to get a unique solution?

63. Solve Example 7.14 except that the left edge is insulated so that no heat is gained or lost.

64. Solve Example 7.14 except that the top edge is held at 20°C and along the right edge $\partial u/\partial x = 15°C/cm$.

Section 7.8

▶ **65.** Solve Exercise 50 by the A.D.I. method, using $\rho = 1.0$. Begin with the initial u-values equal to the arithmetic average of the boundary temperatures. Compare the number of iterations needed to those required with Liebmann's method (Exercise 54) and S.O.R. using the optimum overrelaxation factor (Exercise 55).

66. Repeat Exercise 56, but use the A.D.I. method. Vary the value of ρ to find the optimum value.

67. A cube is 7 cm along each edge. Two opposite faces are held at 100°, the other four faces are held at 0°. Find the interior temperatures at the nodes of a 1-cm network. Use the A.D.I. method

68. Repeat Exercise 67, but now the two opposite edges have a mixed condition: the outward normal gradient equals $0.23(u - 20)$, where u is the surface temperature.

Section 7.9

▶ **69.** Construct a pictorial operator for Figure 7.8 (Eq. (7.20)).

70. A coaxial cable has a circular outer conductor 10 cm in diameter and a concentric inner square conductor that is 3 cm on a side. The outer conductor is at zero volts, the inner is at 100 volts. Find the potential at points in the space between. Compare the results when a grid of 1-cm squares is used, first when the "irregular star" is used at the outer conductor, then when the "distorted boundary" technique is employed.

71. A hollow shaft has an outer diameter of 12 cm and an inner concentric hole that is 3 cm in diameter. What is the value of the torsion function at nodes within the material? (Space the nodes so there are at least 12 of them.) Use whichever of the three techniques seems most appropriate.

72. Repeat Exercise 71, but with the inner hole a square that is 3 cm on a side.

73. Use the Laplacian in polar coordinates to set up the equations to solve

$$\nabla^2 u = x^2 y$$

on a semicircular region whose radius is 4. Take $\Delta r = 0.5$ and $\Delta\theta = \pi/8$ radians. The value of u is 10 in. along the straight edge and $u = 40$ on the curved boundary.

74. Solve Exercise 73, but now use a square mesh with $h = 0.5$. Compare the results with those of Exercise 73 for nodes where $x = 0$. Do this first with "irregular stars" at the curved boundary, then by distorting that boundary.

Section 7.10

75. Which of these boundary-value problems meet the conditions for a unique solution? For those that do not, find a special case for the boundary values for which a solution does exist.

a. $y'' = (x - 1)y' + (x + 1)y + (x^2 - 1)$,
  $y(-1) = -1$, $y(1) = 1$.
b. $y'' = (x + 1)y' + (x - 1)y + (x^2 + 1)$,
  $y(-1) = -1$, $y(1) = 1$.
c. $y'' = y' + q * y + x$,
$$q = \begin{cases} -1, & -1 < x < 0 \\ 0, & x = 0 \\ 1, & 0 < x < 1 \end{cases}$$
  $y(-1) = -1$, $y(1) = 1$.

▶ **76.** For each of the equations in Exercise 75, set up the difference equations to solve with $h = 0.5$. What is the determinant of each matrix?

77. The equation in Exercise 9 is nonlinear. Solve with three assumed values of $y'(-1)$, and then use Muller's method to get the next value.

78. Continue Exercise 77 until you converge on $y(1)$ within 0.0001. Are fewer trials needed than if you linearly interpolate from two values of $y'(-1)$?

79. Show that the power method does not converge if there are two eigenvalues of the same largest magnitude.

Section 7.10

▶ **80.** Matrix A is defined as

$$A = \begin{bmatrix} 4 & 3 & 2 \\ 2 & 3 & 4 \\ 2 & 4 & a \end{bmatrix}.$$

What is the smallest value for a in matrix A for which convergence will be obtained with

a. the Jacobi method?
b. the Gauss–Seidel method?

81. Let T be

$$T = \begin{bmatrix} L & 0 \\ x & L \end{bmatrix}.$$

Show that

$$T^n = \begin{bmatrix} L^n & 0 \\ nL^{n-1}x & L^n \end{bmatrix}.$$

From this infer that $T^n \to$ 0-matrix if $|L| < 1$.

82. Let matrices S and T be

$$S = \begin{bmatrix} Z_1 & 0 & 0 \\ a & Z_2 & 0 \\ b & c & Z_3 \end{bmatrix}, \qquad T = \begin{bmatrix} Z & 0 & 0 \\ a & Z & 0 \\ b & c & Z \end{bmatrix},$$

where $Z = \text{Max}\{|Z_1|, |Z_2|, |Z_3|\}$.
   Write $U = L + D$, where

$$L = \begin{bmatrix} 0 & 0 & 0 \\ a & 0 & 0 \\ b & c & 0 \end{bmatrix}, \qquad D = \begin{bmatrix} Z & 0 & 0 \\ 0 & Z & 0 \\ 0 & 0 & Z \end{bmatrix}.$$

a. Show that $L^3 = $ 0-matrix.
b. Because $LD = DL$, show that

$$T^n = (L + D)^n$$

$$= \binom{n}{n-2} L^2 D^{n-2} + \binom{n}{n-1} LD^{n-1} + D^n,$$

   from the binomial formula. From this, we see that $T^n \to$ 0-matrix and also S^n, if $Z < 1$.
c. Prove this for any general triangular matrix T where $|T_{i,i}| < 1$ for $i = 1, 2, \ldots, n$.

▶ 83. For each matrix, find the optimum overrelaxation factor by the technique of Section 7.10, then compare to the value predicted by Eq. (7.15) or experimentally.

a. $\begin{bmatrix} -4 & 1 \\ 1 & -4 \end{bmatrix}$ b. $\begin{bmatrix} 4 & 3 \\ 3 & 4 \end{bmatrix}$

c. $\begin{bmatrix} -4 & 1 & 0 \\ 1 & -4 & 1 \\ 0 & 1 & -4 \end{bmatrix}$

84. The system of equations given here (as an augmented matrix) can be solved by either Gauss–Jacobi or Gauss–Seidel. Both can be speeded by applying overrelaxation. Make trials with varying values of the factor to find the optimum value. (In this case you will probably find this to be less than unity, meaning it is underrelaxed.)

$$\begin{bmatrix} 8 & 1 & -1 & \vdots & 8 \\ 1 & -7 & 2 & \vdots & -4 \\ 2 & 1 & 9 & \vdots & 12 \end{bmatrix}$$

85. What are the eigenvalues of the iteration matrices of Exercise 84 when the optimum relaxation factor is applied?

In Exercises 86 and 87, the region is a square with four evenly spaced interior nodes. The boundaries are everywhere zero except $u = 10$ on the right side.

86. Demonstrate the maximum–minimum principle for Laplace's equation by solving the problem with $h = 2$,
a. starting with $u = 20$ at the interior nodes.
b. starting with $u = -10$ at the interior nodes.

87. The superposition principle works somewhat differently for numerical methods. Show that solutions of $u_{xx} + u_{yy} = 0$ and $u_{xx} + u_{yy} = f(x, y)$ differ by a value that depends on $f(x, y)$ and h.
a. Solve analytically.
b. Demonstrate by comparing the solutions of both equations for several values of h and several definitions of $f(x, y)$.

Section 7.11

88. Reproduce the values in Table 7.1 with the ode23 routine from MATLAB.

89. Repeat Exercise 88 but now use ode45. Are the results the same?

90. Solve Example 7.2 with MATLAB, using ode45. Paraphrase Table 7.2 in doing this.

In solving Exercises 91 to 96, use appropriate routines from MATLAB.

91. Solve the first set of equations of Example 7.3
a. through getting the inverse matrix.
b. through a "solve" routine.
c. by doing row-reduction.

▶ 92. Get the eigenvalues of this matrix:

$$\begin{bmatrix} 5 & 3 & 2 \\ -2 & 6 & 3 \\ 3 & 2 & 4 \end{bmatrix}$$

93. Add 0.1 to each element of the matrix of Exercise 92 and then get the eigenvalues. Compare the sum of these to the sum of those in Exercise 92.

94. Find the eigenvectors for the matrix of Exercise 92.

95. Find the characteristic polynomial of the matrix of Exercise 92.

96. Do similarity transformations on the matrix

$$\begin{bmatrix} 1 & 0 & 0 \\ 0 & 2 & 0 \\ 0 & 0 & 3 \end{bmatrix}$$

so that another matrix is obtained that has no nonzero elements. Then show that they have the same eigenvalues. Do they have the same eigenvectors?

Applied Problems and Projects

97. If a cantilever beam of length L, which bends due to a uniform load of w lb/ft, is also subject to an axial force P at its free end (see Fig. 7.16), the equation of its elastic curve is

$$EI\frac{dy^2}{dx^2} = Py - \frac{wx^2}{2}.$$

For this equation, the origin O has been taken at the free end. I is the moment of inertia; here $I = bh^3/12$. At the point $x = L$, $dy/dx = 0$; at $(0, 0)$, $y = 0$. Solve this boundary-value problem by the shooting method for a 2 in. $\times$ 4 in. $\times$ 10-ft wooden beam for which $E = 12 \times 10^5$ lb/in². Find y versus x when the beam has the 4-in. dimension vertical with $w = 25$ lb/ft and a tension force of $P = 500$ lb. Also solve for the deflections if the beam is turned so that the 4-in. dimension is horizontal.

98. Solve Problem 97 by replacing the differential equation by difference equations. Compare the solutions by each method.

99. In Problem 97 y'' is used although the radius of curvature should be employed in the equation:

$$EI\frac{y''}{[1 + (y')^2]^{3/2}} = Py - \frac{wx^2}{2}.$$

If the deflection of the beam is small, the difference is negligible, but in the second part of Problem 97 at least, this is not true. Furthermore, if there is considerable bending of the beam, the horizontal distance from the origin to the wall is less than L, the original length of the beam. Solve Problem 97 taking these factors into account, and determine by how much the deflections differ from those previously calculated.

100. A cylindrical pipe has a hot fluid flowing through it. Because the pressure is very high, the walls of the pipe are thick. For such a situation, the differential equation that relates temperatures in the metal wall to radial distance is

$$r\frac{d^2u}{dr^2} + \frac{du}{dr} = 0,$$

where

$$r = \text{radial distance from the centerline,}$$

$$u = \text{temperature.}$$

Solve for the temperatures within a pipe whose inner radius is 1 cm and whose outer radius is 2 cm if the fluid is at 540°C and the temperature of the outer circumference is 20°C.

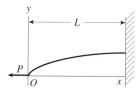

Figure 7.16

101. The pipe in Problem 100 is insulated to reduce the heat loss. The insulation used has properties such that the gradient du/dr at the outer circumference is proportional to the difference in temperatures from the outer wall to the surroundings:

$$\left. \frac{du}{dr} \right|_{r=2} = 0.083[u(2) - 20].$$

Solve Problem 100 with this boundary condition.

102. A simple spring–mass system obeys the equation

$$\frac{d^2y}{dt^2} + \alpha^2 y = 0,$$

where the positive constant α^2 equals k/m, the ratio of the spring constant to the mass. If it is known that $y = 0$ at $t = 0$ and again at $t = 1.26$ sec (so that its period is 2.52 sec), this is a characteristic-value problem that has a solution for only certain values of α. These character-istic values are discussed in Section 7.5 for this equation. Suppose, however, that the spring is not an ideal one with constant k, but that the spring force and elongation vary with y according to the equation $k = b(1 - y^{0.1})$. If it is still true that $y = 0$ at $t = 0$ and also at $t = 1.26$, is this a characteristic-value problem requiring that the ratio k/m be at certain fixed values in order to have a nontrivial solution? If it is, find these values.

103. A classic problem in elliptic partial-differential equations is to solve $\nabla^2 u = 0$ on a region defined by $0 \le x \le \pi$, $0 \le y \le \infty$, with boundary condition of $u = 0$ at $x = 0$, at $x = \pi$, and at $y = \infty$. The boundary at $y = 0$ is held at $u = F(x)$. This can be quite readily solved by the method of separation of variables, to give the series solution

$$u = \sum_{n=1}^{\infty} B_n e^{-ny} \sin nx,$$

with

$$B_n = 2 \int_0^\pi F(x)\sin nx \, dx.$$

Solve this equation numerically for various definitions of $F(x)$. (You will need to redefine the region so that $0 \le y \le M$, where M is large enough that changes in u with y at $y = M$ are negligible.) Compare your results to the series solution. You might try

$$F(x) = 100 \sin(x); \qquad F(x) = 4x(\pi - x)/\pi^2; \qquad F(x) = 100(\pi - |2x - \pi|).$$

104. The equation

$$2\frac{\partial^2 u}{\partial x^2} + \frac{\partial^2 u}{\partial y^2} - \frac{\partial u}{\partial x} = 2$$

is an elliptic equation. Solve it on the unit square, subject to $u = 0$ on the boundaries. Approxi-mate the first derivative by a central-difference approximation. Investigate the effect of size of Δx on the results, to determine at what size reducing it does not have further effect.

105. A second-order boundary-value problem can have one or both boundary conditions specified at infinity. (Of course, the solution must approach zero very rapidly as the boundary at infinity is approached). We have not discussed this. Research ways to handle such problems. (*Hint:* Acton (1970) is a place to start). Apply these techniques to this problem:

$$y'' = 2y^3, \qquad y(1) = 1, \qquad y(\infty) = 0.$$

106. a. Section 7.5 described getting the eigenvalues of a matrix by the QR method and emphasized that it is computationally expensive unless the matrix is first reduced to upper Hessenberg by similarity transformations. An alternative to this first step is to employ similarity transformations to produce a tridiagonal matrix. Do research to find methods that do this and test them with the matrix of Example 7.13.

   b. Suppose that a matrix is symmetric as is true in many practical situations. Can you take advantage of this in getting its eigenvalues? If it is both symmetric and banded, is this an advantage?

107. a. If you write out the equations for Example 7.14, you will find that the coefficient matrix is symmetric and banded. How can you take advantage of this in solving the equations by Gaussian elimination? Would Gauss–Jordan be preferred?

   b. Suppose the nodes in Example 7.14 were numbered by columns instead of by rows. Is the coefficient matrix still symmetric and banded?

108. If we want to improve the accuracy of the solution to Example 7.17, there are several alternative strategies, including

   a. Recompute with nodes more closely spaced but still in a uniform grid.
   b. Use a higher-order approximation, such as Eq. (7.13).
   c. Add additional nodes only near the right and left sides because the gradient is large there (see Table 7.7) and errors will be greater.

   Discuss the pros and cons of each of these choices. Be sure to consider how boundary conditions will be handled. In part (c), how should equations be written where the nodal spacing changes?

109. Investigate the effect of different values for ρ in Example 7.18. What value is optimal? How do the starting values that are used affect this?

8

Parabolic and Hyperbolic Partial-Differential Equations

Contents of This Chapter

Chapter 7 discussed the solution of partial-differential equations that are time-independent. Such steady-state problems are described by a kind of partial-differential equations that are called *elliptic*. This chapter describes the solution of partial-differential equations that are time-dependent. Such unsteady-state problems are described by partial-differential equations that are called *parabolic* or *hyperbolic*.

The use of finite-difference approximations to replace the derivatives has been the customary way to solve these latter two types of problems and it is the way that this chapter will solve the problems. The finite-difference approach is similar to what was used for parabolic equations, but there is now concern for stability and convergence. A more modern procedure is the finite-element method, but we defer a discussion of that method until the next chapter.

Available commercial software to solve the problems discussed in this chapter include PDECOL, distributed by IMSL, which also solves boundary-value problems.

This chapter differs from the others in that we do not discuss the application of computer algebra systems because the procedures here are so similar to those of Chapter 7 that nothing new is involved.

8.1 Types of Partial-Differential Equations

Explains how partial-differential equations are classified into three types: elliptic, parabolic, and hyperbolic. These names suggest a relationship to the equations for conic sections and it is true that the criterion is similar to the discriminant of a second-order algebraic equation. When there are two or three dimensions for the object under consideration, the Laplacian is seen in the equations.

8.2 The Heat Equation and the Wave Equation

Derives the equations for two typical instances where parabolic and hyperbolic partial-differential equations are involved. The first of these finds how temperatures change with time within a rod, beginning from some initial state and proceeding toward the steady state as time progresses. The second (hyperbolic) situation uses a vibrating string as the model and finds how displacements of points along the string vary with time, again beginning from a known initial state. The vibrating string (in the absence of forces that damp the motion) does not reach a steady state but continues to vibrate.

The names *heat equation* and *wave equation* are given to the equations that govern these two examples. Both of these simpler situations are extended to time-dependent problems where there are two or three spatial dimensions. Equations of the same form also apply to many other important real-world applications, not just to flow of heat and vibrations.

8.3 Solution Techniques for the Heat Equation in One Dimension

Discusses three solution methods for parabolic partial-differential equations. All three use finite-difference approximations to replace the derivatives in the equation. The first of these, called the *explicit method,* must limit the value of a parameter in the equations to avoid instability. The second, the *Crank–Nicolson method,* has no such limitation. The third, the *theta method* is a generalization of the Crank–Nicolson technique.

8.4 Solving the Vibrating String Problem

Shows how the displacements of a string vary with time from a given initial position. The method replaces the derivatives in the partial-differential equation with finite-difference approximations. It is found that the solution can exactly match the analytical solution; this is given by a special technique called the *D'Alembert solution.* When the initial velocities are nonzero, integration should be used to get the displacements at the end of the first time step.

8.5 Parabolic Equations in Two or Three Dimensions

Extends the procedures described in Section 8.3 to problems that involve two or three spatial variables. The coefficient matrix for the equations is tridiagonal if an adaptation of the A.D.I. method is used. (This technique was introduced in Chapter 7.) The formulation of the partial-differential equation in polar or spherical coordinates is briefly discussed; this applies when the region is circular or spherical.

8.6 The Wave Equation in Two Dimensions

Shows that solving the problem of a vibrating membrane through using finite-difference approximations to the derivatives does not lead to a solution that is close to the analytical.

8.7 Theoretical Matters

Discusses in some detail the questions of stability and convergence when parabolic equations are solved numerically. It demonstrates that the numerical solution to the vibrating string is stable if the values of Δt and Δx are restricted.

Chapter Summary

Reviews the topics of the chapter in the style of previous chapters.

Computer Programs

Provides examples of computer programs for the methods of the chapter.

Parallel Processing

Chapter 8 is similar to Chapter 7 in that sets of equations must be solved. In the present chapter, these solutions are repeated many times, so the discussion in Chapter 2 is particularly applicable.

8.1   Types of Partial-Differential Equations

Second-order partial-differential equations are classified into three categories depending on the values of the coefficients in this general formulation:

$$A\,\frac{\partial^2 u}{\partial x^2} + B\,\frac{\partial^2 u}{\partial x \partial y} + C\,\frac{\partial^2 u}{\partial y^2} + f = 0, \tag{8.1}$$

where f can be a function of x, y, u, $\partial u/\partial x$, and $\partial u/\partial y$.

The value of $B^2 - 4AC$ determines the type:

$$
\begin{aligned}
&\text{If } B^2 - 4AC < 0, &&\text{it is called } \textit{elliptic,}\\
&\text{If } B^2 - 4AC = 0, &&\text{it is called } \textit{parabolic,}\\
&\text{If } B^2 - 4AC > 0, &&\text{it is called } \textit{hyperbolic.}
\end{aligned}
$$

The terms are the same as for the classification of second-degree polynomial equations based on which of the three types of conic sections they represent.

The partial-differential equations discussed in Chapter 7 were all of the form

$$\frac{\partial^2 u}{\partial x^2} + \frac{\partial^2 u}{\partial y^2} + f\left(x,\ y,\ u,\ \frac{\partial u}{\partial x},\ \frac{\partial u}{\partial y}\right) = 0.$$

Their coefficients had these values: $A = 1$, $B = 0$, and $C = 1$, so they were elliptic equations. In this chapter we discuss the solution of equations of the other two types. Here we will deal with situations where (for second-order equations) the independent variables are a distance, x, and time, t. The general formulation with x and t as variables is then

$$\frac{A\partial^2 u}{\partial x^2} + \frac{B\partial^2 u}{\partial x \partial t} + \frac{C\partial^2 u}{\partial t^2} + f\left(x,\ t,\ u,\ \frac{\partial u}{\partial x},\ \frac{\partial u}{\partial t}\right) = 0.$$

In the next section, we will develop the *heat equation,*

$$\frac{\partial^2 u}{\partial x^2} - \frac{c\rho}{k} * \frac{\partial u}{\partial t} = 0, \tag{8.2}$$

in which $A = 1$, $B = 0$, and $C = 0$, so that $B^2 - 4AC = 0$ and the equation is parabolic. When applied to the molecular diffusion of matter, Eq. (8.2) becomes

$$\frac{\partial^2 u}{\partial x^2} - \frac{1}{D} * \frac{\partial u}{\partial t} = 0,$$

in which the quantity D is the *diffusion coefficient.* Because the quantity $k/c\rho$ appears in Eq. (8.2) exactly as the quantity D does in the diffusion equation, that fraction is called the *thermal diffusivity.* The two quantities have precisely the same units.

Another equation that we will develop is the *wave equation,*

$$\frac{\partial^2 u}{\partial x^2} - \frac{Tg}{w} \frac{\partial^2 u}{\partial t^2} = 0, \tag{8.3}$$

in which $A = 1$, $B = 0$, and $C = -(Tg/w)$ (and T, g, and w are always positive numbers), so that the quantity $B^2 - 4AC > 0$ and the equation is hyperbolic.

Both of these equations can be applied where the region has two or three dimensions. In that case, Eqs. (8.2) and (8.3) become

$$k\nabla^2 u = (c\rho)\frac{\partial u}{\partial t} \qquad \text{and} \qquad \nabla^2 u = \frac{Tg}{w}\frac{\partial^2 u}{\partial t^2},$$

where the Laplacian represents the sum of the second partials with respect to the space variables.

8.2   The Heat Equation and the Wave Equation

The equations that govern flow of heat within an object as well as the equations for vibrations of a string or a membrane will be developed in this section. We begin with flow of heat. What we are calling the heat equation is often called the diffusion equation because it also describes the molecular diffusion of matter. We prefer to use heat flow for the discussion because that phenomenon is more familiar to most people.

To see how the temperatures change with time along a rod, consider the flow of heat within a rod of length L with boundary conditions specified at each end, at $x = 0$ and at $x = L$. Initially the temperatures are given as some function of x at points along the rod. Call this initial temperature distribution $f(x)$.

In the previous chapter we computed the temperatures after equilibrium is reached, the steady-state temperatures. Now we desire to see how the temperatures change from the initial state to that final steady state as a function of time. Figure 8.1 shows a portion of the rod that includes an element of the rod of length dx, located at a distance x from the

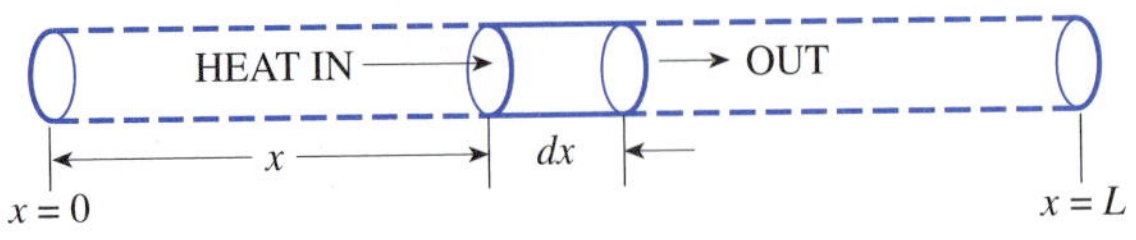

Figure 8.1

left end. As described in Chapter 7, heat flows into the element from the left at a rate, measured in cal/sec, of

$$-kA\left(\frac{du}{dx}\right).$$

The minus sign is required because du/dx expresses how rapidly temperatures increase with x, whereas the heat always flows from high temperature to low.

The rate at which heat leaves the element is given by a similar equation, but now the temperature gradient must be at the point $x + dx$:

$$-kA\left[\frac{du}{dx} + \frac{d}{dx}\left(\frac{du}{dx}\right)dx\right]$$

in which the gradient term is the gradient at x plus the change in the gradient between x and $x + dx$.

These two relations are precisely those of Section 7.1. Now, however, we do not assume that these two rates are equal, but that their difference is the rate at which heat is stored within the element. This heat that is stored within the element raises its temperature. The rate of increase in the amount of heat that is stored is related to the rate of change in temperature of the element by an equation that involves the volume of the element ($A * dx$, measured in cm^3), the density of the material (ρ, measured in cal/gm), and a property of the material called the heat capacity, (c, measured in cal/(gm $* \,^\circ$C)):

$$\text{rate of increase of heat stored} = c\rho(A\ dx)\frac{du}{dt}.$$

We equate this increase in the rate of heat storage to the difference between the rates at which heat enters and leaves:

$$-kA\left(\frac{\partial u}{\partial x}\right) - \left(-kA\left[\frac{\partial u}{\partial x} + \frac{\partial}{\partial x}\left(\frac{\partial u}{\partial x}\right)dx\right]\right) = c\rho(A\ dx)\frac{\partial u}{\partial t}, \tag{8.4}$$

where the derivatives are now partial derivatives because there are two independent variables, x and t. We can simplify Eq. (8.4) to

$$k\left(\frac{\partial^2 u}{\partial x^2}\right) = c\rho\frac{\partial u}{\partial t}, \tag{8.5}$$

which is the same as Eq. (8.2) but rearranged.

If the region is a slab or a three-dimensional object, we have the analogous equation

$$k \; \nabla^2 u \; = \; c\rho \frac{\partial u}{\partial t},$$ (8.6)

in which the Laplacian appears.

It may be that the material is not homogenous and its thermal properties may vary with position. Also, there could be heat generation within the element equal to Q cal/(sec * sec * cm^3). In this more general case we have, in three dimensions,

$$\frac{\partial}{\partial x}\left[(k(x, \; y, \; z)\frac{\partial u}{\partial x}\right] \; + \; \frac{\partial}{\partial y}\left[(k(x, \; y, \; z)\frac{\partial u}{\partial y}\right] \; + \; \frac{\partial}{\partial z}\left[(k(x, \; y, \; z)\frac{\partial u}{\partial z}\right] \; + \; Q(x, \; y, \; z)$$

$$= \; c(x, \; y, \; z) * \rho(x, \; y, \; z)\frac{\partial u}{\partial t}.$$

Our illustrations will stay with the simpler cases represented by Eqs. (8.5) and (8.6).

In order to solve these equations for unsteady-state heat flow (and they apply as well to diffusion or to any problem where the potential is proportional to the gradient), we need to make the solution agree with specified conditions along the boundary of the region of interest. In addition, because the problems are time-dependent, we must begin with specified initial conditions (at $t = 0$) at all points within the region. We might think of these problems as both boundary-value problems with respect to the space variables and as initial-value problems with respect to time.

The Vibrating String

We can develop the one-dimensional wave equation, an example of hyperbolic partial-differential equations, by considering the oscillations of a taut string stretched between two fixed endpoints. Figure 8.2 shows the string with displacements from the straight line between the endpoints greatly exaggerated. The figure shows an element of the string of length dx between points A and B. We use u for the displacements, measured perpendicularly from the straight line between the ends of the string. We focus our attention on the element of the string in Figure 8.2. It is shown enlarged in Figure 8.3, which also shows the angles, α_A and α_B, between the ends of element and the horizontal. (The bending of

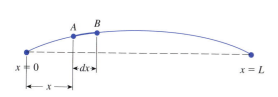

Figure 8.2

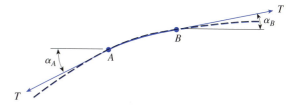

Figure 8.3

the element between points A and B is exaggerated as are the displacements.) The figure also indicates that the tension in the stretched string is a force T. Taking the upward direction as positive, we can write, for the upward forces at each end of the element (these are the vertical components of the tensions),

Upward force at Point $A = -T \sin(\alpha_A)$,

Upward force at Point $B = T \sin(\alpha_B)$.

Remembering that Figure 8.3 has displacements and angles greatly exaggerated, the tangents of these angles are essentially equal to the sines. We then can write

$$\text{Upward force at Point } A = -T \tan(\alpha_A) = -T\left(\frac{\partial u}{\partial x}\right)_A,$$

$$\text{Upward force at Point } B = T \tan(\alpha_B) = T\left(\frac{\partial u}{\partial x}\right)_B = T\left[\left(\frac{\partial u}{\partial x}\right)_A + \frac{\partial}{\partial x}\left(\frac{\partial u}{\partial x}\right)dx\right].$$

The net force acting on the element then is

$$T\left(\frac{\partial^2 u}{\partial x^2}\right)dx.$$

Now, using Newton's law, we equate the force to mass $\times$ acceleration (in the vertical direction). Our simplifying assumptions permit us to use $w\,dx$ as the weight (w is the weight per unit length), so

$$\left(T\frac{\partial^2 u}{\partial x^2}\right)dx = \left(w\frac{dx}{g}\right)\left(\frac{\partial^2 u}{\partial t^2}\right) \quad \text{or} \quad \frac{\partial^2 u}{\partial t^2} = \left(\frac{Tg}{w}\right)\frac{\partial^2 u}{\partial x^2}. \tag{8.7}$$

As pointed out in Section 8.1, when Eq. (8.7) is compared to the general form of second-order partial-differential equation, we see that $A = 1$, $B = 0$, and $C = -Tg/w$, and so this falls in the class of hyperbolic equations.

If we have a stretched membrane (like a drum head) instead of a string, the governing equation is

$$\frac{\partial^2 u}{\partial t^2} = \left(\frac{Tg}{w}\right)\nabla^2 u. \tag{8.8}$$

The solution to Eq. (8.7) or Eq. (8.8) must satisfy given boundary conditions along the boundary of the region of interest as well as given initial conditions at $t = 0$. Because the problem is of second order with respect to t, these initial conditions must include both the initial velocity and the initial displacements at all points within the region.

8.3   Solution Techniques for the Heat Equation in One Dimension

In this section we discuss three methods for solving for the temperatures as they vary with time at points along a rod whose initial temperature is given and whose ends satisfy given boundary conditions. In all three techniques, we will replace the derivatives in the equation with finite-difference quotients just as was done in Chapter 7. The names given to these methods (which are all just variations on the same procedure) are (1) the *explicit method,* (2) the *Crank–Nicolson method,* and (3) the *theta method.*

All three of these techniques work with sets of equations that are obtained by replacing the derivatives in Eq. (8.5) with finite-difference quotients. Consider the heat equation that was derived in Section 8.2, which we restate as:

$$k\left(\frac{\partial^2 u}{\partial x^2}\right) = c\rho \frac{\partial u}{\partial t}. \tag{8.9}$$

The equation applies to a one-dimensional region (a rod of length L) whose initial temperature is specified and whose ends are at given boundary conditions. We shall subdivide the rod into equal subintervals of length Δx and use equal subintervals of time of duration Δt.

The Explicit Method

In the explicit method we replace the time derivative with this approximation:

$$\frac{\partial u}{\partial t} = \frac{u_i^{j+1} - u_i^j}{\Delta t} \quad \text{(at point } x_i \text{ and time } t_j), \tag{8.10}$$

where we use subscripts to indicate the location and superscripts to indicate the time.* For the derivative with respect to x, we use (at point x_i and time t_j):

$$\frac{\partial^2 u}{\partial x^2} = \frac{u_{i+1}^j - 2u_i^j + u_{i-1}^j}{(\Delta x)^2}. \tag{8.11}$$

Observe that we are using a forward difference in Eq. (8.10) but a central difference in Eq. (8.11). From the discussion in Chapter 3, we know that the first has an error of order $O(\Delta t)$ whereas the second has an error of order $O(\Delta x)^2$. This difference in orders has an important consequence as will be seen.

Substituting these approximations into Eq. (8.9) and solving for u_i^{j+1}, we get

$$u_i^{j+1} = r * (u_{i+1}^j + u_{i-1}^j) + (1 - 2r) * u_i^j, \tag{8.12}$$

where

$$r = \frac{k\,\Delta t}{c\rho(\Delta x)^2}.$$

Equation (8.12) is a way that we can march through time one Δt at a time. For $t = t_1$, we have the u's at t_0 from the initial conditions. At each subsequent time interval, we have the values for the previous time from the last computations. We apply the equation at each point along the rod where the temperature is unknown. (If an end-condition involves a temperature gradient, that endpoint is included.)

_____

*The x_i are locations of evenly spaced nodes. The t_j are times spaced apart by Δt.

The use of Eq. (8.12) to compute temperatures as a function of position and time is called the *explicit method* because each subsequent computation is explicitly given from the previous *u*-values.

An example will clarify the procedure.

EXAMPLE 8.1   Solve for the temperatures as a function of time within a large steel plate that is 2-cm thick. For steel, $k = 0.13$ cal/(sec $*$ cm $*$ °C), $c = 0.11$ cal/g $*$ °C), and $\rho = 7.8$ g/cm^3. Because the plate is large, neglect lateral flow of heat and consider only the flow perpendicular to the faces of the plate.

Initially, the temperatures within the plate, measured from the top face (where $x = 0$) to the bottom (where $x = 2$) are given by this relation:

$$u(x) = 100x, \quad 0 \le x \le 1; \quad u(x) = 200 - 100x, \quad 1 \le x \le 2.$$

The boundary conditions, both at $x = 0$ and at $x = 2$, are $u = 0°$. Use $\Delta x = 0.25$ so there are eight subdivisions. Number the interior nodes from 1 to 7 so that node 0 is on the top face and node 8 is at the bottom.

The value that we use for Δt depends on the value that we choose for r, the ratio $(k \Delta t)/[c\rho(\Delta x)^2]$. Let us use $r = 0.5$ for a first trial. Doing so greatly simplifies Eq. (8.12). It becomes:

$$u_i^{j+1} = 0.5(u_{i+1}^j + u_{i-1}^j). \tag{8.13}$$

(We shall compare the results of this first trial to other trials with different values for r.) With $r = 0.5$, the value of Δt is $rc\rho(\Delta x)^2/k = 0.5(0.11)(7.8)(0.25)^2/0.13 = 0.206$ sec.

We use Eq. (8.13) to compute temperatures at each node for several time steps. When this is done, the results shown in Table 8.1 are obtained. Because the values are symmetrical

Table 8.1   Computed and analytical temperatures for Example 8.1

Time steps	t	0.25 (computed)	0.50 (comp)	0.50 (anal)	0.75 (computed)	1.00 (comp)	1.00 (anal)
0	0	25.00	50.00	50.00	75.00	100.00	100.00
1	0.206	25.00	50.00	49.58	75.00	75.00	80.06
2	0.413	25.00	50.00	47.49	62.50	75.00	71.80
3	0.619	25.00	43.75	44.68	62.50	62.50	65.46
4	0.825	21.88	43.75	41.71	53.13	62.50	60.11
5	1.031	21.88	37.50	38.79	53.13	53.13	55.42
6	1.237	18.75	37.50	35.99	45.31	53.13	51.18
7	1.444	18.75	32.03	33.37	45.31	45.31	47.33
8	1.650	16.02	32.03	30.91	38.67	45.31	43.79
9	1.856	16.02	27.34	28.63	38.67	38.67	40.52
10	2.062	13.67	27.34	26.51	33.01	38.67	37.51
11	2.269	13.67	23.34	24.55	33.01	33.01	34.72
12	2.475	11.67	23.34	22.73	28.17	33.01	32.15
13	2.681	11.67	19.92	21.04	28.17	28.17	29.76
14	2.887	9.96	19.92	19.48	24.05	28.17	27.55

The column group header is "x value" spanning the four x-value columns (0.25, 0.50, 0.75, 1.00).

about the center of the rod, only those for the left half are tabulated, and the values for $x = 0$, which are all $u = 0$, are omitted. Table 8.1 also shows values from the "analytical" solution at $x = 0.5$ and at $x = 1$ from the series solution given by a classical method for solving the problem.   ▲

It is apparent from the conditions for this example that the temperatures will eventually reach the steady-state temperatures; at $t = \infty$, u will be $0°$ everywhere. The values in Table 8.1 are certainly approaching this equilibrium temperature. (All temperatures are within 0.1 of 0.0 after 85 time steps.)

The computed values generally follow the analytical but oscillate above and below successive values. This is shown more clearly in Fig. 8.4, where the computed temperatures at the center node and at $x = 0.5$ cm are plotted. The curves represent the analytical solution. If the computations are repeated but with two other values of r ($r = 0.4$ and $r = 0.6$) we find an interesting phenomenon. Of course, the values of Δt will change as well. With the smaller value for r, 0.4, the computed results are much more accurate, and the differences from the analytical values are about half as great during the early time steps and become only one-tenth as great after ten time steps. We would expect somewhat better agreement because the time steps are smaller, but the improvement is much greater than this change would cause.

On the other hand, using a value of 0.6 for r results in extremely large errors. In fact, after only eight time steps, some of the calculated values for u are negative, a patently impossible result. Figure 8.5 illustrates this quite vividly. The open circles in the figure are

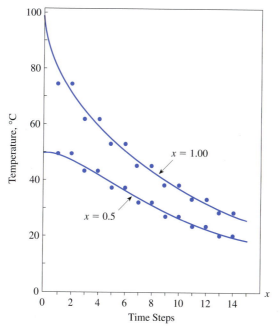

Figure 8.4

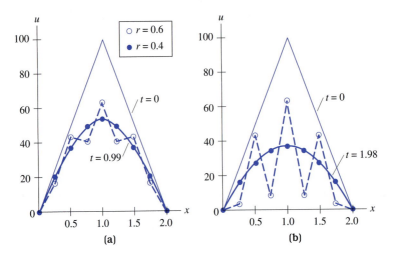

Figure 8.5

results with $r = 0.6$; the solid points are for $r = 0.4$. As discussed in Section 8.7, "Theoretical Matters," the explanation for this behavior is *instability*. The maximum value for r to avoid instability (which is particularly evident for $r = 0.6$) is $r = 0.5$. The oscillation of points about the analytical curve in Fig. 8.4 shows incipient instability. Even this value for r is too large if the boundary conditions involve a gradient.

The Crank–Nicholson Method

The reason why there is instability when r is greater than 0.5 in the explicit method is the difference in orders of the finite-difference approximations for the spatial derivative and the time derivative. The *Crank–Nicolson method* is a technique that makes these finite-difference approximations of the same order.

The difference quotient for the time derivative, $(u_i^{j+1} - u_i^j)/\Delta t$, can be considered a central-difference approximation at the midpoint of the time step. If we do take this as a central-difference approximation, we will need to equate it to a central-difference approximation of the spatial derivative at the same halfway point in the time step, and this we can hope to obtain by averaging two approximations for $\partial^2 u/\partial x^2$, one computed at the start and the other at the end of the time step. So, we write, for

$$\frac{\partial u}{\partial t} = \frac{k}{c\rho} \frac{\partial^2 u}{\partial x^2},$$

this approximation:

$$\frac{u_i^{j+1} - u_i^j}{\Delta t} = \frac{1}{2} \frac{k}{c\rho} \left[\frac{u_{i+1}^j - 2u_i^j + u_{i-1}^j}{(\Delta x)^2} + \frac{u_{i+1}^{j+1} - 2u_i^{j+1} + u_{i-1}^{j+1}}{(\Delta x)^2} \right],$$

which we solve for the u-values at the end of the time step to give

$$-ru_{i+1}^{j+1} + (2 + r)u_i^{j+1} - ru_{i-1}^{j+1} = ru_{i+1}^j + (2 - r)u_i^j + ru_{i-1}^j,$$

$$r = \frac{k\Delta t}{c\rho(\Delta x)^2}.$$ (8.14)

Equation (8.14) is the Crank–Nicolson formula, and using it involves solving a set of simultaneous equations because the equation for u_i^{j+1} includes two adjacent u-values at $t = t^{j+1}$. Hence this is an *implicit method*. Fortunately, the coefficient matrix is tridiagonal. A most important advantage of the method is that it is stable for any value for r, although smaller values usually give better accuracy. This next example illustrates the method.

EXAMPLE 8.2   Solve Example 8.1 but now use the Crank–Nicolson method. Compare the results with $r = 0.5$ and with $r = 1.0$ to the analytical values.

Employing Eq. (8.14) gives the results shown in Table 8.2 for the centerline temperatures with $r = 0.5$ and in Table 8.3 for the centerline temperatures with $r = 1.0$. The error columns are the differences between the computed temperatures and those from the series solution. In Table 8.2, these range from 2.0% to 2.7% of the analytical values whereas in Table 8.3, they range from 1.0% to 2.5%. One would expect the errors with $r = 0.5$ to be smaller but this is not the case. Both sets of computations are more accurate than those in Table 8.1, where the explicit method was used with $r = 0.5$.

Table 8.2   Centerline temperatures with Crank–Nicolson Method, $r = 0.5$

Time steps	t	u-values	Error
0	0	100.00	—
1	0.206	82.32	2.26
2	0.413	73.48	1.68
3	0.619	66.86	1.40
4	0.825	61.34	1.23
5	1.031	56.52	1.10
6	1.237	52.21	1.03
7	1.444	48.30	0.97
8	1.650	44.71	0.92
9	1.856	41.40	0.88
10	2.062	38.36	0.85

Table 8.3   Centerline temperatures with Crank–Nicolson Method, $r = 1.0$

Time steps	t	u-values	Error
0	0	100.00	—
1	0.413	71.13	0.67
2	0.825	61.53	1.42
3	1.237	51.97	0.79
4	1.650	44.67	0.88
5	2.062	38.29	0.78
6	2.475	32.88	0.73
7	2.887	28.23	0.68
8	3.300	24.23	0.61

▲

The Theta Method—A Generalization

In the Crank–Nicolson method, we interpret the finite-difference approximation to the time derivative as a central difference at the midpoint of the time interval. In the *theta method* we make a more general statement by interpreting this approximation to apply at some other point within the time interval. If we interpret it to apply at a fraction θ of Δt, we then equate the time-derivative approximation to a weighted average of the spatial derivatives at the beginning and end of the time interval, giving this relation:

$$\frac{u_i^{j+1} - u_i^j}{\Delta t} = \left(\frac{k}{c\rho}\right)\left[\frac{(1-\theta)(u_{i+1}^j - 2u_i^j + u_{i-1}^j)}{(\Delta x)^2} + \frac{\theta(u_{i+1}^{j+1} - 2u_i^{j+1} + u_{i-1}^{j+1})}{(\Delta x)^2}\right].$$

Observe that using $\theta = 0.5$ gives the Crank–Nicolson method whereas using $\theta = 0$ gives the explicit method. If we use $\theta = 1$, the theta method is often called the *implicit method*. For $\theta = 1$, the analog of Eq. (8.14) is

$$-ru_{i+1}^{j+1} + (1+2r)u_i^{j+1} - ru_{i-1}^{j+1} = u_i^j, \qquad r = \frac{k\Delta t}{c\rho(\Delta x)^2}.$$

For any value of θ, the typical equation is

$$-ru_{i+1}^{j+1} + (1+2r\theta)u_i^{j+1} - ru_{i-1}^{j+1}$$
$$= r(1-\theta)u_{i+1}^j + [1 - 2r(1-\theta)]u_i^j + r(1-\theta)u_{i-1}^j. \quad \textbf{(8.15)}$$

What value is best for θ? Burnett (1987) suggests that $\theta = \frac{2}{3}$ is nearly optimum but he points out that a case can be made for using $\theta = 0.878$. This next example compares the use of these two values.

EXAMPLE 8.3   Solve Example 8.1 with $r = 0.5$ by the theta method with $\theta = \frac{2}{3}$, 0.878, and 1.0. Compare these to results from the Crank–Nicolson and explicit methods.

Using Eq. (8.15), computations were made for ten time steps. Table 8.4 shows how the values at the centerline, $x = 1.0$, differ from the analytical values. It is interesting to observe that, for this problem, the Crank–Nicolson results ($\theta = 0.5$) have smaller errors than those with larger values for θ. Even the results from the explicit method ($\theta = 0$) are better than

Table 8.4   Comparisons of results from the theta method, $r = 0.5$

Time steps	Errors in computed centerline temperatures θ-value				
	2/3	0.878	1.0	0.5	0.0
1	3.57	4.88	5.51	2.26	5.06
2	2.48	3.55	4.15	1.68	3.20
3	1.98	2.79	3.28	1.40	2.96
4	1.71	2.37	2.78	1.23	2.39
5	1.53	2.11	2.46	1.10	2.29
6	1.43	1.97	2.29	1.03	1.95
7	1.35	1.86	2.16	0.97	2.02
8	1.30	1.80	2.09	0.92	1.52
9	1.27	1.76	2.04	0.88	1.85
10	1.23	1.72	2.00	0.85	1.16

those with $\theta = 1.0$ (though the explicit values oscillate around the analytical). This suggests that there is an optimum value for θ less than $\frac{2}{3}$ and greater than zero. We leave this determination as an exercise, as well as the comparison at other values for x. We also leave as an exercise to find if there is an optimum value in other problems.   ▲

8.4   Solving the Vibrating String Problem

If a string is stretched tightly between two supports and displaced a small amount at some point on the string (like plucking a violin string with the finger), it will vibrate. The vibrations cause a sound to be produced as the air vibrates in unison with the string. As shown in Section 8.2, the partial-differential equation that governs the vibrating string problem is an example of hyperbolic differential equations. We repeat that equation

$$\frac{\partial^2 u}{\partial t^2} = \frac{Tg}{w}\frac{\partial^2 u}{\partial x^2}. \tag{8.16}$$

Observe that Eq. (8.16) is second order with respect to both x and t. To solve Eq. (8.16) we need boundary values (u-values and/or derivative values at the two supports) and two initial conditions (u-values and velocities, $\partial u/\partial t$, at $t = 0$).

We can solve this numerically by replacing the derivatives with finite-difference approximations, preferring to use central differences in both cases. If we do this, we get

$$\frac{u_{i+1}^j - 2u_i^j + u_{i-1}^j}{(\Delta x)^2} = \frac{Tg}{w}\frac{u_i^{j+1} - 2u_i^j + u_i^{j-1}}{(\Delta t)^2}$$

where the subscripts indicate x-values and the superscripts indicate t-values.* (If the boundary conditions involve derivatives, we will approximate them with central differences in the way that we are accustomed.) If we solve for the displacement at the end of the current time step, u_i^{j+1}, we get

$$u_i^{j+1} = \frac{Tg(\Delta t)^2}{w(\Delta x)^2}(u_{i+1}^j + u_{i-1}^j) - u_i^{j-1} + 2\left(1 - \frac{Tg(\Delta t)^2}{w(\Delta x)^2}\right)u_i^j.$$

If we make $Tg(\Delta t)^2/w(\Delta x)^2$ equal to 1, the maximum value that avoids instability, there is considerable simplification:

$$u_i^{j+1} = u_{i+1}^j + u_{i-1}^j - u_i^{j-1}, \qquad \Delta t = \frac{\Delta x}{\sqrt{(Tg/w)}}. \tag{8.17}$$

Equation (8.17) shows how one can march through time: To get the new value for u at node i, we add the two u-values last computed at nodes to the right and left and subtract the value at node i at the time step before that. That is fine for the second time step; we have the initial u-values (at $t = 0$) and those for step 1 (at $t = \Delta t$). We also have the necessary

---

*We again assume evenly spaced nodes and evenly spaced time intervals.

information for all subsequent computations. But how do we get the value for the first time step? We seem to need the values of u one time step before the start!

That really is no problem if we recognize that the oscillation of the vibrating string is a periodic function and that the "starting point" is just an arbitrary instant of time at which we happen to know the displacement and the velocity. That suggests that we can get the u-values at $t = -\Delta t$ from the specified initial velocities. If we use a central difference approximation:

$$\frac{u_i^1 - u_i^{-1}}{2\Delta t} \Delta t = \frac{\partial u}{\partial t} \qquad \text{at } x_i \text{ and } t = 0,$$

$\partial u / \partial t$ at $t = 0$ is known; it is one of the initial conditions, call it $g(x)$. So we can write

$$u_i^{-1} = u_i^1 - 2g(x) \, \Delta t. \tag{8.18}$$

If we substitute Eq. (8.18) into Eq. (8.17), we have (but for $t = 0$ only),

$$u_i^1 = \frac{1}{2}(u_{i+1}^0 + u_{i-1}^0) + g(x) \, \Delta t. \tag{8.19}$$

Our procedure then is to use Eq. (8.19) for the first time step, then use Eq. (8.17) to march on through time after that first step.* As we will see, Eq. (8.17) is not only stable but also can give exact answers. It is interesting that using a value for $Tg(\Delta t)^2/w(\Delta x)^2$ less than 1, while stable, gives results that are less accurate.

An example will illustrate the technique.

EXAMPLE 8.4   A banjo string is 80-cm long and weighs 1.0 gm. It is stretched with a tension of 40,000 g. At a point 20 cm from one end it is pulled 0.6 cm from the equilibrium position and then released. Find the displacements along the string as a function of time. Use $\Delta x = 10$ cm. How long does it take to complete one cycle of motion? From this, compute the frequency of the vibrations.

If Eq. (8.19) is used to begin the calculations and Eq. (8.17) thereafter, the results are as shown in Table 8.5. The initial velocities are zero because the string is just released after being displaced. Observe that the displacements are reproduced every 16 time steps.   ▲

Figure 8.6 illustrates how the displacements change with time; it also shows that, after 16 Δt's, the original u-values are reproduced, which will be true for every 16 time steps. Because the original displacements are reproduced every 16 time steps, we can compute the frequency of the vibrations. Each time step is

$$\Delta t = \sqrt{\frac{w}{Tg}} * \Delta x = \sqrt{\frac{1/80}{40000 * 980}} * 10 = 0.000179 \text{ sec,}$$

---

*There is a more accurate way to start the computations that we discuss a little later.

Table 8.5 Results for vibrating string example

Time steps	\multicolumn{9}{c}{u-values at x =}
	0	10	20	30	40	50	60	70	80
0	0.00	0.30	0.60	0.50	0.40	0.30	0.20	0.10	0.00
1	0.00	0.30	0.40	0.50	0.40	0.30	0.20	0.10	0.00
2	0.00	0.10	0.20	0.30	0.40	0.30	0.20	0.10	0.00
3	0.00	−0.10	0.00	0.10	0.20	0.30	0.20	0.10	0.00
4	0.00	−0.10	−0.20	−0.10	0.00	0.10	0.20	0.10	0.00
5	0.00	−0.10	−0.20	−0.30	−0.20	−0.10	0.00	0.10	0.00
6	0.00	−0.10	−0.20	−0.30	−0.40	−0.30	−0.20	−0.10	0.00
7	0.00	−0.10	−0.20	−0.30	−0.40	−0.50	−0.40	−0.30	0.00
8	0.00	−0.10	−0.20	−0.30	−0.40	−0.50	−0.60	−0.30	0.00
9	0.00	−0.10	−0.20	−0.30	−0.40	−0.50	−0.40	−0.30	0.00
10	0.00	−0.10	−0.20	−0.30	−0.40	−0.30	−0.20	−0.10	0.00
11	0.00	−0.10	−0.20	−0.30	−0.20	−0.10	0.00	0.10	0.00
12	0.00	−0.10	−0.20	−0.10	0.00	0.10	0.20	0.10	0.00
13	0.00	−0.10	0.00	0.10	0.20	0.30	0.20	0.10	0.00
14	0.00	0.10	0.20	0.30	0.40	0.30	0.20	0.10	0.00
15	0.00	0.30	0.40	0.50	0.40	0.30	0.20	0.10	0.00
16	0.00	0.30	0.60	0.50	0.40	0.30	0.20	0.10	0.00
17	0.00	0.30	0.40	0.50	0.40	0.30	0.20	0.10	0.00
18	0.00	0.10	0.20	0.30	0.40	0.30	0.20	0.10	0.00
19	0.00	−0.10	0.00	0.10	0.20	0.30	0.20	0.10	0.00
20	0.00	−0.10	−0.20	−0.10	0.00	0.10	0.20	0.10	0.00

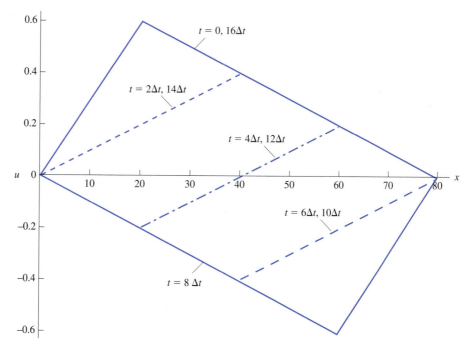

Figure 8.6

and the frequency is

$$f = \frac{1}{16 * 0.000179} = 350 \text{ cycles/sec.}$$

The standard formula from physics is

$$f = \left(\frac{1}{2L}\right)\sqrt{\frac{Tg}{w}} = \left(\frac{1}{160}\right)\sqrt{\frac{40000 * 980}{1.0/80}} = 350 \text{ cycles/sec,}$$

precisely the same!

It seems remarkable that we get exactly the correct frequency, but what about the accuracy of the displacements? We will find that these too are precisely correct as the next discussion shows. It is also apparent that the computations are stable when $Tg(\Delta t)^2/w(\Delta x)^2$ equals 1.

The D'Alembert Solution

The simple vibrating string problem is one where the analytical solution is readily obtained. This analytical solution is called the *D'Alembert solution.* Consider this expression for $u(x, t)$

$$u(x, t) = F(x + ct) + G(x - ct), \tag{8.20}$$

where F and G are arbitrary functions.

If we substitute this into the vibrating string equation, which we repeat,

$$\frac{\partial^2 u}{\partial t^2} = \frac{Tg}{w}\frac{\partial^2 u}{\partial x^2}, \tag{8.21}$$

we find that the partial-differential equation is satisfied, because

$$\frac{\partial u}{\partial t} = F'\frac{\partial(x + ct)}{\partial t} + G'\frac{\partial(x - ct)}{\partial t} = cF' - cG',$$
$$\frac{\partial^2 u}{\partial t^2} = c^2 F'' + c^2 G''; \tag{8.22}$$

$$\frac{\partial u}{\partial x} = F'\frac{\partial(x + ct)}{\partial x} + G'\frac{\partial(x - ct)}{\partial x} = F' + G',$$
$$\frac{\partial^2 u}{\partial x^2} = F'' + G''. \tag{8.23}$$

In Eqs. (8.22) and (8.23), the primes indicate derivatives of the arbitrary functions. Now, substituting these expressions for the second partials into Eq. (8.21), we see that the equation for the vibrating string is satisfied when $c^2 = (Tg/w)$. This means that we can get the solution to Eq. (8.21) if we can find functions F and G that satisfy the initial conditions and the boundary conditions. That too is not difficult. Suppose we are given the initial conditions

$$u(x,\ 0) = f(x), \qquad \frac{\partial u}{\partial t}(x,\ 0) = g(x).$$

The combination

$$u(x,\ t) = \left(\frac{1}{2}\right)[f(x + ct) + f(x - ct)] + \left(\frac{1}{2c}\right)\int_{x-ct}^{x+ct} g(v)\ dv \qquad \textbf{(8.24)}$$

is of the same form as Eq. (8.20). It certainly fulfills the boundary conditions, for substituting $t = 0$ in Eq. (8.24) gives $u(x,\ 0) = f(x)$ and differentiating with respect to t gives

$$\frac{1}{2}[f' * c + f' * (-c)] = 0$$

for the first term of Eq. (8.24), and

$$\left(\frac{1}{2c}\right)\{(c)[g(x + ct)] - (-c)[g(x - ct)]\} = g(x)$$

(when $t = 0$) for the second term.

We have thus shown that the solution to the vibrating string problem is exactly that given by Eq. (8.24). Now we ask "Does the simple algorithm of Eq. (8.17) match Eq. (8.24) for the example problem?" We can show that the answer to the question is "Yes" in the following way.

First, for $Tg(\Delta t)^2/w(\Delta x)^2$ equal to 1, $\Delta x = c\ \Delta t$. Recalling that u_i^j represents the u-value at $x = x_i = i\ \Delta x$ and at $t = t_j = j\ \Delta t$, we see that $ct_j = cj\ \Delta t = j\ \Delta x$. If we write $u(x_i,\ t_j)$ using our subscript/superscript notation, it becomes

$$u_i^j = F(x_i + ct_j) + G(x_i - ct_j) = F(i\ \Delta x + j\ \Delta x) + G(i\ \Delta x - j\ \Delta x) \qquad \textbf{(8.25)}$$
$$= F[(i + j)\ \Delta x] + G[(i - j)\ \Delta x].$$

Now let us use Eq. (8.25) to write each term on the right-hand side of Eq. (8.17), the algorithm that we used in the example.

$$u_{i+1}^j = F[(i + 1 + j)\ \Delta x] + G[(i + 1 - j)\ \Delta x],$$
$$u_{i-1}^j = F[(i - 1 + j)\ \Delta x] + G[(i - 1 - j)\ \Delta x],$$
$$u_i^{j-1} = F[(i + j - 1)\ \Delta x] + G[(i - j + 1)\ \Delta x].$$

In the example, both F and G are linear functions of x, so that $F(a) + F(b) = F(a + b)$, and the same is true for G. If we combine these terms in Eq. (8.17),

$$u_{i+1}^j + u_{i-1}^j - u_i^{j-1} = F\,[(i + 1 + j)\ \Delta x + (i - 1 + j)\ \Delta x - (i + j - 1)\ \Delta x]$$
$$+ G[(i + 1 - j)\ \Delta x + (i - 1 - j)\ \Delta x - (i - j + 1)\ \Delta x]$$
$$= F\{[i + (j + 1)]\ \Delta x\} + G\{[i - (j + 1)]\ \Delta x]$$
$$= u_i^{j+1},$$

and the validity of Eq. (8.17) is proved. The important implication from this is that, if we have correct values for the u's at two successive time steps, all subsequent computed values will be correct.

When the Initial Velocity Is Not Zero

Example 8.4 had the string starting with zero velocities. What if the initial velocity is not zero? Equation (8.19) was a very simple way to begin the computations, but it gave correct results only because $g(x)$ was zero in Eq. (8.24). This next example shows that Eq. (8.19) is inadequate when $g(x) \neq 0$ and that there is a better way to begin.

EXAMPLE 8.5   A string is 9 units long. Initially it is in its equilibrium position (just a straight line between the supports). It is set into motion by striking it so that it has an initial velocity given by $\partial u/\partial t = 3 \sin(\pi x/L)$. Take $\Delta x = 1$ unit and let $c^2 = Tg/w = 4$. When the ratio $c^2(\Delta t)^2/(\Delta x)^2 = 1$, the value of Δt is 0.5 time units. Find the displacements at the end of one Δt.

Because $\Delta x = 1$ and the length is 9, the string is divided into nine intervals; there are eight interior nodes. We are to compute the u-values at $t = \Delta t = 0.5$.

As we have seen, Eq. (8.19) is one way to get these starting values. However, looking at Eq. (8.24), we see that there is an alternative technique. If we substitute $t = \Delta t$ in that equation and remember that $c \, \Delta t = \Delta x$, we get for $u(x_i, \Delta t)$

$$u(x_i, \Delta t) = \frac{1}{2}[f(x_i + \Delta x) + f(x_i - \Delta x)] + \left(\frac{1}{2c}\right) \int_{x - \Delta x}^{x + \Delta x} g(v) \, dv \qquad (8.26)$$

$$= \frac{1}{2}[u^0_{i+1} + u^0_{i-1}] + \left(\frac{1}{2c}\right) \int_{x - \Delta x}^{x + \Delta x} g(v) \, dv.$$

Equation (8.26) differs from Eq. (8.19) only in the last term. If $g(x) = $ a constant, the last terms are equal, but if $g(x)$ is not constant, we should do the integration in Eq. (8.26). Table 8.6 compares the results of both techniques and also gives the answers from the analytical solution. Only values for x between 1 and 4 are given as the displacements for the right half of the string are the same as for the left half. Simpson's rule was used to do the integrations. We see from the tabulated results that the values using Eq. (8.26) are almost exactly the same as the analytical values (they are the same within 1 in the fourth decimal place) but

Table 8.6   Comparison of ways to begin the wave equation at $t = \Delta t$ with $\Delta x = 1$

	u = values from		
x	Eq. (8.19)	Eq. (8.26)	Analytical
1	0.5130	0.5027	0.50267
2	0.9642	0.9448	0.94472
3	1.2990	1.2729	1.27282
4	1.4772	1.4475	1.44740

that the results from Eq. (8.19) are less accurate (they each differ by 2.0% from the analytical). We could improve the accuracy with Eq. (8.19) by decreasing the size of Δx (and reducing Δt correspondingly). By making $\Delta x = 0.5$, the errors are reduced fourfold as expected. ▲

8.5   Parabolic Equations in Two or Three Dimensions

In Section 8.2 the equations for the one-dimensional heat equation and for a vibrating string were extended to two and three spatial dimensions. We restate these equations

$$\frac{\partial u}{\partial t} = \frac{k}{c\rho} \nabla^2 u, \tag{8.27}$$

$$\frac{\partial^2 u}{\partial t^2} = \frac{Tg}{w} \nabla^2 u, \tag{8.28}$$

where the Laplacian in both Eqs. (8.27) and (8.28) represents the sum of the second derivatives of u with respect to the spatial variables.

In this section we apply finite-difference approximations to the derivatives to obtain a set of equations that solve parabolic partial-differential equations in two and three dimensions. We show how this is done through examples.

Heat Flow Within a Two-Dimensional Region

Suppose we have a rectangular region whose edges fit to evenly spaced nodes. If we replace the right-hand side of Eq. (8.27) with central difference approximations, where $\Delta x = \Delta y = h$, and $r = k\,\Delta t/(c\rho h^2)$, the explicit scheme becomes

$$u_{i,j}^{k+1} - u_{i,j}^k = r(u_{i+1,j}^k - 2u_{i,j}^k + u_{i-1,j}^k + u_{i,j+1}^k - 2u_{i,j}^k + u_{i,j-1}^k)$$

or

$$u_{i,j}^{k+1} = r(u_{i+1,j}^k + u_{i-1,j}^k + u_{i,j+1}^k + u_{i,j-1}^k) + (1 - 4r)u_{i,j}^k.$$

In this scheme, stability requires that the value of r be $\frac{1}{4}$ or less in the simple case of Dirichlet boundary conditions. (Note that this corresponds again to the numerical value that gives a particularly simple formula.) In the more general case with $\Delta x \neq \Delta y$, the criterion is

$$\frac{k\,\Delta t}{c\rho[(\Delta x)^2 + (\Delta y)^2]} \leq \frac{1}{8}.$$

The analogous equation in three dimensions, with equal grid spacing each way, has the coefficient $(1 - 6r)$, and $r \leq \frac{1}{6}$ is required for convergence and stability.

The difficulty with the use of the explicit scheme is that the restrictions on Δt require inordinately many rows of calculations. One then looks for a method in which Δt can be made larger without loss of stability. In one dimension, the Crank–Nicolson method was such a method. In the two-dimensional case, using averages of central-difference approximations to give $\partial^2 u/\partial x^2$ and $\partial^2 u/\partial y^2$ at the midvalue of time, we get

$$u_{i,j}^{k+1} - u_{i,j}^k = \frac{r}{2}[u_{i+1,j}^{k+1} - 2u_{i,j}^{k+1} + u_{i-1,j}^{k+1} + u_{i+1,j}^k - 2u_{i,j}^k + u_{i-1,j}^k$$
$$+ u_{i,j+1}^{k+1} - 2u_{i,j}^{k+1} + u_{i,j-1}^{k+1} + u_{i,j+1}^k - 2u_{i,j}^k + u_{i,j-1}^k].$$

The problem now is that a set of $(M)(N)$ simultaneous equations must be solved at each time step, where M is the number of unknown values in the x-direction and N in the y-direction. Furthermore, the coefficient matrix is no longer tridiagonal, so the solution to each set of equations is slower and memory space to store the elements of the matrix may be exorbitant.

The advantage of a tridiagonal matrix is retained in the alternating-direction-implicit scheme (A.D.I.) proposed by Peaceman and Rachford (1955). It is widely used in modern computer programs for the solution of parabolic partial-differential equations. We discussed the A.D.I. method in Chapter 7 as applied to elliptic equations. For parabolic equations, we approximate $\nabla^2 u$ by adding a central-difference approximation to $\partial^2 u/\partial x^2$ written at the beginning of the time interval to a similar expression for $\partial^2 u/\partial y^2$ written at the end of the time interval. We will use subscripts L, R, A, and B to indicate nodes to the left, right, above, and below the central node, respectively, where $u = u_0$. We then have

$$u_0^{j+1} - u_0^j = r[u_L^j - 2u_0^j + u_R^j] + r[u_A^{j+1} - 2u_0^{j+1} + u_B^{j+1}], \quad (8.29)$$

where $r = k\,\Delta t/c\rho\Delta^2$ and $\Delta = \Delta x = \Delta y$. The obvious bias in this formula is balanced by reversing the order of the second derivative approximations in the next time span:

$$u_0^{j+2} - u_0^{j+1} = r[u_L^{j+2} - 2u_0^{j+2} + u_R^{j+2}] + r[u_A^{j+1} - 2u_0^{j+1} + u_B^{j+1}]. \quad (8.30)$$

Observe that in using Eq. (8.29), we make a vertical traverse through the nodes, computing new values for each column of nodes. Similarly, in using Eq. (8.30) we make a horizontal traverse, computing new values row by row. In effect, we consider u_L and u_R as fixed when we do a vertical traverse; we consider u_A and u_B as fixed for horizontal traverses.

EXAMPLE 8.6   A square plate of steel is 8 in. wide and 6 in. high. Initially all points on the plate are at 50°. The edges are suddenly brought to the temperatures shown in Fig. 8.7 and held at these temperatures. Trace the history of temperatures at nodes spaced 2 in. apart using the A.D.I. method, assuming that heat flows only in the x- and y-directions.

Figure 8.7 shows a numbering system for the internal nodes, all of which start at 50°, as well as the temperatures at boundary nodes.

Using Eq. (8.29), the typical equation for a vertical traverse is

$$-ru_A^{j+1} + (1 + 2r)u_0^{j+1} - ru_B^{j+1} = (ru_L + (1 - 2r)u_0 + ru_R)^j.$$

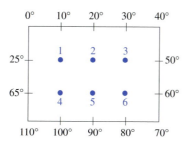

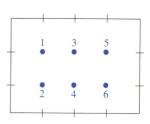

Figure 8.7   Figure 8.8

If we use this equation and the numbering system of Fig. 8.7 to set up the equations for a vertical traverse, we do not get the tridiagonal system that we desire, but we do if the nodes are renumbered as shown in Fig. 8.8. To keep track of the different numbering systems, we will use v for temperatures when a vertical traverse is made (numbered as in Fig. 8.8) and u when a horizontal traverse is made (numbered as in Fig. 8.7).

This is the set of equations for a vertical traverse:

$$
\begin{bmatrix}
(1+2r) & -r & & & & \\
-r & (1+2r) & & & & \\
& & (1+2r) & -r & & \\
& & -r & (1+2r) & & \\
& & & & (1+2r) & -r \\
& & & & -r & (1+2r)
\end{bmatrix}
\begin{Bmatrix} v_1 \\ v_2 \\ v_3 \\ v_4 \\ v_5 \\ v_6 \end{Bmatrix}
=
\begin{Bmatrix}
r(25) + (1-2r)u_1 + ru_2 + r(10) \\
r(65) + (1-2r)u_4 + ru_5 + r(100) \\
ru_1 + (1-2r)u_2 + ru_3 + r(20) \\
ru_4 + (1-2r)u_5 + ru_6 + r(90) \\
ru_2 + (1-2r)u_3 + r(50) + r(30) \\
ru_5 + (1-2r)u_6 + r(60) + r(80)
\end{Bmatrix}.
$$

When we apply Eq. (8.29) to get a set of equations for a horizontal traverse, we get (the dashed lines show they break into subsets)

$$
\begin{bmatrix}
(1+2r) & -r & & & & \\
-r & (1+2r) & -r & & & \\
& -r & (1+2r) & & & \\
& & & (1+2r) & -r & \\
& & & -r & (1+2r) & -r \\
& & & & -r & (1+2r)
\end{bmatrix}
\begin{Bmatrix} u_1 \\ u_2 \\ u_3 \\ u_4 \\ u_5 \\ u_6 \end{Bmatrix}
=
\begin{Bmatrix}
r(10) + (1-2r)v_1 + rv_2 + r(25) \\
r(20) + (1-2r)v_3 + rv_4 \\
r(30) + (1-2r)v_5 + rv_6 + r(50) \\
rv_1 + (1-2r)v_2 + r(100) + r(65) \\
rv_3 + (1-2r)v_4 + r(90) \\
rv_5 + (1-2r)v_6 + r(80) + r(60)
\end{Bmatrix}.
$$

A value must be specified for r. Small r's give better accuracy but smaller Δt's, so more time steps are required to compute the history. If we take $r = 1$, Δt is 26.4 sec.

The first vertical traverse gives results for $t = 26.4$ sec. We get the first set of v's from

$$
\begin{bmatrix}
3 & -1 & & & & \\
-1 & 3 & & & & \\
& & 3 & -1 & & \\
& & -1 & 3 & & \\
& & & & 3 & -1 \\
& & & & -1 & 3
\end{bmatrix}
\begin{Bmatrix} v_1 \\ v_2 \\ v_3 \\ v_4 \\ v_5 \\ v_6 \end{Bmatrix}
=
\begin{Bmatrix}
25 - (1)50 + 50 + 10 \\
65 - (1)50 + 50 + 100 \\
50 - (1)50 + 50 + 20 \\
50 - (1)50 + 50 + 40 \\
50 - (1)50 + 50 + 30 \\
50 - (1)50 + 60 + 80
\end{Bmatrix}
=
\begin{Bmatrix} 35 \\ 165 \\ 70 \\ 140 \\ 80 \\ 140 \end{Bmatrix}.
$$

Solving, we get these values:

$$\{33.75 \quad 66.25 \quad 43.75 \quad 61.25 \quad 47.5 \quad 62.5\}.$$

These values are used to build the right-hand sides for the next computations, a horizontal traverse, getting these equations for $t = 52.8$ sec:

$$
\begin{bmatrix}
3 & -1 & & & & \\
-1 & 3 & -1 & & & \\
 & -1 & 3 & & & \\
 & & & 3 & -1 & \\
 & & & -1 & 3 & -1 \\
 & & & & -1 & 3
\end{bmatrix}
\begin{Bmatrix}
u_1 \\ u_2 \\ u_3 \\ u_4 \\ u_5 \\ u_6
\end{Bmatrix}
=
\begin{Bmatrix}
10 & - (1)33.75 + & 66.25 + 25 \\
20 & - (1)43.75 + & 61.25 \\
30 & - (1)47.5 + & 62.5 + 50 \\
33.75 - (1)66.25 + & 100 & + 65 \\
43.75 - (1)61.25 + & 90 & \\
47.5 - (1)62.5 + & 80 & + 60
\end{Bmatrix}
=
\begin{Bmatrix}
67.5 \\ 37.5 \\ 95 \\ 132.5 \\ 72.5 \\ 125
\end{Bmatrix},
$$

which have the solution (a set of u's)

$$\{35.595 \quad 39.286 \quad 44.762 \quad 66.786 \quad 67.857 \quad 64.286\}.$$

We continue by alternating between vertical and horizontal traverses to get the results shown in Table 8.7. This also shows the steady-state temperatures that are reached after a long time. The steady-state temperatures could have been computed by the methods of Chapter 7. We observe that the A.D.I. algorithm for steady-state temperatures of Chapter 7 is essentially identical to what we have seen here. ▲

The compensation of errors produced by this alternation of direction gives a scheme that is convergent and stable for all values of r, although accuracy requires that r not be too large. The three-dimensional analog alternates three ways, returning to each of the three formulas after every third step. (Unfortunately the three-dimensional case is not stable for all fixed values of $r > 0$. A variant due to Douglas (1962) is unconditionally stable, however.) When the nodes are renumbered, in each case tridiagonal coefficient matrices result.

Note that the equations can be broken up into two independent subsets, corresponding to the nodes in each column or row. (See the first set of equations of Example 8.6.) This is always true in the A.D.I. method; each row gives a set independent of the equations from the other rows. For columns, the same thing occurs. For very large problems, this is important, because it permits the ready overlay of main memory in solving the independent sets. Observe also that each subset can be solved at the same time by parallel processors.

Nonrectangular Regions

When the region in which the heat-flow equation is to be satisfied is not rectangular, we can perturb the boundary to make it agree with a square mesh, or we can interpolate from boundary points to estimate u at adjacent mesh points, as discussed in Chapter 7 for elliptic equations.

The frequency with which circular or spherical regions occur makes it worthwhile to mention the heat equation in polar and spherical coordinates. The basic equation

Table 8.7   Results for Example 8.6 using the A.D.I. method

```
AT START, TEMPS ARE
    0.0000         10.0000        20.0000        30.0000        40.0000
   25.0000         50.0000        50.0000        50.0000        50.0000
   65.0000         50.0000        50.0000        50.0000        60.0000
  110.0000        100.0000        90.0000        80.0000        70.0000

AFTER ITERATION 1 TIME =  26.4  - VALUES ARE
    0.0000         10.0000        20.0000        30.0000        40.0000
   25.0000         33.7500        43.7500        47.5000        50.0000
   65.0000         66.2500        61.2500        62.5000        60.0000
  110.0000        100.0000        90.0000        80.0000        70.0000

AFTER ITERATION 2 TIME =  52.8  - VALUES ARE
    0.0000         10.0000        20.0000        30.0000        40.0000
   25.0000         35.5952        39.2857        44.7619        50.0000
   65.0000         66.7857        67.8571        64.2857        60.0000
  110.0000        100.0000        90.0000        80.0000        70.0000

AFTER ITERATION 3 TIME =  79.2  - VALUES ARE
    0.0000         10.0000        20.0000        30.0000        40.0000
   25.0000         35.2679        42.0536        45.8929        50.0000
   65.0000         67.1131        65.0893        63.1548        60.0000
  110.0000        100.0000        90.0000        80.0000        70.0000

AFTER ITERATION 4 TIME = 105.6  - VALUES ARE
    0.0000         10.0000        20.0000        30.0000        40.0000
   25.0000         36.2443        41.8878        46.3832        50.0000
   65.0000         66.1366        65.2551        62.6644        60.0000
  110.0000        100.0000        90.0000        80.0000        70.0000



STEADY-STATE TEMPERATURES:
    0.0000         10.0000        20.0000        30.0000        40.0000
   25.0000         35.8427        41.8323        46.1760        50.0000
   65.0000         66.5383        65.3106        62.8716        60.0000
  110.0000        100.0000        90.0000        80.0000        70.0000
```

$$\frac{\partial u}{\partial t} = \frac{k}{c\rho}\nabla^2 u$$

becomes, in polar coordinates (r, θ),

$$\frac{\partial u}{\partial t} = \frac{k}{c\rho}\left(\frac{\partial^2 u}{\partial r^2} + \frac{1}{r}\frac{\partial u}{\partial r} + \frac{1}{r^2}\frac{\partial^2 u}{\partial \theta^2}\right),$$

and in spherical coordinates (r, ϕ, θ),

$$\frac{\partial u}{\partial t} = \frac{k}{c\rho}\left(\frac{\partial^2 u}{\partial r^2} + \frac{2}{r}\frac{\partial u}{\partial r} + \frac{1}{r^2}\frac{\partial^2 u}{\partial \theta^2} + \frac{\cot\phi}{r}\frac{\partial u}{\partial \theta} + \frac{1}{r^2\sin^2\theta}\frac{\partial^2 u}{\partial \phi^2}\right).$$

Using finite-difference approximations to convert these to difference equations is straight-forward except at the origin where $r = 0$. For this point, consider $\nabla^2 u$ in rectangular coordinates, so that, in two dimensions,

$$\partial^2 u = \frac{u_{i+1,j} + u_{i-1,j} + u_{i,j+1} + u_{i,j-1} - 4u_{i,j}}{(\Delta r)^2}, \qquad u_{ij} \text{ at } r = 0.$$

This is exactly the same as the expression for the Laplacian in Chapter 7, Eq. (7.12). The expression for $\nabla^2 u$ is obviously independent of the orientation of the axes. In polar coordinates we get the best value by using the average of values at all nodes that are a distance Δr from the origin, so that for $r = 0$,

$$\nabla^2 u = \frac{4(u_{av} - u_0)}{(\Delta r)^2} \qquad \text{at } r = 0.$$

The corresponding relation for spherical coordinates is

$$\nabla^2 u = \frac{6(u_{av} - u_0)}{(\Delta r)^2} \qquad \text{at } r = 0.$$

8.6   The Wave Equation in Two Dimensions

The finite-difference method can be applied to hyperbolic partial-differential equations in two or more space dimensions. A typical problem is the vibrating membrane. Consider a thin flexible membrane stretched over a rectangular frame and set to vibrating. As we have seen, (Eqs. 8.8 and 8.28), the equation is

$$\frac{\partial^2 u}{\partial t^2} = \frac{Tg}{w}\left(\frac{\partial^2 u}{\partial x^2} + \frac{\partial^2 u}{\partial y^2}\right),$$

in which u is the displacement, t is the time, x and y are the space coordinates, T is the uniform tension per unit length, g is the acceleration of gravity, and w is the weight per unit area. For simplification, let $Tg/w = c^2$. Replacing each derivative by its central-difference approximation, and using $h = \Delta x = \Delta y$, gives (we recognize the Laplacian on the right-hand side)

$$\frac{u_{i,j}^{k+1} - 2u_{i,j}^{k} + u_{i,j}^{k-1}}{(\Delta t)^2} = c^2 \frac{u_{i+1,j}^{k} + u_{i-1,j}^{k} + u_{i,j+1}^{k} + u_{i,j-1}^{k} - 4u_{i,j}^{k}}{h^2}. \quad (8.31)$$

Solving for the displacement at time t_{k+1}, we obtain

$$u_{i,j}^{k+1} = \frac{c^2(\Delta t)^2}{h^2} \left\{ \begin{matrix} 1 \\ 1 \quad 0 \quad 1 \\ 1 \end{matrix} \right\} u_{i,j}^{k} - u_{i,j}^{k-1} + \left(2 - 4\frac{c^2(\Delta t)^2}{h^2} \right) u_{i,j}^{k}. \quad (8.32)$$

In Eqs. (8.31) and (8.32), we use superscripts to denote the time. If we let $c^2(\Delta t)^2/h^2 = \frac{1}{2}$, the last term vanishes and we get

$$u_{i,j}^{k+1} = \frac{1}{2} \left\{ \begin{matrix} 1 \\ 1 \quad 0 \quad 1 \\ 1 \end{matrix} \right\} u_{i,j}^{k} - u_{i,j}^{k-1}. \quad (8.33)$$

For the first time step, we get displacements from Eq. (8.34), which is obtained by approximating $\partial u/\partial t$ at $t = 0$ by a central-difference approximation involving $u_{i,j}^{1}$ and $u_{i,j}^{-1}$.

$$u_{i,j}^{1} = \frac{1}{4} \left\{ \begin{matrix} 1 \\ 1 \quad 0 \quad 1 \\ 1 \end{matrix} \right\} u_{i,j}^{0} + (\Delta t)g(x_i, y_j). \quad (8.34)$$

In Eq. (8.34), $g(x, y)$ is the initial velocity.

It should not surprise us to learn that this ratio $c^2(\Delta t)^2/h^2 = \frac{1}{2}$ is the maximum value for stability, in view of our previous experience with explicit methods. However, in contrast with the wave equation in one space dimension, we do not get exact answers from the numerical procedure of Eq. (8.33), and we further observe that we must use smaller time steps in relation to the size of the space interval. Therefore we advance in time more slowly. However, the numerical method is straightforward, as the following example will show.

EXAMPLE 8.7   A membrane for which $c^2 = Tg/w = 3$ is stretched over a square frame that occupies the region $0 \le x \le 2, 0 \le y \le 2$, in the xy-plane. It is given an initial displacement described by

$$u = x(2 - x)y(2 - y),$$

and has an initial velocity of zero. Find how the displacement varies with time.

We divide the region with $h = \Delta x = \Delta y = \frac{1}{2}$, obtaining nine interior nodes. Initial displacements are calculated from the initial conditions: $u^0(x, y) = x(2 - x)y(2 = y)$; Δt is taken at its maximum value for stability, $h/(\sqrt{2}\,c) = 0.2041$. The values at the end of one time step are given by

$$u^1_{i,j} = \frac{1}{4}\left\{\begin{matrix} & 1 & \\ 1 & 0 & 1 \\ & 1 & \end{matrix}\right\} u^0_{i,j}$$

because $g(x, y)$ in Eq. (8.34) is everywhere zero. For succeeding time steps, Eq. (8.33) is used. Table 8.8 gives the results of our calculations. Also shown in Table 8.8 (in parentheses) are analytical values, computed from the double infinite series:

$$u(x,\ y,\ t) = \sum_{m=1}^{\infty} \sum_{n=1}^{\infty} B_{mn} \sin\frac{m\pi x}{a} \sin\frac{n\pi y}{b} \cos\left(c\pi t \sqrt{\frac{m^2}{a^2} - \frac{n^2}{b^2}}\right),$$

$$B_{mn} = \frac{16a^2b^2A}{\pi^6 m^3 n^3}(1 - \cos m\pi)(1 - \cos n\pi),$$

which gives the displacement of a membrane fastened to a rectangular framework, $0 \le x \le a$, $0 \le y \le b$, with initial displacements of $Ax(a - x)y(b - y)$.

We observe that the finite-difference results do not agree exactly with the analytical calculations. The finite-difference values are symmetrical with respect to position and repeat themselves with a regular frequency. The very regularity of the values itself indicates that the finite-difference computations are in error, because they predict that the membrane could emit a musical note. We know from experience that a drum does not give a musical

Table 8.8   Displacements of a vibrating membrane—finite-difference method: $\Delta t = h/(\sqrt{2}\alpha)$

	Grid location								
t	(0.5, 0.5)	(1.0, 0.5)	(1.5, 0.5)	(0.5, 1.0)	(1.0, 1.0)	(1.5, 1.0)	(0.5, 1.5)	(1.0, 1.5)	(1.5, 1.5)
0	0.5625 (0.5625)	0.750 (0.750)	0.5625	0.750	1.000 (1.000)	0.750	0.5625	0.750	0.5625
0.204	0.375 (0.380)	0.531 (0.536)	0.375	0.531	0.750 (0.755)	0.531	0.375	0.531	0.375
0.408	−0.031 (−0.044)	0.000 (−0.009)	−0.031	0.000	0.062 (0.083)	0.000	−0.031	0.000	−0.031
0.612	−0.375 (−0.352)	−0.531 (−0.539)	−0.375	−0.531	−0.750 (−0.813)	−0.531	−0.375	−0.531	−0.375
0.816	−0.500 (−0.502)	−0.750 (−0.746)	−0.500	−0.750	−1.125 (−1.114)	−0.750	−0.500	−0.750	−0.500
1.021	−0.375 (−0.407)	−0.531 (−0.535)	−0.375	−0.531	−0.750 (−0.691)	−0.531	−0.375	−0.531	−0.375
1.225	−0.031 (−0.015)	0.000 (0.008)	−0.031	0.000	0.062 (0.030)	0.000	−0.031	0.000	−0.031
1.429	0.375 (0.410)	0.531 (0.534)	0.375	0.531	0.750 (0.688)	0.531	0.375	0.531	0.375

Note: Analytical values are in parentheses.

tone when struck; therefore the vibrations do not have a cycle pattern of constant frequency, as exhibited by our numerical results.

Decreasing the ratio of $c^2(\Delta t)^2/h^2$ and using Eq. (8.32) gives little or no improvement in the average accuracy; to approach closely to the analytical results, $h = \Delta x = \Delta y$ must be made smaller. When this is done, Δt will need to decrease in proportion, requiring many time steps and leading to many repetitions of the algorithm and extravagant use of computer time. One remedy is the use of implicit methods, which allow the use of larger ratios of $\alpha^2(\Delta t)^2/h^2$. However, with many nodes in the xy-grid, this requires large, sparse matrices similar to the Crank–Nicolson method for parabolic equations in two space dimensions. A.D.I. methods have been used for hyperbolic equations—tridiagonal systems result. We do not discuss these methods. ▲

8.7 Theoretical Matters

Although there is much to say about the theory behind parabolic equations, we will discuss only stability and convergence of our numerical procedures. We already know that, to ensure stability and convergence in the explicit method, the ratio $r = k\,\Delta t/c\rho(\Delta x)^2$ must be 0.5 or less. The Crank–Nicolson and the implicit methods, however, have no such limitations.

By *convergence*, we mean that the results of the method approach the analytical values as Δt and Δx both approach zero. By *stability,* we mean that errors made at one stage of the calculations do not cause increasingly large errors as the computations are continued, but rather will eventually damp out.

We will first discuss convergence, limiting ourselves to the simple case of the unsteady-state heat-flow equation in one dimension: *

$$\frac{\partial U}{\partial t} = \frac{k}{c\rho}\frac{\partial^2 U}{\partial x^2}. \qquad (8.35)$$

Let us use the symbol U to represent the exact solution to Eq. (8.35), and u to represent the numerical solution. At the moment we assume that u is free of round-off errors, so the only difference between U and u is the error made by replacing Eq. (8.35) by the difference equation. Let $e_i^j = U_i^j - u_i^j$, at the point $x = x_i,\ t = t_j$. By the explicit method, Eq. (8.35) becomes

$$u_i^{j+1} = r(u_{i+1}^j + u_{i-1}^j) + (1 - 2r)u_i^j, \qquad (8.36)$$

where $r = k\,\Delta t/c\rho(\Delta x)^2$. Substituting $u = U - e$ into Eq. (8.36), we get

$$e_i^{j+1} = r(e_{i+1}^j + e_{i-1}^j) + (1 - 2r)e_i^j - r(U_{i+1}^j + U_{i-1}^j) - (1 - 2r)U_i^j + U_i^{j+1}. \qquad (8.37)$$

---

*We could have treated the simpler equation $\partial U/\partial T = \partial^2 U/\partial X^2$ without loss of generality, because with the change of variables—$X = \sqrt{c\rho}\,x,\ T = kt$—the two equations are seen to be identical.

By using Taylor-series expansions, we have

$$
U^j_{i+1} = U^j_i + \left(\frac{\partial U}{\partial x}\right)_{i,j} \Delta x + \frac{(\Delta x)^2}{2} \frac{\partial^2 U(\xi_1, t_j)}{\partial x^2}, \qquad x_i < \xi_1 < x_{i+1},
$$

$$
U^j_{i-1} = U^j_i - \left(\frac{\partial U}{\partial x}\right)_{i,j} \Delta x + \frac{(\Delta x)^2}{2} \frac{\partial^2 U(\xi_2, t_j)}{\partial x^2}, \qquad x_{i-1} < \xi_2 < x_i,
$$

$$
U^{j+1}_i = U^j_i + \Delta t \frac{\partial U(x_i, \eta)}{\partial t}, \qquad t_j < \eta < t_{j+1}.
$$

Substituting these into Eq. (8.37) and simplifying, remembering that $r(\Delta x)^2 = k\,\Delta t/c\rho$, we get

$$
e^{j+1}_i = r(e^j_{i+1} + e^j_{i-1}) + (1 - 2r)e^j_i + \Delta t \left[\frac{\partial U(x_i, \eta)}{\partial t} - \frac{k}{c\rho} \frac{\partial^2 U(\xi, t_j)}{\partial x^2}\right],
$$

$$
t_j \le \eta \le t_{j+1}, \qquad x_{i-1} \le \xi \le x_{i+1}. \tag{8.38}
$$

Let E^j be the magnitude of the maximum error in the row of calculations for $t = t_j$, and let $M > 0$ be an upper bound for the magnitude of the expression in brackets in Eq. (8.38). If $r \le \frac{1}{2}$, all the coefficients in Eq. (8.38) are positive (or zero) and we may write the inequality

$$
|e^{j+1}_i| \le 2rE^j + (1 - 2r)E^j + M\,\Delta t = E^j + M\,\Delta t.
$$

This is true for all the e^{j+1}_i at $t = t_{j+1}$, so

$$
E^{j+1} \le E^j + M\,\Delta t.
$$

This is true at each time step,

$$
E^{j+1} \le E^j + M\,\Delta t \le E^{j-1} + 2M\,\Delta t \le \cdots \le E^0 + (j + 1)M\,\Delta t = E^0 + Mt_{j+1}
$$

$$
= Mt_{j+1},
$$

because E^0, the errors at $t = 0$, are zero, as U is given by the initial conditions.

Now, as $\Delta x \to 0$, $\Delta t \to 0$ if $k\,\Delta t/c\rho(\Delta x)^2 \le \frac{1}{2}$, and $M \to 0$, because, as both Δx and Δt get smaller,

$$
\left[\frac{\partial U(x_i, \eta)}{\partial t} - \frac{k}{c\rho} \frac{\partial^2 U(\xi, t_j)}{\partial x^2}\right] \to \left(\frac{\partial U}{\partial t} - \frac{k}{c\rho} \frac{\partial^2 U}{\partial x^2}\right)_{i,j} = 0.
$$

This last is by virtue of Eq. (8.35), of course. Consequently, we have shown that the explicit method is convergent for $r \le \frac{1}{2}$, because the errors approach zero as Δt and Δx are made smaller.

Table 8.9   Propagation of errors—explicit method

t	Endpoint x_1	x_2	x_3	x_4	Endpoint x_5
t_0	0	0	0	0	0
t_1	0	e	0	0	0
t_2	0	0	$0.50e$	0	0
t_3	0	$0.25e$	0	$0.25e$	0
t_4	0	0	$0.25e$	0	0
t_5	0	$0.125e$	0	$0.125e$	0
t_6	0	0	$0.125e$	0	0
t_7	0	$0.062e$	0	$0.062e$	0
t_8	0	0	$0.062e$	0	0

For the solution to the heat-flow equation by the Crank–Nicolson method, the analysis of convergence may be made by similar methods. The treatment is more complicated, but it can be shown that each E^{j+1} is no greater than a finite multiple of E^j plus a term that vanishes as both Δx and Δt become small, and this is independent of r. Hence, because the initial errors are zero, the finite-difference solution approaches the analytical solution as $\Delta t \to 0$ and $\Delta x \to 0$, requiring only that r stay finite. This is also true for the θ method whenever $0.5 \leq \theta \leq 1$.

Let us begin our discussion of stability with a numerical example. Because the heat-flow equation is linear, if two solutions are known, their sum is also a solution. We are interested in what happens to errors made in one line of the computations as the calculations are continued, and because of the additivity feature, the effect of a succession of errors is just the sum of the effects of the individual errors. We follow, then, a single error,* which most likely occurred due to round-off. If this single error does not grow in magnitude, we will call the method *stable,* because then the cumulative effect of all errors affects the later calculations no more than a linear combination of the previous errors would. (Because round-off errors are both positive and negative, we can expect some cancellation.)

Table 8.9 illustrates the principle. We have calculated for the simple case where the boundary conditions are fixed, so that the errors at the endpoints are zero. We assume that a single error of size e occurs at $t = t_1$ and $x = x_2$. The explicit method, $k \, \Delta t / c \rho (\Delta x)^2 = \frac{1}{2}$, was used. The original error quite obviously dies out. As an exercise, it is left to the student to show that with $r > 0.5$, errors have an increasingly large effect on later computations. Table 8.10 shows that errors damp out for the Crank–Nicolson method with $r = 1$ even more rapidly than in the explicit method with $r = 0.5$. The errors with the implicit method also die out with $r = 1$, more rapidly than with the explicit method but less rapidly than with Crank–Nicolson.

---

*A computation made assuming that each of the interior points has an error equal to e at $t = t_1$ demonstrates the effect more rapidly.

Table 8.10   Propagation of errors—Crank–Nicolson method

t	x_1	x_2	x_3	x_4	x_5
t_0	0	0	0	0	0
t_1	0	e	0	0	0
t_2	0	$0.071e$	$0.286e$	$0.071e$	0
t_3	0	$0.092e$	$0.082e$	$0.092e$	0
t_4	0	$0.036e$	$0.064e$	$0.036e$	0
t_5	0	$0.024e$	$0.030e$	$0.024e$	0
t_6	0	$0.012e$	$0.018e$	$0.012e$	0
t_7	0	$0.007e$	$0.009e$	$0.007e$	0
t_8	0	$0.004e$	$0.005e$	$0.004e$	0

A More Analytical Argument

To discuss stability in a more analytical sense, we need some material from linear algebra. In Chapter 7 we discussed eigenvalues and eigenvectors of a matrix. We recall that for the matrix A and vector x, if

$$Ax = \lambda x,$$

then the scalar λ is an eigenvalue of A and x is the corresponding eigenvector. If the N eigenvalues of the $N \times N$ matrix A are all different, then the corresponding N eigenvectors are linearly independent, and any N-component vector can be written uniquely in terms of them.

Consider the unsteady-state heat-flow problem with fixed boundary conditions. Suppose we subdivide into $N + 1$ subintervals so there are N unknown values of the temperature being calculated at each time step. Think of these N values as the components of a vector. Our algorithm for the explicit method (Eq. 8.12) can be written as the matrix equation*

$$
\begin{bmatrix} u_1^{j+1} \\ u_2^{j+1} \\ \vdots \\ u_N^{j+1} \end{bmatrix}
=
\begin{bmatrix} (1 - 2r) & r & & \\ r & (1 - 2r) & r & \\ & & \vdots & \\ & & r & (1 - 2r) \end{bmatrix}
\begin{bmatrix} u_1^{j} \\ u_2^{j} \\ \vdots \\ u_N^{j} \end{bmatrix},
\qquad (8.39)
$$

or

$$u^{j+1} = Au^j,$$

where A represents the coefficient matrix and u^j and u^{j+1} are the vectors whose N components are the successive calculated values of temperature. The components of u^0 are the

---

*A change of variable is required to give boundary conditions of $u = 0$ at each end. This can always be done for fixed end conditions.

initial values from which we begin our solution. The successive rows of our calculations are

$$u^1 = Au^0,$$
$$u^2 = Au^1 = A^2u^0,$$
$$\vdots$$
$$u^j = Au^{j-1} = A^2u^{j-2} = \cdots = A^ju^0.$$

(Here the superscripts on the A are exponents; on the vectors they indicate time.)

Suppose that errors are introduced into u^0, so that it becomes $\bar{u}^0$. We will follow the effects of this error through the calculations. The successive lines of calculation are now

$$\bar{u}^j = A\bar{u}^{j-1} = \cdots = A^j\bar{u}^0.$$

Let us define the vector e^j as $u^j - \bar{u}^j$ so that e^j represents the errors in u^j caused by the errors in $\bar{u}^0$. We have

$$e^j = u^j - \bar{u}^j = A^ju^0 - A^j\bar{u}^0 = A^je^0. \qquad \textbf{(8.40)}$$

This shows that errors are propagated by using the same algorithm as that by which the temperatures are calculated, as was implicitly assumed earlier in this section.

Now the N eigenvalues of A are distinct (see below) so that its N eigenvectors x_1, $x_2, \ldots, x_N$ are independent, and

$$Ax_1 = \lambda_1 x_1,$$
$$Ax_2 = \lambda_2 x_2,$$
$$\vdots$$
$$Ax_N = \lambda_N x_N.$$

We now write the error vector e^0 as a linear combination of the x_i:

$$e^0 = c_1 x_1 + c_2 x_2 + \cdots + c_N x_N,$$

where the c's are constants. Then e^1 is, in terms of the x_i,

$$e^1 = Ae^0 = \sum_{i=1}^{N} Ac_i x_i = \sum_{i=1}^{N} c_i Ax_i = \sum_{i=1}^{N} c_i \lambda_i x_i,$$

and for e^2,

$$e^2 = Ae^1 = \sum_{i=1}^{N} Ac_i \lambda_i x_i = \sum_{i=1}^{N} c_i \lambda_i^2 x_i.$$

(Again, the superscripts on vectors indicate time; on λ they are exponents.) After j steps, Eq. (8.40) can be written

$$e^j = \sum_{i=1}^{N} c_i \lambda_i^j x_i.$$

If the magnitudes of all of the eigenvalues are less than or equal to unity, errors will not grow as the computations proceed; that is, the computational scheme is stable. This then is the analytical condition for stability: that the largest eigenvalue of the coefficient matrix for the algorithm be one or less in magnitude.

The eigenvalues of matrix A (Eq. 8.39) can be shown to be

$$1 - 4r \sin^2 \frac{n\pi}{2(N+1)}, \qquad n = 1, 2, \ldots, N$$

(note that they are all distinct). We will have stability for the explicit scheme if

$$-1 \leq 1 - 4r \sin^2 \frac{n\pi}{2(N+1)} \leq 1.$$

The limiting value of r is given by

$$-1 \leq 1 - 4r \sin^2 \frac{n\pi}{2(N+1)}$$

$$r \leq \frac{\dfrac{1}{2}}{\sin^2 \left(\dfrac{n\pi}{2(N+1)} \right)}.$$

Hence, if $r \leq \frac{1}{2}$, the explicit scheme is stable.

The Crank–Nicolson scheme, in matrix form, is

$$\begin{bmatrix} (2+2r) & -r & & \\ -r & (2+2r) & -r & \\ & & \ddots & \\ & & -r & (2+2r) \end{bmatrix} \begin{bmatrix} u_1^{j+1} \\ u_2^{j+1} \\ \vdots \\ u_N^{j+1} \end{bmatrix} = \begin{bmatrix} (2-2r) & r & & \\ r & (2-2r) & r & \\ & & \ddots & \\ & & r & (2-2r) \end{bmatrix} \begin{bmatrix} u_1^{j} \\ u_2^{j} \\ \vdots \\ u_N^{j} \end{bmatrix},$$

or

$$Au^{j+1} = Bu^{j}.$$

We can write

$$u^{j+1} = (A^{-1}B)u^{j}, \tag{8.41}$$

so that stability is given by the magnitudes of the eigenvalues of $A^{-1}B$. These are

$$\frac{2 - 4r \sin^2 \left(\dfrac{n\pi}{2(N-1)} \right)}{2 + 4r \sin^2 \left(\dfrac{n\pi}{2(N-1)} \right)}, \qquad n = 1, 2, \ldots, N.$$

Clearly, all the eigenvalues are no greater than one in magnitude for any positive value of r. A similar argument shows that the implicit method is also unconditionally stable.

With derivative boundary conditions, a similar analysis shows that the Crank–Nicolson method is stable for any positive value of r. For the explicit scheme, $r = \frac{1}{2}$ leads to instability with a finite surface coefficient. Smith (1978) shows that the limitation on r for stability is

$$r \le \frac{1}{2 + P\,\Delta x},$$

where P is the ratio of surface coefficient to conductivity, H/k.

The Wave Equation

In this section we will demonstrate only that our finite-difference method is stable when applied to the wave equation in one dimension. Because we will ordinarily solve the one-dimensional wave equation numerically only with

$$\frac{Tg(\Delta t)^2}{w(\Delta x)^2} = 1,$$

it is sufficient to demonstrate stability for that scheme. We assume that a set of errorless computations have been made when a single error of size 1 occurs; we trace the effects of the single error. If the error does not have increasingly great effect on subsequent calculations, we call the method *stable*.

This simple procedure is adequate because, since the problem is linear, the principle of superposition lets us total the effects of all errors and lets us add these errors to the true solution to obtain the actual results. Table 8.11 demonstrates the principle, assuming that the displacement of the endpoints is specified so that these are always free of error.

Table 8.11 Propagation of single error in numerical solution to wave equation

Initially error-free values	0.0	0.0	0.0	0.0	0.0	0.0	0.0
	0.0	0.0	0.0	0.0	0.0	0.0	0.0
Error made here →	0.0	0.0	1.0	0.0	0.0	0.0	0.0
	0.0	1.0	0.0	1.0	0.0	0.0	0.0
	0.0	0.0	1.0	0.0	1.0	0.0	0.0
	0.0	0.0	0.0	1.0	0.0	1.0	0.0
	0.0	0.0	0.0	0.0	1.0	0.0	0.0
	0.0	0.0	0.0	0.0	0.0	0.0	0.0
	0.0	0.0	0.0	0.0	-1.0	0.0	0.0
	0.0	0.0	0.0	-1.0	0.0	-1.0	0.0
	0.0	0.0	-1.0	0.0	-1.0	0.0	0.0
	0.0	-1.0	0.0	-1.0	0.0	0.0	0.0
	0.0	0.0	-1.0	0.0	0.0	0.0	0.0
	0.0	0.0	0.0	0.0	0.0	0.0	0.0
	0.0	0.0	1.0	0.0	0.0	0.0	0.0
	0.0	1.0	0.0	1.0	0.0	0.0	0.0
	0.0	0.0	1.0	0.0	1.0	0.0	0.0

As the arrows indicate, the wave equation propagates disturbances in opposite directions, with reflections occurring at fixed ends, with a reversal of sign on reflection. Stability is demonstrated because the original error does not grow in size.

Chapter Summary

Test your understanding of this chapter by asking yourself if you can

1. Classify second-order partial-differential equations with two independent variables into these types: elliptic, parabolic, hyperbolic.

2. Outline the derivation of both the heat equation and the wave equation.

3. Solve a parabolic partial-differential equation in one dimension by any of these methods: explicit, Crank–Nicolson, theta.

4. Explain how implicit and explicit methods differ.

5. Discuss how finite-difference methods can be applied to both parabolic and hyperbolic partial-differential equations.

6. Solve for the displacements of a vibrating string given its initial conditions and tell why integration may be involved in beginning the solution.

7. Show that the finite-difference technique gives the correct answers to a vibrating string problem by comparing to the D'Alembert solution.

8. Apply the A.D.I. method to parabolic and hyperbolic equations and tell when it is preferred.

9. Discuss the questions of stability and convergence in obtaining solutions to the equations of this chapter.

Computer Programs

There are two programs in this chapter. The first, a C program, solves a parabolic equation for a one-dimensional object. It implements the theta method. The second, written in Fortran 90, uses the A.D.I. method to solve a parabolic equation over a rectangular region.

Program 8.1

The C program that uses the theta method to solve for unsteady heat flow within a rod is listed in Figure 8.9. The output from the program for a test run follows the listing. The program requests the parameters of the problem from the user: values for length, c, k, rho, number of interior nodes, the initial temperatures at these nodes, the Dirichlet boundary conditions, and values for theta and r. Comments within the code tell the steps used in solving the problem. The LU technique is employed to solve the equations.

```
/* * * * * * * * * * * * * * * * * * * * * * * * * * * * * * * * * * * * * * * * * * * * * *
 *                                                                                        *
 *   Chapter 8          Temperatures in a Rod by Theta Method                             *
 *                                                                                        *
 *   Gerald/Wheatley, APPLIED NUMERICAL ANALYSIS (sixth edition)                          *
 *                   Addison Wesley Longman, 1999                                         *
 *                                                                                        *
 * * * * * * * * * * * * * * * * * * * * * * * * * * * * * * * * * * * * * * * * * * * * * *
 *                                                                                        *
 *   This program uses the theta method to solve for changes in                           *
 *   temperatures along a rod from its initial temperatures as                            *
 *   time progresses. The thermal properties and length of the                            *
 *   rod are first obtained from the user, then the initial                               *
 *   temperatures and end conditions. Next, the user inputs values                        *
 *   for theta and r.                                                                     *
 *                                                                                        *
 *   Using these parameters of the problem, the program sets up                           *
 *   the coefficient matrix for the equations, displays this,                             *
 *   finds its LU decomposition and displays it. With this                                *
 *   preliminary work done, temperatures at five successive time                          *
 *   steps are computed. The user can request additional sets of                          *
 *   five computations until no more are desired.                                         *
 *                                                                                        *
 * * * * * * * * * * * * * * * * * * * * * * * * * * * * * * * * * * * * * * * * * * * * * */

#include <stdio.h>
#include <math.h>
#include <conio.h>/*

/* main begins */
void main()  {

/* Declare variables */
  float len, c, k, rho, u[20], ul, ur, th, dx, dt, r, t;
  float a[20][4], b[20], y[20];
  int nInt, i, j, quit;
  char inbuf[20];

clrscr();              /* Clear the output screen */

/* Get rod data from user */
  printf ("\n Enter values for length, c, k, rho: ");
  gets (inbuf);
  sscanf (inbuf,"%f %f %f %f", &len, &c, &k, &rho);

/* Get initial u-values from user: nInt intervals, in u[ ] */
  printf ("\n Enter number of intervals: ");
  gets (inbuf);
  sscanf(inbuf,"%d", &nInt);
  printf ("\n Enter the initial u-values:\n");
  for (i=0; i<=nInt; i++)
```

Figure 8.9

```
        {
          printf("u[%d]]: ",i);
          gets (inbuf);
          sscanf(inbuf,"%f", &u[i]);
        }                               /* end of for */

    /* Get end-point values */
        printf ("\n Enter end-values, UL and UR: ");
        gets (inbuf);
        sscanf (inbuf, "%f %f", &ul, &ur);

    /* Get theta and r values from user */
        printf ("\n Enter values for theta and r: ");
        gets (inbuf);
        sscanf (inbuf, "%f %f", &th, &r);

    /* Compute dx, dt */
        dx = len / nInt;
        dt = r * c * rho * dx * dx / k;

        printf ("\n dx and dt are %6.3f %6.3f",dx, dt);
        getch();

    /* Set up the A matrix */
        for (i=1; i<nInt; i++)
        {
          a[i][1] = a[i][3] = -r * th;
          a[i][2] = 1 + 2 * r * th;
        }                               /* end of for */

    /* Print the A matrix */
        printf ("\n The A matrix is :\n");
        for (i=1; i<nInt; i++)
        printf ("%6.3f %6.3f %6.3f\n", a[i][1], a[i][2], a[i][3]);

    /* Get the LU equivalent to A */
        for (i=2; i<nInt; i++)
        {
          a[i-1][3] = a[i-1][3] / a[i-1][2];
          a[i][2] = a[i][2] - a[i][1] * a[i-1][3];
        }                               /* end of for */

    /* Print the LU matrix (it is in A) */
        printf ("\n The LU matrix is :\n");
        for (i=1; i<nInt;i++)
          printf ("%6.3f %6.3f %6.3f\n", a[i][1], a[i][2], a[i][3]);
        getch();

    /* Do a heading */
        printf ("\n        t               u-values:\n");

    /* Print the initial values */
        t = 0;
```

Figure 8.9 *Continued*

```
    printf ("\n%8.3f   ",t);
    for (i=0; i<=nInt; i++)
      printf ("%6.3f ", u[i]);

/* Do computations until user enters ^C
   Repeat five times: set up rhs; adjust b1, bn;
   solve Ly=b; solve Uu=y; print t, u's
*/
    do                              /* do loop begins */
    {
      u[0] = ul;
      u[nInt] = ur;
      for (j=1; j<=5; j++)
      {                             /* a long for loop begins */

/* set up the b vector */
        for (i=1; i<nInt; i++)
          b[i] = r*(1-th)*u[i-1] + (1-2*r*(1-th))*u[i] + r*(1-th)*u[i+1];

/* adjust the end b's */
        b[1] = b[1] - a[1][1]*u[0];
        b[nInt-1] = b[nInt-1] - a[nInt-1][3]*u[nInt];

/* Solve for y from Ly = b */
        y[1] = b[1] / a[1][2];
        for (i=2; i<nInt; i++)
          y[i] = (b[i] - y[i-1]*a[i][1])/a[i][2];

/* Solve for u from Uu = y */
        u[nInt-1] = y[nInt-1];
        for (i=nInt-2; i>0; --i)
          u[i] = y[i] - a[i][3]*u[i+1];

/* Print new values */
        t = t + dt;
        printf ("\n%8.3f ",t);
        for (i=0; i<=nInt; i++)
          printf ("%6.3f ",u[i]);

    }                              /* End of for j */
    printf ("\nEnter 0 to quit, else press any key");
    gets (inbuf);
    sscanf (inbuf, "%d", &quit);
    }                              /* End of do/while */
  while (quit != 0);
}   /* End of main */

     *************** Output for the Theta Program **************

  Enter values for length, c, k, rho: 2 .11 .13 7.8

  Enter number of intervals: 8
```

Figure 8.9 *Continued*

```
Enter the initial u-values:
 u[0]]: 0
 u[1]]: 25
 u[2]]: 50
 u[3]]: 75
 u[4]]: 100
 u[5]]: 75
 u[6]]: 50
 u[7]]: 25
 u[8]]: 0

Enter end-values, UL and UR: 0 0

Enter values for theta and r: .5 .5

dx and dt are  0.250  0.206

   The A matrix is :
   -0.250  1.500  -0.250
   -0.250  1.500  -0.250
   -0.250  1.500  -0.250
   -0.250  1.500  -0.250
   -0.250  1.500  -0.250
   -0.250  1.500  -0.250
   -0.250  1.500  -0.250

   The LU matrix is :
   -0.250  1.500  -0.167
   -0.250  1.458  -0.171
   -0.250  1.457  -0.172
   -0.250  1.457  -0.172
   -0.250  1.457  -0.172
   -0.250  1.457  -0.172
   -0.250  1.457  -0.250


   t          u-values:
 0.000    0.000 25.000 50.000 75.000 100.000 75.000 50.000 25.000 0.000
 0.206    0.000 24.913 49.480 71.967 82.322 71.967 49.480 24.913 0.000
 0.413    0.000 24.510 47.752 66.162 73.484 66.162 47.752 24.510 0.000
 0.619    0.000 23.649 45.125 60.924 66.856 60.924 45.125 23.649 0.000
 0.825    0.000 22.446 42.251 56.236 61.339 56.236 42.251 22.446 0.000
 1.031    0.000 21.087 39.377 51.994 56.523 51.994 39.377 21.087 0.000
Enter 0 to quit, else press any key

 1.237    0.000 19.693 36.608 48.118 52.212 48.118 36.608 19.693 0.000
 1.444    0.000 18.330 33.985 44.556 48.295 44.556 33.985 18.330 0.000
 1.650    0.000 17.028 31.526 41.271 44.707 41.271 31.526 17.028 0.000
 1.856    0.000 15.802 29.231 38.235 41.404 38.235 29.231 15.802 0.000
 2.062    0.000 14.656 27.097 35.426 38.355 35.426 27.097 14.656 0.000
Enter 0 to quit, else press any key
```

Figure 8.9 *Continued*

```
2.269     0.000 13.587 25.115 32.826 35.536 32.826 25.115 13.587 0.000
2.475     0.000 12.594 23.276 30.418 32.927 30.418 23.276 12.594 0.000
2.681     0.000 11.672 21.570 28.186 30.510 28.186 21.570 11.672 0.000
2.887     0.000 10.817 19.989 26.119 28.272 26.119 19.989 10.817 0.000
3.094     0.000 10.025 18.524 24.204 26.198 24.204 18.524 10.025 0.000
Enter 0 to quit, else press any key


                                   .
                                   .
                                   .


7.425     0.000  2.025  3.743  4.890  5.293  4.890  3.743  2.025 0.000
7.631     0.000  1.877  3.468  4.531  4.905  4.531  3.468  1.877 0.000
7.838     0.000  1.739  3.214  4.199  4.545  4.199  3.214  1.739 0.000
8.044     0.000  1.612  2.978  3.891  4.212  3.891  2.978  1.612 0.000
8.250     0.000  1.494  2.760  3.606  3.903  3.606  2.760  1.494 0.000
Enter 0 to quit, else press any key
```

Figure 8.9 *Continued*

Program 8.2

Figure 8.10 lists the Fortran program. Output from a sample run follows.

```
Program ADI_Parabolic
```

```
!!!!!!!!!!!!!!!!!!!!!!!!!!!!!!!!!!!!!!!!!!!!!!!!!!!!!!!!!!!!!!!!!!!!!!!!
!                                                                    !
!    Chapter 8                                                       !
!                                                                    !
!   Gerald/Wheatley, APPLIED NUMERICAL ANALYSIS(Sixth Edition)       !
!                 Addison Wesley Longman, 1999                       !
!                                                                    !
!                                                                    !
!!!!!!!!!!!!!!!!!!!!!!!!!!!!!!!!!!!!!!!!!!!!!!!!!!!!!!!!!!!!!!!!!!!!!!!!
!                                                                    !
!   This program solves the unsteady-state heat flow equation in     !
!   two space dimensions using the A.D.I. method.                    !
!                                                                    !
!                                                                    !
!!!!!!!!!!!!!!!!!!!!!!!!!!!!!!!!!!!!!!!!!!!!!!!!!!!!!!!!!!!!!!!!!!!!!!!!
!                                                                    !
```

Figure 8.10

```
!        Parameters are :                                          !
!                                                                  !
!     U      - vector of temperatures after odd traverses          !
!     V      - temperatures after even traverses                   !
!     U_Coefficient - matrix of coefficients for the U'S           !
!     V_Coefficient - matrix for the V'S                           !
!     U_Bndary - vector of boundary values for the U equations     !
!     V_Bndary - same for the V'S                                  !
!     Top    - vector to hold boundary values across top of region !
!     Bottom   - same for the values across the bottom             !
!     Right    - hold the right hand side values                   !
!     Left   - hold values for the left Hand edge                  !
!     m      - number of rows of nodes in the grid                 !
!     n      - number of columns of nodes                          !
!     diff   - thermal diffusivity = k/c*density                   !
!     H      - the value of delta X = delta Y                      !
!     time   - the time variable                                   !
!     tMAX   - maximum value of time for which computations are     !
!               desired                                            !
!     DT        - time step size, related to delta-x through r     !
!                                                                  !
!                                                                  !
!!!!!!!!!!!!!!!!!!!!!!!!!!!!!!!!!!!!!!!!!!!!!!!!!!!!!!!!!!!!!!!!!!!!!
!                                                                  !
      Implicit None
      Real(Kind=8), Dimension(500,3) :: U_coefficient,V_coefficient
      Real(Kind=8), Dimension(500) :: U,V,U_Bndary, V_Bndary
      Real(Kind=8), Dimension(500) :: Top,Bottom,Left,Right
      Real(Kind=8) :: h = 0.125, diff = 0.152, tMAX = 20.0, r = 0.5
      Real(Kind=8) :: time, dt
      Integer ::= m = 7, n = 15, i, j,k,l,jROW, mSIZE
!
! ----------------------------------------------------------
!
!
      Do j = 1,3
         Do i = 1,500
            U_Coefficient(i,j) = -1.0
            V_Coefficient(i,j) = -1.0
            End Do
         End Do
!
!  We initialize the temperatures everywhere to zero.
!
      Do i = 1,500
         V(i) = 0.0
         Top(i) = 0.0
         Bottom(i) = 0.0
         Right(i) = 100.0
         Left(i) = 0.0
         U_Bndary(i) = 0.0
         V_Bndary(i) = 0.0
         End Do
```

Figure 8.10 *Continued*

```
!
! -----------------------------------------------------------
!
!  Set up the coefficient matrices by over-writing on the
!  diagonal and certain off-diagonal terms.
!
      mSIZE == m * n
      Do i = 1,mSIZE
         U_coefficient(i,2) = 1.0/r + 2.0
         V_coefficient(i,2) = 1.0/r + 2.0
         End Do
!
      Do i = n,mSIZE-1,n
         U_coefficient(i,3) = 0.0
         U_coefficient(i+1,1) = 0.0
         End Do
      Do i = m,mSIZE-1,m
         V_coefficient(i,3) =0.0
         V_coefficient(i+1,1) = 0.0
         End Do
!
! -----------------------------------------------------------
!
!  Now get values into the boundary vectors
!
      Do i = 1,n
         U_Bndary(i) = Top(i)
         j = mSIZE - n + i
         U_Bndary(j) = Bottom(i)
         End Do
      Do  i =1,m
         j = (i-1)*n + 1
         U_Bndary(j) = U_Bndary(j) + Left(i)
         j = I * N
         U_Bndary(j) = U_Bndary(j) + Right(i)
         End Do
      Do  i = 1,m
         V_Bndary(i) = Left(i)
         j = mSIZE - m + i
         V_Bndary(j) = Right(i)
         End Do
      Do  i = 1,n
         j = (i-1)*m + 1
         V_Bndary(j) = V_Bndary(j) + Top(i)
         j = i * m
         V_Bndary(j) = V_Bndary(j) + Bottom(i)
         End Do
!
! -----------------------------------------------------------
!
!  We Now get the LU decompositions of U_coefficient and V_coefficient
!
```

Figure 8.10 *Continued*

```
      Do   i = 2,mSIZE
          U_coefficient(i-1,3) = U_coefficient(i-1,3)/U_coefficient(i-1,2)
          U_coefficient(i,2) = U_coefficient(i,2) - &
                  U_coefficient(i,1)*U_coefficient(i-1,3)
          V_coefficient(i-1,3) = V_coefficient(i-1,3)/V_coefficient(i-1,2)
          V_coefficient(i,2) = V_coefficient(i,2) - &
                  V_coefficient(i,1)*V_coefficient(i-1,3)
      End Do
!
! ------------------------------------------------------------
!
!  Now we do the iterations until time equals tMAX.
!
      time = 0.0
      dt = R / diff*H*H
!
!
!
!
    Write(*,*) '   ************************************************ '
    Write(*,*) '                 Output for Adiun.f90               '
    Write(*,*)
!
    Do
    IF ( time > tMAX ) Stop
!
! ------------------------------------------------------------
!
!  Compute the rhs for the U equations and store in the U
!  vector. First do the top and bottom rows.
!
      Do   i = 1,n
          j = (i-1)*m + 1
          U(i) = ( 1.0/r - 2.0 )*V(j) + V(j+1) + U_Bndary(i)
          k = mSIZE - N + I
          j = i * m
          U(k) = V(j-1) + ( 1.0/r - 2.0 )*V(j) + U_Bndary(k)
      End Do
!
! ------------------------------------------------------------
!
!  Now for the other ones
!
      Do   i = 2,m-1
          Do j = 1,n
          K = (i-1)*n + J
          L = I + (j-1)*m
          U(k) = V(L-1) + (1.0/r-2.0)*V(L) + V(L+1) + U_Bndary(k)
          End Do
      End Do
!
! ------------------------------------------------------------
!
```

Figure 8.10 *Continued*

```
!   Now get the solution for the odd traverse
!

    U(1) = U(1) / U_coefficient(1,2)
    Do   i = 2,mSIZE
        U(i) = ( U(i) - U_coefficient(i,1)*U(i-1)) / U_coefficient(i,2)
        End Do
    Do i = 1,mSIZE-1
        jROW = mSIZE - I
        U(jROW) = U(jROW) - U_coefficient(jROW,3)*U(jROW+1)
        End Do
!
! ------------------------------------------------------------------
!
!   Compute the R.H.S. for the even traverse, store in V. Do the
!   Top and Bottom ones.
!
    Do   i = 1,m
        j = (i-1)*n + 1
        V(i) = ( 1.0/r - 2.0 )*U(j) + U(j+1) + V_Bndary(i)
        K = mSIZE - M + I
        j = i * n
        V(k) = U(j-1) + ( 1.0/r - 2.0 )*U(j) + V_Bndary(k)
        End Do
!
! ------------------------------------------------------------------
!
!   Now the rest of the rows
!
    Do   i = 2,n-1
        Do j = 1,m
            K = (i-1)*m + J
            L = I + (j-1)*n
            V(k) = U(L-1) + (1.0/r-2.0)*U(L) + U(L+1) + V_Bndary(k)
            End Do
        End Do
!
! ------------------------------------------------------------------
!
!   Get the solution for the even traverse
!
    V(1) = V(1) / V_coefficient(1,2)
    Do   i = 2,mSIZE
        V(i) = ( V(i) - V_coefficient(i,1)*V(i-1)) / V_coefficient(i,2)
        End Do
    Do   i = 1,mSIZE-1
        jROW = mSIZE - I
        V(jROW) = V(jROW) - V_coefficient(jROW,3)*V(jROW+1)
        End Do
        time = time + 2.0*dt
!
! ------------------------------------------------------------------
!
```

Figure 8.10 *Continued*

```
!    Print out the last result.
!
     Write(*,200) time, ( V(i), I = 1,mSIZE )
200  format(//' When T = ',F6.3,' V values are:' /(15F8.3))
     End Do
     Stop
 End  Program ADI_Parabolic
```

```
*************** Output for the Adiun Program **************

When T =  0.103 V values are:
 0.000    0.000    0.000    0.000    0.000    0.000    0.000    0.000    0.000
0.000    0.000    0.000    0.000    0.000    0.000
 0.000    0.000    0.000    0.000    0.000    0.000    0.000    0.000    0.000
0.000    0.000    0.000    0.000    0.000    0.000
 0.000    0.000    0.000    0.000    0.000    0.000    0.000    0.000    0.000
0.000    0.000    0.000    0.001    0.001    0.001
 0.001    0.001    0.001    0.001    0.004    0.005    0.005    0.005    0.005
0.005    0.004    0.015    0.018    0.019    0.020
 0.019    0.018    0.015    0.054    0.069    0.072    0.073    0.072    0.069
0.054    0.202    0.256    0.271    0.273    0.271
 0.256    0.202    0.755    0.957    1.010    1.020    1.010    0.957    0.755
2.816    3.570    3.768    3.808    3.768    3.570
 2.816   10.510   13.323   14.063   14.211   14.063   13.323   10.510   39.226
49.723   52.485   53.037   52.485   49.723   39.226


When T =  0.206 V values are:
 0.000    0.000    0.000    0.000    0.000    0.000    0.000    0.000    0.000
0.000    0.000    0.000    0.000    0.000    0.000
 0.000    0.000    0.000    0.000    0.000    0.000    0.000    0.000    0.000
0.000    0.000    0.000    0.000    0.001    0.001
 0.001    0.001    0.001    0.001    0.001    0.002    0.003    0.004    0.004
0.004    0.003    0.002    0.007    0.011    0.013
 0.014    0.013    0.011    0.007    0.022    0.038    0.045    0.047    0.045
0.038    0.022    0.074    0.125    0.147    0.152
 0.147    0.125    0.074    0.239    0.402    0.470    0.487    0.470    0.402
0.239    0.758    1.258    1.464    1.516    1.464
 1.258    0.758    2.328    3.790    4.385    4.535    4.385    3.790    2.328
6.816   10.770   12.346   12.740   12.346   10.770
 6.816   18.442   27.603   31.075   31.924   31.075   27.603   18.442   42.723
56.019   59.986   60.848   59.986   56.019   42.723


When T =  0.308 V values are:
 0.000    0.000    0.000    0.000    0.000    0.000    0.000    0.000    0.000
0.000    0.000    0.000    0.000    0.000    0.000
 0.001    0.001    0.001    0.001    0.001    0.000    0.001    0.002    0.002
0.002    0.002    0.002    0.001    0.003    0.005
 0.006    0.007    0.006    0.005    0.003    0.009    0.016    0.020    0.021
0.020    0.016    0.009    0.027    0.048    0.060
 0.064    0.060    0.048    0.027    0.080    0.141    0.177    0.188    0.177
```

Figure 8.10 *Continued*

```
0.141    0.080    0.230    0.404    0.506    0.537
    0.506    0.404    0.230    0.640    1.120    1.393    1.476    1.393    1.120
0.640    1.705    2.952    3.643    3.851    3.643
    2.952    1.705    4.280    7.285    8.879    9.351    8.879    7.285    4.280
9.965    16.409    19.595    20.509    19.595    16.409
    9.965    21.341    32.794    37.750    39.078    37.750    32.794    21.341    45.063
60.209    65.374    66.622    65.374    60.209    45.063


    When T =  0.411 V values are:
    0.000    0.000    0.000    0.000    0.000    0.000    0.000    0.000    0.001
0.001    0.001    0.001    0.001    0.000    0.001
    0.002    0.003    0.003    0.003    0.002    0.001    0.003    0.006    0.008
0.008    0.008    0.006    0.003    0.009    0.017
    0.022    0.024    0.022    0.017    0.009    0.027    0.049    0.062    0.067
0.062    0.049    0.027    0.073    0.132    0.170
    0.183    0.170    0.132    0.073    0.193    0.349    0.447    0.480    0.447
0.349    0.193    0.490    0.884    1.128    1.209
    1.128    0.884    0.490    1.189    2.130    2.702    2.891    2.702    2.130
1.189    2.723    4.827    6.074    6.479    6.074
    4.827    2.723    5.851    10.178    12.631    13.406    12.631    10.178    5.851
11.832    19.848    24.056    25.329    24.056    19.848
    11.832    23.214    36.244    42.224    43.913    42.224    36.244    23.214    46.133
62.180    67.929    69.383    67.929    62.180    46.133


    When T =  0.514 V values are:
    0.000    0.001    0.001    0.001    0.001    0.001    0.000    0.001    0.002
0.003    0.003    0.003    0.002    0.001    0.003
    0.006    0.008    0.009    0.008    0.006    0.003    0.009    0.017    0.022
0.023    0.022    0.017    0.009    0.024    0.044
    0.057    0.061    0.057    0.044    0.024    0.061    0.112    0.145    0.157
0.145    0.112    0.061    0.151    0.277    0.358
    0.386    0.358    0.277    0.151    0.360    0.657    0.848    0.914    0.848
0.657    0.360    0.817    1.487    1.915    2.060
    1.915    1.487    0.817    1.765    3.193    4.090    4.392    4.090    3.193
1.765    3.611    6.464    8.210    8.788    8.210
    6.464    3.611    7.016    12.326    15.434    16.436    15.434    12.326    7.016
13.140    22.259    27.202    28.731    27.202    22.259
    13.140    24.338    38.317    44.929    46.838    44.929    38.317    24.338    46.814
63.435    69.567    71.155    69.567    63.435    46.814


    When T =  0.617 V values are:
    0.001    0.002    0.002    0.003    0.002    0.002    0.001    0.003    0.006
0.007    0.008    0.007    0.006    0.003    0.008
    0.015    0.019    0.020    0.019    0.015    0.008    0.020    0.036    0.048
0.051    0.048    0.036    0.020    0.048    0.089
    0.116    0.125    0.116    0.089    0.048    0.113    0.209    0.271    0.293
0.271    0.209    0.113    0.257    0.472    0.613
    0.662    0.613    0.472    0.257    0.558    1.023    1.326    1.431    1.326
1.023    0.558    1.158    2.116    2.736    2.948
```

Figure 8.10 *Continued*

```
    2.736    2.116    1.158    2.291    4.165    5.359    5.765    5.359    4.165
 2.291    4.334    7.800    9.955   10.677    9.955
    7.800    4.334    7.892   13.943   17.545   18.722   17.545   13.943    7.892
14.038   23.919   29.369   31.077   29.369   23.919
   14.038   25.096   39.717   46.757   48.816   46.757   39.717   25.096   47.244
64.230   70.605   72.278   70.605   64.230   47.244


When T =  0.720 V values are:
    0.002    0.004    0.005    0.006    0.005    0.004    0.002    0.006    0.012
 0.015    0.017    0.015    0.012    0.006    0.016
    0.029    0.038    0.041    0.038    0.029    0.016    0.037    0.067    0.088
 0.095    0.088    0.067    0.037    0.083    0.152
    0.199    0.215    0.199    0.152    0.083    0.180    0.333    0.433    0.468
 0.433    0.333    0.180    0.379    0.697    0.908
    0.981    0.908    0.697    0.379    0.764    1.404    1.824    1.970    1.824
 1.404    0.764    1.478    2.707    3.508    3.785
    3.508    2.707    1.478    2.744    5.000    6.451    6.947    6.451    5.000
 2.744    4.911    8.866   11.347   12.184   11.347
    8.866    4.911    8.545   15.150   19.122   20.428   19.122   15.150    8.545
14.681   25.107   30.922   32.758   30.922   25.107
   14.681   25.615   40.676   48.009   50.172   48.009   40.676   25.615   47.536
64.770   71.311   73.042   71.311   64.770   47.536
```

Figure 8.10 *Continued*

Exercises

Section 8.1

In the exercises of this chapter, subscripts indicate a partial derivative, for example, $u_{xx} = \partial^2 u/\partial x^2$.

1. Classify the following as elliptic, parabolic, or hyperbolic.

   a. $(Tw_x)_x = p * g.$

   b. $(xu_x)_x + u_y = \dfrac{(2 + x + y)}{(1 - x)}.$

   c. $kU_{tt} + mU_{xt} - (au_x)_x + bU = f(x, t).$

   d. $(TW_x)_x - k^2 W_t = 0, \; W(0) = 0, \; W(L) = 0.$

▶ 2. For what values of x and y is this equation elliptic, parabolic, hyperbolic?

   $(1 + y)u_{xx} + 2(1 - x)u_{xy} - (1 - y)u_{yy} = f(x, y).$

3. Divide the (x, y)-plane into regions where this equation is elliptic, parabolic, hyperbolic:

   $x^3 u_{xx} - 2x^2 y u_{xy} + x u_{yy} = x^2 - u_x + u_y.$

Section 8.2

4. The parameters of the basic equation for unsteady-state heat flow are dimensional. If it is desired to measure u in °F and x in feet, how must the units of k, c, and ρ be chosen in

   $$\frac{\partial^2 u}{\partial x^2} = \frac{c\rho}{k} \frac{\partial u}{\partial t}?$$

5. In Section 8.2, the heat equation is shown for a case where there is heat generation (or absorption) and where k, c, and ρ vary. Rederive this equation for the one-dimensional case.

▶ 6. Suppose that the rod sketched in Fig. 8.1 is tapered, with the diameter varying linearly from 2 in. at the left end to 1.25 in. at the right end; the rod is 14 in. long and is made of steel. If 200 BTU/hr of heat flows from left to right (the flow is the same at each x-value along the rod—steady state), what are the values of the gradient at

a. the left end?

b. the right end?

c. $x = 3$ in.?

7. What would be the equation equivalent to Eq. (8.7) if W, the weight per unit length, is not constant but varies along the string, $w = w(x)$?

Section 8.3

8. Solve for the temperatures at $t = 2.06$ sec in the 2-cm-thick steel slab of Example 8.1 if the initial temperatures are given by

$$u(x, t) = 100 \sin\left(\frac{\pi x}{2}\right).$$

Use the explicit method with $\Delta x = 0.25$ cm. Compare to the analytical solution: $100e^{-0.3738t} \sin(\pi x/2)$.

▶ 9. Repeat Exercise 8, but now use the Crank–Nicolson method.

10. Repeat Exercise 8, but now use the theta method, with

a. $\theta = \frac{2}{3}$.

b. $\theta = 0.878$.

c. $\theta = 1.0$.

11. Solve for the temperatures in a cylindrical copper rod that is 8 in. long and whose curved outer surface is insulated so that heat flows only in one direction. The initial temperature is linear from $0°C$ at one end to $100°C$ at the other, when suddenly the hot end is brought to $0°C$ and the cold end is brought to $100°C$. Use $\Delta x = 1$ in. and an appropriate value of Δt so that $k \, \Delta t / cp(\Delta x)^2 = \frac{1}{2}$. Look up values for k, c, and ρ in a handbook. Carry out the solution for 10 time steps.

12. Repeat Exercise 11, but with $\Delta x = 0.5$ in., and compare the temperature at points 1 in., 3 in., and 6 in. from the cold end with those of the previous exercise. You will need to compute more time steps to match the 10 steps done previously.

   You will find it instructive to graph the temperatures for both sets of computations.

13. Repeat Exercise 11, but with $\Delta x = 1$ in. and Δt such that $k \, \Delta t / cp(\Delta x)^2 = \frac{1}{4}$. Compare results with Exercises 11 and 12.

Section 8.4

14. If the banjo string of Example 8.4 is tightened or if it is shortened (as by holding it against one of the frets with a finger), the frequency of vibration is raised and the pitch of the sound is higher. What would the frequency be if the tension is made 42,500 g and the effective length is 65 cm? Determine this by finding the number of time steps for the original displacement to be repeated. Compare this to

$$f = \left(\frac{1}{2L}\right) \sqrt{\frac{Tg}{w}}.$$

15. A vibrating string has $Tg/w = 4$ cm^2/sec^2 and is 48 cm long. Divide the length into subintervals so that $\Delta x = L/8$. Find the displacement for $t = 0$ to $t = L$ if both ends are fixed and the initial conditions are

a. $y = x(x - L)/L^2$, $y_t = 0$. (y_t is the velocity.)

b. the string is displaced $+2$ units at $L/4$ and -1 unit at $5L/8$, $y_t = 0$.

c. $y = 0$, $y_t = x(L - x)/L^2$. (Use Eq. 8.19.)

d. the string is displaced 1 unit at $L/2$, $y_t = -y$.

e. Compare part (a) to the analytical solution,

$$y = \frac{8}{\pi^3} \sum_{n=1}^{\infty} \frac{1}{(2n-1)^3} \sin\left[(2n-1)\frac{\pi x}{L}\right] \cos\left[(4n-2)\frac{\pi t}{L}\right].$$

▶ 16. The function u satisfies the equation

$$u_{xx} = u_{tt},$$

with boundary conditions of $u = 0$ at $x = 0$ and $u = 0$ at $x = 1$, and with initial conditions

$$u = \sin(\pi x), \quad u_t = 0, \quad \text{for } 0 \le x \le 1.$$

Solve by the finite-difference method and show that the results are the same as the analytical solution,

$$u(x, t) = \sin(\pi x)\cos(\pi t).$$

17. The ends of the vibrating string do not have to be fixed. Solve the equation $u_{xx} = u_{tt}$ with $y(x, 0) = 0$, $y_t(x, 0) = 0$ for $0 \le x \le 1$, and end conditions of

$$y(0, t) = 0, \quad y(1, t) = \sin\left(\frac{\pi t}{4}\right), \quad y_x(1, t) = 0.$$

▶ 18. Equation (8.19) is inaccurate when the initial velocity is not zero or a constant, so the solutions to parts (c) and (d) of Exercise 15 are not exact. Solve again, but use Eq. (8.26) to begin the solution. (Use Simpson's $\frac{1}{3}$ rule as in Example 8.5. How different are these from the original computations?

19. Repeat Exercise 18 but now use more points between x_{i-1} and x_{i+1} in Simpson's rule. Does this make much difference?

20. A string that weighs w lb/ft is tightly stretched between $x = 0$ and $x = L$ and is initially at rest. Each point is given an initial velocity of

$$y_t(x, 0) = v_0 \sin^3\left(\frac{\pi x}{L}\right).$$

The analytical solution is

$$y(x, t) =$$

$$\frac{v_0 L}{12a\pi}\left(9 \sin\frac{\pi x}{L}\sin\frac{a\pi t}{L} - \sin\frac{3\pi x}{L}\sin\frac{3a\pi t}{L}\right),$$

where $a = \sqrt{Tg/w}$, with T the tension and g the acceleration due to gravity. When $L = 3$ ft, $w = 0.02$ lb/ft, and $T = 5$ lb, with $v_0 = 1$ ft/sec, the analytical formula predicts $y = 0.081$ in. at the midpoint when $t = 0.01$ sec. Solve the problem numerically to confirm this. Does your solution agree with the analytical solution at other values of x and t?

Section 8.5

▶ **21.** A rectangular plate 3 in. × 4 in. is initially at 50°. At $t = 0$, one 3-in. edge is suddenly raised to 100°, and one 4-in. edge is suddenly cooled to 0°. The temperature on these two edges is held constant at these temperatures. The other two edges are perfectly insulated. Use a 1-in. grid to subdivide the plate and write the A.D.I. equations for each of the six nodes where unknown temperatures are involved. Use $r = 2$, and solve the equations for four time steps.

22. A cube of aluminum is 4 in. on each side. Heat flows in all three directions. Three adjacent faces lose heat by conduction to a flowing fluid; the other faces are held at a constant temperature different from that of the fluid. Set up the equations that can be solved for the temperature at nodes using the explicit method with a 1-in. spacing between all nodes. How many time steps are needed to reach 15.12 sec using the maximum r-value for stability? (Look up the properties of aluminum in a handbook). How many equations must be solved at each time step?

23. Repeat Exercise 22 for the Crank–Nicolson method, $r = 1$.

24. Repeat Exercise 22 for the implicit method, $r = 1$.

25. Repeat Exercise 22 for the A.D.I. method, $r = 1$.

Section 8.6

26. Solve the vibrating membrane of Example 8.7, but with initial conditions of

$$u(x, y) = 0 \qquad \text{at } t = 0,$$
$$u_t(x, y) = x^2(2 - x)y^2(2 - y) \qquad \text{at } t = 0.$$

27. Solve the vibrating membrane of Example 8.7, but with

$$u(x, y) = x^2(2 - x)y^2(2 - y) \qquad \text{at } t = 0,$$
$$u_t(x, y) = 0 \qquad \text{at } t = 0.$$

28. A membrane is stretched over a frame that occupies the region in the xy-plane bounded by

$$x = 0, \qquad x = 3, \qquad y = 0, \qquad y = 2.$$

At $t = 0$, the point on the membrane at $(1, 1)$ is lifted 1 unit above the xy-plane and then released. If $T = 6$ lb/in. and $w = 0.55$ lb/in.^{2}, find the displacement of the point $(2, 1)$ as a function of time.

29. How do the vibrations of Exercise 28 change if $w = 0.055$, the other parameters being unchanged?

30. The frame holding the membrane of Exercise 28 is distorted by lifting the corner at $(3, 2)$ 1 unit above the xy-plane. (The members of the frame elongate so that the corner moves vertically.) The membrane is set to vibrating in the same way as in Exercise 28. Follow the vibrations through time. [Assume that the rest positions of points on the membrane lie on the two planes defined by the adjacent edges that meet at $(0, 0)$ and at $(3, 2)$.]

Section 8.7

31. Demonstrate by performing calculations similar to those in Table 8.9 that the explicit method is unstable for $r = 0.6$.

32. Demonstrate by performing calculations similar to those in Table 8.9 that the explicit method is stable for $r = 0.25$ but that the errors damp out less rapidly.

33. Suppose that the end conditions are not $u = $ a constant as in Table 8.9, but rather $u_x = 0$. Demonstrate by performing calculations similar to those in Table 8.9 that the explicit method is still stable for $r = 0.5$ but that the errors damp out much more slowly. Observe that the errors at a later stage become a linear combination of earlier errors.

▶ **34.** Demonstrate by performing calculations similar to those in Table 8.9 that the Crank–Nicolson method is stable even if $r = 10$. You will need to solve a system of equations in this exercise.

35. Repeat Exercise 34 but for the implicit method.

▶ **36.** Compute the largest eigenvalue of the coefficient matrix in Eq. (8.39) for $r = 0.5$, then for $r = 0.6$. Do you find that the statements in the text relative to eigenvalues are confirmed?

37. Repeat Exercise 36 but for Eq. (8.41) using $A^{-1}B$. Use
$r = 1.0$ and $r = 2.0$.

38. Starting with the matrix form of the implicit method,
show that for $A^{-1}B$ none of the eigenvalues exceed 1 in
magnitude.

39. Paraphrase Table 8.11 to investigate the stability of the
solution when the equivalent of Eq. (8.17) is used but
when the ratio $Tg(\Delta t)^2/w(\Delta x)^2$ is not 1 but is

a. equal to 0.5.
b. equal to 2.0.

▶ **40.** Redo Table 8.11 [with the ratio $Tg(\Delta t)^2/w(\Delta x)^2 = 1$]
but for end conditions given in Exercise 17. Is stability
still indicated?

Applied Problems and Projects

41. A vibrating string, with a damping force opposing its motion that is proportional to the velocity,
follows the equation

$$\frac{\partial^2 y}{\partial t^2} = \frac{Tg}{w}\frac{\partial^2 y}{\partial x^2} - B\frac{\partial y}{\partial t},$$

where B is the magnitude of the damping force. Solve the problem if the length of the string is
5 ft with $T = 24$ lb, $w = 0.1$ lb/ft, and $B = 2.0$. Initial conditions are

$$y(x)\big|_{t=0} = \frac{x}{3}, \qquad 0 \le x < 3,$$

$$y(x)\big|_{t=0} = \frac{5}{2} - \frac{x}{2}, \qquad 3 \le x \le 5,$$

$$\frac{\partial y}{\partial t}\bigg|_{t=0} = x(x - 5).$$

Compute a few points of the solution by difference equations.

42. When steel is forged, billets are heated in a furnace until the metal is of the proper temperature,
between $2000°$F and $2300°$F. It can then be formed by the forging press into rough shapes that
are later given their final finishing operations. To produce a certain machine part, a billet of size
4 in. $\times$ 4 in. $\times$ 20 in. is heated in a furnace whose temperature is maintained at $2350°$F. You
have been requested to estimate how long it will take all parts of the billet to reach a temperature
above $2000°$F. Heat transfers to the surface of the billet at a very high rate, principally through
radiation. It has been suggested that you can solve the problem by assuming that the surface
temperature becomes $2250°$F instantaneously and remains at that temperature. Using this assumption, find the required heating time.

   Because the steel piece is relatively long compared to its width and thickness, it may not
introduce significant error to calculate as if it were infinitely long. This will simplify the problem, permitting a two-dimensional treatment rather than a three-dimensional one. Such a calculation should also give a more conservative estimate of heating time. Compare the estimates
from two- and three-dimensional approaches.

43. After you have calculated the answers to Problem 42, your results have been challenged on the
basis of assuming constant surface temperature of the steel. Radiation of heat flows according to
the equation

$$q = E\sigma(u_F^4 - u_S^4)\ \text{Btu/hr} \cdot \text{ft}^2,$$

where E = emissivity (use 0.80), σ is the Stefan–Boltzmann constant $(0.171 \times 10^{-8}$ Btu/ hr $\cdot$ ft^2 $\cdot$ °F^4), u_F and u_S are the furnace and surface absolute temperatures, respectively (°F + 460°).

The heat radiating to the surface must also flow into the interior of the billet by conduction, so

$$ q = -k \frac{\partial u}{\partial x}, $$

where k is the thermal conductivity of steel (use 26.2 Btu/hr $\cdot$ ft^2 $\cdot$ °F/ft) and $(\partial u / \partial x)$ is the temperature gradient at the surface in a direction normal to the surface. Solve the problem with this boundary condition, and compare your solution to that of Problem 42. (Observe that this is now a nonlinear problem. Think carefully how your solution can cope with it.)

44. A horizontal elastic rod is initially undeformed and is at rest. One end, at $x = 0$, is fixed, and the other end, at $x = L$ (when $t = 0$), is pulled with a steady force of F lb/ft^2. It can be shown that the displacements $y(x, t)$ of points originally at the point x are given by

$$ \frac{\partial^2 y}{\partial t^2} = a^2 \frac{\partial^2 y}{\partial x^2}, \qquad y(0, t) = 0, \qquad \left.\frac{\partial y}{\partial t}\right|_{x=L} = \frac{F}{E}, $$

$$ y(x, 0) = 0, \qquad \left.\frac{\partial y}{\partial t}\right|_{t=0} = 0, $$

where $a^2 = Eg/\rho$; E = Young's modulus (lb/ft^2); g = acceleration of gravity; ρ = density (lb/ft^3). Find y versus t for the midpoint of a 2-ft-long piece of rubber for which $E = 1.8 \times 10^6$ and $\rho = 70$ if $F/E = 0.7$.

45. A circular membrane, when set to vibrating, obeys the equation (in polar coordinates)

$$ \frac{1}{r} \frac{\partial}{\partial r}\left(r \frac{\partial u}{\partial r}\right) + \frac{1}{r^2} \frac{\partial^2 u}{\partial \theta^2} = \frac{w}{Tg} \frac{\partial^2 y}{\partial t^2}. $$

A 3-ft-diameter kettledrum is started to vibrating by depressing the center $\frac{1}{2}$ in. If $w = 0.072$ lb/ ft^2 and $T = 80$ lb/ft, find how the displacements at 6 in. and 12 in. from the center vary with time. The problem can be solved in polar coordinates, or it can be solved in rectangular coordinates using the methods of Section 7.9 to approximate $\nabla^2 u$ near the boundaries.

46. A flexible chain hangs freely, as shown in Fig. 8.11. For small disturbances from its equilibrium position (hanging vertically), the equation of motion is

$$ x \frac{\partial^2 y}{\partial x^2} + \frac{\partial y}{\partial x} = \frac{1}{g} \frac{\partial^2 y}{\partial t^2}. $$

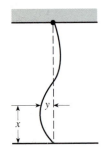

In this equation, x is the distance from the end of the chain, y is the displacement from the equilibrium position, t is the time, and g is the acceleration of gravity. A 10-ft-long chain is originally hanging freely. It is set into motion by striking it sharply at its midpoint, imparting a velocity there of 1 ft/sec. Find how the chain moves as a result of the blow. If you find you need additional information at $t = 0$, make reasonable assumptions.

Figure 8.11

47. Shipment of liquefied natural gas by refrigerated tankers to industrial nations may become an important means of supplying the world's energy needs. It must be stored at the receiving port, however. [A. R. Duffy and his co-workers (1967) discuss the storage of liquefied natural gas in underground tanks.] A commercial design, based on experimental verification of its feasibility, contemplated a prestressed concrete tank 270 ft in diameter and 61 ft deep, holding some 600,000 bbl of liquefied gas at −258°F. Convection currents in the liquid were shown to keep the temperature uniform at this value, the boiling point of the liquid.

Important considerations of the design are the rate of heat gained from the surroundings (causing evaporation of the liquid gas) and variation of temperatures in the earth below the tank (relating to the safety of the tank, which could be affected by possible settling or frost-heaving.)

The tank itself is to be made of concrete 6 in. thick, covered with 8 in. of insulation (on the liquid side). (A sealing barrier keeps the insulation free of liquid; otherwise its insulating capacity would be impaired.) The experimental tests showed that there is a very small temperature drop through the concrete: 12°F. This observed 12°F temperature difference seems reasonable in light of the relatively high thermal conductivity of concrete. We expect then that most of the temperature drop occurs in the insulation or in the earth below the tank.

Because the commercial-design tank is very large, if we are interested in ground temperatures near the center of the tank (where penetration of cold will be a maximum), it should be satisfactory to consider heat flowing in only one dimension, in a direction directly downward from the base of the tank. Making this simplifying assumption, compute how long it will take for the temperature to decrease to 32°F (freezing point of water) at a point 8 ft away from the tank wall. The necessary thermal data are

	Insulation	**Concrete**	**Earth**
Thermal conductivity (Btu/hr · ft · °F)	0.013	0.90	2.6
Density (lb/ft^3)	2.0	150	132
Specific heat (Btu/lb · °F)	0.195	0.200	0.200

Assume the following initial conditions: temperature of liquid, −258°F; temperature of insulation, −258°F to 72°F (inner surface to outer); temperature of concrete, 72°F to 60°F; temperature of earth, 60°F.

48. XYZ Metallurgical has a problem. A slab of steel, 6-ft long, 12-in. wide, and 3-in. thick must be heat-treated and it is a rush job. Unfortunately, their large furnace is down for repairs and the only furnace that can be used will hold just three feet of the slab. It has been proposed that it would be possible to use this furnace if the three feet of the slab that protrude from the furnace are well insulated. (See the figure.) The heat treating requires that all of the slab be held between 950°F and 900°F for at least an hour. The portion that is outside the furnace is covered with a 1-in. thickness of insulation whose thermal conductivity, k, is 0.027 Btu/(hr ∗ ft ∗ °F). Even though you are a new employee, the manager has asked you to determine three things:

(1) Is one inch of this insulation sufficient for all of the slab to reach 900°F with the furnace at 950°F?

(2) If it is, how long will it take for the end of the slab to reach that temperature?

(3) If one inch is insufficient, how much of this same insulation should be used?

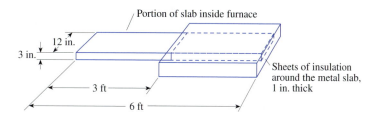

The Finite-Element Method

Contents of this Chapter

The previous two chapters described methods for solving boundary-value problems. These problems were of several types: an ordinary-differential equation of order two or more that must satisfy conditions at the boundaries of some domain, and three types of partial-differential equations that also must satisfy boundary conditions. These latter were classified as (1) elliptic, (2) parabolic, or (3) hyperbolic. The last two types of partial-differential equations are time-dependent. All of these boundary-value problems can be solved by the *finite-difference method* where derivatives are replaced by finite-difference approximations.

When the finite-difference method was used, we usually placed evenly spaced nodes within the region. If nodes do not lie exactly on the boundaries, special adjustments are required for nodes adjacent to the boundary and errors are greater when that is done. (Circular and spherical regions can be accommodated by reformulating the underlying partial-differential equation.)

In this chapter we describe a different approach that is especially valuable for irregular regions—the *finite-element method*. In this method, nodes can be unevenly spaced. This not only solves the problem of matching nodes to a boundary but it also facilitates the placing of nodes closer together in parts of the region where the solution to the problem varies rapidly and this improves the accuracy. The finite-element method is now widely used by applied mathematicians and engineers.

We develop a background for the finite-element method by describing three other techniques that are based on the *calculus of variations*.

9.1 The Rayleigh–Ritz Method

Details a method based on an elegant branch of mathematics, the *calculus of variations*. This method optimizes a so-called *functional*, and by so doing, can approximate the solution to a boundary-value problem.

9.2 The Collocation and Galerkin Methods

Gives alternative techniques for solving a boundary-value problem that avoid having to find the functional, which is sometimes difficult. The Galerkin method is the basis for much of the rest of the chapter.

9.3 Finite Elements for Ordinary-Differential Equations

Applies the Galerkin method to small subregions ("elements") of the region of interest. This provides a way to subdivide the region without requiring that nodes be evenly spaced. The finite-element method (FEM) finds the solution of a boundary-value problem through a system of linear equations. A variety of boundary conditions can be handled. The development is done in several successive steps.

9.4 Finite Elements for Elliptic Partial-Differential Equations

Extends the finite-element procedure to regions of two and three dimensions. In this section, a different approach is taken in developing the system of equations that solves a partial-differential equation. Again, the boundary condition can be one of three types.

9.5 Finite Elements for Parabolic and Hyperbolic Equations

Considers how the finite-element method can be applied to solve time-dependent equations. It turns out that a combination of finite elements and finite differences is employed.

9.6 Theoretical Matters

Discusses some aspects of the finite-element method that relate to accuracy and rate of convergence.

Chapter Summary

Provides a list of the topics that are learned by studying this chapter.

Computer Program

Gives a program that solves a boundary-value problem.

9.1 The Rayleigh–Ritz Method

As a basis for the finite-element method that is the subject of this chapter, you should know something about the Rayleigh–Ritz method. It is based on an elegant branch of mathematics, the *calculus of variations*. With this method we solve a boundary-value problem by approximating the solution with a finite linear combination of simple basis functions that are chosen to fulfill certain criteria, including meeting the boundary conditions.

The calculus of variations seeks to optimize (often minimize) a special class of functions called *functionals*. The usual form for the functional (in problems of one independent variable) is

$$I[y] = \int_a^b F\left(x, y, \frac{dy}{dx}\right) dx. \tag{9.1}$$

Observe that $I[y]$ is not a function of x because x disappears when the definite integral is evaluated. The argument y of $I[y]$ is not a simple variable but a function, $y = y(x)$. The square brackets in $I[y]$ emphasize this fact. A functional can be thought of as a "function of functions." The value of the right-hand side of Eq. (9.1) will change as the function $y(x)$ is varied, but when $y(x)$ is fixed, it evaluates to a scalar quantity (a constant). We seek the $y(x)$ that minimizes $I[y]$.

Let us illustrate this concept by a very simple example where the solution is obvious in advance—find the function $y(x)$ that minimizes the distance between two points. Although we know what $y(x)$ must be, let's pretend we don't. Figure 9.1 suggests that we are to choose from among the set of curves $y_i(x)$ of which $y_1(x)$, $y_2(x)$, and $y_3(x)$ are representative. In this simple case, the functional is the integral of the distance along any of these curves:

$$I[y] = \int_{x_1}^{x_2} \sqrt{(dx)^2 + (dy)^2} = \int_{x_1}^{x_2} \sqrt{1 + \left(\frac{dy}{dx}\right)^2}\, dx.$$

To minimize $I[y]$, just as in calculus, we set its derivative to zero. There are certain restrictions on all the curves $y_i(x)$. Obviously, each must pass through the points (x_1, y_1) and (x_2, y_2). In addition, for the optimal trajectory, the Euler–Lagrange equation must be satisfied:

$$\frac{d}{dx}\left[\frac{\partial}{\partial y'} F(x, y, y')\right] = \frac{\partial}{\partial y} F(x, y, y'). \tag{9.2}$$

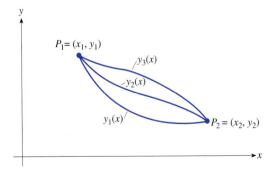

Figure 9.1

Applying this to the functional for shortest distance, we have

$$F(x, y, y') = (1 + (y')^2)^{1/2},$$

$$\frac{\partial F}{\partial y} = 0,$$

$$\frac{\partial F}{\partial y'} = \frac{1}{2}(1 + (y')^2)^{-1/2}(2y'),$$

$$\frac{d}{dx}\frac{\partial F}{\partial y'} = \frac{d}{dx}\left(\frac{y'}{\sqrt{1 + (y')^2}}\right) = \frac{\partial F}{\partial y} = 0.$$

(The last comes from Eq. 9.2)

From this, it follows that

$$\frac{y'}{\sqrt{1 + (y')^2}} = c.$$

Solving for y' gives

$$y' = \sqrt{\frac{c^2}{1 - c^2}} = \text{a constant} = b,$$

and, on integrating,

$$y = bx + a.$$

As stated, $y(x)$ must pass through P_1 and P_2; this condition is used to evaluate the constants a and b.

Let us advance to a less trivial case. Consider this second-order linear boundary-value problem over $[a, b]$: *

$$y'' + Q(x)y = F(x), \qquad y(a) = y_0, \qquad y(b) = y_n. \tag{9.3}$$

(An equation that has $y = $ constant at the endpoints is said to be subject to Dirichlet conditions.) It turns out that the functional that corresponds to Eq. (9.3) is

$$I[u] = \int_a^b \left[\left(\frac{du}{dx}\right)^2 - Qu^2 + 2Fu\right] dx. \tag{9.4}$$

(If the boundary equations involve a derivative of y, the functional must be modified.)

We can transform Eq. (9.4) to Eq. (9.3) through the Euler–Lagrange conditions, so optimizing Eq. (9.4) gives the solution to Eq. (9.3). Observe carefully the benefit of operating with the functional rather than the original equation: We now have only first-order instead of second-order derivatives. This not only simplifies the mathematics but also permits us to find solutions even when there are discontinuities that cause y not to have sufficiently high derivatives.

---

*This equation is a prototype of many equations in applied mathematics. Equations for heat conduction, elasticity, electrostatics, and so on in a one-dimensional situation are of this form.

If we know the solution to our differential equation, substituting it for u in Eq. (9.4) will make $I[u]$ a minimum. If the solution isn't known, perhaps we can approximate it by some (almost) arbitrary function and see whether we can minimize the functional by a suitable choice of the parameters of the approximation. The Rayleigh–Ritz method is based on this idea. We let $u(x)$, which is the approximation to $y(x)$ (the exact solution), be a sum:

$$u(x) = c_0v_0(x) + c_1v_1(x) + \cdots + c_nv_n(x) = \sum_{i=0}^{n} c_iv_i(x). \tag{9.5}$$

There are two conditions on the v's in Eq. (9.5): They must be chosen such that $u(x)$ meets the boundary conditions, and the individual v's must be linearly independent (meaning that no one v can be obtained by a linear combination of the others). We call the v's *trial functions;* the c's and v's are to be chosen to make $u(x)$ a good approximation to the true solution to Eq. (9.3).

If we have some prior knowledge of the true function, $y(x)$, we may be able to choose the v's to closely resemble $y(x)$. Most often we lack such knowledge, and the usual choice then is to use polynomials. We must find a way of getting values for the c's to force $u(x)$ to be close to $y(x)$. We will use the functional of Eq. (9.4) to do this.

If we substitute $u(x)$ as defined by Eq. (9.5) into the functional, Eq. (9.4), we get

$$I(c_0, c_1, \ldots, c_n) = \int_a^b \left[\left(\frac{d}{dx} \Sigma c_iv_i \right)^2 - Q(\Sigma c_iv_i)^2 + 2F\Sigma c_iv_i \right] dx. \tag{9.6}$$

We observe that I is an ordinary function of the unknown c's after this substitution, as reflected in our notation. To minimize I, we take its partial derivatives with respect to each unknown c and set to zero, resulting in a set of equations in the c's that we can solve. This will define $u(x)$ in Eq. (9.5).

We now substitute the $u(x)$ of Eq. (9.5) into the functional. If we partially differentiate with respect to, say, c_i where this is one of the unknown c's, we will get

$$\frac{\partial I}{\partial c_i} = \int_a^b 2 \left(\frac{du}{dx} \right) \frac{\partial}{\partial c_i} \left(\frac{du}{dx} \right) dx - \int_a^b 2Qu \left(\frac{\partial u}{\partial c_i} \right) dx + 2 \int_a^b F \left(\frac{\partial u}{\partial c_i} \right) dx, \tag{9.7}$$

where we have broken the integral into three parts.

An example will clarify the procedure.

EXAMPLE 9.1   Solve the equation $y'' + y = 3x^2$, with boundary points $(0, 0)$ and $(2, 3.5)$. (Here $Q = 1$ and $F = 3x^2$.) Use polynomial trial functions up to degree 3. If we define $u(x)$ as

$$u(x) = \frac{7x}{4} + c_2(x)(x - 2) + c_3(x^2)(x - 2), \tag{9.8}$$

we have linearly independent v's. The boundary conditions are met by the first term, and because the other terms are zero at the boundaries, $u(x)$ also meets the boundary conditions. [It is customary to match the boundary conditions with the initial term(s) of $u(x)$ and then make the succeeding terms equal zero at the boundaries, as we have done here.]

Examination of Eq. (9.7) shows that we need these quantities:

$$\frac{du}{dx} = \frac{7}{4} + c_2(2x - 2) + c_3(3x^2 - 4x),$$

$$\frac{\partial}{\partial c_2}\left(\frac{du}{dx}\right) = 2x - 2, \qquad \frac{\partial}{\partial c_3}\left(\frac{du}{dx}\right) = 3x^2 - 4x, \tag{9.9}$$

$$\frac{\partial u}{\partial c_2} = x(x - 2), \qquad \frac{\partial u}{\partial c_3} = x^2(x - 2).$$

We now substitute from Eqs. (9.9) into Eqs. (9.7). Note that we have two equations, one for the partial with respect to c_2 and the other from the partial with respect to c_3. The results from this step are:

$$\frac{\partial I}{\partial c_2}: \quad 0 = \int_0^2 2\left[\frac{7}{4} + c_2(2x - 2) + c_3(3x^2 - 4x)\right](2x - 2)\, dx$$

$$- \int_0^2 2(1)\left[\frac{7x}{4} + c_2(x^2 - 2x) + c_3(x^3 - 2x^2)\right](x^2 - 2x)\, dx \tag{9.10}$$

$$+ 2\int_0^2 (3x^2)(x^2 - 2x)\, dx,$$

$$\frac{\partial I}{\partial c_3}: \quad 0 = \int_0^2 2\left[\frac{7}{4} + c_2(2x - 2) + c_3(3x^2 - 4x)\right](3x^2 - 4x)\, dx$$

$$- \int_0^2 2(1)\left[\frac{7x}{4} + c_2(x^2 - 2x) + c_3(x^3 - 2x^2)\right](x^3 - 2x^2)\, dx \tag{9.11}$$

$$+ 2\int_0^2 (3x^2)(x^3 - 2x^2)\, dx.$$

We now carry out the integrations. Though there are quite a few of them, all are quite simple in our example. With a more complicated $Q(x)$ and $F(x)$, this might require numerical integrations. The result of this step is the pair of equations

$$\frac{16}{5}c_2 + \frac{16}{5}c_3 = \frac{74}{15},$$

$$\frac{16}{5}c_2 + \frac{128}{21}c_3 = \frac{36}{5}, \tag{9.12}$$

which we solve to get the coefficients in our $u(x)$. On expanding, we find that

$$u(x) = \left(\frac{119}{152}\right)x^3 - \left(\frac{46}{57}\right)x^2 + \left(\frac{53}{228}\right)x. \tag{9.13}$$

Figure 9.2 shows that our $u(x)$ agrees well with the exact solution, which is $6\cos(x) + 3(x^2 - 2)$ over the interval $[0, 2]$. Table 9.1 compares computed values and the error of $u(x)$. The largest errors occur near the middle of the interval.

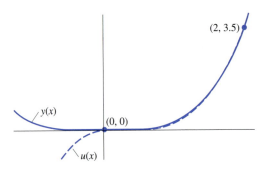

Figure 9.2

Table 9.1

x	$y(x)$	$u(x)$	Error	x	$y(x)$	$u(x)$	Error
0.00	0.000	0.000	0.000	1.10	0.352	0.321	0.031
0.10	0.000	0.016	−0.016	1.20	0.494	0.470	0.024
0.20	0.000	0.020	−0.020	1.30	0.675	0.658	0.017
0.30	0.002	0.018	−0.016	1.40	0.900	0.892	0.008
0.40	0.006	0.014	−0.008	1.50	1.174	1.175	−0.001
0.50	0.015	0.012	−0.003	1.60	1.505	1.513	−0.008
0.60	0.032	0.018	0.014	1.70	1.897	1.909	−0.012
0.70	0.059	0.036	0.023	1.80	2.357	2.370	−0.013
0.80	0.100	0.070	0.030	1.90	2.890	2.898	−0.008
0.90	0.160	0.126	0.034	2.00	3.500	3.500	0.000
1.00	0.242	0.208	0.034				

▲

9.2   The Collocation and Galerkin Methods

There are other ways to approximate $y(x)$ in Example 9.1. The *collocation method* is what is called a "residual method." We begin by defining the residual, $R(x)$, as equal to the left-hand side of Eq. (9.3) minus the right-hand side:

$$R(x) = y'' + Qy - F. \tag{9.14}$$

We approximate $y(x)$ again with $u(x)$ equal to a sum of trial functions, usually chosen as linearly independent polynomials, just as for the Rayleigh–Ritz method. We substitute $u(x)$ into $R(x)$ and attempt to make $R(x) = 0$ by a suitable choice of the coefficients in $u(x)$. Of course, normally we cannot do this everywhere in the interval $[a, b]$, so we select several points at which we make $R(x) = 0$. [The number of points where we do this must equal the number of unknown coefficients in $u(x)$.] An example will clarify the procedure.

EXAMPLE 9.2   Solve the same equation as in Example 9.1, but this time use collocation.
The equation we are to solve is

$$y'' + y = 3x^2, \qquad y(0) = 0, \qquad y(2) = 3.5. \tag{9.15}$$

We take $u(x)$ as before to satisfy the boundary conditions:

$$u(x) = \frac{7x}{4} + c_2(x)(x - 2) + c_3(x^2)(x - 2). \tag{9.16}$$

The residual is, after substituting $u(x)$ for $y(x)$,

$$R(x) = u'' + u - 3x^2, \tag{9.17}$$

which becomes, when we differentiate u twice to get u'',

$$R(x) = c_2(2) + c_3(6x - 4) + \frac{7x}{4} + c_2(x^2 - 2x) + c_3(x^3 - 2x^2) - 3x^2. \tag{9.18}$$

Because there are two unknown constants, we can force $R(x)$ to be zero at two points in
$[0, 2]$. We do not know which two points will be the best choices, so we arbitrarily take
them as $x = 0.7$ and $x = 1.3$. (These points are more or less equally spaced in the interval.)
Setting $R(x) = 0$ for these choices gives a pair of equations in the c's:

$$\text{From } x = 0.7: \qquad \frac{1090c_2 - 437c_3 - 245}{1000} = 0,$$

$$\text{From } x = 1.3: \qquad \frac{1090c^2 + 2617c_3 - 2795}{1000} = 0. \tag{9.19}$$

When these are solved for the c's, we get, for $u(x)$,

$$u(x) = \left(\frac{425}{509}\right)x^3 - \left(\frac{61607}{55481}\right)x^2 + \left(\frac{140023}{221924}\right)x, \tag{9.20}$$

in which the coefficients are quite different than in Eq. (9.13). Figure 9.3 shows that this
approximation is not as good as that obtained by the Rayleigh–Ritz technique. (But the

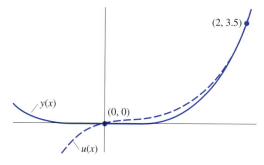

Figure 9.3

Table 9.1

x	$y(x)$	$u(x)$	Error	x	$y(x)$	$u(x)$	Error
0.00	0.00	0.00	0.00	1.10	0.35	0.32	0.03
0.10	0.00	0.05	−0.05	1.20	0.49	0.60	−0.11
0.20	0.00	0.09	−0.09	1.30	0.67	0.78	−0.10
0.30	0.00	0.11	−0.11	1.40	0.90	1.00	−0.10
0.40	0.01	0.13	−0.12	1.50	1.17	1.27	−0.09
0.50	0.02	0.14	−0.13	1.60	1.50	1.59	−0.08
0.60	0.03	0.16	−0.13	1.70	1.90	1.97	−0.07
0.70	0.06	0.18	−0.12	1.80	2.36	2.41	−0.05
0.80	0.10	0.22	−0.12	1.90	2.89	2.92	−0.03
0.90	0.16	0.28	−0.12	2.00	3.50	3.50	0.00
1.00	0.24	0.36	−0.11				

amount of arithmetic is certainly less! We could improve the approximation by using more terms in $u(x)$.) Table 9.2 compares the approximation with the exact solution. ▲

The Galerkin Method

The Galerkin method is widely used, especially in the very popular technique that we will describe in Section 9.3. It is important to know the Galerkin method because of its widespread application.

Like collocation, Galerkin is a "residual method" that uses the $R(x)$ of Eq. (9.14), except that now we multiply $R(x)$ by weighting functions, $W_i(x)$. The $W_i(x)$ can be chosen in many ways, but Galerkin showed that using the individual trial functions, v_i, of Eq. (9.5) is an especially good choice.

Once we have selected the v's for Eq. (9.5), we compute the unknown coefficients by setting the integral over $[a, b]$ of the weighted residual to zero:

$$\int_a^b W_i(x)R(x) \, dx = 0, \qquad i = 0, 1, \ldots, n, \tag{9.21}$$

where $W_i(x) = v_i$. (Observe that using Dirac delta functions for the $W_i(x)$ gives the collocation method.)

Let us use the Galerkin method on the same example as before.

EXAMPLE 9.3   Solve

$$y'' + y = 3x^2, \qquad y(0) = 0, \qquad y(2) = 3.5$$

by the Galerkin method. Use the same $u(x)$ as before:

$$u(x) = \frac{7x}{4} + c_2(x)(x - 2) + c_3(x^2)(x - 2),$$

so that $v_2 = x(x - 2)$ and $v_3 = x^2(x - 2)$.

The residual is

$$R(x) = y'' + y - 3x^2,$$

which becomes, after substituting u'' and u for y'' and y, respectively,

$$R(x) = c_2(2) + c_3(6x - 4) + \frac{7x}{4} + c_2(x)(x - 2) + c_3(x^2)(x - 2) - 3x^2. \quad \textbf{(9.22)}$$

We now carry out two integrations (because there are two unknown c's):

Using v_2 as a W_i: $\displaystyle\int_0^2 [x(x - 2)] * R(x)\ dx = 0,$

Using v_3 as a W_i: $\displaystyle\int_0^2 [x^2(x - 2)] * R(x)\ dx = 0,$

which gives two equations in the c's:

$$-\frac{24c_2 + 24c_3 - 37}{15} = 0,$$

$$-\frac{2(84c_2 + 160c_3 - 180)}{105} = 0. \quad \textbf{(9.23)}$$

Solving Eqs. (9.23) for c_2 and c_3 gives

$$u(x) = \left(\frac{101}{152}\right)x^3 - \left(\frac{103}{228}\right)x^2 - \left(\frac{1}{128}\right)x. \quad \textbf{(9.24)}$$

Although Eq. (9.24) looks different from Eq. (9.13), between $x = 0$ and $x = 3.5$ it gives values for $u(x)$ that are almost identical to those from the Rayleigh–Ritz technique. Equation (9.24) differs from the analytical solution by little more than does the Rayleigh–Ritz equation. (The maximum error of Eq. (9.24) is 0.058; for Eq. (9.13), it is 0.034.)   ▲

Although the Rayleigh–Ritz method is slightly more accurate in this example, the Galerkin method is much easier and we never have to find the variational form.

9.3 Finite Elements for Ordinary-Differential Equations

The disadvantages of the methods of the previous section are twofold: Finding good trial functions (the v's in Eq. (9.5) is not easy, and polynomials [the usual choice when we have no prior knowledge of the behavior of $y(x)$] may interpolate poorly. (We can think of $u(x)$ as an interpolation function between the boundary conditions that also obeys the differential equation.) This is especially true when the interval $[a, b]$ is large.

The remedy to our problem is based on the observations of Chapters 3 and 5 that even low-degree polynomials can reflect the behavior of a function if based on values that are closely spaced. Accordingly, we hope to successfully apply a technique of Section 9.2

(specifically the Galerkin method) using low-degree polynomials if we subdivide $[a, b]$ into smaller subintervals. It will turn out that our hope is fulfilled.

Applying this technique, we have what is known as the *finite-element method.* Our strategy is as follows:

1. Subdivide $[a, b]$ into n subintervals, called *elements,* that join at $x_1, x_2, \ldots, x_{n-1}$. Add to this array $x_0 = a$ and $x_n = b$. We call the x_i the *nodes* of the interval. Number the elements from 1 to n where element (i) runs from x_{i-1} to x_i. The x_i need not be evenly spaced.
2. Apply the Galerkin method to each element separately to interpolate (subject to the differential equation) between the end nodal values, $u(x_{i-1})$ and $u(x_i)$, where these u's are approximations to the $y(x_i)$'s that are the true solution to the differential equation. [These nodal values are actually the c's in our adaptation of Eq. (9.5), the equation for $u(x)$.]
3. Use a low-degree polynomial for $u(x)$. Our development will use a first-degree polynomial, although quadratics or cubics are often used. (The development for these higher-degree polynomials parallels what we will do but is more complicated.)
4. The result of applying Galerkin to element (i) is a pair of equations in which the unknowns are the nodal values at the ends of element (i), the c's. When we have done this for each element, we have equations that involve all the nodal values, which we combine to give a set of equations that we can solve for the unknown nodal values. (The process of combining the separate *element equations* is called *assembling the system.*)
5. These equations are adjusted for the boundary conditions and solved to get approximations to $y(x)$ at the nodes; we get intermediate values for $y(x)$ by linear interpolation.

We now begin the development. Although it involves several steps, each step is straightforward. The differential equation that we will solve is

$$y'' + Q(x)y = F(x) \qquad \text{subject to boundary conditions at } x = a \text{ and } x = b. \quad \textbf{(9.25)}$$

(We will specify the boundary conditions later.)

Step 1:   Subdivide $[a, b]$ into n elements, as discussed. Focus attention on element (i) that runs between x_{i-1} and x_i. To simplify the notation, call the left node L and the right node R.

Step 2:   Write $u(x)$ for element (i):

$$u(x) = c_L N_L + c_R N_R = c_L \frac{x - R}{L - R} + c_R \frac{x - L}{R - L}$$

$$= c_L \frac{x - R}{-h_i} + c_R \frac{x - L}{h_i}, \qquad \textbf{(9.26)}$$

where $h_i = R - L$. In Eq. (9.26) we have used the symbol N for the trial functions, rather than the v's that we have used before, to agree with the literature on finite elements. The N's are called *shape functions* in that literature.

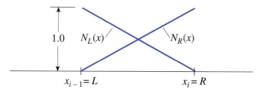

Figure 9.4 N_L and N_R within element (i)

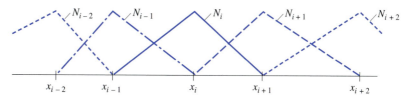

Figure 9.5

Recognize that the N's in Eq. (9.26) are really first-degree Lagrangian polynomials. When we use such linear interpolation, the shape functions are often called *hat functions*. (*Chapeau functions*, from the French, is another name.) The reason for this name will become apparent later on.

Figure 9.4 sketches N_L and N_R within element (i). Because the values of the N's vary (from unity to zero) as x goes from x_L to x_R, they are functions of x. Note also that the c's in Eq. (9.26) are independent of x.

The reason that our N's are called "hat functions" is apparent when we look at a sketch of the N's for several adjacent elements in Fig. 9.5. Observe that we combine the $N_R(x)$ and $N_L(x)$ of Fig. 9.4 that join at x_i into a quantity that we call N_i.

Step 3:   Apply the Galerkin method to element (i). The residual is

$$R(x) = y'' + Qy - F = u'' + Qu - F, \qquad (9.27)$$

where we have substituted $u(x)$ for $y(x)$. The Galerkin method sets the integral of R weighted with each of the N's (over the length of the element) to zero:

$$\int_L^R N_L R(x)\, dx = 0,$$
$$\int_L^R N_R R(x)\, dx = 0. \qquad (9.28)$$

Now expand Eqs. (9.28):

$$\int_L^R u'' N_L\, dx + \int_L^R Qu N_L\, dx - \int_L^R F N_L\, dx = 0. \qquad (9.29)$$

$$\int_L^R u'' N_R\, dx + \int_L^R Qu N_R\, dx - \int_L^R F N_R\, dx = 0. \qquad (9.30)$$

Step 4: Transform Eqs. (9.29) and (9.30) by applying integration by parts* to the first integral. In the second integral, we will take Q out from the integrand as Q_{av}, an average value within the element. We also take F outside the third integral. When this is done, Eq. (9.29) becomes

$$-\int_L^R \left(\frac{du}{dx}\right)\left(\frac{dN_L}{dx}\right) dx + Q_{av} \int_L^R uN_L \, dx - F_{av} \int_L^R N_L \, dx + N_L \frac{du}{dx}\Big|_{x=R} - N_L \frac{du}{dx}\Big|_{x=L} = 0. \quad (9.31)$$

In the last two terms of Eq. (9.31), $N_L = 1$ at L and is zero at R, so the equation can be simplified:

$$-\int_L^R \left(\frac{du}{dx}\right)\left(\frac{dN_L}{dx}\right) dx + Q_{av} \int_L^R uN_L \, dx - F_{av} \int_L^R N_L \, dx - \frac{du}{dx}\Big|_{x=L} = 0. \quad (9.32)$$

Doing similarly with Eq. (9.30) gives

$$-\int_L^R \left(\frac{du}{dx}\right)\left(\frac{dN_R}{dx}\right) dx + Q_{av} \int_L^R uN_R \, dx - F_{av} \int_L^R N_R \, dx + \frac{du}{dx}\Big|_{x=R} = 0. \quad (9.33)$$

Step 5: Change signs in Eqs. (9.32) and (9.33); substitute from Eq. (9.26) for u, du/dx, dN_L/dx, and dN_R/dx; and carry out the integrations. We show this separately for each term in Eq. (9.32):

$$\int_L^R \left(\frac{du}{dx}\right)\left(\frac{dN_L}{dx}\right) dx = \int_L^R \left[\frac{-c_L}{h_i} + \frac{c_R}{h_i}\right]\left(\frac{-1}{h_i}\right) dx = \left(\frac{c_L}{h_i^2} - \frac{c_R}{h_i^2}\right)\int_L^R dx$$

$$= \left(\frac{1}{h_i}\right)c_L - \left(\frac{1}{h_i}\right)c_R. \quad (9.34a)$$

$$-Q_{av}\int_L^R (c_L N_L + c_R N_R)N_L \, dx = -c_L Q_{av} \int_L^R N_L^2 \, dx - c_R Q_{av} \int_L^R N_R N_L \, dx$$

$$= -c_L Q_{av} \int_L^R \left(\frac{x-R}{h_i}\right)^2 dx$$

$$-c_R Q_{av} \int_L^R \left(\frac{x-L}{h_i}\right)\left(\frac{x-R}{-h_i}\right) dx \quad (9.34b)$$

$$= -\left(Q_{av} * \frac{h_i}{3}\right)c_L - \left(Q_{av} * \frac{h_i}{6}\right)c_R.$$

$$F_{av}\int_L^R N_L \, dx = F_{av}\int_L^R \frac{(x-R)}{-h_i} dx = F_{av} * \frac{h_i}{2}. \quad (9.34c)$$

---

*From $d(UV) = U \, dV + V \, dU$, we have

$$\int_a^b U \, dV = -\int_a^b V \, dU + UV \Big|_a^b, \quad \text{so} \quad \int_L^R u''N_i \, dx = -\int_L^R u'(dN_i/dx) + N_i u' \Big|_L^R.$$

Doing the same with Eq. (9.33) gives

$$\int_L^R \left(\frac{du}{dx}\right)\left(\frac{dN_R}{dx}\right) dx = -\left(\frac{1}{h_i}\right)c_L + \left(\frac{1}{h_i}\right)c_R. \qquad \textbf{(9.35a)}$$

$$-Q_{av}\int_L^R (c_L N_L + c_R N_R)N_R\, dx = -\left(Q_{av}*\frac{h_i}{6}\right)c_L - \left(Q_{av}*\frac{h_i}{3}\right)c_R. \qquad \textbf{(9.35b)}$$

$$F_{av}\int_L^R N_R\, dx = F_{av}*\frac{h_i}{2}. \qquad \textbf{(9.35c)}$$

Step 6:   Substitute the result of Step 5 (Eqs. 9.34 and 9.35) into Eqs. (9.32) and (9.33), and rearrange to give two linear equations in the unknown c_L and c_R:

$$\left(\frac{1}{h_i} - \frac{Q_{av}h_i}{3}\right)c_L + \left(\frac{-1}{h_i} - \frac{Q_{av}h_i}{6}\right)c_R = \frac{-F_{av}h_i}{2} - \frac{du}{dx}\bigg|_{x=L},$$

$$\left(\frac{-1}{h_i} - \frac{Q_{av}h_i}{6}\right)c_L + \left(\frac{1}{h_i} - \frac{Q_{av}h_i}{3}\right)c_R = \frac{-F_{av}h_i}{2} + \frac{du}{dx}\bigg|_{x=R}. \qquad \textbf{(9.36)}$$

We call the pair of equations in (9.36) the *element equations*. We can do the same for each element to get n such pairs.

Step 7:   Combine (assemble) all the element equations together to form a system of linear equations for the problem. We now recognize that point R in element (i) is precisely the same as point L in element $(i + 1)$. Renumber the c's as $c_0, c_1, \ldots, c_n$. Also notice that the gradient (du/dx) must be the same on either side of the join of the elements—that is, $(du/dx)_{x=R}$ in element (i) equals $(du/dx)_{x=L}$ in element $(i + 1)$. This means that these terms cancel when we do the assembling except in the first and last equations. (On rare occasions this is not true, but in that case the difference in the two gradients is a known value.)

The result of this step is this set of $n + 1$ equations (numbered from 0 to n),

$$[K]\{c\} = \{b\}, \qquad \textbf{(9.37)}$$

where the diagonal elements of $[K]$ are

$$\left(\frac{1}{h_1} - Q_{av,1}*\frac{h_1}{3}\right) \qquad \text{in row 0,}$$

$$\left(\frac{1}{h_i} - Q_{av,i}*\frac{h_i}{3}\right) + \left(\frac{1}{h_{i+1}} - Q_{av,i+1}*\frac{h_{i+1}}{3}\right) \qquad \text{in rows 1 to } n - 1,$$

$$\left(\frac{1}{h_n} - Q_{av,n}*\frac{h_n}{3}\right) \qquad \text{in row } n;$$

and elements above and to the left of the diagonal in rows 1 to n are

$$\left(\frac{-1}{h_i} - Q_{av,i}*\frac{h_i}{6}\right).$$

The elements of $\{c\}$ are c_i, $i = 0$ to n.

The elements of $\{b\}$ are:

$$-F_{av,1} * \frac{h_1}{2} - \left(\frac{du}{dx}\right)_{x=a} \qquad \text{in row 0,}$$

$$-F_{av,i} * \frac{h_i}{2} - F_{av,i+1} * \frac{h_{i+1}}{2} \qquad \text{in rows 1 to } n-1,$$

$$-F_{av,n} * \frac{h_n}{2} + \left(\frac{du}{dx}\right)_{x=b} \qquad \text{in row } n.$$

In the preceding equations, $Q_{av,i}$ and $F_{av,i}$ are values of Q and F at the midpoints of element (i).

Step 8:   Adjust the set of equations from step 6 for the boundary conditions. We will handle two cases: Case (1), a Dirichlet condition is specified—$y(a)$ = constant [and/or $y(b)$ = constant]. Case (2), a Neumann condition is specified—dy/dx = constant at $x = a$ and/or $x = b$. (If $Q = 0$, we cannot have a Neumann condition at both ends, because the solution would be known only to within an additive constant.) (We leave case (3), mixed conditions, as an exercise; it is a modification of case (2).)

   Case (1): Dirichlet condition.   In this case, c is known at the end node. Suppose this is $y(a) = A$. Then the equation in row 0 is redundant, and so we remove it from the set of equations of step 6. In the next row, we move $k_{10} * A$ to the right-hand side (subtracting this from the element computed in step 6). If the condition is $y(b) = B$, we do the same but with the last and next to last equations.

   Case (2): Neumann condition.   In this case, c is not known at the end node. Suppose the condition is $dy/dx = A$ at $x = a$. We retain the equation in row 0 and substitute the given value of dy/dx into the right-hand side. If the condition is $dy/dx = B$ at $x = b$, we do the same with the last equation.

Step 9:   Solve the set of equations for the unknown c's after adjusting, in step 8, for the boundary conditions. The c's are approximations to $y(x)$ at the nodes. If intermediate values of y are needed between the nodes, we obtain them by linear interpolation.

   Examples will clarify the procedure.

EXAMPLE 9.4   Solve $y'' + y = 3x^2$, $y(0) = 0$, $y(2) = 3.5$. (We solved this same equation in Section 9.2.) Subdivide into seven elements that join at $x = 0.4, 0.7, 0.9, 1.1, 1.3,$ and 1.6.

   Table 9.3 shows the values we need to build the system of equations.

Table 9.3

Element	L	R	Midpoint	h_i	Q_{av}	F_{av}
1	0	0.4	0.2	0.4	1	0.12
2	0.4	0.7	0.55	0.3	1	0.9075
3	0.7	0.9	0.8	0.2	1	1.92
4	0.9	1.1	1.0	0.2	1	3
5	1.1	1.3	1.2	0.2	1	4.32
6	1.3	1.6	1.45	0.3	1	6.3075
7	1.6	2.0	1.8	0.4	1	9.72

The augmented matrix of the set of equations from step 6 is

$$\left[\begin{array}{cccccccc|c} 2.367 & -2.567 & 0.000 & 0.000 & 0.000 & 0.000 & 0.000 & 0.000 & -0.024 \\ -2.567 & 5.600 & -3.383 & 0.000 & 0.000 & 0.000 & 0.000 & 0.000 & -0.160 \\ 0.000 & -3.383 & 8.167 & -5.033 & 0.000 & 0.000 & 0.000 & 0.000 & -0.328 \\ 0.000 & 0.000 & -5.033 & 9.867 & -5.033 & 0.000 & 0.000 & 0.000 & -0.492 \\ 0.000 & 0.000 & 0.000 & -5.033 & 9.867 & -5.033 & 0.000 & 0.000 & -0.732 \\ 0.000 & 0.000 & 0.000 & 0.000 & -5.033 & 8.167 & -3.383 & 0.000 & -1.378 \\ 0.000 & 0.000 & 0.000 & 0.000 & 0.000 & -3.383 & 5.600 & -2.567 & -2.890 \\ 0.000 & 0.000 & 0.000 & 0.000 & 0.000 & 0.000 & -2.567 & 2.367 & -1.944 \end{array}\right] \quad (9.38)$$

To adjust for the boundary conditions, we eliminate the first and last equations and subtract $(0)(-2.567) = 0$ from the right-hand side of the top row and subtract $(3.50)(-2.567) = -8.9845$ from the right-hand side of the bottom row to get

$$\left[\begin{array}{cccccc|c} 5.600 & -3.383 & 0.000 & 0.000 & 0.000 & 0.000 & -0.160 \\ -3.383 & 8.167 & -5.033 & 0.000 & 0.000 & 0.000 & -0.328 \\ 0.000 & -5.033 & 9.867 & -5.033 & 0.000 & 0.000 & -0.492 \\ 0.000 & 0.000 & -5.033 & 9.867 & -5.033 & 0.000 & -0.732 \\ 0.000 & 0.000 & 0.000 & -5.033 & 8.167 & -3.383 & -1.378 \\ 0.000 & 0.000 & 0.000 & 0.000 & -3.383 & 5.600 & 6.094 \end{array}\right] . \quad (9.39)$$

When we solve the set of Eqs. (9.39) we get the solution tabulated in Table 9.4. Observe that we always have a tridiagonal (and symmetrical) system and that this system can be solved quickly.

Table 9.4

x	u(x)	Anal.	Error
0.000	0.0000	0.0000	0
0.400	-0.0024	0.0064	0.0088
0.700	0.0433	0.0591	0.0458
0.900	0.1371	0.1597	0.0831
1.100	0.3232	0.3516	0.0892
1.300	0.6419	0.6750	0.0462
1.600	1.4759	1.5048	0.0081
2.000	3.5000	3.5031	0.0031

▲

EXAMPLE 9.5   Solve $y'' - (x + 1)y = -e^{-x}(x^2 - x + 2)$ subject to Neumann conditions of

$$y'(2) = 0, \qquad y'(4) = -0.036631.$$

Use four elements of equal lengths. Compare to the analytical solution

$$y(x) = e^{-x}(x - 1).$$

Table 9.5 gives values that we need to set up the equations.

Table 9.5

Element	L	R	Midpoint	h_i	Q_{av}	F_{av}
1	2	2.5	2.25	0.5	−3.25	−0.5072
2	2.5	3.0	2.75	0.5	−3.75	−0.4355
3	3.0	3.5	3.25	0.5	−4.25	−0.3611
4	3.5	4.0	3.75	0.5	−4.75	−0.2896

The initial matrix of equations is

$$
\begin{bmatrix}
2.542 & -1.729 & 0.000 & 0.000 & 0.000 & \vdots & 0.127 \\
-1.729 & 5.167 & -1.688 & 0.000 & 0.000 & \vdots & 0.236 \\
0.000 & -1.688 & 5.333 & -1.646 & 0.000 & \vdots & 0.199 \\
0.000 & 0.000 & -1.646 & 5.500 & -1.604 & \vdots & 0.163 \\
0.000 & 0.000 & 0.000 & -1.604 & 2.792 & \vdots & 0.072
\end{bmatrix}.
$$

After adjusting for boundary conditions, we get

$$
\begin{bmatrix}
2.542 & -1.729 & 0.000 & 0.000 & 0.000 & \vdots & 0.127 \\
-1.729 & 5.167 & -1.688 & 0.000 & 0.000 & \vdots & 0.236 \\
0.000 & -1.688 & 5.333 & -1.646 & 0.000 & \vdots & 0.199 \\
0.000 & 0.000 & -1.646 & 5.500 & -1.604 & \vdots & 0.163 \\
0.000 & 0.000 & 0.000 & -1.604 & 2.792 & \vdots & 0.036
\end{bmatrix}
$$

and the solution is

x	$u(x)$	Anal.	Error
2.000	0.1334	0.1353	1.896E-03
2.500	0.1228	0.1231	3.230E-04
3.000	0.0996	0.0996	−2.031E-05
3.500	0.0758	0.0755	−3.280E-04
4.000	0.0564	0.0549	−1.432E-03

▲

9.4 Finite Elements for Elliptic Partial-Differential Equations

When an elliptic partial-differential equation is solved by replacing derivatives with finite-difference approximations, there are serious difficulties if the region is irregular. Analytical methods are also very awkward to apply in such cases.

The finite-element method has no such problems. As we saw in Section 9.3 for one-dimensional boundary-value problems, nodes can be placed wherever the problem solver desires with the finite-element method. This is also true for two- and three-dimensional regions. They can be placed along any boundary so as to approximate it closely. It is the method of choice for solving elliptic partial-differential equations on regions of arbitrary shape.

Although setting up the equations that solve partial-differential equations is no easy task, computer programs are available that do so. It is important to understand how this method works, although this text cannot give everything that today's scientists and engineers might want to know. Our treatment will give a basic knowledge.

The introduction to finite elements in Sections 9.1 and 9.2 is important background for what we shall do here. Recall that two ways of applying variational methods to subdivisions of the region of interest were presented: Rayleigh–Ritz and Galerkin. The first of these minimized the functional for the problem by setting partial derivatives to zero; the second set integrals of a weighted residual to zero. The two methods are equivalent for most problems, and both can be used for elliptic equations. We choose the former, in part to provide variety from the presentation in Section 9.3.

The elliptic equation that we will solve in this section is

$$u_{xx} + u_{yy} + Q(x, y)u = F(x, y) \qquad (9.40)$$

on region R that is bounded by curve L, with boundary conditions

$$u(x, y) = u_0 \quad \text{on } L_1, \qquad \frac{\partial u}{\partial n} = \alpha u + \beta \quad \text{on } L_2,$$

where $\partial u/\partial n$ is the outward normal gradient.

Observe that we have Dirichlet conditions on some parts of the boundary and mixed boundary conditions on other parts. For our notation, we will use $u(x, y)$ as the exact solution to Eq. (9.40) and $v(x, y)$ as our approximation to $u(x, y)$. Although the finite-element method is most often used when the region is three dimensional, we will simplify the development by doing it in only two dimensions.

Here is our plan of attack:

Step 1. Find the functional that corresponds to the partial-differential equation. This is well known for a large class of problems.

Step 2. Subdivide the region into subregions (elements). Although many kinds of elements can be used, our treatment will consider only triangular elements. The elements must span the entire region and approximate the boundary relatively closely. Every node (the vertices of our triangular elements) and every side of the triangles must be common with adjacent elements except for sides on the boundaries.

Step 3. Write an interpolating relation that gives values for the dependent variable within an element based on the values at the nodes (the vertices of the triangles). We will use linear interpolation from the three nodal values for the element. We will write the interpolation function as the sum of three terms; each term involves a quantity c_i, the value of $v(x, y)$ at a node.

Step 4. Substitute the interpolating relation into the functional, and set the partial derivatives of the functional with respect to each c to zero. This gives three equations, with the c's as unknowns for each element.

Step 5. Combine together (assemble) the element equations of step 4 to get a set of system equations. Adjust these for the boundary conditions of the problem, then solve. This will give the values for the unknown nodal values, the c's, that are approximations to $u(x, y)$ at the nodes. We can get approximations to $u(x, y)$ at intermediate points in the region by using the interpolating relations.

We will discuss each step in turn.

Step 1. Find the Functional

For Eq. (9.40) the functional is well known:

$$I[u] = \iint\limits_{\text{Region}} \left[\left(\frac{\partial u}{\partial x}\right)^2 + \left(\frac{\partial u}{\partial y}\right)^2 - Qu^2 + 2Fu \right] dx\ dy - \int_{L_2} [\alpha u^2 + 2\beta u] d_L. \quad \textbf{(9.41)}$$

It is possible to develop Eq. (9.41) using the Galerkin technique. Workers in the field of structural analysis usually derive it from the principle of virtual work. We will take it as a given.

Step 2. Subdivide the Region

As stipulated, we will use triangular elements, which will be defined by our choice of nodes. The placement of nodes is, in part, an art. In general, we place nodes close together in subregions where the solution is expected to vary rapidly. It is advantageous to make the sides run in the direction of the largest gradient. Along the curved parts of the boundary, nodes should be placed so that a side of the triangle closely approximates the boundary.

Some of these recommendations depend on knowing the nature of the solution in advance. Often, however, a better placement for the nodes can be accomplished after some preliminary computations or after preliminary trials using the finite-element method with nodes placed arbitrarily.

The chore of defining the nodes' coordinates is facilitated by computer programs that allow the user to place nodes with a pointing device on a graphical display of the region. These programs even permit rotating three-dimensional regions or looking at cross sections. Once the nodes have been located, the program connects them to create the elements.

Computer routines are available that can divide any given planar region into triangles automatically, but they usually do not have the expertise of an experienced engineer.

Step 3. Write the Interpolating Relations

This part of the development is longer than the previous one. As stated, we will use a linear relation. Figure 9.6(a) is a sketch of typical element (i) whose nodes are numbered r, s, and t in counterclockwise direction. The nodal values are c_r, c_s, and c_t.

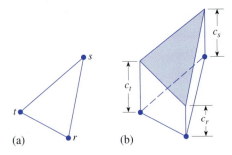

(a)   (b)

Figure 9.6

Within typical element (i), we write

$$v(x, y) = N_r c_r + N_s c_s + N_t c_t = \sum_{j=r,s,t} N_j c_j$$

$$= (N_r \ N_s \ N_t) \begin{Bmatrix} c_r \\ c_s \\ c_t \end{Bmatrix} = (N)\{c\},$$

(9.42)

where the N's (called *shape functions*) will be defined so that $v(x, y)$ at an interior point is a linear interpolation from the nodal values, the c's. We have shown in Eq. (9.42) that $v(x, y)$ can be expressed as the product of vectors (N) and $\{c\}$. (We use parentheses to enclose a row vector and curly brackets to enclose a column vector.) Vector and matrix notation will be useful in this section. We will indicate matrix M by $[M]$.

Figure 9.6(b) suggests that $v(x, y)$ lies on the plane above the element that passes through the nodal values. Equation (9.42) does not define $v(x, y)$ outside of element (i); there will be similar expressions for the other elements, but their N's and c's will differ.

A sketch of the entire region would not show $v(x, y)$ as a plane. Instead, it would be a surface composed of planar facets, each in a plane above an element. $v(x, y)$ for the entire region is continuous, but $v'(x, y)$ is not. (This is one of the flaws in our choice of element. Some other element definitions do not have this flaw.)

Another name for the N's of Eq. (9.42) is *pyramid function*. The reason for this name is illustrated in Fig. 9.7, where N_s of Fig. 9.6 is drawn. Its height at node s is unity and zero at the other nodes. It looks like an unsymmetrical pyramid whose base is the element with its apex directly above node s. The other two N's are similar. It is obvious that the N's are functions of x and y and that the c's are independent of x and y. We now develop expressions for the N's.

Because $v(x, y)$ varies linearly with position within the element, an alternative way to write the linear relation is

$$v(x, y) = a_1 + a_2 x + a_3 y = (1 \ \ x \ \ y)\{a\},$$

(9.43)

which must agree with the nodal values when $(x, y) = (x_j, y_j), j = r, s, t$. Hence

$$
\begin{array}{ll}
v \text{ at } r: & c_r = a_1 + a_2 x_r + a_3 y_r, \\
v \text{ at } s: & c_s = a_1 + a_2 x_s + a_3 y_s, \\
v \text{ at } t: & c_t = a_1 + a_2 x_t + a_3 y_t.
\end{array}
$$

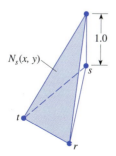

$N_s(x, y)$

1.0

Figure 9.7

This is a system of equations

$$[M]\{a\} = \{c\}, \qquad \text{(Curly brackets show a column vector)} \tag{9.44}$$

where

$$[M] = \begin{bmatrix} 1 & x_r & y_r \\ 1 & x_s & y_s \\ 1 & x_t & y_t \end{bmatrix}, \qquad \{a\} = \begin{Bmatrix} a_1 \\ a_2 \\ a_3 \end{Bmatrix}, \qquad \{c\} = \begin{Bmatrix} c_r \\ c_s \\ c_t \end{Bmatrix}.$$

Solving for $\{a\}$:

$$\{a\} = [M^{-1}]\{c\}.$$

The inverse of M is not difficult to find:

$$[M^{-1}] = \frac{1}{2(\text{Area})} \begin{bmatrix} (x_s y_t - x_t y_s) & (x_t y_r - x_r y_t) & (x_r y_s - x_s y_r) \\ (y_s - y_t) & (y_t - y_r) & (y_r - y_s) \\ (x_t - x_s) & (x_r - x_t) & (x_s - x_r) \end{bmatrix}, \tag{9.45}$$

with $2(\text{Area}) = \det(M)$. The value of the determinant is the sum of the elements in row 1 of Eq. (9.45) within the brackets. Area is the area of the triangular element.* You should verify that $[M^{-1}][M] = [I]$ to ensure that Eq. (9.45) truly gives the inverse matrix.

To apply the interpolating function to the minimization of the quadratic functional, Eq. (9.41), we prefer to write $v(x, y)$ in terms of the shape functions of Eq. (9.42). This task is easy. We have, from Eqs. (9.42) and (9.43),

$$v(x, y) = a_1 + a_2 x + a_3 y = (1 \quad x \quad y)\{a\}$$
$$= (1 \quad x \quad y)[M^{-1}]\{c\}.$$

However, in terms of N (from Eq. 9.42),

$$v(x, y) = (N)\{c\}. \tag{9.46}$$

Comparing the two expressions, we have

$$(N) = (1 \quad x \quad y)[M^{-1}], \tag{9.47}$$

where M^{-1} is given by Eq. (9.45). Observe carefully that Eq. (9.47) says that each N is a linear function of x and y of the form

$$N_j = A_j + B_j x + C_j y, \qquad j = r, s, t \tag{9.48}$$

and that the coefficients are in column j of $[M^{-1}]$.

We have found the expressions for the N's. Before we go on, we digress to show an example that will clarify this step.

EXAMPLE 9.6   For the triangular element shown in Fig. 9.8 with nodes r, s, and t in counterclockwise order, find $\{a\}$, $\{N\}$, and $v(0.8, 0.4)$.

---

*That Area $= \frac{1}{2}\det(M)$ is shown in most books on vectors where the cross product is explained.

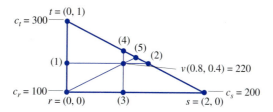

Figure 9.8

Node	x	y	c
r	0	0	100
s	2	0	200
t	0	1	300

Before we do any computations, we can find $v(0.8, 0.4)$ by inspection. (See Fig. 9.8.) Point (1) is at $(0, 0.4)$, so v there is 180 by linear interpolation between nodes r and t. Similarly, v at point (2) is 240. The point $(0.8, 0.4)$ is $\frac{2}{3}$ of the distance from points (1) and (2), so $v(0.8, 0.4) = 180 + \frac{2}{3}(240 - 180) = 220$. We get the same result by interpolating between points (3) and (4), and between node r and (5).

To get $\{a\}$ we first compute $[M^{-1}]$:

$$[M] = \begin{bmatrix} 1 & 0 & 0 \\ 1 & 2 & 0 \\ 1 & 0 & 1 \end{bmatrix}, \qquad [M^{-1}] = \begin{bmatrix} 1 & 0 & 0 \\ -0.5 & 0.5 & 0 \\ -1 & 0 & 1 \end{bmatrix}.$$

Then we compute

$$\{a\} = [M^{-1}]\{c\} = \begin{bmatrix} 1 & 0 & 0 \\ -0.5 & 0.5 & 0 \\ -1 & 0 & 1 \end{bmatrix} \begin{Bmatrix} 100 \\ 200 \\ 300 \end{Bmatrix} = \begin{Bmatrix} 100 \\ 50 \\ 200 \end{Bmatrix},$$

giving $v(x, y) = 100 + 50x + 200y$. (You should confirm that this gives the correct values at each of the nodes.) If we substitute $x = 0.8$, $y = 0.4$, we get $v = 220$, as we should.

From Eq. (9.47),

$$(N) = (1 \quad x \quad y)[M^{-1}] = (1 - 0.5x - y, \quad 0.5x, \quad y).$$

In other words, we have

$$N_r = 1 - 0.5x - y,$$
$$N_s = 0.5x,$$
$$N_t = y.$$

(You should confirm that these also have the proper values at each of the nodes.) It is important to notice that the coefficients of the N's (the A_i, B_i, and C_i of Eq. 9.48) can be read directly from the columns of $[M^{-1}]$. ▲

In what follows we will need the partial derivatives of the N's with respect to x and to y. From $N_j = A_j + B_j x + C_j y$, we see that these are constants that can be read from rows 2 and 3 of $[M^{-1}]$ in column j.

At this point we know how to write $v(x, y)$ within the single triangular element (i) as $v(x, y) = (N^{(i)})\{c^{(i)}\}$. (The superscripts (i) tell which element is being considered whenever this is necessary.) We now stipulate that $(N^{(i)}) \equiv (0)$ everywhere outside of element (i). Therefore we can write

$$v(x, y) = \sum_{i=\text{all elements}} (N^{(i)})\{c^{(i)}\}.$$

This is a mathematical statement of the previous observation that $v(x, y)$ is a surface composed of joined planar facets.

We are now ready for step 4 of our plan. This too is lengthy but each portion is easy.

Step 4. Substitute $v(x, y)$ into the Functional and Minimize

We continue to work with typical element (i) whose nodes are r, s, and t. Repeating Eq. (9.46), our $v(x, y)$ is

$$v(x, y) = (N)\{c\} = N_r c_r + N_s c_s + N_t c_t,$$

where the N's are given by Eqs. (9.47) and (9.45). Recall that each N is $A_j + B_j x + C_j y$ with the coefficients given by the elements in column j of $[M^{-1}]$.

Our objective is to develop a set of three equations for element (i), which is, in matrix form,

$$[K]\{c\} = \{b\},$$

and which is a prototype of similar equations for all other elements.

When we substitute $v(x, y)$ for element (i) into the functional of Eq. (9.41), we get

$$I(c_r, c_s, c_t) = \iint\limits_{\text{element}(i)} \left[\left(\frac{\partial v}{\partial x}\right)^2 + \left(\frac{\partial v}{\partial y}\right)^2 - Qv^2 + 2Fv \right] dx\, dy$$

$$-\oint_{L_2} [\alpha v^2 + 2\beta v^2]\, dL.$$

(9.49)

(I is now an ordinary function of the c's. The integral is only over the area of element (i) because the N's that define $v(x, y)$ in element (i) are zero outside of (i). The last term appears only if element (i) has a side on the boundary. Actually, we will postpone handling this last term for now and handle it as an adjustment to the equations after they have been developed.)

We minimize I by setting the three partials (with respect to each of the three c's) to zero. We now develop expressions for these partials. First consider $\partial I/\partial c_r$.

For the first term in the integrand:

$$\frac{\partial}{\partial c_r}\left(\frac{\partial v}{\partial x}\right)^2 = 2\left(\frac{\partial v}{\partial x}\right)\left(\frac{\partial}{\partial c_r}\left(\frac{\partial v}{\partial x}\right)\right).$$

However,

$$\frac{\partial v}{\partial x} = \left(\frac{\partial N_r}{\partial x}\right)c_r + \left(\frac{\partial N_s}{\partial x}\right)c_t + \left(\frac{\partial N_t}{\partial x}\right)c_t$$

$$= B_r c_r + B_s c_s + B_t c_t,$$

by virtue of Eq. (9.48). (The B's come from row 2 of $[M^{-1}]$.)

Also, $\partial/\partial c_r(\partial v/\partial x) = B_r$ because c_s and c_t are independent of c_r. Hence

$$\frac{\partial}{\partial c_r}\left[\left(\frac{\partial v}{\partial x}\right)^2\right] = 2(B_r^2 + B_r B_s + B_r B_t).$$

The result for the second term is similar:

$$\frac{\partial}{\partial c_r}\left[\left(\frac{\partial v}{\partial y}\right)^2\right] = 2(C_r^2 + C_r C_s + C_r C_t),$$

where the C's come from row 3 of $[M^{-1}]$.

We next consider the Q term. Q is independent of c_r, so

$$\frac{\partial}{\partial c_r}(-Qv^2) = -Q\left[2v\left(\frac{\partial v}{\partial c_r}\right)\right] - 2Q(N_r c_r + N_s c_s + N_t c_t)(N_r).$$

Finally we work with the F term. F is independent of c_r, so

$$\frac{\partial}{\partial c_r}(2Fv) = 2F\left(\frac{\partial v}{\partial c_r}\right) = 2F(N_r).$$

Putting all this together, we have

$$\frac{\partial I}{\partial c_r} = 0 = \iint\limits_{(i)} 2[B_r^2 c_r + B_r B_s c_s + B_r B_t c_t]\ dx\ dy$$

$$+ \iint\limits_{(i)} 2[C_r^2 c_r + C_r C_s c_s + C_r C_t c_t]\ dx\ dy$$

$$- \iint\limits_{(i)} 2Q[N_r^2 c_r + N_r N_s c_s + N_r N_t c_t]\ dx\ dy$$

$$+ \iint\limits_{(i)} 2FN_r\ dx\ dy.$$

(9.50)

Equation (9.50) really is a formulation with the c's unknown:

$$K_{rr} c_r + K_{rs} c_s + K_{rt} c_t = b_r,$$

(9.51)

where

$$K_{rr} = \iint_{(i)} 2B_r^2 \, dx \, dy + \iint_{(i)} 2C_r^2 \, dx \, dy - \iint_{(i)} 2QN_r^2 \, dx \, dy,$$

$$K_{rs} = \iint_{(i)} 2B_r B_s \, dx \, dy + \iint_{(i)} 2C_r C_s \, dx \, dy - \iint_{(i)} 2QN_r N_s \, dx \, dy,$$

$$K_{rt} = \iint_{(i)} 2B_r B_t \, dx \, dy + \iint_{(i)} 2C_r C_t \, dx \, dy - \iint_{(i)} 2QN_r N_t \, dx \, dy,$$

$$b_r = -\iint_{(i)} 2FN_r \, dx \, dy.$$

(Remember, we postpone handling the last part of Eq. (9.49).)

Now we recognize that the B's and C's of Eq. (9.51) are constants, so we can bring them out from under the integral sign. If we use average values for Q and F within the element, we can also bring these out as their average values. (The best average value to use is the value of Q and F at the centroid of the triangular element.)

This means that we have to evaluate these five integrals:

$$I_1: \quad \iint_{(i)} dx \, dy$$

$$I_2: \quad \iint_{(i)} N_r^2 \, dx \, dy \qquad I_3: \quad \iint_{(i)} N_r N_s \, dx \, dy \qquad I_4: \quad \iint_{(i)} N_r N_t \, dx \, dy$$

$$I_5: \quad \iint_{(i)} N_r \, dx \, dy.$$

The first of these is easy: $I_1 =$ Area of the element, which we already know from having computed $[M^{-1}]$. The other integrals are laborious to compute directly, but there is a useful formula for the integral of the product of powers of linear functions over a triangle:

$$\iint_{(\text{triangle})} N_r^\ell N_s^m N_t^n \, dx \, dy = \frac{2\ell! m! n!}{(\ell + m + n + 2)!}(\text{Area}).$$

Using this with the proper values for the exponents, ℓ, m, and n, gives:

$$I_2 = \frac{(\text{Area})}{6},$$

$$I_3 = \frac{(\text{Area})}{12},$$

$$I_4 = \frac{(\text{Area})}{12},$$

$$I_5 = \frac{(\text{Area})}{3}.$$

The terms in Eq. (9.51) are then

$$K_{rr}c_r + K_{rs}c_s + K_{rt}c_t + b_r, \tag{9.52}$$

where

$$K_{rr} = 2(\text{Area})\left(B_r^2 + C_r^2 - \frac{Q_{av}}{6}\right),$$

$$K_{rs} = 2(\text{Area})\left(B_r B_s + C_r C_s - \frac{Q_{av}}{12}\right),$$

$$K_{rt} = 2(\text{Area})\left(B_r B_t + C_r C_t - \frac{Q_{av}}{12}\right),$$

$$b_r = -2(\text{Area})\left(\frac{F_{av}}{3}\right).$$

If we do the same with $\partial I/\partial c_s$ and $\partial I/\partial c_t$, we get two more equations in the c's for element (i). All together we have three equations, which we call the *element equations*. We simplify these equations somewhat by omitting the common factor of 2 for each of them to get

$$[K]\begin{Bmatrix} c_r \\ c_s \\ c_t \end{Bmatrix} = \begin{Bmatrix} b_r \\ b_s \\ b_t \end{Bmatrix},$$

where

(diagonals) $\qquad K_{jj} = \text{Area}\left[B_j^2 + C_j^2 - \frac{Q_{av}}{6}\right], \qquad j = r,\ s,\ t;$

(off-diagonals) $\qquad K_{jk} = \text{Area}\left[B_j B_k + C_j C_k - \frac{Q_{av}}{12}\right], \qquad \begin{cases} j \neq k, \\ j = r,\ s,\ t, \\ k = r,\ s,\ t; \end{cases}$

(rhs) $\qquad b_j = -\text{Area}\left[\frac{F_{av}}{3}\right], \qquad j = r,\ s,\ t.$

Observe that $[K]$ is symmetrical: $K_{ij} = K_{ji}$.

Here is an example to clarify the formation of the element equations.

EXAMPLE 9.7   Find the element equations for the element of Example 9.6 if $Q(x, y) = (xy)/2$ and $F(x, y) = x + y$.

The nodes are $(x, y) = (0, 0)$, $(2, 0)$, and $(0, 1)$. We had, for $[M^{-1}]$,

$$[M^{-1}] = \begin{bmatrix} 1 & 0 & 0 \\ -0.5 & 0.5 & 0 \\ -1 & 0 & 1 \end{bmatrix}.$$

Area = 1. Centroid is at $x = (0 + 2 + 0)/3 = \frac{2}{3}$, $y = (0 + 0 + 1)/3 = \frac{1}{3}$. $Q_{av} = (\frac{2}{3})(\frac{1}{3})/2 = \frac{1}{9}$. $F_{av} = \frac{2}{3} + \frac{1}{3} = 1$.

Using Eq. (9.52), we find that the element equations are

$$\begin{bmatrix} 1.2315 & -0.2592 & -1.0092 \\ -0.2592 & 0.2315 & -0.0093 \\ -1.0092 & -0.0093 & 0.9815 \end{bmatrix} \begin{Bmatrix} c_r \\ c_s \\ c_t \end{Bmatrix} = \begin{Bmatrix} -0.3333 \\ -0.3333 \\ -0.3333 \end{Bmatrix}.$$

▲

We are now ready for step 5 of the plan.

Step 5. Assemble the Equations, Adjust for Boundary Conditions, Solve

There are three separate operations in step 5: (i) assemble the equations, (ii) adjust for boundary conditions, and (iii) solve the equations.

(i) Do the Assembly.   As we have seen, there are three equations for every element. However, some or all nodes of element (i) are shared with other elements; the c-value for a shared node then appears in the equations of all elements that share the node. Combining all of the element equations will create a global system coefficient matrix with as many rows and columns as there are nodes in the system. We combine (assemble) the system matrix in the following way.

Suppose there are n nodes in the system. Number the nodes in order, from 1 to n. Associate the number of each node with the row and column of every element matrix where the c for that node appears on the diagonal. Also associate the node numbers with the rows and columns of the system matrix in the same way.

We get the entry in row (i) and column (j) of the system matrix by adding the values from row (i) of every element matrix that has row (i), then adding these in the columns where the column-node numbers match. We also add the b_i's from these rows to get the b_i of the system matrix. An example will clarify this operation.

EXAMPLE 9.8   Suppose there are five nodes that define three elements, as shown in Fig. 9.9 with the element matrices of Eqs. (9.53) below. Construct the system matrix without adjusting for boundary conditions.

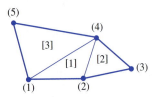

Figure 9.9

$$
\text{Element [1]} \quad
\begin{matrix} (1) \to \\ (2) \to \\ (4) \to \end{matrix}
\begin{bmatrix} K_{11} & K_{12} & K_{13} \\ K_{21} & K_{22} & K_{23} \\ K_{31} & K_{32} & K_{33} \end{bmatrix}
\begin{Bmatrix} c_1 \\ c_2 \\ c_4 \end{Bmatrix}
\begin{matrix} = \\ = \\ = \end{matrix}
\begin{Bmatrix} b_1 \\ b_2 \\ b_3 \end{Bmatrix}
\qquad \textbf{(9.53a)}
$$

$$
\begin{matrix} \uparrow & \uparrow & \uparrow \\ (1) & (2) & (4) \end{matrix}
$$

$$
\text{Element [2]} \quad
\begin{matrix} (2) \to \\ (3) \to \\ (4) \to \end{matrix}
\begin{bmatrix} K_{11} & K_{12} & K_{13} \\ K_{21} & K_{22} & K_{23} \\ K_{31} & K_{32} & K_{33} \end{bmatrix}
\begin{Bmatrix} c_2 \\ c_3 \\ c_4 \end{Bmatrix}
\begin{matrix} = \\ = \\ = \end{matrix}
\begin{Bmatrix} b_1 \\ b_2 \\ b_3 \end{Bmatrix}
\qquad \textbf{(9.53b)}
$$

$$
\begin{matrix} \uparrow & \uparrow & \uparrow \\ (2) & (3) & (4) \end{matrix}
$$

$$
\text{Element [3]} \quad
\begin{matrix} (4) \to \\ (5) \to \\ (1) \to \end{matrix}
\begin{bmatrix} K_{11} & K_{12} & K_{13} \\ K_{21} & K_{22} & K_{23} \\ K_{31} & K_{32} & K_{33} \end{bmatrix}
\begin{Bmatrix} c_4 \\ c_5 \\ c_1 \end{Bmatrix}
\begin{matrix} = \\ = \\ = \end{matrix}
\begin{Bmatrix} b_1 \\ b_2 \\ b_3 \end{Bmatrix}
\qquad \textbf{(9.53c)}
$$

$$
\begin{matrix} \uparrow & \uparrow & \uparrow \\ (4) & (5) & (1) \end{matrix}
$$

[The rows and columns of Eqs. (9.53) could have been in a different order, although we always go counterclockwise around the element in selecting the nodes.]

We construct the system matrix as follows, where the superscripts indicate the element number that provides the value:

For row 1	For row 2
col 1: $K_{11}^{[1]} + K_{33}^{[3]}$	col 1: $K_{21}^{[1]}$
col 2: $K_{12}^{[1]}$	col 2: $K_{22}^{[1]} + K_{11}^{[2]}$
col 3: 0	col 3: $K_{12}^{[2]}$
col 4: $K_{13}^{[1]} + K_{31}^{[3]}$	col 4: $K_{23}^{[1]} + K_{13}^{[2]}$
col 5: $K_{32}^{[3]}$	col 5: 0
b: $b_1^{[1]} + b_3^{[3]}$	b: $b_2^{[1]} + b_1^{[2]}$

and so forth.

[A zero appears in column 3 of row 1 because node (3) is not in any element that includes node (1). A zero appears in column 5 of row 2 because node (5) is not in any element that includes node (2).] ▲

Once the system matrix has been assembled, we make the adjustments for boundary conditions.

(ii) Adjust for Boundary Conditions. There are two types of boundary conditions: non-Dirichlet conditions on some parts of the boundary (L_2) and Dirichlet conditions on other parts (L_1). We will always select nodes such that only one of the two types of conditions pertains to any side of the element. Hence there will always be a node at the point where the two types join. These two types of boundary conditions require two separate adjustments. We prefer to apply the adjustment for a boundary condition that involves the outward normal derivative to the system equations first and then do the adjustment for Dirichlet conditions.

Adjusting for Non-Dirichlet Conditions. Non-Dirichlet conditions (those that involve the outward normal derivative) are associated, not with the nodes, but with sides of the triangular elements, sides that correspond to part of L_2 of Eq. (9.41). Consider an element that has a non-Dirichlet condition on one side that lies between nodes r and s. The effect of the boundary condition on the equations comes from differentiating the last term of Eq. (9.49) with respect to the c's. However, if we take α and β out from the integrand as average values, we see from Eq. (9.49) that they are of the same form as the Q and F terms except they are line integrals rather than area integrals. That similarity lets us immediately write the result of the differentiation with respect to c_r as

$$-2\alpha_{\text{av}} \oint (N_r^2 c_r + N_r N_s c_s)\, dL - 2\beta_{\text{av}} \oint N_r\, dL, \qquad (9.54)$$

where the line integrals are along the side between nodes r and s. It is important to note that we have not included c_t because node t is not on the side we are considering. (The average values of α and β should be taken at the midpoint of the side.) When we integrate, we have

$$-2\alpha_{\text{av}} L \left(\frac{c_r}{3} + \frac{c_s}{6} \right) - 2\beta_{\text{av}} \frac{L}{2}. \qquad (9.55)$$

(It is easy to evaluate the integrals when we remember that the N's are linear from 1 to 0 between the two nodes.)

Precisely the same relations result when we differentiate with respect to node c_s, except the roles of r and s are interchanged in Eqs. (9.54) and (9.55). The net result would be to add $2\beta L/2$ to the right-hand sides of the rows for c_r and c_t. We also would subtract the multipliers of c_r and c_s in Eq. (9.55) from the coefficients for c_r and c_s in row r. The similar equations from the partials with respect to c_s provide subtractions from the coefficients in row s.

Recall, however, that we canceled a 2 factor when we constructed the element equations and so we must do so here. We make this adjustment to the element equations for every element that has a derivative condition on a side.

Adjusting for Dirichlet Conditions. For every node that appears on the boundary where there is a Dirichlet condition, the u-value is specified. We insert this known value in place of the c of that node in every equation where it appears and transpose to the right-hand side. (Actually, if the node number is m, all entries in column m of the matrix are multiplied by the value and subtracted from the right-hand side of the corresponding row.) We also remove the row corresponding to the number of the known node from the set of equations. (The column for this node has already been "removed" by being transferred to the right-hand side.)

Removing the rows for those nodes with a Dirichlet condition is simplified in a computer program if these rows are at the top or the bottom of the matrix. There are other ways to handle Dirichlet conditions that avoid having to remove the rows.

This completes our construction of the system equations.

(iii) Getting the Solution.   We solve the system in the usual way, perhaps preferring an iterative procedure if the system is large.

An example, intentionally simple, follows.

EXAMPLE 9.9   The region shown in Fig. 9.10 has four nodes. It is divided into just two elements. The values for u are specified at nodes (3) and (4), and the outward normal gradient is specified on three sides as indicated. The equation we are to solve is

$$u_{xx} + u_{yy} - \left(\frac{y}{10}\right) u = \frac{x}{4} + y - 12.$$

Find the solution by the finite-element method. The element-matrix inverses are:

M inverse for element 1			**M inverse for element 2**		
area is 11.5			**area is 6**		

$$\begin{bmatrix} 1.000 & 0.000 & 0.000 \\ -0.043 & 0.174 & -0.130 \\ -0.261 & 0.043 & 0.217 \end{bmatrix} \qquad \begin{bmatrix} 1.000 & 0.000 & 0.000 \\ -0.250 & 0.250 & 0.000 \\ 0.083 & -0.417 & 0.333 \end{bmatrix}$$

$$\begin{matrix} (2) & (4) & (1) & \qquad & (2) & (3) & (4) \end{matrix}$$

$$Q_{av} = -0.233 \quad F_{av} = -9.333 \qquad Q_{av} = -0.1 \quad F_{av} = -10.25$$

From these we get these element equations:

Element equations for element 1

$$\begin{matrix} 2 \rightarrow \\ 4 \rightarrow \\ 1 \rightarrow \end{matrix} \begin{bmatrix} 1.2516 & 0.0062 & -0.3633 & 35.7778 \\ 0.0062 & 0.8168 & 0.0714 & 35.7778 \\ -0.3633 & 0.0714 & 1.1864 & 35.7778 \end{bmatrix}$$

$$\begin{matrix} (2) & (4) & (1) \end{matrix}$$

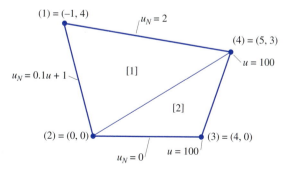

(1) = (−1, 4)

$u_N = 2$

(4) = (5, 3)

$u = 100$

$u_N = 0.1u + 1$

[1]

[2]

(2) = (0, 0)

(3) = (4, 0)

$u_N = 0$

$u = 100$

Figure 9.10

Element equations for element 2

$$
\begin{array}{c}
2 \rightarrow \\
3 \rightarrow \\
4 \rightarrow
\end{array}
\left[
\begin{array}{cccc}
0.5167 & -0.5333 & 0.2167 & 20.5000 \\
-0.5333 & 1.5167 & -0.7833 & 20.5000 \\
0.2167 & -0.7833 & 0.7667 & 20.5000
\end{array}
\right]
$$
$$
\quad\; (2) \qquad\;\; (3) \qquad\;\; (4)
$$

These equations assemble to give this unadjusted system matrix:

$$
\left[
\begin{array}{cccc|c}
1.186 & -0.363 & 0.000 & 0.071 & 35.778 \\
-0.363 & 1.768 & -0.533 & 0.223 & 56.278 \\
0.000 & -0.533 & 1.517 & -0.783 & 20.500 \\
0.071 & 0.223 & -0.783 & 1.583 & 56.278
\end{array}
\right].
$$

We need the lengths of sides $4{-}1$, $1{-}2$, and $2{-}3$. They are:

$$
\text{Side } 4{-}1{:}\;\; 6.083, \qquad \text{Side } 1{-}2{:}\;\; 4.123, \qquad \text{Side } 2{-}3{:}\;\; 4.
$$

We now adjust for the derivative conditions to get the modified system:

$$
\left[
\begin{array}{cccc|c}
1.049 & -0.432 & 0.000 & 0.003 & 43.922 \\
-0.432 & 1.631 & -0.533 & 0.154 & 64.422 \\
0.000 & -0.533 & 1.517 & -0.783 & 20.500 \\
0.003 & 0.154 & -0.783 & 1.446 & 64.422
\end{array}
\right].
$$

The second adjustment is for the known values at nodes (3) and (4), giving

$$
\left[
\begin{array}{cc|c}
1.049 & -0.432 & 43.650 \\
-0.432 & 1.631 & 102.339
\end{array}
\right],
$$

which we solve to get these estimates of u_1 and u_2:

$$
u_1 = 75.71, \qquad u_2 = 82.80. \qquad\qquad \blacktriangle
$$

9.5 Finite Elements for Parabolic and Hyperbolic Equations

As we have just seen, the finite-element method is often preferred for solving boundary-value problems. It is also the preferred method for solving the heat equation when the region of interest is not regular. You should know something about this application of finite elements, but we do not give a full treatment.

Consider the heat-flow equation in two dimensions with heat generation given by $F(x, y)$:

$$
\frac{\partial u}{\partial t} = \frac{k}{c\rho}\left(\frac{\partial^2 u}{\partial x^2} + \frac{\partial^2 u}{\partial y^2}\right) + F(x, y), \tag{9.56}
$$

which is subject to initial conditions at $t = 0$ and boundary conditions that may be Dirichlet or may involve the outward normal gradient. Although this is really a three-variable problem (in x, y, and t), it is customary to approximate the time derivative with a finite difference and apply finite elements only to the spatial region. Doing so, we can rewrite Eq. (9.56) as

$$\frac{u^{m+1} - u^m}{\Delta t} = \frac{k}{c\rho}\left(\frac{\partial^2 u}{\partial x^2} + \frac{\partial^2 u}{\partial y^2}\right) + F(x,\ y), \tag{9.57}$$

where we have used a forward difference as in the explicit method. (We might prefer Crank–Nicolson or the implicit method, but we will keep things simple.)

To apply finite elements to the region, we do exactly as described previously—cover the region with joined elements, write element equations for the right-hand side of Eq. (9.56), assemble these, adjust for boundary conditions, and solve. However, we must also consider the time variable. We do so by considering Eq. (9.57) to apply at a fixed point in time, t_m. Because we know the values of u everywhere within the region at $t = t_0$, we surely know the initial nodal values. We then can solve Eq. (9.57) for the u-values at $t = t_0 + \Delta t$, where the size of Δt is chosen small enough to ensure stability.

We will use the Galerkin procedure to derive the element equations to provide some variety from Section 9.4. In this procedure you will remember that we integrate the residual weighted with each of the shape functions and set them to zero. (The integrations are done over the element area.) If we stay with linear triangular elements, there are three shape functions, N_r, N_s, and N_t, where the subscripts denote the three vertices (nodes) of the element taken in counterclockwise order.

The residual for Eq. (9.56) is

$$\text{Residual} = u_t - \alpha(u_{xx} + u_{yy}) - F, \tag{9.58}$$

where we have used the subscript notation for derivatives and have abbreviated $k/c\rho$ with α.

As stated, we will use linear triangular elements; within each element we approximate u with

$$u(x,\ y) \approx v(x,\ y) = N_r c_r + N_s c_s + N_t c_t. \tag{9.59}$$

This means that Galerkin integrals are

$$\iint N_j[v_t - \alpha(v_{xx} + v_{yy}) + F]\ dx\ dy = 0, \qquad j = r,\ s,\ t. \tag{9.60}$$

If we apply integration by parts (as we did in Section 9.3) to the second derivatives of Eq. (9.60), we can reduce the order of these derivatives. Doing so and replacing v from Eq. (9.59) gives a set of three equations for each element, which we write in matrix form:

$$[C]\left\{\frac{\partial c}{\partial t}\right\} + [K]\{c\} = \{b\}. \tag{9.61}$$

The components of $\{c\}$ are the nodal temperatures of the element, of course; those of $\{\partial c/\partial t\}$ are the time derivatives. The components of the matrices of Eq. (9.61) are

$$C_{ij} = \iint N_i\ dx\ dy, \qquad i = r,\ s,\ t,$$

$$K_{ij} = \alpha_{av} \iint \left[\left(\frac{\partial N_i}{\partial x}\right)\left(\frac{\partial N_j}{\partial x}\right) + \left(\frac{\partial N_i}{\partial y}\right)\left(\frac{N_j}{\partial y}\right)\right] dx\ dy, \qquad i,\ j = r,\ s,\ t, \tag{9.62}$$

$$b_i = \iint FN_i\ dx\ dy + \oint u_N N_i\ dL, \qquad i = r,\ s,\ t.$$

In Eqs. (9.62), the line integral in the b's is present only along a side of an element on the boundary of the region where the outward normal gradient u_N is specified as a boundary condition.

From the development of Section 9.4, we know how to evaluate all of the integrals of Eqs. (9.62) when the elements are triangles. [See, for example, Burnett (1987) for the evaluations for other types of elements.]

As stated, we will use a finite-difference approximation for $\partial c/\partial t$. If this is a forward difference as suggested, we get the explicit formula

$$\frac{1}{\Delta t}[C]\{c^{m+1}\} = \frac{1}{\Delta t}[C]\{c^m\} - [K]\{c^m\} + \{b\}, \qquad (9.63)$$

where all the c's on the right are nodal temperatures at $t = t_m$ and the nodal temperatures on the left in $\{c^{m+1}\}$ are at $t = t_{m+1}$.

We can put Eq. (9.63) into a more familiar iterating form by multiplying through by $\Delta t[C]^{-1}$:

$$\{c^{m+1}\} = \{c^m\} - \Delta t[C]^{-1}[K]\{c^m\} + \Delta t[C]^{-1}\{b\}. \qquad (9.64)$$

[We can make Eq. (9.64) more compact by combining the multipliers of $\{c^m\}$.]

In principle, we have solved the heat-flow problem by finite elements. We construct the equations for every element from Eq. (9.64) and assemble them to get the global matrix, then adjust for boundary conditions just as in Section 9.4. This gives a set of equations in the unknown nodal values that we use to step forward in time from the initial point. With the explicit method illustrated here, each time step is just a matrix multiplication of the current nodal temperatures (and a vector addition) to get the next set of values. If we had used an implicit method such as Crank–Nicolson, we would have had to solve a set of equations at each step, but, unfortunately, they are not tridiagonal. We might hope for some equivalent to the A.D.I. method, but A.D.I. requires that the nodes be uniformly spaced. The conclusion is that the finite-element method in two or three dimensions is a problem that is expensive to solve. In one dimension, however, the system is tridiagonal, so that situation is not bad.

Finite Elements and the Wave Equation

We will only outline how finite elements are applied to the wave equation, because this topic is too complex for full coverage here. Just as for the heat equation, finite elements are used for the space region and finite differences for time derivatives. We will develop only the vibrating string case (one dimension); two or three space dimensions are handled analogously but are harder to follow.

The equation that is usually solved is a more general case of the simple wave equation we have been discussing. In engineering applications, damping forces that serve to decrease the amplitude of the vibrations are important, and external forces that excite the system are usually involved. We therefore use, for a one-dimensional case, this equation for the displacement of points on the vibrating string, $y(x, t)$:

$$\frac{\partial}{\partial x}\left[T(x)\frac{\partial y}{\partial x}\right] - h(x)\frac{\partial y}{\partial t} + F(x, t) = \frac{w(x)}{g}\frac{\partial^2 y}{\partial t^2}.$$ **(9.65)**

Here T represents the tension, which is allowed to vary with x; h represents a damping coefficient that opposes motion in proportion to the velocity; F is the external force; and w/g is the mass density. There are boundary conditions (at $x = a$ and $x = b$) as well as initial conditions that specify initial displacements and velocities.

The approach is essentially identical to that used for unsteady-state heat flow: Apply finite elements to x and finite differences to the time derivatives. We will use linear one-dimensional elements, so we subdivide $[a, b]$ into portions (elements) that join at points that we call nodes. Within each element, we approximate $y(x, t)$ with $v(x, t)$,

$$y(x, t) \approx v(x, t) = N_L c_L + N_R c_R,$$ **(9.66)**

where c_L and c_R are the approximations to the displacements at the nodes at the left and right ends of a typical linear element. The N's are shape functions (in this one-dimensional case, we have called them "hat functions").

By using the Galerkin procedure, we can get this integral equation, which we will eventually transform into the element equations:

$$\int_L^R N_i\left(\frac{w}{g}\right)y_{tt}\ dx + \int_L^R N_i'(T)y_x\ dx + \int_L^R N_i(h)y_t\ dx$$

$$= \int_L^R N_i(F)\ dx + N_i(T)y_x\Bigg]_{x=R} - N_i(T)y_x\Bigg]_{x=L}, \quad i = L, R \quad \textbf{(9.67)}$$

In Eq. (9.67) we have used subscript notation for the partial derivatives of y with respect to t and x and primes to represent the derivatives of the N's with respect to x (because the N's are functions of x only).

We now use Eq. (9.66) to find substitutions for y and its derivatives:

$$y(x, t) \approx N_L c_L + N_R c_R,$$

$$\frac{\partial y}{\partial x} \approx N_L' c_L + N_R' c_R,$$

$$\frac{\partial y}{\partial t} \approx N_L \dot{c}_L + N_R \dot{c}_R,$$ **(9.68)**

$$\frac{\partial^2 y}{\partial t^2} \approx N_L \ddot{c}_L + N_R \ddot{c}_R.$$

Here we employ the dot notation for time derivatives. (The c's vary with time, of course, but the N's do not.)

We now substitute from Eqs. (9.68) into Eq. (9.67) to get a pair of equations for each element (we write them in matrix form):

$$[M]\{\ddot{c}\} + [C]\{\dot{c}\} + [K]\{c\} = \{b\},$$

$$M_{ij} = \int_L^R N_i\left(\frac{w}{g}\right)N_j \, dx,$$

$$\left. C_{ij} = \int_L^R N_i(h)N_j \, dx, \right\} \quad i, j = \text{L, R,} \tag{9.69}$$

$$K_{ij} = \int_L^R N_i'(T)N_j' \, dx,$$

$$b_i = \int_L^R N_i(F) \, dx + N_i(T)\frac{\partial y}{\partial x}\bigg]_{x=R} - N_i(T)\frac{\partial y}{\partial x}\bigg]_{x=L}, \qquad i = \text{L, R.}$$

We will replace the time derivatives with finite differences, selecting central differences because they worked so well in the finite-difference solution to the simple wave equation. Thus we get

$$[M]\frac{\{c^{m+1} - 2c^m + c^{m-1}\}}{(\Delta t)^2} + [C]\frac{\{c^{m+1} - c^{m-1}\}}{2\Delta t} + [K]\{c^m\} = \{b^m\}. \tag{9.70}$$

Now we solve Eq. (9.70) for $\{c^{m+1}\}$:

$$\left(\frac{1}{(\Delta t)^2}[M] + \frac{1}{2\Delta t}[C]\right)\{c^{m+1}\} = \left(\frac{2}{(\Delta t)^2}[M] - [K]\right)\{c^m\} - \left(\frac{1}{(\Delta t)^2}[M] - \frac{1}{2\Delta t}[C]\right)\{c^{m-1}\} + \{b^m\}. \tag{9.71}$$

Notice that we need two previous sets of displacements to advance to the new time, t_{m+1}. We faced this identical problem when we solved the simple wave equation with finite differences, and we solve it in the same way. We use the initial velocities (given as one of the initial conditions) to get $\{c^{-1}\}$ to start the solution:

$$\{c^{-1}\} = \{c^1\} - 2\Delta t\{g(x)\}, \tag{9.72}$$

where $\{g(x)\}$ is the vector of initial velocities. (In view of our earlier work, we expect improved results if we use a weighted average of the g-values if the $g(x)$'s are not constants.)

We have not specifically developed the formulas for the components of the matrices and vector of Eqs. (9.69), but they are identical to those we derived when we applied finite elements to boundary-value problems in Section 9.3, because we will take out w, h, T, and F as average values within the elements. So we just copy from Section 9.3:

$$M_{11} = M_{22} = \left(\frac{w}{g}\right)\frac{\Delta}{3}, \qquad M_{12} = M_{21} = \left(\frac{w}{g}\right)\frac{\Delta}{6},$$

$$C_{11} = C_{22} = h\frac{\Delta}{3}, \qquad C_{12} = C_{21} = h\frac{\Delta}{6},$$

$$K_{11} = K_{22} = \frac{T}{\Delta}, \qquad K_{12} = K_{21} = -\frac{T}{\Delta}, \tag{9.73}$$

$$b_1 = F\frac{\Delta}{2} - \left[T\frac{\partial y}{\partial x}\right]_{x=L}, \qquad b_2 = F\frac{\Delta}{2} + \left[T\frac{\partial y}{\partial x}\right]_{x=R}.$$

In this set, Δ represents the length of the element.

We now have everything we need to construct the element equations. Except for the end elements (and then only if the boundary conditions involve the gradient), the gradient terms in Eqs. (9.73) cancel between adjacent elements. Assembly in this case is very simple because there are always two elements that share each node (except at the ends).

What advantage is there to finite elements over finite differences? The major one is that we can use nodes that are unevenly spaced without having to modify the procedure. The advantage becomes really significant in two- and three-dimensional situations, but the other side of the coin is that solving the equations for each time step is not easy.

9.6 Theoretical Matters

Our treatment of the finite-element method has been brief—one might say that it is superficial. An entire course and a separate book are really needed to do justice to this important method. We discuss the FEM (finite-element method) only to alert the reader to this most important area of applied mathematics. Although we have used heat flow as the application for our description, FEM is also used for stress analysis, vibration modeling, analysis of electrostatic fields, eigenproblems, in fact, for any phenomenon where the potential, u, is governed by an equation of the form

$$\frac{\partial u}{\partial t} = k[\nabla^2 u + Q].$$

Most engineers and scientists use a finite-element program to carry out the computations. These almost always have facilities to make it easy to describe the region of interest and for the placement of nodes through a *graphical user interface* (GUI). Some even place nodes automatically, but these placements may be less appropriate than when an experienced user is involved. Two widely used programs that run on personal computers are MSC/pal (from the MacNeal–Schwendler Corporation) and ViziCad Plus (from Algor Interactive Systems).

One of the computer algebra systems that we describe in this book, MATLAB, has a facility for solving a partial-differential equation by the finite-element method. A very powerful "toolbox," the PDE Toolbox, solves equation of all three types. Because the PDE Toolbox requires the professional version of MATLAB and we describe only the student edition, we do not give examples of its use. However, it is important for the reader to be aware of the MATLAB facility. The description of a two-dimensional region and the placement of nodes is made easy for the user with a GUI.

Kinds of Elements

We have simplified things by describing only two types of elements: a "hat" function for the one-dimensional problem and a triangular element in two dimensions. In both of these, the value is assumed to vary linearly within the element. In many problems, this is not adequate. The reason for this inadequacy is readily seen if we look at a plot of the solution to a typical one-dimensional problem:

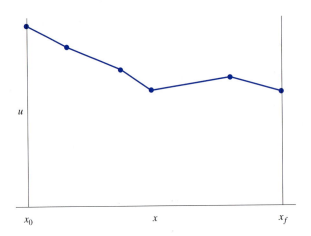

The dots in the figure represent the computed u-values at the nodes and the straight lines that connect the dots are the supposed intermediate values (because the value of u is assumed to vary linearly within the elements). This certainly does not correctly describe the function $u(x)$! (Of course, if the elements are much smaller, the broken line would be a better approximation.) How can we remedy this defect in the procedure?

The obvious way is to use a shape function that better approximates the true function. A quadratic shape function would involve three nodes: two at the ends and an intermediate one. This will force the solution to behave like a parabola that passes through the three nodes. A cubic shape function might be used; it would involve four nodes. Using such higher-order shape functions adds some modest complications to the development of the procedure, but the general approach is identical to what we have shown. When such higher-order shape functions are used, the integration of the analogs of Eqs. (9.31) or (9.5) is usually done by numerical methods. The resulting system of equations is no longer tridiagonal, but the nonzero elements of the coefficient matrix are still clustered about the main diagonal.

There is still a flaw with such higher-order elements in the one-dimensional problem—the curve for u is not continuous in slope at the juncture of the elements. This flaw could be eliminated if the shape function were splines, but this is not often done because that complicates the procedure significantly.

Many users of the finite-element method are interested not only in the u-values but also in the values of the *flux*, a term that means the value of $k\partial u/\partial x$ (in one dimension; in two or three dimensions, it is $k\nabla u$). A linear shape function produces values for the flux that have even greater discontinuities than for $u(x)$. The use of higher-order shape functions helps here.

In two-dimensional problems, a shape function other than triangular may be helpful. We could still assume linearity within the element with a quadrilateral shape function that might better fit the geometry of the region. However, both triangular and quadrilateral elements result in discontinuous slopes where the elements join. This objection is alleviated if we use higher-order shape functions within the elements. These require additional nodes along the edges. A quadratic-triangular element will have one additional node between the vertices of the triangle. Doing this distorts the sides into parabolas; obviously these curved sides can fit better to a curved boundary of the region. Only the more sophisticated finite-element method programs provide for such higher-order elements. The solution is more difficult because the coefficients in the resulting set of equations are no longer as tightly banded.

Convergence Rates

A numerical analyst is always greatly concerned about the accuracy of the numerical solutions. For finite-element-method procedures, the question is "How do the errors decrease when we put nodes closer together?" It can be shown that, with linear elements, errors are of order $O(h^2)$, where h is a measure of the nodal spacing. Quadratic elements give an $O(h^3)$ accuracy; higher orders than two give even better accuracy as the mesh is refined. As we have said, the rate of decrease is a limit value that is achieved only as the h-value gets very small. (The rate of decrease in the errors with quadratic or higher-order shape functions also depends on the integration method used in formulating the system of equations.) Also, a very interesting phenomenon has been observed in studies of the effect of smaller h-values on accuracy—errors may not always decrease uniformly as the spacing is made closer. As a mesh is gradually refined, anomalous behavior can occur.

It is frequently the case that nodes are not uniformly spaced—in fact, this is one of the major advantages of the finite-element method; we can put nodes closer together where the solution $u(x)$ varies most rapidly to get better accuracy in that subregion. This imposes a problem about how best to define "h" in the order of convergence. We shall not pursue this but only remark that if a mesh is refined to improve the accuracy of the numerical solution, we must refine it everywhere, not just in selected parts of the region.

The errors in the flux do not decrease as rapidly with smaller spacing of the nodes. For a linear shape function, errors decrease as $O(h)$.

Burnett (1987) is an excellent reference.

Chapter Summary

When you have learned the material in this chapter, you can:

1. Tell a friend what is meant by the "calculus of variations." You can write the functional for a one-dimensional problem and outline the steps used to solve the problem using the Rayleigh–Ritz method.

2. Use the collocation and Galerkin methods to solve a boundary-value problem. You realize that the solution from the Galerkin technique is essentially the same as that from the Rayleigh–Ritz method.

3. List the steps that are used in solving a boundary-value problem in one dimension through the finite-element method. You can apply this to a simple example.

4. Know what is meant by a "shape function" and can sketch the one used in Section 9.4. You can obtain the "pyramid function" for a triangular element and use this to interpolate for values within the element.

5. Adjust a set of equations for both Dirichlet and mixed boundary conditions to prepare them for solving an elliptic partial-differential equation.

6. Explain how the finite-element method can be applied to parabolic and hyperbolic partial-differential equations. You realize that this uses a combination of finite elements and finite-difference approximations.

Computer Program

In this chapter we present one program, written in C. It solves second-order boundary-value problems that match to

$$u'' + Qu = F, \qquad (Q \text{ and } F \text{ are functions of } x)$$

and with boundary conditions at each end of the region that can be described by $au + bu' = c$. This is the same as for the examples of Section 9.3. The boundary conditions that are permitted are more general than described in Section 9.3 in that a mixed boundary condition is also allowed.

After defining the functions for Q and F, the main procedure defines variables, including the boundary-condition parameters, a matrix for the equations (this has four columns because the equations are tridiagonal), and arrays for the x-values, the h-values, and average values for Q and F within each element. The program begins by echoing the boundary conditions, then it prints the midpoints of the intervals and the h-values (the Δx's), and the averaged Q- and F-values for verification by the user. The equation-matrix is set up as described in step 7 and then is adjusted for the boundary conditions. These matrices are also printed. The equations are then solved using the LU equivalent of the coefficient matrix, and the solution displayed.

Figure 9.11 lists the program. The parameters in it are those that solve Example 9.5. The output is shown at the end of the program.

```
/* * * * * * * * * * * * * * * * * * * * * * * * * * * * * * * * * * * * * * * * * * *
*                                                                                    *
*    Chapter 9    Finite Elements for Boundary-Value Problems                        *
*                                                                                    *
*    Gerald/Wheatley, APPLIED NUMERICAL ANALYSIS (sixth edition)                     *
*                 Addison Wesley Longman, 1999                                       *
*                                                                                    *
* * * * * * * * * * * * * * * * * * * * * * * * * * * * * * * * * * * * * * * * * * * *
*                                                                                    *
*    This program uses the finite element method to solve                            *
*    a second-order boundary-value problem of the form:                              *
*          u" + Qu = F, (Q and F are functions of x).                                *
*    The boundary conditions are described by au + bu' = c,                          *
*    so all types of boundary conditions are accommodated.                           *
*    The coefficients for the boundary conditions are defined                        *
*    within the program as are the x values at the nodes                             *
```

Figure 9.11

```
*                                                                      *
*   The program begins by displaying the boundary conditions and       *
*   then the midpoints of the elements, the h values, and q and        *
*   Q and F at the midpoints are calculated and displayed.             *
*   Following these preliminaries, the tridiagonal system of           *
*   equations is set up, its LU decomposition is found, and            *
*   these are displayed. Use is made of the LU matrix to solve         *
*   for the values of u at the nodes.                                  *
*                                                                      *
* * * * * * * * * * * * * * * * * * * * * * * * * * * * * * * * * * * * * * * * * * * * * * */

#include <stdio.h>
#include <math.h>
#include <conio.h>

/* Define functions for Q and F */
float Q(float x)
{
  return (-(x+1));
}

float F(float x)
{
  return (-exp(-x)*(x*x - x + 2));
}

void main()                   /* beginning of main */
{

/* Define variables */
float aL=0, bL=1, cL=0, aR=0, bR=1, cR=-0.036631;
float x[] = {2.0, 2.5, 3.0, 3.5, 4.0};
float xmid[20];
float m[20][4], h[20], q[20], f[20];
int n = 4;
int i, nstart, nend;

  clrscr();

/* Print the boundary conditions */
  printf ("Boundary conditions \n");
  printf ("At left end: %6.3f %6.3f %6.3f \n", aL, bL, cL);
  printf ("At right end: %6.3f %6.3f %6.3f \n", aR, bR, cR);

/* Get h-values for each element and then Qav, Fav values */
  for (i=1; i<=n; i++)
  {
    h[i] = x[i] - x[i-1];
    xmid[i] = (x[i-1]+x[i])/2;
    q[i] = Q(xmid[i]);
    f[i] = F(xmid[i]);
  }                              /* end of for */
```

Figure 9.11 *Continued*

```
/* Print the xmid, h, q, f */
  printf ("xmid, h, q, f-values \n");
  for (i=1; i<=n; i++)
    printf ("%6.3f %6.3f %6.3f %6.3f \n", xmid[i], h[i], q[i], f[i]);
  printf ("Press a key\n");
  getch();                         /*To freeze the screen */

/*
  Set up equation matrix, m, for a tridiagonal system,
  (m[i][2] are diagonals). First get diagonals in rows 0 and n.
*/
  m[0][2] = 1/h[1] - q[1]*h[1]/3;
  m[n][2] = 1/h[n] - q[n]*h[n]/3;

/* Now the diagonals in rows 1 to (n-1) */
  for (i = 1; i<n; i++)
    m[i][2] = 1/h[i] - q[i]*h[i]/3 + 1/h[i+1] - q[i+1]*h[i+1]/3;

/* And the off-diagonals */
  for (i=1; i<=n; i++)
    m[i-1][3] = m[i][1] = -1/h[i] - q[i]*h[i]/6;

/* Do the rhs before adjustment. First rows 0 and n */
  m[0][4] = -f[1]*h[1]/2;
  m[n][4] = -f[n]*h[n]/2;

/* Now the rest of the rhs */
  for (i=1; i<n; i++)
    m[i][4] = -f[i]*h[i]/2 - f[i+1]*h[i+1]/2;

/* Fill in row 0 and row n */
  m[0][1] = m[n][3] = 0;

/* Print the matrix */
  printf ("The matrix is: \n");
  for (i=0; i<=n; i++)
    printf ("%6.3f %6.3f %6.3f %6.3f \n",
            m[i][1], m[i][2], m[i][3], m[i][4]);
  printf ("Press a key\n");
  getch();

/* Adjust for b.c. First at the left end */
  if (bL == 0)                   /* Dirichlet */
  {
    nstart = 1;
    m[0][4] = cL/aL;
    m[1][4] -= m[1][1]*cL/aL;
  }
  else if (aL == 0)              /* Neumann */
  {
    nstart = 0;
    m[0][4] -= cL/bL;
  }
```

Figure 9.11 *Continued*

```
      else                     /* Mixed condition */
      {
        nstart = 0;
        m[0][4] -= cL/bL;
        m[0][2] += cL/aL;
      }

 /* Do the same for the right end */
    if (bR == 0)                /* Dirichlet */
    {
      nend = n-1;
      m[n][4] = cR/aR;
      m[n-1][4] -= m[n-1][3]*cR/aR;
    }
    else if (aR == 0)          /* Neumann */
    {
      nend = n;
      m[n][4] += cR/bR;
    }
    else                       /* Mixed condition */
    {
      nend = n;
      m[n][4] -= cR/bR;
      m[n][2] += cR/aR;
    }

 /* Print the adjusted matrix */
   printf ("After adjusting, matrix is: \n");
   for (i=nstart; i<=nend; i++)
     printf ("%d %6.3f %6.3f %6.3f %6.3f \n",
            i, m[i][1], m[i][2], m[i][3], m[i][4]);
   getch();

 /* Get the LU of the matrix */
   for (i=(nstart+1); i<=nend; i++)
   {
     m[i][2] = m[i][2] - m[i][1]/m[i-1][2]*m[i-1][3];
     m[i][4] = m[i][4] - m[i][1]/m[i-1][2]*m[i-1][4];
   }

 /* Print the LU matrix */
   printf ("The LU matrix:\n");
   for (i=nstart; i<=nend; i++)
     printf ("%d %6.3f %6.3f %6.3f %6.3f \n",
            i, m[i][1], m[i][2], m[i][3], m[i][4]);
   getch();

 /* Solve the system by back substitution */
   m[nend][4] = m[nend][4]/m[nend][2];
   for (i=(nend-1); i>=nstart; --i)
     m[i][4] = (m[i][4] - m[i][3]*m[i+1][4])/m[i][2];
```

Figure 9.11 *Continued*

```
/* Print the solution */
  printf ("The results are: \n");
  printf ("    node    u-value \n");
  for ( i=0; i<=n; i++)
    printf ("    %d      %f \n", i, m[i][4]);

/* Freeze the output */
  getch();

}       /* End of main */
```

```
        ***************   Output   ****************

        Boundary conditions
        At left end:  0.000    1.000    0.000
        At right end:  0.000    1.000   -0.037
        xmid, h, q, f-values
         2.250    0.500   -3.250   -0.507
         2.750    0.500   -3.750   -0.436
         3.250    0.500   -4.250   -0.361
         3.750    0.500   -4.750   -0.290
        Press a key
        The matrix is:
         0.000    2.542   -1.729    0.127
        -1.729    5.167   -1.688    0.236
        -1.688    5.333   -1.646    0.199
        -1.646    5.500   -1.604    0.163
        -1.604    2.792    0.000    0.072
        Press a key
        After adjusting, matrix is:
        0    0.000    2.542   -1.729    0.127
        1   -1.729    5.167   -1.688    0.236
        2   -1.688    5.333   -1.646    0.199
        3   -1.646    5.500   -1.604    0.163
        4   -1.604    2.792    0.000    0.036

        The LU matrix:
        0    0.000    2.542   -1.729    0.127
        1   -1.729    3.990   -1.688    0.322
        2   -1.688    4.620   -1.646    0.335
        3   -1.646    4.914   -1.604    0.282
        4   -1.604    2.268    0.000    0.128
        The results are:
            node    u-value
            0       0.133439
            1       0.122804
            2       0.099594
            3       0.075821
            4       0.056378
```

Figure 9.11 *Continued*

Exercises

Section 9.1

1. Show that the integrand of Eq. (9.4) is equivalent to Eq. (9.3) if the Euler–Lagrange condition is used. This means that Eq. (9.4) is the functional for any second-order boundary-value problem of the form

$$y'' + Q(x)y = F(x),$$

subject to Dirichlet boundary conditions

$$y(a) = A, \qquad y(b) = B,$$

where A and B are constants.

▶ **2.** Use the Rayleigh–Ritz method to approximate the solution of

$$y'' = 3x + 1, \qquad y(0) = 0, \qquad y(1) = 0,$$

using a quadratic in x as the approximating function. Compare to the analytical solution by graphing the approximation and the analytical solution.

3. Repeat Exercise 2, but this time, for the approximating function, use

$$ax(x - 1) + bx^2(x - 1).$$

Show that this reproduces the analytical solution.

4. Another approximating function that meets the boundary condition of Exercise 3 is

$$ax(x - 1) + bx(x - 1)^2.$$

Use this to solve by the Rayleigh–Ritz technique.

5. Suppose that the boundary conditions in Exercise 3 are $y(0) = 1$, $y(1) = 3$. Modify the procedure of Exercise 3 to get a solution.

Section 9.2

▶ **6.** Solve Exercise 2 by collocation, setting the residual to zero at $x = \frac{1}{3}$ and $x = \frac{2}{3}$. Compare this solution to that from Exercise 2.

7. Repeat Exercise 6, except now use different points within [0, 1] for setting the residual to zero. Are some pairs of points better than others?

8. Repeat Exercise 3, but now use collocation. Does it matter where within [0, 1] you set the residual to zero?

▶ **9.** Use Galerkin's technique to solve Exercise 2. Is the same solution obtained?

10. Repeat Exercise 3, but now use Galerkin.

Section 9.3

11. Suppose that $Q(x) = \sin(x)$ and $F(x) = x^2 + 2$ in Eq. (9.25). For an element that occupies [0.33, 0.45],
   a. find N_L and N_R (Eq. 9.26).
   b. set up the Galerkin integrals (Eq. 9.28), writing out $R(x)$ in full.
   c. write the element equations (Eq. 9.36).
   d. compute the correct average values for Q and F.

12. Repeat Exercise 11 for the two adjacent elements. These occupy [0.21, 0.33] and [0.45, 0.71].

13. Assemble the three pairs of element equations of Exercises 11 and 12 to form a set of four equations with the nodal values at $x = 0.21$, $x = 0.33$, $x = 0.45$, and $x = 0.71$ as unknowns.

▶ **14.** Solve by the finite-element method of Section 9.3:

$$y'' + xy = x^3 - \frac{4}{x^3}, \qquad y(1) = -1, \qquad y(2) = 3.$$

Put nodes at $x = 1.2, 1.5$, and 1.75 as well as at the ends of [1, 2]. Compare your solution to the analytical solution, which is $y = x^2 - 2/x$.

15. Repeat Exercise 14 except for the end condition at $x = 1$ of $y'(1) = 4$.

16. Repeat Exercise 14 but with more nodes. Place added nodes at $x = 1.1, 1.3, 1.4, 1.65$, and 1.9. Compare the errors with those of Exercise 14.

Section 9.4

17. Confirm that Eq. (9.45) is in fact the inverse of matrix M in Eq. (9.44).

18. Find M^{-1}, a, N, and $u(x, y)$ for these triangular elements:
   a. Nodes: (1.2, 3.1), (−0.2, 4), (−2, −3); u-values at these nodes: 5, 20, 7; point where u is to be determined: (−1, 0)
   b. Nodes: (20, 40), (50, 10), (5, 10); u-values at these nodes: 12.5, 6.2, 10.1; point where u is to be determined: (20, 20)
   ▶ c. Nodes: (12.1, 11.3), (8.6, 9.3), (13.2, 9.3); u-values at these nodes: 121, 215, 67; point where u is to be determined: (10.6, 9.6)

19. Confirm that the sum of the entries in the first row of M^{-1} is equal to twice the area for each of the elements in Exercise 18.

▶ **20.** Find the element equations for the element in part (c) of Exercise 18 if $Q = x^2y$ and $F = -x/y$ (these refer to Eq. 9.40). There are no derivative conditions on any of the element boundaries.

21. Solve Example 7.14 (Section 7.7) by finite elements. Place nodes at each corner and at the midpoints of the top and bottom edges, also at points 9, 12, and 14. Draw triangular elements whose vertices are at these nodes. Compare the answers at each node to those obtained with finite-difference approximations to the derivatives.

22. In Exercise 21, the temperatures in the top half of the slab are the same as those in the bottom half because of symmetry in the boundary conditions. Solve the problem for the top half only of the slab with the same nodes as in Exercise 21. (Along the horizontal midline, the gradient will be zero).

Section 9.5

▶ **23.** For a triangular element that has nodes at points (1.2, 3.2), (4.3, 2.7), and (2.4, 4.1), find the components of each matrix in the element equations (Eqs. 9.61 and 9.62) if the material is aluminum.

24. Consider heat flow in one dimension; it will be governed by Eq. (8.9). For this situation, repeat the development that leads to the analog of Eq. (9.62).

25. Use the equations that you derived in Exercise 24 to solve Exercise 11 of Chapter 8. Place the nodes exactly as those used in the finite-difference solution. Are the resulting equations the same?

26. Use finite elements to solve Exercise 21 of Chapter 8. Place interior nodes at three arbitrarily selected points (but do not make these symmetrical). Create triangular elements with these nodes and the four corner points. Set up the element equations, assemble, and solve for four time steps. Use the resulting nodal temperatures to estimate the same set of temperatures that were computed by finite differences. Compare the two methods of solving the problem.

27. Solve Example 7.17 (Section 7.7) by finite elements. Place nodes strategically along the edges and within the slab so there are a total of 14 or 15 nodes. Use triangular elements. Compare the solution to that obtained with finite-difference approximations. (You may want to take advantage of symmetry in the boundary conditions to solve the problem with fewer elements.)

▶ **28.** Rederive Eq. (9.64), but now for the Crank–Nicolson method.

29. Repeat Exercise 28, but now for the theta method.

30. Set up the finite-element equations for advancing the solution to part (a) of Exercise 15 of Chapter 8.

31. Set up the finite-element equations for starting the solution to part (a) of Exercise 15 of Chapter 8. Do this first for the analog of Eq. (8.19) and then for the analog of Eq. (8.26).

▶ **32.** If we were to solve part (c) of Exercise 15 of Chapter 8, would there be an advantage to using shorter elements near the middle of the string where the displacements depart more from linearity?

33. Solve, using finite elements, Example 8.7 of Section 8.6, except with initial conditions of

$$u(x, y) = 0, \qquad u_t(x, y) = x^2(2 - x)y^2(2 - y).$$

34. Repeat Exercise 33, but with these initial conditions:

$$u(x, y) = x^2(2 - x)y^2(2 - y), \qquad u_t(x, y) = 0.$$

35. Solve Exercise 28 of Chapter 8 using finite elements. Where do you think interior nodes should be placed if there are

a. 6 of them?

b. if there are 12?

Compare the solutions from these two cases to that from the finite-difference method.

36. Solve Exercise 30 of Chapter 8 by finite elements, placing five interior nodes at points that you think are best. Justify your choice of nodal positions.

Applied Problems and Projects

37. Use the Internet to find software that solves both ordinary- and partial-differential equations. Can you find any that use the finite-element method? (Hint: try http://gams.nist.gov/ and search the topic: partial differential equations.)

38. Repeat Exercise 37, but now for Eq. (9.56). Test the program by solving Exercises 26 and 27.

39. Write a computer program that solves the vibrating string problem through finite elements.

40. Write and test a program that solves the vibrating membrane problem using the finite-element method.

41. Write a computer program (using your favorite language) to solve a two-dimensional elliptic partial-differential equation. Allow for both Dirichlet and non-Dirichlet boundary conditions. Have the program read in the required data from a file. Provide function procedures to compute the values for $f(x, y)$ and $q(x, y)$. Here is a suggested data structure:

NN = the total number of nodes

NK = the number of boundary nodes with Dirichlet conditions. (NN − NK = number of nodes whose values are not specified, that is, the interior nodes and those boundary nodes whose values are not specified)

VX (NN) = an array to hold the x-values for all nodes in the order that nodes are numbered. There is an advantage if the nodes whose u-values are specified are numbered so as to follow those nodes where the u-values must be computed.

VY (NN) = an array to hold the corresponding y-values for all nodes

M (NE, 4, 3) = an array to hold the element matrices. The first subscript indicates the element number. The second and third subscripts indicate the row and column of the matrix. The fourth row holds the node numbers for nodes in this element in counterclockwise order. There is an advantage if the unspecified nodes come before the nodes whose u-values are known.

UU (NN) = an array to hold unknown and known u-values at nodes in order of the node number. Zeros may be used as fillers for unknown u-values.

AE (NE) = an array to hold areas of the elements

F (NE) = an array to hold average f-values for each element

Q (NE) = an array to hold average q-values for each element

A (NN, NN + 1) = the system matrix

Here is what your logic might look like:

1. Read in NN, NE, NU.
2. Read in (x, y) values for the nodes, storing in VX and VY.
3. Read in node numbers for each element in turn (nodes should be in counterclockwise order), storing in the fourth row of the element matrices.
4. Read in the unknown and known u-values for each node.
5. Compute average values for f and q in each element. (You may prefer to evaluate these at the centroid of the element.) Store in F and Q.
6. Read in the known u-values, storing in UK.
7. Compute the area for each element and its inverse (Eq. 9.49, the area from the first row elements).
8. Find the element equations and add the appropriate values to the system matrix.
9. Adjust the system matrix for non-Dirichlet boundary conditions. (You may want to have the user input the a and b values for these and the node numbers at the ends of the element boundary where this applies. Alternatively, these could have been read in with the other parts of the data.)
10. Adjust the system matrix for Dirichlet conditions using values from the UU array.

11. Solve the system.
12. Display the u-values for each node.

42. In developing the element equations, a number of integrals must be evaluated (see Eq. 9.51). For triangular elements, these are very easy to get: each is just the area divided by a number. These simple triangular elements that we have discussed are called C^0-*linear elements.*

    Other types of elements besides these simple triangles are sometimes useful. For example, connecting the nodes with lines that form quadrilateral elements can cut the number of elements almost in half. For these, the integrals are not so readily evaluated.

    Even if we stay with triangular elements, the accuracy of the solution is improved if we add one node within each of the three sides. Such additional nodes can even permit the "triangle" to have curved sides. Such a more elaborate triangular element is called a C^0-*quadratic element.* This idea can be extended to add more than three nodes to the triangle, and additional nodes are sometimes added to quadrilateral elements.

    For all of these more elaborate elements, the shape functions no longer have a "flat top" like that sketched in Fig. 9.7. The normal procedure for these is to employ Gaussian quadrature in which a weighted sum of the integrand at certain points, called *Gauss-points,* approximates the integral quite well.

    For a square region with opposite corners at $(-1, -1)$ and $(1, 1)$, these Gauss-points are at $x = \pm\sqrt{3}/3$, $y = \pm\sqrt{3}/3$, as given in Table 5.14. For a region that is a triangle with vertices at $(0, 0)$, $(1, 0)$, $(0, 1)$, there are three Gauss-points at $\left(\frac{1}{6}, \frac{1}{6}\right)$, $\left(\frac{2}{3}, \frac{1}{6}\right)$, and $\left(\frac{1}{6}, \frac{2}{3}\right)$, each weighted with $\frac{1}{3}$. For elements that do not conform to these basic cases, they must be mapped to coincide with them. Where are the Gauss-points for

    a. a triangle whose vertices are $(-1, 3)$, $(7, 1)$, and $(2, 7)$?
    b. a quadrilateral whose vertices are $(1, 2)$, $(5, -1)$, $(6, 3)$, $(3, 5)$?

43. The various C programs of this book exhibit different styles. Make a critical comparison of them and draw conclusions as to which style you prefer. Consider their structure, ways of incorporating comments, and how their data are defined.

APPENDIX

Some Basic Information from Calculus

Because a number of results and theorems from the calculus are frequently used in the text, we collect here a number of these items for ready reference, and to refresh the student's memory.

Open and Closed Intervals

For the open interval $a < x < b$, we use the notation (a, b), and for the closed interval $a \leq x \leq b$, we use the notation $[a, b]$.

Continuous Functions

If a real-valued function is defined on the interval (a, b), it is said to be *continuous* at a point x_0 in that interval if for every $\epsilon > 0$ there exists a positive nonzero number δ such that $|f(x) - f(x_0)| < \epsilon$ whenever $|x - x_0| < \delta$ and $a < x < b$. In simple terms, we can meet any criterion of matching the value of $f(x_0)$ (the criterion is the quantity ϵ) by choosing x near enough to x_0, without having to make x equal to x_0, when the function is continuous.

If a function is continuous for all x-values in an interval, it is said to be continuous on the interval. A function that is continuous on a closed interval $[a, b]$ will assume a maximum value and a minimum value at points in the interval (perhaps the endpoints). It will also assume any value between the maximum and the minimum at some point in the interval.

Similar statements can be made about a function of two or more variables. We then refer to a domain in the space of the several variables instead of to an interval.

Sums of Values of Continuous Functions

When x is in $[a, b]$, the value of a continuous function $f(x)$ must be no greater than the maximum and no less than the minimum value of $f(x)$ on $[a, b]$. The sum of n such values must be bounded by $(n)(m)$ and $(n)(M)$, where m and M are the minimum and maximum values. Consequently, the sum is n times some intermediate value of the function. Hence

$$\sum_{i=1}^{n} f(\xi_i) = nf(\xi) \qquad \text{if } a \le \xi_i \le b, \qquad i = 1, 2, \ldots, n, \qquad a \le \xi \le b.$$

Similarly, it is obvious that

$$c_1 f(\xi_1) + c_2 f(\xi_2) = (c_1 + c_2)f(\xi), \qquad \xi_1, \xi_2, \xi \text{ in } [a, b],$$

for the continuous function f when c_1 and c_2 are both equal to or greater than one. If the coefficients are positive fractions, dividing by the smaller gives

$$c_1 f(\xi_i) + c_2 f(\xi_2) = c_1 \left[f(\xi_1) + \frac{c_2}{c_1} f(\xi_2) \right] = c_1 \left(1 + \frac{c_2}{c_1} \right) f(\xi) = (c_1 + c_2)f(\xi),$$

so the rule holds for fractions as well. If c_1 and c_2 are of unlike sign, this rule does not hold unless the values of $f(\xi_1)$ and $f(\xi_2)$ are narrowly restricted.

Mean-Value Theorem for Derivatives

When $f(x)$ is continuous on the closed interval $[a, b]$, then at some point ξ in the interior of the interval

$$f'(\xi) = \frac{f(b) - f(a)}{b - a}, \qquad a < \xi < b,$$

provided, of course, that $f'(x)$ exists at all interior points. Geometrically this means that the curve has at one or more interior points a tangent parallel to the secant line connecting the ends of the curve (Fig. A.1).

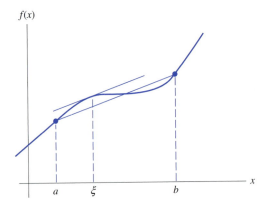

Figure A. 1

Mean-Value Theorems for Integrals

If $f(x)$ is continuous and integrable on $[a, b]$, then

$$\int_a^b f(x) \, dx = (b - a)f(\xi), \qquad a < \xi < b.$$

This says, in effect, that the value of the integral is an average value of the function times the length of the interval. Because the average value lies between the maximum and minimum values, there is some point ξ at which $f(x)$ assumes this average value.

If $f(x)$ and $g(x)$ are continuous and integrable on $[a, b]$, and if $g(x)$ does not change sign on $[a, b]$, then

$$\int_a^b f(x)g(x) \, dx = f(\xi) \int_a^b g(x) \, dx, \qquad a < \xi < b.$$

Note that the previous statement is a special case $[g(x) = 1]$ of this last theorem, which is called the *second theorem of the mean for integrals*.

Taylor Series

If a function $f(x)$ can be represented by a power series on the interval $(-a, a)$, then the function has derivatives of all orders on that interval and the power series is

$$f(x) = f(0) + f'(0)x + \frac{f''(0)}{2!}x^2 + \frac{f'''(0)}{3!}x^3 + \cdots.$$

The preceding power-series expansion of $f(x)$ about the origin is called a *Maclaurin series*. Note that if the series exists, it is unique and any method of developing the coefficients gives this same series.

If the expansion is about the point $x = a$, we have the *Taylor series*

$$f(x) = f(a) + f'(a)(x - a) + \frac{f''(a)}{2!}(x - a)^2 + \frac{f'''(a)}{3!}(x - a)^3 + \cdots.$$

We frequently represent a function by a polynomial approximation, which we can regard as a truncated Taylor series. Usually we cannot represent a function exactly by this means, so we are interested in the error. Taylor's formula with a remainder gives us the error term. The remainder term is usually derived in elementary calculus texts in the form of an integral:

$$f(x) = f(a) + f'(a)(x - a) + \frac{f''(a)}{2!}(x - a)^2 + \cdots$$

$$+ \frac{f^{(n)}(a)}{n!}(x - a)^n + \int_a^x \frac{(x - t)^n}{n!} f^{(n+1)}(t) \, dt.$$

Because $(x - t)$ does not change sign as t varies from a to x, the second theorem of the mean allows us to write the remainder term as

$$\text{Remainder of Taylor series} = \frac{(x - a)^{n+1}}{(n + 1)!} f^{(n+1)}(\xi), \qquad \xi \text{ in } [a, x].$$

The derivative form is the more useful for our purposes. It is occasionally useful to express a Taylor series in a notation that shows how the function behaves at a distance h from a fixed point a. If we call $x = a + h$ in the preceding series, so that $x - a = h$, we get

$$f(a + h) = f(a) + f'(a)h + \frac{f''(a)}{2!} h^2 + \cdots + \frac{f^{(n)}(a)}{n!} h^n + \frac{f^{(n+1)}(\xi)}{(n + 1)!} h^{n+1}.$$

Taylor Series for Functions of Two Variables

For a function of two variables, $f(x, y)$, the rate of change of the function can be due to changes in either x or y. The derivatives of f can be expressed in terms of the partial derivatives. For the expansion in the neighborhood of the point (a, b),

$$f(x, y) = f(a, b) + f_x(a, b)(x - a) + f_y(a, b)(y - b)$$

$$+ \frac{1}{2!} [f_{xx}(a, b)(x - a)^2 + 2f_{xy}(a, b)(x - a)(y - b) + f_{yy}(a, b)(y - b)^2]$$

$$+ \cdots .$$

Descartes' Rule of Signs

Let $p(x)$ be a polynomial with real coefficients and consider the equation $p(x) = 0$. Descartes' rule of signs is a simple method for giving us an estimate of the number of real roots of this equation on both sides of $x = 0$. The rule states:

1. The number of positive real roots is equal to the number of variations in the signs of the coefficients of $p(x)$ or is less than that number by an even integer.
2. The number of negative real roots is determined the same way, but for $p(-x)$. Here also the number of negative roots is equal to the number of variations in the signs of the coefficients of $p(-x)$ or is less than that number by an even integer.

For example, the polynomial equation $p(x) = x^6 - 3x^5 + 2x^4 - 6x^3 - x^2 + 4x - 1 = 0$ will have 5, 3, or 1 positive and 1 negative real root. We can assume then that the number of real roots are at least 2 but can be as many as 6!

Deriving Formulas by the Method of Undetermined Coefficients

In Chapter 5 we developed formulas for integration and differentiation of functions by replacing the actual function with a polynomial that agrees at a number of points (a so-called *interpolating polynomial*), and then integrating or differentiating the polynomial. These formulas are valuable if one wishes to write computer programs for integration and differentiation, but the most important use is for solving differential equations numerically. Because computers calculate their functional values rather than interpolate in a table, there is less interest today in interpolating polynomials than in earlier times. Hence there is reason to present an alternative method of deriving the formulas for derivatives and integrals that are needed to solve differential equations.

We will call the method that we employ in this appendix the *method of undetermined coefficients.* Basically, we impose certain conditions on a formula of desired form and use these conditions to determine values of the unknown coefficients in the formula. Hamming (1973) presents the method in considerable detail.

B.1 Derivative Formulas by the Method of Undetermined Coefficients

Because the derivative of a function is the rate of change of the function relative to changes in the independent variable, we should expect that formulas for the derivative would involve differences between function values in the neighborhood of the point where we wish to evaluate the derivative. It is, in fact, possible to approximate the derivative as a linear combination of such function values. Although one can argue that formulas of greatest accuracy will use function values very near to the point in question, the formulas important in practice impose the restriction that only function values at equally spaced x-values are to be used.

For example, we can write a formula for the first derivative in terms of $n + 1$ equispaced points:

$$f'(x_0) = c_0 f(x_0) + c_1 f(x_1) + c_2 f(x_2) + \cdots + c_n f(x_n),$$

$$x_{i+1} - x_i = h = \text{constant.}$$ **(B.1)**

The more terms we employ, the greater accuracy we shall expect, as more information about the function is being fed in. We will evaluate the coefficients in the equation by requiring that the formula be exact whenever the function is a polynomial of degree n or less.* (Throughout this appendix, we will find that the method of undetermined coefficients uses this criterion to develop formulas. It has validity because any function that is continuous on an interval can be approximated to any specified precision by a polynomial of sufficiently high degree. Using polynomials to replace the function also greatly simplifies the work, in contrast to replacing with other functions.)

Let us illustrate the method of undetermined coefficients with a simple case. We shall simplify the notation by defining

$$f_i = f(x_i).$$

If we write the derivative in terms of only two function values, we have

$$f'_0 = c_0 f_0 + c_1 f_1.$$ **(B.2)**

We require that the formula be exact if $f(x)$ is a polynomial of degree 1 or less. (The maximum degree is always one less than the number of undetermined constants.) Hence, because the formula is to be exact if $f(x)$ is any first-degree polynomial, it must be exact if either $f(x) = x$ or $f(x) = 1$, for these definitions of $f(x)$ are just two special cases of the general first-degree function $ax + b$. We write Eq. (B.2) for both these cases:

$$f(x) = x, \quad f'(x) = 1, \quad 1 = c_0(x_0) + c_1(x_0 + h),$$

$$f(x) = 1, \quad f'(x) = 0, \quad 0 = c_0(1) + c_1(1).$$ **(B.3)**

We solve Eqs. (B.3) simultaneously to get $c_0 = -1/h$, $c_1 = 1/h$. Consequently,

$$f'(x_0) \doteq \frac{f_1 - f_0}{h}.$$ **(B.4)**

The dot over the equal sign in Eq. (B.4) reminds us that the formula is only an approximation. It is exact if $f(x)$ is a polymomial of degree 1, but not exact if $f(x)$ is a polynomial of higher degree, or some transcendental function.

Similarly, a three-term formula can be derived by replacing the function with x^2 and x and 1 in

$$f'_0 = c_0 f_0 + c_1 f_1 + c_2 f_2.$$

---

*Intuitively, it seems reasonable that we can satisfy this criterion, for a polynomial of degree n is determined uniquely by its $n + 1$ coefficients and our formula contains $n + 1$ constants.

The set of equations to solve is

$$2x_0 = c_0(x_0)^2 + c_1(x_0 + h)^2 + c_2(x_0 + 2h)^2,$$
$$1 = c_0(x_0) + c_1(x_0 + h) + c_2(x_0 + 2h),$$
$$0 = c_0(1) + c_1(1) + c_2(1).$$

(B.5)

The arithmetic is simplified by letting $x_0 = 0$. That this is valid is readily seen. Imagine the graph of $f(x)$ versus x. The derivative we desire is the slope of the curve at the point where $x = x_0$, which obviously is unchanged by a translation of the axes. Taking $x_0 = 0$ is the equivalent of a translation of axes so that the origin corresponds to x_0. With this change, Eqs. (B.5) become

$$0 = c_0(0) + c_1(h)^2 + c_2(2h)^2,$$
$$1 = c_0(0) + c_1(h) + c_2(2h),$$
$$0 = c_0(1) + c_1(1) + c_2(1).$$

Solving, we get $c_0 = -3/2h$, $c_1 = 2/h$, $c_2 = -1/2h$; so a three-term formula for the derivative is

$$f'_0 \doteq \frac{-3f_0 + 4f_1 - f_2}{2h}.$$

(B.6)

We could extend these formulas to include more and more terms, but after a little reflection we can conclude that the original form, Eq. (B.1), is not the best to use. It utilizes only functional values to one side of the point in question, whereas it would be better to utilize information from both sides of x. After all, the limit in the definition of the derivative is two-sided, and further, we utilize closer and hence more pertinent information by going to both the left and the right.

We expect an improvement over Eq. (B.6) if we begin with

$$f'_0 = c_{-1}f_{-1} + c_0f_0 + c_1f_1,$$

(B.7)

where $f_{-1} = f(x_0 - h)$. Adopting the simplification of letting $x_0 = 0$ as before, and taking x^2, x, and 1 for $f(x)$, we have to solve

$$0 = c_{-1}(-h)^2 + c_0(0) + c_1(h)^2,$$
$$1 = c_{-1}(-h) + c_0(0) + c_1(h),$$
$$0 = c_{-1}(1) + c_0(1) + c_1(1).$$

(B.8)

Completing the algebra, we get $c_{-1} = -1/2h$, $c_0 = 0$, $c_1 = 1/2h$.

The following equation is particularly important:

$$f'_0 \doteq \frac{f_1 - f_{-1}}{2h}.$$

(B.9)

In the next section we will compare its accuracy with Eqs. (B.4) and (B.6). Because the point where the derivative is evaluated is centered among the function values whose differences appear in the formula, it is called a *central-difference approximation*. Higher-order central-difference approximations to the first derivative, utilizing five or seven or more points, can be derived by this same procedure.

We now apply the method of undetermined coefficients to higher derivatives. We will discuss only the central-difference approximations as they are the more widely used. In terms of values both to the right and left of x_0,

$$f_0'' = c_{-1}f_{-1} + c_0f_0 + c_1f_1.$$

Taking $f(x) = x^2$, x, and 1, we get the relations ($x_0 = 0$),

$$2 = c_{-1}(-h)^2 + c_0(0) + c_1(h)^2,$$
$$0 = c_{-1}(-h) + c_0(0) + c_1(h),$$
$$0 = c_{-1}(1) + c_0(1) + c_1(1).$$

The resulting formula is

$$f_0'' \doteq \frac{f_{-1} - 2f_0 + f_1}{h^2}. \tag{B.10}$$

Equation (B.10), like Eq. (B.9), is particularly useful.

We are not confined to using only functional values in the method of undetermined coefficients.* Suppose, for example, we have values of the first derivative as well as functional values. A formula for the second derivative might then be written as

$$f_0'' = c_0f_0 + c_1f_1 + c_3f_0'.$$

As before, we take x^2, x, and 1 for $f(x)$, with $x_0 = 0$:

$$2 = c_0(0)^2 + c_1(h)^2 + c_3(0),$$
$$0 = c_0(0) + c_1(h) + c_3(1),$$
$$0 = c_0(1) + c_1(1) + c_3(0).$$

Solving, we obtain

$$f_0'' = 2\frac{f_1 - f_0 - hf_0'}{h^2}.$$

In the same way, we can derive formulas for the third and fourth derivatives. Derivatives beyond these do not often appear in applied problems. We must use a minimum of four and five terms, however, because only polynomials of degrees 3 and 4 have nonzero third or fourth derivatives. For the third-degree formula, complete symmetry with four points is impossible; five-term formulas for both derivatives are therefore given. We present the results only:

$$f_0''' = \frac{-f_{-2} + 2f_{-1} - 2f_1 + f_2}{2h^3} \tag{B.11}$$

$$f_0^4 = \frac{f_{-2} - 4f_{-1} + 6f_0 - 4f_1 + f_2}{h^4}. \tag{B.12}$$

———————

*We must be sure that the set of values is sufficient to determine a polynomial uniquely, however. For example, f_{-1}, f_1, and f_0' will not give a formula for f_0''.

B.2 Error Terms for Derivative Formulas

In the previous section, we used the method of undetermined coefficients to derive several formulas for the first derivative of a function, utilizing function values at equispaced x-values. Using the notation that $f_i = f(x_i)$, these were

$$f_0' \doteq \frac{f_1 - f_0}{h}, \tag{B.13}$$

$$f_0' \doteq \frac{-3f_0 + 4f_1 - f_2}{2h}, \tag{B.14}$$

$$f_0' \doteq \frac{f_1 - f_{-1}}{2h}. \tag{B.15}$$

Although each of these formulas is not exact, as suggested by the $\doteq$ symbol, we argued heuristically that the error should decrease in each succeeding one. We now wish to develop expressions for the errors.

Begin with the Taylor-series expansion of $f(x_1) = f(x_0 + h)$ in terms of $x_1 - x_0 = h$:

$$f(x_1) = f(x_0) + hf'(x_0) + \frac{1}{2}h^2 f''(x_0) + \frac{1}{6}h^3 f'''(x_0) + \cdots.$$

Changing to our subscript notation, and truncating after the term in h, with the usual error term, we have

$$f_1 = f_0 + hf_0' + \frac{1}{2}h^2 f''(\xi), \qquad x_0 < \xi < x_0 + h. \tag{B.16}$$

Solving for f_0', we have

$$f_0' = \frac{f_1 - f_0}{h} - \frac{1}{2}hf''(\xi), \qquad x_0 < \xi < x_0 + h, \tag{B.17}$$

so that the error term is $-\frac{1}{2}hf''(\xi)$ for the derivative formula, Eq. (B.13).

In Eq. (B.17), the error term involves the second derivative of the function evaluated at a place that is known only within a certain interval. In many of the applications for numerical differentiation, not only is the point of evaluation uncertain, but the function $f(x)$ is also unknown. If we do not know $f(x)$, we can hardly expect to know its derivatives. We do know, however, that the error involves the first power of $h = x_{i+1} - x_i$, and, in fact, the only way we can change the error is to change h. Making h smaller will decrease h, and in the limit as $h \to 0$ the error will go to zero. Further, as $h \to 0$, $x_1 \to x_0$, and the value of ξ is squeezed into a smaller and smaller interval. In other words, $f''(\xi)$ approaches a fixed value, specifically $f''(x_0)$, as h goes to zero.

The special importance of h in the error term is denoted by a special notation in numerical analysis, the *order relation*. We say the error of Eq. (B.17) is "of order h" and write

$$\text{Error} = O(h), \qquad \text{when } \lim_{h \to 0} (\text{error}) = ch,$$

where c is a fixed value not equal to zero.

To develop the error term for Eq. (B.14), we proceed similarly, except that we write expansions for both f_1 and f_2. It is also necessary to carry terms through h^3 because the second derivatives cancel, resulting in an error term involving the third derivative:

$$f_1 = f_0 + hf_0' + \frac{1}{2}h^2 f_0'' + \frac{1}{6}h^3 f'''(\xi_1), \qquad x_0 < \xi_1 < x_0 + h;$$

$$f_2 = f_0 + 2hf_0' + \frac{1}{2}(2h)^2 f_0'' + \frac{1}{6}(2h)^3 f'''(\xi_2), \qquad x_0 < \xi_2 < x_0 + 2h.$$

Note that the values of ξ_1 and ξ_2 may not be identical. If we multiply the first equation by 4, the second by -1, and add $-3f_0$ to their sum, we get

$$-3f_0 + 4f_1 - f_2 = (-3 + 4 - 1)f_0 + 2hf_0' + (2 - 2)f_0'' + \frac{4h^3}{6}[f'''(\xi_1) - 2f'''(\xi_2)].$$

Solving for f_0', we get

$$f_0' = \frac{-3f_0 + 4f_1 - f_2}{2h} + \frac{1}{3}h^2[2f'''(\xi_2) - f'''(\xi_1)], \quad x_0 < \xi_1 < x_0 + h, \qquad x_0 < \xi_2 < x_0 + 2h. \quad \textbf{(B.18)}$$

The last term of Eq. (B.18) is the error term. As $h \to 0$, the two values of ξ approach the same value. Consequently, the error term approaches $\frac{1}{3}h^2 f'''(x_0)$. We then conclude that the error of Eq. (B.14) is $O(h^2)$.

For the error of Eq. (B.15), we proceeded similarly. It is not hard to show that

$$f_0' = \frac{f_1 - f_{-1}}{2h} - \frac{1}{6}h^2 f'''(\xi), \qquad x_0 - h < \xi < x_0 + h,$$

$$= \frac{f_1 - f_{-1}}{2h} + O(h^2). \qquad \textbf{(B.19)}$$

The error terms of Eqs. (B.18) and (B.19) are both $O(h^2)$, but the coefficient in Eq. (B.19) is only half the magnitude of the coefficient in Eq. (B.18). The progressive increase in accuracy we anticipated is confirmed.

By similar arguments, we can show that the formulas for third and fourth derivatives, Eqs. (B.11) and (B.12), both have errors $O(h^2)$.

B.3 Integration Formulas by the Method of Undetermined Coefficients

A numerical integration formula will estimate the value of the integral of the function by a formula involving the values of the function at a number of points in or near the interval of integration. Figure B.1 illustrates the general situation. If we desire to evaluate $\int_a^b f(x)\,dx$, it is obvious that if we could find some average value of the function on the interval $[a, b]$, the integral would be:

$$\int_a^b f(x)\,dx = (b - a)f_{\text{av}}.$$

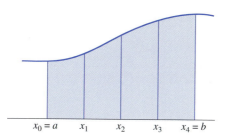

$x_0 = a$ x_1 x_2 x_3 $x_4 = b$

Figure B.1

It seems reasonable to assume that f_{av} could be approximated by a linear combination of function values within or near the interval.* Therefore, we write, similarly to our procedure for derivatives,

$$\int_a^b f(x)\ dx = c_0 f(x_0) + c_1 f(x_1) + \cdots + c_n f(x_n),$$

where the coefficients c_i are to be determined. The points x_i at which the function is to be evaluated can also be left undetermined, but for the formulas that we need to derive, we will again impose the restriction that they are equally spaced with the value of $x_{i+1} - x_i = \Delta x = h = $ constant. We will impose the further restriction that the boundaries of the interval of integration, a and b, coincide with two of the x_i-values.

We start with a simple case. Suppose we wished to express the integral in terms of just two functional values, specifically $f(a)$ and $f(b)$. Our formula takes the form

$$\int_a^b f(x)\ dx = c_0 f(a) + c_1 f(b). \tag{B.20}$$

Obviously, the values of c_0 and c_1 depend on $f(x)$, and probably on the values of a and b as well. The method of undetermined coefficients assumes that $f(x)$ can be approximated by a polynomial over the interval $[a, b]$ and determines c_0 and c_1 so that Eq. (B.20) is exact for all polynomials of a certain maximum degree or less.

Equation (B.20) has only two arbitrary constants, so we would not expect that it could be made exact for polynomials of degree higher than the first, because more than two parameters appear in second- or higher-degree polynomials. We therefore force Eq. (B.20) to be exact when $f(x)$ is replaced by either a first-degree polynomial or one of zero degree (a constant function). If it is to be exact for all first-degree polynomials, it must be exact for the very simple function $f(x) = x$. If it is to be exact when the function is any constant, it must hold if $f(x) = 1$. We express these two relations:

$$\int_a^b x\ dx = c_0(a) + c_1(b),$$
$$\int_a^b (1)\ dx = c_0(1) + c_1(1). \tag{B.21}$$

---

*One could generalize the concept by extending this to include derivatives of the function as well, but we stay with the simpler case.

The integrals of Eqs. (B.21) are easily evaluated, so we have, as conditions that the constants must satisfy,

$$\int_a^b x \, dx = \frac{x^2}{2}\Big|_a^b = \frac{b^2}{2} - \frac{a^2}{2} = c_0 a + c_1 b,$$

$$\int_a^b dx = x\Big|_a^b = b - a = c_0 + c_1.$$

Solving these equations gives $c_0 = (b - a)/2$, $c_1 = (b - a)/2$.

Consequently,

$$\int_a^b f(x) \doteq \frac{b - a}{2}[f(a) + f(b)]. \tag{B.22}$$

This formula is the *trapezoidal rule,* and was derived in Chapter 5 as Eq. (5.29) through an entirely different approach. We have put a dot over the equality sign in Eq. (B.22) to remind ourselves that the relation is only approximately true, for the function $f(x)$ cannot ordinarily be replaced without error by a first-degree polynomial. It is intuitively obvious that unless the interval $[a, b]$ is very small, the error will be considerable.

We can reduce the error in the preceding formula by using a higher-degree polynomial to replace $f(x)$, but we would then need to take a linear combination of more than two function values. In fact, we will have to preserve a balance between the number of undetermined coefficients in the formula and the number of parameters in the polynomial. The number of coefficients hence must be one greater than the degree of the polynomial.

We now look at the next case, a three-term formula corresponding to replacing $f(x)$ by a quadratic. For three terms, using $x_0 = a$ and $x_2 = b$, with x_1 at the midpoint,

$$\int_a^b f(x) \, dx = c_0 f(a) + c_1 f\left(\frac{a + b}{2}\right) + c_2 f(b).$$

The formula must be exact if $f(x) = x^2$, or $f(x) = x$, or $f(x) = 1$, so

$$\int_a^b x^2 \, dx = \frac{b^3}{3} - \frac{a^3}{3} = c_0 a^2 + c_1\left(\frac{a + b}{2}\right)^2 + c_2 b^2,$$

$$\int_a^b x \, dx = \frac{b^2}{2} - \frac{a^2}{2} = c_0 a + c_1\left(\frac{a + b}{2}\right) + c_2 b,$$

$$\int_a^b dx = b - a = c_0 + c_1 + c_2.$$

The solution is $c_0 = c_2 = \frac{1}{6}(b - a)$, $c_1 = \frac{4}{6}(b - a)$. Hence

$$\int_a^b f(x) \, dx \doteq \frac{b - a}{6}\left(f(a) + 4f\left(\frac{a + b}{2}\right) + f(b)\right). \tag{B.23}$$

A more common form of Eq. (B.23) is found by writing $b - a = 2h$ (the interval from a to b is subdivided into two panels) and substituting x_0 for a, x_2 for b, and x_1 for the midpoint. We then have

$$\int_{x_0}^{x_0+2h} f(x) \; dx \doteq \frac{h}{3}(f_0 + 4f_1 + f_2). \tag{B.24}$$

Formula (B.24) is Simpson's $\frac{1}{3}$ rule, a particularly popular formula. Again, it is not exact— as indicated by the dot over the equality sign.

The application of Eqs. (B.22) and (B.24) over an extended interval is straightforward. It is intuitively apparent that the error in these formulas will be large if the interval is not small. (We discuss these errors quantitatively in Section B.5.) To apply these to a large interval of integration and still maintain control over the error, we break the interval into a large number of small subintervals and add the formulas applied to the subintervals. When this is done we get the *extended trapezoidal rule:*

$$\int_a^b f(x) \; dx \doteq \frac{b-a}{2n}[f(x_0) + 2f(x_1) + 2f(x_2) + \cdots + 2f(x_{n-1}) + f(x_n)]; \tag{B.25}$$

and the extended Simpson's $\frac{1}{3}$ rule:

$$\int_a^b f(x) \; dx \doteq \frac{b-a}{3n}[f(x_0) + 4f(x_1) + 2f(x_2) + 4f(x_3) + \cdots$$
$$+ 2f(x_{2n-2}) + 4f(x_{2n-1}) + f(x_{2n})]. \tag{B.26}$$

These forms of the trapezoidal rule (Eq. (B.25)) and of Simpson's $\frac{1}{3}$ rule (Eq. (B.26)) are widely used in computer programs for integration. By applying them with n increasing, the error can be made arbitrarily small.* For Simpson's rule, observe that the number of subintervals must be even.

B.4 Integration Formulas Using Points Outside the Interval

In studying numerical methods to solve differential equations we shall have need for some specific integral formulas that involve function values computed at points outside the interval of integration. Figures B.2, B.3, and B.4 sketch three special cases of importance. In Fig. B.2 the curve that passes through the four points whose abscissas are x_{n-3}, x_{n-2}, x_{n-1}, and x_n is extrapolated to x_{n+1}, and we desire the integral only over the extrapolated interval. Figure B.3 presents the case where the curve passes through four points, and the integral is taken only over the last panel. In Fig. B.4 we consider a case where a curve that fits at three points is extrapolated in both directions, and integration is taken over four panels.

In the derivations of this section it will be convenient to adopt the notation that $f_i = f(x_i)$, so that subscripts on the function indicate the x-value at which it is evaluated.

---

*This is true except for round-off error effects, which eventually will dominate because they are not decreased by small subdivision of the interval and, in fact, may increase as the number of computations increases.

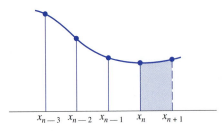

Figure B.2

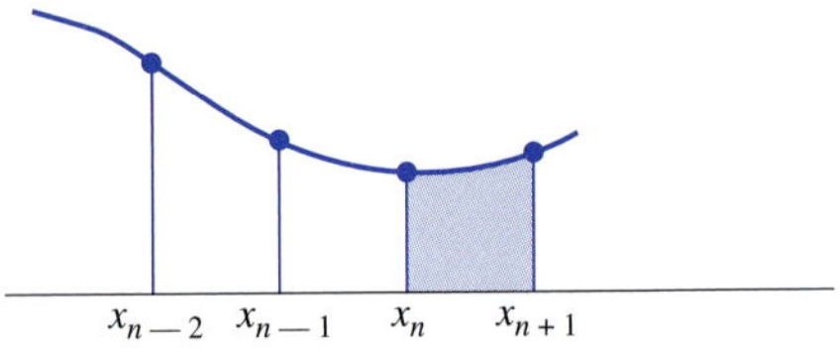

Figure B.3

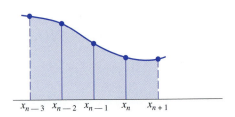

Figure B.4

For the first case (Fig. B.2) we desire a formula of the form

$$\int_{x_n}^{x_{n+1}} f(x) \, dx = c_0 f_{n-3} + c_1 f_{n-2} + c_2 f_{n-1} + c_3 f_n.$$

With four constants, we can make the formula exact when $f(x)$ is any polynomial of degree 3 or less. Accordingly, we replace $f(x)$ successively by x^3, x^2, x, and 1 to evaluate the coefficients.

It is apparent that the formula must be independent of the actual x-values. To simplify the equations, let us shift the origin to the point $x = x_n$; our integral is then taken over the interval from 0 to h, where $h = x_{n+1} - x_n$:

$$\int_0^h f(x) \, dx = c_0 f(-3h) + c_1 f(-2h) + c_2 f(-h) + c_3 f(0).$$

Carrying out the computations by replacing $f(x)$ with the particular polynomials, we have

$$\frac{h^4}{4} = c_0(-3h)^3 + c_1(-2h)^3 + c_2(-h)^3 + c_3(0),$$

$$\frac{h^3}{3} = c_0(-3h)^2 + c_1(-2h)^2 + c_2(-h)^2 + c_3(0),$$

(B.27)

$$\frac{h^2}{2} = c_0(-3h) + c_1(-2h) + c_2(-h) + c_3(0),$$

$$h = c_0(1) + c_1(1) + c_2(1) + c_3(1).$$

After completion of the algebra we get

$$\int_{x_n}^{x_{n+1}} f(x)\, dx \doteq \frac{h}{24}(-9f_{n-3} + 37f_{n-2} - 59f_{n-1} + 55f_n). \tag{B.28}$$

For the second case, illustrated by Fig. B.3, we again translate the origin to x_n to simplify. The set of equations analogous to Eqs. (B.27) is

$$\frac{h^4}{4} = c_0(-2h)^3 + c_1(-h)^3 + c_2(0) + c_3(h)^3,$$

$$\frac{h^3}{3} = c_0(-2h)^2 + c_1(-h)^2 + c_2(0) + c_3(h)^2,$$

$$\frac{h^2}{2} = c_0(-2h) + c_1(-h) + c_2(0) + c_3(h),$$

$$h = c_0(1) + c_1(1) + c_2(1) + c_3(1).$$

These give values of the constants in the integration formula:

$$\int_{x_n}^{x_{n+1}} f(x)\, dx \doteq \frac{h}{24}(f_{n-2} - 5f_{n-1} + 19f_n + 9f_{n+1}). \tag{B.29}$$

The third case, Fig. B.4, where extrapolation of a quadratic in both directions is involved, leads to the equation

$$\int_{x_{n-3}}^{x_{n+1}} f(x)\, dx \doteq \frac{4h}{3}(2f_{n-2} - f_{n-1} + 2f_n). \tag{B.30}$$

The particular methods for solving differential equations that use the formulas derived in this section are known as *Adams' method* and *Milne's method*.

B.5   Error of Integration Formulas

In each of the formulas that we derived so far, we used the symbol $\doteq$ to remind us that the formulas are not exact unless $f(x)$ is in fact a polynomial of a certain degree or less. It is important to derive expressions for the error.

We first determine the error term for the trapezoidal rule over one panel. Eq. (B.22). Begin with a Taylor expansion:

$$F(x_1) = F(x_0) + F'(x_0)h + \frac{1}{2}F''(x_0)h^2 + \frac{1}{6}F'''(\xi_1)h^3, \qquad x_0 < \xi_1 < x_1 = x_0 + h. \tag{B.31}$$

Let us define $F(x) = \int_a^x f(t)\, dt$ so that $F'(x) = f(x)$, $F''(x) = f'(x)$, $F'''(x) = f''(x)$. With this definition of $F(x)$, Eq. (B.31) becomes

$$\int_a^{x_1} f(x)\, dx = \int_a^{x_0} f(x)\, dx + f(x_0)h + \frac{1}{2}f'(x_0)h^2 + \frac{1}{6}f''(\xi_1)h^3.$$

On rearranging, and using subscript notation, we get

$$\int_a^{x_1} f(x)\, dx - \int_a^{x_0} f(x)\, dx = \int_{x_0}^{x_1} f(x)\, dx = f_0 h + \frac{1}{2} f_0' h^2 + \frac{1}{6} f''(\xi_1) h^3. \qquad \textbf{(B.32)}$$

We now replace the term f_0' by the forward-difference approximation with error term from Section B.2:

$$f_0' = \frac{f_1 - f_0}{h} - \frac{1}{2} h f''(\xi_2), \qquad x_0 < \xi_2 < x_1.$$

Equation (B.32) becomes

$$\int_{x_0}^{x_1} f(x)\, dx = f_0 h + \frac{1}{2} h(f_1 - f_0) - \frac{1}{4} h^3 f''(\xi_2) + \frac{1}{6} h^3 f''(\xi_1).$$

We can combine the last two terms into $-\frac{1}{12} h^3 f''(\xi)$, where ξ also lies in $[x_0, x_1]$. Hence

$$\int_{x_0}^{x_1} f(x)\, dx = \frac{h}{2} (f_0 + f_1) - \frac{1}{12} h^3 f''(\xi).$$

The error of the trapezoidal rule over one panel is then $O(h^3)$. This is called the *local error*. Because we normally use a succession of applications of this rule over the interval $[a, b]$ to evaluate $\int_a^b f(x)\, dx$, by subdividing into $n = (b - a)/h$ panels, we need to determine the so-called global error for such intervals of integration. The global error will be the sum of the local errors in each of the n panels:

$$\text{Global error} = -\frac{1}{12} h^3 f''(\xi_1) - \frac{1}{12} h^3 f''(\xi_2) - \cdots - \frac{1}{12} h^3 f''(\xi_n)$$

$$= -\frac{1}{12} h^3 [f''(\xi_1) + f''(\xi_2) + \cdots + f''(\xi_n)].$$

Here the subscripts on ξ indicate the panel in whose interval the value lies. If $f''(x)$ is continuous throughout the interval of integration,

$$\sum_{i=1}^{n} f''(\xi_i) = n f''(\xi), \qquad a < \xi < b.$$

Using $n = (b - a)/h$, we get

$$\text{Global error} = -\frac{1}{12} h^3 \left(\frac{b - a}{h} \right) f''(\xi) = -\frac{b - a}{12} h^2 f''(\xi) = O(h^2).$$

Therefore the global error is $O(h^2)$, even though the local error is $O(h^3)$.
   Similar treatment shows that the local error for Simpson's $\frac{1}{3}$ rule is

$$-\frac{1}{90} h^5 f^4(\xi) = O(h^5),$$

and the global error is

$$-\frac{1}{180}(b - a)h^4 f^4(\xi) = O(h^4).$$

We can also show that the local error of each of the integration formulas in Section B.4 is $O(h^5)$.

APPENDIX

Software Resources

Besides the programs and subroutines that accompany the book, there are many other excellent programs and subroutines available that are geared to professionals who need more reliable and robust software. Our programs are presented primarily to illustrate the particular algorithms presented in the book. The following list is a brief introduction to some of the software available to those who work on numerical problems.

DERIVE is a software package with a menu-driven interface that includes two-dimensional and three-dimensional plotting. In addition to a version for the PC, there is a ROM-card version for the HP 95LX and 100LX Palmtop computers. *Source:* Soft Warehouse, Inc., 3660 Waialae Avenue, Suite 304, Honolulu, HI, 96816-3236, (808)734-5801.

GAMS The *Guide to Available Mathematical Software* currently contains information on some 9000 software modules from about 80 packages found at NIST (National Institute for Standards and Technology) and on netlib (see below). It is very easy to search for software on GAMS using keyword, name, or problem classification. Using the Web, the researcher can retrieve abstracts, documentation, examples, and source code. *Source:* http://gams.nist.gov/.

IBM's ESSL (*Engineering and Scientific Subroutine Library*) Version 2 consists of more than 400 subroutines callable from C, C++, and FORTRAN/Fortran 90. The packages include topics in numerical quadrature, interpolation, random number generation, FFTs, linear systems, and eigenvalue problems. These subroutines are especially tuned for IBM RISC System/6000 workstations running AIX. *Source:* (800)IBM-CALL, e-mail: wyaddow@vnet.ibm.com.

IMSL (*International Mathematical and Statistical Library*) is a library of over 900 useful FORTRAN subroutines that cover both mathematical and statistical problems. Fortran 90 and C versions have been more recently developed. Moreover, this library can be installed on a variety of machines from PCs and up. *Source:* Visual Numerics, Inc. (800) 364-8880.

LAPACK is a library of FORTRAN 77 subroutines for solving problems in numerical linear algebra including eigenvalue problems and singular value problems. This library supersedes LINPACK and EISPACK. *Source:* Individual routines are available through *netlib*. *Sources:* Currently, the e-mail addresses are: netlib@ornl.gov and netlib@research.att.com. For just general information send mail to one of these two addresses with the message: *send index from LAPACK*. The complete package consisting of about 600,000 lines of FORTRAN code can be obtained on magnetic media from NAG (see below).

MACSYMA is a software package that has been around for a long time. Some of its features include simplification of trig functions, differential equation packages, color graphics, etc. It is now available for Workstations, PCs running Windows 95 and NT. *Source:* MACSYMA, Inc., 20 Academy Street, Arlington, MA 02174; Phone: (800) 622-7962; http://www.macsyma.com.

MAPLE V is a very powerful software package that performs both symbolic and numeric computations. It provides both two-dimensional and three-dimensional color graphing capabilities. This software runs on almost all computers including PC's, Macintoshes, workstations, and mainframes. There is an Academic version as well. *Source:* Waterloo Maple Inc. 450 Phillip St., Waterloo, ON, Canada N2L5J2. Phone: (519) 747-2373; email: info@maplesoft.com; http://www.maplesoft.com

MATHEMATICA 3.0 continues to be one of the best known software packages for doing a large variety of mathematical problems. Moreover, it has excellent two-dimensional and three-dimensional graphing capabilities. It does both symbolic and numeric computations. This system is available on a very large variety of computers. *Source:* (800) 976-5301 or on the Web at: http://www.wolfram.com.

MATLAB covers several topics in mathematics such as linear algebra, digital signal processing, graphing, numerical methods, etc. It is available both for the MS-DOS machines as well as the Macintosh computers. There is a student edition of MATLAB 5 that includes three of the several *Toolboxes* that are available with this package. *Source:* The Math Works Inc., 24 Prime Park Way, Natick, MA 01760. Phone: (508) 647-7000; e-mail: info@mathworks.com; http://www.mathworks.com.

NAG (Numerical Algorithms Group) has provided in the past a large library of FORTRAN 77 subprograms. In addition this group now has come out with a Fortran 90, NAG Parallel Library, and a NAG C library. *Source:* Numerical Algorithms Group, Inc. 1400 Opus Place, Suite 200, Downers Grove, IL 60515-5702. Phone (630) 971-2337; http://www.nag.com.

NETLIB mentioned with regard to the LAPACK software is a tremendous resource for numerical libraries. Rather than get a long list, one could send an e-mail message containing the line "help" to one of the Internet addresses: netlib@research.att.com, netlib@ornl.gov, netlib@nec.no for information on how to use netlib and for an overview of the mathematical software.

NUMERICAL RECIPES is a well-known set of over 300 numerical routines that come on disks for a variety of programming languages. There are versions for C, Fortran 77, Fortran 90, Pascal, and BASIC. In addition to the book, the source codes for each version is available on diskette. *Source:* Cambridge University Press, 40 West 20th Street, New York, NY 10011-4211; http://www.cup.org.

Answers to Selected Exercises

Chapter 0 **6.** This begins as a trigonometric problem. If we let H = depth of well, D = diameter of well, W = width of ladder, V = width of side rails, L = length of ladder, and A = angle of inclination from the vertical, we can write

$$\tan(A) = \frac{2\sqrt{\left(\frac{D}{2}\right)^2 - \left(\frac{W}{2}\right)^2} - V * \cos(A)}{H - V * \sin(A)}$$

and

$$L = \frac{H - V * \sin(A)}{\cos(A)}$$

With the given values for H, D, W, and V, MATLAB can solve the first to give A = 0.34965 radian, and from this we get, from the second equation, L = 181.557 in.

9. Endless loop if tolerance = 1E-8. Stop with "BREAK."

11. b. x = 0.453398 in 19 iterations with tolerance = 1E-6.

16. Total numbers = 180,001; largest positive number = 0.9999E5; most negative number = $-0.9999E5$; smallest positive number = 0.1000E-4; smallest negative number = $-0.1000E-4$.

18. a. 0.123E2 (chop), 0.123E2 (round)
b. $-0.319E-1$ (chop), $-0.320E-1$ (round)
c. 0.122E2 (chop), 0.123E2 (round)
d. $-0.288E3$ (chop), $-0.289E3$ (round)
e. 0.130E3 (chop), 0.130E3 (round)
f. $-0.156E5$ (chop), $-0.156E5$ (round)
g. 0.123E-6 (chop), 0.123E-6 (round)

24. Exact = -1.297387
Chopped 3 digits = -1.31, relative error = -0.00972
Round 3 digits = -1.30, relative error = -0.00201

28. The series will converge because with a very large value for N, $1/N$ = zero. Looping is impractical because of the time required. Compute through the largest positive number. (N = 3.402823E38 in QuickBASIC, single precision. Other systems may have a smaller limit.)

31. The answer depends on the spacing between the roots. If evenly spaced, we obtain the middle one, of course. Closely spaced roots act like a multiple root. If $f(x) > 0$ beyond $x = b$, the root found tends to be a larger root than the middle one.

35. Parallel processing is applicable when iterate $n + 1$ does not require that iterate n be available.

Chapter 1

3. A graph indicates a root near -1.5. Beginning from $[-2, -1]$, the root is -1.491644 in 19 iterations, tolerance = 1E-6.

8. The plots intersect near $x = 4.5$, y about 56. Using *regula falsi* from $[4, 5]$, $x = 4.53786$ after 20 iterations, tolerance = 1E-5. The secant method from $[4, 5]$ gets the same value in 4 iterations, tolerance = 1E-5. Substituting this value into either equation gets $y = 55.7978$.

13. $F(x) = x^2 - N = 0$, $F'(x) = 2x$, so

$$x_1 = x_0 - \frac{x_0^2 - N}{2x_0} = \frac{x_0^2 + N}{2x_0} = \frac{1}{2}\left(x_0 + \frac{N}{x_0}\right).$$

18. From $x_0 = 1.1$, the positive root is obtained in 3 iterations, also from $x_0 = 0.9$. From $x_0 = -1.1$ or $x_0 = -0.9$, it takes 18 iterations to get the negative root.

20. From graph, $f'(x) = 0$ at $x = -1.0$ and at $x = 0.5$.

24. For each example of Exercise 4, Muller's method does get the root nearest zero. In some cases this is not true, such as $(x + 0.3)(x - 0.2)(x - 0.3)$, or when there is a root very near to -0.5 or 0.5 in addition to a smaller root. In part d, $[-0.5, 0, 0.5]$ fails but $[-1, 0, 1]$ works.

26. The method will attempt to get the square root of a negative number. Try different starting values.

30. Without acceleration, fixed point iteration approximates the root as $x = 0.618033$ in 13 iterations. With Aitken acceleration, $x = 0.618034$ after 6 iterations.

33. Starting from $x = 1.2$:
(1) $x = ((6 + 4x - 4x^2)/2)^{1/3}$, 26 iterations (6 if accelerated)
(2) $x = ((6 + 4x - 2x^3)/4)^{1/3}$, 23 iterations (6 if accelerated)
(3) $x = ((6 + 4x)/(2x + 4))^{1/2}$, 4 iterations (3 if accelerated)

36. $x = -4.5615528$, $x = -0.43844719$; factor is $2x^2 - 3x + 7$; roots: $0.75 \pm 1.713913i$

39.

Change in coeff	Max change in magnitude of any root
$2.00 \to 2.02$	0.73%
$7.00 \to 7.07$	0.56%
$4.00 \to 4.04$	0.71%
$29.00 \to 29.29$	0.80%
$14.00 \to 14.14$	0.89%

43. $P(x) = (x^2 - 1.5x + 4.3)(x^2 - 4.2x + 16.1)$; roots: $0.75 \pm 1.9333i$, $2.1 \pm 3.4190i$

45. $P(x) = (x^2 - 5x + 2)(2x^2 - 3x + 7)$; roots: -0.438445, -4.56155, $0.75 \pm 1.7139i$

49. a. After convergence, q's approximate the roots: $x = 1.8019$, -1.2470, 0.4450.
b. Two real roots: $x = -3.3480$ and 1.1390; quadratic factor: $x^2 - 1.3090x + 3.1048$

53. If there are complex roots, an initial estimate near their modulus asks for the square root of a negative number. If there are multiple roots, this is true as well.

57. Bairstow gives these factors:

$$x^2 - 2x + 1, \quad x^2 - 5.9999x + 8.9997, \quad x - 3.0001,$$

whose roots are 1, 1; 3, 2.9999; 3.0001. The results depend on the starting value and the tolerance value.

60. $P(x)/P'(x) = (x^2 - 4x + 3)/(5x - 9)$. Starting with $x_0 = 1.1$, successive errors are 0.100000, 0.00826, and 0.00005, and each of these is about the square of the previous one. Starting with $x_0 = 3.3$, errors are 0.30000, 0.02243, and 0.00017; these actually decrease faster than $O(\text{error}^2)$.

64. a. Starting with $x_0 = 2.05$, converges to root at $x = 0$ but there are wide swings at first. Near convergence, errors are -0.71630, -0.15299, and -0.00232. The last two show quadratic convergence.
b. Starting with $x_0 = 2.00$, converges to root at 9.41757. The last errors are 0.03679, 0.00130 showing about quadratic convergence.

67. c. Convergence is very, very slow. There is a linear convergence; each error is 2/3 times the previous value.

71. The soLve command does not find the roots. Using NEWTON (see Section 10.1 of the DERIVE manual), a root at 0.86033 is found after 3 iterations beginning at $x = 1$. (A root at $x = -0.86033$ is found after 3 iterations from -1.)

78. From the graphs:

a. 1.1462718
b. 6.1353472, 1.1737446
c. 3.7333079 (other roots: -0.4589623, 0.91000757)
d. 4.3026887 (there are many other roots)

Chapter 2 **1.**
$$\text{b. } A - B = \begin{bmatrix} -3 & -9 & 0 & -3 \\ 4 & -5 & 0 & -1 \\ 3 & 0 & 1 & -4 \end{bmatrix}, \quad Ax = \begin{bmatrix} 2 \\ 19 \\ 9 \end{bmatrix}, \quad By = \begin{bmatrix} 34 \\ 28 \\ 36 \end{bmatrix}$$

$$\text{c. } x^T y = [-6], \quad xy^T = \begin{bmatrix} 0 & 8 & 4 & 12 \\ 0 & -12 & -6 & -18 \\ 0 & 0 & 0 & 0 \\ 0 & 4 & 2 & 6 \end{bmatrix}$$

2.
$$\text{a. } BA = \begin{bmatrix} -18 & 7 & 9 \\ -15 & -8 & -1 \\ 8 & 11 & 26 \end{bmatrix}, \quad B^3 = \begin{bmatrix} -203 & 45 & 190 \\ -40 & -28 & 55 \\ -150 & 45 & -58 \end{bmatrix}, \quad AA^T = \begin{bmatrix} 14 & -3 & -1 \\ -3 & 13 & 4 \\ -1 & 4 & 14 \end{bmatrix}$$

4. b. $\text{eig}(A) = -1, 7$
$\text{eig}(B) = 5, 6.42442, -3.42442$

6.
$$\begin{bmatrix} 2 & -6 & 1 \\ -5 & 1 & -2 \\ 1 & 2 & 7 \end{bmatrix} \begin{bmatrix} x \\ y \\ z \end{bmatrix} = \begin{bmatrix} 11 \\ -12 \\ 20 \end{bmatrix}$$

7. b. $x_3 = 2$; $x_2 = (3 + 6)/3 = 3$; $x_1 = (7 - 4 + 3)/2 = 3$

11. By using the elementary row operations, we can reduce the augmented matrix to
$$\begin{bmatrix} 3 & 2 & -1 & -4 & 10 \\ 1 & -1 & 3 & -1 & -4 \\ 2 & 1 & -3 & 0 & 16 \\ 0 & -1 & 8 & -5 & 3 \end{bmatrix} \rightarrow \begin{bmatrix} 3 & 2 & -1 & -4 & 10 \\ 0 & 5 & -10 & -1 & 22 \\ 0 & 0 & -45 & 39 & 162 \\ 0 & 0 & 0 & 0 & 435 \end{bmatrix}.$$

The last line would imply that $0 = 435$!

13.   Let $R_1 = (3 \quad 2 \quad -1 \quad -4), \dots, R_4 = (0 \quad -1 \quad 8 \quad -5)$ represent the four rows of the coefficient matrix of Exercise 11. Then we have the relationship $R_1 + R_2 - 2R_3 = R_4$.

15.

a. $\begin{bmatrix} 1 & 1 & -2 & 3 \\ 4 & -2 & 1 & 5 \\ 3 & -1 & 3 & 8 \end{bmatrix} \rightarrow \cdots \rightarrow \begin{bmatrix} 4 & -2 & 1 & 5 \\ \frac{1}{4} & \frac{3}{2} & -\frac{9}{4} & \frac{7}{4} \\ \frac{3}{4} & \frac{1}{3} & 3 & \frac{11}{3} \end{bmatrix}$

Using back-substitution, we get

$$x_3 = \frac{11}{9}; \qquad x_2 = \left(\frac{7}{4} + \frac{11}{4}\right) * \frac{2}{3} = 3; \qquad x_1 = \frac{22}{9}$$

b. $\det(A) = (-1) * 4 * (3/2) * 3 = -18$

c. $\begin{bmatrix} 1 & 0 & 0 \\ \frac{1}{4} & 1 & 0 \\ \frac{3}{4} & \frac{1}{3} & 1 \end{bmatrix} \begin{bmatrix} 4 & -2 & 1 \\ 0 & \frac{3}{2} & -\frac{9}{4} \\ 0 & 0 & 3 \end{bmatrix} = \begin{bmatrix} 4 & -2 & 1 \\ 1 & 1 & -2 \\ 3 & -1 & 3 \end{bmatrix}$

16.   a. $x^T = (1.30, -1.35, -0.275)$
b. $x^T = (1.45, -1.59, -0.276)$
c. Calculated right-hand sides are

$$(0.02, 1.02, -0.21) \qquad \text{and} \qquad (0.04, 1.03, -0.54).$$

19.   c. In general, the number of multiplications/divisions for the Gauss–Jordan method is $O(n^3/2)$ versus $O(n^3/3)$ for Gaussian elimination.

20.   a. $Az = b$ implies $(B + Ci)(x + yi) = (Bx - Cy) + (By + Cx)i = p + qi$. We need to solve

$$\begin{bmatrix} B & -C \\ C & B \end{bmatrix} = \begin{bmatrix} p \\ q \end{bmatrix}.$$

b. $2n^2 + 2n$ if done with complex arithmetic versus $4n^2 + 2n$ if done as in part a.

21.   b. $x_1 = 1, x_2 = 2, y_1 = -1, y_2 = 0$,
$z_1 = 1 - i, z_2 = 2 + 0i$

22.

$$A = \begin{bmatrix} 1 & 0 & 0 \\ 4 & -6 & 0 \\ 3 & -4 & 3 \end{bmatrix} \begin{bmatrix} 1 & 1 & -2 \\ 0 & 1 & -\frac{3}{2} \\ 0 & 0 & 1 \end{bmatrix}$$

First solving for $Ly = b$, we get: $y = (3, 7/6, 11/9)$; then solving $Ux = (3, 7/6, 11/9)^T$ we get the answer: $(22/9, 3, 11/9)$.

29.   b. $a_1 = 13, a_2 = 12, a_3 = -5$ (or any nonzero multiple of these)

30.   b. Solving for the first three equations gives us the unique solution: $(3/2, -1/2, -3/2)$. However, substituting these values into the fourth equation does not produce the correct result. This set of four equations does not have a solution.

31.   a. $\det(H)$ is very small (about 1.65×10^{-6}); one also cannot avoid a very small divisor in solving.
b. $x^T = (1.11, 0.228, 1.95, 0.797)$
c. $x^T = (0.988, 1.42, -0.428, 2.10)$

37.

b. $H^{-1} = \begin{bmatrix} 16 & -120 & 240 & -140 \\ -120 & 1200 & -2700 & 1680 \\ 240 & -2700 & 6480 & -4200 \\ -140 & 1680 & -4200 & 2800 \end{bmatrix}$

38. Gaussian elimination: 25 mult/div; 11 add/subtracts. Gauss–Jordan: 29 mult/div; 15 add/subtracts. The system here is too small to illustrate the true difference between the methods.

39. f. Yes, the triangle inequality holds.
    for 1-norm: $22 + 12 > 17$
    for f-norm: $18.87 + 10.34 > 15.68$
    for inf-norm: $23 + 11 > 17$

40. 25/12, which is the sum of the elements of the first row

42. a. $x^T = (1592.6, -631.911, -493.62)$

44. $x^T = (119.5, -47.14, -36.84)$. This is further evidence of ill-condition, in that small changes in the coefficients make large changes in the solution vector.

48. $\dfrac{1}{\text{C.N.}} \dfrac{\|r\|_2}{\|b\|_2} = 3.95 \times 10^{-5}$,   C.N. $\dfrac{\|r\|_2}{\|b\|_2} = 120{,}640$

    $\dfrac{\|e\|_2}{\|\bar{x}\|_2} = 14.46$, $\dfrac{\|e\|_2}{\|x\|_2} = 1.07$

The relation is verified with either $\|\bar{x}\|$ or $\|x\|$.

51. Using 3-digit arithmetic to compute e:
$$e = \{-0.00272, 0.00126, 0.00103\}, \quad x = \{0.153, 0.144, 0.166\}.$$
$(x + e) = \{0.15029, 0.14526, -0.16497\}$; compare with $x = \{0.15094, 0.14525, -0.16592\}$. The improvement is remarkable even in one iteration.

52. Gauss–Seidel diverges faster.
After 10 iterations, G–S gets $-201551.9, 604659.8$.
After 10 iterations, Jacobi gets $-76.76, -76.76$.

55. Interchange lines 2 and 3. Begin with $(1, 1, 1)$.
a. Using the Jacobi method, we converse to the solution vector $(-2, 1, -3)$ in 9 iterations.
b. With the Gauss–Seidel method there is convergence in 5 iterations.

56. $x^T = (1, -1, 2)$

59. $(0.72595, 0.50295)$ in 5 iterations from $(1, 1)$;
$(-1.6701, 0.34513)$ in 6 iterations from $(-2, 1)$

61. $(1.64304, -2.34978), (-2.07929, -3.16173)$

64. b. Multiply $P_{1,4} * A$ where $P_{1,4}$ is the identity matrix with the first and fourth rows interchanged.
c. Multiply $A * P_{1,2}$ where $P_{1,2}$ is similarly defined.
d. $P_{2,4} * A * P_{2,4}$

Chapter 3   **2.** $P_3(x) = \dfrac{(x-3)(x-5)(x-9)}{-32} + \dfrac{(x-1)(x-5)(x-9)}{8} + \dfrac{(x-1)(x-3)(x-9)}{-32/3}$

    $+ \dfrac{(x-1)(x-3)(x-5)}{24}$

    $P_3(x) = x^3/12 - 9x^2/8 + 53x/12 - 11/8$

6. The Neville table for $f(0.4)$ is

3.0	20.7180	75.4040	46.5355	53.7704	56.0694
5.0	130.0900	−40.0700	10.3609	39.9768	
7.0	470.4100	312.9461	247.2881		
−2.0	−1.9817	94.0861			
−3.0	−17.9930				

Interpolates are 46.5355 for $P_2(4)$,
53.7704 for $P_3(4)$,
56.0694 for $P_4(4)$.

15. For: $f(0.15)$, at $x = 0.0, 0.1, -0.2$ or 0.5,
$f(-0.1)$, at $x = -0.2, 0.0, 0.1$,
$f(1.2)$, at $x = 0.1, 0.5, 0.7$.

18. Interpolate $= 1.22183$, error $= -0.00043$. Bounds on error are -0.00033 and -0.00045, which bracket the error. (If more accurate values for e^x had been used, the error would be -0.00039, so round-off has entered.)

22. Third difference $=$

$$0.3365 - 3(0.3001) + 3(0.2624) - 0.2231 = 0.0003$$

26. $P_2(0.203) = 0.78024$, next term increment $= 1.32\text{E}-3$
$P_3(0.203) = 0.78156$, next term increment $= 7.2\text{E}-5$

30. Fourth degree because all fourth differences are constant.

34. The system is

$$\begin{bmatrix} 1.2200 & 0.490 & 0 & 0 & -1.326 \\ 0.490 & 1.240 & 0.130 & 0 & -1.715 \\ 0 & 0.130 & 0.620 & 0.180 & -0.829 \\ 0 & 0 & 0.180 & 2.440 & -1.865 \end{bmatrix}$$

The S-values are $0, -0.678, -1.018, -0.922, -0.696, 0$.

39. Plot of cubic spline, interpolating polynomial, and original curve. (Evenly spaced points are not the best choice.)

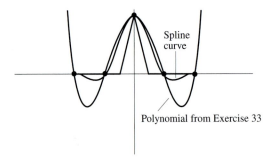

Spline curve

Polynomial from Exercise 33

43.

$$[u^4, u^3, u^2, u, 1] \begin{bmatrix} 1 & -4 & 6 & -4 & 1 \\ -4 & 12 & -12 & 4 & 0 \\ 6 & -12 & 6 & 0 & 0 \\ -4 & 4 & 0 & 0 & 0 \\ 1 & 0 & 0 & 0 & 0 \end{bmatrix} \begin{bmatrix} p_0 \\ p_1 \\ p_2 \\ p_3 \\ p_4 \end{bmatrix}$$

45. $dx/du = -(3/6)(1-u)^2 x_{i-1} + (1/6)(9u^2 - 12u)x_i + (1/6)(-9u^2 + 6u + 3)x_{i+1} + (3/6)(u^2)x_{i+2}$
At $u = 0$, $dx/du = -(3/6)x_{i-1} + 0 + (3/6)x_{i+1} + 0 = (3/6)(x_{i+1} - x_{i-1})$.

Similarly for dy/du, so

$$dy/dx = \frac{y_{i+1} - y_{i-1}}{x_{i+1} - x_{i-1}},$$

which is the slope between points adjacent to p_i. A similar relation holds at the right end.

52. $f(1.6, 0.33) = 1.8328$

54. Interpolating by rows: $f(1.1, 0.71) = 0.70726$, $f(3.0, 0.71) = 5.2614$, $f(3.7, 0.71) = 8.0028$, $f(5.2, 0.71) = 15.8077$. From these, $f(3.32, 0.71) = 6.4435$.

59. $y = 2.908x + 2.02533$

61. $z = 2.8530x - 1.9145y + 1.0399$

67. $\ln(S) = \ln(1.2019) + 0.009603T$, or $S = 1.2019 * e^{0.009603T}$.

73. Fitting polynomials of increasing degree, we find the variances are degree 2, 533.3; degree 3, 85.47; degree 4, 86.73; degree 5, 64.84. This indicates that the optimum degree is 3 and, for this,

$$f(x) = 0.1223x^3 - 3.614x^2 + 24.11x - 6.392.$$

78. Values for $P(0.1)$ from polynomials:

Degree	$P(0.1)$	Error
2	0.99	-0.39
3	0	0.60
4	0.9504	-0.3504
5	0.2256	0.3744

Cannot compare to error bounds because $f'(x)$ is discontinuous and hence requirements for Eq. (3.30) are not met.

84. 1.8407

86. $\ln(s) = 0.183932 + 0.0096028T$

Chapter 4

4. Zeros at 0, ± 0.58778525, ± 0.95105651

8. The ninth-degree Maclaurin series is very accurate near $x = 0$, but the error increases rapidly to 0.04952 at $x = \pm 1$. The error of P_3 has four maxima (counting the endpoints) and has a maximum error of 0.0349 at $x = \pm 1$.

11. $y = (2x - b - a)/(b - a)$

12. a. Cannot find $R_{3,3}$.

$$R_{4,2} = \frac{15x^2 - 3x^4}{15 + 2x^2}, \qquad \text{maximum error} = 0.00219.$$

b. Cannot get Padé for $N = 6$.

c. $R_{33} = \dfrac{1 + x/2 + x^2/10 + x^2/120}{1 - x/2 + x^2/10 - x^3/120}$, maximum error $= 2.81E{-}5$

15. For part (c):

$$-1 + \cfrac{24}{-x + 12 - \cfrac{50}{x + 10/x}}$$

22. $A_0 = 2/3$, $A_n = 4/(n^2\pi^2)$, $n = 1, 2, 3, \ldots,$
$B_n = -4/(n\pi)$, $n = 1, 2, 3, \ldots$

27.   a. A_n values:

$n =$	0	1	2	3	4
$A_n =$	0.097599	−0.250735	−0.306423	−0.0944593	−0.0629500

b. B_n values:

$n =$	1	2	3	4
$B_n =$	0.199033	−0.159817	−0.173484	−0.133883

29.   a. Maxima are 0.12 at $x = 0$ and 0.17 at $x = \pm 1.0$;
       minima are −0.105625 at $x = \pm 0.689202$.
       $T_4/8$ has maxima of 0.125 and minima of −0.125.
      b. Maxima are 0.126 at $x = 0$ and 0.086 at $x = \pm 1.0$;
       minima are −0.1444 at $x = \pm 0.721110$.
       $T_4/8$ has maxima of 0.125 and minima of −0.125.

35. $L_1 = x; \; L_2 = \dfrac{-1 + 3x^2}{2}; \; L_3 = \dfrac{-3x + 5x^3}{2};$

$L_4 = \dfrac{3 - 30x^2 + 35x^4}{8}; \; L_5 = \dfrac{15x - 70x^3 + 63x^5}{8}; \; L_6 = \dfrac{-5 + 105x^2 - 315x^4 - 231x^6}{16}$

38.   a.

x	Error Maclaurin	Error economized
−1.0	−0.00724	−0.00258
−0.5	−0.0000683	0.00591
0.0	0	−0.00046
0.5	0.0000957	−0.00502
1.0	0.014137	−0.00314

b.

x	Error Maclaurin	Error economized
−1.0	−11.7154	−10.6291
−0.5	−0.211435	−0.088626
0.0	0	0.840733
0.5	0.289899	−1.51437
1.0	30.1944	27.4266

c. Cannot economize because the Maclaurin series is just

$$x, \text{ for } x > 0, \; -x \text{ for } x < 0.$$

Chapter 5   4.   Derivative values:

x	Computed	Exact
0.21	2.0904	2.0681
0.22	1.9754	1.9741
0.23	1.8973	1.8882
0.24	1.8191	1.8096
0.25	1.7409	1.7372
0.26	1.6627	1.6704
0.27	1.6230	1.6085

11. a. $f'(0.72)$ from $P_3(x) = 1.2505$ (exact $= 1.25060$)
b. $f'(1.33)$ from $P_2(x) = 1.0790$ (exact $= 1.07949$)
c. $f'(0.50)$ from $P_4(x) = 1.2925$ (exact $= 1.29253$)

14. $f'(0.5) = 1.2905$, exact $= 1.29253$, actual error is 0.0020. Compare to bounds: 0.0021, 0.0017.

22.

No. terms	Value	Actual error	Estimated error
1	−0.404148	−0.007082	−0.007048
2	−0.411988	0.000758	0.000000
3	−0.411988	0.000758	0.000000

(Exact value $= -0.411230$)

26. From Taylor series:

$$f'(x_0) = \frac{f_1 - f_0}{h} - \frac{h}{2} f''(\xi)$$

$$f'(x_0) = \frac{f_0 - f_{-1}}{h} + \frac{h}{2} f''(\xi)$$

31. $f'(0.32) = 0.15728$ after 2 extrapolations, agrees to 5 digits.

33. The order of error is $O(h)$, so Eq. (4.21) applies with $n = 1$:

$$\text{Extrapolated value} = \text{More accurate} + \frac{1}{2 - 1} * (\text{More accurate} - \text{less})$$

Agrees to 5 digits after 7 extrapolations.

38. (We used undetermined coefficients.)

$$\text{Integral from } x_0 \text{ to } x_0 + 4h = \left(\frac{h}{45}\right)(14f_0 + 64f_1 + 24f_2 + 64f_3 + 14f_4) + O(h^7)$$

Integral from x_0 to $x_0 + 5h$

$$= \left(\frac{h}{288}\right)(95f_0 + 375f_1 + 250f_2 + 250f_3 + 375f_4 + 95f_5) + O(h^7)$$

41. Analytical value is 1.7669731. Trapezoidal rule results:

h	No. panels	Value	Error	Error/h^2
0.1	8	1.7684	−0.0014	−0.140
0.2	4	1.7728	−0.0058	−0.145
0.4	2	1.7904	−0.0234	−0.146

45. Estimate of integral is 0.69315; $\ln(2) = 0.693147$

50. With 8 panels ($h = 0.125$), integral is 1.718284, analytical is 1.7182820.

55. Let $P_3(x) = a + bx + cx^2 + dx^3$. By change of variable, integration can be from $-h$ to h with midpoint at $x = 0$. The quadratic that fits at three evenly spaced points is $a + (b + dh^2)x + cx^2$. The integral of each is the same: $2ah + 2ch^3/3$.

59. We want the value of the integral to be $af_0 + bf_1$ where a and b are to be determined. By change of variable, the limits can be from $x = 0$ to $x = h$. Letting $f(x) = 1$ and then $f(x) = x$, we need to solve:

$$\begin{cases} h = a + b, \\ \dfrac{h^2}{2} = bh. \end{cases}$$

which gives $a = h/2, b = h/2$.

62. Three-term formula gives 1.71281, which is accurate to 6 significant figures.

64. Correct value is -0.700943. Even five terms in the Gaussian formula is not enough. Simpson's $\frac{1}{3}$ rule attains 5 digits accurate wth 400 intervals. Extrapolating from Simpson's $\frac{1}{3}$ rule to 7 levels also attains this (requiring 128 intervals).

69. With TOL $= 0.45$, answer is 3.657243, error $= 0.003\%$; with TOL $= 0.5$, answer is 3.666552, error $= 0.257\%$.

73. a. With x-values vertical:

$$\frac{\Delta x}{3} \frac{\Delta y}{2} \begin{Bmatrix} 1 & 2 & 2 & 1 \\ 4 & 8 & 8 & 4 \\ 2 & 4 & 4 & 2 \\ 4 & 8 & 8 & 4 \\ 1 & 2 & 2 & 1 \end{Bmatrix}$$

d. For part (a), even number in x-direction.
   For part (b), even number in both directions.
   For part (c), divisible by 3 in both directions.

78. Analytical value is 2/3.

x	y	Integral	Error	Error/h^2
0.5	0.5	0.75	-0.0833	-0.3333
0.25	0.5	0.7185	-0.05208	-0.3333*
0.5	0.25	0.7185	-0.05208	-0.3333*
0.25	0.25	0.6875	-0.0208	-0.3333
0.125	0.125	0.6719	-0.0052	-0.3333

*Used the average of the squares of the h-values.

82. a. Procedure:
   (1) Locate the Gauss points on the y-axis between $[0, 1]$.
   (2) For each of these y-values, locate the Gauss points for x between $x = 0$ and the x-value on the circle.
   (3) For each y-value in (1), compute the weighted sum of function values at each of the Gauss points in (2); divide the sum by 2.
   (4) Compute the weighted sum of the sums in (3); divide this by 2.
   Result: (a) 0.28208 Analytical value $= 0.28108$

87. Computed values follow. The errors of $f''(x)$ are shown.

x	f'	$f''(x)$	Error in $f''(x)$
1.0	-0.1111	0.0843	-0.0102
1.5	-0.0808	0.0368	0.0098
2.0	-0.0635	0.0326	-0.0014
2.5	-0.0491	0.0250	-0.0030
3.0	-0.0400	0.0115	0.0045

88. Analytical value is 1.30176. From the cubic spline, the value is 1.299188, error = 0.00257. By Simpson's $\frac{1}{3}$ rule (4 panels), value is 1.30160, error = 0.00016.

92. **a.**

Analytical	Simpson's rule
4.00000	4.00000
1.21585	1.21565
0.30396	0.30311
0.13510	0.13296
0.07599	0.07155
−2.15946	−2.15954
−1.22486	−1.22587
−0.83449	−0.83838
−0.63057	−0.64090

b.

Analytical		Simpson's rule
$A_0 = 8/3$	= 2.666667	2.666667
$A_1 = -8/\pi$	= −2.545479	−2.546619
$A_2 = -4/\pi^2$	= −0.405385	−0.404144
$A_3 = 8/(3\pi)$	= 0.848826	0.852982
$A_4 = 1/\pi^2$	= 0.101321	0.095440
$B_1 = -16/\pi^2$	= −1.621139	−1.620870
$B_2 = 4/\pi$	= 1.273240	1.274396
$B_3 = 16/(9/\pi^2)$	= 0.180127	0.177278
$B_4 = -2/\pi$	= −0.636620	−0.647426

c.

Analytical		Simpson's rule
$A_0 =$	0.088992	0.088987
$A_1 =$	−0.133910	−0.133930
$A_2 =$	0.050823	0.050900
$A_3 =$	0.000237	−0.000243
$A_4 =$	−0.022006	−0.022491
$B_1 =$	0.052388	0.052372
$B_2 =$	−0.036334	−0.036378
$B_3 =$	0.036060	0.036298
$B_4 =$	−0.013001	−0.013265

97. The last row is

0 8 4 12 2 10 6 14 1 9 5 13 3 11 7 15 and this matches Fig. 5.8 and also the bit-reversing method.

104. The function used was $f(x) = e^x$. For the local error, two panels were used, starting at $x = 1$:

Limits	h	Integral	Error	Error/h^5
[1, 2]	0.5	4.672349	−0.001575	−0.0504
[1, 1.5]	0.25	1.763445	−0.000037	−0.0379
[1, 1.25]	0.125	0.7720621	−0.000001	−0.0317

which shows that errors are about proportional to h^5.

Chapter 6

1. a. $y(x) = 1 + x + (3/2)x^2 + (5/6)x^3 + (7/12)x^4 + (17/60)x^5$;
    $y(0.1) = 1.11589$, $y(0.5) = 2.0245$

3. $y(0.2) = 1.2015$, $y(0.4) = 1.4128$, $y(0.6) = 1.6471$

5. With $h = 0.01$, $y(0.1) = 1.11418$, error $= 0.00171$. To reduce the error to 0.00005 (34-fold), we must reduce h 34-fold, or to about 0.00029.

7. a. With $h = 0.025$, $y(0.1) = 1.11587$ (four decimals are correct). The simple Euler method will require about 340 steps and 340 function evaluations for similar accuracy, compared to four steps and eight function evaluations with the modified Euler method.

9. | x: | 0.1 | 0.2 | 0.3 | 0.4 | 0.5 |
    |------|-----|-----|-----|-----|-----|
    | y: | 2.2150 | 2.4630 | 2.7473 | 3.0715 | 3.4394 |

13. | x: | 0.2 | 0.4 | 0.6 |
    |------|-----|-----|-----|
    | y: | 2.0933 | 2.1755 | 2.2493 |

18. The exact answer is $y(x) = -5e^{-x} + 2x^2 - 4x + 4$.

x	y(Runge–Kutta–Fehlberg)	Analytical
0.00	−1.000000	−1.000000
0.10	−0.904187	−0.904187
0.20	−0.813654	−0.813654
0.30	−0.724091	−0.724091
.	.	.
.	.	.
.	.	.
1.80	2.453506	2.543506
1.90	2.872158	2.872157
2.00	3.323325	3.323324

20. a. $y(0.5) = -0.28326$, error $= -0.00077$
    b. $y(0.5) = -0.28387$, error $= -0.00016$
    c. $y(0.5) = -0.28396$, error $= -0.00007$

24. $y(4)$ predicted $= 4.1149$, $y(4)$ corrected $= 4.2229$. From these, the error in y_c is estimated as 0.0038; equivalently y_c has three correct digits. (The actual error is −0.0998.) Obviously, the starting values must have at least three digits of accuracy also.

26. | x: | 0.8 | 1.0 | 1.2 | 1.6 | 2.0 |
    |------|-----|-----|-----|-----|-----|
    | y: | 2.3163 | 2.3780 | 2.4350 | 2.5380 | 2.6294 |
    | Error estimate: | 0.0003 | <0.00005 | 0 | −0.00005 | −0.00002 |

    After $x = 1.2$, h was increased to 0.4.

29. By Runge–Kutta:

x:	0	0.2	0.4	0.6
y:	0	0.0004	0.0064	0.0324

By Adams–Moulton:

x:	0.8	0.9	1.0	1.1	1.2	1.25	1.3	1.35	1.4
y:	0.1025	0.1644	0.2513	0.3704	0.5321	0.6340	0.7544	0.8990	1.0772

The step size was halved after $x = 0.8$ and $x = 1.2$ to provide four-decimal accuracy.

31. a. $f_y = \sin(x)$, so $h_{max} = (24/9)/1 = 2.67$.
b. With $h = 0.267$, D cannot exceed 10×10^{-N} for N-decimal-place accuracy.
c. If $D = 14.2 \times 10^{-N}$, h cannot exceed $(1/14.2)h_{max}$.

37. $y' = z$, $y(0) = 0$;
$$z' = \frac{M(x)}{EI}(1 + z^2)^{3/2}, \ z(0) = 0$$

39.
t:	0	0.2	0.4	0.6
y:	1	0.982	0.934	0.865
x:	0	0.022	0.093	0.221

42.

x	y	y'
0	1	−1
0.1	0.8950	−1.0995
0.2	0.7802	−1.1956
0.3	0.6561	−1.2847
0.4	0.5236	−1.3629
0.5	0.3840	−1.4263
0.6	0.2389	−1.4715

46.

t	x	Analytical
0	0	0
0.1	0.0717	0.0717
0.2	0.1999	0.1999
0.3	0.2028	0.2026
0.4	−0.0193	−0.0233
0.5	−0.3543	−0.3784
0.6	−0.6003	−0.5977
0.7	−0.4643	−0.4419
0.8	0.0276	0.0924

49. a. Satisfies the Lipschitz condition, because we have
$$|f(x, y_1) - f(x, y_2)| \le |y_1 + y_2| \, |y_1 - y_2| < 2|y_1 - y_2|.$$
b. Does not satisfy the Lipschitz condition, because
$$|f(x, y_1) - f(x, y_2)| = x^2/|y_1 y_2| \, |y_1 - y_2| \text{ is unbounded at } y = 0.$$

51. Examine $f(x, y) = |x|$ on the unit square. In general, consider the integral of a bounded function with a finite number of discontinuities.

52. Parts (b) and (c) are stable, whereas parts (a) and (d) are unstable.

54.

x	y	f	$1 + hf_y$	$h^2y''/2$	Est'd error	Actual error
1.0	1.000	1.000	1.200	0.015	0	0
1.1	1.100	1.331	1.242	0.022	0.019	0.017
1.2	1.233	1.825	1.296	0.035	0.052	0.049
1.3	1.416	2.605	1.368	0.058	0.115	0.111
1.4	1.676	3.933	1.469	0.106	0.234	0.247
1.5	2.069	6.423	1.621	0.221	0.481	0.597
1.6	2.712	11.765	1.868	0.547	1.091	1.834

59. $-1/(-1 + \exp(1/2 * x^2 - x + C1)) * \exp(1/2 * x^2 - x + C1)$, which looks better when written as

$$\frac{C \exp(x^2/2)}{\exp(x) - C \exp(x^2/2)}.$$

62. $y(x) = 1 - x + 3/2x^2 - 7/6x^3 + 19/24x^4 - 9/24x^5$.

Chapter 7 **2.** The temperatures are linear within each portion. The gradient between $x = 0$ and $x = X$ is proportional to A/k_1; from $x = X$ to $x = L$, it is proportional to A/k_2. Solving for U, the temperature at the junction, gives

$$U = \frac{100k_2X}{k_1(L - X) + k_2X}.$$

5. Required value of $y'(1)$ to give $y(2) = 15.0000$: with modified Euler, 5.48408; with fourth-order Runge–Kutta, 5.49872; with Runge–Kutta–Fehlberg, 5.50012. Analytical solution has $y'(1) = 5.50000$.

7. Modified Euler can match analytical to 6 significant figures if $h = 0.01$. (Runge–Kutta–Fehlberg matches if $h = 0.125$.)

12. a.

θ	y	% error
0	0	0
$\pi/4$	0.77015	0.6250
$\pi/2$	1.42153	0.5176
$3\pi/4$	1.85370	0.3214
π	2	0

b. With $h = \pi/5$. The largest error is 0.4041%.
c. The shooting method with just two intervals gives results within 0.5% (largest error is 0.041%).

15. It requires 32 intervals.

19.

x:	0	$\pi/8$	$\pi/4$	$3\pi/8$	$\pi/2$
y:	1.5000	1.5828	1.4215	1.0410	0.5000
By RKF:	1.5000	1.5772	1.4142	1.0360	0.5000

25. Exact = 2.46166
a. 2.000 ($h = 1/2$)
b. 2.25895 ($h = 1/3$)

c. 2.34774 ($h = 1/4$)

d. 2.4636 (extrapolated)

29. Cannot get a second eigenvalue with $h = 1/2$. From $h = 1/3$, 3.59125; from $h = 1/4$, 4.0000; from $h = 1/5$, 4.19885.

34. Eigenvalues and eigenvectors.

$$
\begin{aligned}
-4.6242; & \quad [0.95724, \quad 0.25143, \quad -0.14308] \\
7.2024; & \quad [0.37006, \quad 0.92577, \quad -0.77473] \\
-9.5782; & \quad [0.41719, \quad -0.90275, \quad -0.10479]
\end{aligned}
$$

40. Upper Hessenberg matrix:

$$
\begin{bmatrix}
-5.0000 & -1.7889 & -1.3416 \\
-2.2361 & 3 & -7 \\
0 & -7 & -5
\end{bmatrix}
$$

From this, it takes only 6 iterations to get the upper-triangular result.

46.
$$
\left\{
\begin{array}{ccccc}
 & & -1 & & \\
 & & 16 & & \\
-1 & 16 & -60 & 16 & -1 \\
 & & 16 & & \\
 & & -1 & &
\end{array}
\right\}
\frac{u_{ij}}{12h^2} = 0
$$

48. The gradient is $100/L$, where L is the width of the plate. Let $h = L/n$, the nodal spacing with points at $x_i = i * h$, ($i = 0, \ldots, n$), measured from the left end. Then $u_i = 100 + i * h(100/L)$; $u_{i-1} + u_{i+1} = 2u_i$, $u_{i-2} + u_{i+2} = 2u_i$.

a.
$$
\left\{
\begin{array}{ccc}
 & u_i & \\
u_{i-1} & -4u_i & u_{i+1} \\
 & u_i &
\end{array}
\right\}/h^2 = (2u_i - 2u_i)/h^2 = 0
$$

$$
\left\{
\begin{array}{ccc}
u_{i-1} & 4u_i & u_{i+1} \\
4u_{i-1} & -20u_i & 4u_{i+1} \\
u_{i-1} & 4u_i & u_{i+1}
\end{array}
\right\}/(6h^2) = (12u_i - 12u_i)/(6h^2) = 0
$$

b.
$$
\left\{
\begin{array}{ccccc}
 & & -u_i & & \\
 & & 16u_i & & \\
-u_{i-2} & 16u_{i-1} & -60u_i & 16u_{i+1} & -u_{i+2} \\
 & & 16u_i & & \\
 & & -u_i & &
\end{array}
\right\}/(12h^2) = (-2u_i + 32u_i - 30u_i)/(12h^2) = 0
$$

For points adjacent to the boundary, assume outside fictitious points with the same gradient as inside.

50. Interior temperatures:

64.21	105.20	146.65	186.41
61.63	89.94	114.99	134.00
52.39	77.93	89.38	85.59

52. $3u_{xx} + 2u_{yy} = 3(u_L - 2u_0 + u_R)/h^2 + 2(u_A - 2u_0 + u_B)/h^2 = \left\{ \begin{array}{ccc} & 2 & \\ 3 & -10 & 3 \\ & 2 & \end{array} \right\} \dfrac{u_{ij}}{h^2} = 0$

56. a. $\phi = 0.444$ at each interior node with $h = 2/3$.

   b. Values at the interior nodes (observe the symmetry)

0.2115	0.3120	0.3419	0.3120	0.2115
0.3120	0.4722	0.5214	0.4722	0.3120
0.3419	0.5214	0.5769	0.5214	0.3419
0.3120	0.4722	0.5214	0.4722	0.3120
0.2115	0.3120	0.3419	0.3120	0.2115

(Compare node value of 0.4722 with 0.444 from part (a)).

60.

| w: | 1.30 | 1.32 | 1.34 | 1.35 | 1.36 | 1.40 | 1.50 |
| Iterations: | 21 | 19 | 16 | 15 | 17 | 18 | 21 |

(Eq. (7.16) does not apply—the region is not a rectangle.)

65. 27 iterations are needed. Exercise 54 needs 27; Exercise 55 needs only 17.

69.

$$2\left\{\frac{1}{hL(hL+hR)} \quad -\frac{hL+hR}{hL*hR(hL+hR)} \begin{array}{c} \dfrac{1}{hA(hA+hB)} \\[6pt] -\dfrac{hA+hB}{hA*hB(hA+hB)} \\[6pt] \dfrac{1}{hB(hA+hB)} \end{array} \quad \frac{1}{hR(hR+hR)}\right\}u0.$$

76.   a. $Ax = b$ is: $(\det(A) = -6.5820)$

$$\begin{bmatrix} -2.125 & 1.375 & 0.000 & 0.4375 \\ 0.750 & -2.250 & 1.250 & -0.250 \\ 0.000 & 0.875 & -2.375 & -1.3125 \end{bmatrix}.$$

b. $Ax = b$ is: $(\det(A) = -3.0352)$

$$\begin{bmatrix} -1.625 & 0.875 & 0.000 & 1.4375 \\ 1.250 & -1.750 & 0.750 & 0.250 \\ 0.000 & 1.375 & -1.875 & -0.3125 \end{bmatrix}.$$

c. $Ax = b$ is: $(\det(A) = -4.336)$

$$\begin{bmatrix} -1.750 & 0.750 & 0.000 & 1.125 \\ 1.250 & -2.000 & 0.750 & 0.0000 \\ 0.000 & 1.250 & -2.250 & 0.625 \end{bmatrix}.$$

80.   a. 19.333
b. 6.001

83.   a. $w_{opt} = 1.01612$
Eigenvalues, $w = 1$: $0, 0.0625$
Eigenvalues at w_{opt}: $0.01524, 0.01524$ (over 4-fold reduction)
Equation (7.15) with $p = 2$, $q = 3$ gives 1.01613

b. $w_{opt} = 1.20377$
Eigenvalues, $w = 1$: $0, 0.5625$
Eigenvalues at w_{opt}: both are 0.20378
Equation (7.15) doesn't apply. Experimentally w_{opt} is about 1.21.

c. The method is difficult to apply. Equation (7.15) with $p = 2$, $q = 4$ gives $w_{opt} = 1.03337$.
Eigenvalues, $w = 1$: $0, 0.125, 0.125$.
Eigenvalues with $w = 1.03337$ all have magnitude 0.0486, a reduction by a factor of 3.75.
Equation (7.15) is confirmed.

92.   3.30798, 3.64310, 8.04892.

Chapter 8 2. The discriminant is $4(1 - x)^2 + 4(1 + y)(1 - y)$. Setting this to zero and rearranging gives an equation whose points are where the equation is parabolic:

$$y^2 - x^2 + 2x = 2.$$

This is a hyperbola with center at $(1, 0)$ and vertices at $(1, 1)$ and $(1, -1)$. It is symmetrical about the line $x = 1$ and has asymptotes of $y = \pm(x - 1)$. At points above the upper curve and below the lower curve, the equation is elliptic; the equation is hyperbolic at points on the other sides of the curves.

6. (Using $k = 2.156$ Btu/(hr $*$ in. $*$ °F))
   a. -29.53°F/in.
   b. -75.59°F/in.
   c. -34.91°F/in.

9. At $t = 2.06$ sec and with $r = 1$,

x:	0	0.25	0.50	0.75	1.00	1.25
u:	0	17.85	32.98	43.09	46.64	43.09
anal:	0	17.72	32.74	42.78	46.30	42.78

$u(x)$ is symmetrical about $x = 1.00$. The errors are about 1/3 those of Exercise 8.

16. The computed values agree exactly with the analytical solution. There is symmetry about $x = 0.5$. Some values:

Time steps	0	1/8	2/8	3/8	4/8	5/8
1	0	0.354	0.653	0.854	0.924	0.854
2	0	0.271	0.500	0.653	0.707	0.653
8	0	-0.383	-0.707	-0.924	-1.000	-0.924
9	0	-0.354	-0.653	-0.854	-0.924	-0.854

(Eight time steps is one-half the period)

18. For both, the full period is 16 time steps = 24 sec. Displacements at the center point:

t, sec:	0	3	6	9	12	15	18
Part (c): Eq. (8.26):	0	0.688	1.281	1.688	1.828	1.688	1.281
Eq. (8.19):	0	0.750	1.406	1.875	2.063	1.875	1.406
Part (d): Eq. (8.26):	1.000	-2.000	-4.000	-5.500	-6.000	-6.000	-5.000
Eq. (8.19):	1.000	-2.250	-4.000	-5.750	-6.000	-6.250	-5.000

21. Let the right edge be a 3-in. edge held at 100° and the bottom edge be a 4-in. edge held at 0°. In addition to the six interior nodes, we need three nodes along the left edge and three more along the top edge making a total of 12 nodes. The layout of node values is (u for the horizontal traverse, v for the vertical traverse):

$$u_1(v_1) \quad u_2(v_4) \quad u_3(v_7) \quad u_4(v_{10})$$
$$u_5(v_2) \quad u_6(v_5) \quad u_7(v_8) \quad u_8(v_{11})$$
$$u_9(v_3) \quad u_{10}(v_6) \quad u_{11}(v_9) \quad u_{12}(v_{12})$$

We imagine seven fictitious nodes, three outside the left edge and four above the top edge. Because there is perfect insulation on these edges, the gradients there are all zero and the values at each fictitious node is equal to the value at the opposite interior node. Some typical equations:

$$\text{For } u_1: \qquad (1 + 2r)u_1 - 2ru_2 = \qquad (1 - 2r)v_1 + 2rv_2$$
$$\text{For } u_6: -ru_5 + (1 + 2r)u_6 - ru_7 = rv_4 + (1 - 2r)v_5 + rv_6$$
$$\text{For } u_{11}: -ru_{10} + (1 + 2r)u_{11} - ru_{12} = rv_8 + (1 - 2r)v_9 + 0$$
$$\text{For } u_{12}: -ru_{11} + (1 + 2r)u_{12} \qquad = rv_{11} + (1 - 2r)v_{12} + 0 + 100r$$

$$
\begin{aligned}
\text{For } v_1: &\qquad (1 + 2r)v_1 - 2rv_2 = \qquad (1 - 2r)u_1 + 2ru_2 \\
\text{For } v_5: &\ -rv_4 + (1 + 2r)v_5 - rv_6 = ru_5 + (1 - 2r)u_6 + ru_7 \\
\text{For } v_9: &\ -rv_8 + (1 + 2r)v_9 \qquad = ru_{10} + (1 - 2r)u_{11} + ru_{12} - 0 \\
\text{For } v_{12}: &\ -rv_{11} + (1 + 2r)v_{12} \qquad = ru_{11} + (1 - 2r)u_{12} + 100r + 0
\end{aligned}
$$

25.   There are 125 equations to solve at each time step, but they can be decomposed into 25 subsets of five tridiagonal equations each. If $r = 1$, only two time steps are needed to reach $t = 15.12$.

28.   Time steps are 0.00544 sec. There appears to be no repetitive pattern. Some displacements at (2, 1):

Time steps:	0	1	2	4	6
u:	0.5	0.375	0.125	−0.328	−0.586
Steps:	8	10	14	18	
u:	0.333	0.668	−0.815	0.565	

34.   A single error of magnitude 1 becomes less than 0.0005 after 56 time steps. The line of the table for t_8 is

$$0 \qquad 0.199 \qquad -0.137 \qquad -0.002 \qquad 0.$$

36.   For $N = 4$; $r = 0.5, 0.8090$; $r = 0.6, -1.1708$
      For $N = 5$; $r = 0.5, 0.8660$; $r = 0.6, -1.2392$

40.   The table resembles Table 8.11 except the errors are reflected earlier.

Chapter 9   2.   Letting $u(x) = cx(x - 1)$, the Rayleigh–Ritz integral gives

$$\frac{2c}{3} + 0 = 2\left(\frac{5}{12}\right), \qquad c = \frac{5}{4},$$

so $u(x) = (5/4)x(x - 1)$; $y(x)$ anal. $= x^3/2 + x^2/2 - x$.

x:	0	0.2	0.4	0.6	0.8	1
$u(x)$:	0	−0.200	−0.300	−0.300	−0.200	0
$y(x)$:	0	−0.176	−0.288	−0.312	−0.224	0

6. $R(x) = y'' = 3x - 1$. Let $u(x) = cx(x - 1)$, so $u'' = 2c$. Setting $R(x) = 0$ at $x = 0.5$ gives $2c - 3(0.5) - 1 = 0$, $c = 5/4$. The answer is identical to Exercise 2.

9.   Galerkin integral is

$$\int_0^1 [x(x - 1)][2c - 3x - 1]\, dx = 0,$$

which gives $(5 - 4c)/12 = 0$, $c = 5/4$. The answer is identical to those for Exercises 2 and 6.

14.   In the following $u(x)$ is the approximation, $y(x)$ is exact.

x:	1.0	1.2	1.5	1.75	2.0
$u(x)$:	−1	−0.2307	0.9174	1.9197	3
$y(x)$:	−1	−0.2267	0.9167	1.9196	3

18.

$$
\text{c. } M^{-1} = \begin{bmatrix} -4.650 & 3.982 & 1.668 \\ 0.000 & -0.217 & 0.217 \\ 0.500 & -0.120 & -0.380 \end{bmatrix}
$$

$$a = [405.16, -32.174, 9.3044]$$

$$
N = \begin{bmatrix} -4.650 & + 0.500y \\ 3.982 & - 0.217x - 0.120y \\ 1.688 & + 0.217x - 0.380y \end{bmatrix}
$$

$$u(10.6, 9.6) = 153.444$$

20. The augmented matrix:

$$\begin{bmatrix} -974.54 & -488.12 & -488.72 & 1.7385 \\ -488.12 & -975.41 & -487.83 & 1.7385 \\ -488.72 & -487.83 & -974.81 & 1.7385 \end{bmatrix}$$

23.
$$c_{ij} = \begin{cases} 0.2825, & i = j \\ 0.1412 & i \neq j \end{cases}$$

$$[K * \text{area}] = \frac{k}{c\rho} \begin{bmatrix} 0.484 & 0.089 & -0.573 \\ 0.089 & 0.196 & -0.285 \\ -0.573 & -0.285 & 0.857 \end{bmatrix}$$

$$b_i = F_{av} * 0.565$$

28. Beginning from Eq. (9.61), replace $\{\partial c/\partial t\}$ with a forward difference approximation but consider this to apply at the time halfway between t_m and t_{m+1}. Equate this to the average of values for $-[C^{-1}][K]\{c\}$ at t_m and t_{m+1}. On doing this and rearranging, we get

$$(2 + r)\{c_{m+1}\} = (2 - r)\{c_m\} + 2r[K^{-1}]\{b\},$$

where $r = \Delta t[C^{-1}][K]$.

32. Yes, this will always be true. In this example, the curvature near the center is greatest when $t = 2.25$ sec (and at every 4.5 sec thereafter).

References

Acton, F. S. (1970). *Numerical Methods That Work.* New York: Harper and Row.

Aki, S. G. (1989). *The Design and Analysis of Parallel Algorithms.* Englewood Cliffs, NJ: Prentice-Hall.

Allaire, P. W. (1985). *Basics of the Finite Element Method.* Dubuque, IA: Brown.

Allen, D. N. (1954). *Relation Methods.* New York: McGraw-Hill.

Anderson, E., Z. Gai, C. Bishof, J. Demmel, J. Dongarra, et al. (1996). *LAPACK Users' Guide.* 2nd ed. Philadelphia: SIAM.

Andrews, G., and R. Olsson (1993). *The SR Programming Language.* Redwood City, CA: Benjamin Cummings.

Andrews, Larry C. (1985). *Elementary Partial Differential Equations with Boundary Value Problems.* Philadelphia: Saunders College Publishing.

Arney, David C. (1987). *The Student Edition of DERIVE, Manual.* Reading, MA: Addison-Wesley — Benjamin Cummings.

Atkinson, K. E. (1978). *An Introduction to Numerical Analysis.* New York: Wiley.

Bartels, Richard, J. Beatty, and B. Barsky (1987). *An Introduction to Splines for Use in Computer Graphics and Geometric Modeling.* Los Altos, CA: Morgan Kaufmann.

Bertsekas, Dimitri, and Tsitsiklis, John (1989): *Parallel and Distributed Computation: Numerical Methods.* Englewood Cliffs, NJ: Prentice-Hall.

Birkhöff, Garrett, Richard Varga, and David Young (1962). Alternating direction implicit methods. *Advances in Computers* 3 : 187–273.

Boisvert, R. (1994) NIST's GAMS: A "Card Catalog" for the Computer User. *SIAM NEWS* Volume 27/Number 8.

Borse, G. J. (1997). *Numerical Methods with MATLAB.* Boston: ITP.

Bracewell, Ronald N. (1986). *The Hartley Transform.* New York: Oxford University Press.

Brigham, E. Oron (1974). *The Fast Fourier Transform.* Englewood Cliffs, NJ: Prentice-Hall.

Burnett, David S. (1987). *Finite Element Analysis: From Concepts to Applications.* Reading, MA: Addison-Wesley.

Campbell, Leon, and Laizi Jacchia (1941). *The Story of Variable Stars.* Philadelphia: Blakiston.

Carnahan, Brice (1964). *Radiation Induced Cracking of Pentanes and Dimethylbutanes.* Ph.D. dissertation, University of Michigan.

Carnahan, Brice, et al. (1969). *Applied Numerical Methods.* New York: Wiley.

Carslaw, H. S., and J. C. Jaeger (1959). *Conduction of Heat in Solids.* 2nd ed. London: Oxford University Press.

Chandy, K. M., and S. Taylor (1992). *An Introduction to Parallel Programming.* Boston: Jones and Bartlett.

Char, B., K. Geddes, G. Gonnet, B. Leong, M. Monagan, and S. Watt (1992). *First Leaves: A Tutorial Introduction to Maple V.* New York: Springer-Verlag.

Condon, Edward, and Hugh Odishaw, eds. (1967). *Handbook of Physics.* New York: McGraw-Hill.

Conte, S. D., and C. de Boor (1980). *Elementary Numerical Analysis.* 3rd ed. New York: McGraw-Hill.

Cooley, J. W., and J. W. Tukey (1965). An algorithm for the machine calculations of complex Fourier series. *Mathematics of Computation* 19:297–301.

Corliss, G., and Y. F. Chang (1982). Solving ordinary differential equations using Taylor series. *ACM Transactions on Mathematical Software* 8:114–144.

Crow, Frank (1987). Origins of a teapot. *IEEE Computer Graphics and Applications* 7(1):8–19.

Datta, B. N. (1995). *Numerical Linear Algebra and Applications.* Pacific Grove, CA: Brooks/Cole.

Davis, Alan J. (1980). *The Finite Element Method.* Oxford: Clarendon Press.

Davis, Phillip J., and Phillip Rabinowitz (1967). *Numerical Integration.* Waltham, MA: Blaisdell.

de Boor, C. (1978). *A Practical Guide to Splines.* New York: Springer-Verlag.

De Santis, R., F. Gironi, and L. Marelli (1976). Vector-liquid equilibrium from a hard-sphere equation of state. *Industrial and Engineering Chemistry Fundamentals* 15(3):183–189.

Dongarra, J., I. Duff, D. Sorensen, and H. van der Vorst (1991). *Solving Linear Systems on Vector and Shared Memory Computers.* Philadelphia: SIAM.

Dongarra, J. J., J. R. Bunch, C. B. Moler, and G. W. Stewart. (1979). *LINPACK User's Guide.* Philadelphia: SIAM.

Douglas, J. (1962). Alternating direction methods for three space variables. *Numerical Mathematics* 4:41–63.

Duffy, A. R., J. E. Sorenson, and R. E. Mesloh (1967). Heat transfer characteristics of belowground LNG storage. *Chemical Engineering Progress* 63(6):55–61.

Etter, D. M. (1993). *Quattro Pro—A Software Tool for Engineers and Scientists.* Redwood City, CA: Benjamin/Cummings.

Fike, C. T. (1968). *Computer Evaluation of Mathematical Functions.* Englewood Cliffs, NJ: Prentice-Hall.

Forsythe, G. E., M. A. Malcolm, and C. B. Moler (1977). *Computer Methods for Mathematical Computation.* Englewood Cliffs, NJ: Prentice-Hall.

Forsythe, G. E., and C. B. Moler (1967). *Computer Solution of Linear Algebraic Systems.* Englewood Cliffs, NJ: Prentice-Hall.

Fox, L., (1965). *An Introduction to Numerical Linear Analysis.* New York: Oxford University Press.

Gear, C. W. (1967). The numerical integration of ordinary differential equations. *Mathematics of Computation* 21:146–156.

Gear, C. W. (1971). *Numerical Initial Value Problems in Ordinary Differential Equations.* Englewood Cliffs, NJ: Prentice-Hall.

Hageman, L. A., and D. M. Young (1981). *Applied Iterative Methods.* New York: Academic Press.

Hamming, R. W. (1971). *Introduction to Applied Numerical Analysis.* New York: McGraw-Hill.

Hamming, R. W. (1973). *Numerical Methods for Scientists and Engineers.* 2nd ed. New York: McGraw-Hill.

Harrington, Steven (1987). *Computer Graphics: A Programming Approach.* New York: McGraw-Hill.

Henrici, P. H. (1964). *Elements of Numerical Analysis.* New York: Wiley.

Hornbeck, R. W. (1975). *Numerical Methods.* New York: Quantum.

Householder, A. S. (1970). *The Numerical Treatment of a Single Nonlinear Equation.* New York: McGraw-Hill.

IEEE Standard for Binary Floating-Point Arithmetic (1985). Institute of Electrical and Electronics Engineers, Inc., New York.

JaJa, J. (1992). *An Introduction to Parallel Algorithms.* Reading, MA: Addison-Wesley.

Jones, B. (1982). A note on the T transformation. *Nonlinear Analysis, Theory, Methods and Applications* 6:303–305.

Kahaner, D., C. Moler, S. Nash (1989). *Numerical Methods and Software.* Englewood Cliffs, NJ: Prentice-Hall.

Lee, Peter, and Geoffrey Duffy (1976). Relationships between velocity profiles and drag reduction in turbulent fiber suspension flow. *Journal of the American Institute of Chemical Engineering* 22(4):750–753.

Love, Carl H. (1966). *Abscissas and Weights for Gaussian Quadrature.* National Bureau of Standards, Monograph 98.

Muller, D. E. (1956). A method of solving algebraic equations using an automatic computer. *Math Tables and Other Aids to Computation* 10:208–215.

O'Neill, Mark A. (1988). Faster Than Fast Fourier. *BYTE* 13(4):293–300.

Orvis, William J. (1987). *1-2-3 for Scientists and Engineers.* San Francisco: Sybex.

Peaceman, D. W., and H. H. Rachford (1955). The numerical solution of parabolic and elliptic differential equations. *Journal of the Society for Industrial and Applied Mathematics* 3:28–41.

Penrod, E. B., and K. V. Prasanna (1962). Design of a flat-plate collector for a solar earth heat pump. *Solar Energy* 6(1):9–22.

Pinsky, Mark A. (1991). *Partial Differential Equations and Boundary Value Problems with Applications.* 2nd ed. New York: McGraw-Hill.

Pizer, Stephen J. (1975). *Numerical Computing and Mathematical Analysis.* Chicago: Science Research Associates.

Pokorny, C., and C. Gerald (1989). *Computer Graphics: The Principles Behind the Art and Science.* Irvine, CA: Franklin, Beedle, and Associates.

Prenter, P. M. (1975). *Splines and Variational Methods.* New York: Wiley.

Press, W., B. Flannery, S. Teudolsky, and W. Vetterling (1992). *Numerical Recipes in C: The Art of Scientific Computing.* 2nd ed. New York: Cambridge University Press.

Press, W., B. Flannery, S. Teudolsky, and W. Vetterling (1992). *Numerical Recipes in FORTRAN: The Art of Scientific Computing.* 2nd ed. New York: Cambridge University Press.

Press, W., B. Flannery, S. Teudolsky, and W. Vetterling (1996). *Numerical Recipes in FORTRAN 90: The Art of Parallel Scientific Computing.* 2nd ed. New York: Cambridge University Press.

Rall, L. B. (1981). *Automatic Differentiation: Techniques and Applications.* Springer-Verlag.

Ralston, Anthony (1965). *A First Course in Numerical Analysis.* New York: McGraw-Hill.

Ramirez, Robert W. (1985). *The FFT, Fundamentals and Concepts.* Englewood Cliffs, NJ: Prentice-Hall.

Rice, John R. (1983). *Numerical Methods, Software, and Analysis.* New York: McGraw-Hill.

Richtmyer, R. D. (1957). *Difference Methods for Initial Value Problems.* New York: Wiley Interscience.

Sabot, G. W. ed. (1995). *High Performance Computing:* Reading, MA: Addison-Wesley.

Sedgwick, R. (1992). *Algorithms in C++.* Reading, MA: Addison-Wesley. [Other versions available: *in C, in Pascal.*]

Shampine, L., and R. Allen (1973). *Numerical Computing.* Philadelphia: Saunders.

Smith, G. D. (1978). *Numerical Solution of Partial Differential Equations.* 2nd ed. London: Oxford University Press.

Stallings, William (1990). *Computer Organization and Architecture.* New York: Macmillan.

Stewart, G. W. (1973). *Introduction to Matrix Computations.* New York: Academic Press.

Stoer, J., and R. Burlirsch (1993). *Introduction to Numerical Analysis.* 2nd edition. New York: Springer-Verlag.

Traub, J. F. (1964). *Iterative Methods for the Solution of Equations.* Englewood Cliffs, NJ: Prentice-Hall.

Van Loan, C. F. (1997). *Introduction to SCIENTIFIC COMPUTING: A Matrix-Vector Approach Using MATLAB.* Englewood Cliffs, NJ: Prentice-Hall.

Varga, Richard (1959). *p*-Cyclic matrices: A generalization of the Young-Frankel successive over-relaxation scheme. *Pacific Journal of Mathematics* 9:617–628.

Vichnevetsky, R. (1981). *Computer Methods for Partial Differential Equations. Vol. 1: Elliptic Equations and the Finite Element Method.* Englewood Cliffs, NJ: Prentice-Hall.

Walker, D. W., and J. J. Dongarra. (1996). MPI: A Standard Message-Passing Interface. *SIAM NEWS,* Volume 29/Number 1.

Waser, S., and M. J. Flynn (1982). *Introduction to Arithmetic for Digital Systems Designers.* New York: Holt, Rinehart and Winston.

Wilkinson, J. H. (1963). *Rounding Errors in Algebraic Processes.* Englewood Cliffs, NJ: Prentice-Hall.

Wilkinson, J. H. (1965). *The Algebraic Eigenvalue Problem.* London: Oxford University Press.

Wolfram, Stephen (1988). *Mathematics: A System for Doing Mathematics by Computer.* Reading, MA: Addison-Wesley.

Index

Date Due

PRINTED IN U.S.A.   CAT. NO. 24 161

**OVERDUE FINES ARE
$0.25 PER DAY**